HUGHES

ELECTRICAL AND ELECTRONIC
TECHNOLOGY

PEARSON
Education

We work with leading authors to develop the
strongest educational materials in engineering,
bringing cutting-edge thinking and best learning
practice to a global market.

Under a range of well-known imprints, including
Prentice Hall, we craft high quality print and
electronic publications which help readers to
understand and apply their content,
whether studying or at work.

To find out more about the complete range of our
publishing please visit us on the World Wide Web at:
www.pearsoned.co.uk

HUGHES

ELECTRICAL AND ELECTRONIC TECHNOLOGY

ninth edition

EDWARD HUGHES

Revised by John Hiley, Keith Brown and

Ian McKenzie Smith

PEARSON

Prentice
Hall

Harlow, England • London • New York • Boston • San Francisco • Toronto • Sydney • Singapore • Hong Kong
Tokyo • Seoul • Taipei • New Delhi • Cape Town • Madrid • Mexico City • Amsterdam • Munich • Paris • Milan

Pearson Education Limited
Edinburgh Gate
Harlow
Essex CM20 2JE
England

and Associated Companies throughout the world

Visit us on the World Wide Web at:
www.pearsoned.co.uk

First published under the Longman imprint 1960
Ninth edition 2005

© Pearson Education Limited 1960, 2005

ISBN 978-0-13-114397-5

British Library Cataloguing-in-Publication Data
A catalogue record for this book is available from the British Library

Library of Congress Cataloging-in-Publication Data
Hughes, Edward, 1888–
      Hughes electrical and electronic technology / Edward Hughes.—9th ed. / revised by
Ian McKenzie Smith with John Hiley and Keith Brown.
          p. cm.
      Rev. ed. of: Hughes electrical technology.
      Includes index.
      ISBN 0-13-114397-2
      1. Electric engineering—Textbooks.   2. Electronics—Textbooks.   I. Title: Electrical and
electronic technology.   II. Smith, Ian McKenzie.   III. Hiley, John.   IV. Brown, Keith, 1962–
V. Hughes, Edward, 1888–. Hughes electrical technology.   VI. Title.

TK146.H9 2005
621.3—dc22

2004058680

10   9   8   7   6   5   4
09   08   07
Typeset in 10/11pt Ehrhardt MT by 35
Printed and bound by Ashford Colour Press, Gosport, Hants.

Short contents

Contents

Supporting resource

Visit **www.pearsoned.co.uk/hughes** to find a valuable online resource

For instructors
• Complete, downloadable Instructor's Manual

For more information please contact your local Pearson Education sales representative or visit
www.pearsoned.co.uk/hughes

Preface to the Ninth Edition

The electrical and electronic technology upon which this textbook seeks to shed light continues to develop apace. Accordingly, the ninth edition has been updated and new material added, reflecting the need to provide new text to support the understanding of some of those advances. New sections on passive filters, nodal analysis both for d.c. and a.c. circuits and Karnaugh maps have been added and the topic of resonance in a.c. circuits revised and extended. We have also taken the opportunity to revise the logical flow of the material of the book. At the same time, the authors have continued the excellent tradition, established by Edward Hughes and continued by Ian McKenzie Smith, of illustrating the new material with a large number of numerical examples.

Indeed, we wholeheartedly endorse the sentiments expressed by Edward Hughes in the Preface to the First Edition. His advice remains just as valuable today as it was in 1959. Then he exhorted students of the subject to test the thoroughness of their understanding by working out as many numerical problems as possible. He also pointed out the importance of being able to communicate to others both the method of solution and the solution itself. The worked examples continue to provide guidance in this regard. In addition, he underlined the need to attach the names of units to solutions. We would suggest that this is done wherever possible during the solution process too. Dimensional analysis frequently guides that process and aids overall understanding. It helps too in the cross-checking of calculator answers which are all too often beset by input errors! Here too the value of common sense and an approximation of the appropriate order of magnitude for a solution should not be underestimated.

We would encourage all teachers who use this book to use the solutions to the end of chapter exercises, which are available on the Pearson Education website for this book: www.pearsoned.co.uk/hughes. There too you can leave not only your comments on this edition but also your suggestions for new material and improvements to old. Feedback, from students and teachers alike, is most welcome and is invaluable for ensuring the usefulness of this text for our vast subject area. We thank you, in anticipation of your comments.

Finally, we would like to acknowledge the past and present support of our colleagues, the academic, secretarial and technical staff of the Department of Electrical, Electronic and Computer Engineering at Heriot–Watt University. This book is dedicated to them.

Keith Brown
John Hiley
Heriot–Watt University, Edinburgh
May 2004

Preface to the First Edition

This volume covers the electrical engineering syllabuses of the Second and Third Year Courses for the Ordinary National Certificate in Electrical Engineering and of the First Year Course leading to a Degree of Engineering.

The rationalized M.K.S. system of units has been used throughout this book. The symbols, abbreviations and nomenclature are in accordance with the recommendations of the British Standards Institution; and, for the convenience of students, the symbols and abbreviations used in this book have been tabulated in the Appendix.

It is impossible to acquire a thorough understanding of electrical principles without working out a large number of numerical problems; and while doing this, students should make a habit of writing the solutions in an orderly manner, attaching the name of the unit wherever possible. When students tackle problems in examinations or in industry, it is important that they express their solutions in a way that is readily intelligible to others, and this facility can only be acquired by experience. Guidance in this respect is given by the 106 worked examples in the text, and the 670 problems afford ample opportunity for practice.

Most of the questions have been taken from examination papers; and for permission to reproduce these questions I am indebted to the University of London, the East Midland Educational Union, the Northern Counties Technical Examination Council, the Union of Educational Institutions and the Union of Lancashire and Cheshire Institutes.

I wish to express my thanks to Dr F. T. Chapman, C.B.E., M.I.E.E., and Mr E. F. Piper, A.M.I.E.E., for reading the manuscript and making valuable suggestions.

Edward Hughes
Hove
April 1959

Section one   Electrical Principles

Chapter one

International System of Measurement

Objectives

When you have studied this chapter, you should
- be familiar with the International System of Measurement
- be familiar with a variety of derived SI units
- be aware of the concepts of torque and turning moment
- be capable of analysing simple applications of the given SI units
- have an understanding of work, energy and power
- be capable of analysing simple applications involving work, energy and power
- have an understanding of efficiency and its relevance to energy and power
- be capable of analysing the efficiency of simple applications
- have an understanding of temperature and its units of measurement

Contents

Electrical technology is a subject which is closely related to the technologies of mechanics, heat, light and sound. For instance, we use electrical motors to drive machines such as cranes, we use electric heaters to keep us warm, we use electric lamp bulbs perhaps to read this book and we use electric radios to listen to our favourite music.

At this introductory stage, let us assume that we have some understanding of physics in general and, in particular, let us assume that we have some understanding of the basic mechanics which form part of any study of physics. It is not necessary to have an extensive knowledge and, in this chapter we shall review the significant items of which you should have an understanding. We shall use these to develop an appreciation of electrical technology.

In particular, we shall be looking at the concepts of work, energy and power since the underlying interest that we have in electricity is the delivery of energy to a point of application. Thus we drive an electric train yet the power source is in a generating station many kilometres away, or we listen to a voice on the phone speaking with someone possibly on the other side of the world. It is electricity which delivers the energy to make such things happen.

1.1 The International System

The International System of Units, known as SI in every language, was formally introduced in 1960 and has been accepted by most countries as their only legal system of measurement.

One of its most important advantages over its predecessors is that it is a coherent system wherever possible. A system is coherent if the product or quotient of any two quantities is the unit of the resultant quantity. For example, unit area results when unit length is multiplied by unit length. Similarly unit velocity results when unit length or distance is divided by unit time.

The SI is based on the measures of six physical quantities:

> Mass
> Length
> Time
> Electric current
> Absolute temperature
> Luminous intensity

All other units are derived units and are related to these base units by definition.

If we attempt to analyse relationships between one unit and another, this can be much more readily achieved by manipulating symbols, e.g. A for areas, W for energy and so on. As each quantity is introduced, its symbol will be highlighted as follows:

Energy   Symbol: W

Capital letters are normally used to represent constant quantities – if they vary, the symbols can be made lower case, i.e. W indicates constant energy while w indicates a value of energy which is time varying.

The names of the SI units can be abbreviated for convenience. Thus the unit for energy – the joule – can be abbreviated to J. This will be highlighted as follows:

Energy   Symbol: W Unit: **joule (J)**

Here the unit is given the appropriate unit abbreviation in brackets. These are only used after numbers, e.g. 16 J. By comparison, we might refer to a few joules of energy.

Now let us consider the six base quantities.

The *kilogram* is the mass of a platinum–iridium cylinder preserved at the International Bureau of Weights and Measures at Sèvres, near Paris, France.

Mass   Symbol: m Unit: **kilogram (kg)**

It should be noted that the megagram is also known as the tonne (t).

The *metre* is the length equal to 1 650 763.73 wavelengths of the orange line in the spectrum of an internationally specified krypton discharge lamp.

Length   Symbol: l Unit: **metre (m)**

Length and distance are effectively the same measurement but we use the term distance to indicate a length of travel. In such instances, the symbol d may be used instead of l. In the measurement of length, the centimetre is additional to the normal multiple units.

The *second* is the interval occupied by 9 192 631 770 cycles of the radiation corresponding to the transition of the caesium–133 atom.

Time Symbol: t Unit: second (s)

Although the standard submultiples of the second are used, the multiple units are often replaced by minutes (min), hours (h), days (d) and years (a).
The *ampere* is defined in section 2.7.

Electric current Symbol: I Unit: ampere (A)

The *kelvin* is 1/273.16 of the thermodynamic temperature of the triple point of water. On the Celsius scale the temperature of the triple point of water is 0.01 °C,

hence 0 °C = 273.15 K

A temperature interval of 1 °C = a temperature interval of 1 K.
The *candela* is the unit of luminous intensity.

1.2 SI derived units

Although the physical quantities of area, volume, velocity, acceleration and angular velocity are generally understood, it is worth noting their symbols and units.

Area Symbol: A Unit: square metre (m²)

Volume Symbol: V Unit: cubic metre (m³)

Velocity Symbol: u Unit: metre per second (m/s)

Acceleration Symbol: a Unit: metre per second squared (m/s²)

Angular velocity Symbol: ω Unit: radian per second (rad/s)

The unit of force, called the newton, is that force which, when applied to a body having a mass of one kilogram, gives it an acceleration of one metre per second squared.

Force Symbol: F Unit: newton (N)

$$F = ma \qquad [1.1]$$

F [newtons] = m [kilograms] × a [metres per second²]

Weight The weight of a body is the gravitational force exerted by the earth on that body. Owing to the variation in the radius of the earth, the gravitational force on a given mass, at sea-level, is different at different latitudes, as shown in Fig. 1.1. It will be seen that the weight of a 1 kg mass at sea-level in the London area is practically 9.81 N. For most purposes we can assume

$$\text{The weight of a body} \simeq 9.81m \text{ newtons} \qquad [1.2]$$

where m is the mass of the body in kilograms.

Fig. 1.1 Variation of weight with latitude

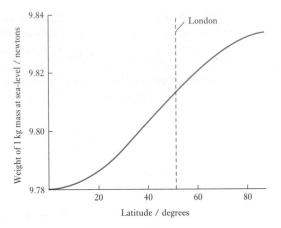

Example 1.1 A force of 50 N is applied to a mass of 200 kg. Calculate the acceleration.

Substituting in expression [1.1], we have

$$50 \, [\text{N}] = 200 \, [\text{kg}] \times a$$

$$\therefore \qquad a = \mathbf{0.25 \ m/s^2}$$

Example 1.2 A steel block has a mass of 80 kg. Calculate the weight of the block at sea-level in the vicinity of London.

Since the weight of a 1 kg mass $\simeq$ 9.81 N,

$$\therefore \qquad \text{Weight of the steel block} = 80 \, [\text{kg}] \times 9.81 \, [\text{N/kg}]$$

$$= \mathbf{785 \ N}$$

In the above example, it is tempting to give the answer as 784.8 N but this would be a case of false accuracy. The input information was only given to three figures and therefore the answer should only have three significant numbers, hence 784.8 ought to be shown as 785. Even here, it could be argued that the 80 kg mass was only given as two figures and the answer might therefore have been shown as 780 N. Be careful to show the answer as a reasonable compromise. In the following examples, such adjustments will be brought to your attention.

1.3 **Unit of turning moment or torque**

If a force F, in newtons, is acting at right angles to a radius r, in metres, from a point, the turning moment or torque about that point is

$$Fr \text{ newton metres}$$

Torque   Symbol: T (or M)   Unit: newton metre (N m)

If the perpendicular distance from the line of action to the axis of rotation is r, then

$$T = Fr \qquad\qquad\qquad [1.3]$$

The symbol M is reserved for the torque of a rotating electrical machine.

1.4	Unit of work or energy

The SI unit of energy is the *joule* (after the English physicist, James P. Joule, 1818–89). *The joule is the work done when a force of 1 N acts through a distance of 1 m in the direction of the force.* Hence, if a force F acts through distance l in its own direction

$$\text{Work done} = F \text{ [newtons]} \times l \text{ [metres]}$$

$$= Fl \text{ joules}$$

Work or energy   Symbol: W Unit: joule (J)

$$W = Fl \qquad\qquad [1.4]$$

Note that energy is the capacity for doing work. Both energy and work are therefore measured in similar terms.

If a body having mass m, in kilograms, is moving with velocity u, in metres per second

$$\text{Kinetic energy} = \tfrac{1}{2}mu^2 \text{ joules}$$

$$\therefore \qquad W = \tfrac{1}{2}mu^2 \qquad\qquad [1.5]$$

If a body having mass m, in kilograms, is lifted vertically through height h, in metres, and if g is the gravitational acceleration, in metres per second^2, in that region, the potential energy acquired by the body is

$$\text{Work done in lifting the body} = mgh \text{ joules}$$

$$W \simeq 9.81mh \qquad\qquad [1.6]$$

Example 1.3

A body having a mass of 30 kg is supported 50 m above the earth's surface. What is its potential energy relative to the ground?

If the body is allowed to fall freely, calculate its kinetic energy just before it touches the ground. Assume gravitational acceleration to be 9.81 m/s^2.

$$\text{Weight of body} = 30 \text{ [kg]} \times 9.81 \text{ [N/kg]} = 294.3 \text{ N}$$

$$\therefore \qquad \text{Potential energy} = 294.3 \text{ [N]} \times 50 \text{ [m]} = \mathbf{14\,700\ J}$$

Note – here we carried a false accuracy in the figure for the weight and rounded the final answer to three figures.

If u is the velocity of the body after it has fallen a distance l with an acceleration g.

$$u = \sqrt{(2gl)} = \sqrt{(2 \times 9.81 \times 50)} = 31.32 \text{ m/s}$$

and

$$\text{Kinetic energy} = \tfrac{1}{2} \times 30 \text{ [kg]} \times (31.32)^2 \text{ [m/s]}^2 = \mathbf{14\,700\ J}$$

Hence the whole of the initial potential energy has been converted into kinetic energy. When the body is finally brought to rest by impact with the ground, practically the whole of this kinetic energy is converted into heat.

1.5	Unit of power

Since power is the rate of doing work, it follows that the SI unit of power is the *joule per second* or *watt* (after the Scottish engineer, James Watt, 1736–1819). In practice, the watt is often found to be inconveniently small and so the *kilowatt* is frequently used.

Power   Symbol: P Unit: watt (W)

$$P = \frac{W}{t} = \frac{F \cdot l}{t} = F \cdot \frac{l}{t}$$

$$P = Fu \qquad\qquad [1.7]$$

In the case of a rotating electrical machine:

$$P = M\omega = \frac{2\pi N_r M}{60} \qquad\qquad [1.8]$$

where N_r is measured in revolutions per minute.

Rotational speed   Symbol: N_r Unit: revolution per minute (r/min)

In the SI, the rotational speed ought to be given in revolutions per second but this often leads to rather small numbers, hence it is convenient to give rotational speed in revolutions per minute. The old abbreviation was rev/min and this is still found to be widely in use.

Rotational speed   Symbol: n_r Unit: revolution per second (r/s)

There is another unit of energy which is used commercially – the kilowatt hour (kW h). It represents the work done by working at the rate of one kilowatt for a period of one hour. Once known as the Board of Trade Unit, it is still widely referred to, especially by electricity suppliers, as the unit.

$$1 \text{ kW h} = 1000 \text{ watt hours}$$
$$= 1000 \times 3600 \text{ watt seconds or joules}$$
$$= 3\,600\,000 \text{ J} = 3.6 \text{ MJ}$$

Example 1.4

A stone block, having a mass of 120 kg, is hauled 100 m in 2 min along a horizontal floor. The coefficient of friction is 0.3. Calculate

(a) the horizontal force required;
(b) the work done;
(c) the power.

(a)   Weight of stone $\simeq 120\,[\text{kg}] \times 9.81\,[\text{N/kg}] = 1177.2$ N

∴   force required $= 0.3 \times 1177.2\,[\text{N}] = 353.16$ N $= \mathbf{353}$ N

(b)   Work done $= 353.16\,[\text{N}] \times 100\,[\text{m}] = 35\,316$ J

$$= \mathbf{35.3 \text{ kJ}}$$

(c)   Power $= \dfrac{35\,316\,[\text{J}]}{(2 \times 60)\,[\text{s}]} = \mathbf{294 \text{ W}}$

Example 1.5 An electric motor is developing 10 kW at a speed of 900 r/min. Calculate the torque available at the shaft.

$$\text{Speed} = \frac{900\,[\text{r/min}]}{60\,[\text{s/min}]} = 15\,\text{r/s}$$

Substituting in expression [1.8], we have

$$10\,000\,[\text{W}] = T \times 2\pi \times 15\,[\text{r/s}]$$

$$\therefore \qquad T = 106\,\text{N m}$$

1.6 Efficiency

It should be noted that when a device converts or transforms energy, some of the input energy is consumed to make the device operate. The efficiency of this operation is defined as

$$\text{Efficiency} = \frac{\text{energy output in a given time}}{\text{energy input in the same time}} = \frac{W_o}{W_{in}}$$

$$= \frac{\text{power output}}{\text{power input}} = \frac{P_o}{P_{in}}$$

Efficiency   Symbol: η Unit: none

$$\therefore \qquad \eta = \frac{P_o}{P_{in}} \qquad\qquad [1.9]$$

Example 1.6 A generating station has a daily output of 280 MW h and uses 500 t (tonnes) of coal in the process. The coal releases 7 MJ/kg when burnt. Calculate the overall efficiency of the station.

Input energy per day is

$$W_{in} = 7 \times 10^6 \times 500 \times 1000$$

$$= 35.0 \times 10^{11}\,\text{J}$$

Output energy per day is

$$W_o = 280\,\text{MW h}$$

$$= 280 \times 10^6 \times 3.6 \times 10^3 = 10.1 \times 10^{11}\,\text{J}$$

$$\eta = \frac{W_o}{W_{in}} = \frac{10.1 \times 10^{11}}{35.0 \times 10^{11}} = 0.288$$

Example 1.7 A lift of 250 kg mass is raised with a velocity of 5 m/s. If the driving motor has an efficiency of 85 per cent, calculate the input power to the motor.

Weight of lift is

$$F = mg = 250 \times 9.81$$

$$= 2452\,\text{N}$$

Output power of motor is

$$P_o = Fu = 2452 \times 5 = 12\ 260 \text{ W}$$

Input power to motor is

$$P_{in} = \frac{P_o}{\eta} = \frac{12\ 260}{0.85} = 14\ 450 \text{ W} = \textbf{14.5 kW}$$

1.7 Temperature

Some mention is required about temperature measurement, which is in the Celsius scale. Absolute temperature is measured in kelvin, but for most electrical purposes at an introductory stage it is sufficient to measure temperature in degrees Celsius.

It should be remembered that both degrees of temperature represent the same change in temperature – the difference lies in the reference zero.

Temperature   Symbol: θ Unit: degree Celsius (°C)

A useful constant to note is that it takes 4185 J to raise the temperature of 1 litre of water through 1 °C.

Example 1.8

An electric heater contains 40 litres of water initially at a mean temperature of 15 °C; 2.5 kW h is supplied to the water by the heater. Assuming no heat losses, what is the final mean temperature of the water?

$$W_{in} = 2.5 \times 3.6 \times 10^6 = 9 \times 10^6 \text{ J}$$

Energy to raise temperature of 40 litres of water through 1 °C is

$$40 \times 4185 \text{ J}$$

Therefore change in temperature is

$$\Delta\theta = \frac{9 \times 10^6}{40 \times 4185} = 53.8 \text{ °C}$$

$$\theta_2 = \theta_1 + \Delta\theta = 15 + 53.8 = \textbf{68.8 °C}$$

Summary of important formulae

F [newtons] $= m$ [kilograms] $\times a$ [metres per second squared]	[1.1]
i.e. $F = ma$	
Torque $T = Fr$ (newton metres)	[1.3]
Work $W = Fl$ (joules)	[1.4]
Work = Energy	
Kinetic energy $W = \frac{1}{2}mu^2$	[1.5]
Power $P = Fu$ (watts)	[1.7]
$= T\omega = M\omega = 2\pi nT$	[1.8]
Efficiency $\eta = P_o/P_{in}$	[1.9]

Terms and concepts

Force, when applied to a body, causes the body to accelerate.

Weight is the gravitational force exerted by the earth on a body.

Torque, when applied to a body, causes the body to rotationally accelerate.

Energy is the capacity to do work. When selling energy, it is measured in kilowatt hours rather than joules.

Power is the rate of working.

Efficiency is the ratio of output power to input power. The difference between output and input is usually due to wastage.

Exercises 1

1. A force of 80 N is applied to a mass of 200 kg. Calculate the acceleration in metres per second squared.
2. Calculate the force, in kilonewtons, required to give a mass of 500 kg an acceleration of 4 m/s².
3. What is the weight, in newtons, of a body of mass 10 kg?
4. A ball falls off the top of a wall. Determine its downward velocity 1 s, 2 s and 3 s after commencing its fall.
5. A body of mass 10 tonnes is acted upon by a force of 1 kN. How long will it take the body to reach a speed of 5 m/s?
6. A 10 000 tonne ship when slowing down with its engines stopped is found to slow from 3 m/s to 2 m/s in a distance of 40 m. Determine the average resistance to motion.
7. A body of mass 10 kg rests on a surface travelling upwards with uniform velocity 3 m/s. Determine the apparent weight of the body that it exerts on the surface. If the surface accelerated at 3 m/s², what would be the new value of the apparent weight?
8. A body of true weight 10 N appears to weigh 9 N when its weight is measured by means of a spring balance in a moving lift. What is the acceleration of the lift at the time of weighing?
9. A train having a mass of 300 Mg is hauled at a constant speed of 90 km/h along a straight horizontal track. The track resistance is 5 mN per newton of train weight. Calculate (a) the tractive effort in kilonewtons, (b) the energy in megajoules and in kilowatt hours expended in 10 minutes, (c) the power in kilowatts and (d) the kinetic energy of the train in kilowatt hours (neglecting rotational inertia).
10. The power required to drive a certain machine at 350 r/min is 600 kW. Calculate the driving torque in newton metres.

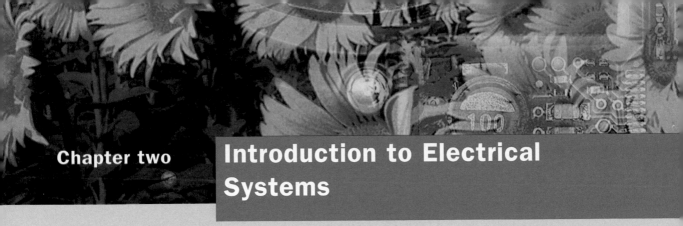

Chapter two

Introduction to Electrical Systems

Objectives

When you have studied this chapter, you should

- have an understanding of the importance electricity has for engineers
- be familiar with the constituent parts of an electric circuit
- be familiar with electric charge
- recognize that an electric current is the rate of flow of electric charge
- have an understanding of the effect that electromotive force has on a circuit
- be capable of differentiating between electromotive force and volt drop
- be familiar with basic electrical units of measurement
- have an understanding of Ohm's law
- be capable of applying Ohm's law to the analysis of simple circuits
- be familiar with resistors and their coding
- be aware of the difference between conductors and insulators

Contents

Electrical systems involve the use of circuits. This chapter introduces you to the construction of a circuit and classifies the principal parts which are to be found in every circuit, and this will lead to an understanding of circuit diagrams. We shall also address what happens to bring about the action of an electric circuit. The principal activity involves electric charge – when we arrange for electric charge to move in a predetermined way, we achieve an electric current. To produce this effect, we require to enlist the aid of an electromotive force.

Georg Ohm related the electromotive force to the current in his simple law and by applying Ohm's law, we can find out about resistance which is an important physical property associated with all circuits. This will lead us to discover that circuits can have conductors, insulators and resistors depending on the way in which we regard the resistance of the component parts. And most significantly, we find that current passing through a resistor produces heat – and this is important in practice since it determines whether a cable can pass a small current or a large one.

2.1 Electricity and the engineer

Electricity can be considered from two points of view. The scientist is concerned with what happens in an electric system and seeks to explain its mysteries. The engineer accepts that electricity is there and seeks to make use of its properties without the need to fully understand them.

Because this book is written for engineers, let us concentrate on the features of electricity which are most significant – and the most significant is that an electrical system permits us easily to transmit energy from a source of supply to a point of application.

In fact, electrical engineering could be summarized into four categories:

1. The production of electrical energy.
2. The transmission of electrical energy.
3. The application of electrical energy.
4. The control of electrical energy.

Most electrical engineers concern themselves with electronic control systems which involve not only computers but also all forms of communications. Transmission systems are varied and include the electronic communications systems as well as the power systems which appear as tower lines. For the electronics engineer, the source, which produces the energy, and the load to which the energy is applied, are less significant; for the power engineer, they are the most significant.

To understand an electrical system better, let us consider a simple situation with which we are familiar – the electric light in our room.

2.2 An electrical system

A basic electrical system has four constituent parts as shown in Fig. 2.1.

1. The source.   The function of the source is to provide the energy for the electrical system. A source may usually be thought of as a battery or a generator, although for simplicity we might even think of a socket outlet as a source.

2. The load.   The function of the load is to absorb the electrical energy supplied by the source. Most domestic electrical equipment constitutes loads. Common examples include lamps and heaters, all of which accept energy from the system.

3. The transmission system.   This conducts the energy from the source to the load. Typically the transmission system consists of insulated wire.

4. The control apparatus.   As the name suggests, its function is to control. The most simple control is a switch which permits the energy to flow or else interrupts the flow.

Fig. 2.1 Parts of an electrical system

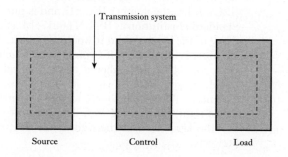

Transmission system

Source   Control   Load

Fig. 2.2 Simple lamp system

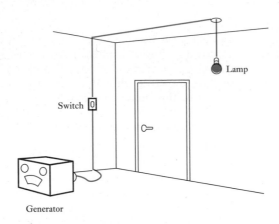

Fig. 2.3 Simple lamp circuit

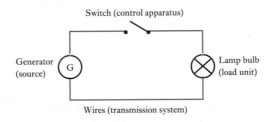

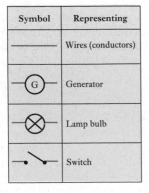

Symbol	Representing
——————	Wires (conductors)
—Ⓖ—	Generator
—⊗—	Lamp bulb
—•⁄—	Switch

Fig. 2.4 Symbols used in Fig. 2.3

A simple system is shown in Fig. 2.2; a generator supplies a lamp bulb, while a switch is included to put the lamp on and off. This example serves to show two points. First, it illustrates the fundamental function of any electric system which is to transport energy from the input source to the energy-converting load. The generator could well be a long distance away from the point of application to the lamp, and this transport of energy is called transmission, i.e. the energy has been transmitted. Secondly, the sketch of the system arrangement is difficult to interpret. The system used to take the energy from the generator to the lamp is almost impossible to follow. However, the alternative form of diagram shown in Fig. 2.3 is easy to follow because symbols have replaced detailed sketches of the components.

Such symbolic diagrams do not take long to draw, but they involve a new means of communication; this new means is the use of the symbols which are separately shown in Fig. 2.4.

To obtain the best use of these symbols, it is necessary that everyone should use the same system of symbols, and such a system is published in a specification drawn up by the International Electrotechnical Commission (IEC). It has the number IEC 617 and is published in the UK by the British Standards Institution as BS EN 60617. Most engineers become familiar with many of the symbols and it would be unusual to require to remember them all. Most symbols are self-explanatory as each diagram is introduced.

It should be remembered that electrical circuit diagrams, as they are called, are generally drawn to show a clear sequence of events; in particular, the energy flows from source to load. Normally this flow should read from left to right; thus, in Fig. 2.3, the generator was drawn at the left-hand side and the lamp bulb at the right-hand side, with the controlling switch in between.

Electricity permits the source of energy to be remote from the point of application. Electrical engineering is concerned with the study of how this energy transmission takes place, but, before getting down to applying electric current to our use, it is necessary to become familiar with some of the basic electrical terms.

2.3	Electric charge

An electrical system generally transmits energy due to the movement of electric charge. Although we need not study electric charge in depth, we need to have some understanding in order to develop a system of measurement of electrical quantities and also to relate these to the measurements which we have reviewed in Chapter 1.

Electricity appears in one of two forms which, by convention, are called negative and positive electricity. Electric charge is the excess of negative or positive electricity on a body or in space. If the excess is negative, the body is said to have a negative charge and vice versa.

An electron is an elementary particle charged with a small and constant quantity of negative electricity. A proton is similarly defined but charged with positive electricity while the neutron is uncharged and is therefore neutral. In an atom the number of electrons normally equals the number of protons; it is the number of protons that determines to which element type the atom belongs. An atom can have one or more electrons added to it or taken away. This does not change its elemental classification but it disturbs its electrical balance. If the atom has excess electrons, it is said to be negatively charged. A charged atom is called an ion.

A body containing a number of ionized atoms is also said to be electrically charged. It can be shown that positively and negatively charged bodies are mutually attracted to one another while similarly charged bodies repel one another.

2.4	Movement of electrons

All electrons have a certain potential energy. Given a suitable medium in which to exist, they move freely from one energy level to another and this movement, when undertaken in a concerted manner, is termed an electric current flow. Conventionally it is said that the current flows from a point of high energy level to a point of low energy level. These points are said to have high potential and low potential respectively. For convenience the point of high potential is termed the positive and the point of low potential is termed the negative, hence conventionally a current is said to flow from positive to negative.

This convention was in general use long before the nature of electric charge was discovered. Unfortunately it was found that electrons move in the other direction since the negatively charged electron is attracted to the positive potential. Thus conventional current flows in the opposite direction to that of electron current. Normally only conventional current is described by the term current and this will apply throughout the text.

The transfer of electrons takes place more readily in a medium in which atoms can readily release electrons, e.g. copper, aluminium, silver, etc. Such a material is termed a conductor. A material that does not readily permit electron flow is termed an insulator, e.g. porcelain, nylon, rubber, etc. There is also a family of materials termed semiconductors which have certain characteristics that belong to neither of the other groups.

For most practical applications it is necessary that the current flow continues for as long as it is required; this will not happen unless the following conditions are fulfilled:

1. There must be a complete circuit around which the electrons may move. If the electrons cannot return to the point of starting, then eventually they will all congregate together and the flow will cease.

2. There must be a driving influence to cause the continuous flow. This influence is provided by the source which causes the current to leave at a high potential and to move round the circuit until it returns to the source at a low potential. This circuit arrangement is indicated in Fig. 2.5.

The driving influence is termed the electromotive force, hereafter called the e.m.f. Each time the charge passes through the source, more energy is provided by the source to permit it to continue round once more. This is a continuous process since the current flow is continuous. It should be noted that the current is the rate of flow of charge through a section of the circuit.

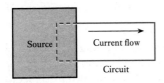

Fig. 2.5 Elementary circuit

The e.m.f. represents the driving influence that causes a current to flow. The e.m.f. is not a force, but represents the energy expended during the passing of a unit charge through the source; an e.m.f. is always connected with energy conversion.

The energy introduced into a circuit is transferred to the load unit by the transmission system, and the energy transferred due to the passage of unit charge between two points in a circuit is termed the potential difference (p.d.). If all the energy is transferred to the load unit, the p.d. across the load unit is equal to the source e.m.f.

It will be observed that both e.m.f. and p.d. are similar quantities. However, an e.m.f. is always active in that it tends to produce an electric current in a circuit while a p.d. may be either passive or active. A p.d. is passive whenever it has no tendency to create a current in a circuit.

Unless it is otherwise stated, it is usual to consider the transmission system of a circuit to be ideal, i.e. it transmits all the energy from the source to the load unit without loss. Appropriate imperfections will be considered later.

Certain conventions of representing the e.m.f. and p.d. in a circuit diagram should be noted. Each is indicated by an arrow as shown in Fig. 2.6. In each case, the arrowhead points towards the point of high (or assumed higher) potential. It is misleading to show an arrowhead at each end of

Fig. 2.6 Circuit diagram conventions

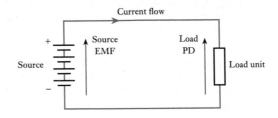

the line as if it were a dimension line. An arrowhead is drawn on the transmission system to indicate the corresponding direction of conventional current flow.

It will be seen that the current flow leaves the source at the positive terminal and therefore moves in the same direction as indicated by the source e.m.f. arrow. The current flow enters the load at the positive terminal, and therefore in the opposite direction to that indicated by the load p.d. arrow. Energy is converted within the load unit and, depending on the nature of this conversion, the p.d. may be constituted in a variety of ways. It is sufficient at first to consider the p.d. as the change in energy level across the terminals of the load unit. This is termed a volt drop since the p.d. (and e.m.f.) are measured in volts.

In Fig. 2.6, the source indicated consists of a battery which delivers direct current, i.e. current which flows in one direction. The source in Fig. 2.3 was shown as a circle which indicates that a rotating machine provided the current. A general symbol for any type of source of direct current is shown in Fig. 2.7.

Fig. 2.7 General symbol for d.c. source

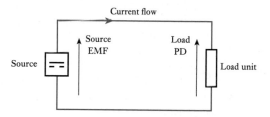

2.7   **Electrical units**

The unit of current is the *ampere* and is one of the SI base units mentioned in section 1.1.

(a) Current

The ampere is defined as that *current which, if maintained in two straight parallel conductors of infinite length, of negligible circular cross-section, and placed 1 metre apart in a vacuum, would produce between these conductors a force of 2 × 10⁻⁷ newtons per metre of length.* The conductors are attracted towards each other if the currents are in the same direction, whereas they repel each other if the currents are in opposite directions.

Current   Symbol: I Unit: ampere (A)

This definition is outstanding for its complexity. However, by using such a definition, most of the electrical units take on suitable magnitudes. The figure of 2×10^{-7} is therefore one of convenience and the definition will be explained in section 7.3.

The value of the current in terms of this definition can be determined by means of a very elaborately constructed balance in which the force between fixed and moving coils carrying the current is balanced by the force of gravity acting on a known mass.

(b) Quantity of electricity

The unit of electrical quantity is the *coulomb*, namely the *quantity of electricity passing a given point in a circuit when a current of 1 ampere is maintained for 1 second*. Hence

$$Q \, [\text{coulombs}] = I \, [\text{amperes}] \times t \, [\text{seconds}]$$

$\therefore \qquad Q = It$ [2.1]

Charge   Symbol: Q Unit: coulomb (C)

From equation [2.1], it can be seen that the coulomb is an ampere second. Batteries are used to hold charge but they are usually rated in ampere hours.

$$1 \text{ ampere hour} = 3600 \text{ coulombs}$$

The rate of charge passing a point is the current but it has become common practice to describe a flow of charge as a current. We have already met this misuse in the last paragraph of (a) above when it was said that the coils were carrying a current. Thus we shall find the term 'current' being used to indicate both a flow of charge and also the rate of flow of charge. It sounds confusing but fortunately it rarely gives rise to difficulties.

Example 2.1

If a charge of 25 C passes a given point in a circuit in a time of 125 ms, determine the current in the circuit.

From equation [2.1]

$$Q = It$$

$$\therefore \qquad I = \frac{Q}{t} = \frac{25}{125 \times 10^{-3}} = 200 \text{ A}$$

(c) Potential difference

The unit of potential difference is the *volt*, namely *the difference of potential between two points of a conducting wire carrying a current of 1 ampere, when the power dissipated between these points is equal to 1 watt*.

The term *voltage* originally meant a difference of potential expressed in volts, but it is now used as a synonym for potential difference irrespective of the unit in which it is expressed. For instance, the voltage between the lines of a transmission system may be 400 kV, while in communication and electronic circuits the voltage between two points may be 5 μV. The term *potential difference* is generally abbreviated to *p.d.*

Electric potential   Symbol: V Unit: volt (V)

Electromotive force has the symbol E but has the same unit. Because p.d.s are measured in volts, they are also referred to as volt drops or voltages. By experiment, it can be shown that the relation corresponding to the definition is

$$V = \frac{P}{I}$$

This is better expressed as

$$P = VI$$ [2.2]

It also follows that

$$V = \frac{P}{I} = \frac{W}{t} \cdot \frac{t}{Q}$$

$\therefore$
$$V = \frac{W}{Q}$$ [2.3]

That is, the p.d. is equal to the energy per unit charge.

Example 2.2 A circuit delivers energy at the rate of 20 W and the current is 10 A. Determine the energy of each coulomb of charge in the circuit.

From [2.2]

$$V = \frac{P}{I} = \frac{20}{10} = 2 \text{ V}$$

From [2.3]

$$W = VQ = 2 \times 1 = \mathbf{2\,J}$$

(d) Resistance

The unit of electric resistance is the *ohm*, namely *the resistance between two points of a conductor when a potential difference of 1 volt, applied between these points, produces in this conductor a current of 1 ampere, the conductor not being a source of any electromotive force.*

Alternatively, the *ohm* can be defined as *the resistance of a circuit in which a current of 1 ampere generates heat at the rate of 1 watt.*

Electric resistance   Symbol: R Unit: ohm (Ω)

If V represents the p.d., in volts, across a circuit having resistance R, in ohms, carrying a current I, in amperes, for time t, in seconds,

$$V = IR$$ [2.4]

or

$$I = \frac{V}{R}$$

Power $$P = IV = I^2 R$$ [2.5]

$$= \frac{V^2}{R}$$

Also the energy dissipated is given by

$$W = Pt = I^2 Rt = IVt$$

Example 2.3	A current of 5 A flows in a resistor of resistance 8 Ω. Determine the rate of heat dissipation and also the heat dissipated in 30 s.

$$P = I^2 R = 5^2 \times 8 = \mathbf{200\ W}$$

$$W = Pt = 200 \times 30 = \mathbf{6000\ J}$$

(e) Electromotive force

An electromotive force is that which tends to produce an electric current in a circuit, and the *unit of e.m.f.* is the *volt*.

Electromotive force   Symbol: E Unit: volt (V)

The principal sources of e.m.f. are as follows:

1. The electrodes of dissimilar materials immersed in an electrolyte, as in primary and secondary cells, i.e. batteries.
2. The relative movement of a conductor and a magnetic flux, as in electric generators; this source can, alternatively, be expressed as the variation of magnetic flux linked with a coil (sections 6.10 and 8.3).
3. The difference of temperature between junctions of dissimilar metals, as in thermo-junctions.

2.8 **Ohm's law**	

One of the most important steps in the analysis of the circuit was undertaken by Georg Ohm, who found that the p.d. across the ends of many conductors is proportional to the current flowing between them. This, he found, was a direct proportionality, provided that temperature remained constant. Since the symbol for current is I, this relationship may be expressed as

$$V \propto I \qquad\qquad\qquad [2.6]$$

Relation [2.6] is the mathematical expression of what is termed Ohm's law.

Subsequent experimental evidence has shown that many other factors affect this relationship, and that in fact few conduction processes give a direct proportionality between p.d. and current. However, this relationship is almost constant for many electrical circuits and it is convenient at this introductory stage to consider only circuits in which the relationship is constant. The corresponding characteristic is shown in Fig. 2.8.

Fig. 2.8 Constant potential difference/current characteristic and the circuit from which it is obtained

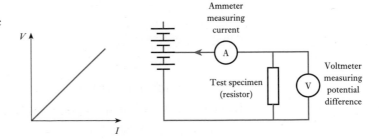

Since the relationship is assumed constant, then

$$\frac{V}{I} = R$$

where R is a constant termed the resistance of the conductor. The boxes used in Figs 2.6 and 2.7 are generally used to represent a load with resistance properties. The expression involving R is usually expressed as

$$V = IR \qquad\qquad [2.4]$$

It should be noted that this relationship is derived from Ohm's law and is not a symbolic expression for it. Ohm's law only notes the constancy of p.d. to current provided that other physical factors remain unchanged, i.e. for a given p.d. the current will vary in consequence of variation of external physical factors.

Example 2.4

A motor gives an output power of 20 kW and operates with an efficiency of 80 per cent. If the constant input voltage to the motor is 200 V, what is the constant supply current?

$$P_o = 20\,000 \text{ W}$$

$$P_{in} = \frac{P_o}{\eta} = \frac{20\,000}{0.8} = 25\,000 \text{ W} = VI$$

$$I = \frac{25\,000}{200} = \textbf{125 A}$$

Example 2.5

A 200 t train experiences wind resistance equivalent to 62.5 N/t. The operating efficiency of the driving motors is 0.87 and the cost of electrical energy is 8 p/kW h. What is the cost of the energy required to make the train travel 1 km?

If the train is supplied at a constant voltage of 1.5 kV and travels with a velocity of 80 km/h, what is the supply current?

In moving 1 km

$$W_o = Fl$$

$$= 200 \times 62.5 \times 1000 = 12.5 \times 10^6 \text{ J}$$

$$W_{in} = \frac{W_o}{\eta} = \frac{12.5 \times 10^6}{0.87} = 14.4 \times 10^6 \text{ J}$$

But 1 kW h $= 3.6 \times 10^6$ J, hence

$$W_{in} = \frac{14.4 \times 10^6}{3.6 \times 10^6} = 4.0 \text{ kW h}$$

$\therefore \qquad$ Cost of energy $= 8.0 \times 4.0 = \textbf{32 p}$

Work done in 1 h when moving with a velocity of 80 km/h is

$$(14.4 \times 10^6 \times 80) \text{ J}$$

Work done in 1 s (equivalent to the input power) is

$$\frac{14.4 \times 10^{6} \times 80}{60 \times 60} = 320 \times 10^{3} \text{ W} = P_{\text{in}}$$

But $P_{\text{in}} = VI$

$\therefore$ $I = \dfrac{P_{\text{in}}}{V} = \dfrac{320 \times 10^{3}}{1.5 \times 10^{3}} = \mathbf{214 \text{ A}}$

2.9   Resistors

A resistor is a device which provides resistance in an electrical circuit. The resistance of a resistor is said to be linear if the current through the resistor is proportional to the p.d. across its terminals. If the resistance were to vary with the magnitude of either the voltage or the current, the resistor is said to be non-linear. Resistors made from semiconductor materials (see Chapter 20) are examples of non-linear resistors.

In this book resistors may be assumed linear, i.e. their resistance will be assumed to remain constant when the temperature is maintained constant. We cannot ignore the effect of temperature since all resistors dissipate heat when operating, i.e. if a resistor passes a current I, then energy is brought into the resistor at the rate $I^{2}R$. In order to release this heat energy, the resistor must become warmer than its surroundings until it can release the heat energy at the same rate as it is arriving. Therefore we have to assume that the effect of this temperature rise is negligible.

All resistors have a power rating which is the maximum power that can be dissipated without the temperature rise being such that damage occurs to the resistor. Thus a 1 W resistor with a resistance of 100 Ω can pass a current of 100 mA whereas a $\frac{1}{4}$ W resistor with the same resistance could only handle a current of 50 mA. In either case, if the current level were exceeded for any length of time, the resistor would overheat and might burn out.

Conductor wires and cables are similarly rated. Although we like to assume that a conductor has no resistance, in fact all have some resistance. The passing of a current therefore causes the conductor to heat, and if the heating effect is too great the insulant material can be damaged. The rating is therefore determined by the temperature which the insulant can withstand.

Therefore if we wish to purchase a resistor, we require to specify not only the required resistance but also the power rating. In electronic circuits, the common standard rating are $\frac{1}{4}$ W, $\frac{1}{2}$ W, 1 W and 2 W. Thus if we required a resistor to dissipate 1.1 W, we would select a 2 W resistor since the rating must exceed the operational value. In power circuits, much higher ratings up to a megawatt and more can be experienced, but such operational conditions are expensive since we would be paying for energy which we are throwing away. Thus power engineers avoid the use of resistors as much as possible, seeking to utilize energy with minimal loss.

For the purpose of this book we will assume that the ratings have been correctly specified, and therefore resistors will only have their resistances given. Resistors can be made in a variety of ways but all fall into the following categories given in Fig. 2.9. Most fixed resistors (also called non-variable resistors) are used in electronic circuits and are made from carbon mouldings or from metal-oxide film. These will be considered further in section 2.10, but they have the common feature of having low power ratings. When ratings

Symbol	Representing
▭	Fixed resistor
⌁⌁⌁	Resistor symbol found in old diagrams, no longer used
or	Variable resistor (or rheostat)
	Potentiometer

Fig. 2.9 Resistor types and symbols

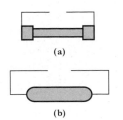

Fig. 2.10 Wire-wound resistors. (a) Cement coated on a ceramic former; (b) vitreous enamel coated on a ceramic former

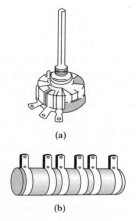

Fig. 2.11 Variable resistors. (a) Wire-wound; (b) mains 'dropper' resistor with fixed tappings; cement coated on a ceramic former

of more than 1 W are required, this often involves the winding of a thin nichrome wire on a ceramic former. As we shall see in section 3.6, the thinner and the longer then the greater is its resistance. Usually the wire is of a very fine gauge and the coil of wire is coated with a cement or a vitreous enamel. Such resistors can operate up to 10 or 20 W and are shown in Fig. 2.10. At the top end of the scale, when hundreds of watts and more are involved, the resistors resemble lumps of cast iron or bent metal bars held in cages so that the air can circulate and take away the waste heat.

Wire-wound resistors may be wound on what looks like a large washer made from an insulant material such as card. An arm is mounted through the centre of the washer. By rotating the arm the length of wire between the point of contact and the end of the coil is varied, hence the resistance is varied. Usually such variable resistors have three connections, being each end of the coil plus the connection from the wiper arm. When all three connections are used, the device is said to be a potentiometer, as shown in Fig. 2.11.

In some instances, fixed connections are made to the wire so that we know the resistance offered between any two terminals. Such intermediate connections are known as tappings. A typical tapped resistor, still often referred to by its old name – the rheostat – is shown in Fig. 2.11.

2.10 Resistor coding

We have already noted that there are resistors made from carbon mouldings or from metal-oxide film. Both are small, if not very small, and therefore we would find it almost impossible to mark them with a rating such as 47 000 Ω, $\frac{1}{4}$ watt.

In the case of carbon resistors, it is usual to identify the ratings by means of rings painted around the resistors, as shown in Fig. 2.12. One of the bands is always placed near to the end of the resistor and should be taken as the first band. The first, second and third bands are used to indicate the resistance of the resistor by means of a colour code which is also given in Fig. 2.12.

The application of this code is best explained by the example shown in Fig. 2.12. Here the first two bands are orange and blue which, from the table, are 3 and 6 respectively. Therefore we are being told that the resistance has a numerical value of 36. The third band tells us how many zeros to put after that number. In this case, the third band is green and there should be five zeros, i.e. the resistance is 3 600 000 Ω.

Fig. 2.12 Colour coding of resistors

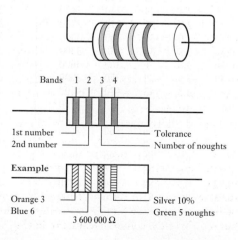

Digit	Colour
0	Black
1	Brown
2	Red
3	Orange
4	Yellow
5	Green
6	Blue
7	Violet
8	Grey
9	White

Tolerance	Colour
5%	Gold
10%	Silver
20%	No colour band

Let us look at similar examples. If the first three bands were yellow, violet and red, the resistance would be 4700 Ω. However, if the colours were orange, black and gold then the resistance would be 3.0 Ω.

The fourth band is the tolerance which is the extent to which the actual value of the resistance can vary relative to nominal value indicated by the first three bands. Thus for the resistor shown in Fig. 2.12, the fourth band is silver which indicates that the resistance has a value between 3 600 000 + 10 per cent and 3 600 000 − 10 per cent. Ten per cent of 3 600 000 is 360 000. Therefore the resistance will be somewhere between 3 960 000 and 3 240 000 Ω.

This may seem an unexpected range of values, but it reflects the problems in manufacturing resistors with specific resistances. Fortunately this is not a problem in electronic circuits, and later we will see that the variation of resistance values can be compensated with little detriment to circuit performance.

Some resistors have a fifth band which indicates a reliability factor which is the percentage of failure per 1000 h of use. For instance, a 1 per cent failure rate would suggest that one from every hundred resistors will not remain with tolerance after 1000 h of use.

Band 5 colours indicate the following percentages.

1	Brown
0.1	Red
0.01	Orange
0.001	Yellow

A list of readily available standard values appears in Table 2.1. All are available with 5 per cent tolerance while those in bold type are also available with 10 per cent tolerance.

Table 2.1 Standard values of available resistors

Ohms (Ω)					Kilohms (kΩ)		Megohms (MΩ)	
0.10	**1.0**	**10**	**100**	**1000**	**10**	**100**	**1.0**	**10.0**
0.11	1.1	11	110	1100	11	110	1.1	11.0
0.12	**1.2**	**12**	**120**	**1200**	**12**	**120**	**1.2**	**12.0**
0.13	1.3	13	130	1300	13	130	1.3	13.0
0.15	**1.5**	**15**	**150**	**1500**	**15**	**150**	**1.5**	**15.0**
0.16	1.6	16	160	1600	16	160	1.6	16.0
0.18	**1.8**	**18**	**180**	**1800**	**18**	**180**	**1.8**	**18.0**
0.20	2.0	20	200	2000	20	200	2.0	20.0
0.22	**2.2**	**22**	**220**	**2200**	**22**	**220**	**2.2**	**22.0**
0.24	2.4	24	240	2400	24	240	2.4	
0.27	**2.7**	**27**	**270**	**2700**	**27**	**270**	**2.7**	
0.30	3.0	30	300	3000	30	300	3.0	
0.33	**3.3**	**33**	**330**	**3300**	**33**	**330**	**3.3**	
0.36	3.6	36	360	3600	36	360	3.6	
0.39	**3.9**	**39**	**390**	**3900**	**39**	**390**	**3.9**	
0.43	4.3	43	430	4300	43	430	4.3	
0.47	**4.7**	**47**	**470**	**4700**	**47**	**470**	**4.7**	
0.51	5.1	51	510	5100	51	510	5.1	
0.56	**5.6**	**56**	**560**	**5600**	**56**	**560**	**5.6**	
0.62	6.2	62	620	6200	62	620	6.2	
0.68	**6.8**	**68**	**680**	**6800**	**68**	**680**	**6.8**	
0.75	7.5	75	750	7500	75	750	7.5	
0.82	**8.2**	**82**	**820**	**8200**	**82**	**820**	**8.2**	
0.91	9.1	91	910	9100	91	910	9.1	

Many resistors, especially the metal-oxide resistors, can be so small or of the wrong shape so that the colour coding is difficult to apply. Instead a letter code is used and this is best explained by examination of the following list:

Resistance	Marking
0.47 Ω	R47
4.7 Ω	4R7
47 Ω	47R
470 Ω	470R
4.7 kΩ	4K7
47 kΩ	47K
4.7 MΩ	4M7

Example 2.6

A resistor is marked
1st band   Brown
2nd band   Black
3rd band   Orange
No other band
What is its resistance and between what values does it lie?

Brown   (1) = 1   first unit
Black   (0) = 0   second unit
Orange   (3) =   000   number of zeros
                10000 = 10 kΩ

Since no further band is given the tolerance is ±20 per cent. The resistance lies between 10 000 + 2000 and 10 000 − 2000, i.e. **12 kΩ** and **8 kΩ**.

2.11   Conductors and insulators

So far, we have assumed that the current in a circuit will move around the circuit as desired. But why should the electric charge pass along the wires and not leak away as water might leak out from a faulty pipe? Or, if you cut open an electrical insulated wire, why are there copper strands in the middle surrounded by plastic and not plastic surrounded by copper?

In part, the answers to these questions have already been given in the consideration of the structure of atoms and their electron shells. However, more direct answers to the questions can be found by looking at a simple experiment.

Make a number of rods of different materials so that each is identical in size and shape. Connect each in turn to a battery which provides a source of constant e.m.f., and measure the resulting current by a device called an ammeter. The circuit arrangement is shown in Fig. 2.13.

By trying rods of different materials, it can soon be seen that those rods made from metals permit quite reasonable currents to flow while those made from non-metallic materials permit virtually no current to flow. Not

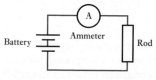

Fig. 2.13   Experimental circuit

Table 2.2 Typical electrical materials

Conductors	Insulators
Copper	Glass
Aluminium	Rubber
Silver	Plastic
Platinum	Air
Bronze	Varnish
Gold	Paper
	Wood
	Mica
	Ceramic
	Certain oils

all the metals conduct as well as each other, copper being better than steel, aluminium being better than zinc and so on.

In this simple experiment, the non-metallic materials permit so little current to flow that no comparison can be made between them, but nevertheless the observation may be made that there are certain materials which permit current to flow and others which do not. Those materials which permit current to flow are the conductors, while those that do not permit current to flow are the insulators. Common examples of each are given in Table 2.2.

This classification is rather over-simplified because no material completely stops the flow of current, just as no material permits the passage of charge without some opposition. However, recalling the insulated wire, the copper provides an easy path and the charge which would leak away through the insulating plastic covering is negligible by comparison. Just how negligible will become apparent in later studies, but in these initial stages it is reasonable to consider the current as remaining within the conductors.

Therefore the function of the conductors is to provide a complete circuit at all points where there is material with free electrons. If at any part of the circuit free electrons are not available, and if they cannot readily be introduced into the material, then current will not flow in the circuit. Materials with no energy gap readily provide the free electrons and are used to make up a circuit, but those materials with sizeable energy gaps between the valence and conduction bands are used to insulate the circuit and to contain the current within the conductors.

2.12   The electric circuit in practice

Before finishing this introduction to electricity, there are certain practical comments that require to be made. For instance, where do we get this electricity from?

The most common sources of electricity are the generating stations, most of which are operated by electricity suppliers. However a surprising amount of electrical energy is generated by other commercial and private enterprises for their own use and possibly as much as 20 per cent of all the electrical energy generated in the UK comes from such sources.

Generating stations contain rotating electric machines which generate the electricity and these are driven either by steam turbines or by water turbines. This latter form is described as hydroelectric.

Generated electricity is transmitted from one place to another by power lines, in which the conductors are held up in the air clear of all activities on the ground, or by cables, which are buried in the ground. In this way, the electricity is brought to our industries and to our homes.

The need for conductor systems limits our ability to move electrical devices about, i.e. to make them portable. In such cases, we use batteries. Batteries can be used for torches, transistor radios, hearing aids, cameras, watches and cars. In most of these examples we use a form of battery that provides an amount of energy after which the battery is thrown away. Such batteries are called primary cells. As a method of purchasing energy, they are extremely expensive, but we are prepared to pay the appropriate costs for the sake of convenience.

Primary cells can only provide small amounts of energy, and an alternative form of battery is required when more demanding tasks are to be undertaken, e.g. the starting of a car engine. In such instances, we use a form of battery made from a group of secondary cells and these have the advantages both of being able to give larger amounts of energy and to be recharged with energy. However, such devices are heavy and can therefore only be used in such application as cars.

Whether we use a generator or a battery, we nevertheless require a circuit in which to utilize the available energy, and there are two circuit conditions that are of extreme importance. These occur when the resistance is at its lowest value of $0\ \Omega$ and at its highest value of infinity. In the first instance there is no limit to the current that flows, the volt drop being zero ($V = IR = I \times 0$), and the circuit condition is termed a short-circuit. It follows that if two points are connected by a conductor of zero resistance they are said to be short-circuited.

In the second instance, no current can flow through an infinite resistance ($I = V/R = V/\infty = 0$). This circuit condition is termed an open-circuit, and if two points are connected by a conductor of infinite resistance they are said to be open-circuited. Effectively this means that there is no connection between them. For example, if the wire of a circuit is broken, there is no connection across the break and the circuit is open-circuited.

Summary of important formulae	Electric charge $Q = It$ (coulombs)	[2.1]
	Voltage $V = P/I$ (volts)	[2.2]
	$V = W/Q$	[2.3]
	$V = IR$	[2.4]
	Power $P = IV = I^2R = V^2/R$ (watts)	[2.5]

Terms and concepts

Current is the rate of flow of electric charge in a circuit. The term is often used to describe the flow of electric charge, e.g. 'a current is flowing in a circuit'; this is ambiguous but is so common that we have to accept it.

A **source** supplies energy to a system.

A **load** accepts energy from a system.

Electric charge may be either positive or negative. Negative electrons are free to move around a circuit thus transporting energy from source to load.

To maintain a current, the source must provide a driving force called the **electromotive force** (e.m.f.).

The **potential difference** across a load indicates in volts the energy lost per coulomb of charge passing through the load.

Since the current is the rate of flow, its product with the voltage gives the rate of energy transmission, i.e. the **power**.

Resistance is a measure of the opposition to the flow of charge through a load.

Ohm's law states that the ratio of voltage to current is constant, provided other physical factors such as temperature remain unchanged.

The resistances of resistors can be identified by a code system.

Exercises 2

1. What is a simpler way of expressing 0.000 005 A?
2. What is a simpler way of expressing 3 000 000 V?
3. A p.d. of 6 V causes a current of 0.6 A to flow in a conductor. Calculate the resistance of the conductor.
4. Find the p.d. required to pass a current of 5 A through a conductor of resistance 8 Ω.
5. A 960 Ω lamp is connected to a 240 V supply. Calculate the current in the lamp.
6. A p.d. of 1.35 V causes a current of 465 μA to flow in a conductor. Calculate the resistance of the conductor.
7. An accidental short-circuit to a 240 V supply is caused by the connection of a component of 8.5 mΩ across the supply terminals. What will be the short-circuit current?
8. A p.d. of 24 V is applied to a 4.7 kΩ resistor. Calculate the circuit current.
9. What is the voltage across an electric heater of resistance 5 Ω through which passes a current of 22 A?
10. Calculate the current in a circuit due to a p.d. of 10 V applied to a 10 kΩ resistor. If the supply voltage is doubled while the circuit resistance is trebled, what is the new current in the circuit?

11. A current in a circuit is due to a p.d. of 10 V applied to a resistor of resistance 100 Ω. What resistance would permit the same current to flow if the supply voltage were 100 V?
12. For the V/I characteristic shown in Fig. A, calculate the resistance of the load.

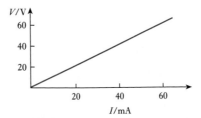

Fig. A

13. Plot the V/I characteristic for a 4.7 Ω resistor, given that the applied voltage range is 0–5 V.
14. From the V/I characteristic shown in Fig. B, derive the R/I characteristic of the load.
15. What is the charge transferred in a period of 8 s by current flowing at the rate of 2.5 A?

Exercises 2 continued

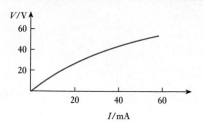

Fig. B

16. A p.d. of 12 V is applied to a 7.5 Ω resistor for a period of 5 s. Calculate the electric charge transferred in this time.

17. A voltage of 20 V is required to cause a current of 2 A to flow in a resistor of resistance 10 Ω. What voltage would be required to make the same current flow if the resistance were 40 Ω?

18. A d.c. motor connected to a 240 V supply is developing 20 kW at a speed of 900 r/min. Calculate the useful torque.

19. If the motor referred to in Q. 18 has an efficiency of 0.88, calculate (a) the current and (b) the cost of the energy absorbed if the load is maintained constant for 6 h. Assume the cost of electrical energy to be 8.0 p/kW h.

20. (a) An electric motor runs at 600 r/min when driving a load requiring a torque of 200 Nm. If the motor input is 15 kW, calculate the efficiency of the motor and the heat lost by the motor per minute, assuming its temperature to remain constant.

(b) An electric kettle is required to heat 0.5 kg of water from 10 °C to boiling point in 5 min, the supply voltage being 230 V. If the efficiency of the kettle is 0.80, calculate the resistance of the heating element. Assume the specific heat capacity of water to be 4.2 kJ/kg K.

21. A pump driven by an electric motor lifts 1.5 m³ of water per minute to a height of 40 m. The pump has an efficiency of 90 per cent and the motor an efficiency of 85 per cent. Determine: (a) the power input to the motor; (b) the current taken from a 480 V supply; (c) the electrical energy consumed when the motor runs at this load for 8 h. Assume the mass of 1 m³ of water to be 1000 kg.

22. An electric kettle is required to heat 0.6 litre of water from 10 °C to boiling point in 5 min, the supply voltage being 240 V. The efficiency of the kettle is 78 per cent. Calculate: (a) the resistance of the heating element; (b) the cost of the energy consumed at 8.0 p/kW h. Assume the specific heat capacity of water to be 4190 J/kg K and 1 litre of water to have a mass of 1 kg.

23. An electric furnace is to melt 40 kg of aluminium per hour, the initial temperature of the aluminium being 12 °C. Calculate: (a) the power required, and (b) the cost of operating the furnace for 20 h, given that aluminium has the following thermal properties: specific heat capacity, 950 J/kg K; melting point, 660 °C; specific latent heat of fusion, 450 kJ/kg. Assume the efficiency of the furnace to be 85 per cent and the cost of electrical energy to be 8.0 p/kW h.

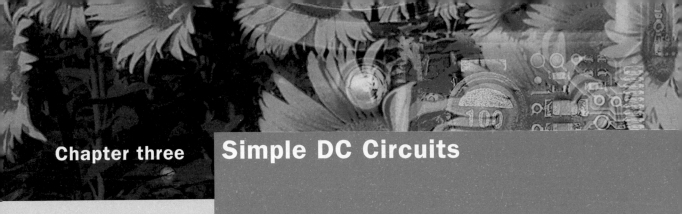

Chapter three

Simple DC Circuits

Objectives

When you have studied this chapter, you should
- recognize series- and parallel-connected loads
- have an understanding that series-connected loads all pass the same current
- have an understanding that parallel-connected loads all share the same applied voltage
- be familiar with Kirchhoff's laws
- be capable of analysing relatively simple circuits and networks containing series- and parallel-connected loads
- have an understanding of the power and energy associated with electric circuits and networks
- be capable of analysing the power and energy associated with loads passing current
- be familiar with the temperature coefficient of resistance
- be aware of the effects of temperature rise in electrical components

Contents

Life is rarely simple and so it is with electrical technology. Our simple circuit with a single source and a single load seldom exists except in battery torches. It follows that we should look to systems with an increased number of loads. This does not require a huge increase – it is generally sufficient to be able to handle two or three loads at a time but this will let us appreciate the problems encountered in practice.

To handle two or more loads, we need to become adept at recognizing series-connected loads and parallel-connected loads. This is best undertaken by first considering a considerable number of worked problems and then trying more problems until such operations become second nature. To do this, we are helped by two principles termed Kirchhoff's laws. We shall find that we use them almost all the time whenever we analyse electrical circuits so they are very important to us.

We have already noted that resistors produce heat when passing current and this activity requires further investigation. So there is a lot to be done in this chapter.

3.1   Series circuits

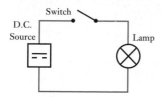

Fig. 3.1   Simple lamp circuit

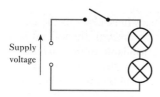

Fig. 3.2   Incorrect lamp connection

So far we have been introduced to the most simple circuit possible. In particular, the circuit contained only one load unit or resistor. We now need to consider what happens when there are two or three load units.

Let us consider what might happen if the load unit were a lamp bulb as shown in Fig. 3.1. In practice, we might find that the light output was insufficient and decide to add in a second lamp to give more light. Let us wire in this second lamp as shown in Fig. 3.2. It looks as though it should work satisfactorily because the current from the first lamp now passes on to the second. But when we switch on, both lamps give out very little light. What has gone wrong?

We have already noted that the passing of a current through the lamp filament wire can make it so hot that it gives out light. This only happens if there is sufficient current to bring the temperature of the filament up to about 3000 °C. However, if the current is insufficient then there is only a dull glow from the filament because the temperature has not risen sufficiently. This is what has happened with the circuit shown in Fig. 3.2 – but why then is the current too small?

In the original circuit the lamp operated at its normal brightness by passing a certain current. This current was determined by the supply voltage in conjunction with the resistance of the filament wire, the volt drop across the filament being equal to that of the applied voltage. Since these two voltages are equal, then, when the second lamp is inserted into the circuit, there will be no voltage 'left over' to permit the current to even flow through the lamp. In a complete circuit, it would not be possible for the current to flow in part of the circuit and not the remainder. Some compromise must therefore be reached whereby the first lamp 'uses up' less of the supply voltage and leaves some for the second lamp. This compromise can be achieved by having less current in the circuit; in this way, there is less voltage drop across the first lamp and what remains can be used to pass the current on through the second lamp. The current must be just the right amount to obtain the correct balance in each lamp.

Owing to the lack of certain technical statements that can be made about electric circuits, this explanation has been made in a rather non-technical manner. It has also involved assumptions that can be shown only by experiment. For instance, it has been assumed that the current in the first lamp is the same as the current in the second lamp, i.e. that the current must be the same in all parts of the circuit. This is illustrated by the experimental circuit shown in Fig. 3.3.

In the circuit, one ammeter is connected to measure the current in lamp 1 while another ammeter measures the current in lamp 2. By observation, it can be seen that no matter what voltage is applied to the input, the current registered on each ammeter is the same. The current in each lamp is

Fig. 3.3   Observation of a current in a circuit

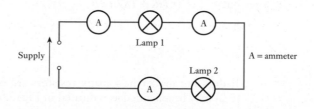

Fig. 3.4 Investigation of volt drops in a series circuit

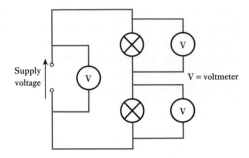

therefore the same, which is reasonable because where else can the current flow having passed through the first lamp but on through the second lamp, there being no other conduction path shown as available to the current?

Further experiments can be carried out to show that the same result is obtained no matter how many lamps are connected in this way. When lamps, or any other forms of load units, are connected in this manner, they are said to be connected in *series*. In series circuits, the current must pass through each and every one of the components so connected.

A further assumption was that the volt drops across each of the lamps add up to give the total supply voltage. Again a simple experiment may be arranged to illustrate this point as shown in Fig. 3.4. A voltmeter is connected across each of the lamps while a further voltmeter is connected across the supply. Different supply voltages are applied to the circuit, and it is observed that no matter what supply voltage is applied, it is always equal to the sum of the voltmeter readings across the loads. As with the investigation of current, it is of no consequence how many load units are connected in series: in each case the sum of volt drops across each of the components is equal to the total supply voltage.

Replacing the lamps with simple resistive loads, as shown in Fig. 3.5, and by using the notation shown, it can be observed that

$$V = V_1 + V_2 + V_3 \qquad [3.1]$$

Since, in general, $V = IR$, then $V_1 = IR_1$, $V_2 = IR_2$ and $V_3 = IR_3$, the current I being the same in each resistor. Substituting in equation [3.1],

$$V = IR_1 + IR_2 + IR_3$$

For the complete circuit, the effective resistance of the load R represents the ratio of the supply voltage to the circuit current whence

$$V = IR$$

but $\qquad V = IR_1 + IR_2 + IR_3$

hence $\qquad IR = IR_1 + IR_2 + IR_3$

and $\qquad R = R_1 + R_2 + R_3 \qquad [3.2]$

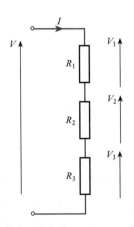

Fig. 3.5 Volt drops in a series circuit

It does not matter how many resistors are connected in series, as relation [3.2] may be amended as indicated in Fig. 3.6.

Fig. 3.6 Series–connected resistors

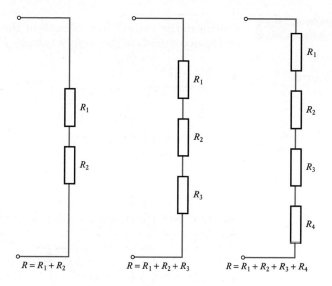

$$R = R_1 + R_2 \qquad R = R_1 + R_2 + R_3 \qquad R = R_1 + R_2 + R_3 + R_4$$

Example 3.1

Calculate for each of the circuits shown in Fig. 3.7 the current flowing in the circuit given that $R = 3 \text{ k}\Omega$.

Fig. 3.7 Circuit diagrams for Example 3.1

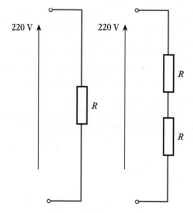

220 V 220 V

In the first case

$$I = \frac{V}{R} = \frac{220}{3 \times 10^3} = 0.073 \text{ A} = 73 \text{ mA}$$

In the second case the circuit resistance is given by

$$R = R_1 + R_2 = 3 \times 10^3 + 3 \times 10^3 = 6000 \ \Omega$$

$$I = \frac{V}{R} = \frac{220}{6000} = 0.037 \text{ A} = 37 \text{ mA}$$

Notice that doubling the circuit resistance has halved the current. This is similar to the effect with the lamps in Fig. 3.2 where the two lamps in series halved the current with the resulting diminution of the output of light.

| **Example 3.2** | Calculate the voltage across each of the resistors shown in Fig. 3.8 and hence calculate the supply voltage V. |

Fig. 3.8   Circuit diagram for Example 3.2

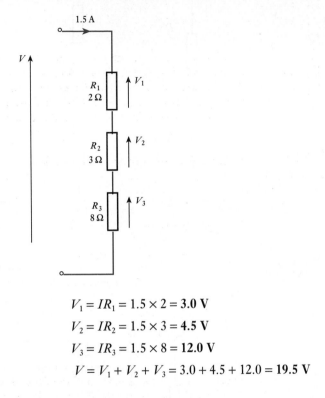

$$V_1 = IR_1 = 1.5 \times 2 = \textbf{3.0 V}$$
$$V_2 = IR_2 = 1.5 \times 3 = \textbf{4.5 V}$$
$$V_3 = IR_3 = 1.5 \times 8 = \textbf{12.0 V}$$
$$V = V_1 + V_2 + V_3 = 3.0 + 4.5 + 12.0 = \textbf{19.5 V}$$

| **Example 3.3** | For the circuit shown in Fig. 3.9, calculate the circuit current, given that the supply is 100 V. |

Fig. 3.9   Circuit diagram for Example 3.3

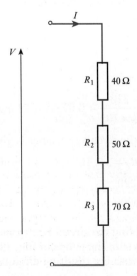

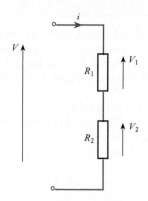

Fig. 3.10 Voltage division between two resistors

Total resistance

$$R = R_1 + R_2 + R_3 = 40 + 50 + 70 = 160 \ \Omega$$

$$I = \frac{V}{R} = \frac{100}{160} = 0.625 \ \text{A}$$

While there are many problems that can be set concerning series-connected resistors, there is one form of application which is especially useful. This involves the division of voltage between only two resistors connected in series, as shown in Fig. 3.10. Given the supply voltage V, it is required to determine the volt drop across R_1. The total resistance of the circuit is

$$R = R_1 + R_2$$

and therefore the current in the circuit is

$$I = \frac{V}{R_1 + R_2}$$

The volt drop across R_1 is given by

$$IR_1 = \frac{V}{R_1 + R_2} \times R_1 = V_1$$

whence

$$\frac{V_1}{V} = \frac{R_1}{R_1 + R_2} \qquad\qquad [3.3]$$

The ratio of the voltages therefore depends on the ratio of the resistances. This permits a rapid determination of the division of volt drops in a simple series circuit and the arrangement is called a voltage divider.

Example 3.4

A voltage divider is to give an output voltage of 10 V from an input voltage of 30 V as indicated in Fig. 3.11. Given that $R_2 = 100 \ \Omega$, calculate the resistance of R_1.

$$\frac{V_2}{V} = \frac{R_2}{R_1 + R_2}$$

$$\frac{10}{30} = \frac{100}{R_1 + 100}$$

$$R_1 + 100 = 3 \times 100 = 300$$

$$R_1 = 200 \ \Omega$$

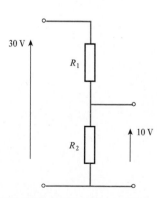

Fig. 3.11 Circuit diagram for Example 3.4

Let us return to the problem of connecting in an extra lamp bulb. We still appear to be in the dark although at least we have learned that extra resistance in a circuit reduces the current. And we have learned that reducing the current also reduces the light output from a bulb so we have to seek an alternative method of connection. Such a method must ensure that the full supply voltage appears across each bulb, thereby ensuring that each bulb passes the required current. Such a method is connection in *parallel*.

3.2 Parallel networks

Owing to the lack of current in the series-connected lamps, each gave out only a dull light. Yet when there had been only one lamp, there were no difficulties because the full supply voltage had been applied to the lamp. In the alternative method of connecction, the full voltage is applied to each lamp as shown in Fig. 3.12.

This arrangement requires double wiring, which in this case is not necessary since both wires at the top of the diagram are at the same high potential while both at the foot of the diagram are at the same low potential. One wire will do in place of two, as shown in the modified arrangement in Fig. 3.13.

The arrangement shown in Fig. 3.13 is very practical since we can now use the wires previously installed to supply the first lamp bulb. Connecting in the second lamp bulb as shown causes both the lamps to operate with full brilliance just as we had intended.

The connection arrangements shown in Fig. 3.13 are termed a *parallel* network. It will be noted that the current may pass from the top conductor to the bottom by means of two paths which run side by side or in parallel with one another. Each of these parallel paths is called a branch, so in this case there are two branches in parallel.

Also it can be seen that there is more than one circuit, as indicated in Fig. 3.14. In any arrangement in which there is more than one circuit, it is appropriate to call the system a network. A network consists of two or more circuits. An investigation of the relations in this simple network may be made in a similar manner to that applied to the series circuit. Again the currents may be investigated using the network shown in Fig. 3.15.

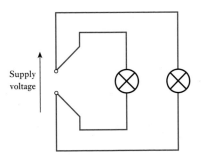

Fig. 3.12 Two lamps connected directly to a source

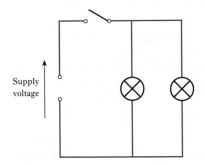

Fig. 3.13 Correct lamp connection

Fig. 3.14 Circuits in a simple network

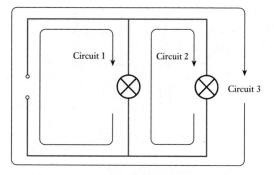

Fig. 3.15 Observation of current in a network

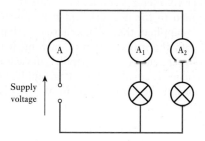

It may be observed that no matter what supply voltage is applied, the sum of the currents as indicated by ammeter A_1 and ammeter A_2 is always equal to the supply current as indicated by ammeter A. It makes no difference whether identical or dissimilar lamps are used – the results are always the same and

$$I = I_1 + I_2 \qquad\qquad [3.4]$$

This observation is reasonable because where else can the currents in each of the lamps come from but from the source, there being no other conduction path available to the currents in the lamps?

Further experiments can be carried out to show that the same results can be obtained no matter how many lamps are connected in parallel. When lamps, or any other forms of load units, are connected in parallel, the sum of the currents in the load units is equal to the supply current.

By transforming the arrangement shown in Fig. 3.12 into the network shown in Fig. 3.13, a further assumption is made that the volt drops across each of the lamps are the same as that of the supply voltage. Again a simple experiment may be carried out to illustrate this point as shown in Fig. 3.16.

Fig. 3.16 Investigation of volt drops in a parallel network

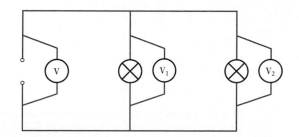

Fig. 3.17   Currents in a parallel network

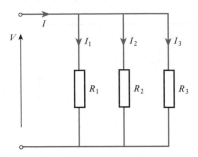

In this investigation, voltmeters are connected across each of the lamps while a further voltmeter is connected across the supply. Different supply voltages are applied to the network and it is observed that no matter what supply voltage is applied, the voltages across each of the loads are equal to it.

As with the investigation of current, it is of no consequence how many load units are connected in parallel; in each case, the volt drop across each of the branches is equal to the voltage applied to the network.

Replacing the lamps with simple resistive loads, as shown in Fig. 3.17, and by using the notation shown, it may be observed that

$$I = I_1 + I_2 + I_3 \qquad\qquad\qquad\qquad [3.5]$$

Since in general

$$I = \frac{V}{R}, \quad \text{then } I_1 = \frac{V}{R_1}, I_2 = \frac{V}{R_2} \quad \text{and} \quad I_3 = \frac{V}{R_3}$$

the voltage across each branch being the same. Substituting in equation [3.5], we get

$$I = \frac{V}{R_1} + \frac{V}{R_2} + \frac{V}{R_3}$$

For the complete network, the effective resistance of the load R represents the ratio of the supply voltage to the supply current, whence

$$I = \frac{V}{R}$$

but $\qquad I = \frac{V}{R_1} + \frac{V}{R_2} + \frac{V}{R_3}$

hence

$$\frac{V}{R} = \frac{V}{R_1} + \frac{V}{R_2} + \frac{V}{R_3}$$

and

$$\frac{1}{R} = \frac{1}{R_1} + \frac{1}{R_2} + \frac{1}{R_3} \qquad\qquad\qquad [3.6]$$

It does not matter how many resistors are connected in parallel, as relation [3.6] may be amended as indicated in Fig. 3.18.

Fig. 3.18 Parallel-connected resistors

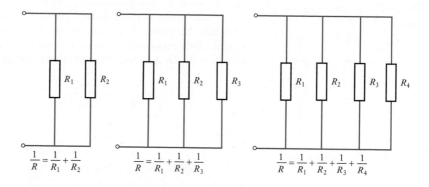

$$\frac{1}{R} = \frac{1}{R_1} + \frac{1}{R_2} \qquad \frac{1}{R} = \frac{1}{R_1} + \frac{1}{R_2} + \frac{1}{R_3} \qquad \frac{1}{R} = \frac{1}{R_1} + \frac{1}{R_2} + \frac{1}{R_3} + \frac{1}{R_4}$$

Example 3.5

Calculate the supply current to the network shown in Fig. 3.19.

Let the currents be I, I_1 and I_2 as indicated on the network.

$$I_1 = \frac{V}{R_1} = \frac{110}{22} = 5.0\,\text{A}$$

$$I_2 = \frac{V}{R_2} = \frac{110}{44} = 2.5\,\text{A}$$

$$I = I_1 + I_2 = 5.0 + 2.5 = \textbf{7.5 A}$$

Fig. 3.19 Circuit diagram for Example 3.5

Example 3.6

For the network shown in Fig. 3.20, calculate the effective resistance and hence the supply current.

$$\frac{1}{R} = \frac{1}{R_1} + \frac{1}{R_2} + \frac{1}{R_3} = \frac{1}{6.8} + \frac{1}{4.7} + \frac{1}{2.2}$$

$$= 0.147 + 0.213 + 0.455 = 0.815$$

hence

$$R = \frac{1}{0.815} = 1.23\,\Omega$$

$$I = \frac{V}{R} = \frac{12}{1.23} = \textbf{9.76 A}$$

Fig. 3.20 Circuit diagram for Example 3.6

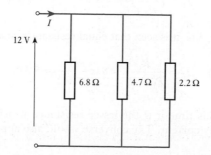

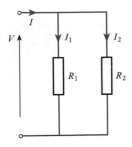

Fig. 3.21 Circuit with two resistors in parallel

As with series circuits, so with parallel networks there are many problems that may be posed, but there is one form of problem which is especially useful. This involves the combination of two resistors in parallel, as shown in Fig. 3.21. In this case, the effective resistance R is given by

$$\frac{1}{R} = \frac{1}{R_1} + \frac{1}{R_2} = \frac{R_1 + R_2}{R_1 R_2}$$

hence

$$R = \frac{R_1 R_2}{R_1 + R_2} \qquad [3.7]$$

Thus the total effective resistance in the case of two resistors connected in parallel is given by the product of the resistances divided by the sum of the resistances. This cannot be extended to the case of three or more parallel resistors and must only be applied in the two-resistor network.

It may also be convenient to determine the manner in which two parallel resistors share a supply current. With reference again to Fig. 3.21

$$V = IR = I\frac{R_1 R_2}{R_1 + R_2}$$

also $V = I_1 R_1$, hence

$$I_1 = I\frac{R_2}{R_1 + R_2} \qquad [3.8]$$

So, to find how the current is shared, the current in one resistor is that portion of the total given by the ratio of the other resistance to the sum of the resistances. This permits a rapid determination of the division of the currents in a simple parallel network.

| Example 3.7 |

A current of 8 A is shared between two resistors in the network shown in Fig. 3.22. Calculate the current in the 2 Ω resistor, given that

(a) $R_1 = 2\ \Omega$;
(b) $R_1 = 4\ \Omega$.

(a) $I_2 = I\dfrac{R_1}{R_1 + R_2} = 8 \times \dfrac{2}{2 + 2} = 4.0\ \text{A}$

From this, it is seen that equal resistances share the current equally.

(b) $I_2 = \dfrac{R_1}{R_1 + R_2} = 8 \times \dfrac{4}{4 + 2} = 5.3\ \text{A}$

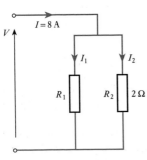

Fig. 3.22 Circuit diagram for Example 3.7

This time it is the lesser resistance which takes the greater part of the supply current. The converse would also apply.

3.3 Series circuits versus parallel networks

It can take a little time to sort out one circuit arrangement from the other when they are being introduced, so the comparisons given in Table 3.1 should be useful.

Table 3.1

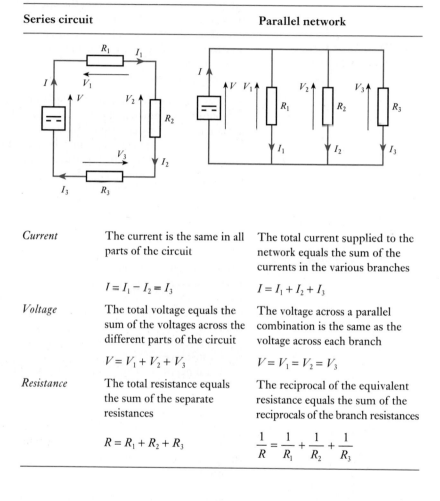

Series circuit		Parallel network
Current	The current is the same in all parts of the circuit	The total current supplied to the network equals the sum of the currents in the various branches
	$I = I_1 - I_2 = I_3$	$I = I_1 + I_2 + I_3$
Voltage	The total voltage equals the sum of the voltages across the different parts of the circuit	The voltage across a parallel combination is the same as the voltage across each branch
	$V = V_1 + V_2 + V_3$	$V = V_1 = V_2 = V_3$
Resistance	The total resistance equals the sum of the separate resistances	The reciprocal of the equivalent resistance equals the sum of the reciprocals of the branch resistances
	$R = R_1 + R_2 + R_3$	$\dfrac{1}{R} = \dfrac{1}{R_1} + \dfrac{1}{R_2} + \dfrac{1}{R_3}$

Here are some points to remember:

1. In a series circuit, the total resistance is always greater than the greatest resistance in the circuit. This serves as a check when combining series resistances.

2. In a parallel network, the total resistance is always less than the smallest resistance in the network. Again this serves as a useful check especially as it is easy to forget to invert the term $1/R$ during the evaluations of R.

3.   To tell the difference between series and parallel, if in doubt imagine being an electron faced with the problem of passing through the circuits. If the electron has no choice but to pass through all the load units, then they are in series. If the electron has the choice of which load unit through which to pass, then the load units are in parallel.

4.   Finally there are certain practical points to note about the two systems. For instance, with the parallel lamp arrangement, either lamp could fail to operate without affecting the operating of the other lamp. Lamp bulbs only last a certain length of time after which the filament breaks and the circuit is interrupted. This only interrupts the current flow in one branch and the remaining branch (or branches if there are more than two lamps) continues to pass current as before. However, if lamps are connected in series and one fails then all the lamps are extinguished.

This is a problem which we sometimes see with Christmas tree lights. One bulb fails and all (or at least a number of) the lights go out. The advantage of series-connected bulbs is that they share the supply voltage, hence cheap low voltage lamps may be used. For most lighting purposes, this advantage is far outweighed by the unreliability of having most of the lamps giving out light should one fail and greater reliability is provided by the parallel arrangement. Therefore the parallel arrangement is highly preferable in practical terms.

3.4   Kirchhoff's laws

From our consideration of series and of parallel connections of resistors, we have observed certain conditions appertaining to each form of connection. For instance, in a series circuit, the sum of the voltages across each of the components is equal to the applied voltage; again the sum of the currents in the branches of a parallel network is equal to the supply current.

Gustav Kirchhoff, a German physicist, observed that these were particular instances of two general conditions fundamental to the analysis of any electrical network. These conditions may be stated as follows:

First (current) law. At any instant the algebraic sum of the currents at a junction in a network is zero. Different signs are allocated to currents held to flow towards the junction and to those away from it.

Second (voltage) law. At any instant in a closed loop, the algebraic sum of the e.m.f.s acting round the loop is equal to the algebraic sum of the p.d.s round the loop.

Stated in such words, the concepts are difficult to understand but we can easily understand them by considering some examples. In Fig. 3.23, the currents flowing towards the junction have been considered positive and those flowing away from the junction are negative, hence the equation given below the diagram. Had we said that currents out were positive and those in were negative, we would have obtained a similar equation except that the polarities would be reversed – but multiplying both sides of the equation by −1 would bring us to that given in the diagram.

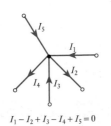

$$I_1 - I_2 + I_3 - I_4 + I_5 = 0$$

Fig. 3.23 Kirchhoff's first (current) law

| Example 3.8 | For the network junction shown in Fig. 3.24, calculate the current I_3, given that $I_1 = 3$ A, $I_2 = -4$ A and $I_4 = 2$ A. |

$$I_1 - I_2 + I_3 - I_4 = 0$$

$$I_3 = -I_1 + I_2 + I_4 = -3 - 4 + 2 = -5\,\text{A}$$

Fig. 3.24 Circuit diagram for Example 3.8

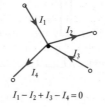

$$I_1 - I_2 + I_3 - I_4 = 0$$

| Example 3.9 | With reference to the network shown in Fig. 3.25, determine the relationship between the currents I_1, I_2, I_4 and I_5. |

Fig. 3.25 Circuit diagram for Example 3.9

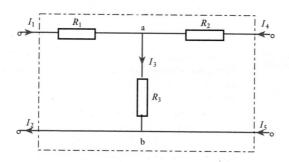

For junction a:

$$I_1 + I_4 - I_3 = 0$$

hence $\qquad I_3 = I_1 + I_4$

For junction b:

$$I_3 + I_5 - I_2 = 0$$

$\therefore \qquad\qquad I_3 = I_2 - I_5$

$$I_1 + I_4 = I_2 - I_5 \quad \text{and} \quad I_1 - I_2 + I_4 + I_5 = 0$$

From the result of this example, it may be noted that Kirchhoff's first law need not only apply to a junction but may also apply to a section of a network. The result of the above problem indicates the application of this law to the dotted box indicated in Fig. 3.25. It follows that knowledge of the performance of quite a complicated network may not be required if only the input and output quantities are to be investigated. This is illustrated by the following problem.

| Example 3.10 | For the network shown in Fig. 3.26, $I_1 = 2.5$ A and $I_2 = -1.5$ A. Calculate the current I_3. |

Fig. 3.26 Circuit diagram for Example 3.10

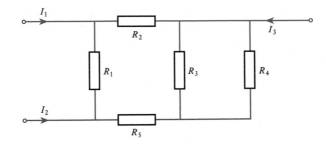

By Kirchhoff's law:

$$I_1 + I_2 + I_3 = 0$$

$$I_3 = -I_1 - I_2 = -2.5 + 1.5 = -1.0 \text{ A}$$

Kirchhoff's first law may be applied at any point within a network. This is illustrated by Example 3.11.

| Example 3.11 | Write down the current relationships for junctions a, b and c of the network shown in Fig. 3.27 and hence determine the currents I_2, I_4 and I_5. |

Fig. 3.27 Circuit diagram for Example 3.11

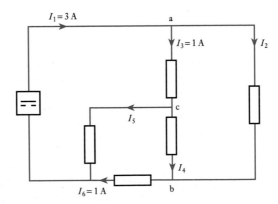

For junction a:

$$I_1 - I_2 - I_3 = 0$$

$$I_2 = I_1 - I_3 = 3 - 1 = 2 \text{ A}$$

For junction b:

$$I_2 + I_4 - I_6 = 0$$

$$I_4 = I_6 - I_2 = 1 - 2 = -1 \text{ A}$$

This answer shows that the current is flowing from b to c. So far, we have drawn all our diagrams so that the current flows from the top conductor to the bottom conductor so it might be strange that here we have a current going upwards. The reason is that we draw a diagram *assuming* that the current flow is downwards. If subsequently we find that the current is flowing in the reverse direction then it only shows that our assumption was incorrect. If it were important then we should redraw the diagram so that b appears above c.

For junction c:

$$I_3 - I_4 - I_5 = 0$$

$$I_5 = I_3 - I_4 = 1 + 1 = \textbf{2 A}$$

The examples chosen so far have permitted the addition and subtraction of the currents at junctions. Parallel arrangements require the division of currents, a point that has already been noted. However, it may also have been observed from the examples given that with more than two resistors, it is possible to make considerably more complicated networks. And these networks need not fit into either the series or the parallel classifications. The network shown in Fig. 3.28 illustrates this observation.

Fig. 3.28   Series–parallel network

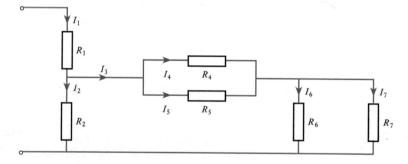

Starting with the points that may be readily observed, it can be seen that R_4 is in parallel with R_5. Also R_6 is in parallel with R_7. In each case the current divides between the two components and then comes together again.

What of R_2? It is tempting to think that it is in parallel with R_6, probably because both I_2 and I_6 are derived from I_1 and, after passing through their respective resistors, the currents immediately come together again. However, I_6 is not derived immediately from I_1 and instead there is the intervening network comprising R_4 and R_5. Because the currents are not immediately derived from I_1, then their respective branches are not in parallel.

Here R_2 is in parallel with the network comprising R_4, R_5, R_6 and R_7. In this case it should be remembered that this specified network takes the current I_3, and I_3 and I_2 are directly derived from I_1.

Finally it may be observed that the network comprising R_4 and R_5 is in series with the network comprising R_6 and R_7. In this case, the current in the one network has no choice but to then pass through the other network, this being the condition of series connection. It is, however, the networks that are in series and not the individual resistors, thus you cannot describe R_6 alone as being in series with, say, R_5. This would only apply if you could be sure that only the current in R_5 then passed to R_6, which cannot be said in the given arrangement.

| **Example 3.12** | For the network shown in Fig. 3.29, determine I_1 and I_2. |

$$I_3 = \frac{R_2}{R_2 + R_3} \cdot I_1$$

$$I_1 = \frac{R_2 + R_3}{R_2} \cdot I_3 = \frac{60 + 30}{60} \times 1 = \textbf{1.5 A}$$

$$0 = I_2 + I_3 - I_1$$

$$I_2 = I_1 - I_3 = 1.5 - 1 = \textbf{0.5 A}$$

Fig. 3.29   Circuit diagram for Example 3.12

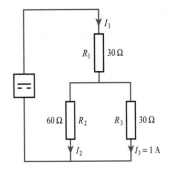

This example illustrates the third arrangement of connection of three resistors, the other arrangements being the three resistors all in series or all in parallel. The network shown in Fig. 3.29 is termed a series–parallel network, i.e. R_1 is in series with the network comprising R_2 in parallel with R_3.

Kirchhoff's second (voltage) law is most readily exemplified by consideration of a simple series circuit as shown in Fig. 3.30.

In this circuit

$$E = V_1 + V_2 + V_3$$

In even the most simple parallel network, there are three possible loops that may be considered. Figure 3.31 shows a reasonably simple arrangement in which the three loops are indicated.

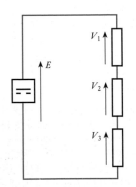

Fig. 3.30   Kirchhoff's second (voltage) law

| **Example 3.13** | For the network shown in Fig. 3.31, determine the voltages V_1 and V_3. |

For loop A:

$$E = V_1 + V_2$$

$$V_1 = E - V_2 = 12 - 8 = \textbf{4 V}$$

For loop B:

$$0 = -V_2 + V_3 + V_4$$

$$V_3 = V_2 - V_4 = 8 - 2 = \textbf{6 V}$$

Fig. 3.31 Circuit diagram for Example 3.13

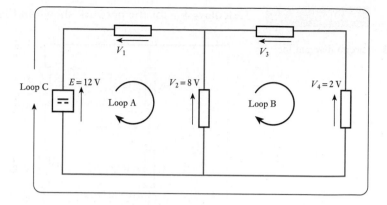

For loop C:

$$E = V_1 + V_3 + V_4$$
$$12 = 4 + 6 + 2 = 12$$

thus confirming the results obtained.

It is important to note that there need not be an e.m.f. in a given loop and this was instanced by loop B. Also it is important to note that p.d.s acting in a clockwise direction round a loop are taken to be negative, which compares with the treatment of currents flowing out from a junction.

The application of Kirchhoff's second law need not be restricted to actual circuits. Instead, part of a circuit may be imagined, as instanced by Fig. 3.32. In this case, we wish to find the total p.d. across three series-connected resistors, i.e. to determine V. Let V be the p.d. across the imaginary section shown by dotted lines and apply Kirchhoff's second law to the loop thus formed.

$$0 = -V + V_1 + V_2 + V_3 \quad \text{and} \quad V = V_1 + V_2 + V_3$$

This is a result that was observed when first investigating series circuits but now we may appreciate it as yet another instance of the principle described as Kirchhoff's second law.

Fig. 3.32 Potential difference across series-connected resistors

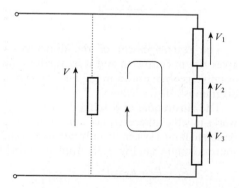

Example 3.14   Calculate V_{AB} for the network shown in Fig. 3.33.

Fig. 3.33   Circuit diagram for Example 3.14

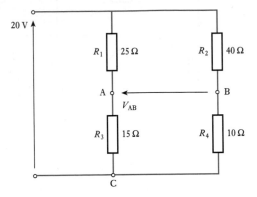

For branch A, let V_{AC} be the voltage at A with respect to C:

$$V_{AC} = \frac{R_3}{R_1 + R_3} \cdot V = \frac{15}{25 + 15} \times 20 = 7.5 \text{ V}$$

For branch B:

$$V_{BC} = \frac{R_4}{R_2 + R_4} \cdot V = \frac{10}{40 + 10} \times 20 = 4.0 \text{ V}$$

Applying Kirchhoff's second law to loop ABC:

$$0 = V_{AB} + V_{BC} + V_{CA} = V_{AB} + V_{BC} - V_{AC}$$

$$V_{AB} = V_{AC} - V_{BC} = 7.5 - 4.0 = \mathbf{3.5 \text{ V}}$$

Fig. 3.34   Alternative circuit layout for Fig. 3.33

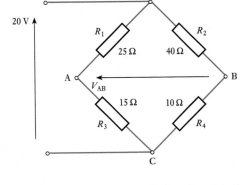

The rearrangement of the drawing layout of a network sometimes gives rise to confusion and it is worth noting that the network used in this example is often drawn in the form shown in Fig. 3.34. This form of circuit diagram is called a bridge.

The illustrations of Kirchhoff's second law have so far only dealt with networks in which there has been only one source of e.m.f. However, there is no reason to limit a system to only one source and a simple circuit involving three sources is shown in Fig. 3.35. Applying Kirchhoff's second law to this circuit,

$$E_1 + E_2 - E_3 = V$$

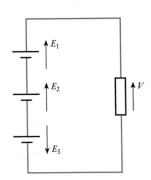

Fig. 3.35   Circuit loop with three sources

| Example 3.15 | Figure 3.36 shows a network with two sources of e.m.f. Calculate the voltage V_1 and the e.m.f. E_2. |

Fig. 3.36 Circuit diagram for Example 3.15

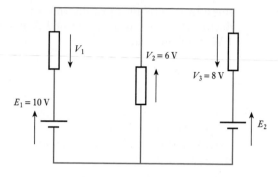

Applying Kirchhoff's second law to the left-hand loop,

$$E_1 = V_1 + V_2$$
$$V_1 = E_1 - V_2 = 10 - 6 = \mathbf{4\ V}$$

The right-hand loop gives

$$-E_2 = -V_2 - V_3$$
$$E_2 = V_2 + V_3 = 6 + 8 = \mathbf{14\ V}$$

These results may be checked by considering the outside loop

$$E_1 - E_2 = V_1 - V_3$$
$$10 - 14 = 4 - 8$$

which confirms the earlier results.

Finally it may be observed that this section has merely stated Kirchhoff's laws and illustrated each in terms of its isolated application. Every time a problem requires that currents be added then the addition conforms to the principle described by the first law while all voltage additions conform to the principle described by the second law. Kirchhoff's laws need not be complicated affairs and nine times out of ten they apply to two currents or two voltages being added together. Nevertheless, the laws may be applied jointly to the solution of complicated networks.

| 3.5 | **Power and energy** | By consideration of the problems of our wiring, it has been seen that a reduction in the voltage and in the current to the lamp bulbs causes their light output to be reduced. The light output is the rate at which the light energy is given out, i.e. the power of the lamp. It can be shown that |

$$P \propto V$$

and $\qquad P \propto I$

whence

$$P \propto VI \qquad\qquad\qquad [3.9]$$

But it may be recalled that the volt is that potential difference across a conductor when passing a current of 1 A and dissipating energy at the rate of 1 W. It follows that

$$P = VI \qquad\qquad\qquad [3.10]$$

Example 3.16

A 230 V lamp is rated to pass a current of 0.26 A. Calculate its power output. If a second similar lamp is connected in parallel to the lamp, calculate the supply current required to give the same power output in each lamp.

$$P = VI = 230 \times 0.26 = 60 \text{ W}$$

With the second lamp in parallel:

$$P = 60 + 60 = \textbf{120 W}$$

$$I = \frac{P}{V} = \frac{120}{230} = \textbf{0.52 A}$$

Example 3.17

Assuming the lamps in Example 3.16 to have reasonably constant resistance regardless of operating conditions, estimate the power output if the lamps are connected in series.

For one lamp

$$R = \frac{V}{I} = \frac{230}{0.26} = 885 \ \Omega$$

With both lamps connected in series, the circuit resistance is

$$R = 885 + 885 = 1770 \ \Omega$$

$$I = \frac{V}{R} = \frac{230}{1770} = 0.13 \text{ A}$$

$$P = VI = 230 \times 0.13 = \textbf{30 W}$$

This is the combined power output and therefore each lamp has an output of 15 W. No wonder we could not get much light when we connected the lamps in series. Instead of 60 W, we were only developing 15 W, a quarter of what we expected from each bulb. Remember that this is only an estimate since no allowance has been made for the effect of the different operating conditions due to temperature rise as the lamp gives out more light.

Consider again relation [3.10]. In a simple load $V = IR$, and substituting in equation [3.10],

$$P = (IR)I = I^2R$$

$$\therefore \qquad P = I^2R \qquad\qquad\qquad [3.11]$$

Seen in this form, the expression emphasizes the power-dissipation effect of a current which creates heat in a conductor. This is generally known as the heating effect of a current and is termed the conductor or I^2R loss since the energy transferred in this way is lost to the electrical system. In the heating bar of an electric radiator or the heating element of a cooker, the I^2R loss is beneficial, but in other cases it may be simply energy lost to the surroundings.

It should be noted that the expression I^2R represents a power, i.e. the rate at which energy is transferred or dissipated, and not the energy itself. Thus, strictly speaking, the previous paragraph is somewhat misleading, but the method of expression is that commonly used. So the power loss describes the energy lost! This is due to the electrical engineer's preoccupation with power and is a consequence of current being the rate of flow and not the flow itself. Once again we meet with this misuse of terminology and again you should remember that it rarely seems to cause difficulties. It is sufficient simply to bear in mind that this double use of 'current' appears throughout electrical engineering.

Should the energy be required, then

$$W = Pt = VIt \qquad [3.12]$$

and $\qquad W = I^2Rt \qquad [3.13]$

This energy is measured in joules, but for the purposes of electricity supply, the joule is too small a unit. From relation [3.11], it can be seen that the unit could also be the watt second. By taking a larger power rating (a kilowatt) and a longer period of time (an hour) then a larger unit of energy (the kilowatt hour) is obtained. Electricity suppliers call this simply a unit of electricity, and it forms the basis of electricity supply measurement whereby consumers are charged for the energy supplied to them.

Example 3.18

A current of 3 A flows through a 10 Ω resistor.

Find:
  (a) the power developed by the resistor;
  (b) the energy dissipated in 5 min.

  (a) $P = I^2R = 3^2 \times 10 = \textbf{90 W}$

  (b) $W = Pt = 90 \times (5 \times 60) = \textbf{2700 J}$.

Example 3.19

A heater takes a current of 8 A from a 230 V source for 12 h. Calculate the energy consumed in kilowatt hours.

$$P = VI = 230 \times 8 = 1840 \text{ W} = 1.84 \text{ kW}$$

$$W = 1.84 \times 12 = \textbf{22 kW h}$$

Example 3.20

For the network shown in Fig. 3.37, calculate the power developed by each resistor.

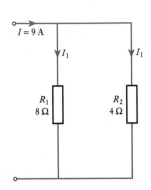

$$I_1 = I\frac{R_2}{R_1 + R_2} = 9 \times \frac{4}{8 + 4} = 3\,\text{A}$$

$$P_1 = I_1^2 R_1 = 3^2 \times 8 = \textbf{72 W}$$

$$I_2 = I - I_1 = 9 - 3 = 6\,\text{A}$$

$$P_2 = I_2^2 R_2 = 6^2 \times 4 = \textbf{140 W}$$

Note: it is tempting to leave this answer as 144 W but the accuracy of the input information is only to one significant figure. Even assuming a second figure, the answer can only be given to two figures, hence the 144 should be rounded to the second significant figure which is 140. Therefore it would be more consistent to give the answer as **140 W** and the other power as **70 W**. However, at this stage the solution answers are sufficient but beware the pitfalls of false accuracy!

Fig. 3.37 Circuit diagram for Example 3.20

3.6   Resistivity

Certain materials permit the reasonably free passage of electric charge and are termed conductors, while others oppose such a free passage and are termed insulators. These abilities are simply taken relative to one another and depend on the material considered. However, other factors also have to be taken into account.

Consider a conductor made of a wire which has a resistance of 1 Ω for every 10 cm of its length. If the wire is made 20 cm long then effectively there are two sections each of 10 cm connected in series. This being the case, the resistance will be 2 Ω. This form of argument may be continued so that 30 cm of wire will have a resistance of 3 Ω and so on, resulting in the length/resistance characteristic shown in Fig. 3.38. Since the graph has the form of a straight line, then the resistance is proportional to the length of wire, i.e.

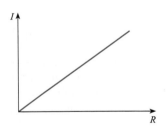

Fig. 3.38 Length/resistance characteristic of a conductor

$$R \propto l \qquad\qquad [3.14]$$

Example 3.21

A cable consists of two conductors which, for the purposes of a test, are connected together at one end of the cable. The combined loop resistance measured from the other end is found to be 100 Ω when the cable is 700 m long. Calculate the resistance of 8 km of similar cable.

$$R \propto l$$

$$\therefore \qquad \frac{R_1}{R_2} = \frac{l_1}{l_2}$$

$$R_2 = \frac{R_1 l_2}{l_1} = \frac{100 \times 8000}{700} = 1143\,\Omega$$

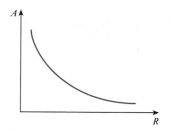

Fig. 3.39 Area/resistance
characteristic of a conductor

Again considering 10 cm pieces of conductor of resistance $1\,\Omega$, if two such pieces are connected in parallel then the resistance is $0.5\,\Omega$. Equally if three such pieces are connected in parallel, the total resistance is $0.33\,\Omega$, and so on. The addition of each piece of wire increases the area of conductor available to the passage of current and hence the area/resistance characteristic of Fig. 3.39 is obtained. The form of this characteristic is such that the resistance is inversely proportional to the area available, i.e.

$$R \propto \frac{1}{A} \qquad\qquad [3.15]$$

Example 3.22

A conductor of 0.5 mm diameter wire has a resistance of $300\,\Omega$. Find the resistance of the same length of wire if its diameter were doubled.

$$R \propto \frac{l}{A}$$

$$\frac{R_1}{R_2} = \frac{A_2}{A_1} = \frac{d_2^2}{d_1^2}$$

$$\frac{300}{R_2} = \frac{1.0^2}{0.5^2}$$

$$R_2 = 75\,\Omega$$

Combining relations [3.14] and [3.15] we get

$$R \propto \frac{l}{A}$$

Rather than deal in proportionality, it is better to insert a constant into this relation, thereby taking into account the type of material being used. This constant is termed the *resistivity* of the material.

Resistivity is measured in ohm metres and is given the symbol ρ (ρ is the Greek letter rho)

$$R = \rho\frac{l}{A} \qquad\qquad [3.16]$$

Resistivity   Symbol: ρ Unit: $\Omega\,m$ (ohm metre)

The value of the resistivity is that resistance of a unit cube of the material measured between opposite faces of the cube. Typical values of resistivity are given in Table 3.2.

Table 3.2   Typical values of resistivity

Material	ρ (Ω m) at 0 °C
Aluminium	2.7×10^{-8}
Brass	7.2×10^{-8}
Copper	1.59×10^{-8}
Eureka	49.00×10^{-8}
Manganin	42.00×10^{-8}
Carbon	6500.00×10^{-8}
Tungsten	5.35×10^{-8}
Zinc	5.37×10^{-8}

Example 3.23

A coil is wound from a 10 m length of copper wire having a cross-sectional area of 1.0 mm². Calculate the resistance of the coil.

$$R = \rho \frac{l}{A} = \frac{1.59 \times 10^{-8} \times 10}{1 \times 10^{-6}} = 0.159 \,\Omega$$

3.7   Temperature coefficient of resistance

The resistance of all pure metals increases with increase of temperature, whereas the resistance of carbon, electrolytes and insulating materials decreases with increase of temperature. Certain alloys, such as manganin, show practically no change of resistance for a considerable variation of temperature. For a moderate range of temperature, such as 100 °C, the change of resistance is usually proportional to the change of temperature; the ratio of the change of resistance per degree change of temperature to the resistance at some definite temperature, adopted as standard, is termed *the temperature coefficient of resistance* and is represented by the Greek letter α.

Temperature coefficient of resistance

Symbol: α Unit: /°C (reciprocal degree centigrade)

The variation of resistance of copper for a range over which copper conductors are usually operated is represented by the graph in Fig. 3.40. If this graph is extended backwards, the point of intersection with the horizontal

Fig. 3.40   Variation of resistance of copper with temperature

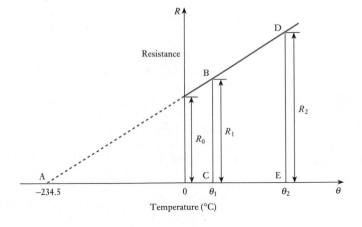

Temperature (°C)

axis is found to be –234.5 °C. Hence, for a copper conductor having a resistance of 1 Ω at 0 °C, the change of resistance for 1 °C change of temperature is (1/234.5) Ω, namely 0.004 264 Ω,

$$\therefore \quad \alpha_0 = \frac{0.004\ 264\ [\Omega/°C]}{1\ [\Omega]} = 0.004\ 264/°C$$

In general, if a material having a resistance R_0 at 0 °C, taken as the standard temperature, has a resistance R_1 at θ_1 and R_2 at θ_2, and if α_0 is the temperature coefficient of resistance at 0 °C,

$$R_1 = R_0(1 + \alpha_0\theta_1) \quad \text{and} \quad R_2 = R_0(1 + \alpha_0\theta_2)$$

$$\therefore \quad \frac{R_1}{R_2} = \frac{1 + \alpha_0\theta_1}{1 + \alpha_0\theta_2} \tag{3.17}$$

In some countries the standard temperature is taken to be 20 °C, which is roughly the average atmospheric temperature. This involves using a different value for the temperature coefficient of resistance, e.g. the temperature coefficient of resistance of copper at 20 °C is 0.003 92/°C. Hence, for a material having a resistance R_{20} at 20 °C and temperature coefficient of resistance α_{20} at 20 °C, the resistance R_t at temperature θ is given by:

$$R_t = R_{20}\{1 + \alpha_{20}(\theta - 20)\} \tag{3.18}$$

Hence if the resistance of a coil, such as a field winding of an electrical machine, is measured at the beginning and at the end of a test, the mean temperature rise of the whole coil can be calculated.

The value of α depends on the initial temperature of the conductor and usually the values of the coefficient are given relative to 0 °C, i.e. α_0. Typical values of the temperature coefficient of resistance are given in Table 3.3.

Table 3.3 Typical temperature coefficients of resistance referred to 0 °C

Material	$\alpha_0(/°C)$ at 0 °C
Aluminium	0.003 81
Copper	0.004 28
Silver	0.004 08
Nickel	0.006 18
Tin	0.004 4
Zinc	0.003 85
Carbon	−0.000 48
Manganin	0.000 02
Constantan	0
Eureka	0.000 01
Brass	0.001

Example 3.24

A coil of copper wire has a resistance of 200 Ω when its mean temperature is 0 °C. Calculate the resistance of the coil when its mean temperature is 80 °C.

$$R_1 = R_0(1 + \alpha_0\theta_1) = 200(1 + 0.004\ 28 \times 80) = \mathbf{268.5\ \Omega}$$

Example 3.25

When a potential difference of 10 V is applied to a coil of copper wire of mean temperature 20 °C, a current of 1.0 A flows in the coil. After some time the current falls to 0.95 A yet the supply voltage remains unaltered. Determine the mean temperature of the coil given that the temperature coefficient of resistance of copper is $4.28 \times 10^{-3}/°C$ referred to 0 °C.

At 20 °C

$$R_1 = \frac{V_1}{I_1} = \frac{10}{1} = 10.0 \ \Omega$$

At θ_2

$$R_2 = \frac{V_2}{I_2} = \frac{10}{0.95} = 10.53 \ \Omega$$

$$\frac{R_1}{R_2} = \frac{(1 + \alpha_0 \theta_1)}{(1 + \alpha_0 \theta_2)}$$

$$\frac{10.0}{10.53} = \frac{(1 + 0.004\ 28 \times 20)}{(1 + 0.004\ 28 \times \theta_2)}$$

whence

$$\theta_2 = 33.4 \ °C$$

Most materials that are classified as conductors have a positive temperature coefficient of resistance, i.e. their resistances increase with increase of temperature. This would have a considerable effect on the estimate of what happened to the lamps when connected in series, as indicated in Example 3.25, especially when it is noted that the working temperature of a tungsten lamp filament is 2500 °C.

At the other extreme, when the temperature falls to absolute zero, 0 K, (−273 °C), the resistance falls to zero and there will be no I^2R losses. Conductors close to these conditions are known as superconductors.

Some alloys are made with a zero temperature coefficient of resistance, thus their resistance does not vary with temperature. This is convenient in the manufacture of measuring instruments which may thus operate effectively without reference to temperature correction in their indications.

Some materials such as carbon have a negative temperature coefficient of resistance, i.e. their resistances fall with increase in temperature. This gives rise to certain difficulties with a group of materials termed semiconductors. The heat created in these materials, if not effectively dissipated by cooling, causes the resistance to fall and the current to increase. This causes the rate of heat creation to increase, the temperature to rise still further and the resistance to continue falling. If unchecked, this process continues until there is sufficient heat to destroy the structure of the semiconductor completely. The process is termed thermal runaway.

3.8   Temperature rise

The maximum power which can be dissipated as heat in an electrical circuit is limited by the maximum permissible temperature, and the latter depends upon the nature of the insulating material employed. Materials such as paper and cotton become brittle if their temperature is allowed to exceed about 100 °C, whereas materials such as mica and glass can withstand a much higher temperature without any injurious effect on their insulating and mechanical properties.

When an electrical device is *loaded* (e.g. when supplying electrical power in the case of a generator, mechanical power in the case of a motor or acting as an amplifier in the case of a transistor), the temperature rise of the device is largely due to the I^2R loss in the conductors; and the greater the load, the greater is the loss and therefore the higher the temperature rise. The *full load* or *rated output* of the device is the maximum output obtainable under specified conditions, e.g. for a specified temperature rise after the device has been loaded continuously for a period of minutes or hours.

The temperature of a coil can be measured by the following means:

1. A thermometer.
2. The increase of resistance of the coil.
3. Thermo-junctions embedded in the coil.

The third method enables the distribution of temperature throughout the coil to be determined, but is only possible if the thermo-junctions are inserted in the coil when the latter is being wound. Since the heat generated at the centre of the coil has to flow outwards, the temperature at the centre may be considerably higher than that at the surface.

The temperature of an electronic device, especially one incorporating a semiconductor junction, is of paramount importance, since even a small rise of temperature above the maximum permissible level rapidly leads to a catastrophic breakdown.

Summary of important formulae

In a series circuit

voltage $V = V_1 + V_2 + V_3$ (volts)   [3.1]

resistance $R = R_1 + R_2 + R_3$ (ohms)   [3.2]

In a parallel network

current $I = I_1 + I_2 + I_3$ (amperes)   [3.5]

$$\frac{1}{R} = \frac{1}{R_1} + \frac{1}{R_2} + \frac{1}{R_3}$$ [3.6]

Effective resistance of two parallel resistors:

$$R = \frac{R_1 R_2}{R_1 + R_2}$$ [3.7]

Current division rule for two resistors:

$$I_1 = \frac{R_2}{R_1 + R_2} \cdot I$$ [3.8]

Energy $W = I^2 R t$ [3.13]

Resistance $R = \rho l / A$ [3.16]

Using the temperature coefficient of resistance:

$$\frac{R_1}{R_2} = \frac{1 + \alpha_0 \theta_1}{1 + \alpha_0 \theta_2}$$ [3.17]

Terms and concepts

Loads are **series** connected when the same current flow passes through each of them.

Loads are connected in **parallel** when the same potential difference is applied to each of them.

Kirchhoff's laws state that the sum of the currents entering a junction is equal to the sum of the currents leaving and the sum of the volt drops round any loop is equal to the sum of the e.m.f.s.

The most common application of Kirchhoff's current law is to two branches in parallel, i.e. one current in and two out (or vice versa).

The most common application of Kirchhoff's voltage law is to a single circuit with one source and one load.

Resistivity is a constant for a material relating its resistance to its length and cross-sectional area.

Resistivity generally varies with change of temperature.

The **temperature coefficient of resistance** relates the changes of resistance to change of temperature according to the initial temperature.

Temperature rise can damage insulation and hence is the basis of rating electrical equipment.

Exercises 3

1. Three resistors of 2 Ω, 3 Ω and 5 Ω are connected in series and a current of 2 A flows through them. Calculate the p.d. across each resistor and the total supply voltage.
2. The lamps in a set of Christmas tree lights are connected in series; if there are 20 lamps and each lamp has a resistance of 25 Ω, calculate the total resistance of the set of lamps, and hence calculate the current taken from a 230 V supply.
3. Three lamps are connected in series across a 120 V supply and take a current of 1.5 A. If the resistance of two of the lamps is 30 Ω each, what is the resistance of the third lamp?
4. The field coil of a d.c. generator has a resistance of 60 Ω and is supplied from a 240 V source. Given that the current in the coil is to be limited to 2 A, calculate the resistance of the resistor to be connected in series with the coil
5. Given that V_{LN} is the p.d. of L with respect to N, calculate, for the circuit shown in Fig. A, the values of: (a) V_{AB}, (b) V_{BC}, (c) V_{AC}, (d) V_{BN}.
6. Given that V_{NL} is the p.d. of N with respect to L, then, for the circuit shown in Fig. B, $V_{LN} = -50$ V. What are the corresponding values of (a) V_{WX}, (b) V_{YZ}, (c) V_{XW}, (d) V_{ZW}, (e) V_{YX}?

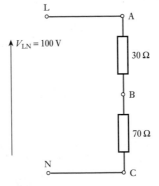

Fig. A

7. Three resistors of 6 Ω, 9 Ω and 15 Ω are connected in parallel to a 9 V supply. Calculate: (a) the current in each branch of the network; (b) the supply current; (c) the total effective resistance of the network.
8. The effective resistance of two resistors connected in parallel is 8 Ω. The resistance of one of the resistors is 12 Ω. Calculate: (a) the resistance of the other resistor; (b) the effective resistance of the two resistors connected in series.

Exercises 3 continued

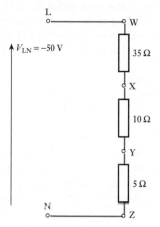

Fig. B

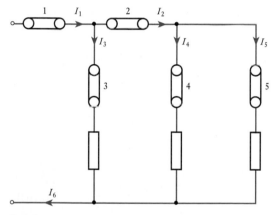

Fig. C

9. A parallel network consists of three resistors of 4 Ω, 8 Ω and 16 Ω. If the current in the 8 Ω resistor is 3 A, what are the currents in the other resistors?

10. With respect to Fig. C, which of the following statements are correct:
(a) I_1 is the total supply current;
(b) $I_1 = I_2 + (-I_3) + I_4$; (c) $I_1 = I_2 + I_3 + I_4$;
(d) $I_2 = I_3 + I_4 + I_5$;
(e) $I_1 = I_3 + I_4 + I_5$?

11. With reference to Fig. C, if link 3 is removed, which of the following statements is correct:
(a) $I_1 = I_2 + I_3$;
(b) $I_1 = I_4 + I_5$;
(c) $I_1 = I_2$;
(d) $I_1 = I_3$;
(e) $I_1 = I_3 + I_4 + I_5$?

12. Two coils having resistances of 5 Ω and 8 Ω respectively are connected across a battery having an e.m.f. of 6 V and an internal resistance of 1.5 Ω. Calculate: (a) the terminal voltage and (b) the energy in joules dissipated in the 5 Ω coil if the current remains constant for 4 min.

13. A coil of 12 Ω resistance is in parallel with a coil of 20 Ω resistance. This combination is connected in series with a third coil of 8 Ω resistance. If the whole circuit is connected across a battery having an e.m.f. of 30 V and an internal resistance of 2 Ω, calculate (a) the terminal voltage of the battery and (b) the power in the 12 Ω coil.

14. A coil of 20 Ω resistance is joined in parallel with a coil of R Ω resistance. This combination is then joined in series with a piece of apparatus A, and the whole circuit connected to 100 V mains. What must be the value of R so that A shall dissipate 600 W with 10 A passing through it?

15. Two circuits, A and B, are connected in parallel to a 25 V battery, which has an internal resistance of 0.25 Ω. Circuit A consists of two resistors, 6 Ω and 4 Ω, connected in series. Circuit B consists of two resistors, 10 Ω and 5 Ω, connected in series. Determine the current flowing in, and the potential difference across, each of the four resistors. Also, find the power expended in the external circuit.

16. A load taking 200 A is supplied by copper and aluminium cables connected in parallel. The total length of conductor in each cable is 200 m, and each conductor has a cross-sectional area of 40 mm². Calculate: (a) the voltage drop in the combined cables; (b) the current carried by each cable; (c) the power wasted in each cable. Take the resistivity of copper and aluminium as 0.018 $\mu\Omega$ m and 0.028 $\mu\Omega$ m respectively.

17. A circuit, consisting of three resistances 12 Ω, 18 Ω and 36 Ω respectively, joined in parallel, is connected in series with a fourth resistance. The whole is supplied at 60 V and it is found that the power dissipated in the 12 Ω resistance is 36 W. Determine the value of the fourth resistance and the total power dissipated in the group.

18. A coil consists of 2000 turns of copper wire having a cross-sectional area of 0.8 mm². The mean length per turn is 80 cm and the resistivity of copper is 0.02 $\mu\Omega$ m at normal working temperature. Calculate the resistance of the coil and the power dissipated when the coil is connected across a 110 V d.c. supply.

Exercises 3 continued

19. An aluminium wire 7.5 m long is connected in parallel with a copper wire 6 m long. When a current of 5 A is passed through the combination, it is found that the current in the aluminium wire is 3 A. The diameter of the aluminium wire is 1.0 mm. Determine the diameter of the copper wire. Resistivity of copper is 0.017 $\mu\Omega$ m; that of aluminium is 0.028 $\mu\Omega$ m.

20. The field winding of a d.c. motor is connected directly across a 440 V supply. When the winding is at the room temperature of 17 °C, the current is 2.3 A. After the machine has been running for some hours, the current has fallen to 1.9 A, the voltage remaining unaltered. Calculate the average temperature throughout the winding, assuming the temperature coefficient of resistance of copper to be 0.004 26/°C at 0 °C.

21. Define the term *resistance–temperature coefficient*.

    A conductor has a resistance of R_1 Ω at θ_1 °C, and consists of copper with a resistance–temperature coefficient α referred to 0 °C. Find an expression for the resistance R_2 of the conductor at temperature θ_2 °C.

    The field coil of a motor has a resistance of 250 Ω at 15 °C. By how much will the resistance increase if the motor attains an average temperature of 45 °C when running? Take $\alpha = 0.004\,28$/°C referred to 0 °C.

22. Explain what is meant by the *temperature coefficient of resistance* of a material.

    A copper rod, 0.4 m long and 4.0 mm in diameter, has a resistance of 550 $\mu\Omega$ at 20 °C. Calculate the resistivity of copper at that temperature. If the rod is drawn out into a wire having a uniform diameter of 0.8 mm, calculate the resistance of the wire when its temperature is 60 °C. Assume the resistivity to be unchanged and the temperature coefficient of resistance of copper to be 0.004 26/°C.

23. A coil of insulated copper wire has a resistance of 150 Ω at 20 °C. When the coil is connected across a 230 V supply, the current after several hours is 1.25 A. Calculate the average temperature throughout the coil, assuming the temperature coefficient of resistance of copper at 20 °C to be 0.0039/°C.

Chapter four

Network Theorems

Objectives

When you have studied this chapter, you should

- be familiar with the relevance of Kirchhoff's laws to the analysis of networks
- be capable of analysing networks by the applications of Kirchhoff's laws
- be capable of analysing networks by the application of Mesh analysis
- be capable of analysing networks by the application of Nodal analysis
- be capable of analysing networks by the application of Thévenin's theorem
- have an understanding of the constant-current generator
- be capable of analysing networks by the application of Norton's theorem
- recognize star and delta connections
- be capable of transforming a star-connected load into a delta-connected load and vice versa
- be familiar with the condition required for maximum power to be developed in a load
- have an understanding of the significance the maximum power condition has in practice

Contents

Many practical circuits can be understood in terms of series and parallel circuits. However, some electrical engineering applications, especially in electronic engineering, involve networks with large numbers of components. It would be possible to solve many of them using the techniques introduced in Chapter 3 but these could be lengthy and time-consuming procedures. Instead, in this chapter, we shall develop a variety of techniques such as Nodal analysis, the Superposition theorem, Thévenin's theorem and Norton's theorem, which will speed up the analysis of the more complicated networks. It is always a good idea to make life as easy as possible!

Not all loads are connected in series or in parallel. There are two other arrangements known as star and delta. They are not so common but, because they are interchangeable, we can readily find a solution to any network in which they appear – so long as we can transform the one into the other.

We have seen that the function of a circuit is to deliver energy or power to a load. It may have crossed your mind – what is the condition for the greatest power to be developed? Well we shall answer that later in this chapter. It is a question which is important to the electronic engineer.

4.1 New circuit analysis techniques

A direct application of Kirchhoff's current and voltage laws using the principles discussed in Chapter 3 can solve many circuit problems. However, there are a variety of techniques, all based on these two laws, that can simplify circuit analysis. The main techniques, to be introduced in this chapter, are:

- Mesh analysis.
- Nodal analysis.
- Thévenin's theorem.
- Norton's theorem.

Each of these techniques has particular strengths aimed at solving particular types of circuit problem. In this way, what would be laborious using one method can be straightforward using another. Familiarization with all the different methods will enable you to choose the method which best suits a particular problem. This simplifies circuit solution and makes less work overall!

4.2 Kirchhoff's laws and network solution

Kirchhoff's laws can be applied to network solution in any of the following ways:

1. By direct application to the network in conjunction with Ohm's law.
2. By indirect application to the network in conjunction with the manipulation of the component resistances.
3. By direct application to the network resulting in solution by simultaneous equations.

These statements appear to be most complicated, but the following series of examples will illustrate the forms of application of the laws to network solution. The form that ought to be most obvious is the first form, in which the laws are directly applied; curiously this form of solution tends to be so obvious that it is all too often neglected, as will be illustrated.

Example 4.1

For the network shown in Fig. 4.1, determine the supply current and the source e.m.f.

Since R_3 and R_4 are in parallel

$$V_3 = I_4 R_4 = 3 \times 8 = 24 \text{ V} = I_3 R_3 = I_3 \times 16$$

$$I_3 = \frac{24}{16} = 1.5 \text{ A}$$

Fig. 4.1 Circuit diagram for Example 4.1

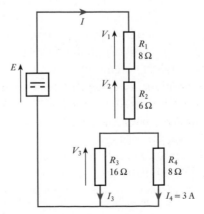

By Kirchhoff's first law

$$I = I_3 + I_4 = 1.5 + 3 = \mathbf{4.5\ A}$$

Also $\quad V_1 = IR_1 = 4.5 \times 8 = 36\ V$

$$V_2 = IR_2 = 4.5 \times 6 = 27\ V$$

By Kirchhoff's second law

$$E = V_1 + V_2 + V_3 = 36 + 27 + 24 = \mathbf{87\ V}$$

This is not the only form of solution to the given problem. For instance, the supply current could have been derived directly from I_3 by applying the current–sharing rule, or the source e.m.f. could have been derived from the product of the supply current and the total effective resistance which could have been determined – but the direct solution is readily available without the need to resort to such devices. The following two examples illustrate again the availability of a direct approach to network problems.

Example 4.2 Given the network shown in Fig. 4.2, determine I_1, E, I_3 and I.

Fig. 4.2 Circuit diagram for Example 4.2

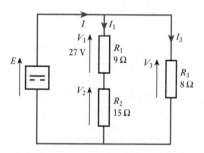

$$I_1 = \frac{V_1}{R_1} = \frac{27}{9} = \mathbf{3\ A}$$

$$V_2 = I_1 R_2 = 3 \times 15 = 45\ V$$

$$E = V = V_1 + V_2 = 27 + 45 = \mathbf{72\ V}$$

$$I_3 = \frac{V}{R_3} = \frac{72}{8} = \mathbf{9\ A}$$

$$I = I_1 + I_3 = 3 + 9 = \mathbf{12\ A}$$

Example 4.3 For the network shown in Fig. 4.3, the power dissipated in R_3 is 20 W. Calculate the current I_3 and hence evaluate R_1, R_3, I_1, I_2 and V.

Potential difference across the 10 Ω resistor is $1 \times 10 = 10$ V. For resistor R_3,

$$P = 20\ W = 10 \times I_3$$

Hence $\quad I_3 = \dfrac{20}{10} = 2\ A$

$$P = I_3^2 R_3 = 20$$

Fig. 4.3 Circuit diagram for
Example 4.3

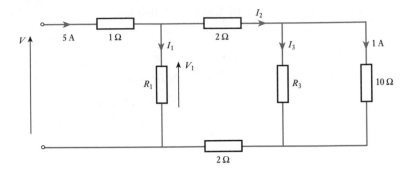

hence $20 = 2^2 \times R_3$

$R_3 = \textbf{5 } \boldsymbol{\Omega}$

$I_2 = 2 + 1 = \textbf{3 A}$

Potential difference across each of the two 2 Ω resistors is $3 \times 2 = 6$ V. Thus

$V_1 = 6 + 10 + 6 = \textbf{22 V}$

$I_1 = 5 - 3 = \textbf{2 A}$

$R_1 = \dfrac{V_1}{I_1} = \dfrac{22}{2} = \textbf{11 } \boldsymbol{\Omega}$

Potential difference across the 1 Ω resistor is $5 \times 1 = 5$ V, hence

$V = 5 + 22 = \textbf{27 V}$

This last example in particular illustrates that a quite complicated network can readily be analysed by this direct approach. However, it is not always possible to proceed in this way, either because most of the information given relates to the resistances or because there is insufficient information concerning any one component of the network.

An instance of the information being presented mainly in terms of resistance is given in Example 4.4 and it also brings us to the second form of application of Kirchhoff's laws.

Example 4.4

For the network shown in Fig. 4.4, determine the supply current and current I_4.

Essentially this network consists of three parts in series, but one of them comprises R_3 and R_4 in parallel. These can be replaced by an equivalent resistance, thus

$R_e = \dfrac{R_3 R_4}{R_3 + R_4} = \dfrac{16 \times 8}{16 + 8} = 5.33 \ \Omega$

Replacing R_3 and R_4 by R_e, the network becomes that shown in Fig. 4.5.

Now that the network has been reduced to a simple series circuit the total effective resistance is

$R = R_1 + R_2 + R_e = 8 + 6 + 5.33 = 19.33 \ \Omega$

$I = \dfrac{V}{R} = \dfrac{87}{19.33} = \textbf{4.5 A}$

Fig. 4.4 Circuit diagram for Example 4.4

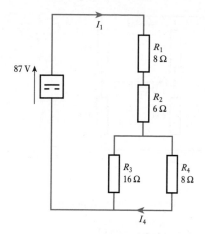

Fig. 4.5 Circuit diagram for Example 4.4

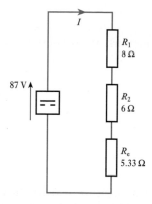

Reverting now to the original network,

$$I_4 = \frac{R_3}{R_3 + R_4} \cdot I = \frac{16}{16 + 8} \times 4.5 = 3\,\text{A}$$

This example compares with Example 4.1 and the figures are in fact the same. However, in this second instance the given voltage and current information stemmed from the source and not from the load, hence the emphasis of the calculation lay in dealing with the resistances of the network. The calculation was based on network reduction, i.e. by replacing two or more resistors by one equivalent resistor. A further example of this approach is given below, in which two instances of network reduction transform the problem into a form that can be readily analysed.

Example 4.5 Determine V_{AB} in the network shown in Fig. 4.6.

This is quite a complex network. However, there are two instances of parallel resistors that may be replaced by equivalent resistors. For the 10 Ω and 15 Ω resistors

$$R = \frac{10 \times 15}{10 + 15} = 6\,\Omega$$

Fig. 4.6 Circuit diagram for Example 4.5

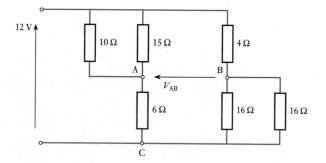

Fig. 4.7 Circuit diagram for
Example 4.5

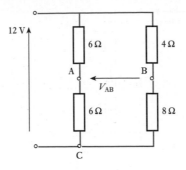

For the two 16 Ω resistors in parallel

$$R = \frac{16 \times 16}{16 + 16} = 8\,\Omega$$

If these equivalent values are inserted into the network, the network transforms into that shown in Fig. 4.7. Thus

$$V_{AC} = \frac{6}{6 + 6} \times 12 = 6\text{ V}$$

and $$V_{BC} = \frac{8}{4 + 8} \times 12 = 8\text{ V}$$

$$V_{AB} = V_{AC} - V_{BC} = 6 - 8 = -2\text{ V}$$

   Having now observed the two methods of analysis being demonstrated,
you may well wonder how to tell when each should be used. As a general
rule, if the information given concerns the voltage or the current associated
with one or more components of the network, then you would apply the first
form of approach. However, if the information given concerns the supply
voltage or current, then you would try to apply the second form of approach
by network reduction. This is not always possible because resistors may be
connected in a manner that is neither series nor parallel – such an arrangement is shown in Fig. 4.8.

Example 4.6   For the network shown in Fig. 4.8, calculate the currents in each of
the resistors.

Fig. 4.8 Circuit diagram for
Example 4.6

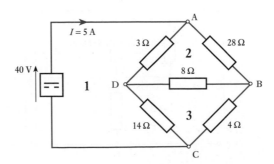

Fig. 4.9 Circuit diagram for
Example 4.7

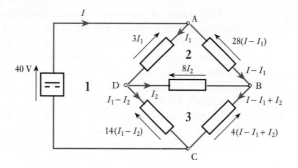

In this network the resistors are neither in series nor in parallel and therefore a more difficult method of analysis must be employed. Let the current in the 3 Ω resistor be I_1 and therefore by Kirchhoff's first law, the current in the 28 Ω resistor is $I - I_1$. Further, let the current in the 8 Ω resistor flowing from D to B be I_2. It follows that the current in the 14 Ω resistor is $I_1 - I_2$ while that in the 4 Ω resistor is $I - I_1 + I_2$. The resulting volt drops are shown in Fig. 4.9.

Applying Kirchhoff's second law to loop 1 (comprising source to ADC):

$$40 = 3I_1 + 14(I_1 - I_2)$$
$$40 = 17I_1 - 14I_2 \tag{a}$$

Applying Kirchhoff's second law to loop 2 (ABD):

$$0 = 28(I - I_1) - 8I_2 - 3I_1$$
$$= 28I - 31I_1 - 8I_2$$

But $I - 5$ A

Therefore

$$140 = 31I_1 + 8I_2 \tag{b}$$

(a) × 4 $160 = 68I_1 - 56I_2$ $\tag{c}$

(b) × 7 $980 = 217I_1 + 56I_2$ $\tag{d}$

(c) + (d) $1140 = 285I_1$

$$I_1 = \textbf{4 A in 3 Ω resistor}$$

Substituting in (b),

$$140 = 124 + 8I_2$$

$$I_2 = \textbf{2 A in 8 Ω resistor}$$

Hence current in 28 Ω resistor is

$$5 - 4 = \textbf{1 A}$$

current in 14 Ω resistor is

$$4 - 2 = \textbf{2 A}$$

and current in 4 Ω resistor is

$$5 - 4 + 2 = \textbf{3 A}$$

This form of solution requires that you proceed with great caution, otherwise it is a simple matter to make mistakes during the mathematical processes. However, in the instance given, it is necessary to involve such an analysis; had a different current been given in this example, such a solution would not have been required since it would then have been possible to achieve a solution by applying the first approach, i.e. directly applying Kirchhoff's laws.

If two parallel e.m.f.s appear in a network as exemplified by Fig. 4.10, it might again be necessary to employ the approach using simultaneous equations resulting from the application of Kirchhoff's laws.

Example 4.7   Calculate the currents in the network shown in Fig. 4.10.

Fig. 4.10   Circuit diagram for Example 4.7

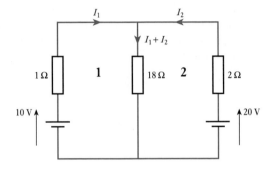

Applying Kirchhoff's second law to loop 1:

$$10 = 1I_1 + 18(I_1 + I_2)$$

$$10 = 19I_1 + 18I_2 \qquad\qquad\text{(a)}$$

Applying Kirchhoff's second law to loop 2:

$$20 = 2I_2 + 18(I_1 + I_2)$$

$$20 = 18I_1 + 20I_2 \qquad\qquad\text{(b)}$$

(a) × 10 $100 = 190I_1 + 180I_2$ (c)

(b) × 9 $180 = 162I_1 + 180I_2$ (d)

(d) − (c) $80 = -28I_1$

$$I_1 = -2.85 \text{ A}$$

Substituting in (a)

$$10 = -54.34 + 18I_2$$

$$I_2 = 3.57 \text{ A}$$

Current in 18 Ω resistor is

$$3.57 - 2.85 = 0.72 \text{ A}$$

This form of solution is fraught with the danger of mathematical mistakes and therefore should only be employed when all else fails. This section commenced by stating that the obvious solution is all too easily ignored. Thus if the 2 Ω resistor were removed from the network shown in Fig. 4.10, it might be overlooked that the 20 V battery is now directly applied to the 18 Ω resistor and so, knowing the voltage drop across one of the components, it is possible to revert to the first form of analysis as shown in Example 4.8.

Example 4.8   Calculate the currents in the network shown in Fig. 4.11.

Fig. 4.11   Circuit diagram for Example 4.8

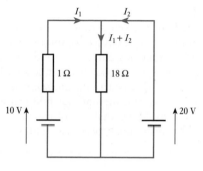

Current in 18 Ω resistor is

$$\frac{20}{18} = 1.1\,\text{A}$$

Applying Kirchhoff's second law to the outside loop:

$$20 - 10 = -I_1 \times 1$$

$$I_1 = -\mathbf{10}\,\textbf{A}$$

$$I_2 = -(-10) + 1.1 = \mathbf{11.1}\,\textbf{A}$$

Example 4.9   Three similar primary cells are connected in series to form a closed circuit as shown in Fig. 4.12. Each cell has an e.m.f. of 1.5 V and an internal resistance of 30 Ω. Calculate the current and show that points A, B and C are at the same potential.

In Fig. 4.12, E and R represent the e.m.f. and internal resistance respectively of each cell.

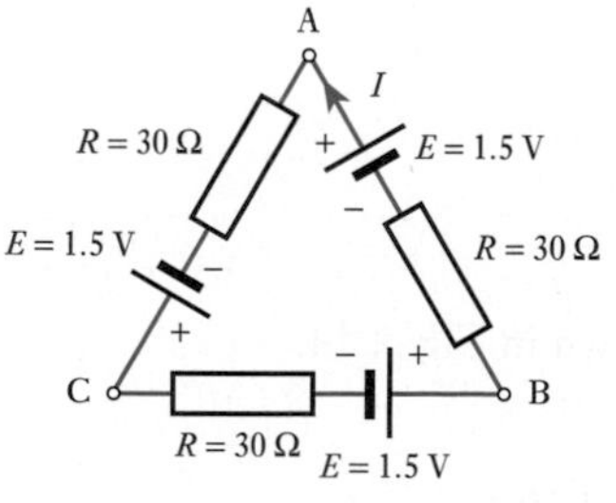

Total e.m.f. = $1.5 \times 3 = 4.5$ V

total resistance = $30 \times 3 = 90$ Ω

$\therefore$ current = $\dfrac{4.5}{90} = 0.05$ A

The volt drop due to the internal resistance of each cell is 0.05×30, namely 1.5 V. Hence the e.m.f. of each cell is absorbed in sending the current through the internal resistance of that cell, so that there is no difference of potential between the two terminals of the cell. Consequently the three junctions A, B and C are at the same potential.

Fig. 4.12   Circuit diagram for Example 4.9

To summarize, therefore, the approach to network analysis should be to determine whether component voltages and currents are known, in which case a direct approach to the analysis may be made using the principles observed by Kirchhoff's laws. If this is not possible then network reduction should be tried in order that the network is sufficiently simplified that it becomes manageable. Should all else fail, then Kirchhoff's laws should be applied to derive loop simultaneous equations from which the solution will be obtained.

| 4.3 | Mesh analysis |

This method is given a number of different names – all of which are an indication of the analysis technique employed. It is variously known as Maxwell's circulating current method, loop analysis or Mesh current analysis. The terminology is chosen to distinguish it from the familiar 'branch current' technique, in which currents are assigned to individual branches of a circuit. The branch current method was first introduced in Chapter 3 and has been used hitherto. Mesh analysis, of course, relies on Kirchhoff's laws just the same. The technique proceeds as follows:

- Circulating currents are allocated to closed loops or meshes in the circuit rather than to branches.
- An equation for each loop of the circuit is then obtained by equating the algebraic sum of the e.m.f.s round that loop to the algebraic sum of the potential differences (in the direction of the loop, mesh or circulating current), as required by Kirchhoff's voltage (second) law.
- Branch currents are found thereafter by taking the algebraic sum of the loop currents common to individual branches.

| Example 4.10 |

Calculate the current in each branch of the network shown in Fig. 4.13.

Fig. 4.13 Circuit diagram for Example 4.10

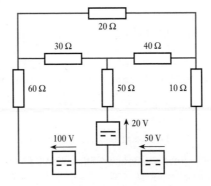

Let the circulating loop currents be as shown in Fig. 4.14.

In loop ①:

$$100 - 20 = I_1(60 + 30 + 50) - I_2 50 - I_3 30$$

$$\therefore \qquad 80 = 140I_1 - 50I_2 - 30I_3 \qquad\qquad\qquad (a)$$

Fig. 4.14 Circuit diagram for
Example 4.10

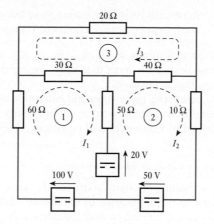

In loop ②:

$$50 + 20 - I_2(50 + 10 + 10) - I_1 50 - I_3 10$$

$$\therefore \qquad 70 = -50I_1 + 100I_2 - 40I_3 \qquad \qquad \text{(b)}$$

In loop ③:

$$0 = I_3(30 + 20 + 40) - I_1 30 - I_2 40$$

$$\therefore \qquad 0 = -30I_1 - 40I_2 + 90I_3 \qquad \qquad \text{(c)}$$

Solving for these equations gives

$$I_1 = 1.65 \text{ A} \qquad I_2 = 2.16 \text{ A} \qquad I_3 = 1.50 \text{ A}$$

Current in 60 $\Omega = I_1 = $ **1.65 A** in direction of I_1
Current in 30 $\Omega = I_1 - I_3 = $ **0.15 A** in direction of I_1
Current in 50 $\Omega = I_2 - I_1 = $ **0.51 A** in direction of I_2
Current in 40 $\Omega = I_2 - I_3 = $ **0.66 A** in direction of I_2
Current in 10 $\Omega = I_2 = $ **2.16 A** in direction of I_2
Current in 20 $\Omega = I_3 = $ **1.50 A** in direction of I_3.

In Example 4.10 all the circulating loop currents have been taken in the
same direction (i.e. clockwise). This is not essential when using this method,
but if the same direction is adopted for the loop currents then the equations
will always be of the form:

$$E_1 = R_{11}I_1 - R_{12}I_2 - R_{13}I_3 \ldots - R_{1n}I_n$$

$$E_2 = -R_{21}I_1 + R_{22}I_2 - R_{23}I_3 \ldots - R_{2n}I_n$$

$$E_3 = -R_{31}I_1 - R_{32}I_2 + R_{33}I_3 \ldots - R_{3n}I_n$$

$$E_n = -R_{n1}I_1 - R_{n12}I_2 - R_{n3}I_3 \ldots + R_{nn}I_n$$

where $E_1 = $ the algebraic sum of the e.m.f.s in loop ① in the direction of I_1;
$\qquad E_2 = $ the algebraic sum of the e.m.f.s in loop ② in the direction of I_2,
$\qquad \qquad$ etc.;
$\qquad R_{11} = $ sum of resistances in loop ①;
$\qquad R_{22} = $ sum of resistances in loop ②, etc.;

R_{12} = total resistance common to loops ① and ②;
R_{23} = total resistance common to loops ② and ③, etc.

By their definitions $R_{12} = R_{21}$, $R_{23} = R_{32}$, etc. Note that in the equation derived from each loop it is only the term in the loop's own circulating current that is positive.

By observing these rules the equations necessary for the solution of the circuit problem can be written down by inspection of the circuit. This can be confirmed by examination of equations (a), (b) and (c) in Example 4.10.

4.4   Nodal analysis

This technique of circuit solution, also known as the Node Voltage method, is based on the application of Kirchhoff's first (current) law at each junction (node) of the circuit, to find the node voltages. It should be noted that, in contrast, both the branch current and Mesh current techniques of circuit analysis are based on the applications of Kirchhoff's second (voltage) law, often to find unknown currents.

The Node Voltage method generally proceeds as follows:

Step 1: Choose a reference node to which all node voltages can be referred. Label all the other nodes with (unknown) values of voltage, V_1, V_2, etc.
Step 2: Assign currents in each connection to each node, except the reference node, in terms of the node voltages, V_1, V_2, etc.
Step 3: Apply Kirchhoff's current law at each node, obtaining as many equations as there are unknown node voltages.
Step 4: Solve the resulting equations to find the node voltages.

Example 4.11

Using Nodal analysis, calculate the voltages V_1 and V_2, in the circuit of Fig. 4.15.

Fig. 4.15 Network for Example 4.11

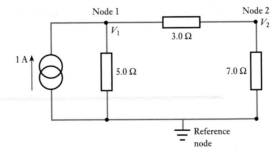

Refer to the four steps previously indicated:

Step 1: Reference node chosen. Voltages V_1 and V_2 assigned to the other two nodes.
Step 2: Assign currents in each connection to each node (Fig. 4.16).

Fig. 4.16 Part of Example 4.11

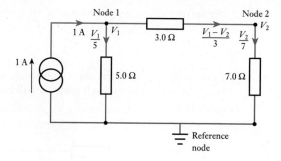

Step 3: Apply Kirchhoff's current law to sum the currents at each node.

At node 1:

$$\frac{V_1}{5} + \left(\frac{V_1 - V_2}{3}\right) = 1$$

which can be simplified to

$$V_1\left(\frac{1}{5} + \frac{1}{3}\right) - \frac{V_2}{3} = 1 \tag{a}$$

At node 2:

$$\frac{V_1 - V_2}{3} = \frac{V_2}{7}$$

which simplifies to

$$\frac{V_1}{3} - V_2\left(\frac{1}{3} + \frac{1}{7}\right) = 0 \tag{b}$$

Step 4: Solve node voltage equations (a) and (b).

From equation (b), by multiplying each term by 21,

$$7V_1 - V_2(7 + 3) = 0$$

∴ $$7V_1 = 10V_2$$

so $$V_2 = \frac{7}{10}V_1 \tag{c}$$

From equation (a), by multiplying each term by 15,

$$8V_1 - 5V_2 = 15 \tag{d}$$

Substitute for V_2, from equation (c), in equation (d):

$$8V_1 - \frac{35V_1}{10} = 15$$

so $$4.5V_1 = 15$$

$$V_1 = \frac{10}{3} \text{ V}$$

From (c)

$$V_2 = \frac{7}{3} \text{ V}$$

To check the accuracy of the calculation, see for yourself if Kirchhoff's current law is obeyed for each node. It will be seen that the currents are as in the circuit of Fig. 4.17.

Fig. 4.17　Part of Example 4.11

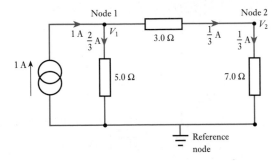

Example 4.12　Using the Node Voltage method calculate the voltages V_1 and V_2 in Fig. 4.18 and hence calculate the currents in the 8 Ω resistor.

Fig. 4.18　Network for Example 4.12

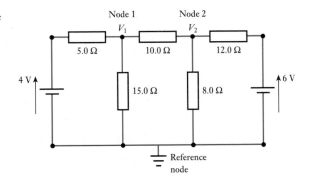

Step 1: Reference node shown. Voltages V_1 and V_2 assigned.
Step 2: Assign currents in each connection to each node (Fig. 4.19).

Fig. 4.19　Part of Example 4.12

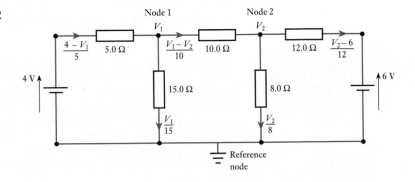

Step 3: Apply Kirchhoff's current law at each node.

At node 1:

$$\frac{4 - V_1}{5} = \frac{V_1 - V_2}{10} + \frac{V_1}{15}$$

Multiply each term by 30:

$$24 - 6V_1 = 3V_1 - 3V_2 + 2V_1$$

$$11V_1 - 3V_2 = 24 \qquad\qquad\qquad\qquad (a)$$

At node 2:

$$\frac{V_1 - V_2}{10} = \frac{V_2 - 6}{12} + \frac{V_2}{8}$$

Multiply each term by 120:

$$12V_1 - 12V_2 = 10V_2 - 60 + 15V_2$$

$$12V_1 - 37V_2 = -60 \qquad\qquad\qquad\qquad (b)$$

Step 4: Solve for V_1 and V_2.

Equation (a) $\times \dfrac{12}{11}$ gives:

$$12V_1 - \frac{36V_2}{11} = \frac{24 \times 12}{11} \qquad\qquad\qquad (c)$$

Equation (c) − equation (b) gives

$$33.37V_2 = 86.18$$

$$V_2 = \textbf{2.55 V}$$

From (a)

$$11V_1 = 24 + 3 + 2.55 = 31.65$$

$$V_1 = \textbf{2.88 V}$$

Hence the current in the 8 Ω resistor is

$$\frac{V_2}{8} = \textbf{0.32 A}$$

A second method of solving this problem by Nodal analysis, using source conversion techniques, is shown in section 4.7.

4.5 Superposition theorem

The Superposition theorem states that in any network containing more than one source, the current in, or the p.d. across, any branch can be found by considering each source separately and adding their effects: omitted sources of e.m.f. are replaced by resistances equal to their internal resistances.

This sounds very complicated, but is really quite simple when demonstrated by example. Example 4.13 illustrates the manner in which Example 4.7 would be solved by means of the Superposition theorem.

Example 4.13	By means of the Superposition theorem, calculate the currents in the network shown in Fig. 4.20(a).

Fig. 4.20   Circuit diagrams for Example 4.11

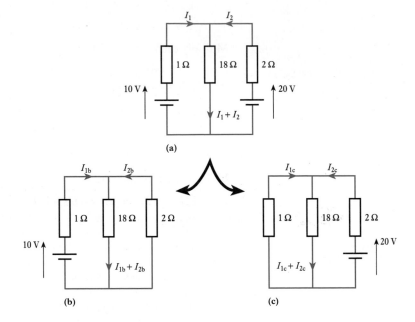

Because there are two sources of e.m.f. in the network, then two separate networks need to be considered, each having one source of e.m.f. Figure 4.20(b) shows the network with the 20 V source replaced by a short-circuit, there being zero internal resistance indicated. Also Fig. 4.20(c) shows the network with the 10 V source similarly replaced.

For the (b) arrangement, the total resistance is

$$1 + \frac{2 \times 18}{2 + 18} = 2.8 \ \Omega$$

thus $I_{1b} = \dfrac{10}{2.8} = 3.57 \ A$

and $I_{2b} = -\dfrac{18}{2 + 18} \times 3.57 = -3.21 \, A$

also $I_{1b} + I_{2b} = 3.57 - 3.21 = 0.36 \ A$

N.B. The current I_{2b} is negative due to the direction in which it has been shown.

For the (c) arrangement, the total resistance is

$$2 + \frac{1 \times 18}{1 + 18} = 2.95 \ \Omega$$

thus $I_{2c} = \dfrac{20}{2.95} = 6.78 \ A$

and $I_{1c} = -\dfrac{18}{1 + 18} \times 6.78 = -6.42 \ A$

$$I_{2c} + I_{1c} = 6.78 - 6.42 = 0.36 \text{ A}$$

Thus $\quad I_1 = I_{1b} + I_{1c} = 3.57 - 6.42 = -2.85 \text{ A}$

and $\quad I_2 = I_{2b} + I_{2c} = -3.21 + 6.78 = 3.57 \text{ A}$

also $\quad I_1 + I_2 = -2.85 + 3.57 = 0.72 \text{ A}$

4.6 Thévenin's theorem

The current through a resistor R connected across any two points A and B of an active network [i.e. a network containing one or more sources of e.m.f.] *is obtained by dividing the p.d. between A and B, with R disconnected, by (R + r), where r is the resistance of the network measured between points A and B with R disconnected and the sources of e.m.f. replaced by their internal resistances.*

An alternative way of stating Thévenin's theorem is as follows: *An active network having two terminals A and B can be replaced by a constant-voltage source having an e.m.f. E and an internal resistance r. The value of E is equal to the open-circuit p.d. between A and B, and r is the resistance of the network measured between A and B with the load disconnected and the sources of e.m.f. replaced by their internal resistances.*

Suppose A and B in Fig. 4.21(a) to be the two terminals of a network consisting of resistors having resistances R_2 and R_3 and a battery having an e.m.f. E_1 and an internal resistance R_1. It is required to determine the current through a load of resistance R connected across AB. With the load disconnected as in Fig. 4.21(b)

$$\text{Current through } R_3 = \frac{E_1}{R_1 + R_3}$$

and

$$\text{p.d. across } R_3 = \frac{E_1 R_3}{R_1 + R_3}$$

Fig. 4.21 Networks to illustrate Thévenin's theorem

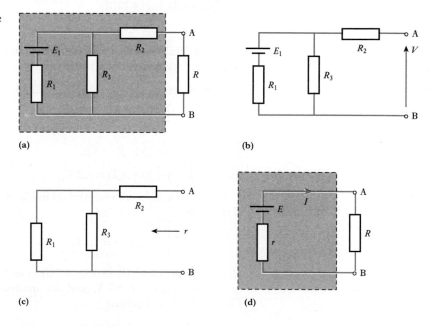

(a)

(b)

(c)

(d)

Since there is no current through R_2, p.d. across AB is

$$V = \frac{E_1 R_3}{R_1 + R_3}$$

Figure 4.21(c) shows the network with the load disconnected and the battery replaced by its internal resistance R_1. Resistance of network between A and B is

$$r = R_2 + \frac{R_1 R_3}{R_1 + R_3}$$

Thévenin's theorem merely states that the active network enclosed by the dotted line in Fig. 4.21(a) can be replaced by the very simple circuit enclosed by the dotted line in Fig. 4.21(d) and consisting of a source having an e.m.f. E equal to the open-circuit potential difference V between A and B, and an internal resistance r, where V and r have the values determined above. Hence

$$\text{Current through } R = I = \frac{E}{r + R}$$

Thévenin's theorem – sometimes referred to as Helmholtz's theorem – is an application of the Superposition theorem. Thus, if a source having an e.m.f. E equal to the open-circuit p.d. between A and B in Fig. 4.21(b) were inserted in the circuit between R and terminal A in Fig. 4.21(a), the positive terminal of the source being connected to A, no current would flow through R. Hence, this source could be regarded as circulating through R a current superimposed upon but opposite in direction to the current through R due to E_1 *alone*. Since the resultant current is zero, it follows that a source of e:m.f. E connected in series with R and the equivalent resistance r of the network, as in Fig. 4.21(d), would circulate a current I having the same value as that through R in Fig. 4.21(a), but in order that the direction of the current through R may be from A towards B, the polarity of the source must be as shown in Fig. 4.21(d).

| Example 4.14 | In Fig. 4.22(a) C and D represent the two terminals of an active network. Calculate the current through R_3. |

With R_3 disconnected, as in Fig. 4.22(b),

$$I_1 = \frac{6 - 4}{2 + 3} = 0.4 \text{ A}$$

and p.d. across CD is $E_1 - I_1 R_1$,

i.e. $E = 6 - (0.4 \times 2) = 5.2 \text{ V}$

When the e.m.f.s are removed, as in Fig. 4.22(c), total resistance between C and D is

$$\frac{2 \times 3}{2 + 3}, \quad \text{i.e.} \quad r = 1.2 \ \Omega$$

Hence the network AB in Fig. 4.22(a) can be replaced by a single source having an e.m.f. of 5.2 V and an internal resistance of 1.2 Ω, as in Fig. 4.22(d); consequently

Fig. 4.22 Circuit diagrams for Example 4.14

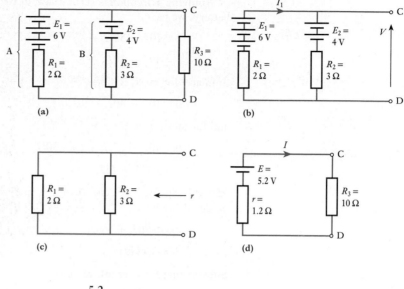

(a)

(b)

(c)

(d)

$$I = \frac{5.2}{1.2 + 10} = 0.46\,\text{A}$$

Example 4.15 The resistances of the various arms of a bridge are given in Fig. 4.23. The battery has an e.m.f. of 2.0 V and a negligible internal resistance. Determine the value and direction of the current in BD, using:

(a) Kirchhoff's laws;
(b) Thévenin's theorem.

(a) *By Kirchhoff's laws.* Let I_1, I_2 and I_3 be the currents in arms AB, AD and BD respectively, as shown in Fig. 4.23. Then by Kirchhoff's first law

current in BC $= I_1 - I_3$

and   current in DC $= I_2 + I_3$

Fig. 4.23 Network for Example 4.15

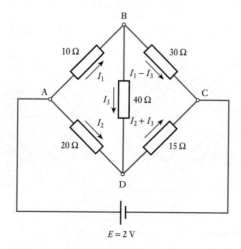

Applying Kirchhoff's second law to the mesh formed by ABC and the battery, we have

$$2 = 10I_1 + 30(I_1 - I_3)$$

$$= 40I_1 - 30I_3 \tag{a}$$

Similarly for mesh ABDA,

$$0 = 10I_1 + 40I_3 - 20I_2 \tag{b}$$

and for mesh BDCB,

$$0 = 40I_3 + 15(I_2 + I_3) - 30(I_1 - I_3)$$

$$= -30I_1 + 15I_2 + 85I_3 \tag{c}$$

Multiplying (b) by 3 and (c) by 4, and adding the two expressions thus obtained, we have

$$0 = -90I_1 + 460I_3$$

$$\therefore \qquad I_1 = 5.111I_3$$

Substituting for I_1 in (a), we have

$$I_3 = 0.0115 \text{ A} = \textbf{11.5 mA}$$

Since the value of I_3 is positive, the direction of I_3 is that assumed in Fig. 4.23, namely from B and D.

(b) *By Thévenin's theorem.* Since we require to find the current in the 40 Ω resistor between B and D, the first step is to remove this resistor, as in Fig. 4.24(a). Then p.d. between A and B is

$$2 \times \frac{10}{10 + 30} = 0.5 \text{ V}$$

and p.d. between A and D is

$$2 \times \frac{20}{20 + 15} = 1.143 \text{ V}$$

therefore p.d. between B and D is

$$1.143 - 0.5 = 0.643 \text{ V}$$

Fig. 4.24 Diagrams for solution of Example 4.15 by Thévenin's theorem

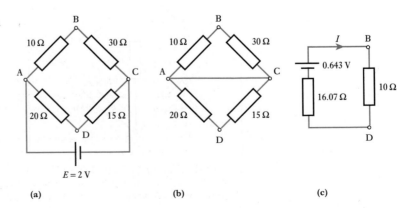

(a) (b) (c)

B being positive relative to D. Consequently, current in the 40 Ω resistor, when connected between B and D, will flow from B to D.

The next step is to replace the battery by a resistance equal to its internal resistance. Since the latter is negligible in this problem, junctions A and C can be short-circuited as in Fig. 4.24(b). Equivalent resistance of BA and BC is

$$\frac{10 \times 30}{10 + 30} = 7.5\,\Omega$$

and equivalent resistance of AD and CD is

$$\frac{20 \times 15}{20 + 15} = 8.57\,\Omega$$

therefore total resistance of network between B and D = 16.07 Ω. Hence the network of Fig. 4.24(a) is equivalent to a source having an e.m.f. of 0.643 V and an internal resistance of 16.07 Ω as in Fig. 4.24(c).

$$\therefore \qquad \text{Current through BD} = \frac{0.643}{16.07 + 40} = 0.0115\,\text{A}$$

$$= \textbf{11.5 mA from B to D}$$

4.7 The constant-current generator

It was shown in section 4.6 that a source of electrical energy could be represented by a source of e.m.f. in series with a resistance. This is not, however, the only form of representation. Consider such a source feeding a load resistor R_L as shown in Fig. 4.25.

From this circuit:

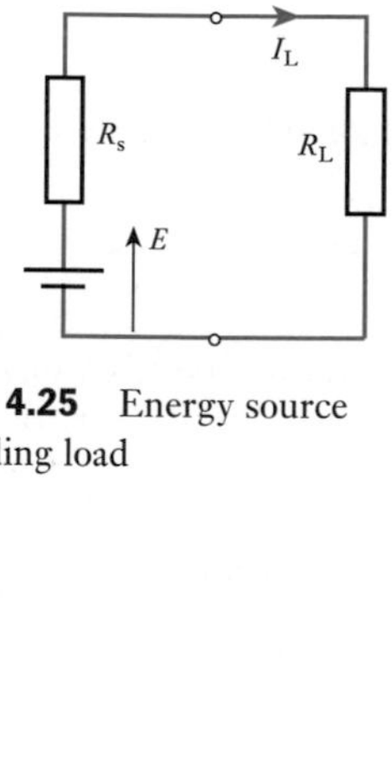

$$I_L = \frac{E}{R_s + R_L} = \frac{\frac{E}{R_s}}{\frac{R_s + R_L}{R_s}}$$

$$\therefore \qquad I_L = \frac{R_s}{R_s + R_L} \times I_s \qquad\qquad [4.1]$$

Fig. 4.25 Energy source feeding load

where $I_s = E/R_s$ is the current which would flow in a short-circuit across the output terminals of the source.

It can be seen from relation [4.1] that, when viewed from the load, the source appears as a source of current (I_s) which is dividing between the internal resistance (R_s) and the load resistor (R_L) connected in parallel. For the solution of problems, either form of representation can be used. In many practical cases an easier solution is obtained using the current form. Figure 4.26 illustrates the equivalence of the two forms.

The resistance of the constant-current generator must be taken as infinite, since the resistance of the complete source must be R_s as is obtained with the constant-voltage form.

The ideal constant-voltage generator would be one with zero internal resistance so that it would supply the same voltage to all loads. Conversely, the ideal constant-current generator would be one with infinite internal resistance so that it supplied the same current to all loads. These ideal conditions can be approached quite closely in practice.

Fig. 4.26 Equivalence of constant-voltage generator and constant-current generator forms of representation

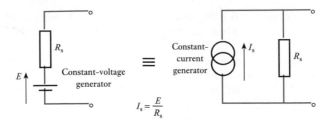

$$I_s = \frac{E}{R_s}$$

| Example 4.16 | Represent the network shown in Fig. 4.27 by one source of e.m.f. in series with a resistance. |

Potential difference across output terminals is

$$V_o = 1 \times 15 = 15 \text{ V}$$

Resistance looking into output terminals is

$$5 + 15 = 20 \; \Omega$$

therefore the circuit can be represented as shown in Fig. 4.28.

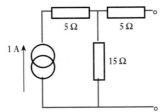

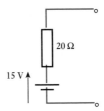

Fig. 4.27   Network for Example 4.15

Fig. 4.28   Part of Example 4.14

| Example 4.17 | The Node Voltage technique (Nodal analysis) lends itself to circuit models having current instead of voltage sources. To illustrate the technique, we will convert all the voltage sources of Fig. 4.18 to current sources and replace them in the circuit of Fig. 4.29. This produces Fig. 4.30 to which we apply the rules of Nodal analysis. |

At node 1:

$$0.8 = \frac{V_1}{5} + \frac{V_1}{15} + \frac{V_1 - V_2}{10}$$

$$0.8 = V_1\left(\frac{1}{5} + \frac{1}{15} + \frac{1}{10}\right) - \frac{V_2}{10}$$

Multiply by 30:

$$24 = V_1(6 + 2 + 3) - 3V_2$$

$$24 = 11V_1 - 3V_2 \tag{a}$$

Fig. 4.29 Source conversions
for Example 4.17

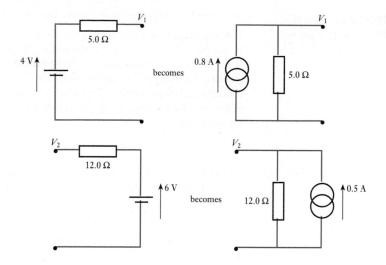

Fig. 4.30 New circuit for
Example 4.17

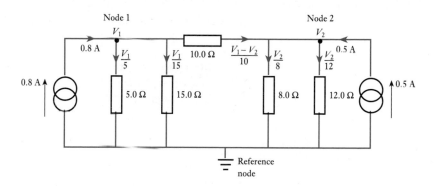

At node 2:

$$0.5 = \frac{V_2}{8} + \frac{V_2}{12} - \left(\frac{V_1 - V_2}{10}\right)$$

$$0.5 = \frac{-V_1}{10} + V_2\left(\frac{1}{8} + \frac{1}{10} + \frac{1}{12}\right)$$

Multiply by 120:

$$60 = -12V_1 + V_2(15 + 12 + 10)$$

$$60 = -12V_1 + 37V_2 \qquad\qquad\qquad\qquad\qquad \text{(b)}$$

(a) $\times \dfrac{12}{11}$ $\qquad 26.8 = 12V_1 - 3.273V_2 \qquad\qquad\qquad$ (c)

(c) + (b) $\qquad 86.8 = 33.727V_2$

$$V_2 = 2.55 \text{ V}$$

Hence the current in the 8 Ω resistor $(V_2/8) = \textbf{0.32 A}$ as before.

4.8   **Norton's theorem**

When a branch in a circuit is open-circuited the remainder of the circuit can be represented by one source of e.m.f. in series with a resistor; it follows from what has been said in section 4.7 that it could equally well be represented by a source of current in parallel with the same resistor. Norton's theorem is therefore a restatement of Thévenin's theorem using an equivalent current-generator source instead of the equivalent voltage-generator source. It can therefore be stated that:

> The current which flows in any branch of a network is the same as that which would flow in the branch if it were connected across a source of electrical energy, the short-circuit current of which is equal to the current that would flow in a short-circuit across the branch, and the internal resistance of which is equal to the resistance which appears across the open-circuited branch terminals.

Norton's theorem is illustrated in Fig. 4.31.

Fig. 4.31   Norton's theorem

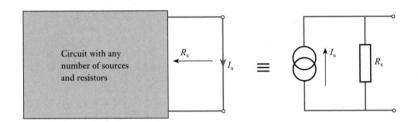

Example 4.18

Calculate the potential difference across the 2.0 Ω resistor in the network shown in Fig. 4.32.

Short-circuiting the branch containing the 2.0 Ω resistor gives the network shown in Fig. 4.33.

Fig. 4.32   Network for Example 4.15

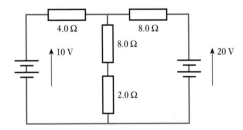

Fig. 4.33   Part of Example 4.15

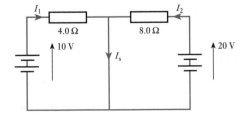

Fig. 4.34 Part of Example 4.18

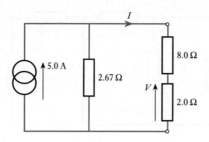

$$10 = 4.0I_1 \quad \therefore \quad I_1 = \frac{10}{4.0} = 2.5 \text{ A}$$

$$20 = 8.0I_2 \quad \therefore \quad I_2 = \frac{20}{8.0} = 2.5 \text{ A}$$

$$\therefore \qquad I_s = I_1 + I_2 = 5.0 \text{ A}$$

Resistance across open–circuited branch is

$$\frac{4.0 \times 8.0}{4.0 + 8.0} = 2.67 \, \Omega$$

therefore the circuit reduces to that shown in Fig. 4.34.

$$I = \frac{2.67}{2.67 + 10.0} \times 5.0 = 1.06 \text{ A}$$

$$\therefore \qquad V = 1.06 \times 2.0 = \mathbf{2.1 \ V}$$

Example 4.19 Calculate the current in the 5.0 Ω resistor in the network shown in Fig. 4.35.

Short-circuiting the branch containing the 5.0 Ω resistor gives the circuit shown in Fig. 4.36. Since the branch containing the 4 Ω and 6 Ω is short-circuited, we can ignore it, thus the source current is divided between the 8 Ω branch and the short-circuit branch which still has the 2 Ω resistance in series with the short circuit.

$$I_s = \frac{8.0}{8.0 + 2.0} \times 10 = 8.0 \text{ A}$$

Fig. 4.35 Network for Example 4.19

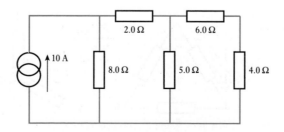

Fig. 4.36   Part of Example 4.19

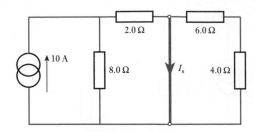

When obtaining the equivalent source resistance seen from the terminal of the open-circuit (i.e. when the 5.0 Ω has been removed) the current generator is replaced by an open-circuit, hence the resistance looking into the output terminals is

$$\frac{(2.0 + 8.0)(6.0 + 4.0)}{(2.0 + 8.0) + (6.0 + 4.0)} = \frac{10 \times 10}{20} = 5.0\,\Omega$$

therefore the circuit reduces to that shown in Fig. 4.37.

$$I = \frac{5.0}{5.0 + 5.0} \times 8.0 = \mathbf{4.0\,A}$$

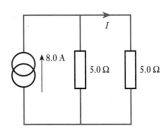

Fig. 4.37   Part of Example 4.19

4.9   Delta–star transformation

Figure 4.38(a) shows three resistors R_1, R_2 and R_3 connected in a closed mesh or *delta* to three terminals A, B and C, *their numerical subscripts 1, 2 and 3 being opposite to the terminals* A, B *and* C *respectively*. It is possible to replace these delta-connected resistors by three resistors R_a, R_b and R_c connected respectively between the same terminals A, B and C and a common point S, as in Fig. 4.38(b). Such an arrangement is said to be *star-connected*. It will be noted that the letter subscripts are now those of the terminals to which the respective resistors are connected. If the star-connected network is to be equivalent to the delta-connected network, the resistance between any two terminals in Fig. 4.38(b) must be the same as that between the same two terminals in Fig. 4.38(a). Thus, if we consider terminals A and B in Fig. 4.38(a), we have a circuit having a resistance R_3 in parallel with a circuit having resistances R_1 and R_2 in series; hence

$$R_{AB} = \frac{R_3(R_1 + R_2)}{R_1 + R_2 + R_3} \qquad\qquad [4.2]$$

Fig. 4.38   Delta–star transformation

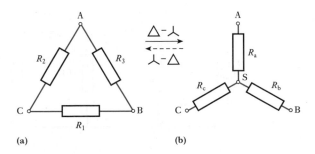

(a)   (b)

For Fig. 4.38(b), we have

$$R_{AB} = R_a + R_b \qquad [4.3]$$

In order that the networks of Fig. 4.38(a) and (b) may be equivalent to each other, the values of R_{AB} represented by expressions [4.2] and [4.3] must be equal.

$$\therefore \qquad R_a + R_b = \frac{R_1 R_3 + R_2 R_3}{R_1 + R_2 + R_3} \qquad [4.4]$$

Similarly

$$R_b + R_c = \frac{R_1 R_2 + R_1 R_3}{R_1 + R_2 + R_3} \qquad [4.5]$$

and

$$R_a + R_c = \frac{R_1 R_2 + R_2 R_3}{R_1 + R_2 + R_3} \qquad [4.6]$$

Subtracting equation [4.5] from [4.6], we have

$$R_a - R_c = \frac{R_2 R_3 - R_1 R_2}{R_1 + R_2 + R_3} \qquad [4.7]$$

Adding equations [4.6] and [4.7] and dividing by 2, we have

$$R_a = \frac{R_2 R_3}{R_1 + R_2 + R_3} \qquad [4.8]$$

Similarly

$$R_b = \frac{R_3 R_1}{R_1 + R_2 + R_3} \qquad [4.9]$$

and

$$R_c = \frac{R_1 R_2}{R_1 + R_2 + R_3} \qquad [4.10]$$

These relationships may be expressed thus: *the equivalent star resistance connected to a given terminal is equal to the product of the two delta resistances connected to the same terminal divided by the sum of the delta resistances.*

4.10 Star–delta transformation

Let us next consider how to replace the star-connected network of Fig. 4.38(b) by the equivalent delta-connected network of Fig. 4.38(a). Dividing equation [4.8] by equation [4.9], we have

$$\frac{R_a}{R_b} = \frac{R_2}{R_1}$$

$$\therefore \qquad R_2 = \frac{R_1 R_a}{R_b}$$

Similarly, dividing equation [4.8] by equation [4.10], we have

$$\frac{R_a}{R_c} = \frac{R_3}{R_1}$$

$$\therefore \qquad R_3 = \frac{R_1 R_a}{R_c}$$

Substituting for R_2 and R_3 in equation [4.8], we have

$$R_1 = R_b + R_c + \frac{R_b R_c}{R_a} \qquad\qquad [4.11]$$

Similarly

$$R_2 = R_c + R_a + \frac{R_c R_a}{R_b} \qquad\qquad [4.12]$$

and

$$R_3 = R_a + R_b + \frac{R_a R_b}{R_c} \qquad\qquad [4.13]$$

These relationships may be expressed thus: *the equivalent delta resistance between two terminals is the sum of the two star resistances connected to those terminals plus the product of the same two star resistances divided by the third star resistance.*

4.11 Maximum power transfer

Let us consider a source, such as a battery or a d.c. generator, having an e.m.f. E and an internal resistance r, as shown enclosed by the dotted rectangle in Fig. 4.39. A variable resistor is connected across terminals A and B of the source. If the value of the load resistance is R, then

$$I = \frac{E}{r + R}$$

and power transferred to load is

$$I^2 R = \frac{E^2 R}{(r + R)^2} = \frac{E^2 R}{r^2 + 2rR + R^2}$$

$$I^2 R = \frac{E^2}{(r^2/R) + 2r + R} \qquad\qquad [4.14]$$

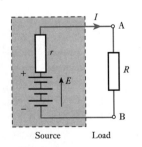

Fig. 4.39 Resistance matching

This power is a maximum when the denominator of [4.14] is a minimum, i.e. when

$$\frac{\mathrm{d}}{\mathrm{d}R}\left(\frac{r^2}{R} + 2r + R\right) = 0$$

$$\therefore \qquad -\frac{r^2}{R^2} + 1 = 0$$

or $\qquad R = r \qquad\qquad [4.15]$

To check that this condition gives the minimum and not the maximum value of the denominator in expression [4.14], expression $\{-(r^2/R^2) + 1\}$ should be differentiated with respect to R, thus

$$\frac{\mathrm{d}}{\mathrm{d}R}\left\{1 - \frac{r^2}{R^2}\right\} = 2r^2/R^3$$

Since this quantity is positive, expression [4.15] is the condition for the denominator of equation [4.14] to be a minimum and therefore the output power to be a maximum. Hence the power transferred from the source to the load is a maximum when the resistance of the load is equal to the internal resistance of the source. This condition is referred to as *resistance matching*.

Resistance matching is of importance in communications and electronic circuits where the source usually has a relatively high resistance and where it is desired to transfer the largest possible amount of power from the source to the load. In the case of power sources such as generators and batteries, the internal resistance is so low that it is impossible to satisfy the above condition without overloading the source.

Summary of important formulae

For delta–star transformation

$$R_a = \frac{R_2 R_3}{R_1 + R_2 + R_3}$$ [4.8]

For star–delta transformation

$$R_1 = R_b + R_c + \frac{R_b R_c}{R_a}$$ [4.11]

For maximum power transfer

$$R = r$$ [4.15]

Terms and concepts

Most circuit problems can be solved by applying Kirchhoff's laws to produce simultaneous equations; the solution of these equations is often unnecessarily difficult.

In **Mesh analysis**, circulating currents are allocated to closed loops or meshes in the circuit rather than to branches.

Nodal analysis is based on the application of Kirchhoff's first (current) law at each junction (node) of a circuit, to find the node voltages.

The **Superposition theorem** states that we can solve a circuit problem one source at a time, finally imposing the analyses one on another.

Thévenin's theorem states that any network supplying a load can be replaced by a constant-voltage source in series with an internal resistance.

Norton's theorem states that any network supplying a load can be replaced by a constant-current source in parallel with an internal resistance.

The **delta–star transformation** permits us to replace any three loads connected in delta by an equivalent three loads connected in star. The star–delta transformation permits the converse transfer.

The **maximum-power transfer theorem** states that maximum power is dissipated by a load when its resistance is equal to the equivalent internal resistance of the source.

Exercises 4

1. State Kirchhoff's laws and apply them to the solution of the following problem.

    Two batteries, A and B, are connected in parallel, and an 80 Ω resistor is connected across the battery terminals. The e.m.f. and the internal resistance of battery A are 100 V and 5 Ω respectively, and the corresponding values for battery B are 95 V and 3 Ω respectively. Find (a) the value and direction of the current in each battery and (b) the terminal voltage.

2. State Kirchhoff's laws for an electric circuit, giving an algebraic expression for each law.

    A network of resistors has a pair of input terminals AB connected to a d.c. supply and a pair of output terminals CD connected to a load resistor of 120 Ω. The resistances of the network are AC = BD = 180 Ω, and AD = BC = 80 Ω. Find the ratio of the current in the load resistor to that taken from the supply.

3. State Kirchhoff's laws as applied to an electrical circuit.

    A secondary cell having an e.m.f. of 2 V and an internal resistance of 1 Ω is connected in series with a primary cell having an e.m.f. of 1.5 V and an internal resistance of 100 Ω, the negative terminal of each cell being connected to the positive terminal of the other cell. A voltmeter having a resistance of 50 Ω is connected to measure the terminal voltage of the cells. Calculate the voltmeter reading and the current in each cell.

4. State and explain Kirchhoff's laws relating to electric circuits. Two storage batteries, A and B, are connected in parallel for charging from a d.c. source having an open–circuit voltage of 14 V and an internal resistance of 0.15 Ω. The open–circuit voltage of A is 11 V and that of B is 11.5 V; the internal resistances are 0.06 Ω and 0.05 Ω respectively. Calculate the initial charging currents.

    What precautions are necessary when charging batteries in parallel?

5. State Kirchhoff's laws as applied to an electrical circuit.

    Two batteries A and B are joined in parallel. Connected across the battery terminals is a circuit consisting of a battery C in series with a 25 Ω resistor, the negative terminal of C being connected to the positive terminals of A and B. Battery A has an e.m.f. of 108 V and an internal resistance of 3 Ω, and the corresponding values for battery B are 120 V and 2 Ω. Battery C has an e.m.f. of 30 V and a negligible internal resistance. Determine (a) the value and direction of the current in each battery and (b) the terminal voltage of battery A.

6. A network is arranged as shown in Fig. A. Calculate the value of the current in the 8 Ω resistor by (a) the Superposition theorem, (b) Kirchhoff's laws, (c) Thévenin's theorem and (d) Nodal analysis.

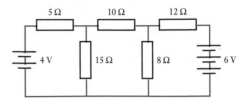

Fig. A

7. Find the voltage across the 4 Ω resistor in Fig. B. using (a) Nodal analysis, (b) the Superposition theorem and (c) Thévenin's theorem.

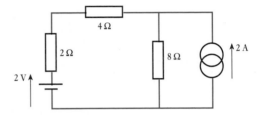

Fig. B

8. A network is arranged as in Fig. C. Calculate the equivalent resistance between (a) A and B, and (b) A and N.

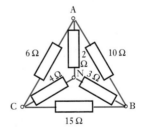

Fig. C

9. A network is arranged as in Fig. D, and a battery having an e.m.f. of 2 V and negligible internal resistance is connected across AC. Determine the value and direction of the current in branch BE.

10. Calculate the value of the current through the 40 Ω resistor in Fig. E.

11. Using Thévenin's theorem, calculate the current through the 10 Ω resistor in Fig. F.

12. A certain generator has an open–circuit voltage of 12 V and an internal resistance of 40 Ω. Calculate: (a) the load resistance for maximum power transfer; (b) the corresponding values of the terminal voltage and of the power supplied to the load.

Exercises 4 continued

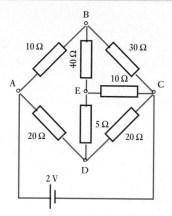

Fig. D

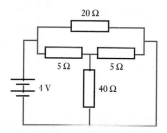

Fig. E

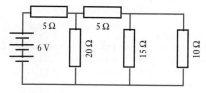

Fig. F

10 Ω respectively and the corresponding values for the other are 150 V and 20 Ω. A resistor of 50 Ω is connected across the battery terminals. Calculate (a) the current through the 50 Ω resistor, and (b) the value and direction of the current through each battery. If the 50 Ω resistor were reduced to 20 Ω resistance, find the new current through it.

15. Three resistors having resistances 50 Ω, 100 Ω and 150 Ω are star-connected to terminals A, B and C respectively. Calculate the resistances of equivalent delta-connected resistors.

16. Three resistors having resistance 20 Ω, 80 Ω and 30 Ω are delta-connected between terminals AB, BC and CA respectively. Calculate the resistances of equivalent star-connected resistors.

17. With the aid of delta and star connection diagrams, state the basic equations from which the delta–star and star–delta conversion equations can be derived.

A star network, in which N is the star point, is made up as follows: A–N = 70 Ω, B–N = 100 Ω and C–N = 90 Ω. Find the equivalent delta network. If the above star and delta networks were superimposed, what would be the measured resistance between terminals A and C?

18. Calculate the current in the 10 Ω resistor in the network shown in Fig. G.

19. For the network shown in Fig. H, calculate the potential difference V_{NO}. Calculate the resistance of a resistor connected across NO that would draw a current of 1.0 A.

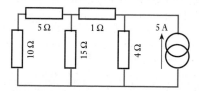

Fig. G

If the load resistance were increased to twice the value for maximum power transfer, what would be the power absorbed by the load?

13. A battery having an e.m.f. of 105 V and an internal resistance of 1 Ω is connected in parallel with a d.c. generator of e.m.f. 110 V and internal resistance of 0.5 Ω to supply a load having a resistance of 8 Ω. Calculate: (a) the currents in the battery, the generator and the load; (b) the potential difference across the load.

14. State your own interpretation of Thévenin's theorem and use it to solve the following problem.

Two batteries are connected in parallel. The e.m.f. and internal resistance of one battery are 120 V and

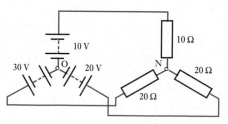

Fig. H

Chapter five

Capacitance and Capacitors

A capacitor is a device which can store electric charge for short periods of time. Like resistors, capacitors can be connected in series and in parallel and therefore we can analyse them after the fashion which we have developed in previous chapters.

We know that a resistor makes the passage of electric charge difficult, hence the production of heat, but otherwise we do not bother too much about what happens in the resistor. However, the effect of storing charge in a capacitor has much more significance not only within the capacitor but also in the space surrounding it. The effect in such space is termed the electric field and this requires that we investigate it some detail.

If we wished to fill a container with water, we know that it takes time to pour in that water. In much the same way, it takes time to pour charge into a capacitor and again this speed of action is something with which we need to become familiar since it has a great deal of influence on the application of capacitors.

Capacitors are widely used in all branches of electrical engineering and the effect of capacitance is to be found wherever there is an electric circuit. We shall find that capacitors are one of the three main components in any electrical system.

5.1   Capacitors

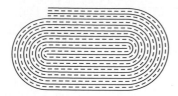

Fig. 5.1 Paper-insulated capacitor

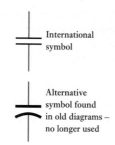

Fig. 5.2 Circuit symbols for a capacitor

In Chapter 2 we considered the presence of electric charge. The next step requires that we introduce a device which can hold a reasonable amount of charge. Such a device is called a *capacitor*, although some people still use the old term condenser.

Early experimenters found that conductors would hold much greater electric charges provided that they were held in close proximity to one another yet kept apart. They also found that the greater the surface area of the conductors then the greater the stored charge.

A simple capacitor can be made from two strips of metal foil sandwiched with two thin layers of insulation. Waxed paper is a suitable insulant; the wax is needed to keep damp out of the paper which otherwise would quickly cease being an insulator. The foil and paper is rolled as shown in Fig. 5.1. Thus we have a device bringing two conductors of large area into very close proximity with one another yet which are insulated, and this would provide a practical capacitor which can be used to hold electric charge.

A capacitor's ability to hold electric charge is measured in *farads*. This is a very large unit and most capacitors are rated in microfarads or less. In circuit diagrams, there are two common symbols for a capacitor as shown in Fig. 5.2. In subsequent diagrams the international symbol will be used.

A charged capacitor may be regarded as a reservoir of electricity and its action can be demonstrated by connecting a capacitor of, say, 20 μF in series with a resistor R, a centre-zero microammeter A and a two-way switch S, as in Fig. 5.3. A voltmeter V is connected across C. If R has a resistance of, say, 1 MΩ, it is found that when switch S is closed on position a, the ammeter A shows a deflection rising immediately to its maximum value and then falling off to zero. This means that initially there has been a significant current due to the inrush of electric charge into the uncharged capacitor, subsequently reducing to zero once the capacitor was fully charged. This change of current is indicated by curve G in Fig. 5.4.

At the same time the voltmeter indicates a rise in voltage across the capacitor C. This rise of voltage is indicated by curve M.

When the switch S is moved over to position b, the ammeter again performs as before except that the indication is in the reverse direction. The reverse deflection is due to the charge rushing out from the capacitor. The current is indicated by curve H.

At the same time the voltmeter indicates a fall in voltage across the capacitor C. This fall in voltage is indicated by curve N.

If the experiment is repeated with a resistance of, say, 2 MΩ, it is found that the initial current, both on charging and on discharging, is halved, but

Fig. 5.3 Capacitor charged and discharged through a resistor

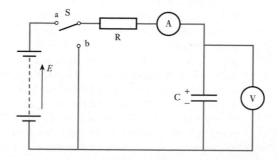

Fig. 5.4 Charging and
discharging currents and p.d.s

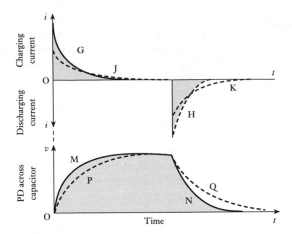

it takes about twice as long to rise up and to fall off, as shown by the dotted
curves J and K. Curves P and Q represent the corresponding variation of the
p.d. across C during charge and discharge respectively.

The shaded area between curve G and the horizontal axis in Fig. 5.4
represents the product of the charging current (in amperes) and the time (in
seconds), namely the quantity of electricity (in coulombs) required to charge
the capacitor to a p.d. of V volts. Similarly the shaded area enclosed by curve
H represents the same quantity of electricity obtainable during discharge.

5.2   **Hydraulic
analogy**

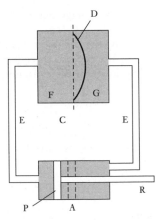

Fig. 5.5 Hydraulic analogy of a
capacitor

The operation of charging and discharging a capacitor may be more easily
understood if we consider the hydraulic analogy given in Fig. 5.5, where P
represents a piston operated by a rod R and D is a rubber diaphragm
stretched across a cylindrical chamber C. The cylinders are connected by
pipes E and are filled with water.

When no force is being exerted on P, the diaphragm is flat, as shown dotted,
and the piston is in position A. If P is pushed towards the left, water is withdrawn from G and forced into F and the diaphragm is in consequence distended, as shown by the full line. The greater the force applied to P, the greater
is the amount of water displaced. But the rate at which this displacement takes
place depends upon the resistance offered by pipes E; thus the smaller the
cross-sectional area of the pipes the longer is the time required for the steady
state to be reached. The force applied to P is analogous to the e.m.f. of the
battery, the quantity of water displaced corresponds to the charge, the rate at
which the water passes any point in the pipes corresponds to the current and
the cylinder C with its elastic diaphragm is the analogue of the capacitor.

When the force exerted on P is removed, the distended diaphragm forces
water out of F back into G; and if the frictional resistance of the water in the
pipes exceeds a certain value, it is found that the piston is merely pushed
back to its original position A. The strain energy stored in the diaphragm due
to its distension is converted into heat by the frictional resistance. The effect
is similar to the discharge of the capacitor through a resistor.

No water can pass from F to G through the diaphragm so long as it
remains intact; but if it is strained excessively it bursts, just as the insulation
in a capacitor is punctured when the p.d. across it becomes excessive.

5.3 Charge and voltage

The experiment described in section 5.1 shows that charge is transferred but it is unsuitable for accurate measurement of the charge. A suitable method is to discharge the capacitor through a ballistic galvanometer G, since the deflection of the latter is proportional to the charge.

Let us charge a capacitor C (Fig. 5.6) to various voltages by means of a slider on a resistor R connected across a battery B, S being at position a; and for each voltage, note the deflection of G when C is discharged through it by moving S over to position b. Thus, if θ is the first deflection or 'throw' observed when the capacitor, charged to a p.d. of V volts, is discharged through G, and if k is the ballistic constant of G in coulombs per unit of first deflection, then discharge through G is

$$Q = k\theta \text{ coulombs}$$

It is found that, for a given capacitor,

$$\frac{\text{Charge on C [coulombs]}}{\text{p.d. across C [volts]}} = \text{a constant} \qquad [5.1]$$

Fig. 5.6 Measurement of charge by ballistic galvanometer

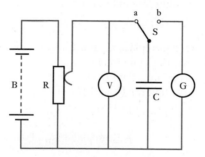

5.4 Capacitance

The property of a capacitor to store an electric charge when its plates are at different potentials is referred to as its *capacitance*.

The unit of capacitance is termed the *farad* (abbreviation F) which may be defined as *the capacitance of a capacitor between the plates of which there appears a potential difference of 1 volt when it is charged by 1 coulomb of electricity*.

Capacitance   **Symbol:** C **Unit:** farad (F)

It follows from expression [5.1] and from the definition of the farad that

$$\frac{\text{Charge [coulombs]}}{\text{Applied p.d. [volts]}} = \text{capacitance [farads]}$$

or in symbols

$$\frac{Q}{V} = C$$

∴ $Q = CV \quad \text{coulombs}$ [5.2]

In practice, the farad is found to be inconveniently large and the capacitance is usually expressed in *microfarads* (μF) or in *picofarads* (pF), where

$$1 \ \mu F = 10^{-6} \ F$$

and $$1 \ pF = 10^{-12} \ F$$

Example 5.1

A capacitor having a capacitance of 80 μF is connected across a 500 V d.c. supply. Calculate the charge.

From equation [5.2]

$$Q = CV$$

$\therefore$ charge $= (80 \times 10^{-6}) \ [F] \times 500 \ [V]$

$$= 0.04 \ C = \textbf{40 mC}$$

5.5   **Capacitors in parallel**

Suppose two capacitors, having capacitances C_1 and C_2 farads respectively, to be connected in parallel (Fig. 5.7) across a p.d. of V volts. The charge on C_1 is Q_1 coulombs and that on C_2 is Q_2 coulombs, where

$$Q_1 = C_1 V \quad \text{and} \quad Q_2 = C_2 V$$

If we were to replace C_1 and C_2 by a single capacitor of such capacitance C farads that the same total charge of $(Q_1 + Q_2)$ coulombs would be produced by the same p.d., then $Q_1 + Q_2 = CV$.

Substituting for Q_1 and Q_2, we have

$$C_1 V + C_2 V = CV$$

$$\boxed{C = C_1 + C_2 \quad \text{farads}} \qquad [5.3]$$

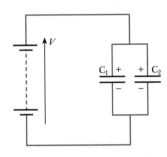

Fig. 5.7 Capacitors in parallel

Hence the *resultant capacitance of capacitors in parallel is the arithmetic sum of their respective capacitances.*

5.6   **Capacitors in series**

Suppose C_1 and C_2 in Fig. 5.8 to be two capacitors connected in series with suitable centre-zero ammeters A_1 and A_2, a resistor R and a two-way switch S. When S is put over to position a, A_1 and A_2 are found to indicate exactly the same charging current, each reading decreasing simultaneously from a maximum to zero, as already shown in Fig. 5.4. Similarly, when S is put over to position b, A_1 and A_2 indicate similar discharges. It follows that during charge the displacement of electrons from the positive plate of C_1 to the negative plate of C_2 is exactly the same as that from the upper plate (Fig. 5.8) of C_2 to the lower plate of C_1. In other words the displacement of Q coulombs of electricity is the same in every part of the circuit, and the charge on each capacitor is therefore Q coulombs.

If V_1 and V_2 are the corresponding p.d.s across C_1 and C_2 respectively, then from equation [5.2]:

$$Q = C_1 V_1 = C_2 V_2$$

so that

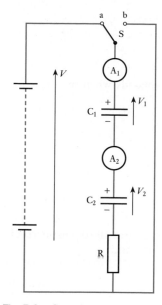

Fig. 5.8 Capacitors in series

$$V_1 = \frac{Q}{C_1} \quad \text{and} \quad V_2 = \frac{Q}{C_2} \qquad [5.4]$$

If we were to replace C_1 and C_2 by a single capacitor of capacitance C farads such that it would have the same charge Q coulombs with the same p.d. of V volts, then

$$Q = CV \quad \text{or} \quad V = \frac{Q}{C}$$

But it is evident from Fig. 5.8 that $V = V_1 + V_2$. Substituting for V, V_1 and V_2, we have

$$\frac{Q}{C} = \frac{Q}{C_1} + \frac{Q}{C_2}$$

$$\therefore \qquad \frac{1}{C} = \frac{1}{C_1} + \frac{1}{C_2} \qquad [5.5]$$

Hence *the reciprocal of the resultant capacitance of capacitors connected in series is the sum of the reciprocals of their respective capacitances.*

5.7 Distribution of voltage across capacitors in series

From expression [5.4]

$$\frac{V_2}{V_1} = \frac{C_1}{C_2} \qquad [5.6]$$

But $V_1 + V_2 = V$

$$\therefore \qquad V_2 = V - V_1$$

Substituting for V_2 in equation [5.6], we have

$$\frac{V - V_1}{V_1} = \frac{C_1}{C_2}$$

$$\therefore \qquad V_1 = V \times \frac{C_2}{C_1 + C_2} \qquad [5.7]$$

and

$$V_2 = V \times \frac{C_1}{C_1 + C_2} \qquad [5.8]$$

Example 5.2 Three capacitors have capacitances of 2, 4 and 8 μF respectively. Find the total capacitance when they are connected

(a) in parallel;
(b) in series.

(a) From equation [5.3]:

$$C = C_1 + C_2 + C_3$$

Total capacitance $= 2 + 4 + 8 = \textbf{14} \, \boldsymbol{\mu}\textbf{F}$

(b) If C is the resultant capacitance in microfarads when the capacitors are in series, then from equation [5.5]:

$$\frac{1}{C} = \frac{1}{C_1} + \frac{1}{C_2} + \frac{1}{C_3} = \frac{1}{2} + \frac{1}{4} + \frac{1}{8}$$

$$= 0.5 + 0.25 + 0.125 = 0.875$$

$\therefore \qquad C = \textbf{1.14} \, \boldsymbol{\mu}\textbf{F}$

| Example 5.3 | If two capacitors having capacitances of 6 μF and 10 μF respectively are connected in series across a 200 V supply, find |

(a) the p.d. across each capacitor,
(b) the charge on each capacitor.

(a) Let V_1 and V_2 be the p.d.s. across the 6 μF and 10 μF capacitors respectively; then, from expression [5.7],

$$V_1 = 200 \times \frac{10}{6 + 10} = \textbf{125 V}$$

and $\qquad V_2 = 200 - 125 = \textbf{75 V}$

(b) Charge on each capacitor

$$Q = \text{charge on } C_1$$

$$= 6 \times 10^{-6} \times 125 = 0.000\,75 \text{ C} = \textbf{750} \, \boldsymbol{\mu}\textbf{C}$$

5.8 Capacitance and the capacitor

It follows from expression [5.3] that if two similar capacitors are connected in parallel, the capacitance is double that of one capacitor. But the effect of connecting two similar capacitors in parallel is merely to double the area of each plate. In general, we may therefore say that the capacitance of a capacitor is proportional to the area of the plates.

On the other hand, if two similar capacitors are connected in series, it follows from expression [5.5] that the capacitance is halved. We have, however, doubled the thickness of the insulation between the plates that are connected to the supply. Hence we may say in general that the capacitance of a capacitor is inversely proportional to the distance between the plates; and the above relationships may be summarized thus:

$$\text{Capacitance} \propto \frac{\text{area of plates}}{\text{distance between plates}}$$

In order to clarify this relationship, we now need to consider the space between the charged plates of a capacitor. In this space, the charges set up electric fields. The study of such electric fields is known as electrostatics.

5.9   Electric fields

The space surrounding a charge can be investigated using a small charged body. This investigation is similar to that applied to the magnetic field surrounding a current-carrying conductor. However, in this case the charged body is either attracted or repelled by the charge under investigation. The space in which this effect can be observed is termed the electric field of the charge and the force on the charged body is the electric force.

The lines of force can be traced out and they appear to have certain properties:

1. In an electric field, each line of force emanates from or terminates in a charge. The conventional direction is from the positive charge to the negative charge.
2. The direction of the line is that of the force experienced by a positive charge placed at a point in the field, assuming that the search charge has no effect on the field which it is being used to investigate.
3. The lines of force never intersect since the resultant force at any point in the field can have only one direction.

The force of attraction or of repulsion acts directly between two adjacent charges. All points on the surface of a conductor may be assumed to be at the same potential, i.e. equipotential, and the lines of force radiate out from equipotential surfaces at right angles. The most simple case is that of the isolated spherical charge shown in Fig. 5.9. However, most electric fields exist between two conductors. The two most important arrangements are those involving parallel plates (as in a simple capacitor) and concentric cylinders (as in a television aerial cable). The resulting fields are shown in Fig. 5.10.

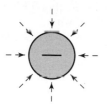

Fig. 5.9   Electric field about an isolated spherical charge

Fig. 5.10   Electric fields between oppositely charged surfaces. (a) Parallel plates; (b) concentric cylinders (cable)

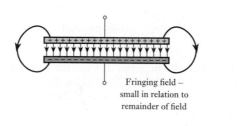

Fringing field – small in relation to remainder of field

(a)

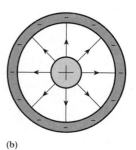

(b)

It should not be overlooked that the space between the conductors needs to be filled with an insulator, otherwise the charges would move towards one another and therefore be dissipated. The insulant is called a dielectric.

5.10   Electric field strength and electric flux density

We can investigate an electric field by observing its effect on a charge. In the SI method of measurement this should be a unit charge, i.e. a coulomb. In practice this is such a large charge that it would disrupt the field being investigated. Therefore our investigation is a matter of pure supposition.

The magnitude of the force experienced by this unit charge at any point in a field is termed the electric field strength at that point. Electric field

Fig. 5.11   A parallel–plate capacitor

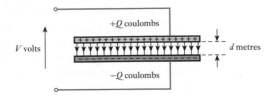

strength is sometimes also known as electric stress. It can be measured in newtons per unit charge and represented by the symbol E. (Since E can also represent e.m.f., we use a bold type for E when representing electric field strength and later we will meet D representing electric flux density.)

It should be recalled that 1 J of work is necessary to raise the potential of 1 C of charge through 1 V. When a charge moves through an electric field, the work done against or by the electric field forces is indicated by the change in potential of the charge. Therefore to move a unit charge through a field so that its potential changes by V volts requires V joules of work.

The most simple field arrangement which we can investigate is that between parallel charged plates as shown in Fig. 5.11. Let us suppose that the plates are very large and that the distance between them is very small. By doing this, we can ignore any fringing effects of the type shown in Fig. 5.10 and assume that all the field exists between the plates. Let us also assume that there is free space between the plates.

There is a potential difference of V volts between the plates, therefore the work in transferring 1 C of charge between the plates is V joules. But work is the product of force and distance, and in this case the distance is d metres. Therefore the force experienced by the charge is the electric field strength E given by

$$E = \frac{V}{d} \qquad \text{volts per metre} \qquad [5.9]$$

The total electric effect of a system as described by the lines of electric force is termed the electric flux linking the system. Flux is measured in the same units as electric charge, hence a flux of Q coulombs is created by a charge of Q coulombs.

The electric flux density is the measure of the electric flux passing at right angles through unit area, i.e. an area of 1 m^2. It follows that if the area of the plates in the capacitor of Fig. 5.11 is A then the electric flux density D is given by

$$D = \frac{Q}{A} \qquad \text{coulombs per square metre} \qquad [5.10]$$

From expressions [5.9] and [5.10]

$$\frac{\text{Electric flux density}}{\text{Electric field strength}} = \frac{D}{E} = \frac{Q}{A} \div \frac{V}{d} = \frac{Q}{V} \times \frac{d}{A} = \frac{Cd}{A}$$

In electrostatics, the ratio of the electric flux density in a vacuum to the electric field strength is termed the *permittivity of free space* and is represented by ϵ_0. Hence,

$$\epsilon_0 = \frac{Cd}{A}$$

or $$C = \frac{\epsilon_0 A}{d} \quad \text{farads}$$ [5.11]

Permittivity of free space Symbol: ϵ_0 Unit: farad per metre (F/m)

The value of ϵ_0 can be determined experimentally by charging a capacitor, of known dimensions and with vacuum dielectric, to a p.d. of V volts and then discharging it through a ballistic galvanometer having a known ballistic constant k coulombs per unit deflection. If the deflection is θ divisions,

$$Q = CV = k\theta$$

$$\therefore \qquad \epsilon_0 = C \cdot \frac{d}{A} = \frac{k\theta}{V} \cdot \frac{d}{A}$$

From carefully conducted tests it has been found that the value of ϵ_0 is 8.85×10^{-12} F/m.

Hence the capacitance of a parallel-plate capacitor with vacuum or air dielectric is given by

$$C = \frac{(8.85 \times 10^{-12}) \, [\text{F/m}] \times A \, [\text{m}^2]}{d \, [\text{m}]} \quad \text{farads}$$ [5.12]

5.11 Relative permittivity

If the experiment described in section 5.9 is performed with a sheet of glass filling the space between plates as shown in Fig. 5.12, it is found that the value of the capacitance is greatly increased; the ratio of the capacitance of a capacitor having a given material as dielectric to the capacitance of that capacitor with vacuum (or air) dielectric is termed the *relative permittivity* of that material and is represented by the symbol ϵ_r. Values of the relative permittivity of some of the most important insulating materials are given in Table 5.1: note that some of these vary with frequency.

Table 5.1 Important insulating materials

Material	Relative permittivity
Vacuum	1.0
Air	1.0006
Paper (dry)	2–2.5
Polythene	2–2.5
Insulating oil	3–4
Bakelite	4.5–5.5
Glass	5–10
Rubber	2–3.5
Mica	3–7
Porcelain	6–7
Distilled water	80
Barium titanate	6000+

Fig. 5.12   A parallel-plate capacitor with a glass dielectric

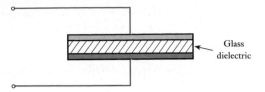

Glass
dielectric

Relative permittivity   Symbol: ϵ_r Unit: none

From expression [5.11], it follows that if the space between the metal plates of the capacitor in Fig. 5.12 is filled with a dielectric having a relative permittivity ϵ_r, capacitance

$$C = \frac{\epsilon_0 \epsilon_r A}{d} \quad \text{farads} \tag{5.13}$$

$$= \frac{(8.85 \times 10^{-12})\,[\text{F/m}] \times \epsilon_r \times A\,[\text{m}^2]}{d\,[\text{m}]} \quad \text{farads}$$

and charge due to a p.d. of V volts is

$$Q = CV$$

$$= \frac{\epsilon_0 \epsilon_r A V}{d} \quad \text{coulombs}$$

$$\therefore \quad \frac{\text{Electric flux density}}{\text{Electric field strength}} = \frac{D}{E} = \frac{Q}{A} \div \frac{V}{d} = \frac{Qd}{VA} = \epsilon_0 \epsilon_r$$

Let $\epsilon_0 \epsilon_r = \epsilon$ \hfill [5.14]

where ϵ is the absolute permittivity

$$\therefore \quad \text{Absolute permittivity } \epsilon = \epsilon_0 \epsilon_r = \frac{C\,[\text{farads}] \times d\,[\text{metres}]}{A\,[\text{metres}^2]}$$

$$= \frac{Cd}{A} \quad \text{farads per metre}$$

hence the units of absolute permittivity are *farads per metre*, e.g.

$$\epsilon_0 = 8.85 \times 10^{-12} \text{ F/m}$$

5.12   Capacitance of a multi-plate capacitor

Suppose a capacitor to be made up of n parallel plates, alternate plates being connected together as in Fig. 5.13. Let

A = area of *one* side of each plate in square metres

d = thickness of dielectric in metres

and ϵ_r = relative permittivity of the dielectric

Figure 5.13 shows a capacitor with seven plates, four being connected to A and three to B. It will be seen that each side of the three plates connected to B is in contact with the dielectric, whereas only one side of each of the

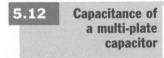

Fig. 5.13   Multi-plate capacitor

outer plates is in contact with it. Consequently, the useful surface area of each set of plates is $6A$ square metres. For n plates, the useful area of each set is $(n - 1)A$ square metres.

$$\therefore \qquad \text{Capacitance} = \frac{\epsilon_0\epsilon_r(n - 1)A}{d} \quad \text{farads}$$

$$\text{Capacitance} = \frac{8.85 \times 10^{-12}\,\epsilon_r(n - 1)A}{d} \quad \text{farads} \qquad [5.15]$$

Example 5.4   A capacitor is made with seven metal plates connected as in Fig. 5.13 and separated by sheets of mica having a thickness of 0.3 mm and a relative permittivity of 6. The area of one side of each plate is 500 cm². Calculate the capacitance in microfarads.

Using expression [5.15], we have $n = 7$, $A = 0.05$ m², $d = 0.0003$ m and $\epsilon_r = 6$.

$$\therefore \qquad C = \frac{8.85 \times 10^{-12} \times 6 \times 6 \times 0.05}{0.0003} = 0.0531 \times 10^{-6}\text{ F}$$

$$= 0.053\ \mu\text{F}$$

Example 5.5   A p.d. of 400 V is maintained across the terminals of the capacitor of Example 5.4. Calculate

(a) the charge;
(b) the electric field strength or potential gradient;
(c) the electric flux density in the dielectric.

(a) Charge

$$Q = CV = 0.0531\,[\mu\text{F}] \times 400\,[\text{V}] = \textbf{21.2}\ \mu\text{C}$$

(b) Electric field strength or potential gradient

$$E = V/d = 400\,[\text{V}]/0.0003\,[\text{m}] = 1\,333\,000\text{ V/m}$$

$$= \textbf{1330 kV/m}$$

(c) Electric flux density

$$D = Q/A = 21.24\,[\mu\text{C}]/(0.05 \times 6)\,[\text{m}^2]$$

$$= \textbf{70.8}\ \mu\text{C/m}^2$$

5.13 Composite-dielectric capacitors

Suppose the space between metal plates M and N to be filled by dielectrics 1 and 2 of thickness d_1 and d_2 metres respectively, as shown in Fig. 5.14(a). Let

Q = charge in coulombs due to p.d. of V volts

and A = area of each dielectric in square metres

then $D = Q/A$

which is the electric flux density, in coulombs per metre squared, in A and B.

Fig. 5.14   Parallel-plate capacitor with two dielectrics

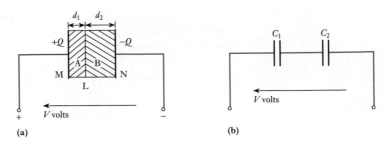

(a)   (b)

Let E_1 and E_2 = electric field strengths in 1 and 2 respectively; then if the relative permittivities of 1 and 2 are ϵ_1 and ϵ_2 respectively, electric field strength in A is

$$E_1 = \frac{D}{\epsilon_1 \epsilon_0} = \frac{Q}{\epsilon_1 \epsilon_0 A}$$

and electric field strength in B is

$$E_2 = \frac{D}{\epsilon_2 \epsilon_0} = \frac{Q}{\epsilon_2 \epsilon_0 A}$$

Hence $\dfrac{E_1}{E_2} = \dfrac{\epsilon_2}{\epsilon_1}$ [5.16]

i.e. for dielectrics having the same cross-sectional area in series, the electric field strengths (or potential gradients) are inversely proportional to their relative permittivities. Potential drop in a dielectric is

electric field strength × thickness

Therefore p.d. between plate M and the boundary surface L between 1 and 2 is $E_1 d_1$. Hence all points on surface L are at the same potential, i.e. L is an *equipotential surface* and is at right angles to the direction of the electric field strength. It follows that if a very thin metal foil were inserted between 1 and 2, it would not alter the electric field in the dielectrics. Hence the latter may be regarded as equivalent to two capacitances, C_1 and C_2, connected in series as in Fig. 5.14(b), where

$$C_1 = \frac{\epsilon_1 \epsilon_0 A}{d_1} \quad \text{and} \quad C_2 = \frac{\epsilon_2 \epsilon_0 A}{d_2}$$

and total capacitance between plates M and N is

$$\frac{C_1 C_2}{C_1 + C_2}$$

Example 5.6   A capacitor consists of two metal plates, each 400 × 400 mm, spaced 6 mm apart. The space between the metal plates is filled with a glass plate 5 mm thick and a layer of paper 1 mm thick. The relative permittivities of the glass and paper are 8 and 2 respectively. Calculate

(a) the capacitance, neglecting any fringing flux, and
(b) the electric field strength in each dielectric in kilovolts per millimetre due to a p.d. of 10 kV between the metal plates.

Fig. 5.15 Diagrams for Example 5.6

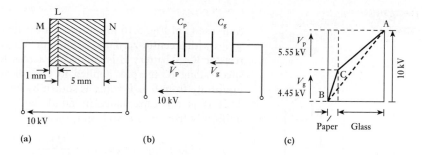

(a)   (b)   (c)

(a) Figure 5.15(a) shows a cross-section (not to scale) of the capacitor; and in Fig. 5.15(b), C_p represents the capacitance of the paper layer between M and the equipotential surface L and C_g represents that of the glass between L and N. From expression [5.13] we have

$$C_p = \frac{8.85 \times 10^{-12} \times 2 \times 0.4 \times 0.4}{0.001} = 2.83 \times 10^{-9}\,\text{F}$$

and

$$C_g = \frac{8.85 \times 10^{-12} \times 8 \times 0.4 \times 0.4}{0.005} = 2.265 \times 10^{-9}\,\text{F}$$

If C is the resultant capacitance between M and N

$$\frac{1}{C} = \frac{10^9}{2.83} + \frac{10^9}{2.265} = 0.7955 \times 10^9$$

$$\therefore \qquad C = 1.257 \times 10^{-9}\,\text{F} = 0.001\,257\,\mu\text{F}$$

$$\equiv \textbf{1260 pF}$$

(b) Since C_p and C_g are in series across 10 kV, it follows from expression [5.7] that the p.d., V_p, across the paper is given by

$$V_p = \frac{10 \times 2.265}{2.83 + 2.265} = 4.45\,\text{kV}$$

and $\quad V_g = 10 - 4.45 = 5.55\,\text{kV}$

These voltages are represented graphically in Fig. 5.15(c). Electric field strength in the paper dielectric is

$$4.45/1 = \textbf{4.45 kV/mm}$$

and electric field strength in the glass dielectric is

$$5.55/5 = \textbf{1.11 kV/mm}$$

These electric field strengths are represented by the slopes of AC and CB for the glass and paper respectively in Fig. 5.15(c). Had the dielectric between plates M and N been homogeneous, the electric field strength would have been 10/6 = 1.67 kV/mm, as represented by the slope of the dotted line AB in Fig. 5.15(c).

From the result of Example 5.6 it can be seen that the effect of using a composite dielectric of two materials having different relative permittivities is to increase the electric field strength in the material having the lower relative permittivity. This effect has very important applications in high-voltage work.

5.14   Charging and discharging currents

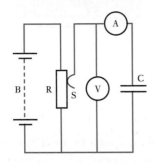

Fig. 5.16 Charging and discharging of a capacitor

Fig. 5.17 Voltage and current during charging and discharging of a capacitor

Suppose C in Fig. 5.16 to represent a capacitor of, say, 30 μF connected in series with a centre-zero microammeter A across a slider S and one end of a resistor R. A battery B is connected across R. If S is moved at a uniform speed along R, the p.d. applied to C, indicated by voltmeter V, increases uniformly from 0 to V volts, as shown by line OD in Fig. 5.17.

If C is the capacitance in farads and if the p.d. across C increases uniformly from 0 to V volts in t_1 seconds

$$\text{Charging current} = i_1 = \frac{Q\,[\text{coulombs or ampere seconds}]}{t_1\,[\text{seconds}]}$$

$$= CV/t_1 \text{ amperes}$$

i.e. charging current in amperes is equal to rate of change of charge in coulombs per second and is

$$C\,[\text{farads}] \times \text{rate of change of p.d. in volts per second}$$

Since the p.d. across C increases at a uniform rate, the charging current, i_1, remains constant and is represented by the dotted line LM in Fig. 5.17.

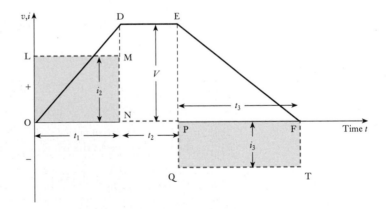

Suppose the p.d. across C to be maintained constant at V volts during the next t_2 seconds. Since the rate of change of p.d. is now zero, the current (apart from a slight leakage current) is zero and is represented by the dotted line NP. If the p.d. across C is then reduced to zero at a uniform rate by moving slider S backwards, the microammeter indicates a current i_3 flowing in the reverse direction, represented by the dotted line QT in Fig. 5.17. If t_3 is the time in seconds for the p.d. to be reduced from V volts to zero, then

$$Q = -i_3 t_3 \text{ coulombs}$$

$$\therefore \qquad i_3 = -Q/t_3 = -C \times V/t_3 \text{ amperes}$$

i.e. discharge current in amperes is equal to rate of change of charge in coulombs per second and is

$$C\,[\text{farads}] \times \text{rate of change of p.d. in volts per second}$$

Since $Q = i_1 t_1 = -i_3 t_3$ (assuming negligible leakage current through C),

$$\therefore \qquad \text{areas of rectangles OLMN and PQTF are equal}$$

In practice it is seldom possible to vary the p.d. across a capacitor at a constant rate, so let us consider the general case of the p.d. across a capacitor of C farads being increased by dv volts in dt seconds. If the corresponding increase of charge is dq coulombs

$$dq = C \cdot dv$$

If the charging current at that instant is i amperes

$$dq = i \cdot dt$$

$$\therefore \quad i \cdot dt = C \cdot dv$$

and $\qquad i = C \cdot dv/dt$

$$i = C \times \text{rate of change of p.d.} \qquad [5.17]$$

If the capacitor is being discharged and if the p.d. falls by dv volts in dt seconds, the discharge current is given by

$$i = \frac{dq}{dt} \quad i = C \cdot \frac{dv}{dt} \qquad [5.18]$$

Since dv is now negative, the current is also negative.

5.15 Growth and decay

In section 5.1 we derived the curves of the voltage across a capacitor during charging and discharging from the readings on a voltmeter connected across the capacitor. We will now consider how these curves can be derived graphically from the values of the capacitance, the resistance and the applied voltage. At the instant when S is closed on position a (Fig. 5.3), there is no p.d. across C; consequently the whole of the voltage is applied across R and the initial value of the charging current $= I = V/R$.

The growth of the p.d. across C is represented by the curve in Fig. 5.18. Suppose v to be the p.d. across C and i to be the charging current t seconds after S is put over to position a. The corresponding p.d. across R $= V - v$, where V is the terminal voltage of the battery. Hence

$$iR = V - v$$

and $\qquad i = \dfrac{V - v}{R} \qquad [5.19]$

If this current remained *constant* until the capacitor was fully charged, and if the time taken was x seconds, the corresponding quantity of electricity is

$$ix = \frac{V - v}{R} \times x \text{ coulombs}$$

With a constant charging current, the p.d. across C would have increased uniformly up to V volts, as represented by the tangent LM drawn to the curve at L.

But the charge added to the capacitor also equals increase of p.d. $\times C$ which is

Fig. 5.18 Growth of p.d. across a capacitor in series with a resistor

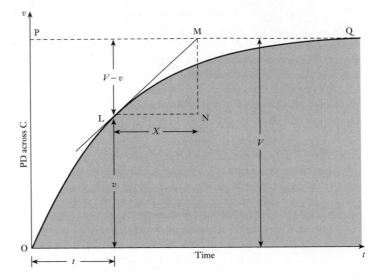

Fig. 5.19 Growth of p.d. across a capacitor in series with a resistor

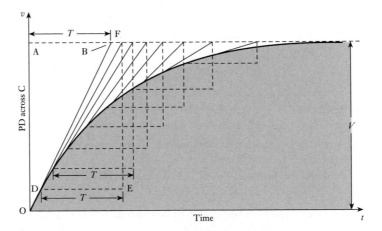

$$(V - v) \times C$$

Hence $\dfrac{V - v}{R} \times x = C(V - v)$

and $x = CR =$ the time constant, T, of the circuit

i.e. $\boxed{T = CR}$ seconds   [5.20]

The construction of the curve representing the growth of the p.d. across a capacitor is therefore similar to that described in section 8.6 for the growth of current in an inductive circuit. Thus, OA in Fig. 5.19 represents the battery voltage V, and AB the time constant T. Join OB, and from a point D fairly near the origin draw DE = T seconds and draw EF perpendicularly. Join DF, etc. Draw a curve such that OB, DF, etc. are tangents to it.

From expression [5.19] it is evident that the instantaneous value of the charging current is proportional to $(V - v)$, namely the vertical distance between the curve and the horizontal line PQ in Fig. 5.18. Hence the shape of the curve representing the charging current is the inverse of that of the p.d. across the capacitor and is the same for both charging and discharging currents (assuming the resistance to be the same), and its construction is illustrated by the following example.

Example 5.7

A 20 μF capacitor is charged to a p.d. of 400 V and then discharged through a 100 000 Ω resistor. Derive a curve representing the discharge current.

From equation [5.20]:

$$\text{Time constant} = 100\,000\,[\Omega] \times \frac{20}{1\,000\,000}\,[\text{F}] = 2\text{ s}$$

Initial value of discharge current is

$$\frac{V}{R} = \frac{400}{100\,000} = 0.004\text{ A} = 4\text{ mA}$$

Hence draw OA in Fig. 5.20 to represent 4 mA and OB to represent 2 s. Join AB. From a point C corresponding to, say, 3.5 mA, draw CD equal to 2 s and DE vertically. Join CE. Repeat the construction at intervals of, say, 0.5 mA and draw a curve to which AB, CE, etc. are tangents. This curve represents the variation of discharge current with time.

Fig. 5.20 Discharge current, Example 5.7

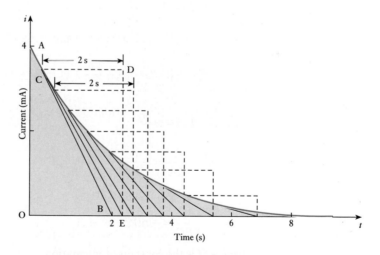

5.16   **Analysis of growth and decay**

Suppose the p.d. across capacitor C in Fig. 5.3, t seconds after S is switched over to position a, to be v volts, and the corresponding charging current to be i amperes, as indicated in Fig. 5.21. Also, suppose the p.d. to increase from v to $(v + \mathrm{d}v)$ volts in $\mathrm{d}t$ seconds, then, from expression [5.17],

$$i = C \cdot \frac{\mathrm{d}v}{\mathrm{d}t}$$

Fig. 5.21 Variation of current and p.d. during charging

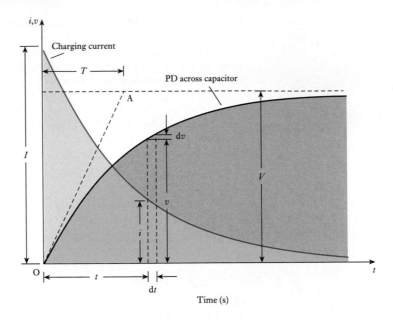

and corresponding p.d. across R is

$$Ri = RC \cdot \frac{dv}{dt}$$

But V = p.d. across C + p.d. across R

$$V = v + RC \cdot \frac{dv}{dt}$$ [5.21]

$\therefore \qquad V - v = RC \cdot \dfrac{dv}{dt}$

so that $\dfrac{dt}{RC} = \dfrac{dv}{V - v}$

Integrating both sides, we have

$$\frac{t}{RC} = -\ln(V - v) + A$$

where A is the constant of integration.
When $t = 0$, $v = 0$,

$\therefore \qquad A = \ln V$

so that

$$\frac{t}{RC} = \ln \frac{V}{V - v}$$

$\therefore \qquad \dfrac{V}{V - v} = e^{\frac{t}{RC}}$

and $\qquad v = V(1 - e^{-\frac{t}{RC}})$ volts [5.22]

Also $\qquad i = C \cdot \dfrac{dv}{dt} = CV \cdot \dfrac{d}{dt}(1 - e^{-\frac{t}{RC}})$

$\therefore \qquad i = \dfrac{V}{R} e^{-\frac{t}{RC}}$ [5.23]

At the instant of switching on, $t = 0$ and $e^{-0} = 1$,

$\therefore \qquad$ initial value of current $= \dfrac{V}{R} = (\text{say})I$

This result is really obvious from the fact that at the instant of switching on there is no charge on C and therefore no p.d. across it. Consequently the whole of the applied voltage must momentarily be absorbed by R.

Substituting for V/R in expression [5.23], we have instantaneous charging current

$\qquad i = I e^{-\frac{t}{RC}}$ [5.24]

If the p.d. across the capacitor continued increasing at the initial rate, it would be represented by OA, the tangent drawn to the initial part of the curve. If T is the *time constant* in seconds, namely the time required for the p.d. across C to increase from zero to its final value if it continued increasing at its initial rate, then initial rate of increase of p.d. =

$\qquad \dfrac{V}{T}$ volts per second [5.25]

Time constant $\qquad$ Symbol: T $\qquad$ Unit: second (s)

But it follows from equation [5.21] that at the instant of closing the switch on position a $v = 0$, then

$\qquad V = RC\dfrac{dv}{dt}$

Therefore initial rate of change of p.d. is

$\qquad \dfrac{dv}{dt} = \dfrac{V}{RC}$ [5.26]

Equating [5.25] and [5.26], we have

$\qquad \dfrac{V}{T} = \dfrac{V}{RC}$

$\therefore \qquad T = RC$ seconds [5.27]

Hence we can rewrite equations [5.22] and [5.24] thus:

$\qquad v = V(1 - e^{-\frac{t}{T}})$ [5.28]

and $\qquad i = I e^{-\frac{t}{T}}$ [5.29]

| 5.17 | Discharge of a capacitor through a resistor |

Having charged capacitor C in Fig. 5.3 to a p.d. of V volts, let us now move switch S over to position b and thereby discharge the capacitor through R. The pointer of microammeter A is immediately deflected to a maximum value in the negative direction, and then the readings on both the microammeter and the voltmeter (Fig. 5.3) decrease to zero as indicated in Fig. 5.22.

Fig. 5.22 Variation of current and p.d. during discharge

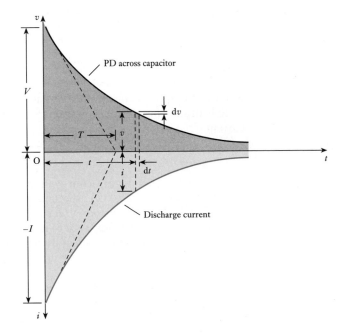

Suppose the p.d. across C to be v volts t seconds after S has been moved to position b, and the corresponding current to be i amperes, as in Fig. 5.22, then

$$i = -\frac{v}{R}$$ [5.30]

The negative sign indicates that the direction of the discharge current is the reverse of that of the charging current.

Suppose the p.d. across C to change by dv volts in dt seconds,

$$\therefore \qquad i = C \cdot \frac{dv}{dt}$$ [5.31]

Since dv is now negative, i must also be negative, as already noted. Equating [5.30] and [5.31], we have

$$-\frac{v}{R} = C \cdot \frac{dv}{dt}$$

so that

$$\frac{\mathrm{d}t}{RC} = -\frac{\mathrm{d}v}{v}$$

Integrating both sides, we have

$$\frac{t}{RC} = -\ln v + A$$

When $t = 0$, $v = V$, so that $A = \ln V$. Hence

$$\frac{t}{RC} = \ln V/v$$

so that

$$\frac{V}{v} = \mathrm{e}^{\frac{t}{RC}}$$

and

$$v = V\,\mathrm{e}^{-\frac{t}{RC}} = V\,\mathrm{e}^{-\frac{t}{T}} \qquad\qquad [5.32]$$

Also

$$i = -\frac{v}{R} = -\frac{V}{R}\mathrm{e}^{-\frac{t}{RC}} = -I\,\mathrm{e}^{-\frac{t}{T}} \qquad\qquad [5.33]$$

$$\therefore \qquad i = -I\,\mathrm{e}^{-\frac{t}{RC}} \qquad\qquad [5.34]$$

where I = initial value of the discharge current = V/R.

Example 5.8 An 8 μF capacitor is connected in series with a 0.5 MΩ resistor across a 200 V d.c. supply. Calculate:

  (a) the time constant;
  (b) the initial charging current;
  (c) the time taken for the p.d. across the capacitor to grow to 160 V;
  (d) the current and the p.d. across the capacitor 4.0 s after it is connected to the supply.

(a) From equation [5.27]

$$\text{time constant} = 0.5 \times 10^6 \times 8 \times 10^{-6} = \textbf{4.0 s}$$

(b) Initial charging current is

$$\frac{V}{R} = \frac{200}{0.5 \times 10^6}\,\text{A}$$

$$= \textbf{400 } \mu\textbf{A}$$

(c) From equation [5.28]

$$160 = 200(1 - \mathrm{e}^{-\frac{t}{4}})$$

$$\therefore \qquad \mathrm{e}^{-\frac{t}{4}} = 0.2$$

From mathematical tables

$$\frac{t}{4} = 1.61$$

$$\therefore \qquad t = 6.44 \text{ s}$$

Or alternatively

$$e^{\frac{t}{4}} = \frac{1}{0.2} = 5$$

$$\therefore \qquad \left(\frac{t}{4}\right) \log_e = \log 5$$

But $e = 2.718$

$$t = \frac{4 \times 0.699}{0.4343} = \mathbf{6.44 \text{ s}}$$

(d) From equation [5.28]

$$v = 200(1 - e^{-\frac{4}{4}}) = 200(1 - 0.368)$$

$$= 200 \times 0.632 = 126.4 \text{ V}$$

It will be seen that the time constant can be defined as the time required for the p.d. across the capacitor to grow from zero to 63.2 per cent of its final value.

From equation [5.29]

$$\text{corresponding current} = i = 400 \cdot e^{-1} = 400 \times 0.368$$

$$= \mathbf{147 \ \mu A}$$

5.18 Transients in CR networks

We have considered the charging and discharging of a capacitor through a resistor. In each case, the arrangement has involved a network containing capacitance and resistance, hence such a network is known as a *CR* network.

In practice, we are not likely to come across such simple arrangements as the connection of a battery to a *CR* network, but it is not much of a progression to an exceedingly common situation. Most communications circuits now involve the use of short pulses of voltage being applied to a variety of circuits, some of which are quite similar to the *CR* networks which we have considered. A pulse basically consists of the sudden application of the voltage source followed almost immediately by its being switched off.

This first of all begs the question – what do we mean by switching off almost immediately? Let us therefore again consider expression [5.24], i.e.

$$i = I\,e^{-\frac{t}{CR}}$$

Such a relation is an exponential expression of the form e^{-x}. Let us consider the way in which this expression changes for increasing values of x; there is no point in considering negative values since these would relate to a situation prior to switching which does not fall within our period of interest. It would be convenient to establish Table 5.2 by programming a computer or calculator.

Table 5.2

Variation of e^{-x}	
$x = 0$	$e^{-0} = 1.0000$
$x = 1$	$e^{-1} = 0.3679$
$x = 2$	$e^{-2} = 0.1353$
$x = 3$	$e^{-3} = 0.0479$
$x = 4$	$e^{-4} = 0.0183$
$x = 5$	$e^{-5} = 0.0067$
$x = 10$	$e^{-10} = 0.000\ 05$

Fig. 5.23 The exponential function

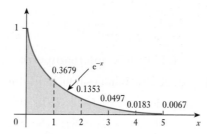

If we draw the corresponding characteristic, it takes the form shown in Fig. 5.23. This is known as an exponential decay and it has the same form as all the curves which have appeared in this chapter.

Returning to expression [5.24] above, when $t = CR$, i.e. when the time after switching is equal to the time constant, then

$$i = I\,e^{-1} = 0.368I$$

However when $t = 5CR$, then

$$i = I\,e^{-5} = 0.007I$$

Effectively the decay has ended at this point. This lets us interpret the term 'almost immediately'. If a supply is switched on and off then, provided the period between switching is less than five times the time constant of the network, this can be considered as almost immediately. If the period between switching is longer, the action of the first operation is effectively independent of the second.

In practice, many networks responding to pulsed switching experience a rate of switching which causes the second transient change to commence before the first has finished. The effects are best demonstrated by means of the following examples.

Example 5.9

For the network shown in Fig. 5.24:

(a) determine the mathematical expressions for the variation of the voltage across the capacitor and the current through the capacitor following the closure of the switch at $t = 0$ on to position 1;

(b) the switch is closed on to position 2 when $t = 100$ ms: determine the new expressions for the capacitor voltage and current;

(c) plot the voltage and current waveforms for $t = 0$ to $t = 200$ ms.

Fig. 5.24   Network for
Example 5.9

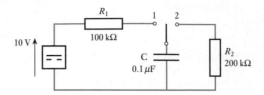

(a) For the switch in position 1, the time constant is

$$T_1 = CR_1 = 0.1 \times 10^{-6} \times 100 \times 10^3 \equiv 10 \text{ ms}$$

$$\therefore \qquad v_c = V(1 - e^{-\frac{t}{T_1}}) = 10(1 - e^{-\frac{t}{10 \times 10^{-3}}}) \text{ volts}$$

and $\qquad i_c = I e^{-\frac{t}{T_1}} = \dfrac{10}{100 \times 10^3} e^{-\frac{t}{10 \times 10^{-3}}}$

$$\equiv 100 \, e^{-\frac{t}{10 \times 10^{-3}}} \text{ microamperes}$$

(b) For the switch in position 2, the time constant is

$$T_2 = CR_2 = 0.1 \times 10^{-6} \times 200 \times 10^3 \equiv 20 \text{ ms}$$

In the transient expressions, t has to be measured from the second switching and not from the initial switching. Hence

$$v_c = V \, e^{-\frac{t}{T_2}} = 10 \, e^{-\frac{t}{20 \times 10^{-3}}} \text{ volts}$$

and

$$i_c = I e^{-\frac{t}{T_2}} = \frac{10}{200 \times 10^3} e^{-\frac{t}{20 \times 10^{-3}}}$$

$$\equiv 50 \, e^{-\frac{t}{20 \times 10^{-3}}} \text{ microamperes}$$

(c) The current and voltage waveforms are shown in Fig. 5.25.

It will be noted that in the first switching period, five times the time constant is 50 ms. The transient has virtually finished at the end of this time and it would not have mattered whether the second switching took place then or later. However, during the second period, the transient takes the full 100 ms.

Fig. 5.25   Voltage and current
waveforms for Example 5.9

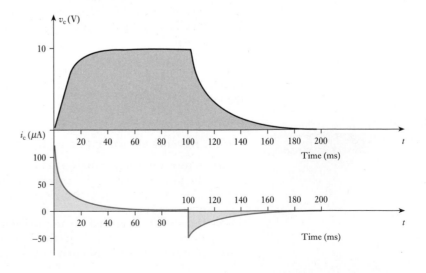

Example 5.10   For the network shown in Fig. 5.24, the switch is closed on to position 1 as in Example 5.9. However, it is closed on to position 2 when $t = 10$ ms. Again determine the voltage and current expressions and hence plot the voltage and current waveforms.

For the switch in position 1, the time constant is 10 ms as in Example 5.9, and the voltage and current expressions are again as before. However, the switch is moved to position 2 while the transient is proceeding.

When $t = 10$ ms

$$v_c = 10(1 - e^{-\frac{t}{10 \times 10^{-3}}}) = 10(1 - e^{-1}) = 6.32 \text{ V}$$

The second transient commences with an initial voltage across the capacitor of 6.32 V. The voltage decay is therefore

$$v_c = V e^{-\frac{t}{T_2}} = \mathbf{6.32} \, e^{-\frac{t}{20 \times 10^{-3}}} \text{ volts}$$

and $$i_c = \frac{6.32}{200 \times 10^{-3}} e^{-\frac{t}{20 \times 10^{-3}}} \equiv \mathbf{31.6} \, e^{-\frac{t}{20 \times 10^{-3}}} \text{ microamperes}$$

The current and voltage waveforms are shown in Fig. 5.26.

It would be possible to extend such an example by repeatedly switching the supply on and off. The analysis would be a repetition of either Example 5.9 or 5.10 depending on the rate of switching.

We can also analyse more complex networks by combining the application of network theorems with exponential expressions. Again this can more readily be illustrated by means of an example.

Fig. 5.26   Voltage and current waveforms for Example 5.10

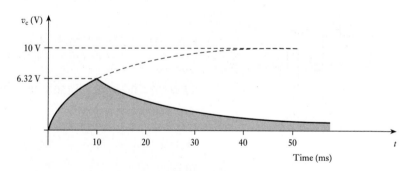

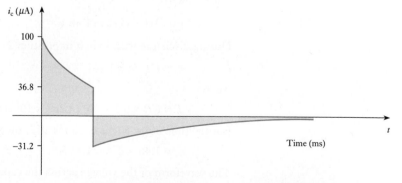

Example 5.11   For the network shown in Fig. 5.27, the switch is closed on to position 1 when $t = 0$ and then moved to position 2 when $t = 20$ ms. Determine the voltage across the capacitor when $T = 30$ ms.

Fig. 5.27   Network for
Example 5.11

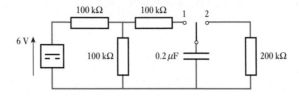

In order to analyse the transient effect, it is necessary to simplify the supply network to the capacitor by means of Thévenin's theorem. The supply voltage is shown in Fig. 5.28. The equivalent resistance is given by

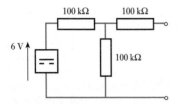

Fig. 5.28   For Example 5.11

$$R_e = 100 + \frac{100 \times 100}{100 + 100} = 150 \text{ k}\Omega$$

$$V_{o/c} = 6 \times \frac{100}{100 + 100} = 3 \text{ V}$$

The network hence can be replaced as shown in Fig. 5.29.
During charging

$$v_c = V(1 - e^{-\frac{t}{CR}}) = 3(1 - e^{-\frac{t}{T}})$$

where

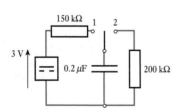

Fig. 5.29   For Example 5.11

$$T = CR = 0.2 \times 10^{-6} \times 150 \times 10^3 \equiv 30 \text{ ms}$$

hence

$$v_c = 3(1 - e^{-\frac{t}{30 \times 10^{-3}}})$$

For $t = 20$ ms,

$$v_c = 3(1 - e^{-\frac{20}{30}}) = 1.46 \text{ V}$$

During discharge with switch in position 2

$$v_c = V e^{-\frac{t}{CR}} = 1.46 e^{-\frac{t}{T}}$$

where

$$T = CR = 0.2 \times 10^{-6} \times 200 \times 10^3 \equiv 40 \text{ ms}$$

But the time of $t = 30$ ms is 10 ms after the second switching action, hence

$$v_c = 1.46 \times e^{-\frac{10}{40}} = \mathbf{1.14 \text{ V}}$$

The waveform of the voltage across the capacitor is shown in Fig. 5.30.

Fig. 5.30 Waveform of voltage across capacitor in Example 5.11

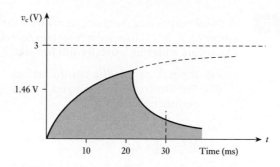

5.19 Energy stored in a charged capacitor

Suppose the p.d. across a capacitor of capacitance C farads is to be increased from v to $(v + dv)$ volts in dt seconds. From equation [5.18], the charging current, i amperes, is given by

$$i = C \frac{dv}{dt}$$

Instantaneous value of power to capacitor is

$$iv \text{ watts} = vC \cdot \frac{dv}{dt} \text{ watts}$$

and energy supplied to capacitor during interval dt is

$$vC \cdot \frac{dv}{dt} \cdot dt = Cv \cdot dv \text{ joules}$$

Hence total energy supplied to capacitor when p.d. is increased from 0 to V volts is

$$\int_0^V Cv \cdot dv = \tfrac{1}{2}C\left[v^2\right]_0^V = \tfrac{1}{2}CV^2 \text{ joules}$$

$$\therefore \qquad W = \tfrac{1}{2}CV^2 \qquad\qquad [5.35]$$

also $\qquad W = \tfrac{1}{2}C\left[\dfrac{Q}{C}\right]^2 = \tfrac{1}{2} \cdot \dfrac{Q^2}{C}$

For a capacitor with dielectric of thickness d metres and area A square metres, energy per cubic metre is

$$\frac{1}{2} \cdot \frac{CV^2}{Ad} = \frac{1}{2} \cdot \frac{\epsilon A}{d} \cdot \frac{V^2}{Ad}$$

$$= \frac{1}{2}\epsilon\left(\frac{V}{d}\right)^2 = \frac{1}{2}\epsilon E^2$$

$$\frac{1}{2} \cdot \frac{CV^2}{Ad} = \tfrac{1}{2} \cdot \epsilon E^2 = \frac{1}{2}DE = \frac{1}{2}\frac{D^2}{\epsilon} \text{ joules} \qquad [5.36]$$

These expressions are similar to expression [8.22] for the energy stored per cubic metre of a magnetic field.

Example 5.12

A 50 μF capacitor is charged from a 200 V supply. After being disconnected it is immediately connected in parallel with a 30 μF capacitor which is initially uncharged. Find:

(a) the p.d. across the combination;
(b) the electrostatic energies before and after the capacitors are connected in parallel.

(a) From equation [5.2]

$$Q = CV$$

$$\text{charge} = (50 \times 10^{-6})\,[\text{F}] \times 200\,[\text{V}] = 0.01\ \text{C}$$

When the capacitors are connected in parallel, the total capacitance is 80 μF, and the charge of 0.01 C is divided between the two capacitors:

$$Q = CV$$

$$\therefore \qquad 0.01\,[\text{C}] = (80 \times 10^{-6})\,[\text{F}] \times \text{p.d.}$$

$$\therefore \qquad \text{p.d. across capacitors} = \mathbf{125\ V}$$

(b) From equation [5.35] it follows that when the 50 μF capacitor is charged to a p.d. of 200 V:

$$W = \tfrac{1}{2}CV^2$$

$$\text{Electrostatic energy} = \tfrac{1}{2} \times (50 \times 10^{-6})\,[\text{F}] \times (200)^2\,[\text{V}^2]$$

$$= \mathbf{1.0\ J}$$

With the capacitors in parallel:

$$\text{Total electrostatic energy} = \tfrac{1}{2} \times 80 \times 10^{-6} \times (125)^2 = \mathbf{0.625\ J}$$

It is of interest to note that there is a reduction in the energy stored in the capacitors. This loss appears as heat in the resistance of the circuit by the current responsible for equalizing the p.d.s in the spark that may occur when the capacitors are connected in parallel, and in electromagnetic radiation if the discharge is oscillatory.

5.20 Force of attraction between oppositely charged plates

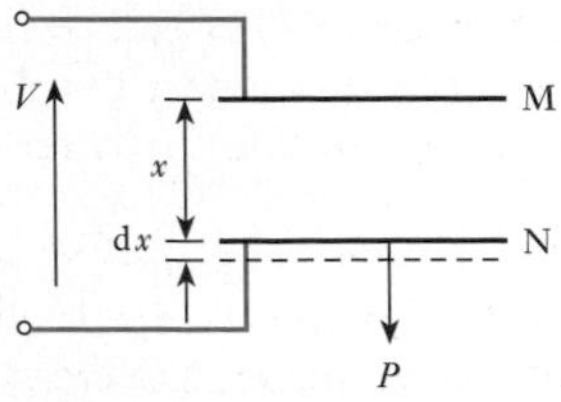

Fig. 5.31 Attraction between charged parallel plates

Let us consider two parallel plates M and N (Fig. 5.31) immersed in a homogeneous fluid, such as air or oil, having an absolute permittivity ϵ. Suppose the area of the dielectric to be A square metres and the distance between M and N to be x metres. If the p.d. between the plates is V volts, then from [5.35], energy per cubic metre of dielectric is

$$\frac{1}{2}\epsilon\left(\frac{V}{x}\right)^2 \text{ joules}$$

Suppose plate M to be fixed and N to be movable, and let F be the force of attraction, in newtons, between the plates. Let us next disconnect the charged capacitor from the supply and then pull plate N outwards through a distance dx metres. If the insulation of the capacitor is perfect, the charge on the plates remains constant. This means that the electric flux density and therefore the potential gradient in the dielectric must remain unaltered, the constancy of the potential gradient being due to the p.d. between plates M and N increasing in proportion to the distance between them. It follows from expression [5.36] that the energy per cubic metre of the dielectric remains constant. Consequently, all the energy in the *additional* volume of the dielectric must be derived from the work done when the force F newtons acts through distance dx metres, namely $F \cdot dx$ joules, i.e.

$$F \cdot dx = \frac{1}{2}\epsilon \left(\frac{V}{x}\right)^2 \cdot A \cdot dx$$

$$\therefore \qquad F = \frac{1}{2}\epsilon A \left(\frac{V}{x}\right)^2 \qquad \text{newtons} \qquad\qquad [5.37]$$

$$= \tfrac{1}{2}\epsilon A \times (\text{potential gradient in volts per metre})^2$$

Example 5.13

Two parallel metal discs, each 100 mm in diameter, are spaced 1.0 mm apart, the dielectric being air. Calculate the force, in newtons, on each disc when the p.d. between them is 1.0 kV.

Area of one side of each plate

$$A = \frac{\pi}{4}d^2 = 0.7854 \times (0.1)^2$$

$$= 0.007\ 854 \text{ m}^2$$

Potential gradient

$$E = \frac{V}{d} = 1000 \text{ [V]}/0.001 \text{ [m]}$$

$$= 10^6 \text{ V/m}$$

From expression [5.37] force

$$F = \frac{1}{2}\epsilon A \left(\frac{V}{x}\right)^2$$

$$= \frac{1}{2} \times (8.85 \times 10^{-12}) \text{ [F/m]} \times 0.007\ 854 \text{ [m]}^2 \times (10^6)^2 \text{ [V/m]}^2$$

$$= \textbf{0.035 N}$$

5.21   **Dielectric strength**

If the p.d. between the opposite sides of a sheet of solid insulating material is increased beyond a certain value, the material breaks down. Usually this results in a tiny hole or puncture through the dielectric so that the latter is then useless as an insulator.

The potential gradient necessary to cause breakdown of an insulating medium is termed its *dielectric strength* and is usually expressed in megavolts per metre. The value of the dielectric strength of a given material decreases with increase of thickness, and Table 5.3 gives the approximate dielectric strengths of some of the most important materials.

Table 5.3

Material	Thickness (mm)	Dielectric strength (MV/m)
Air (at normal pressure and temperature)	0.2	5.75
	0.6	4.92
	1	4.46
	6	3.27
	10	2.98
Mica	0.01	200
	0.1	176
	1.0	61
Glass (density 2.5)	1	28.5
	5	18.3
Ebonite	1	50
Paraffin-waxed paper	0.1	40–60
Transformer oil	1	200
Ceramics	1	50

5.22 Leakage and conduction currents in capacitors

When considering dielectric strength, we noted that the flow of electrons in a dielectric could be due to breakdown. However, it would be incorrect to consider that there is no flow of electrons when the applied voltage is at a value less than breakdown. No dielectric is perfect; instead every dielectric has a few free electrons (partly due to impurities) and therefore effectively acts as an insulator of very high resistance between the plates of a capacitor.

It follows that when we apply a voltage across the plates of a capacitor, a small leakage current passes between the plates due to the free electrons in the dielectric. For most practical purposes this can be neglected because the leakage current is so small. The effect can be represented as shown in Fig. 5.32, but in most instances the resistance has a value in excess of 100 MΩ.

If we were to charge a capacitor and then switch off the supply voltage, the capacitor would remain charged. However, it would be found that after, say, a few hours some of the charge would have disappeared. The reason is that the equivalent resistance would give rise to the decay situation which we have analysed, but with a very large time constant.

We should not confuse the leakage current with the charging current which is in the conductors connecting the voltage source to the plates of a capacitor. If we consider the circuit shown in Fig. 5.33, closure of the switch

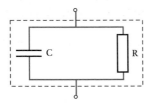

Fig. 5.32 Equivalent circuit of a practical capacitor

Fig. 5.33 Conduction current

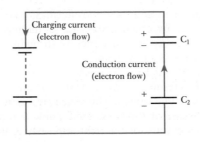

causes electrons to flow from the positive plate of C_1 via the battery to the negative plate of C_2.

At the same time, the negative plate of C_1 receives electrons from the positive plate of C_2. This flow of electrons which does not pass through the battery is referred to as *conduction current*.

5.23	Displacement current in a dielectric

Let us consider the capacitor in Fig. 5.34 with a vacuum between the plates. There are no electrons in the space between the plates and therefore there cannot be any movement of electrons in this space when the capacitor is being charged. We know, however, that an electric field is being set up and that energy is being stored in the space between the plates; in other words, the space between the plates of a charged capacitor is in a state of electrostatic strain.

We do not know the exact nature of this strain (any more than we know the nature of the strain in a magnetic field), but James Clerk Maxwell, in 1865, introduced the concept that any *change* in the electric flux in any region is equivalent to an electric current in that region, and he called this electric current a *displacement* current, to distinguish it from the *conduction* current referred to above.

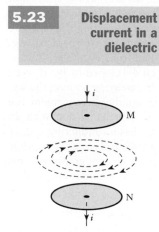

Fig. 5.34 Magnetic field due to displacement current

5.24	Types of capacitor and capacitance

We have already noted that capacitors are devices which promote capacitance, i.e. they are designed to have a high ability to hold electric charge. Capacitors are generally made to have a fixed value of capacitance, but some are variable. The symbols for fixed and variable capacitors are shown in Fig. 5.35.

(a) Fixed capacitors

The fixed capacitors come in a variety of groups depending on the type of dielectric used.

Paper capacitors

This type has already been considered in Fig. 5.1, the electrodes of the capacitor being layers of metal foil interleaved with paper impregnated with wax or oil. Such capacitors are commonly used in the power circuits of household appliances.

Electrolytic capacitors

The type most commonly used consists of two aluminium foils, one with an oxide film and one without, the foils being interleaved with a material such as paper saturated with a suitable electrolyte; for example, ammonium borate.

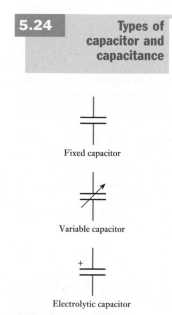

Fixed capacitor

Variable capacitor

Electrolytic capacitor

Fig. 5.35 Circuit symbols for capacitors

The aluminium oxide film is formed on the one foil by passing it through an electrolytic bath of which the foil forms the positive electrode. The finished unit is assembled in a container – usually of aluminium – and hermetically sealed. The oxide film acts as the dielectric, and as its thickness in a capacitor suitable for a working voltage of 100 V is only about 0.15 μm, a very large capacitance is obtainable in a relatively small volume.

The main disadvantages of this type of capacitor are: (a) the insulation resistance is comparatively low, and (b) it is only suitable for circuits where the voltage applied to the capacitor never reverses its direction. Electrolytic capacitors are mainly used where very large capacitances are required, e.g. for reducing the ripple in the voltage wave obtained from a rectifier.

Solid types of electrolytic capacitors have been developed to avoid some of the disadvantages of the wet electrolytic type. In one arrangement, the wet electrolyte is replaced by manganese dioxide. In another arrangement the anode is a cylinder of pressed sintered tantalum powder coated with an oxide layer which forms the dielectric. This oxide has a conducting coat of manganese dioxide which acts as an electron conductor and replaces the ionic conduction of the liquid electrolyte in the wet type. A layer of graphite forms the connection with a silver or copper cathode and the whole is enclosed in a hermetically sealed steel can.

Mica capacitors

This type consists either of alternate layers of mica and metal foil clamped tightly together, or of thin films of silver sputtered on the two sides of a mica sheet. Owing to its relatively high cost, this type is mainly used in high-frequency circuits when it is necessary to reduce to a minimum the loss in the dielectric.

Polyester capacitors

Polyester is relatively new as a dielectric when used in capacitors. It is manufactured in very thin films of thickness as little as 2 μm and is metallized on one side. Two films are then rolled together rather like the paper-insulated capacitor.

Such capacitors can be very small so that there is insufficient outside surface on which to print the ratings and other data. For this reason, they often come with a colour coding after the fashion used with resistors. Usually a black band is printed near the lead connected to the outer metal foil electrode. This lead should be kept at the lower working potential.

These capacitors can operate at high voltages, i.e. a few thousand volts, and the leakage resistance is high, say 100 MΩ.

Ceramic capacitors

The ceramic capacitor is manufactured in many forms, but all are basically the same. A thin ceramic dielectric is coated on both sides with a metal. The capacitor is made up by making a stack of these ceramic layers, each layer being separated from the next by more ceramic. The plates are connected by electrodes to the supply leads and a coating of ceramic is then applied to the outside of the stack. The arrangement is then fired to give a solid device.

Such capacitors generally have small capacitance values from 1 pF to about 1 μF. Like the polyester capacitors, working voltages can be up to a few thousand volts, but the leakage resistance can be even higher, say 1000 MΩ. Ceramic capacitors are useful in high-temperature situations.

Ceramic materials include compounds of barium titanate which, it will be recalled, has an exceptionally high relative permittivity (6000 +). This permits very small separation between the plates and gives rise to high values of capacitance from relatively small capacitors.

Tantalum electrolytic capacitors

These capacitors are much smaller than the corresponding aluminium electrolytic capacitors. The construction may take the form indicated in Fig. 5.36, in which one plate consists of pressed, sintered tantalum powder coated with an oxide layer which is the dielectric. The case of brass, copper or even silver forms the other plate. Layers of manganese dioxide and graphite form the electrolyte.

Fig. 5.36 Sintered tantalum capacitor

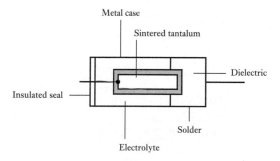

(b) Variable capacitors

These require two sets of rigid plates which can be moved between one another as indicated in Fig. 5.37. The plates must be rigid so that they can move between each other without touching. It follows that the dielectric between the plates is air. Normally one set of plates is fixed and the other made to rotate. The greater the insertion of the movable plates then the greater the capacitance. Most of us know this type of capacitor because it is the device used to tune radios.

Fig. 5.37 A variable capacitor

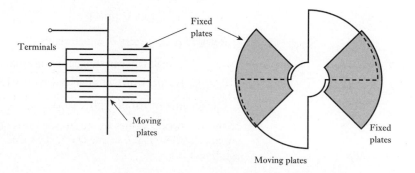

(c)   Other capacitance

It is worth noting that capacitance exists between any two conductors. This means that capacitance exists in every circuit. However, normally the conductors or wires are so small and so far apart that the capacitance between them can be ignored. In long transmission lines or in high-frequency communications systems this is not always the case. At a time when we are being introduced to circuit theory, it is sufficient to ignore circuit capacitance except when capacitors are part of the circuit.

Summary of important formulae

$$Q = CV$$

$$Q \, [\text{coulombs}] = C \, [\text{farads}] \times V \, [\text{volts}] \qquad [5.2]$$

$$1 \, \mu\text{F} = 10^{-6} \, \text{F}$$

$$1 \, \text{pF} = 10^{-12} \, \text{F}$$

For capacitors in parallel

$$C = C_1 + C_2 + \ldots \qquad [5.3]$$

For capacitors in series

$$\frac{1}{C} = \frac{1}{C_1} + \frac{1}{C_2} + \ldots \qquad [5.5]$$

For C_1 and C_2 in series

$$V_1 = V \cdot \frac{C_2}{C_1 + C_2} \qquad [5.7]$$

$$\text{Electric field strength in dielectric} = E = \frac{V}{d} \qquad [5.9]$$

$$\text{Electric flux density} = D = \frac{Q}{A} \qquad [5.10]$$

$$\text{Capacitance } C = \frac{\epsilon_0 \epsilon_r A}{d} \qquad [5.13]$$

$$\text{Absolute permittivity} = \frac{D}{E} = \epsilon = \epsilon_0 \epsilon_r \qquad [5.14]$$

Permittivity of free space is

$$\epsilon_0 = 8.85 \times 10^{-12} \, \text{F/m}$$

Relative permittivity of a material is

$$\frac{\text{capacitance of capacitor with that material as dielectric}}{\text{capacitance of same capacitor with vacuum dielectric}}$$

Capacitance of parallel-plate capacitor with n plates is

$$\frac{\epsilon(n-1)A}{d} \qquad [5.15]$$

Summary of important formulae continued

For two dielectrics, A and B, of same areas, in series

$$\frac{\text{electric field strength or potential gradient in A}}{\text{electric field strength or potential gradient in B}}$$

$$= \frac{\text{relative permittivity of B}}{\text{relative permittivity of A}} \qquad [5.16]$$

Charging current of capacitor i is

$$\frac{\mathrm{d}q}{\mathrm{d}t} = C \cdot \frac{\mathrm{d}v}{\mathrm{d}t} \qquad [5.18]$$

For R and C in series across d.c. supply,

$$v = V(1 - e^{-\frac{t}{RC}}) \qquad [5.22]$$

and $\quad i = I\, e^{-\frac{t}{RC}} \qquad [5.24]$

Time constant is

$$T = RC \qquad [5.27]$$

For C discharged through R

$$v = V\, e^{-\frac{t}{RC}} \qquad [5.32]$$

and $\quad i = -I\, e^{-\frac{t}{RC}} \qquad [5.33]$

Energy stored in capacitor is

$$W = \tfrac{1}{2}CV^2 \text{ joules} \qquad [5.35]$$

Energy per cubic metre of dielectric is

$$\frac{1}{2}\epsilon E^2 = \frac{1}{2}DE = \frac{1}{2}\frac{D^2}{\epsilon} \text{ joules} \qquad [5.36]$$

Electrostatic attraction between parallel plates is

$$f = \frac{1}{2}\epsilon A\left(\frac{V}{x}\right)^2 \text{ newtons} \qquad [5.37]$$

Terms and concepts

Capacitance is a measure of the ability to store electric charge.

Capacitance is also a measure of the ability to store energy in an electric field.

Charging is the process of increasing the charge held in a capacitor.

Discharging is the process of reducing the charge held in a capacitor.

Farad is the capacitance of a capacitor which has a p.d. of 1 V when maintaining a charge of 1 C.

Leakage current is the rate of movement of charge through a dielectric.

Permittivity is the ratio of electric flux density to electric field strength measured in farads per metre.

Exercises 5

1. A 20 μF capacitor is charged at a constant current of 5 μA for 10 min. Calculate the final p.d. across the capacitor and the corresponding charge in coulombs.

2. Three capacitors have capacitances of 10 μF, 15 μF and 20 μF respectively. Calculate the total capacitance when they are connected (a) in parallel, (b) in series.

3. A 9 μF capacitor is connected in series with two capacitors, 4 μF and 2 μF respectively, which are connected in parallel. Determine the capacitance of the combination. If a p.d. of 20 V is maintained across the combination, determine the charge on the 9 μF capacitor and the energy stored in the 4 μF capacitor.

4. Two capacitors, having capacitances of 10 μF and 15 μF respectively, are connected in series across a 200 V d.c. supply. Calculate: (a) the charge on each capacitor; (b) the p.d. across each capacitor. Also find the capacitance of a single capacitor that would be equivalent to these two capacitors in series.

5. Three capacitors of 2, 3 and 6 μF respectively are connected in series across a 500 V d.c. supply. Calculate: (a) the charge on each capacitor; (b) the p.d. across each capacitor; and (c) the energy stored in the 6 μF capacitor.

6. A certain capacitor has a capacitance of 3 μF. A capacitance of 2.5 μF is required by combining this capacitance with another. Calculate the capacitance of the second capacitor and state how it must be connected to the first.

7. A capacitor A is connected in series with two capacitors B and C connected in parallel. If the capacitances of A, B and C are 4, 3 and 6 μF respectively, calculate the equivalent capacitance of the combination. If a p.d. of 20 V is maintained across the whole circuit, calculate the charge on the 3 μF capacitor.

8. Three capacitors, A, B and C, are connected in series across a 200 V d.c. supply. The p.d.s across the capacitors are 40, 70 and 90 V respectively. If the capacitance of A is 8 μF, what are the capacitances of B and C?

9. Two capacitors, A and B, are connected in series across a 200 V d.c. supply. The p.d. across A is 120 V. This p.d. is increased to 140 V when a 3 μF capacitor is connected in parallel with B. Calculate the capacitances of A and B.

10. Show from first principles that the total capacitance of two capacitors having capacitances C_1 and C_2 respectively, connected in parallel, is $C_1 + C_2$.

   A circuit consists of two capacitors A and B in parallel connected in series with another capacitor C. The capacitances of A, B and C are 6 μF, 10 μF and 16 μF

respectively. When the circuit is connected across a 400 V d.c. supply, calculate: (a) the potential difference across each capacitor; (b) the charge on each capacitor.

11. On what factors does the capacitance of a parallel-plate capacitor depend?

   Derive an expression for the resultant capacitance when two capacitors are connected in series.

   Two capacitors, A and B, having capacitances of 20 μF and 30 μF respectively, are connected in series to a 600 V d.c. supply. Determine the p.d. across each capacitor. If a third capacitor C is connected in parallel with A and it is then found that the p.d. across B is 400 V, calculate the capacitance of C and the energy stored in it.

12. Derive an expression for the energy stored in a capacitor of C farads when charged to a potential difference of V volts.

   A capacitor of 4 μF capacitance is charged to a p.d. of 400 V and then connected in parallel with an uncharged capacitor of 2 μF capacitance. Calculate the p.d. across the parallel capacitors and the energy stored in the capacitors before and after being connected in parallel. Explain the difference.

13. Derive expressions for the equivalent capacitance of a number of capacitors: (a) in series; (b) in parallel.

   Two capacitors of 4 μF and 6 μF capacitance respectively are connected in series across a p.d. of 250 V. Calculate the p.d. across each capacitor and the charge on each. The capacitors are disconnected from the supply p.d. and reconnected in parallel with each other, with terminals of similar polarity being joined together. Calculate the new p.d. and charge for each capacitor. What would have happened if, in making the parallel connection, the connections of one of the capacitors had been reversed?

14. Show that the total capacitance of two capacitors having capacitances C_1 and C_2 connected in series is $C_1C_2/(C_1 + C_2)$.

   A 5 μF capacitor is charged to a potential difference of 100 V and then connected in parallel with an uncharged 3 μF capacitor. Calculate the potential difference across the parallel capacitors.

15. Find an expression for the energy stored in a capacitor of capacitance C farads charged to a p.d. of V volts.

   A 3 μF capacitor is charged to a p.d. of 200 V and then connected in parallel with an uncharged 2 μF capacitor. Calculate the p.d. across the parallel capacitors and the energy stored in the capacitors before and after being connected in parallel. Account for the difference.

Exercises 5 continued

16. Explain the terms *electric field strength* and *permittivity*.

    Two square metal plates, each of size 400 cm^2, are completely immersed in insulating oil of relative permittivity 5 and spaced 3 mm apart. A p.d. of 600 V is maintained between the plates. Calculate: (a) the capacitance of the capacitor; (b) the charge stored on the plates; (c) the electric field strength in the dielectric; (d) the electric flux density.

17. A capacitor consists of two metal plates, each having an area of 900 cm^2, spaced 3.0 mm apart. The whole of the space between the plates is filled with a dielectric having a relative permittivity of 6. A p.d. of 500 V is maintained between the two plates. Calculate: (a) the capacitance; (b) the charge; (c) the electric field strength; (d) the electric flux density.

18. Describe with the aid of a diagram what happens when a battery is connected across a simple capacitor comprising two metal plates separated by a dielectric.

    A capacitor consists of two metal plates, each having an area of 600 cm^2, separated by a dielectric 4 mm thick which has a relative permittivity of 5. When the capacitor is connected to a 400 V d.c. supply, calculate: (a) the capacitance; (b) the charge; (c) the electric field strength; (d) the electric flux density.

19. Define: (a) the farad; (b) the relative permittivity.

    A capacitor consists of two square metal plates of side 200 mm, separated by an air space 2.0 mm wide. The capacitor is charged to a p.d. of 200 V and a sheet of glass having a relative permittivity of 6 is placed between the metal plates immediately they are disconnected from the supply. Calculate: (a) the capacitance with air dielectric; (b) the capacitance with glass dielectric; (c) the p.d. across the capacitor after the glass plate has been inserted; (d) the charge on the capacitor.

20. What factors affect the capacitance that exists between two parallel metal plates insulated from each other?

    A capacitor consists of two similar, square, aluminium plates, each 100 mm × 100 mm, mounted parallel and opposite each other. Calculate the capacitance when the distance between the plates is 1.0 mm and the dielectric is mica of relative permittivity 7.0. If the plates are connected to a circuit which provides a constant current of 2 μA, how long will it take the potential difference of the plates to change by 100 V, and what will be the increase in the charge?

21. What are the factors which determine the capacitance of a parallel-plate capacitor? Mention how a variation in each of these factors will influence the value of capacitance.

    Calculate the capacitance in microfarads of a capacitor having 11 parallel plates separated by mica sheets 0.2 mm thick. The area of one side of each plate is 1000 mm^2 and the relative permittivity of mica is 5.

22. A parallel-plate capacitor has a capacitance of 300 pF. It has 9 plates, each 40 mm × 30 mm, separated by mica having a relative permittivity of 5. Calculate the thickness of the mica.

23. A capacitor consists of two parallel metal plates, each of area 2000 cm^2 and 5.0 mm apart. The space between the plates is filled with a layer of paper 2.0 mm thick and a sheet of glass 3.0 mm thick. The relative permittivities of the paper and glass are 2 and 8 respectively. A potential difference of 5 kV is applied between the plates. Calculate: (a) the capacitance of the capacitor; (b) the potential gradient in each dielectric; (c) the total energy stored in the capacitor.

24. Obtain from first principles an expression for the capacitance of a single-dielectric, parallel-plate capacitor in terms of the plate area, the distance between plates and the permittivity of the dielectric.

    A sheet of mica, 1.0 mm thick and of relative permittivity 6, is interposed between two parallel brass plates 3.0 mm apart. The remainder of the space between the plates is occupied by air. Calculate the area of each plate if the capacitance between them is 0.001 μF. Assuming that air can withstand a potential gradient of 3 MV/m, show that a p.d. of 5 kV between the plates will not cause a flashover.

25. Explain what is meant by electric field strength in a dielectric and state the factors upon which it depends.

    Two parallel metal plates of large area are spaced at a distance of 10 mm from each other in air, and a p.d. of 5000 V is maintained between them. If a sheet of glass, 5.0 mm thick and having a relative permittivity of 6, is introduced between the plates, what will be the maximum electric field strength and where will it occur?

26. Two capacitors of capacitance 0.2 μF and 0.05 μF are charged to voltages of 100 V and 300 V respectively. The capacitors are then connected in parallel by joining terminals of corresponding polarity together. Calculate: (a) the charge on each capacitor before being connected in parallel; (b) the energy stored on each capacitor before being connected in parallel; (c) the charge on the combined capacitors; (d) the p.d. between the terminals of the combination; (e) the energy stored in the combination.

27. A capacitor consists of two metal plates, each 200 mm × 200 mm, spaced 1.0 mm apart, the dielectric being air. The capacitor is charged to a p.d. of 100 V and then discharged through a ballistic galvanometer having a ballistic constant of 0.0011 microcoulombs per scale division. The amplitude of the first deflection is 32 divisions. Calculate the value of the absolute permittivity of air. Calculate also the electric field strength and the electric flux density in the air dielectric when the terminal p.d. is 100 V.

28. When the capacitor of Q. 27 is immersed in oil, charged to a p.d. of 30 V and then discharged through the same galvanometer, the first deflection is 27 divisions. Calculate: (a) the relative permittivity of the oil; (b) the electric field strength and the electric flux density in the oil when the terminal p.d. is 30 V; (c) the energy stored in the capacitor.

29. A 20 μF capacitor is charged and discharged thus:

Steady charging current of 0.02 A   from 0 to 0.5 s
Steady charging current of 0.01 A   from 0.5 to 1.0 s
Zero current   from 1.0 to 1.5 s
Steady discharging current of 0.01 A from 1.5 to 2.0 s
Steady discharging current of 0.005 A   from 2.0 to 4.0 s

Draw graphs to scale showing how the current and the capacitor voltage vary with time.

30. Define the *time constant* of a circuit that includes a resistor and capacitor connected in series.

A 100 μF capacitor is connected in series with an 800 Ω resistor. Determine the time constant of the circuit. If the combination is connected suddenly to a 100 V d.c. supply, find: (a) the initial rate of rise of p.d. across the capacitor; (b) the initial charging current; (c) the ultimate charge in the capacitor; and (d) the ultimate energy stored in the capacitor.

31. A 10 μF capacitor connected in series with a 50 kΩ resistor is switched across a 50 V d.c. supply. Derive graphically curves showing how the charging current and the p.d. across the capacitor vary with time.

32. A 2 μF capacitor is joined in series with a 2 MΩ resistor to a d.c. supply of 100 V. Draw a current–time graph and explain what happens in the period after the circuit is made, if the capacitor is initially uncharged. Calculate the current flowing and the energy stored in the capacitor at the end of an interval of 4 s from the start.

33. Derive an expression for the current flowing at any instant after the application of a constant voltage V to a circuit having a capacitance C in series with a resistance R.

Determine, for the case in which $C = 0.01$ μF, $R = 100\,000$ Ω and $V = 1000$ V, the voltage to which the capacitor has been charged when the charging current has decreased to 90 per cent of its initial value, and the time taken for the current to decrease to 90 per cent of its initial value.

34. Derive an expression for the stored electrostatic energy of a charged capacitor.

A 10 μF capacitor in series with a 10 kΩ resistor is connected across a 500 V d.c. supply. The fully charged capacitor is disconnected from the supply and discharged by connecting a 1000 Ω resistor across its terminals. Calculate: (a) the initial value of the charging current; (b) the initial value of the discharge current; and (c) the amount of heat, in joules, dissipated in the 1000 Ω resistor.

35. A 20 μF capacitor is found to have an insulation resistance of 50 MΩ, measured between the terminals. If this capacitor is charged off a d.c. supply of 230 V, find the time required after disconnection from the supply for the p.d. across the capacitor to fall to 60 V.

36. A circuit consisting of a 6 μF capacitor, an electrostatic voltmeter and a resistor in parallel, is connected across a 140 V d.c. supply. It is then disconnected and the reading on the voltmeter falls to 70 V in 127 s. When the test is performed without the resistor, the time taken for the same fall in voltage is 183 s. Calculate the resistance of the resistor.

37. A constant direct voltage of V volts is applied across two plane parallel electrodes. Derive expressions for the electric field strength and flux density in the field between the electrodes and the charge on the electrodes. Hence, or otherwise, derive an expression for the capacitance of a parallel-plate capacitor.

An electronic flash tube requires an energy input of 8.5 J which is obtained from a capacitor charged from a 2000 V d.c. source. The capacitor is to consist of two parallel plates, 11 cm in width, separated by a dielectric of thickness 0.1 mm and relative permittivity 5.5. Calculate the necessary length of each capacitor plate.

38. An electrostatic device consists of two parallel conducting plates, each of area 1000 cm^2. When the plates are 10 mm apart in air, the attractive force between them is 0.1 N. Calculate the potential difference between the plates. Find also the energy stored in the system. If the device is used in a container filled with a gas of relative permittivity 4, what effect does this have on the force between the plates?

39. The energy stored in a certain capacitor when connected across a 400 V d.c. supply is 0.3 J. Calculate: (a) the capacitance; and (b) the charge on the capacitor.

40. A variable capacitor having a capacitance of 800 pF is charged to a p.d. of 100 V. The plates of the capacitor are then separated until the capacitance is reduced to 200 pF. What is the change of p.d. across the capacitor? Also, what is the energy stored in the capacitor when its capacitance is: (a) 800 pF; (b) 200 pF? How has the increase of energy been supplied?

41. A 200 pF capacitor is charged to a p.d. of 50 V. The dielectric has a cross-sectional area of 300 cm^2 and a relative permittivity of 2.5. Calculate the energy density (in J/m^3) of the dielectric.

42. A parallel–plate capacitor, with the plates 20 mm apart, is immersed in oil having a relative permittivity of 3. The plates are charged to a p.d. of 25 kV. Calculate the force between the plates (in newtons per square metre of plate area) and the energy density (in J/m^3) within the dielectric.

43. A capacitor consists of two metal plates, each 600 mm × 500 mm, spaced 1.0 mm apart. The space between the metal plates is occupied by a dielectric having a relative permittivity of 6, and a p.d. of 3 kV is maintained between the plates. Calculate: (a) the capacitance in picofarads; (b) the electric field strength and the electric flux density in the dielectric; and (c) the force of attraction, in newtons, between the plates.

Chapter six

Electromagnetism

Objectives

When you have studied this chapter, you should
- have an understanding of magnetic fields and be able to draw maps using lines of flux
- be familiar with the magnetic fields associated with conductors and solenoids
- be capable of determining the force experienced by a current-carrying conductor lying in a magnetic field
- have an understanding of Fleming's rules and Lenz's law
- be capable of determining the e.m.f. induced in a conductor moving in a magnetic field

Contents

Most of us have seen permanent magnets which can pick up pins and other small steel objects. In this chapter, we shall find that the passage of an electric current in a conductor produces a similar magnetic field and that, by winding the conductor into a coil, the magnetic field can be made quite strong. Taking this a stage further, we can also observe that if we introduce a current-carrying conductor into such a magnetic field, it experiences a force. If we develop this observation, we can make an electric motor which can drive things.

In this chapter, we shall meet the essential principle of a generator in that if we move a conductor through a magnetic field, we find that an e.m.f. is induced in it. This e.m.f. causes current to flow and so provides us with most of the electric current which we meet in practice.

6.1   Magnetic field

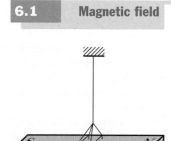

Fig. 6.1   A suspended permanent magnet

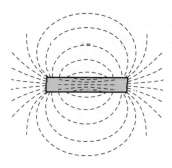

Fig. 6.2   Use of steel filings for determining distribution of magnetic field

The first known recognition of magnetism was made by the Chinese in 2637 BC. An emperor, Hoang-ti, is reputed to have had a chariot with a figure-head which pointed south no matter in what direction the chariot was moving. This arrangement was developed into the compass and it is even suggested that it was King Solomon who invented it. The ore from which the magnet was produced was called magnesian stone, hence the name magnet.

Nowadays we make compasses from steel, but the action remains the same. The pointer of a compass is called a permanent magnet because it always retains its peculiar properties, i.e. if a permanent magnet is suspended in a horizontal plane, as shown in Fig. 6.1, it takes up a position such that one end points to the earth's North Pole. That end is said to be the north-seeking end of the magnet while the other end is called the south-seeking end. These are called the north (or N) and south (or S) poles respectively of the magnet.

Let us place a permanent magnet on a table, cover it over with a sheet of smooth cardboard and sprinkle steel filings uniformly over the sheet. Slight tapping of the latter causes the filings to set themselves in curved chains between the poles, as shown in Fig. 6.2. The shape and density of these chains enable one to form a mental picture of the magnetic condition of the space or 'field' around a bar magnet and lead to the idea of *lines of magnetic flux*. It should be noted, however, that these lines of magnetic flux have no physical existence; they are purely imaginary and were introduced by Michael Faraday as a means of visualizing the distribution and density of a magnetic field. It is important to realize that the magnetic flux permeates the whole of the space occupied by that flux. This compares with the electric field lines introduced in Chapter 5.

6.2   Direction of magnetic field

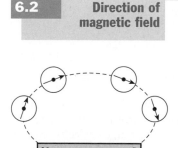

Fig. 6.3   Use of compass needles for determining direction of magnetic field

The direction of a magnetic field is taken as that in which the north-seeking pole of a magnet points when the latter is suspended in the field. Thus, if a bar magnet rests on a table and four compass needles are placed in positions indicated in Fig. 6.3, it is found that the needles take up positions such that their axes coincide with the corresponding chain of filings (Fig. 6.2) and their N poles are all pointing along the dotted line from the N pole of the magnet to its S pole. The lines of magnetic flux are assumed to pass through the magnet, emerge from the N pole and return to the S pole.

6.3   Characteristics of lines of magnetic flux

In spite of the fact that lines of magnetic flux have no physical existence, they do form a very convenient and useful basis for explaining various magnetic effects and for calculating their magnitudes. For this purpose, lines of magnetic flux are assumed to have the following properties:

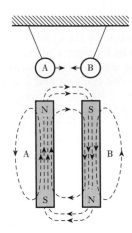

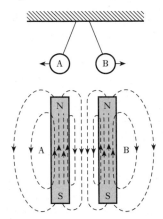

Fig. 6.4 Attraction between magnets

Fig. 6.5 Repulsion between magnets

1.   *The direction of a line of magnetic flux at any point in a non-magnetic medium, such as air, is that of the north-seeking pole of a compass needle placed at that point.*

2.   *Each line of magnetic flux forms a closed loop*, as shown by the dotted lines in Figs 6.4 and 6.5. This means that a line of flux emerging from any point at the N-pole end of a magnet passes through the surrounding space back to the S-pole end and is then assumed to continue through the magnet to the point at which it emerged at the N-pole end.

3.   *Lines of magnetic flux never intersect.* This follows from the fact that if a compass needle is placed in a magnetic field, its north-seeking pole will point in one direction only, namely in the direction of the magnetic flux at that point.

4.   *Lines of magnetic flux are like stretched elastic cords, always trying to shorten themselves.* This effect can be demonstrated by suspending two permanent magnets, A and B, parallel to each other, with their poles arranged as in Fig. 6.4. The distribution of the resultant magnetic field is indicated by the dotted lines. The lines of magnetic flux passing between A and B behave as if they were in tension, trying to shorten themselves and thereby causing the magnets to be attracted towards each other. In other words, unlike poles attract each other.

5.   *Lines of magnetic flux which are parallel and in the same direction repel one another.* This effect can be demonstrated by suspending the two permanent magnets, A and B, with their N poles pointing in the same direction, as in Fig. 6.5. It will be seen that in the space between A and B the lines of flux are practically parallel and are in the same direction. These flux lines behave as if they exerted a lateral pressure on one another, thereby causing magnets A and B to repel each other. Hence like poles repel each other.

6.4	Magnetic field due to an electric current

When a conductor carries an electric current, a magnetic field is produced around that conductor – a phenomenon discovered by Oersted at Copenhagen in 1820. He found that when a wire carrying an electric current was placed above a magnetic needle (Fig. 6.6) and in line with the normal direction of the latter, the needle was deflected clockwise or anticlockwise, depending upon the direction of the current. Thus it is found that if we look along the conductor and if the current is flowing away from us, as shown by the cross inside the conductor in Fig. 6.7, the magnetic field has a clockwise direction and the lines of magnetic flux can be represented by concentric circles around the wire.

We should note the interesting convention for showing the direction of current flow in a conductor. In Fig. 6.8, we have a conductor in which we have drawn an arrow indicating the direction of conventional current flow. However, if we observe the conductor end on, the current would be flowing either towards us or away from us. If the current is flowing towards us, we

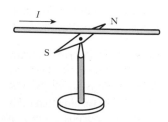

Fig. 6.6 Oersted's experiment

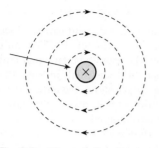

Fig. 6.7 Magnetic flux due to current in a straight conductor

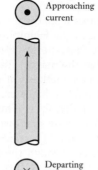

Approaching current

Departing current

Fig. 6.8 Current conventions

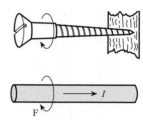

Fig. 6.9 Right-hand screw rule

indicate this by a dot equivalent to the approaching point of the arrow, and if the current is flowing away then it is represented by a cross equivalent to the departing tail feathers of the arrow.

A convenient method of representing the relationship between the direction of a current and that of its magnetic field is to place a corkscrew or a woodscrew (Fig. 6.9) alongside the conductor carrying the current. In order that the screw may travel in the same direction as the current, namely towards the right in Fig. 6.9, it has to be turned clockwise when viewed from the left-hand side. Similarly, the direction of the magnetic field, viewed from the same side, is clockwise around the conductor, as indicated by the curved arrow F.

An alternative method of deriving this relationship is to grip the conductor with the *right* hand, with the thumb outstretched parallel to the conductor and pointing in the direction of the current; the fingers then point in the direction of the magnetic flux around the conductor.

6.5 Magnetic field of a solenoid

If a coil is wound on a steel rod, as in Fig. 6.10, and connected to a battery, the steel becomes magnetized and behaves like a permanent magnet. The magnetic field of the electromagnet is represented by the dotted lines and its direction by the arrowheads.

The direction of the magnetic field produced by a current in a solenoid may be deduced by applying either the **screw** or the **grip rule**.

If the axis of the **screw** is placed along that of the solenoid and if the screw is turned in the direction of the current, it travels in the direction of the magnetic field *inside* the solenoid, namely towards the right in Fig. 6.10.

The **grip rule** can be expressed thus: if the solenoid is gripped with the *right* hand, with the fingers pointing in the direction of the current, i.e. conventional current, then the thumb outstretched parallel to the axis of the solenoid points in the direction of the magnetic field *inside* the solenoid.

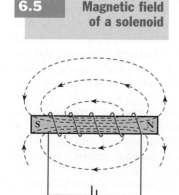

Fig. 6.10 Solenoid with a steel core

In section 6.4 it was shown that a conductor carrying a current can produce a force on a magnet situated in the vicinity of the conductor. By Newton's third law of motion, namely that to every force there must be an equal and opposite force, it follows that the magnet must exert an equal force on the conductor. One of the simplest methods of demonstrating this effect is to take a copper wire, about 2 mm in diameter, and bend it into a rectangular loop as represented by BC in Fig. 6.11. The two tapered ends of the loop dip into mercury contained in cups, one directly above the other, the cups being attached to metal rods P and Q carried by a wooden upright rod D. A current of about 5 A is passed through the loop and the N pole of a permanent magnet NS is moved towards B. If the current in this wire is flowing downwards, as indicated by the arrow in Fig. 6.11, it is found that the loop, when viewed from above, turns anticlockwise, as shown in plan in Fig. 6.12. If the magnet is reversed and again brought up to B, the loop turns clockwise.

Fig. 6.11 Force on conductor carrying current across a magnetic field

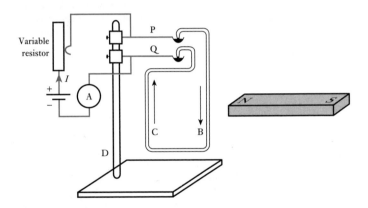

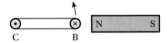

Fig. 6.12 Direction of force on conductor in Fig. 6.11

If the magnet is placed on the other side of the loop, the latter turns clockwise when the N pole of the magnet is moved near to C, and anticlockwise when the magnet is reversed. This may seem to be a most awkward method of demonstrating the interaction between a current-carrying conductor and a magnetic field. However, it is important to recognize that the action did not come from two pieces of magnetized steel. By means of the experiment which had only one piece of steel, we can be certain that the action arose from the current-carrying conductor.

Being convinced of this observation, we can introduce extra steel components in an experiment to explain the effects which we have noted. A suitable experimental apparatus is shown in elevation and plan in Fig. 6.13. Two permanent magnets NS rest on a sheet of paper or glass G, and steel pole-pieces P are added to increase the area of the magnetic field in the gap between them. Midway between the pole-pieces is a wire W passing vertically downwards through glass G and connected through a switch to a 6 V battery capable of giving a very large current for a short time.

With the switch open, steel filings are sprinkled over G and the latter is gently tapped. The filings in the space between PP take up the distribution

Fig. 6.13 Flux distribution with and without current

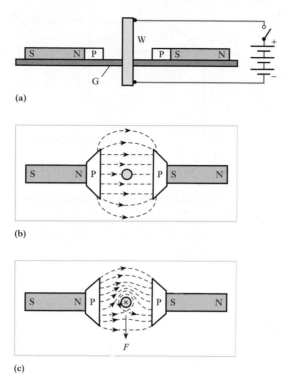

shown in Fig. 6.13(b). If the switch is closed momentarily, the filings rearrange themselves as in Fig. 6.13(c). It will be seen that the lines of magnetic flux have been so distorted that they partially surround the wire. This distorted flux acts like stretched elastic cords bent out of the straight; the lines of flux try to return to the shortest paths between PP, thereby exerting a force F urging the conductor out of the way.

It has already been shown in section 6.4 that a wire W carrying a current downwards in Fig. 6.13(a) produces a magnetic field as shown in Fig. 6.7. If this field is compared with that of Fig. 6.13(b), it is seen that on the upper side the two fields are in the same direction, whereas on the lower side they are in opposition. Hence, the combined effect is to strengthen the magnetic field on the upper side and weaken it on the lower side, thus giving the distribution shown in Fig. 6.13(c).

By combining diagrams similar to Figs 6.13(b) and 6.7, it is easy to understand that if either the current in W or the polarity of magnets NS is reversed, the field is strengthened on the lower side and weakened on the upper side of diagrams corresponding to Fig. 6.13(b), so that the direction of the force acting on W is the reverse of that shown in Fig. 6.13(c).

On the other hand, if both the current through W and the polarity of the magnets are reversed, the *distribution* of the resultant magnetic field and therefore the direction of the force on W remain unaltered.

By observation of the experiments, it can also be noted that the mechanical force exerted by the conductor always acts in a direction perpendicular to the plane of the conductor and the magnetic field direction. The direction is given by the left-hand rule illustrated in Fig. 6.14.

Fig. 6.14   Left-hand rule

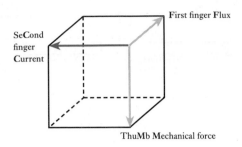

The rule can be summarized as follows:

1. Hold the thumb, first finger and second finger of the left hand in the manner indicated by Fig. 6.14, whereby they are mutually at right angles.
2. Point the **F**irst finger in the **F**ield direction.
3. Point the se**C**ond finger in the **C**urrent direction.
4. The thu**M**b then indicates the direction of the **M**echanical force exerted by the conductor.

By trying this with your left hand, you can readily demonstrate that if either the current or the direction of the field is reversed then the direction of the force is also reversed. Also you can demonstrate that, if both current and field are reversed, the direction of the force remains unchanged.

<table>
<tr><td>6.7</td><td>Force
determination</td></tr>
</table>

With the apparatus of Fig. 6.11, it can be shown qualitatively that the force on a conductor carrying a current at right angles to a magnetic field is increased (a) when the current in the conductor is increased, and (b) when the magnetic field is made stronger by bringing the magnet nearer to the conductor. With the aid of more elaborate apparatus, the force on the conductor can be measured for various currents and various densities of the magnetic field, and it is found that

Force on conductor $\propto$ current $\times$ (flux density)

$\times$ (length of conductor)

If F is the force on conductor in newtons, I the current through conductor in amperes and l the length, in metres, of conductor at right angles to magnetic field

$$F \text{ [newtons]} \propto \text{flux density} \times l \text{ [metres]} \times I \text{ [amperes]}$$

The *unit of flux density* is taken as *the density of a magnetic field such that a conductor carrying 1 ampere at right angles to that field has a force of 1 newton per metre acting upon it.* This unit is termed a *tesla** (T).

Magnetic flux density   Symbol: B Unit: tesla (T)

---

* Nikola Tesla (1857–1943), a Yugoslav who emigrated to the USA in 1884, was a very famous electrical inventor. In 1888 he patented two-phase and three-phase synchronous generators and motors.

For a flux density of B teslas,

force on conductor = BlI newtons

$$F = BlI \qquad\qquad [6.1]$$

For a magnetic field having a cross-sectional area of A square metres and a uniform flux density of B teslas, the *total flux* in *webers*† (Wb) is represented by the Greek capital letter Φ (phi).

Magnetic flux   Symbol: Φ Unit: weber (Wb)

It follows that,

$$\Phi \,[\text{webers}] = B\,[\text{teslas}] \times A\,[\text{metres}^2]$$

$\therefore \qquad \Phi = BA \qquad\qquad [6.2]$

and $\qquad B = \dfrac{\Phi}{A} \qquad\qquad [6.3]$

or $\qquad B\,[\text{teslas}] = \dfrac{\Phi\,[\text{webers}]}{A\,[\text{metres}]}$

i.e. $\qquad 1\ \text{T} = 1\ \text{Wb/m}^2$

Before we define the unit of magnetic flux, i.e. the weber, we need to introduce the concept of electromagnetic induction.

Example 6.1

A conductor carries a current of 800 A at right angles to a magnetic field having a density of 0.5 T. Calculate the force on the conductor in newtons per metre length.

From expression [6.1]

$$F = BlI$$

force per metre length is

$$0.5\,[\text{T}] \times 1\,[\text{m}] \times 800\,[\text{A}] = \textbf{400 N}$$

Example 6.2

A rectangular coil measuring 200 mm by 100 mm is mounted such that it can be rotated about the midpoints of the 100 mm sides. The axis of rotation is at right angles to a magnetic field of uniform flux density 0.05 T. Calculate the flux in the coil for the following conditions:

(a) the maximum flux through the coil and the position at which it occurs;
(b) the flux through the coil when the 100 mm sides are inclined at 45° to the direction of the flux (Fig. 6.15).

---

† Wilhelm Eduard Weber (1804–91), a German physicist, was the first to develop a system of absolute electrical and magnetic units.

Fig. 6.15

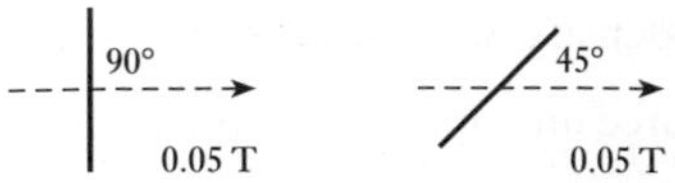

90°
0.05 T

45°
0.05 T

(a) The maximum flux will pass through the coil when the plane of the coil is at right angles to the direction of the flux.

$$\Phi = BA = 0.05 \times 200 \times 10^{-3} \times 100 \times 10^{-3} = 1 \times 10^{-3} \text{ Wb}$$

$$= 1 \text{ mWb}$$

(b) $\Phi = BA \sin \theta = 1 \times 10^{-3} \times \sin 45° = 0.71 \times 10^{-3}$ Wb $= \textbf{0.71 mWb}$

6.8   Electromagnetic induction

In 1831, Michael Faraday made the great discovery of *electromagnetic induction*, namely a method of obtaining an electric current with the aid of magnetic flux. He wound two coils, A and C, on a steel ring R, as in Fig. 6.16 and found that, when switch S was closed, a deflection was obtained on galvanometer G, and that when S was opened, G was deflected in the reverse direction. A few weeks later he found that when a permanent magnet NS was moved relative to a coil C (Fig. 6.17), galvanometer G was deflected in one direction when the magnet was moved towards the coil and in the reverse direction when the magnet was withdrawn; and it was this experiment that finally convinced Faraday that an electric current could be produced by the movement of magnetic flux relative to a coil. Faraday also showed that the magnitude of the induced e.m.f. is proportional to the rate at which the magnetic flux passed through the coil is varied. Alternatively, we can say that when a conductor cuts or is cut by magnetic flux, an e.m.f. is generated in the conductor and the magnitude of the generated e.m.f. is proportional to the rate at which the conductor cuts or is cut by the magnetic flux.

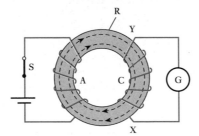

Fig. 6.16   Electromagnetic induction

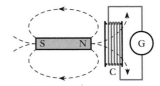

Fig. 6.17   Electromagnetic induction

6.9   Direction of induced e.m.f.

Two methods are available for deducing the direction of the induced or generated e.m.f., namely (a) Fleming's* right-hand rule and (b) Lenz's law. The former is empirical, but the latter is fundamental in that it is based upon electrical principles.

---

* John Ambrose Fleming (1849–1945) was Professor of Electrical Engineering at University College, London.

(a) Fleming's right-hand rule

*If the first finger of the right hand is pointed in the direction of the magnetic flux, as in Fig. 6.18, and if the thumb is pointed in the direction of motion of the conductor **relative** to the magnetic field, then the second finger, held at right angles to both the thumb and the first finger, represents the direction of the e.m.f.*

Fig. 6.18 Fleming's right-hand rule

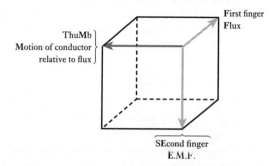

ThuMb
Motion of conductor
relative to flux

First finger
Flux

SEcond finger
E.M.F.

The manipulation of the thumb and fingers and their association with the correct quantity present some difficulty to many students. Easy manipulation can be acquired only by experience; and it may be helpful to associate Field or Flux with First finger, Motion of the conductor relative to the field with the M in thuMb and e.m.f. with the E in sEcond finger. If any two of these are correctly applied, the third is correct automatically.

(b) Lenz's law

In 1834 Heinrich Lenz, a German physicist, enunciated a simple rule, now known as Lenz's law, which can be expressed thus: *The direction of an induced e.m.f. is always such that it tends to set up a current opposing the motion or the change of flux responsible for inducing that e.m.f.*

Let us consider the application of Lenz's law to the ring shown in Fig. 6.16. By applying either the screw or the grip rule given in section 6.5, we find that when S is closed and the battery has the polarity shown, the direction of the magnetic flux in the ring is clockwise. Consequently, the current in C must be such as to try to produce a flux in an anticlockwise direction, tending to oppose the growth of the flux due to A, namely the flux which is responsible for the e.m.f. induced in C. But an anticlockwise flux in the ring would require the current in C to be passing through the coil from X to Y (Fig. 6.16). Hence, this must also be the direction of the e.m.f. induced in C.

6.10   **Magnitude of the generated or induced e.m.f.**

Figure 6.19 represents the elevation and plan of a conductor AA situated in an airgap between poles NS. Suppose AA to be carrying a current, I amperes, in the direction shown. By applying either the screw or the grip rule of section 6.4, it is found that the effect of this current is to strengthen the field on the right and weaken that on the left of A, so that there is a force of BII newtons (section 6.7) urging the conductor towards the left, where B is the flux

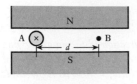

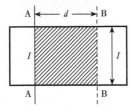

Fig. 6.19 Conductor moved across magnetic field

density in teslas and l is the length in metres of conductor in the magnetic field. Hence, a force of this magnitude has to be applied in the opposite direction to move A towards the right.

The work done in moving conductor AA through a distance d metres to position BB in Fig. 6.19 is $(BlI \times d)$ joules. If this movement of AA takes place at a uniform velocity in t seconds, the e.m.f. induced in the conductor is constant at, say, E volts. Hence the electrical power generated in AA is IE watts and the electrical energy is IEt watt seconds or joules. Since the mechanical energy expended in moving the conductor horizontally across the gap is all converted into electrical energy, then

$$IEt = BlId$$

$$\therefore \qquad E = \frac{Bld}{t}$$

and $\qquad \boxed{E = Blu}$ [6.4]

where u is the velocity in metres per second. But Bld is the total flux, Φ webers, in the area shown shaded in Fig. 6.19. This flux is cut by the conductor when the latter is moved from AA to BB. Hence

$$E\,[\text{volts}] = \frac{\Phi\,[\text{webers}]}{t\,[\text{seconds}]}$$

i.e. the e.m.f., in volts, generated in a conductor is equal to the rate (in webers per second) at which the magnetic flux is cutting or being cut by the conductor; and the *weber* may therefore be defined as *that magnetic flux which, when cut at a uniform rate by a conductor in 1 s, generates an e.m.f. of 1 V.*

In general, if a conductor cuts or is cut by a flux of $d\phi$ webers in dt seconds

e.m.f. generated in conductor $= d\phi/dt$ volts

$$\therefore \qquad \boxed{e = \frac{d\phi}{dt}}$$ [6.5]

Example 6.3   Calculate the e.m.f. generated in the axle of a car travelling at 80 km/h, assuming the length of the axle to be 2 m and the vertical component of the earth's magnetic field to be 40 μT (microteslas).

$$80\,\text{km/h} = \frac{(80 \times 1000)\,[\text{m}]}{3600\,[\text{s}]}$$

$$\therefore \qquad u = 22.2\,\text{m/s}$$

Vertical component of earth's field is

$$40 \times 10^{-6}\,\text{T}$$

$$\therefore \qquad \text{Flux cut by axle} = 40 \times 10^{-6}\,[\text{T}] \times 2\,[\text{m}] \times 22.2\,[\text{m s}]$$

$$= 1776 \times 10^{-6}\,\text{Wb/s}$$

and e.m.f. generated in axle is 1776×10^{-6} V

$$\therefore \qquad e = \mathbf{1780\ \mu V}$$

6.11 **Magnitude of e.m.f. induced in a coil**

Suppose the magnetic flux through a coil of N turns to be increased by Φ webers in t seconds due to, say, the relative movement of the coil and a magnet (Fig. 6.17). Since each of the lines of magnetic flux cuts each turn, one turn can be regarded as a conductor cut by Φ webers in t seconds; hence, from expression [6.5], the average e.m.f. induced in each turn is Φ/t volts. The current due to this e.m.f., by Lenz's law, tries to prevent the increase of flux, i.e. tends to set up an opposing flux. Thus, if the magnet NS in Fig. 6.17 is moved towards coil C, the flux passing from left to right through the latter is increased. The e.m.f. induced in the coil circulates a current in the direction represented by the dot and cross in Fig. 6.20, where – for simplicity – coil C is represented as one turn. The effect of this current is to distort the magnetic field as shown by the dotted lines, thereby tending to push the coil away from the magnet. By Newton's third law of motion, there must be an equal and opposite force tending to oppose the movement of the magnet.

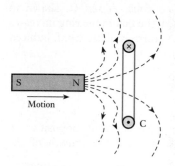

Fig. 6.20 Distortion of magnetic field by induced current

The induced e.m.f. circulates a current tending to oppose the increase of flux through the coil, hence the average e.m.f. induced in one turn is

$$\frac{\Phi}{t} \text{ volts}$$

which is the average rate of *change* of flux in webers per second, and the average e.m.f. induced in coil is

$$\frac{N\Phi}{t} \text{ volts} \qquad\qquad [6.6]$$

which is the average rate of *change* of flux-linkages per second. The term 'flux-linkages' merely means the product of the flux in webers and the number of turns with which the flux is linked. Thus if a coil of 20 turns has a flux of 0.1 Wb through it, the flux-linkages = $0.1 \times 20 = 2$ weber-turns (Wb).

The turn is a dimensionless factor, hence the product of webers and turns is measured only in webers. However, some prefer to retain the term 'turn' although this does not conform to the SI.

Flux linkage   Symbol: Ψ (psi)   Unit: weber (Wb)

$$\Psi = N\Phi$$

From expression [6.5] it follows that instantaneous value of e.m.f., in volts, induced in a coil is the rate of change of flux-linkages, in weber-turns per second, or

$$e = \frac{d}{dt}(N\phi) \text{ volts} \qquad\qquad [6.7]$$

and $$e = \frac{d\psi}{dt} \qquad\qquad [6.8]$$

This relationship is usually known as *Faraday's law*, though it was not stated in this form by Faraday.

From expression [6.6] we can define the *weber as that magnetic flux which, linking a circuit of one turn, induces in it an e.m.f. of 1 V when the flux is reduced to zero at a uniform rate in 1 s.*

Next, let us consider the case of the two coils, A and C, shown in Fig. 6.16. Suppose that when switch S is closed, the flux in the ring increases by Φ webers in t seconds. Then, if coil A has N_1 turns, average e.m.f. induced in A is

$$\frac{N_1 \Phi}{t} \text{ volts}$$

This e.m.f., in accordance with Lenz's law, is acting in opposition to the e.m.f. of the battery, thereby trying to prevent the growth of the current.

If coil C is wound with N_2 turns, and if all the flux produced by coil A passes through C, average e.m.f. induced in C is

$$\frac{N_2 \Phi}{t} \text{ volts}$$

In this case the e.m.f. circulates a current in such a direction as to tend to set up a flux in opposition to that produced by the current in coil A, thereby delaying the growth of flux in the ring.

In general, if the magnetic flux through a coil increases by $d\phi$ webers in dt seconds e.m.f. induced in coil is

$$N \cdot \frac{d\phi}{dt} \text{ volts}$$

$$\therefore \qquad e = N\frac{d\phi}{dt} \qquad\qquad [6.7]$$

Example 6.4　A magnetic flux of 400 μWb passing through a coil of 1200 turns is reversed in 0.1 s. Calculate the average value of the e.m.f. induced in the coil.

The magnetic flux has to decrease from 400 μWb to zero and then increase to 400 μWb in the reverse direction; hence the *increase* of flux in the original direction is 800 μWb.

Substituting in expression [6.6], we have

$$E = \frac{N\Phi}{t}$$

average e.m.f. induced in coil is

$$\frac{1200 \times (800 \times 10^{-6})}{0.1} = \textbf{9.6 V}$$

Summary of important formulae

Force on a conductor

$$F = BlI \quad \text{(newtons)} \qquad [6.1]$$

$$\text{Flux } \Phi = BA \quad \text{(webers)} \qquad [6.2]$$

$$\text{Flux density } B = \Phi/A \quad \text{(teslas)} \qquad [6.3]$$

Induced e.m.f.

$$E = Blu \quad \text{(volts)} \qquad [6.4]$$

hence

$$e = \frac{\mathrm{d}\phi}{\mathrm{d}t} \qquad [6.5]$$

or

$$e = \frac{\mathrm{d}}{\mathrm{d}t}(N\phi) \qquad [6.7]$$

Terms and concepts

A **magnetic field** can be described using **lines of flux**. Such lines form closed loops, do not cross and, when parallel, repel one another.

Magnetic fields have **north and south poles**. Like poles repel one another. Unlike poles attract one another.

A current-carrying conductor lying in a magnetic field experiences a force.

The relative directions of the field, force and current are given by the **left-hand rule**.

When the magnetic flux linking a circuit is varied, an e.m.f. is induced in the circuit. This is known as **Faraday's law**.

The induced e.m.f. opposes the change of condition. This is known as **Lenz's law**.

The relative directions of the field, motion and induced e.m.f. are given by the **right-hand rule** (sometimes known as Fleming's right-hand rule).

Exercises 6

1. A current-carrying conductor is situated at right angles to a uniform magnetic field having a density of 0.3 T. Calculate the force (in newtons per metre length) on the conductor when the current is 200 A.

2. Calculate the current in the conductor referred to in Q. 1 when the force per metre length of the conductor is 15 N.

3. A conductor, 150 mm long, is carrying a current of 60 A at right angles to a magnetic field. The force on the conductor is 3 N. Calculate the density of the field.

4. The coil of a moving-coil loudspeaker has a mean diameter of 30 mm and is wound with 800 turns. It is situated in a radial magnetic field of 0.5 T. Calculate the force on the coil, in newtons, when the current is 12 mA.

5. Explain what happens when a long straight conductor is moved through a uniform magnetic field at constant velocity. Assume that the conductor moves perpendicularly to the field. If the ends of the conductor are connected together through an ammeter, what will happen?

   A conductor, 0.6 m long, is carrying a current of 75 A and is placed at right angles to a magnetic field of uniform flux density. Calculate the value of the flux density if the mechanical force on the conductor is 30 N.

Exercises 6 continued

6. State Lenz's law.

   A conductor, 500 mm long, is moved at a uniform speed at right angles to its length and to a uniform magnetic field having a density of 0.4 T. If the e.m.f. generated in the conductor is 2 V and the conductor forms part of a closed circuit having a resistance of 0.5 Ω, calculate: (a) the velocity of the conductor in metres per second; (b) the force acting on the conductor in newtons; (c) the work done in joules when the conductor has moved 600 mm.

7. A wire, 100 mm long, is moved at a uniform speed of 4 m/s at right angles to its length and to a uniform magnetic field. Calculate the density of the field if the e.m.f. generated in the wire is 0.15 V. If the wire forms part of a closed circuit having a total resistance of 0.04 Ω, calculate the force on the wire in newtons.

8. Give three practical applications of the mechanical force exerted on a current-carrying conductor in a magnetic field.

   A conductor of active length 30 cm carries a current of 100 A and lies at right angles to a magnetic field of density 0.4 T. Calculate the force in newtons exerted on it. If the force causes the conductor to move at a velocity of 10 m/s, calculate (a) the e.m.f. induced in it and (b) the power in watts developed by it.

9. The axle of a certain motor car is 1.5 m long. Calculate the e.m.f. generated in it when the car is travelling at 140 km/h. Assume the vertical component of the earth's magnetic field to be 40 μT.

10. An aeroplane having a wing span of 50 m is flying horizontally at a speed of 600 km/h. Calculate the e.m.f. generated between the wing tips, assuming the vertical component of the earth's magnetic field to be 40 μT. Is it possible to measure this e.m.f.?

11. A copper disc, 250 mm in diameter, is rotated at 300 r/min about a horizontal axis through its centre and perpendicular to its plane. If the axis points magnetic north and south, calculate the e.m.f. between the circumference of the disc and the axis. Assume the horizontal component of the earth's field to be 18 μT.

12. A coil of 1500 turns gives rise to a magnetic flux of 2.5 mWb when carrying a certain current. If this current is reversed in 0.2 s, what is the average value of the e.m.f. induced in the coil?

13. A short coil of 200 turns surrounds the middle of a bar magnet. If the magnet sets up a flux of 80 μWb, calculate the average value of the e.m.f. induced in the coil when the latter is removed completely from the influence of the magnet in 0.05 s.

14. The flux through a 500-turn coil increases uniformly from zero to 200 μWb in 3 ms. It remains constant for the fourth millisecond and then decreases uniformly to zero during the fifth millisecond. Draw to scale a graph representing the variation of the e.m.f. induced in the coil.

15. Two coils, A and B, are wound on the same ferromagnetic core. There are 300 turns on A and 2800 turns on B. A current of 4 A through coil A produces a flux of 800 μWb in the core. If this current is reversed in 20 ms, calculate the average e.m.f.s induced in coils A and B.

Chapter seven

Simple Magnetic Circuits

Objectives

When you have studied this chapter, you should

- have an understanding of magnetic circuits and be able to recognize solenoids and toroids

- be familiar with magnetomotive force and magnetic field strength

- be capable of relating magnetic field strength to magnetic flux density by means of permeability

- have an understanding of the permeability of free space

- be familiar with B/H characteristics and be capable of applying them to the analysis of simple magnetic circuits

- be capable of relating magnetomotive force to magnetic flux by means of reluctance

Contents

Having found that we can produce a magnetic field using a current-carrying conductor and that we can make the field stronger by winding the conductor into a coil, we find that it would be useful if we could produce an even stronger magnetic field. If we place a piece of steel within a coil, we find that the field becomes hundreds of times stronger. This is essential to the manufacture of motors and other electrical devices and therefore we need to understand the effect of the steel.

Our investigation leads to an introduction to permeability, the magnetic field's equivalent to permittivity in electric fields.

7.1   Introduction to magnetic circuits

In Chapter 6, we observed that one of the characteristics of lines of magnetic flux is that each line forms a closed loop. For instance, in Fig. 7.1, the dotted lines represent the flux set up within a ring made of steel. The complete closed path followed by any group of magnetic flux lines is referred to as a *magnetic circuit*. One of the simplest forms of magnetic circuit is the ring shown in Fig. 7.1 where the steel ring provides the space in which the magnetic flux is created. Most rings are made like anchor rings in that their cross-section is circular – such a ring is called a toroid.

Fig. 7.1   A toroid

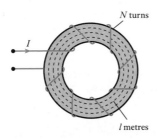

In an electric circuit, the current is due to the existence of an electromotive force. By analogy, we may say that in a magnetic circuit the magnetic flux is due to the existence of a *magnetomotive force* (m.m.f.) caused by a current flowing through one or more turns. The value of the m.m.f. is proportional to the current and to the number of turns, and is descriptively expressed in *ampere-turns*; but for the purpose of dimensional analysis, it is expressed in *amperes*, since the number of turns is dimensionless. Hence the unit of magnetomotive force is the *ampere*.

7.2   Magnetomotive force and magnetic field strength

Magnetomotive force   Symbol: F Unit: ampere (A)

Sometimes the unit of magnetomotive force is shown as ampere turns abbreviated as At. This does not conform to the SI but some people find it helpful.

If a current of I amperes flows through a coil of N turns, as shown in Fig. 7.1, the magnetomotive force F is the *total* current linked with the magnetic circuit, namely IN amperes. If the magnetic circuit is homogeneous and of uniform cross-sectional area, the magnetomotive force per metre length of the magnetic circuit is termed the *magnetic field strength* and is represented by the symbol H. Thus, if the mean length of the magnetic circuit of Fig. 7.1 is l metres,

$$H = IN/l \quad \text{amperes per metre} \tag{7.1}$$

Magnetic field strength   Symbol: H Unit: ampere per metre (A/m)

Again this unit is sometimes given in ampere–turns per metre (At/m)

and $$H = \frac{F}{l}$$

where $$F = NI \quad \text{amperes} \tag{7.2}$$

7.3	**Permeability of free space or magnetic constant**

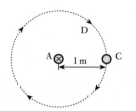

Fig. 7.2   Magnetic field at 1 m radius due to current in a long straight conductor

Suppose conductor A in Fig. 7.2 to represent the cross-section of a long straight conductor, situated in a vacuum and carrying a current of 1 A towards the paper, and suppose the return path of this current to be some considerable distance away from A so that the effect of the return current on the magnetic field in the vicinity of A may be neglected. The lines of magnetic flux surrounding A will, by symmetry, be in the form of concentric circles, and the dotted circle D in Fig. 7.2 represents the path of one of these lines of flux at a radius of 1 m. Since conductor A and its return conductor form one turn, the magnetomotive force acting on path D is 1 A; and since the length of this line of flux is 2π metres, the magnetic field strength, H, at a radius of 1 m is $1/(2\pi)$ amperes per metre.

If the flux density in the region of line D is B teslas, it follows from expression [6.1] that the force per metre length on a conductor C (parallel to A) carrying 1 A at right angles to this flux is given by

$$\text{Force per metre length} = B \, [\text{T}] \times 1 \, [\text{m}] \times 1 \, [\text{A}] = B \text{ newtons}$$

But from the definition of the ampere given in section 2.7 this force is 2×10^{-7} N, therefore flux density at 1 m radius from conductor carrying 1 A is

$$B = 2 \times 10^{-7} \text{ T}$$

Hence

$$\frac{\text{Flux density at C}}{\text{Magnetic field strength at C}} = \frac{B}{H} = \frac{2 \times 10^{-7} \, [\text{T}]}{1/2\pi \, [\text{A/m}]}$$

$$= 4\pi \times 10^{-7} \text{ H/m}$$

The ratio B/H is termed the *permeability of free space* and is represented by the symbol μ_0. Thus

Permeability of free space   **Symbol:** μ_0 **Unit: henry per metre (H/m)**

$$\mu_0 = \frac{B}{H} \qquad\qquad [7.3]$$

The value of this is almost exactly the same whether the conductor A is placed in free space, in air or in any other non-magnetic material such as water, oil, wood, copper, etc.

The unit of permeability is the henry per metre which is abbreviated to H/m. We will explain this unit after having been introduced to inductance in Chapter 8.

Returning to the permeability of free space

$$\mu_0 = \frac{B}{H} \text{ for a vacuum and non-magnetic materials}$$

$$\therefore \qquad \mu_0 = 4\pi \times 10^{-7} \text{ H/m} \qquad\qquad [7.4]$$

and magnetic field strength for non-magnetic materials is

$$H = \frac{B}{\mu_0} = \frac{B}{4\pi \times 10^{-7}} \qquad \text{amperes per metre} \qquad [7.5]$$

It may be mentioned at this point that there is a definite relationship between μ_0, ϵ_0 and the velocity of light and other electromagnetic waves; thus

$$\frac{1}{\mu_0 \epsilon_0} = \frac{1}{4\pi \times 10^{-7} \times 8.85 \times 10^{-12}} \simeq 8.99 \times 10^{16}$$

$$\simeq (2.998 \times 10^8)^2$$

But the velocity of light $= 2.998 \times 10^8$ metres per second

$$\therefore \qquad \left.\begin{array}{r} \text{velocity of light} \\ \text{in metres per second} \end{array}\right\} = \frac{1}{\sqrt{(\mu_0 \epsilon_0)}} \qquad\qquad [7.6]$$

This relationship was discovered by James Clerk Maxwell in 1865 and enabled him to predict the existence of radio waves about twenty years before their effect was demonstrated experimentally by Heinrich Hertz.

Example 7.1

A coil of 200 turns is wound uniformly over a wooden ring having a mean circumference of 600 mm and a uniform cross-sectional area of 500 mm². If the current through the coil is 4.0 A, calculate

(a) the magnetic field strength;
(b) the flux density;
(c) the total flux.

(a) Mean circumference = 600 mm = 0.6 m.

$$\therefore \qquad\qquad H = 4 \times 200/0.6 = \mathbf{1330 \ A/m}$$

(b) From expression [7.3]:

$$\text{Flux density} = \mu_0 H = 4\pi \times 10^{-7} \times 1333$$

$$= 0.001\ 675 \ \text{T} \equiv \mathbf{1680 \ \mu T}$$

(c) Cross-sectional area = 500 mm² = 500×10^{-6} m²

$$\therefore \qquad \text{Total flux} = 1675 \ [\mu\text{T}] \times (500 \times 10^{-6}) \ [\text{m}^2] \equiv \mathbf{0.838 \ \mu Wb}$$

Example 7.2

Calculate the magnetomotive force required to produce a flux of 0.015 Wb across an airgap 2.5 mm long, having an effective area of 200 cm².

$$\text{Area of airgap} = 200 \times 10^{-4} = 0.02 \ \text{m}^2$$

$$\therefore \qquad \text{Flux density} = \frac{0.015 \ [\text{Wb}]}{0.02 \ [\text{m}^2]} = 0.75 \ \text{T}$$

From expression [7.5]:

$$\text{Magnetic field strength for gap} = \frac{0.75}{4\pi \times 10^{-7}} = 597\,000 \ \text{A/m}$$

$$\text{Length of gap} = 2.5 \ \text{mm} = 0.0025 \ \text{m}$$

therefore m.m.f. required to send flux across gap is

$$597\,000 \ [\text{A/m}] \times 0.0025 \ [\text{m}] = \mathbf{1490 \ A}$$

7.4 Relative permeability

It can be shown that the magnetic flux inside a coil is intensified when a steel core is inserted. It follows that if the non-magnetic core of a toroid, such as that shown in Fig. 7.1, is replaced by a steel core, the flux produced by a given m.m.f. is greatly increased; and *the ratio of the flux density produced in a material to the flux density produced in a vacuum* (or in a non-magnetic core) *by the same magnetic field strength* is termed the *relative permeability* and is denoted by the symbol μ_r.

Any form of steel produces an increase in flux density for a given magnetic field strength. The same observation can be made of all ferromagnetic materials, i.e. those which contain iron, cobalt, nickel or gadolinium. Steels contain iron and are the most common form of ferromagnetic material. For air, $\mu_r = 1$, but for some forms of nickel–iron alloys, the relative permeability can be 100 000.

The value of the relative permeability of a ferromagnetic material is not a constant. Instead it varies considerably for different values of the magnetic field strength, and it is usually convenient to represent the relationship between the flux density and the magnetic field strength graphically as in Fig. 7.3; and the curves in Figs 7.4 and 7.5 represent the corresponding values of the relative permeability plotted against the magnetic field strength and the flux density respectively.

From expression [7.3], $B = \mu_0 H$ for a non-magnetic material; hence, for a material having a relative permeability μ_r,

$$B = \mu_r \mu_0 H \qquad\qquad [7.7]$$

Fig. 7.3 Magnetization characteristics of soft-magnetic materials

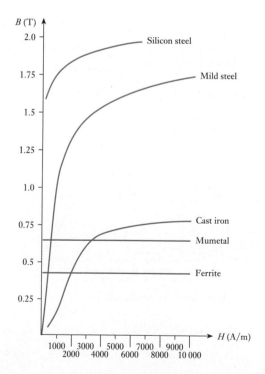

Fig. 7.4 μ_r/H characteristics
for soft–magnetic materials

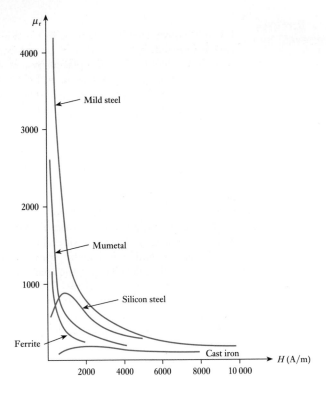

Fig. 7.5 μ_r/B characteristics for
soft–magnetic materials

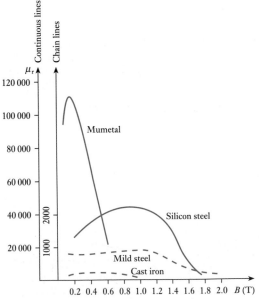

where absolute permeability

$$\mu = \mu_0\mu_r$$ [7.8]

$\therefore$ $B = \mu H$

7.5	**Reluctance**

Let us consider a ferromagnetic ring having a cross-sectional area of A square metres and a mean circumference of l metres (Fig. 7.1), wound with N turns carrying a current I amperes, then total flux (Φ) = flux density × area

$$\therefore \qquad \Phi = BA \qquad\qquad\qquad [7.9]$$

and m.m.f. (F) = magnetic field strength × length.

$$\therefore \qquad F = Hl \qquad\qquad\qquad [7.10]$$

Dividing equation [7.9] by [7.10], we have

$$\frac{\Phi}{F} = \frac{BA}{Hl} = \mu_r\mu_0 \times \frac{A}{l}$$

$$\therefore \qquad \Phi = \frac{F}{\frac{l}{\mu_0\mu_r A}}$$

where

$$\frac{F}{\Phi} = \frac{l}{\mu_0\mu_r A} = S \qquad\qquad\qquad [7.11]$$

S is the reluctance of the magnetic circuit where

$$F = \Phi S \qquad\qquad\qquad [7.12]$$

and

$$S = \frac{l}{\mu_0\mu_r A} \qquad\qquad\qquad [7.13]$$

This expression is similar in form to

$$\frac{E}{I} = \frac{\rho l}{A}$$

for the electric circuit. The expression

$$\frac{l}{\mu_0\mu_r A}$$

in expression [7.11] is similar in form to $\rho l/A$ for the resistance of a conductor except that the absolute permeability, $\mu_r\mu_0$, for the magnetic material corresponds to the reciprocal of the resistivity, namely the conductivity of the electrical material.

Since the m.m.f. is equal to the total number of amperes ($= IN$) acting on the magnetic circuit,

$$\therefore \qquad \text{magnetic flux} = \frac{\text{m.m.f.}}{\text{reluctance}}$$

and

$$\Phi = \frac{F}{S} = \frac{IN}{S}$$

The unit of reluctance is the ampere per weber abbreviated to A/Wb.

Reluctance   Symbol: S Unit: ampere per weber (A/Wb)

7.6   Comparison of electric and magnetic circuits

It is helpful to tabulate side by side the various electric and magnetic quantities and their relationships (Table 7.1).

One important difference between the electric and magnetic circuits is the fact that energy must be supplied to *maintain* the flow of electricity in a circuit, whereas the magnetic flux, once it is set up, does not require any further supply of energy. For instance, once the flux produced by a current in a solenoid has attained its maximum value, the energy subsequently absorbed by that solenoid is all dissipated as heat due to the resistance of the winding. However, the magnetic circuit stores energy in its field, whereas the electric circuit immediately releases its energy as heat.

Table 7.1

Electric circuit		Magnetic circuit	
Quantity	Unit	Quantity	Unit
EMF	volt	MMF	ampere
Electric field strength	volts per metre	Magnetic field strength	ampere per metre
Current	ampere	Magnetix flux	weber
Current density	ampere per square metre	Magnetic flux density	tesla
Resistance $\left(= \rho \cdot \dfrac{l}{A}\right)$	ohm	Reluctance $\left(= \dfrac{1}{\mu_r \mu_0} \cdot \dfrac{l}{A}\right)$	ampere per weber

Current = e.m.f./resistance; flux = m.m.f./reluctance.

Example 7.3

A mild-steel ring having a cross-sectional area of 500 mm^2 and a mean circumference of 400 mm has a coil of 200 turns wound uniformly around it. Calculate:

(a) the reluctance of the ring;

(b) the current required to produce a flux of 800 μWb in the ring.

(a) Flux density in ring is

$$\frac{800 \times 10^{-6}\,[\text{Wb}]}{500 \times 10^{-6}\,[\text{m}^{-2}]} = 1.6\,\text{T}$$

From Fig. 7.5, the relative permeability of mild steel for a flux density of 1.6 T is about 380. Therefore reluctance of ring is

$$\frac{0.4}{380 \times 4\pi \times 10^{-7} \times 5 \times 10^{-4}} = 1.68 \times 10^6\,\text{A/Wb}$$

(b) From expression [7.11]

$$\Phi = \frac{F}{S}$$

$$\therefore \qquad 800 \times 10^{-6} = \frac{\text{m.m.f.}}{1.677 \times 10^6}$$

$$\therefore \qquad \text{m.m.f. } F = 1342\,\text{A}$$

and magnetizing current is

$$\frac{F}{N} = \frac{1342}{200} = 6.7 \, \text{A}$$

Alternatively, from expression [7.5],

$$H = \frac{B}{\mu_r \mu_0} = \frac{1.6}{380 \times 4\pi \times 10^{-7}} = 3350 \, \text{A/m}$$

$$\therefore \qquad \text{m.m.f.} = 3350 \times 0.4 = 1340 \, \text{A}$$

and magnetizing current is

$$\frac{1340}{200} = 6.7 \, \text{A}$$

Magnetic circuits have an equivalent to the potential difference of electric circuits. This is the magnetic potential difference which allows us to apply Kirchhoff's laws to magnetic circuit analysis. This is demonstrated by Example 7.4.

Example 7.4	A magnetic circuit comprises three parts in series, each of uniform cross-sectional area (c.s.a.). They are:

(a) a length of 80 mm and c.s.a. 50 mm^2,
(b) a length of 60 mm and c.s.a. 90 mm^2,
(c) an airgap of length 0.5 mm and c.s.a. 150 mm^2.

A coil of 4000 turns is wound on part (b), and the flux density in the airgap is 0.30 T. Assuming that all the flux passes through the given circuit, and that the relative permeability μ_r is 1300, estimate the coil current to produce such a flux density.

$$\Phi = B_c A_c = 0.3 \times 1.5 \times 10^{-4} = 0.45 \times 10^{-4} \, \text{Wb}$$

$$F_a = \Phi S_a = \Phi \cdot \frac{l_a}{\mu_0 \mu_r A_a}$$

$$= \frac{0.45 \times 10^{-4} \times 80 \times 10^{-3}}{4\pi \times 10^{-7} \times 1300 \times 50 \times 10^{-6}} = 44.1 \, \text{At}$$

$$F_b = \Phi S_b = \Phi \cdot \frac{l_b}{\mu_0 \mu_r A_b}$$

$$= \frac{0.45 \times 10^{-4} \times 60 \times 10^{-3}}{4\pi \times 10^{-7} \times 1300 \times 90 \times 10^{-6}} = 18.4 \, \text{At}$$

$$F_c = \Phi S_c = \Phi \cdot \frac{l_c}{\mu_0 \mu_r A_c}$$

$$= \frac{0.45 \times 10^{-4} \times 0.5 \times 10^{-3}}{4\pi \times 10^{-7} \times 1 \times 150 \times 10^{-6}} = 119.3 \, \text{At}$$

$$F = F_a + F_b + F_c = 44.1 + 18.4 + 119.3 = 181.8 \, \text{At} = IN$$

$$I = \frac{181.8}{4000} = 45.4 \times 10^{-3} \, \text{A} = \textbf{45.4 mA}$$

7.7   Determination of the *B/H* characteristic

(a)   By means of a fluxmeter

Figure 7.6 shows a steel ring of uniform cross-section, uniformly wound with a coil P, thereby eliminating magnetic leakage. Coil P is connected to a battery through a reversing switch RS, an ammeter A and a variable resistor R_1. Another coil S, which need not be distributed around the ring, is connected through a two-way switch K to fluxmeter F which is a special type of permanent-magnet moving-coil instrument. Current is led into and out of the moving coil of F by fine wires or ligaments so arranged as to exert negligible control over the position of the moving coil. When the flux in the ring is varied, the e.m.f. induced in S sends a current through the fluxmeter and produces a deflection that is proportional to the change of flux-linkages in coil S.

Fig. 7.6 Determination of the magnetization curve for a steel ring

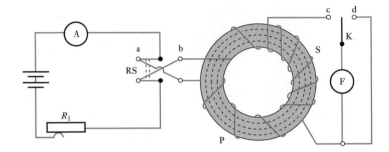

The current through coil P is adjusted to a desired value by means of R_1 and switch RS is then reversed several times to bring the steel into a 'cyclic' condition, i.e. into a condition such that the flux in the ring reverses from a certain value in one direction to the same value in the reverse direction. During this operation, switch K should be on d, thereby short-circuiting the fluxmeter. With switch RS on, say, a, switch K is moved over to c, the current through P is reversed by moving RS quickly over to b and the fluxmeter deflection is noted.

If N_P is the number of turns on coil P, l the mean circumference of the ring, in metres, and I the current through P, in amperes, the magnetic field strength is

$$H = \frac{IN_P}{l} \text{ amperes per metre}$$

If θ is the fluxmeter deflection when current through P is reversed and c the fluxmeter constant = no. of weber-turns per unit of scale deflection,

$$\text{change of flux-linkages with coil S} = c\theta \qquad [7.14]$$

If the flux in the ring changes from Φ to $-\Phi$ when the current through coil P is reversed, and if N_S is the number of turns on S, change of flux-linkages with coil S is

$$\text{change of flux} \times \text{no. of turns on S} = 2\,\Phi N_S \qquad [7.15]$$

Equating [7.14] and [7.15], we have

$$2\Phi N_S = c\theta$$

so that

$$\Phi = \frac{c\theta}{2N_S} \text{ webers}$$

If A is the cross-sectional area of ring in square metres

$$\text{flux density in ring} = B = \frac{\Phi}{A}$$

$$B = \frac{c\theta}{2AN_S} \text{ teslas} \qquad [7.16]$$

The test is performed with different values of the current; and from the data, a graph representing the variation of flux density with magnetic field strength can be plotted, as in Fig. 7.3.

(b) By means of a ballistic galvanometer

A ballistic galvanometer has a moving coil suspended between the poles of a permanent magnet, but the coil is wound on a *non-metallic* former, so that there is very little damping when the coil has a resistor of high resistance in series. The first deflection or 'throw' is proportional to the number of coulombs discharged through the galvanometer if the duration of the discharge is short compared with the time of one oscillation.

If θ is the first deflection or 'throw' of the ballistic galvanometer when the current through coil P is reversed and k the ballistic constant of the galvanometer = quantity of electricity in coulombs per unit deflection, quantity of electricity through galvanometer is

$$k\theta \text{ coulombs} \qquad [7.17]$$

If Φ is the flux produced in ring by I amperes through P and t the time, in seconds, of reversal of flux, average e.m.f. induced in S is

$$N_S \times \frac{2\Phi}{t} \text{ volts}$$

If R is the total resistance of the secondary circuit, the quantity of electricity through a ballistic galvanometer is average current × time which is

$$\frac{2\Phi N_S}{tR} \times t = \frac{2\Phi N_S}{R} \text{ coulombs} \qquad [7.18]$$

Equating [7.17] and [7.18], we have

$$k\theta = \frac{2\Phi N_S}{R}$$

$$\therefore \qquad \Phi = \frac{k\theta R}{2N_S} \text{ webers}$$

The values of the flux density, etc. can then be calculated as already described for method (a).

7.8	Comparison of electromagnetic and electrostatic terms

It may be helpful to compare the terms and symbols used in electrostatics with the corresponding terms and symbols used in electromagnetism (see Table 7.2).

Table 7.2

Electrostatics		Electromagnetism	
Term	Symbol	Term	Symbol
Electric flux	Q	Magnetic flux	Φ
Electric flux density	D	Magnetix flux density	B
Electric field strength	E	Magnetic field strength	H
Electromotive force	E	Magnetomotive force	F
Electric potential difference	V	Magnetic potential difference	—
Permittivity of free space	ϵ_0	Permeability of free space	μ_0
Relative permittivity	ϵ_r	Relative permeability	μ_r
Absolute permittivity		Absolute permeability	

$$= \frac{\text{electric flux density}}{\text{electric field strength}} \qquad\qquad = \frac{\text{magnetic flux density}}{\text{magnetic field strength}}$$

i.e. $\epsilon_0 \epsilon_r = \epsilon = D/E$ i.e. $\mu_0 \mu_r = \mu = B/H$

Summary of important formulae

Magnetomotive force

$$F = NI \quad \text{(amperes or ampere-turns)} \qquad [7.2]$$

Magnetic field strength

$$H = F/l = NI/l \quad \text{(amperes per metre)} \qquad [7.1]$$

Flux density

$$B = \mu \cdot H \quad \text{(teslas)} \qquad [7.7]$$

Permeability of free space

$$\mu_0 = 4\pi \times 10^{-7} \text{ (henrys per metre)} \qquad [7.4]$$

Reluctance of a magnetic circuit

$$S = F/\Phi \qquad [7.12]$$

$$S = \frac{l}{\mu_0 \mu_r A} \qquad [7.13]$$

Terms and concepts	A **magnetic flux** is created by a magnetomotive force.

A **magnetic flux** is created by a magnetomotive force.

The **magnetic field strength** is the m.m.f. gradient at any point in a field.

The **magnetic field strength** and the **flux density** at any point in a field are related by the **permeability** of the material in which the magnetic field is created.

The ratio of the permeability to that of free space is termed the **relative permeability**. For ferromagnetic materials, the relative permeability varies according to the magnetic field strength.

The variation of flux density with magnetic field strength is illustrated by the **magnetization characteristic** (or B/H curve).

The **reluctance** of a magnetic circuit is the ratio of the magnetomotive force to the flux.

Exercises 7

Data of B/H, when not given in question, should be taken from Fig. 7.3.

1. A mild steel ring has a mean diameter of 160 mm and a cross-sectional area of 300 mm². Calculate: (a) the m.m.f. to produce a flux of 400 μWb; and (b) the corresponding values of the reluctance of the ring and of the relative permeability.

2. A steel magnetic circuit has a uniform cross-sectional area of 5 cm² and a length of 25 cm. A coil of 120 turns is wound uniformly over the magnetic circuit. When the current in the coil is 1.5 A, the total flux is 0.3 mWb; when the current is 5 A, the total flux is 0.6 mWb. For each value of current, calculate: (a) the magnetic field strength; (b) the relative permeability of the steel.

3. A mild steel ring has a mean circumference of 500 mm and a uniform cross-sectional area of 300 mm². Calculate the m.m.f. required to produce a flux of 500 μWb. An airgap, 1.0 mm in length, is now cut in the ring. Determine the flux produced if the m.m.f. remains constant. Assume the relative permeability of the mild steel to remain constant at 220.

4. A steel ring has a mean diameter of 15 cm, a cross-section of 20 cm² and a radial airgap of 0.5 mm cut in it. The ring is uniformly wound with 1500 turns of insulated wire and a magnetizing current of 1 A produces a flux of 1 mWb in the airgap. Neglecting the effect of magnetic leakage and fringing, calculate: (a) the reluctance of the magnetic circuit; (b) the relative permeability of the steel.

5. (a) A steel ring, having a mean circumference of 750 mm and a cross-sectional area of 500 mm², is wound with a magnetizing coil of 120 turns. Using the following data, calculate the current required to set up a magnetic flux of 630 μWb in the ring.

Flux density (T)	0.9	1.1	1.2	1.3
Magnetic field strength (A/m)	260	450	600	820

(b) The airgap in a magnetic circuit is 1.1 mm long and 2000 mm² in cross-section. Calculate: (i) the reluctance of the airgap; and (ii) the m.m.f. to send a flux of 700 microwebers across the airgap.

6. A magnetic circuit consists of a cast steel yoke which has a cross-sectional area of 200 mm² and a mean length of 120 mm. There are two airgaps, each 0.2 mm long. Calculate: (a) the m.m.f. required to produce a flux of 0.05 mWb in the airgaps; (b) the value of the relative permeability of cast steel at this flux density. The magnetization curve for cast steel is given by the following:

B (T)	0.1	0.2	0.3	0.4
H (A/m)	170	300	380	460

7. An electromagnet has a magnetic circuit that can be regarded as comprising three parts in series: A, a length of 80 mm and cross-sectional area 60 mm²; B, a length of 70 mm and cross-sectional area 80 mm²; C, an airgap of length 0.5 mm and cross-sectional area 60 mm². Parts A and B are of a material having magnetic characteristics given by the following table:

H (A/m)	100	210	340	500	800	1500
B (T)	0.2	0.4	0.6	0.8	1.0	1.2

Determine the current necessary in a coil of 4000 turns wound on part B to produce in the airgap a flux density of 0.7 T. Magnetic leakage may be neglected.

8. A certain magnetic circuit may be regarded as consisting of three parts, A, B and C in series, each one of which has a uniform cross-sectional area. Part A has a length of 300 mm and a cross-sectional area of 450 mm². Part B has a length of 120 mm and a cross-sectional area of 300 mm². Part C is an airgap 1.0 mm in length and of cross-sectional area 350 mm². Neglecting magnetic leakage and fringing, determine the m.m.f. necessary to produce a flux of 0.35 mWb in the airgap. The magnetic characteristic for parts A and B is given by:

H (A/m)	400	560	800	1280	1800
B (T)	0.7	0.85	1.0	1.15	1.25

9. A magnetic circuit made of silicon steel is arranged as in Fig. A. The centre limb has a cross-sectional area of 800 mm² and each of the side limbs has a cross-sectional area of 500 mm². Calculate the m.m.f. required to produce a flux of 1 mWb in the centre limb, assuming the magnetic leakage to be negligible.

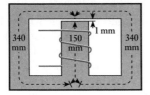

Fig. A

10. A magnetic core made of mild steel has the dimensions shown in Fig. B. There is an airgap 1.2 mm long in one side limb and a coil of 400 turns is wound on the centre limb. The cross-sectional area of the centre limb is 1600 mm² and that of each side limb is 1000 mm². Calculate the exciting current required to produce a flux of 1000 μWb in the airgap. Neglect any magnetic leakage and fringing.

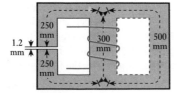

Fig. B

11. An electromagnet with its armature has a core length of 400 mm and a cross-sectional area of 500 mm².

There is a total airgap of 1.8 mm. Assuming a leakage factor of 1.2, calculate the m.m.f. required to produce a flux of 400 μWb in the armature. Points on the B/H curve are as follows:

Flux density (T)	0.8	1.0	1.2
Magnetic field strength (A/m)	800	1000	1600

12. Define the *ampere* in terms of SI units.

A coil P of 300 turns is uniformly wound on a ferromagnetic ring having a cross-sectional area of 500 mm² and a mean diameter of 400 mm. Another coil S of 30 turns wound on the ring is connected to a fluxmeter having a constant of 200 microweber-turns per division. When a current of 1.5 A through P is reversed, the fluxmeter deflection is 96 divisions. Calculate: (a) the flux density in the ring; and (b) the magnetic field strength.

13. The magnetization curve of a ring specimen of steel is determined by means of a fluxmeter. The specimen, which is uniformly wound with a magnetizing winding of 1000 turns, has a mean length of 65 cm and a cross-sectional area of 7.5 cm². A search coil of 2 turns is connected to the fluxmeter, which has a constant of 0.1 mWb-turn per scale division. When the magnetizing current is reversed, the fluxmeter deflection is noted, the following table being the readings which are obtained:

Magnetizing current (amperes)	0.2	0.4	0.6	0.8	1.0
Fluxmeter deflection (divisions)	17.8	13.2	38.0	42.3	44.7

Explain the basis of the method and calculate the values of the magnetic field strength and flux density for each set of readings.

14. A ring specimen of mild steel has a cross-sectional area of 6 cm² and a mean circumference of 30 cm. It is uniformly wound with two coils, A and B, having 90 turns and 300 turns respectively. Coil B is connected to a ballistic galvanometer having a constant of 1.1×10^{-8} coulomb per division. The total resistance of this secondary circuit is 200 000 Ω. When a current of 2.0 A through coil A is reversed, the galvanometer gives a maximum deflection of 200 divisions. Neglecting the damping of the galvanometer, calculate the flux density in the steel and its relative permeability.

15. Two coils of 2000 and 100 turns respectively are wound uniformly on a non-magnetic toroid having a mean circumference of 1 m and a cross-sectional

Exercises 7 continued

area of 500 mm². The 100-turn coil is connected to a ballistic galvanometer, the total resistance of this circuit being 5.1 kΩ. When a current of 2.5 A in the 2000-turn coil is reversed, the galvanometer is deflected through 100 divisions. Neglecting damping, calculate the ballistic constant for the instrument.

16. A long straight conductor, situated in air, is carrying a current of 500 A, the return conductor being far removed. Calculate the magnetic field strength and the flux density at a radius of 80 mm.

17. (a) The flux density in air at a point 40 mm from the centre of a long straight conductor A is 0.03 T. Assuming that the return conductor is a considerable distance away, calculate the current in A.

    (b) In a certain magnetic circuit, having a length of 500 mm and a cross-sectional area of 300 mm², an m.m.f. of 200 A produces a flux of 400 μWb. Calculate: (i) the reluctance of the magnetic circuit; and (ii) the relative permeability of the core.

18. Two long parallel conductors, spaced 40 mm between centres, are each carrying a current of 5000 A. Calculate the force in newtons per metre length of conductor.

19. Two long parallel conductors P and Q, situated in air and spaced 8 cm between centres, carry currents of 600 A in opposite directions. Calculate the values of the magnetic field strength and of the flux density at points A, B and C in Fig. C, where the dimensions are given in centimetres. Calculate also the values of the same quantities at the same points if P and Q are each carrying 600 A in the same direction.

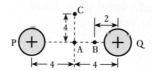

Fig. C

Chapter eight

Inductance in a DC Circuit

Having set up a magnetic circuit, we need some measure of our success. Have we achieved a good magnetic field or is it rather mediocre? Our factor of success is termed the inductance and, in this chapter, we shall explore the applications of inductance. The most important application lies in relating the efficiency of a magnetic circuit to the induction of an e.m.f. in a circuit. Other applications relate the inductance to the dimensions of a coil and to the ferromagnetic nature of that coil.

Also we have seen that an electrostatic field can store energy and that such storage takes time to build up or to decay. We shall find that a similar set of activities can be observed in electromagnetic fields. These observations will be used in later chapters to explain that the response of electric circuits cannot be instantaneous when changes occur and these are important when we investigate alternating currents.

Finally we shall find that two coils can interact with one another giving rise to the concept of mutual inductance. This principle will be developed to explain the action of transformers which are the backbone of the alternating electrical supplies found throughout the country.

8.1 Inductive and non-inductive circuits

Most readers are likely to find that they do not require to give more consideration to the electromagnetic field than has been covered by Chapter 7. Those who have interests in communications systems or in electrical machines will require more informed coverage, some of which appears in Sections Two and Three respectively of this text.

However, every reader will become involved with the induced e.m.f. created by electromagnetic fields in electrical circuits. This chapter is therefore devoted to the circuits affected by electromagnetic fields as distinct from the electromagnetic fields themselves.

Let us consider what happens when a coil L (Fig. 8.1) and a resistor R, connected in parallel, are switched across a battery B; L consists of a large number of turns wound on a steel core D (or it might be the field winding of a generator or motor) and R is connected in series with a *centre-zero* ammeter A_2.

Fig. 8.1 Inductive and non-inductive circuits

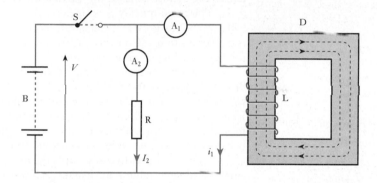

In the analysis which follows, capital symbols are used to indicate quantities which do not vary and lower-case symbols are used to denote variable quantities.

Although ideally a conductor has no resistance, in practice the length of conductor wire used to make a coil is such that the coil has an effective resistance. We will need to recognize this resistance in the experiment which we are now observing.

When switch S is closed, it is found that the current I_2 through R increases almost instantly to its final value. The current i_1 through L takes an appreciable time to grow. This is indicated in Fig. 8.2. Eventually the final current I_1 is given by

$$I_1 = \frac{\text{battery voltage } V}{\text{resistance of coil } L}$$

In Fig. 8.2, I_2 is shown as being greater than I_1 – this is only to keep them apart on the diagram, but they could have been equal or I_1 could have been greater than I_2.

When S is opened, current through L decreases comparatively slowly, but the current through R instantly reverses its direction and becomes the same current as i_1; in other words the current of L is circulating round R.

Let us now consider the reason for the difference in the behaviour of the currents in L and R.

The growth of current in L is accompanied by an increase of flux – shown dotted – in the steel core D. But it has been pointed out in section 6.8 that

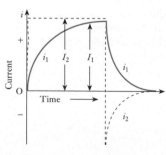

Fig. 8.2 Variation of switch-on and switch-off currents

any change in the flux linked with a coil is accompanied by an e.m.f. induced in that coil, the direction of which – described by Lenz's law – is always such as to oppose the change responsible for inducing the e.m.f., namely the growth of current in L. In other words the induced e.m.f. is acting in opposition to the current and, therefore, to the applied voltage. In circuit R, the flux is so small that its induced e.m.f. is negligible.

When switch S is opened, the currents in both L and R tend to decrease; but any decrease of i_1 is accompanied by a decrease of flux in D and therefore by an e.m.f. induced in L in such a direction as to oppose the decrease of i_1. Consequently the induced e.m.f. is now acting in the same direction as the current. But it is evident from Fig. 8.1 that after S has been opened, the only return path for the current of L is that via R; hence the reason why i_1 and i_2 are now one and the same current.

If the experiment is repeated without R, it is found that the growth of i_1 is unaffected, but when S is opened there is considerable arcing at the switch due to the maintenance of the current across the gap by the e.m.f. induced in L. The more quickly S is opened, the more rapidly does the flux in D collapse and the greater is the e.m.f. induced in L. This is the reason why it is dangerous to break quickly the full excitation of an electromagnet such as the field winding of a d.c. machine.

Any circuit in which a change of current is accompanied by a change of flux, and therefore by an induced e.m.f., is said to be *inductive* or to possess *self-inductance* or merely *inductance*. It is impossible to have a perfectly *non-inductive* circuit, i.e. a circuit in which no flux is set up by a current; but for most purposes a circuit which is not in the form of a coil may be regarded as being practically non-inductive – even the open helix of an electric fire is almost non-inductive. In cases where the inductance has to be reduced to the smallest possible value – for instance, in resistance boxes – the wire is bent back on itself, as shown in Fig. 8.3, so that the magnetizing effect of the current in one conductor is neutralized by that of the adjacent conductor. The wire can then be coiled round an insulator without increasing the inductance.

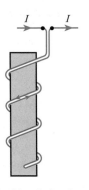

Fig. 8.3 Non-inductive resistor

8.2   **Unit of inductance**

The unit of inductance is termed the *henry*, in commemoration of a famous American physicist, Joseph Henry (1797–1878), who, quite independently, discovered electromagnetic induction within a year after it had been discovered in Britain by Michael Faraday in 1831. *A circuit has an inductance of 1 henry (or 1 H) if an e.m.f. of 1 volt is induced in the circuit when the current varies uniformly at the rate of 1 ampere per second.* If either the inductance or the rate of change of current is doubled, the induced e.m.f. is doubled. Hence if a circuit has an inductance of L henrys and if the current *increases* from i_1 to i_2 amperes in t seconds the average rate of change of current is

$$\frac{i_2 - i_1}{t} \text{ amperes per second}$$

and average induced e.m.f. is

$$L \times \text{rate of change of current} = L \times \frac{i_2 - i_1}{t} \text{ volts} \qquad [8.1]$$

Self-inductance   Symbol: L Unit: henry (H)

Considering instantaneous values, if di = increase of current, in amperes, in time dt seconds, rate of change of current is

$$\frac{\mathrm{d}i}{\mathrm{d}t} \text{ amperes per second}$$

and e.m.f. induced in circuit is

$$L \cdot \frac{\mathrm{d}i}{\mathrm{d}t} \text{ volts}$$

i.e.
$$e = L \cdot \frac{\mathrm{d}i}{\mathrm{d}t}$$
[8.2]

While this term gives the magnitude of the e.m.f., there remains the problem of polarity. When a force is applied to a mechanical system, the system reacts by deforming, or mass-accelerating or dissipating or absorbing energy. A comparable state exists when a force (voltage) is applied to an electric system, which accelerates (accepts magnetic energy in an inductor) or dissipates energy in heat (in a resistor). The comparable state to deformation is the acceptance of potential energy in a capacitor which was dealt with in Chapter 5. In the case of a series circuit containing resistance and inductance then

$$v = Ri + L \cdot \frac{\mathrm{d}i}{\mathrm{d}t}$$

There are now two schools of thought as to how to proceed. One says

$$v = v_R + e_L$$

The other says

$$v = v_R - e_L$$
$$= v_R + v_L$$

This requires that

$$e = -L \cdot \frac{\mathrm{d}i}{\mathrm{d}t}$$

The first method suggests that the only voltage to be measured is a component of the voltage applied and this is

$$v_L = +L \cdot \frac{\mathrm{d}i}{\mathrm{d}t}$$

Both arguments are acceptable and the reader will find both systems have wide application. Although the International Electrotechnical Commission prefer the second method, for the purposes of this text the positive version will be used.

It will be noted that the positive version of the relation appears to be more logical in a circuit diagram. In Fig. 8.4, both versions are considered. If the induced e.m.f. is taken as an effective volt drop, it may be represented by an arrow pointing upwards. If the negative version is used, the arrow must point in the direction of the current flow.

Fig. 8.4 Polarity of e.m.f. in a circuit diagram

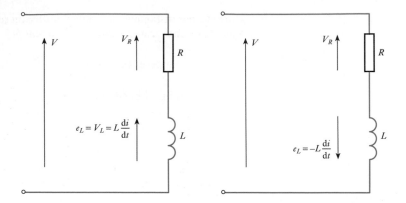

This interpretation of the diagram arises from the intention that we clearly identify active circuit components which are sources of e.m.f. as distinct from inactive circuit components which provide only volt drops. Active components include batteries, generators and (because they store energy) inductors and capacitors. A charged capacitor can act like a battery for a short period of time and a battery possesses an e.m.f. which, when the battery is part of a circuit fed from a source of voltage, acts against the passage of current through it from the positive terminal to the negative terminal, i.e. the process whereby a battery is charged. An inductor opposes the increase of current in it by acting against the applied voltage by means of an opposing e.m.f.

When the e.m.f. opposes the increase of current in the inductor, energy is taken into the inductor. Similarly, when the e.m.f. opposes the decrease of current in the inductor, energy is supplied from the inductor back into the electric circuit.

Example 8.1

If the current through a coil having an inductance of 0.5 H is reduced from 5 A to 2 A in 0.05 s, calculate the mean value of the e.m.f. induced in the coil.

Average rate of change of current is

$$\frac{I_2 - I_1}{t} = \frac{2 - 5}{0.05} = -60 \text{ A/s}$$

From equation [8.1] average e.m.f. induced in coil is

$$0.5 \times (-60) = -30 \text{ V}$$

The direction of the induced e.m.f. is opposite to that of the current, opposing its decrease.

8.3   Inductance in terms of flux-linkages per ampere

Suppose a current of I amperes through a coil of N turns to produce a flux of Φ webers, and suppose the reluctance of the magnetic circuit to remain constant so that the flux is proportional to the current. Also, suppose the inductance of the coil to be L henrys.

If the current is increased from zero to I amperes in t seconds, the average rate of change of current is

$$\frac{I}{t} \text{ amperes per second}$$

$\therefore$ average e.m.f. induced in coil $= \dfrac{LI}{t}$ volts   [8.3]

In section 6.10, it was explained that the value of the e.m.f., in volts, induced in a coil is equal to the rate of change of flux-linkages per second. Hence, when the flux increases from zero to Φ webers in t seconds,

$$\text{average rate of change of flux} = \frac{\Phi}{t} \text{ webers per second}$$

and average e.m.f. induced in coil is

$$\frac{N\Phi}{t} \text{ volts} \tag{8.4}$$

Equating expressions [8.3] and [8.4], we have

$$\frac{LI}{t} = \frac{N\Phi}{t}$$

$\therefore$ $L = \dfrac{N\Phi}{I}$ henrys = flux linkages per ampere   [8.5]

Considering instantaneous values, if $d\phi$ is the increase of flux, in webers, due to an increase di amperes in dt seconds

$$\text{Rate of change of flux} = \frac{d\phi}{dt} \text{ webers per second}$$

and   Induced e.m.f. $= N \cdot \dfrac{d\phi}{dt}$ volts

$$e = N\frac{d\phi}{dt} \tag{8.6}$$

Equating expressions [8.2] and [8.6], we have

$$L \cdot \frac{di}{dt} = N \cdot \frac{d\phi}{dt}$$

$$L = N \cdot \frac{d\phi}{dt} \tag{8.7}$$

$$= \frac{\text{Change of flux linkages}}{\text{Change of current}}$$

For a coil having a magnetic circuit of constant reluctance, the flux is proportional to the current; consequently, $d\phi/di$ is equal to the flux per ampere, so that

$$L = \text{flux-linkages per ampere}$$

$$L = \frac{N\Phi}{I} \text{ henrys} \qquad\qquad [8.5]$$

This expression gives us an alternative method of defining the unit of inductance, namely: *a coil possesses an inductance of 1 henry if a current of 1 ampere through the coil produces a flux-linkage of 1 weber-turn.*

Example 8.2   A coil of 300 turns, wound on a core of non-magnetic material, has an inductance of 10 mH. Calculate:

(a) the flux produced by a current of 5 A;
(b) the average value of the e.m.f. induced when a current of 5 A is reversed in 8 ms (milliseconds).

(a) From expression [8.5]:

$$10 \times 10^{-3} = 300 \times \Phi/5$$

$$\Phi = 0.167 \times 10^{-3} \text{ Wb} = \textbf{167 } \boldsymbol{\mu}\textbf{Wb}$$

(b) When a current of 5 A is reversed, the flux decreases from 167 μWb to zero and then increases to 167 μWb in the reverse direction, therefore change of flux is 334 μWb and average rate of change of flux is

$$\frac{\Delta\Phi}{\Delta t} = \frac{334 \times 10^{-6}}{8 \times 10^{-3}} = 0.041\,75 \text{ Wb/s}$$

therefore average e.m.f. induced in coil is

$$N\frac{\Delta\Phi}{\Delta t} = 0.041\,75 \times 300 = \textbf{12.5 V}$$

Alternatively, since the current changes from 5 to -5 A, average rate of change of current is

$$\frac{\Delta I}{\Delta t} = \frac{5 \times 2}{8 \times 10^{-3}} = 1250 \text{ A/s}$$

Hence, from expression [8.2], average e.m.f. induced in coil is

$$L\frac{\Delta I}{\Delta t} = 0.01 \times (1250) = \textbf{12.5 V}$$

The sign is positive because the e.m.f. is acting in the direction of the original current, at first trying to prevent the current decreasing to zero and then opposing its growth in the reverse direction

8.4 Factors determining the inductance of a coil

Let us first consider a coil uniformly wound on a *non-magnetic* ring of uniform section – similar to that of Fig. 7.1. From expression [7.5], it follows that the flux density, in teslas, in such a ring is $4\pi \times 10^{-7} \times$ the magnetic field strength, in amperes per metre. Consequently, if l is the length of the magnetic circuit in metres and A its cross-sectional area in square metres, then for a coil of N turns with a current I amperes:

From expression [7.1]

$$\text{Magnetic field strength} = \frac{IN}{l}$$

and

$$\text{total flux} = \Phi = BA = \mu_0 HA$$

$$= 4\pi \times 10^{-7} \times \frac{IN}{l} A$$

Substituting for Φ in expression [8.5] we have

$$\boxed{\text{Inductance} = L = 4\pi \times 10^{-7} \times \frac{AN^2}{l} \text{ henrys}} \qquad [8.8]$$

Hence the inductance is proportional to the square of the number of turns and to the cross-sectional area, and is inversely proportional to the length of the magnetic circuit.

If the coil is wound on a closed ferromagnetic core, such as a ring, the problem of defining the inductance of such a coil becomes involved due to the fact that the variation of flux is no longer proportional to the variation of current. Suppose the relationship between the magnetic flux and the magnetizing current to be as shown in Fig. 8.5, then if the core has initially no residual magnetism, an increase of current from zero to OA causes the flux to increase from zero to AC, but when the current is subsequently reduced to zero, the decrease of flux is only DE. If the current is then increased to OG in the reverse direction, the change of flux is EJ. Consequently, we can have an infinite number of inductance values, depending upon the particular variation of current that we happen to consider.

Since we are usually concerned with the effect of inductance in an a.c. circuit, where the current varies from a maximum in one direction to the same maximum in the reverse direction, it is convenient to consider the value of the inductance as being the ratio of the change of flux-linkages to the change of current when the latter is reversed. Thus, for the case shown in Fig. 8.5:

$$\text{Inductance of coil} = \frac{DJ}{AG} \times \text{number of turns}$$

This value of inductance is the same as if the flux varied linearly along the dotted line COH in Fig. 8.5.

If μ_r represents the value of the relative permeability corresponding to the maximum value AC of the flux, then the inductance of the steel-cored coil, as defined above, is μ_r times that of the same coil with a non-magnetic core. Hence, from expression [8.8], we have inductance of a ferromagnetic-cored coil (for reversal of flux) is

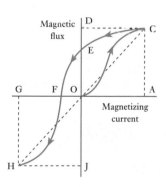

Fig. 8.5 Variation of magnetic flux with magnetizing current for a closed ferromagnetic circuit

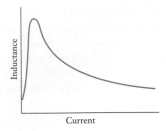

Fig. 8.6   Inductance of a ferromagnetic-cored coil

$$L = 4\pi \times 10^{-7} \times \frac{AN^2}{l} \times \mu_r \text{ henrys} \qquad [8.9]$$

The variations of relative permeability with magnetic field strength for various qualities of steel are shown in Fig. 7.4; hence it follows from expression [8.9] that as the value of an alternating current through a coil having a closed steel circuit is increased, the value of the inductance increases to a maximum and then decreases, as shown in Fig. 8.6. It will now be evident that when the value of the inductance of such a coil is stated, it is also necessary to specify the current variation for which that value has been determined.

From expression [8.9]:

$$L \text{ (henrys)} = \mu_0 \mu_r \times A \text{ (metres}^2) \times \frac{N^2}{l \text{ (metres)}}$$

$$\therefore \qquad \text{Absolute permeability} = \mu_0 \mu_r = \frac{L \text{ (henrys)} \times l \text{ (metres)}}{N^2 \times A \text{ (metres}^2)}$$

hence the units of absolute permeability are *henrys per metre* (or H/m), e.g. $\mu_0 = 4\pi \times 10^{-7}$ H/m.

Example 8.3

A ring of mild steel stampings having a mean circumference of 400 mm and a cross-sectional area of 500 mm^2 is wound with 200 turns. Calculate the inductance of the coil corresponding to a reversal of a magnetizing current of:

(a) 2 A;
(b) 10 A.

(a) $\qquad H = \dfrac{NI}{l} = \dfrac{2 \text{ [A]} \times 200 \text{ [turns]}}{0.4 \text{ [m]}} = 1000 \text{ A/m}$

From Fig. 7.3:

corresponding flux density $= 1.13$ T

$\therefore \qquad$ total flux $\Phi = BA = 1.13 \text{ [T]} \times 0.0005 \text{ [m}^2] = 0.000\ 565$ Wb

From expression [8.5],

$$\text{Inductance } L = \frac{N\Phi}{I} = (0.000\ 565 \times 200)/2 = \textbf{56.6 mH}$$

(b) $\qquad H = \dfrac{NI}{l} = \dfrac{10 \text{ [A]} \times 200 \text{ [turns]}}{0.4 \text{ [m]}} = 5000 \text{ A/m}$

From Fig. 7.3

Corresponding flux density $B = 1.63$ T

$\therefore \qquad$ Total flux $\Phi = 1.63 \times 0.0005 = 0.000\ 815$ Wb

and $\qquad$ Inductance $L = \dfrac{N\Phi}{I} = (0.000\ 815 \times 200)/10 = \textbf{16.3 mH}$

<table>
<tr><td>**Example 8.4**</td><td>If a coil of 200 turns is wound on a non-magnetic core having the same dimensions as the mild steel ring for Example 8.3, calculate its inductance.</td></tr>
</table>

From expression [8.8] we have

$$\text{Inductance} = \frac{(4\pi \times 10^{-7})\,[\text{H/m}] \times 0.0005\,[\text{m}^2] \times (200)^2\,[\text{turns}^2]}{0.4\,[\text{m}]}$$

$$= 0.000\ 062\ 8\ \text{H}$$

$$= 62.8\ \mu\text{H}$$

A comparison of the results from Examples 8.3 and 8.4 for a coil of the same dimensions shows why a ferromagnetic core is used when a large inductance is required.

8.5 Ferromagnetic-cored inductor in a d.c. circuit

An inductor (i.e. a piece of apparatus used primarily because it possesses inductance) is frequently used in the output circuit of a rectifier to smooth out any variation (or ripple) in the direct current. If the inductor was made with a closed ferromagnetic circuit, the relationship between the flux and the magnetizing current would be represented by curve OBD in Fig. 8.7. It will be seen that if the current increases from OA to OC, the flux increases from AB to CD. If this increase takes place in t seconds, then average induced e.m.f. is number of turns × rate of change of flux:

$$E = N\frac{\Delta\Phi}{\Delta t}$$

$$E = N \times \frac{(\text{CD} - \text{AB})}{t}\ \text{volts} \qquad [8.10]$$

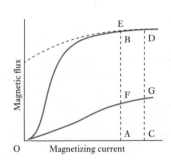

Fig. 8.7 Effect of inserting an airgap in a ferromagnetic core

Let L_δ be the *incremental* inductance of the coil over this range of flux variation, i.e. the effective value of the inductance when the flux is not proportional to the magnetizing current and varies over a relatively small range, then average induced e.m.f. is

$$L_\delta \times \frac{(\text{OC} - \text{OA})}{t}\ \text{volts} \qquad [8.11]$$

Equating expressions [8.10] and [8.11], we have

$$L_\delta \times \frac{(\text{OC} - \text{OA})}{t} = N \times \frac{(\text{CD} - \text{AB})}{t}$$

$$\therefore \qquad L_\delta = N \times \frac{\text{CD} - \text{AB}}{\text{OC} - \text{OA}} \qquad [8.12]$$

$$= N \times \text{average slope of curve BD}$$

From Fig. 8.7 it is evident that the slope is very small when the core is saturated. This effect is accentuated by hysteresis (section 44.7); thus if the current is reduced from OC to OA, the flux decreases from CD only to AE, so that the effective inductance is still further reduced.

If a short radial airgap were made in the ferromagnetic ring, the flux produced by current OA would be reduced to some value AF. For the reduced flux density in the core, the total m.m.f. required for the ferromagnet and the gap is approximately proportional to the flux; and for the same increase of current, AC, the increase of flux = CG − AF. As (CG − AF) may be much greater than (CD − AB), we have the curious result that the effective inductance of a ferromagnetic-cored coil in a d.c. circuit may be increased by the introduction of an airgap.

An alternative method of increasing the flux-linkages per ampere and maintaining this ratio practically constant is to make the core of compressed magnetic dust, such as small particles of ferrite or nickel–iron alloy, bound by shellac. This type of coil is used for 'loading' telephone lines, i.e. for inserting additional inductance at intervals along a telephone line to improve its transmission characteristics.

Expression [8.12] indicates that the inductance L of a magnetic system need not be a constant. It follows that if L can vary with time then expression [8.2] requires to be stated as

$$e = \frac{\mathrm{d}(Li)}{\mathrm{d}t}$$

It can be shown that this expands to give

$$e = L \cdot \frac{\mathrm{d}i}{\mathrm{d}t} + i \cdot \frac{\mathrm{d}L}{\mathrm{d}t}$$

It is for this reason that inductance is no longer defined in terms of the rate of change of current since this presumes that the inductance is constant, which need not be the case in practice.

It is interesting to note that the energy conversion associated with $L \cdot \mathrm{d}i/\mathrm{d}t$ is stored in the magnetic field yet the energy associated with the other term is partially stored in the magnetic field, while the remainder is converted to mechanical energy which is the basis of motor or generator action.

Example 8.5   A laminated steel ring is wound with 200 turns. When the magnetizing current varies between 5 and 7 A, the magnetic flux varies between 760 and 800 μWb. Calculate the incremental inductance of the coil over this range of current variation.

From expression [8.12] we have

$$L_\delta = 200 \times \frac{(800 - 760) \times 10^{-6}}{(7 - 5)} = 0.004 \text{ H}$$

8.6   Growth in an inductive circuit

In section 8.1 the growth of current in an inductive circuit was discussed qualitatively; we shall now consider how to derive the curve showing the growth of current in a circuit of known resistance and inductance (assumed constant).

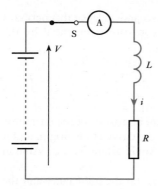

Fig. 8.8 Inductive circuit

Fig. 8.9 Growth of current in an inductive circuit

This derivation can be undertaken either graphically or analytically. It is worth undertaking the graphical approach once and thereafter using the approach used in section 8.7.

When dealing with an inductive circuit it is convenient to separate the effects of inductance and resistance by representing the inductance L as that of an inductor or coil having no resistance and the resistance R as that of a resistor having no inductance, as shown in Fig. 8.8. It is evident from the latter that the current ultimately reaches a steady value I (Fig. 8.9), where $I = V/R$.

Let us consider any instant A during the growth of the current. Suppose the current at that instant to be i amperes, represented by AB in Fig. 8.9. The corresponding p.d. across R is Ri volts. Also at that instant the rate of change of the current is given by the slope of the curve at B, namely the slope of the tangent to the curve at B.

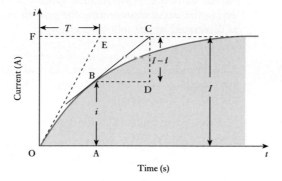

But the slope of BC $= \dfrac{CD}{BD} = \dfrac{I-i}{BD}$ amperes per second.

Hence e.m.f. induced in L at instant A is

$$L \times \text{rate of change of current} = L \times \frac{I-i}{BD} \text{ volts}$$

The total applied voltage V is absorbed partly in providing the voltage drop across R and partly in neutralizing the e.m.f. induced in L, i.e.

$$V = Ri + L \times \frac{I-i}{BD}$$

Substituting RI for V, we have

$$RI = Ri + L \times \frac{I-i}{BD}$$

$$\therefore \qquad R(I-i) = L \times \frac{I-i}{BD}$$

hence

$$BD = \frac{L}{R}$$

In words, this expression means that the rate of growth of current at any instant is such that if the current continued increasing at that rate, it would reach its maximum value of I amperes in L/R seconds. Hence this period is termed the *time constant* of the circuit and is usually represented by the symbol T, i.e.

$$\text{Time constant} = T = \frac{L}{R} \quad \text{seconds} \qquad [8.13]$$

Immediately after switch S is closed, the rate of growth of the current is given by the slope of the tangent OE drawn to the curve at the origin; and if the current continued growing at this rate, it would attain its final value in time FE = T seconds.

From expression [8.13] it follows that the greater the inductance and the smaller the resistance, the larger is the time constant and the longer it takes for the current to reach its final value. Also this relationship can be used to derive the curve representing the growth of current in an inductive circuit, as illustrated by the following example.

Example 8.6

A coil having a resistance of 4 Ω and a constant inductance of 2 H is switched across a 20 V d.c. supply. Derive the curve representing the growth of the current.

From equation [8.13], time constant is

$$T = \frac{L}{R} = \frac{2}{4} = 0.5 \, \text{s}$$

Final value of current is

$$I = \frac{V}{R} = \frac{20}{4} = 5 \, \text{A}$$

With the horizontal and vertical axes suitably scaled, as in Fig. 8.10, draw a horizontal dotted line at the level of 5 A. Along this line mark off a period MN = T = 0.5 s and join ON.

Take any point P relatively near the origin and draw a horizontal dotted line PQ = T = 0.5 s and at Q draw a vertical dotted line QS. Join PS.

Repeat the operation from a point X on PS, Z on XY, etc.

A curve touching OP, PX, XZ, etc. represents the growth of the current. The greater the number of points used in the construction, the more accurate is the curve.

Fig. 8.10 Graph for Example 8.6

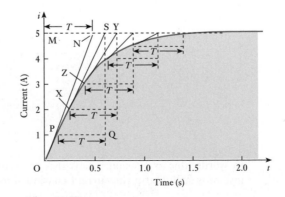

In the above graphical derivation, if you remember nothing else, always keep the relationship between the initial gradient and the time constant in your mind's eye. It helps in understanding the effect of the mathematical analysis which follows.

| 8.7 | Analysis of growth |

Let us again consider the circuit shown in Fig. 8.8 and suppose i amperes to be the current t seconds after the switch is closed, and di amperes to be the increase of current in dt seconds, as in Fig. 8.11. Then rate of change of current is

Fig. 8.11 Growth of current in an inductive circuit

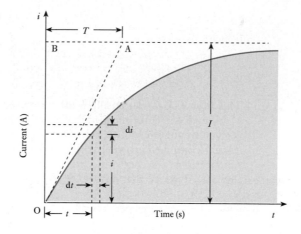

$$\frac{di}{dt} \text{ amperes per second}$$

and induced e.m.f. is

$$L \cdot \frac{di}{dt} \text{ volts}$$

Since total applied voltage is p.d. across R + induced e.m.f.

$$\therefore \qquad V = Ri + L \cdot \frac{di}{dt} \qquad\qquad [8.14]$$

so that

$$V - Ri = L \cdot \frac{di}{dt}$$

and

$$\frac{V}{R} - i = \frac{L}{R} \cdot \frac{di}{dt}$$

But

$$\frac{V}{R} = \text{final value of current} = \text{(say)} \ I$$

$$\therefore \qquad \frac{R}{L} \cdot dt = \frac{di}{I - i}$$

Integrating both sides, we have

$$\frac{Rt}{L} = -\ln(I - i) + A$$

where A is the constant of integration.

At the instant of closing the switch, $t = 0$ and $i = 0$, so that $A = \ln I$,

$$\therefore \qquad \frac{Rt}{L} = -\ln(I - i) + \ln I \qquad\qquad [8.15]$$

$$= \ln \frac{I}{I - i}$$

Hence $\dfrac{I - i}{I} = e^{-\frac{Rt}{L}}$

$$\therefore \qquad i = I(1 - e^{-\frac{Rt}{L}}) \qquad\qquad [8.16]$$

This exponential relationship is often referred to as the Helmholtz equation.

Immediately after the switch is closed, the rate of change of the current is given by the slope of tangent OA drawn to the curve at the origin. If the current continued growing at this initial rate, it would attain its final value, I amperes, in T seconds, the *time constant* of the circuit (section 8.6). From Fig. 8.11 it is seen that initial rate of growth of current is

$$\frac{I}{T} \text{ amperes per second}$$

At the instant of closing the switch $i = 0$; hence, from expression [8.14],

$$V = L \times \text{initial rate of change of current}$$

$$\therefore \qquad \text{initial rate of change of current} = \frac{V}{L}$$

Hence, $\dfrac{I}{T} = \dfrac{V}{L}$

so that

$$T = \frac{LI}{V} = \frac{L}{R} \text{ seconds}$$

$$\therefore \qquad T = \frac{L}{R} \qquad\qquad [8.17]$$

Substituting for R/L in equation [8.16], we have

$$i = I(1 - e^{-\frac{t}{T}}) \qquad\qquad [8.18]$$

For $t = T$

$$i = I(1 - 0.368) = 0.632I$$

hence the time constant is the time required for the current to attain 63.2 per cent of its final value.

| Example 8.7 | A coil having a resistance of 4 Ω and a constant inductance of 2 H is switched across a 20 V d.c. supply. Calculate |

(a) the time constant
(b) the final value of the current
(c) the value of the current 1.0 s after the switch is closed.

(a) Time constant is

$$\frac{L}{R} = \frac{2}{4} = 0.5 \text{ s}$$

(b) Final value of current is

$$\frac{V}{R} = \frac{20}{4} = 5.0 \text{ A}$$

(c) Substituting $t = 1$, $T = 0.5$ s and $I = 5$ A in equation [8.18],

$$i = 5(1 - e^{\frac{1}{0.5}}) = 5(1 - e^{-2})$$
$$e^{-2} = 0.1353$$
$$\therefore \qquad i = 5(1 - 0.1353) = \textbf{4.32 A}$$

| **8.8** | **Analysis of decay** | Figure 8.12 represents a coil, of inductance L henrys and resistance R ohms, in series with a resistor r across a battery. The function of r is to prevent the battery current becoming excessive when switch S is closed. Suppose the steady current through the coil to be I amperes when S is open. Also, suppose i amperes in Fig. 8.13 to be the current through the coil t seconds after S is closed. Since the external applied voltage V of equation [8.14] is zero as far as the coil is concerned, we have |

$$0 = Ri + L \cdot \frac{di}{dt}$$

$$\therefore \qquad Ri = -L \cdot \frac{di}{dt} \qquad\qquad [8.19]$$

In this expression, di is numerically negative since it represents a decrease of current. Hence

$$\left(\frac{R}{L}\right) dt = -\frac{1}{i} \cdot di$$

Integrating both sides, we have

$$\left(\frac{R}{L}\right) t = -\ln i + A$$

where A is the constant of integration.

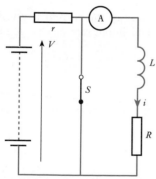

Fig. 8.12 Short-circuited inductive circuit

Fig. 8.13 Decay of current in an inductive circuit

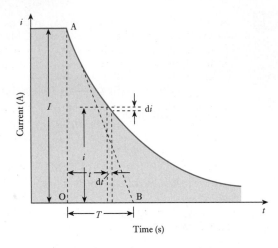

At the instant of closing switch S, $t = 0$ and $i = I$, so that

$$0 = -\ln I + A$$

$$\therefore \qquad \left(\frac{R}{L}\right)t = \ln I - \ln i = \ln \frac{I}{i}$$

Hence $\quad \dfrac{I}{i} = e^{\frac{Rt}{L}}$

and $\qquad i = I\,e^{-\frac{Rt}{L}}$ \hfill [8.20]

Immediately after S is closed, the rate of decay of the current is given by the slope of the tangent AB in Fig. 8.13, and initial rate of change of current is

$$-\frac{I}{T} \text{ amperes per second}$$

Also, from equation [8.19], since initial value of i is I, initial rate of change of current is

$$-\frac{RI}{L} \text{ amperes per second}$$

Hence, $\quad \dfrac{RI}{L} = \dfrac{I}{T}$

so that

$$T = \frac{L}{R} = \text{time constant of circuit}$$

namely, the value already deduced in section 8.6.

The curve representing the decay of the current can be derived graphically by a procedure similar to that used in section 8.6 for constructing the curve representing the current growth in an inductive circuit.

8.9 Transients in LR networks

We have considered the charging and discharging of an inductor through circuit resistance. In each case, the arrangement has involved a network containing inductance and resistance, hence such a network is known as an *LR* network.

As in the *CR* networks which we considered in section 5.18, it is significant to consider the effects of switching on followed by switching off. Once again the difference in time between switching actions is only important if it is not greater than five times the time constant given by expression [8.17]. If the time between switching is greater than five times the time constant, the transient may be considered finished and steady-state conditions can be applied.

In practice, many networks responding to pulsed switching experience a rate of switching which causes the second transient change to commence before the first has finished. The effects are best demonstrated by means of the following examples.

Example 8.8

For the network shown in Fig. 8.14:

(a) determine the mathematical expressions for the variation of the current in the inductor following the closure of the switch at $t = 0$ on to position 1;

(b) when the switch is closed on to position 2 at $t = 100$ ms, determine the new expression for the inductor current and also for the voltage across R;

(c) plot the current waveforms for $t = 0$ to $t = 200$ ms.

(a) For the switch in position 1, the time constant is

$$T_1 = \frac{L}{R_1} = \frac{0.1}{10} \equiv 10 \text{ ms}$$

$$\therefore \quad i_1 = I(1 - e^{-\frac{t}{T_1}}) = \frac{10}{10}(1 - e^{-\frac{t}{10 \times 10^{-3}}})$$

$$= (1 - e^{-\frac{t}{10 \times 10^{-3}}})$$

(b) For the switch in position 2, the time constant is

$$T_2 = \frac{L}{R_2} = \frac{0.1}{10 + 15} \equiv 4 \text{ ms}$$

$$\therefore \quad i_2 = I e^{-\frac{t}{T_2}} = 1 e^{-\frac{t}{4 \times 10^{-3}}}$$

$$= e^{-\frac{t}{4 \times 10^{-3}}} \text{ amperes}$$

The current continues to flow in the same direction as before, therefore the volt drop across R is negative relative to the direction of the arrow shown in Fig. 8.14.

$$v_R = i_2 R = -15 e^{-\frac{t}{4 \times 10^{-3}}} \text{ volts}$$

The current waveforms are shown in Fig. 8.15.

It will be noted that in the first switching period, five times the time constant is 50 ms. The transient has virtually finished at the end of this time and it would not have mattered whether the second switching took place then or later. During the second period, the transient took only 25 ms.

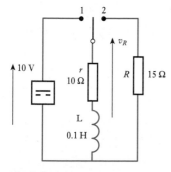

Fig. 8.14 Network for Example 8.8

Fig. 8.15 Current waveforms for Example 8.8

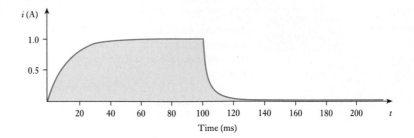

Example 8.9

For the network shown in Fig. 8.14, the switch is closed on to position 1 as in Example 8.8. However, it is closed on to position 2 when $t = 10$ ms. Again determine the current expressions and hence plot the current waveforms.

For the switch in position 1, the time constant is 10 ms as in Example 8.8, and the current expression is as before. However, the switch is moved to position 2 while the transient is proceeding.

When $t = 10$ ms

$$i = (1 - \mathrm{e}^{-\frac{t}{10 \times 10^{-3}}}) = (1 - \mathrm{e}^{-1}) = 0.632 \text{ A}$$

The second transient commences with an initial current in R of 0.632 A. The current decay is therefore

$$i_2 = 0.632 \, \mathrm{e}^{-\frac{t}{4 \times 10^{-3}}} \text{ amperes}$$

Fig. 8.16 Current waveforms for Example 8.9

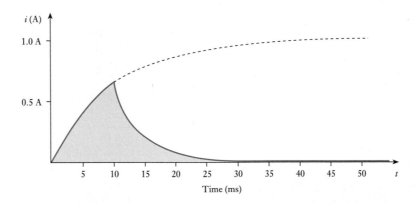

The current waveforms are shown in Fig. 8.16.

It would be possible to extend such an example by repeatedly switching the supply on and off. The analysis would be a repetition of either Example 8.8 or 8.9 depending on the rate of switching.

We can also analyse more complex networks by combining the application of network theorems with exponential expressions. Again, this can be more readily illustrated by means of an example.

Example 8.10

For the network shown in Fig. 8.17, the switch is closed on to position 1 when $t = 0$ and then moved to position 2 when $t = 1.5$ ms. Determine the current in the inductor when $t = 2.5$ ms.

Fig. 8.17 Network for Example 8.10

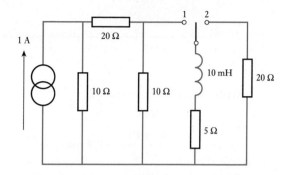

In order to analyse the transient effect, it is necessary to simplify the supply network to the inductor by means of Norton's theorem (Fig. 8.18). The equivalent resistance is given by

$$R_e = \frac{(10 + 20)10}{10 + 20 + 10} = 7.5\,\Omega$$

$$I_{s/c} = \frac{10}{10 + 20} \times 1 = 0.33\,\text{A}$$

Hence the network can be replaced as shown in Fig. 8.19.

The transient current with the switch in position 1 will have a final current of

$$I = \frac{7.5}{5 + 7.5} \times 0.33 = 0.2\,\text{A}$$

At $t = 1.5$ ms, the inductor current is

$$i_1 = 0.2(1 - e^{-\frac{t}{T_1}})$$

where $\quad T_1 = \dfrac{L}{R} = \dfrac{10 \times 10^{-3}}{5} \equiv 2.0\,\text{ms}$

$\therefore \qquad i_1 = 0.2(1 - e^{-\frac{1.5}{2}}) = 0.106\,\text{A}$

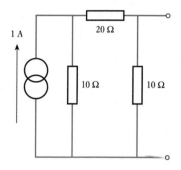

Fig. 8.18 For Example 8.10

Fig. 8.19 For Example 8.10

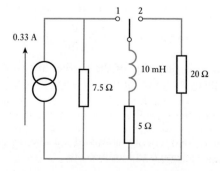

Fig. 8.20 Waveform of current in inductor in Example 8.10

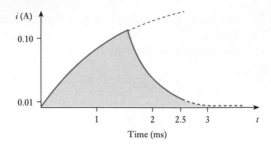

With the switch in position 2,

$$i_2 = 0.106\,e^{-\frac{t}{T_2}}$$

where $\quad T_2 = \dfrac{10 \times 10^{-3}}{5 + 20} \equiv 0.4 \text{ ms}$

$\therefore \qquad i_2 = 0.106\,e^{-\frac{1}{0.4}}$

$$= 0.009 \text{ A} = \mathbf{9 \text{ mA}}$$

The waveform of the current in the inductor is shown in Fig. 8.20.

8.10 Energy stored in an inductor

If the current in a coil having a *constant* inductance of L henrys grows at a uniform rate from zero to I amperes in t seconds, the average value of the current is $\frac{1}{2}I$ and the e.m.f. induced in the coil is $(L \times I/t)$ volts. The product of the current and the component of the applied voltage to neutralize the induced e.m.f. represents the power absorbed by the magnetic field associated with the coil.

Hence average power absorbed by the magnetic field is

$$\tfrac{1}{2}I \times \frac{LI}{t} \text{ watts}$$

and total energy absorbed by the magnetic field is

$$\text{average power} \times \text{time} = \tfrac{1}{2}I \times \frac{LI}{t} \times t$$

$\therefore \qquad \boxed{W_f = \tfrac{1}{2}LI^2} \quad \text{joules} \qquad\qquad\qquad\qquad [8.21]$

Let us now consider the general case of a current increasing at a uniform or a non-uniform rate in a coil having a *constant* inductance L henrys. If the current increases by di amperes in dt seconds

$$\text{Induced e.m.f.} = L \cdot \frac{di}{dt} \text{ volts}$$

and if i is the value of the current at that instant, energy absorbed by the magnetic field during time dt seconds is

$$iL \cdot \frac{di}{dt} \cdot dt = Li \cdot di \text{ joules}$$

Hence total energy absorbed by the magnetic field when the current increases from 0 to I amperes is

$$L \int_0^I i \cdot \mathrm{d}i = L \times \frac{1}{2} \left[i^2 \right]_0^I$$

$$L \int_0^I i \cdot \mathrm{d}i = \tfrac{1}{2} L I^2 \text{ joules}$$

From expression [8.9]

$$L = N^2 \mu \frac{A}{l}$$

for a homogeneous magnetic circuit of uniform cross-sectional area. Therefore energy per cubic metre ω_f is

$$\tfrac{1}{2} I^2 N^2 \frac{\mu}{l^2} = \tfrac{1}{2} \mu H^2$$

$$\therefore \qquad \omega_f = \tfrac{1}{2} H B = \frac{1}{2} \cdot \frac{B^2}{\mu_0 \mu_r} \text{ joules} \qquad\qquad [8.22]$$

This expression has been derived on the assumption that μ_r remains constant. When the coil is wound on a closed ferromagnetic core, the variation of μ_r renders this expression inapplicable and the energy has to be determined graphically as will be explained in section 44.9. For non-magnetic materials, $\mu_r = 1$ and the energy stored per cubic metre is $\tfrac{1}{2} B^2/\mu_0$ joules, which will be developed in section 36.4.

When an inductive circuit is opened, the current has to die away and the magnetic energy has to be dissipated. If there is no resistor in parallel with the circuit the energy is largely dissipated as heat in the arc at the switch. With a parallel resistor, as described in section 8.1, the energy is dissipated as heat generated by the decreasing current in the total resistance of the circuit in which that current is flowing.

| Example 8.11 | A coil has a resistance of 5 Ω and an inductance of 1.2 H. The current through the coil is increased uniformly from zero to 10 A in 0.2 s, maintained constant for 0.1 s and then reduced uniformly to zero in 0.3 s. Plot graphs representing the variation with time of: |

    (a) the current;
    (b) the induced e.m.f.;
    (c) the p.d.s across the resistance and the inductance;
    (d) the resultant applied voltage;
    (e) the power to and from the magnetic field.

Assume the coil to be wound on a non-metallic core.

The variation of current is represented by graph A in Fig. 8.21: and since the p.d. across the resistance is proportional to the current, this p.d. increases from zero to (10 A × 5 Ω), namely 50 V, in 0.2 s, remains constant at 50 V for 0.1 s and then decreases to zero in 0.3 s, as represented by graph B.

During the first 0.2 s, the current is increasing at the rate of 10/0.2, namely 50 A/s,

$$\therefore \qquad \text{Corresponding induced e.m.f.} = 50 \times 1.2 = 60 \text{ V}$$

Fig. 8.21   Graphs for
Example 8.11
A: current
B: p.d. across resistance
D: induced e.m.f.
E: resultant applied voltage
F: (+) energy absorbed
(−) energy returned

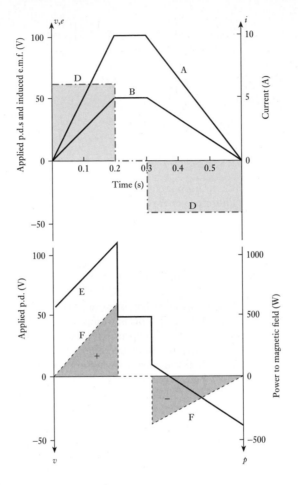

During the following 0.1 s, the induced e.m.f. is zero, and during the last 0.3 s, the current is decreasing at the rate of −10/0.3, namely −33.3 A/s,

$$\therefore \qquad \text{Corresponding induced e.m.f.} = (-33.3 \times 1.2) = -40 \text{ V}$$

The variation of the induced e.m.f. is represented by the uniformly dotted graph D in Fig. 8.21.

The resultant voltage applied to the coil is obtained by adding graphs B and D; thus the resultant voltage increases uniformly from 60 to 110 V during the first 0.2 s, remains constant at 50 V for the next 0.1 s and then changes uniformly from 10 to −40 V during the last 0.3 s, as shown by graph E in Fig. 8.21.

The power supplied to the magnetic field increases uniformly from zero to (10 A × 60 V), namely 600 W, during the first 0.2 s. It is zero during the next 0.1 s. Immediately the current begins to decrease, energy is being returned from the magnetic field to the electric circuit, and the power decreases uniformly from (−40 V × 10 A), namely −400 W, to zero as represented by graph F.

The positive shaded area enclosed by graph F represents the energy ($= \frac{1}{2} \times 600 \times 0.2 = 60$ J) absorbed by the magnetic field during the first 0.2 s; and the negative shaded area represents the energy ($= \frac{1}{2} \times 400 \times 0.3 = 60$ J) returned from the magnetic field to the electric circuit during the last 0.3 s. The two areas are obviously equal in magnitude, i.e. all the energy supplied to the magnetic field is returned to the electric circuit.

8.11 Mutual inductance

If two coils A and C are placed relative to each other as in Fig. 8.22, then, when S is closed, some of the flux produced by the current in A becomes linked with C, and the e.m.f. induced in C circulates a momentary current through galvanometer G. Similarly when S is opened the collapse of the flux induces an e.m.f. in the reverse direction in C. Since a change of current in one coil is accompanied by a change of flux linked with the other coil and therefore by an e.m.f. induced in the latter, the two coils are said to have *mutual inductance*.

Fig. 8.22 Mutual inductance

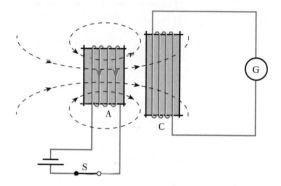

The unit of mutual inductance is the same as for self-inductance, namely the *henry*; and *two coils have a mutual inductance of 1 henry if an e.m.f. of 1 volt is induced in one coil when the current in the other coil varies uniformly at the rate of 1 ampere per second.*

Mutual inductance   Symbol: M Unit: **henry (H)**

If two circuits possess a mutual inductance of M henrys and if the current in one circuit – termed the *primary* circuit – increases by di amperes in dt seconds, e.m.f. induced in *secondary* circuit is

$$M \cdot \frac{di}{dt} \text{ volts}$$

[8.23]

The induced e.m.f. tends to circulate a current in the secondary circuit in such a direction as to oppose the increase of flux due to the increase of current in the primary circuit.

If $d\phi$ webers is the increase of flux linked with the secondary circuit due to the increase of di amperes in the primary, e.m.f. induced in secondary circuit is

$$N_2 \cdot \frac{d\phi}{dt} \text{ volts}$$

[8.24]

where N_2 is the number of secondary turns. From expressions [8.23] and [8.24]

$$M \cdot \frac{\mathrm{d}i}{\mathrm{d}t} = N_2 \cdot \frac{\mathrm{d}\phi}{\mathrm{d}t}$$

$$\therefore \qquad M = N_2 \cdot \frac{\mathrm{d}\phi}{\mathrm{d}i} \qquad\qquad [8.25]$$

$$= \frac{\text{change of flux-linkages with secondary}}{\text{change of current in primary}}$$

If the relative permeability of the magnetic circuit remains constant, the ratio $\mathrm{d}\phi/\mathrm{d}i$ must also remain constant and is equal to the flux per ampere, so that

$$M = \frac{\text{flux-linkages with secondary}}{\text{current in primary}} = \frac{N_2 \Phi_2}{I_1} \qquad\qquad [8.26]$$

where Φ_2 is the flux linked with the secondary circuit due to a current I_1 in the primary circuit.

The mutual inductance between two circuits, A and B, is precisely the same, whether we assume A to be the primary and B the secondary or vice versa; for instance, if the two coils are wound on a non-metallic cylinder, as in Fig. 8.23, then, from expression [8.21], energy in the magnetic field due to current I_A in coil A alone is

$$\tfrac{1}{2}L_A I_A^2 \text{ joules}$$

and energy in the magnetic field due to current I_B in coil B alone is

$$\tfrac{1}{2}L_B I_B^2 \text{ joules}$$

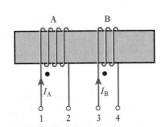

Fig. 8.23   Mutual inductance

Suppose the current in B to be maintained steady at I_B amperes in the direction shown in Fig. 8.23, and the current in A to be increased by $\mathrm{d}i$ amperes in $\mathrm{d}t$ seconds, then

$$\text{e.m.f. induced in B} = M_{12} \cdot \frac{\mathrm{d}i}{\mathrm{d}t} \text{ volts}$$

where M_{12} is the mutual inductance when A is primary.

If the direction of I_A is that indicated by the arrowhead in Fig. 8.23, then, by Lenz's law, the direction of the e.m.f. induced in B is anticlockwise when the coil is viewed from the right-hand end, i.e. the induced e.m.f. is in opposition to I_B and the p.d. across terminals 3 and 4 has to be increased by $M_{12} \cdot \mathrm{d}i/\mathrm{d}t$ volts to maintain I_B constant. Hence the *additional* electrical energy absorbed by coil B in time $\mathrm{d}t$ is

$$I_B M_{12}\left(\frac{\mathrm{d}i}{\mathrm{d}t}\right) \times \mathrm{d}t = I_B M_{12} \cdot \mathrm{d}i \text{ joules}$$

Since I_B remains constant, the I^2R loss in B is unaffected, and there is no e.m.f. induced in coil A apart from that due to the increase of I_A; therefore this additional energy supplied to coil B is absorbed by the magnetic field. Hence, when the current in A has increased to I_A, total energy in magnetic field is

$$\tfrac{1}{2}L_A I_A^2 + \tfrac{1}{2}L_B I_B^2 + \int_0^{I_A} I_B M_{12} \cdot \mathrm{d}i$$

$$= \tfrac{1}{2}L_A I_A^2 + \tfrac{1}{2}L_B I_B^2 + M_{12} I_A I_B \text{ joules}$$

If the direction of either I_A or I_B was reversed, the direction of the e.m.f. induced in B, while the current in A was increasing, would be the same as that of I_B, and coil B would then be acting as a source. By the time the current in A would have reached its steady value I_A, the energy withdrawn from the magnetic field and generated in coil B would be $M_{12} I_A I_B$ joules, and final energy in magnetic field would be

$$\tfrac{1}{2}L_A I_A^2 + \tfrac{1}{2}L_B I_B^2 - M_{12} I_A I_B \text{ joules}$$

Hence, in general, total energy in magnetic field is

$$\tfrac{1}{2}L_A I_A^2 + \tfrac{1}{2}L_B I_B^2 \pm M_{12} I_A I_B \text{ joules} \qquad [8.27]$$

the sign being positive when the ampere-turns due to I_A and I_B are additive, and negative when they are in opposition.

If M_{21} were the mutual inductance with coil B as primary, it could be shown by a similar procedure that the total energy in the magnetic field is

$$\tfrac{1}{2}L_A I_A^2 + \tfrac{1}{2}L_B I_B^2 \pm M_{21} I_A I_B \text{ joules}$$

Since the final conditions are identical in the two cases, the energies must be the same,

$$\therefore \qquad M_{12} I_A I_B = M_{21} I_A I_B$$

or $\qquad M_{12} = M_{21} = \text{(say)} \ M$

i.e. the mutual inductance between two circuits is the same whichever circuit is taken as the primary.

When the two coils are shown on a common core, as in Fig. 8.23, it is obvious that the magnetomotive forces due to I_A and I_B are additive when the directions of the currents are as indicated by the arrowheads. If, however, the coils are drawn as in Fig. 8.24, it is impossible to state whether the magnetomotive forces due to currents I_A and I_B are additive or in opposition; and it is to remove this ambiguity that the dot notation has been adopted. Thus, in Figs 8.23 and 8.24, dots are inserted at ends 1 and 3 of the coils to indicate that when currents *enter both* coils (or *leave both* coils) at these ends, as in Fig. 8.24(a), the magnetomotive forces of the coils are additive,

Fig. 8.24 Application of the dot notation

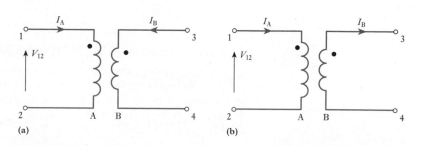

(a)   (b)

and the mutual inductance is then said to be *positive*. But if I_A *enters* coil A at the dotted end and I_B *leaves* coil B at the dotted end, as in Fig. 8.24(b), the m.m.f.s of the coils are in opposition and the mutual inductance is then said to be *negative*.

An application of the dot notation is given at the end of section 8.13. A further application appears in Chapter 13.

8.12 Coupling coefficient

Suppose a ring of non-magnetic material to be wound *uniformly* with two coils, A and B, the turns of one coil being as close as possible to those of the other coil, so that the whole of the flux produced by current in one coil is linked with all the turns of the other coil.

It will be recalled from expression [7.11] that the reluctance S for a magnetic circuit is given by

$$S = \frac{F}{\Phi} = \frac{NI}{\Phi}$$

If coil A has N_1 turns and B has N_2 turns, and if the reluctance of the magnetic circuit is S amperes per weber, then, from expression [8.5], the self-inductances of A and B are

$$L_1 = \frac{N_1\Phi_1}{I_1} = \frac{N_1^2\Phi_1}{I_1N_1} = \frac{N_1^2}{S} \qquad [8.28]$$

and

$$L_2 = \frac{N_2\Phi_2}{I_2} = \frac{N_2^2}{S} \qquad [8.29]$$

where Φ_1 and Φ_2 are the magnetic fluxes due to I_1 in coil A and I_2 in coil B respectively, and

$$S = \frac{I_1N_1}{\Phi_1} = \frac{I_2N_2}{\Phi_2}$$

Since the whole of flux Φ_1 due to I_1 is linked with coil B, it follows from expression [8.26] that

$$M = \frac{N_2\Phi_1}{I_1} = \frac{N_1N_2\Phi_1}{I_1N_1}$$

$$M = \frac{N_1N_2}{S} \qquad [8.30]$$

Hence, from equations [8.28], [8.29] and [8.30],

$$L_1L_2 = \frac{N_1^2N_2^2}{S^2} = M^2$$

so that

$$M = \sqrt{(L_1L_2)} \qquad [8.31]$$

We have assumed that

1. The reluctance remains constant.
2. The magnetic leakage is zero, i.e. that all the flux produced by one coil is linked with the other coil.

The first assumption means that expression [8.31] is strictly correct only when the magnetic circuit is of non-magnetic material. It is, however, approximately correct for a ferromagnetic core if the latter has one or more gaps of air or non-magnetic material, since the reluctance of such a magnetic circuit is approximately constant.

When there is magnetic leakage, i.e. when all the flux due to current in one coil is not linked with the other coil,

$$M = k\sqrt{(L_1 L_2)} \qquad\qquad [8.32]$$

where k is termed the *coupling coefficient*. 'Coupling coefficient' is a term much used in radio work to denote the degree of coupling between two coils; thus, if the two coils are close together, most of the flux produced by current in one coil passes through the other and the coils are said to be *tightly* coupled. If the coils are well apart, only a small fraction of the flux is linked with the secondary, and the coils are said to be *loosely* coupled.

Example 8.12

A ferromagnetic ring of cross-sectional area 800 mm^2 and of mean radius 170 mm has two windings connected in series, one of 500 turns and one of 700 turns. If the relative permeability is 1200, calculate the self-inductance of each coil and the mutual inductance of each assuming that there is no flux leakage.

$$S = \frac{l}{\mu_0 \mu_r A} = \frac{2\pi \times 170 \times 10^{-3}}{4\pi \times 10^{-7} \times 1200 \times 800 \times 10^{-6}}$$

$$= 8.85 \times 10^5 \, \text{H}$$

$$L_1 = \frac{N_1^2}{S} = \frac{500^2}{8.85 \times 10^5} = 0.283 \, \text{H}$$

$$L_2 = \frac{N_2^2}{S} = \frac{700^2}{8.85 \times 10^5} = 0.552 \, \text{H}$$

$$M = k(L_1 L_2)^{\frac{1}{2}} = 1 \times (0.283 \times 0.552)^{\frac{1}{2}} = 0.395 \, \text{H}$$

8.13 Coils connected in series

Figure 8.25(a) shows two coils A and B wound coaxially on an insulating cylinder, with terminals 2 and 3 joined together. It will be evident that the fluxes produced by a current i through the two coils are in the same direction, and the coils are said to be cumulatively coupled. Suppose A and B to have self-inductances L_A and L_B henrys respectively and a mutual inductance M henrys, and suppose the arrowheads to represent the positive direction of the current.

Fig. 8.25 Cumulative and differential coupling of two coils connected in series

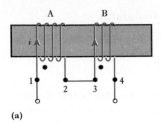 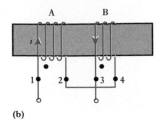

(a)   (b)

If the current increases by $\mathrm{d}i$ amperes in $\mathrm{d}t$ seconds, e.m.f. induced in A due to its self-inductance is

$$L_\mathrm{A} \cdot \frac{\mathrm{d}i}{\mathrm{d}t} \text{ volts}$$

and e.m.f. induced in B due to its self-inductance is

$$L_\mathrm{B} \cdot \frac{\mathrm{d}i}{\mathrm{d}t} \text{ volts}$$

Also, e.m.f. induced in A due to increase of current in B is

$$M \cdot \frac{\mathrm{d}i}{\mathrm{d}t} \text{ volts}$$

and e.m.f. induced in B due to increase of current in A is

$$M \cdot \frac{\mathrm{d}i}{\mathrm{d}t} \text{ volts}$$

Hence total e.m.f. induced in A and B is

$$(L_\mathrm{A} + L_\mathrm{B} + 2M) \cdot \frac{\mathrm{d}i}{\mathrm{d}t}$$

If the windings between terminals 1 and 4 are regarded as a single circuit having a self-inductance L_1 henrys, then for the same increase $\mathrm{d}i$ amperes in $\mathrm{d}t$ seconds e.m.f. induced in the whole circuit is

$$L_1 \cdot \frac{\mathrm{d}i}{\mathrm{d}t} \text{ volts}$$

But the e.m.f. induced in the whole circuit is obviously the same as the sum of the e.m.f.s induced in A and B, i.e.

$$L_1 \cdot \frac{\mathrm{d}i}{\mathrm{d}t} = (L_\mathrm{A} + L_\mathrm{B} + 2M) \cdot \frac{\mathrm{d}i}{\mathrm{d}t}$$

$$\therefore \qquad L_1 = L_\mathrm{A} + L_\mathrm{B} + 2M \qquad\qquad [8.33]$$

Let us next reverse the direction of the current in B relative to that in A by joining together terminals 2 and 4, as in Fig. 8.25(b). With this differential coupling, the e.m.f., $M \cdot \mathrm{d}i/\mathrm{d}t$, induced in coil A due to an increase $\mathrm{d}i$ amperes in $\mathrm{d}t$ seconds in coil B, is in the same direction as the current and is therefore in opposition to the e.m.f. induced in coil A due to its self-inductance. Similarly, the e.m.f. induced in B by mutual inductance is in opposition to that induced by the self-inductance of B. Hence, total e.m.f. induced in A and B is

$$L_A \cdot \frac{di}{dt} + L_B \cdot \frac{di}{dt} - 2M \cdot \frac{di}{dt}$$

If L_2 is the self-inductance of the whole circuit between terminals 1 and 3 in Fig. 8.25(b), then

$$L_2 \cdot \frac{di}{dt} = (L_A + L_B - 2M) \cdot \frac{di}{dt}$$

$$\therefore \qquad \boxed{L_2 = L_A + L_B - 2M} \qquad [8.34]$$

Hence the total inductance of inductively coupled circuits is

$$\boxed{L_A + L_B \pm 2M} \qquad [8.35]$$

The positive sign applies when the coils are cumulatively coupled, the mutual inductance then being regarded as positive; the negative sign applies when they are differentially coupled.

From expressions [8.33] and [8.34], we have

$$\boxed{M = \frac{L_{t1} - L_{t2}}{4}} \qquad [8.36]$$

i.e. the mutual inductance between two inductively coupled coils is a quarter of the difference between the total self-inductance of the circuit when the coils are cumulatively coupled and that when they are differentially coupled.

Example 8.13

When two coils are connected in series, their effective inductance is found to be 10.0 H. However, when the connections to one coil are reversed, the effective inductance is 6.0 H. If the coefficient of coupling is 0.6, calculate the self-inductance of each coil and the mutual inductance.

$$L = L_1 + L_2 \pm 2M = L_1 + L_2 \pm 2k(L_1 L_2)^{1/2}$$

$$\therefore \qquad 10 = L_1 + L_2 + 2k(L_1 L_2)^{1/2}$$

$$\text{and} \qquad 6 = L_1 + L_2 - 2k(L_1 L_2)^{1/2}$$

$$8 = L_1 + L_2$$

$$10 = 8 - L_2 + L_2 + 1.2(8L_2 - L_2^2)^{1/2}$$

$$0 = L_2^2 - 8L_2 + 2.78$$

$$\therefore \qquad L_2 = 7.63 \text{ H or } 0.37 \text{ H}$$

$$\therefore \qquad L_1 = 0.37 \text{ H or } 7.63 \text{ H}$$

$$2M = 10 - 7.63 - 0.37$$

$$M = 1.0 \text{ H}$$

8.14   **Types of inductor and inductance**

We have already noted that inductors are devices which promote inductance, i.e. they are designed to have a great ability to hold magnetic energy. Inductors are generally made to have a fixed value of inductance, but some are variable. The symbols for fixed and variable inductors are shown in Fig. 8.26.

Inductors, unlike resistors and capacitors, cannot be considered as pure elements. Most resistors can be considered to be purely resistive and likewise

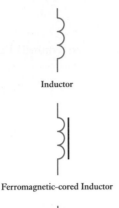

Inductor

Ferromagnetic-cored Inductor

Variable Inductor

Fig. 8.26 Circuit symbols for inductors

most capacitors can be considered to be purely capacitive. Inductors always introduce inductance but also resistance into a circuit.

Inductance is the ratio of flux-linkages to current, i.e. the flux linking the turns through which it appears to pass. Any circuit must comprise at least a single turn, and therefore the current in the circuit sets up a flux which links the circuit itself. It follows that any circuit has inductance. However, the inductance can be negligible unless the circuit includes a coil so that the number of turns ensures high flux-linkage or the circuit is large enough to permit high flux-linkage. The latter infers a transmission line which is effectively long.

Inductors always involve coils of conductor wire. Such conductors are made of wire which cannot be of too large a cross-section. Because the cross-section is small, the coil resistance is at least a few ohms, but can easily be as much as a few thousand ohms.

Inductors fall into two categories – those with an air core and those with a ferromagnetic core. The air core has the advantage that it has a linear B/H characteristic which means that the inductance L is the same no matter what current is in the coil. However, the relative permeability of air being 1 means that the values of inductance attained are very low.

The ferromagnetic core produces very much higher values of inductance, but the B/H characteristic is not linear and therefore the inductance L varies indirectly with the current. However, many of the sintered ferromagnetic materials have almost linear characteristcs and they are therefore almost ideal.

There are variable inductors in which the core is mounted on a screw so that it can be made to move in and out of the coil, thus varying the inductance.

Like capacitors, the weakness of inductors lies in the insulation. In particular if the insulation fails and as a result one or more turns of the coil are short-circuited, the inductance reduces to a value similar to that of an air-cored inductor. The consequence is that there is little back e.m.f. when the coil current is varied.

Summary of important formulae

Induced e.m.f.

$$e = L \cdot \frac{di}{dt} \quad \text{(volts)} \qquad [8.2]$$

$$= N \cdot \frac{d\phi}{dt} \qquad [8.6]$$

Inductance

$$L = N\Phi/I \quad \text{(webers per ampere or henrys)} \qquad [8.5]$$

The time constant of an LR circuit

$$T = L/R \quad \text{(seconds)} \qquad [8.13]$$

Current rise in an LR circuit

$$i = I(1 - e^{-\frac{R}{L}t}) \quad \text{(amperes)} \qquad [8.16]$$

Summary of important formulae continued

Current decay in an LR circuit

$$i = I e^{-\frac{R}{L}t}$$ [8.20]

Energy stored in an inductor

$$W_f = \tfrac{1}{2}LI^2 \quad \text{(joules)}$$ [8.21]

Energy density in a magnetic field

$$w_f = \tfrac{1}{2}BH \quad \text{(joules per cubic metre)}$$ [8.22]

EMF induced by mutual inductance

$$e = M \cdot \frac{\mathrm{d}i}{\mathrm{d}t} \quad \text{(volts)}$$ [8.23]

Mutual inductance

$$M = \frac{N_2\Phi_2}{I_1}$$ [8.26]

$$= \frac{N_1 N_2}{S}$$ [8.30]

Coupling coefficient of a mutual inductor

$$M = k\sqrt{(L_1 L_2)}$$ [8.32]

Effective inductance of a mutual inductor

$$L = L_1 + L_2 \pm 2M$$ [8.35]

Terms and concepts

Inductance is a factor of goodness for a magnetic circuit. The higher the inductance, the better the flux linkage per ampere.

Self-inductance arises when an e.m.f. is induced due to change of flux linkage created by its associated current.

Whether an e.m.f. is positive or negative depends entirely on the assumed direction of action. Self-induced e.m.f.s are assumed to act as though they were load volt drops.

The inductance depends on the number of turns of the energizing coil, the length and cross-sectional area of the magnetic circuit and the material from which the magnetic circuit is made.

Ferromagnetic-cored inductors produce significantly higher inductances than other inductors.

The current in an inductor cannot change instantaneously but has to rise or fall exponentially.

When a magnetic field is set up by an inductor, it stores energy.

When the magnetic field of one coil links with a second coil, the coils are said to be mutually linked and they have **mutual inductance**. How well they are linked is indicated by the **coupling coefficient**.

Exercises 8

1. A 1500-turn coil surrounds a magnetic circuit which has a reluctance of 6×10^6 A/Wb. What is the inductance of the coil?

2. Calculate the inductance of a circuit in which 30 V are induced when the current varies at the rate of 200 A/s.

3. At what rate is the current varying in a circuit having an inductance of 50 mH when the induced e.m.f. is 8 V?

4. What is the value of the e.m.f. induced in a circuit having an inductance of 700 μH when the current varies at a rate of 5000 A/s?

5. A certain coil is wound with 50 turns and a current of 8 A produces a flux of 200 μWb. Calculate: (a) the inductance of the coil corresponding to a reversal of the current; (b) the average e.m.f. induced when the current is reversed in 0.2 s.

6. A toroidal coil of 100 turns is wound uniformly on a non-magnetic ring of mean diameter 150 mm. The circular cross-sectional area of the ring is 706 mm². Estimate: (a) the magnetic field strength at the inner and outer edges of the ring when the current is 2 A; (b) the current required to produce a flux of 0.5 μWb; (c) the self-inductance of the coil. If the ring had a small radial airgap cut in it, state, giving reasons, what alterations there would be in the answers to (a), (b) and (c).

7. A coil consists of two similar sections wound on a common core. Each section has an inductance of 0.06 H. Calculate the inductance of the coil when the sections are connected (a) in series, (b) in parallel.

8. A steel rod, 1 cm diameter and 50 cm long, is formed into a closed ring and uniformly wound with 400 turns of wire. A direct current of 0.5 A is passed through the winding and produces a flux density of 0.75 T. If all the flux links with every turn of the winding, calculate: (a) the relative permeability of the steel; (b) the inductance of the coil; (c) the average value of the e.m.f. induced when the interruption of the current causes the flux in the steel to decay to 20 per cent of its original value in 0.01 s.

9. Explain, with the aid of diagrams, the terms *self-inductance* and *mutual inductance*. In what unit are they measured? Define this unit.

    Calculate the inductance of a ring-shaped coil having a mean diameter of 200 mm wound on a wooden core of diameter 20 mm. The winding is evenly wound and contains 500 turns. If the wooden core is replaced by a ferromagnetic core which has a relative permeability of 600 when the current is 5 A, calculate the new value of inductance.

10. Name and define the unit of *self-inductance*.

    A large electromagnet is wound with 1000 turns. A current of 2 A in this winding produces a flux through the coil of 0.03 Wb. Calculate the inductance of the electromagnet. If the current in the coil is reduced from 2 A to zero in 0.1 s, what average e.m.f. will be induced in the coil? Assume that there is no residual flux.

11. Explain what is meant by the self-inductance of a coil and define the practical unit in which it is expressed.

    A flux of 0.5 mWb is produced in a coil of 900 turns wound on a wooden ring by a current of 3 A. Calculate: (a) the inductance of the coil; (b) the average e.m.f. induced in the coil when a current of 5 A is switched off, assuming the current to fall to zero in 1 ms; (c) the mutual inductance between the coils, if a second coil of 600 turns was uniformly wound over the first coil.

12. Define the *ampere* in terms of SI units.

    A steel ring, having a mean circumference of 250 mm and a cross-sectional area of 400 mm², is wound with a coil of 70 turns. From the following data calculate the current required to set up a magnetic flux of 510 μWb.

B (T)	1.0	1.2	1.4
H (A/m)	350	600	1250

    Calculate also: (a) the inductance of the coil at this current; (b) the self-induced e.m.f. if this current is switched off in 0.005 s. Assume that there is no residual flux.

13. Explain the meaning of *self-inductance* and define the unit in which it is measured.

    A coil consists of 750 turns and a current of 10 A in the coil gives rise to a magnetic flux of 1200 μWb. Calculate the inductance of the coil, and determine the average e.m.f. induced in the coil when this current is reversed in 0.01 s.

14. Explain what is meant by the self-inductance of an electric circuit and define the unit of self-inductance.

    A non-magnetic ring having a mean diameter of 300 mm and a cross-sectional area of 500 mm² is uniformly wound with a coil of 200 turns. Calculate from first principles the inductance of the winding.

15. Two coils, A and B, have self-inductances of 120 μH and 300 μH respectively. When a current of 3 A through coil A is reversed, the deflection on a fluxmeter connected across B is 600 μWb-turns. Calculate: (a) the mutual inductance between the coils; (b) the

average e.m.f. induced in coil B if the flux is reversed in 0.1 s; and (c) the coupling coefficient.

16. A steel ring having a mean diameter of 20 cm and cross-section of 10 cm^2 has a winding of 500 turns upon it. The ring is sawn through at one point, so as to provide an airgap in the magnetic circuit. How long should this gap be, if it is desired that a current of 4 A in the winding should produce a flux density of 1.0 T in the gap? State the assumptions made in your calculation. What is the inductance of the winding when a current of 4 A is flowing through it?

The permeability of free space is $4\pi \times 10^{-7}$ H/m and the data for the B/H curve of the steel are given below:

H (A/m)	190	254	360	525	1020	1530	2230
B (T)	0.6	0.8	1.0	1.2	1.4	1.5	1.6

17. A certain circuit has a resistance of 10 Ω and a constant inductance of 3.75 H. The current through this circuit is increased uniformly from 0 to 4 A in 0.6 s, maintained constant at 4 A for 0.1 s and then reduced uniformly to zero in 0.3 s. Draw to scale graphs representing the variation of (a) the current, (b) the induced e.m.f. and (c) the resultant applied voltage.

18. A coil having a resistance of 2 Ω and an inductance of 0.5 H has a current passed through it which varies in the following manner: (a) a uniform change from zero to 50 A in 1 s; (b) constant at 50 A for 1 s; (c) a uniform change from 50 A to zero in 2 s. Plot the current graph to a time base. Tabulate the potential difference applied to the coil during each of the above periods and plot the graph of potential difference to a time base.

19. A coil wound with 500 turns has a resistance of 2 Ω. It is found that a current of 3 A produces a flux of 500 μWb. Calculate: (a) the inductance and the time constant of the coil; (b) the average e.m.f. induced in the coil when the flux is reversed in 0.3 s. If the coil is switched across a 10 V d.c. supply, derive graphically a curve showing the growth of the current, assuming the inductance to remain constant.

20. Explain the term *time constant* in connection with an inductive circuit.

A coil having a resistance of 25 Ω and an inductance of 2.5 H is connected across a 50 V d.c. supply. Determine graphically: (a) the initial rate of growth of the current; (b) the value of the current after 0.15 s; and (c) the time required for the current to grow to 1.8 A.

21. The field winding of a d.c. machine has an inductance of 10 H and takes a final current of 2 A when connected to a 200 V d.c. supply. Calculate: (a) the initial rate of growth of current; (b) the time constant; and (c) the current when the rate of growth is 5 A/s.

22. A 200 V d.c. supply is suddenly switched across a relay coil which has a time constant of 3 ms. If the current in the coil reaches 0.2 A after 3 ms, determine the final steady value of the current and the resistance and inductance of the coil. Calculate the energy stored in the magnetic field when the current has reached its final steady value.

23. A coil of inductance 4 H and resistance 80 Ω is in parallel with a 200 Ω resistor of negligible inductance across a 200 V d.c. supply. The switch connecting these to the supply is then opened, the coil and resistor remaining connected together. State, in each case, for an instant immediately before and for one immediately after the opening of the switch: (a) the current through the resistor; (b) the current through the coil; (c) the e.m.f. induced in the coil; and (d) the voltage across the coil.

Give rough sketch graphs, with explanatory notes, to show how these four quantities vary with time. Include intervals both before and after the opening of the switch, and mark on the graphs an approximate time scale.

24. A circuit consists of a 200 Ω non-reactive resistor in parallel with a coil of 4 H inductance and 100 Ω resistance. If this circuit is switched across a 100 V d.c. supply for a period of 0.06 s and then switched off, calculate the current in the coil 0.012 s after the instant of switching off. What is the maximum p.d. across the coil?

25. Define the units of: (a) magnetic flux, and (b) inductance.

Obtain an expression for the induced e.m.f. and for the stored energy of a circuit, in terms of its inductance, assuming a steady rise of current from zero to its final value and ignoring saturation.

A coil, of inductance 5 H and resistance 100 Ω, carries a steady current of 2 A. Calculate the initial rate of fall of current in the coil after a short-circuiting switch connected across its terminals has been suddenly closed. What was the energy stored in the coil, and in what form was it dissipated?

26. If two coils have a mutual inductance of 400 μH, calculate the e.m.f. induced in one coil when the

Exercises 8 continued

current in the other coil varies at a rate of 30 000 A/s.

27. If an e.m.f. of 5 V is induced in a coil when the current in an adjacent coil varies at a rate of 80 A/s, what is the value of the mutual inductance of the two coils?

28. If the mutual inductance between two coils is 0.2 H, calculate the e.m.f. induced in one coil when the current in the other coil is increased at a uniform rate from 0.5 to 3 A in 0.05 s.

29. If the toroid of Q. 6 has a second winding of 80 turns wound over the first winding of 100 turns, calculate the mutual inductance.

30. When a current of 2 A through a coil P is reversed, a deflection of 36 divisions is obtained on a fluxmeter connected to a coil Q. If the fluxmeter constant is 150 μWb-turns/div, what is the value of the mutual inductance of coils P and Q?

31. Explain the meaning of the terms *self-inductance* and *mutual inductance* and define the unit by which each is measured.

    A long solenoid, wound with 1000 turns, has an inductance of 120 mH and carries a current of 5 A. A search coil of 25 turns is arranged so that it is linked by the whole of the magnetic flux. A ballistic galvano-meter is connected to the search coil and the combined resistance of the search coil and galvanometer is 200 Ω. Calculate, from first principles, the quantity of electricity which flows through the galvanometer when the current in the solenoid is reversed.

32. Define the unit of mutual inductance.

    A cylinder, 50 mm in diameter and 1 m long, is uniformly wound with 3000 turns in a single layer. A second layer of 100 turns of much finer wire is wound over the first one, near its centre. Calculate the mutual inductance between the two coils. Derive any formula used.

33. A solenoid P, 1 m long and 100 mm in diameter, is uniformly wound with 600 turns. A search-coil Q, 30 mm in diameter and wound with 20 turns, is mounted coaxially midway along the solenoid. If Q is connected to a ballistic galvanometer, calculate the quantity of electricity through the galvanometer when a current of 6 A through the solenoid is reversed. The resistance of the secondary circuit is 0.1 MΩ. Find, also, the mutual inductance between the two coils.

34. When a current of 2 A through a coil P is reversed, a deflection of 43 divisions is obtained on a fluxmeter connected to a coil Q. If the fluxmeter constant is 150 μWb-turns/div, find the mutual inductance of coils P and Q. If the self-inductances of P and Q are 5 mH and 3 mH respectively, calculate the coupling coefficient.

35. Two coils, A and B, have self-inductances of 20 mH and 10 mH respectively and a mutual inductance of 5 mH. If the currents through A and B are 0.5 A and 2 A respectively, calculate: (a) the two possible values of the energy stored in the magnetic field; and (b) the coupling coefficient.

36. Two similar coils have a coupling coefficient of 0.25. When they are connected in series cumulatively, the total inductance is 80 mH. Calculate: (a) the self-inductance of each coil; (b) the total inductance when the coils are connected in series differentially; and (c) the total magnetic energy due to a current of 2 A when the coils are connected in series (i) cumulatively and (ii) differentially.

37. Two coils, with terminals AB and CD respectively, are inductively coupled. The inductance measured between terminals AB is 380 μH and that between terminals CD is 640 μH. With B joined to C, the inductance measured between terminals AD is 1600 μH. Calculate: (a) the mutual inductance of the coils; and (b) the inductance between terminals AC when B is connected to D.

Chapter nine

Alternating Voltage and Current

Objectives

When you have studied this chapter, you should

- have an understanding of alternating currents and voltages
- be familiar with a simple means of generating an alternating e.m.f.
- bo capable of analysing the generated alternating e.m.f.
- have an understanding of the terms 'average' and 'r.m.s.'
- be capable of analysing the average and r.m.s. values of an alternating current whether sinusoidal or non-sinusoidal
- be capable of representing a sinusoidal quantity by means of a phasor
- be capable of adding and subtracting sinusoidal quantities by means of a phasor diagram
- be able to cite practical frequencies and their applications

Contents

Having already developed quite an extensive understanding of circuits in which the current comes from a d.c. source such as a battery, we now need to progress to circuits in which the direction of current flow alternates. In practice most electrical circuits operate in this way and such a current is known as an alternating current. Almost every electrical supply to houses and to industry uses alternating current.

In order to understand such systems, we need to be familiar with the terms used to analyse alternating currents. It cannot be said that investigating such terms is the most exciting activity but when you progress to the application of such terms, you will understand the need to spend some time on the introductory mathematics. Fortunately most alternating systems operate on a sinusoidal basis and this helps to simplify our approach.

Sinusoidal waveforms are tricky things to draw and fortunately we shall find that when we use phasor diagrams we can represent them by straight lines. By joining up such lines, we can undertake apparently difficult additions and subtractions, and this simplifies the later analyses which we shall be considering.

Please remember that this chapter is a means to an end – it is the foundation for things yet to come.

9.1 Alternating systems

In previous chapters we have considered circuits and networks in which the current has remained constant, i.e. direct current systems, or those in which the current has varied for a short period of time, i.e. transient systems. However, there remains another type of system – the alternating system – in which the magnitudes of the voltage and of the current vary in a repetitive manner. Examples of such repetitive currents are shown in Fig. 9.1.

A current which varies after the fashion suggested in Fig. 9.1 is known as an alternating current. It flows first in one direction and then in the other. The cycle of variation is repeated exactly for each direction.

Fig. 9.1 Alternating current waveforms

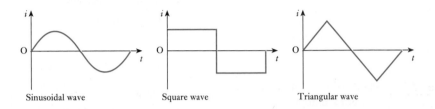

Sinusoidal wave   Square wave   Triangular wave

Alternating current can be abbreviated to a.c., hence a system with such an alternating current is known as an a.c. system. The curves relating current to time are known as *waveforms*. Those shown in Fig. 9.1 are simple waveforms, but waveforms can be quite complicated as shown in Fig. 9.5.

Of the waveforms shown in Fig. 9.1, the sinusoidal example is the most important. At this stage, the most significant reason for giving it further attention is that almost all electrical power supplies involve sinusoidal alternating current which is derived from sinusoidal alternating voltages, although in later chapters we will see that it is also significant in many communications systems.

Before we can set about analysing the performance of a.c. circuits and networks, we need to be introduced to a number of terms which are used to describe the effects of an alternating current. Therefore the remainder of this chapter will be used to analyse alternating waveforms in preparation for a.c. network analysis.

9.2 Generation of an alternating e.m.f.

Figure 9.2 shows a loop AB carried by a spindle DD rotated at a constant speed in an anticlockwise direction in a uniform magnetic field due to poles NS. The ends of the loop are brought out to two slip-rings C_1 and C_2, attached to but insulated from DD. Bearing on these rings are carbon brushes E_1 and E_2, which are connected to an external resistor R.

When the plane of the loop is horizontal, as shown in Fig. 9.3(a), the two sides A and B are moving parallel to the direction of the magnetic flux; it follows that no flux is being cut and no e.m.f. is being generated in the loop. Subsequent diagrams in Fig. 9.3 show the effects which occur as the coil is rotated. In Fig. 9.3(b), the coil sides are cutting the flux and therefore an e.m.f. is induced in the coil sides. Since the coil sides are moving in opposite directions, the e.m.f.s act in opposite directions, as shown by the dot and cross notation. However, they do act in the same direction around the coil so that the e.m.f. which appears at the brushes is twice that which is induced in

Fig. 9.2 Generation of an alternating e.m.f.

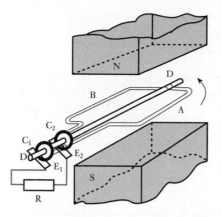

Fig. 9.3 E.m.f. in rotating coil

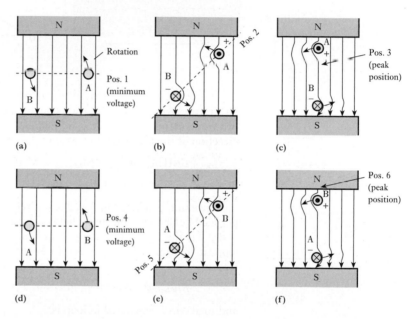

a coil side. Once the coil reaches the position shown in Fig. 9.3(c), the rate of cutting reaches a maximum. Thereafter the e.m.f. falls to zero by the time the coil has rotated to the position shown in Fig. 9.3(d).

The induced e.m.f. in the position shown in Fig. 9.3(e) is of particular interest. At first sight, it appears that the diagram is the same as that of Fig. 9.3(b), but in fact it is side A which bears the cross while side B has the dot. This means that the e.m.f. is of the same magnitude but of the opposite polarity. This observation also applies to Fig. 9.3(f). It follows that the variation of induced e.m.f. during the second half of the cycle of rotation is the same in magnitude as during the first half but the polarity of the e.m.f. has reversed.

We can now analyse the general case shown in Fig. 9.4(a) in which coil AB is shown after it has rotated through an angle θ from the horizontal position, namely the position of zero e.m.f. Suppose the peripheral velocity of each side of the loop to be u metres per second; then at the instant shown in Fig. 9.4, this peripheral velocity can be represented by the length of a line

Fig. 9.4 Instantaneous value of generated e.m.f.

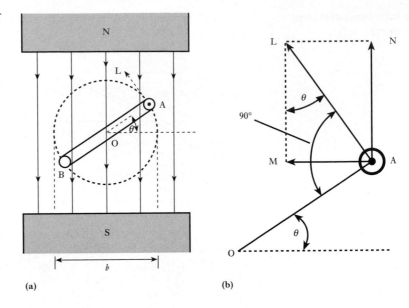

(a)   (b)

AL drawn at right angles to the plane of the loop. We can resolve AL into two components, AM and AN, perpendicular and parallel respectively to the direction of the magnetic flux, as shown in Fig. 9.4(b). Since

$$\angle MLA = 90° - \angle MAL = \angle MAO = \theta$$

$$\therefore \qquad AM = AL \sin \theta = u \sin \theta$$

The e.m.f. generated in A is due entirely to the component of the velocity perpendicular to the magnetic field. Hence, if B is the flux density in teslas and if l is the length in metres of each of the parallel sides A and B of the loop, it follows from expression (6.4) that e.m.f. generated in one side of loop is

$$Blu \sin \theta \text{ volts}$$

and total e.m.f. generated in loop is

$$2Blu \sin \theta \text{ volts}$$

$$\therefore \qquad e = 2Blu \sin \theta \qquad\qquad\qquad [9.1]$$

i.e. the generated e.m.f. is proportional to $\sin \theta$. When $\theta = 90°$, the plane of the loop is vertical and both sides of the loop are cutting the magnetic flux at the maximum rate, so that the generated e.m.f. is then at its maximum value E_m. From expression [9.1], it follows that when $\theta = 90°$, $E_m = 2Blu$ volts.

If b is the breadth of the loop in metres, and n the speed of rotation in revolutions per second, then u is πbn metres per second and

$$E_m = 2Bl \times \pi bn \text{ volts}$$

$$= 2\pi BAn \text{ volts}$$

where

$$A = lb = \text{area of loop in square metres}$$

If the loop is replaced by a coil of N turns in series, each turn having an area of A square metres, maximum value of e.m.f. generated in coil is

$$E_m = 2\pi BAnN \quad \text{volts} \tag{9.2}$$

and instantaneous value of e.m.f. generated in coil is

$$e = E_m \sin\theta = 2\pi BAnN \sin\theta \text{ volts}$$

$\therefore$

$$e = 2\pi BAnN \sin\theta \tag{9.3}$$

Lower-case letters are used to represent instantaneous values and upper-case letters represent definite values such as maximum, average or r.m.s. values. In a.c. circuits, capital I and V without any subscript represent r.m.s. values – we will meet r.m.s. values in section 9.5.

This e.m.f. can be represented by a sine wave as in Fig. 9.5, where E_m represents the maximum value of the e.m.f. and e is the value after the loop has rotated through an angle θ from the position of zero e.m.f. When the loop has rotated through 180° or π radians, the e.m.f. is again zero. When θ is varying between 180° and 360° (π and 2π radians), side A of the loop is moving towards the right in Fig. 9.4 and is therefore cutting the magnetic flux in the opposite direction to that during the first half-revolution. Hence, if we regard the e.m.f. as positive while θ is varying between 0 and 180°, it is negative while θ is varying between 180° and 360°, i.e. when θ varies between 180° and 270°, the value of the e.m.f. increases from zero to $-E_m$ and then decreases to zero as θ varies between 270° and 360°. Subsequent revolutions of the loop merely produce a repetition of the e.m.f. wave.

It is significant that we have concentrated on one cycle of events arising from the single rotation of the coil AB shown in Fig. 9.2. However, alternating e.m.f.s and alternating voltages continue to repeat the cycle as suggested in Fig. 9.6. Further, the effect at each of the situations shown in Fig. 9.3 recurs

Fig. 9.5 Sine wave of e.m.f.

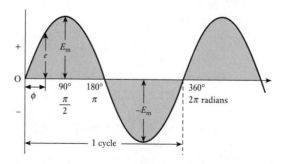

Fig. 9.6 Extended sine wave of e.m.f.

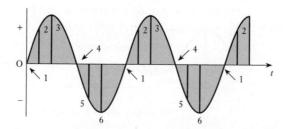

in each subsequent cycle. For instance, the e.m.f. which was induced in position 2 can be seen to recur each time the e.m.f. waveform rises from zero.

9.3 Waveform terms and definitions

Our consideration of alternating systems has already introduced a number of terms and we will find the need of a few more. It will therefore be helpful to consider the terms which we most commonly use:

Waveform. The variation of a quantity such as voltage or current shown on a graph to a base of time or rotation is a waveform.

Cycle. Each repetition of a variable quantity, recurring at equal intervals, is termed a cycle.

Period. The duration of one cycle is termed its period. (Cycles and periods need not commence when a waveform is zero. Figure 9.7 illustrates a variety of situations in which the cycle and period have identical values.)

Instantaneous value. The magnitude of a waveform at any instant in time (or position of rotation). Instantaneous values are denoted by lower-case symbols such as e, v and i.

Peak value. The maximum instantaneous value measured from its zero value is known as its peak value.

Peak-to-peak value. The maximum variation between the maximum positive instantaneous value and the maximum negative instantaneous value is the peak-to-peak value. For a sinusoidal waveform, this is twice the peak value. The peak-to-peak value is E_{pp} or V_{pp} or I_{pp}.

Peak amplitude. The maximum instantaneous value measured from the mean value of a waveform is the peak amplitude. Later we will find how to determine the mean value, but for most sinusoidal alternating voltages and currents the mean value is zero. The peak amplitude is E_m or V_m or I_m. The peak amplitude is generally described as the maximum value, hence the maximum voltage has the symbol V_m.

The relationships between peak value, peak-to-peak value and peak amplitude (maximum value) are illustrated in Fig. 9.8.

Fig. 9.7 Cycles and periods

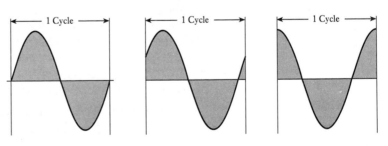

Fig. 9.8 Peak values

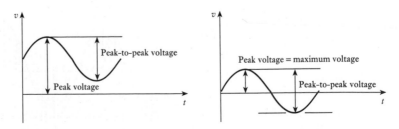

Fig. 9.9 Effect on waveforms by varying frequency

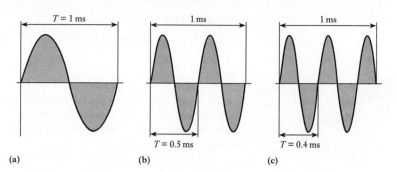

(a)   (b)   (c)

Frequency. The number of cycles that occur in 1 second is termed the frequency of that quantity. Frequency is measured in hertz (Hz) in memory of Heinrich Rudolf Hertz, who, in 1888, was the first to demonstrate experimentally the existence and properties of electromagnetic radiation predicted by Maxwell in 1865. It follows that frequency *f* is related to the period *T* by the relation

$$f = \frac{1}{T} \qquad\qquad [9.4]$$

where *f* is the frequency in hertz and *T* is the period in seconds. Assuming each graph to be drawn to the same scale of time, the effect of increasing the frequency is shown in Fig. 9.9. The diagrams assume frequencies of 1000 Hz (1 kHz), 2000 Hz (2 kHz) and 2500 Hz (2.5 kHz).

Frequency   Symbol: *f*   Unit: hertz (Hz)

Example 9.1 A coil of 100 turns is rotated at 1500 r/min in a magnetic field having a uniform density of 0.05 T, the axis of rotation being at right angles to the direction of the flux. The mean area per turn is 40 cm². Calculate

(a) the frequency;
(b) the period;
(c) the maximum value of the generated e.m.f.;
(d) the value of the generated e.m.f. when the coil has rotated through 30° from the position of zero e.m.f.

(a) Since the e.m.f. generated in the coil undergoes one cycle of variation when the coil rotates through one revolution,

∴   Frequency = no. of cycles per second

= no. of revolutions per second

$$= \frac{1500}{60} = \textbf{25 Hz}$$

(b)   Period = time of 1 cycle

$$= \frac{1}{25} = \textbf{0.04 s}$$

(c) From expression [9.2]

$$E_\mathrm{m} = 2\pi \times 0.05 \times 0.004 \times 100 \times 1500/60 = \textbf{3.14 V}$$

(d) For $\theta = 30°$, $\sin 30° = 0.5$,

$\therefore$ $e = 3.14 \times 0.5 = \mathbf{1.57\ V}$

| 9.4 | Relationship between frequency, speed and number of pole pairs |

The waveform of the e.m.f. generated in an a.c. generator undergoes one complete cycle of variation when the conductors move past a N and a S pole; and the shape of the wave over the negative half is exactly the same as that over the positive half.

The generator shown in Fig. 9.4 has two poles which can also be described as having one pair of poles. Machines can have two or more pairs of poles. For example, if there were N poles placed top and bottom and S poles to either side then the machine would have two pairs of poles.

If an a.c. generator has p *pairs* of poles and if its speed is n revolutions per second, then

Frequency $= f =$ no. of cycles per second

$=$ no. of cycles per revolution

$\times$ no. of revolutions per second

$\therefore$ $\boxed{f = pn}$ hertz   [9.5]

Thus if a two-pole machine is to generate an e.m.f. having a frequency of 50 Hz, then from expression [9.5],

$50 = 1 \times n$

$\therefore$ Speed $= 50$ revolutions per second $= 50 \times 60 = 3000$ r/min

Since it is not possible to have fewer than two poles, the highest speed at which a 50 Hz a.c. generator can be operated is 3000 r/min. Similarly a 60 Hz a.c. generator can only operate with a maximum speed of 3600 r/min. In practice, these are the operating speeds of most generators in power stations other than in hydroelectric generating plant in which lower speeds occur.

| 9.5 | Average and r.m.s. values of an alternating current |

Most electrical energy is provided by rotating a.c. generators operating on the principles already described in this chapter. The e.m.f.s and the resulting voltages and currents are for the most part sinusoidal which is the waveform on which we have concentrated. However, the use of electronic switching has resulted in many circuits operating with waveforms which are anything but sinusoidal; square waveforms are especially common in communication circuits.

Let us first consider the general case of a current the waveform of which cannot be represented by a simple mathematical expression. For instance, the wave shown in Fig. 9.10 is typical of the current taken by a transformer on no load. If n equidistant mid-ordinates, i_1, i_2, etc. are taken over either the positive or the negative half-cycle, then *average* value of current over half a cycle is

$$I_{\mathrm{av}} = \frac{i_1 + i_2 + \ldots + i_n}{n}$$ [9.6]

Fig. 9.10 Average and r.m.s. values

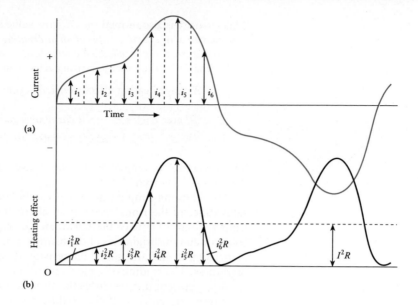

(a)

(b)

Or, alternatively, average value of current is

$$\frac{\text{Area enclosed over half-cycle}}{\text{Length of base over half-cycle}} \qquad [9.7]$$

This method of expressing the average value is the more convenient when we come to deal with sinusoidal waves.

In a.c. work, however, the average value is of comparatively little importance. This is due to the fact that it is the power produced by the electric current that usually matters. Thus, if the current represented in Fig. 9.10(a) is passed through a resistor having resistance R ohms, the heating effect of i_1 is i_1^2R, that of i_2 is i_2^2R, etc. as shown in Fig. 9.10(b). The variation of the heating effect during the second half-cycle is exactly the same as that during the first half-cycle.

$$\therefore \qquad \text{Average heating effect} = \frac{i_1^2R + i_2^2R + \ldots + i_n^2R}{n}$$

Suppose I to be the value of *direct* current through the same resistance R to produce a heating effect equal to the average heating effect of the alternating current, then

$$I^2R = \frac{i_1^2R + i_2^2R + \ldots + i_n^2R}{n}$$

$$\therefore \qquad I = \sqrt{\left(\frac{i_1^2 + i_2^2 + \ldots + i_n^2}{n} \right)} \qquad [9.8]$$

= square *root* of the *mean* of the *squares* of the current

= root-mean-square (or r.m.s.) value of the current

This quantity is also termed the *effective* value of the current. It will be seen that the r.m.s. or *effective value of an alternating current is measured in terms of the* direct *current that produces the same heating effect in the same resistance.*

Alternatively, the average heating effect can be expressed as follows:

Average heating effect over half-cycle

$$= \frac{\text{area enclosed by } i^2R \text{ curve over half-cycle}}{\text{length of base}} \qquad [9.9]$$

This is a more convenient expression to use when deriving the r.m.s. value of a sinusoidal current.

The following simple experiment can be found useful in illustrating the significance of the r.m.s. value of an alternating current. A metal-filament lamp L (Fig. 9.11) is connected to an a.c. supply by closing switch S on contact a and the brightness of the filament is noted. Switch S is then moved to position b and the slider on resistor R is adjusted to give the same brightness. The reading on a moving-coil ammeter A then gives the value of the direct current that produces the same heating effect as that produced by the alternating current. If the reading on ammeter A is, say, 0.3 A when equality of brightness has been attained, the r.m.s. value of the alternating current is 0.3 A.

Fig. 9.11 An experiment to demonstrate the r.m.s. value of an alternating current

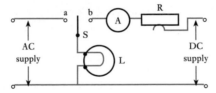

The r.m.s. value is always greater than the average except for a rectangular wave, in which case the heating effect remains constant so that the average and the r.m.s. values are the same.

Form factor of a wave is

$$\frac{\text{r.m.s. value}}{\text{Average value}} \qquad [9.10]$$

Peak or *crest factor* of a wave is

$$\frac{\text{Peak or maximum value}}{\text{r.m.s. value}} \qquad [9.11]$$

9.6   Average and r.m.s. values of sinusoidal currents and voltages

If I_m is the maximum value of a current which varies sinusoidally as shown in Fig. 9.12(a), the instantaneous value i is represented by

$$i = I_m \sin \theta$$

where θ is the angle in radians from instant of zero current.

Fig. 9.12 Average and r.m.s. values of a sinusoidal current

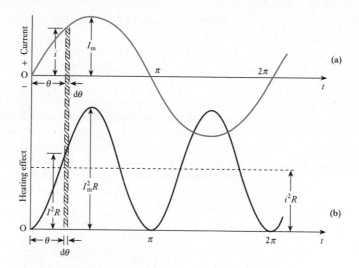

For a very small interval $d\theta$ radians, the area of the shaded strip is $i \cdot d\theta$ ampere radians. The use of the unit 'ampere radian' avoids converting the scale on the horizontal axis from radians to seconds, therefore, total area enclosed by the current wave over half-cycle is

$$\int_0^\pi i \cdot d\theta = I_m \int_0^\pi \sin\theta \cdot d\theta = -I_m \Big[\cos\theta\Big]_0^\pi$$

$$= -I_m[-1 - 1] = 2I_m \text{ ampere radians}$$

From expression [9.7], average value of current over a half-cycle is

$$\frac{2I_m \text{ (ampere radians)}}{\pi \text{ (radians)}}$$

i.e. $\qquad I_{av} = 0.637 I_m \qquad \text{amperes}$ $\hfill$ [9.12]

If the current is passed through a resistor having resistance R ohms, instantaneous heating effect $= i^2 R$ watts.

The variation of $i^2 R$ during a complete cycle is shown in Fig. 9.12(b). During interval $d\theta$ radians, heat generated is $i^2 R \cdot d\theta$ watt radians and is represented by the area of the shaded strip. Hence heat generated during the first half-cycle is area enclosed by the $i^2 R$ curve and is

$$\int_0^\pi i^2 R \cdot d\theta = I_m^2 R \int_0^\pi \sin^2\theta \cdot d\theta$$

$$= \frac{I_m^2 R}{2} \int_0^\pi (1 - \cos 2\theta) \cdot d\theta$$

$$= \frac{I_m^2 R}{2} \Big[\theta - \tfrac{1}{2}\sin 2\theta\Big]_0^\pi$$

$$= \frac{\pi}{2} I_m^2 R \text{ watt radians}$$

From expression [9.9], average heating effect is

$$\frac{(\pi/2)I_{\mathrm{m}}^{2}R \text{ (watt radians)}}{\pi \text{ (radians)}} = \tfrac{1}{2}I_{\mathrm{m}}^{2}R \text{ watts} \qquad [9.13]$$

This result can be observed from the equation $\sin^2\theta = \tfrac{1}{2} - \tfrac{1}{2}\cos 2\theta$. In words, this means that the square of a sine wave may be regarded as being made up of two components: (a) a constant quantity equal to half the maximum value of the $\sin^2\theta$ curve, and (b) a cosine curve having twice the frequency of the $\sin \theta$ curve. From Fig. 9.12 it is seen that the curve of the heating effect undergoes two cycles of change during one cycle of current. The average value of component (b) over a complete cycle is zero; hence the average heating effect is $\tfrac{1}{2}I_m^2R$.

If I is the value of direct current through the same resistance to produce the same heating effect

$$I^2R = \tfrac{1}{2}I_m^2R$$

$\therefore$
$$I = \frac{I_{\mathrm{m}}}{\sqrt{2}} = 0.707I_{\mathrm{m}} \qquad [9.14]$$

Whilst I is I_{RMS} it is normal practice to omit the RMS subscript, as this is the most common current.

Since the voltage across the resistor is directly proportional to the current, it follows that the relationships derived for currents also apply to voltages.

Hence, in general, average value of a sinusoidal current or voltage is

0.637 × maximum value

$\therefore$
$$I_{\mathrm{av}} = 0.637I_{\mathrm{m}} \qquad [9.15]$$

r.m.s. value of a sinusoidal current or voltage is

0.707 × maximum value

$\therefore$
$$I = 0.707I_{\mathrm{m}} \qquad [9.16]$$

From expressions [9.15] and [9.16], form factor of a sine wave is

$$\frac{0.707 \times \text{maximum value}}{0.637 \times \text{maximum value}}$$

$$k_{\mathrm{f}} = 1.11 \qquad [9.17]$$

and peak or crest factor of a sine wave is

$$\frac{\text{maximum value}}{0.707 \times \text{maximum value}}$$

$\therefore$
$$k_{\mathrm{p}} = 1.414 \qquad [9.18]$$

Example 9.2 An alternating current of sinusoidal waveform has an r.m.s. value of 10.0 A. What are the peak values of this current over one cycle?

$$I_m = \frac{I}{0.707} = \frac{10}{0.707} = 14.14 \text{ A}$$

The peak values therefore are **14.14 A** and **−14.14 A**.

Example 9.3 An alternating voltage has the equation $v = 141.4 \sin 377t$; what are the values of:

(a) r.m.s. voltage;
(b) frequency;
(c) the instantaneous voltage when $t = 3$ ms?

The relation is of the form $v = V_m \sin \omega t$ and, by comparison,

(a) $V_m = 141.4 \text{ V} = \sqrt{2}V$

hence $V = \dfrac{141.4}{\sqrt{2}} = \textbf{100 V}$

(b) Also by comparison

$$\omega = 377 \text{ rad/s} = 2\pi f$$

hence $f = \dfrac{377}{2\pi} = \textbf{60 Hz}$

(c) Finally

$$v = 141.4 \sin 377t$$

When $t = 3 \times 10^{-3}$ s

$$v = 141.4 \sin(377 \times 3 \times 10^{-3}) = 141.4 \sin 1.131$$

$$= 141.4 \times 0.904 = \textbf{127.8 V}$$

Note that, in this example, it was necessary to determine the sine of 1.131 rad, which could be obtained either from suitable tables, or from a calculator. Alternatively, 1.131 rad may be converted into degree measurement, i.e.

$$1.131 \text{ rad} \equiv 1.131 \times \frac{180}{\pi} = 64.8°$$

Example 9.4 A moving-coil ammeter, a thermal* ammeter and a rectifier are connected in series with a resistor across a 110 V sinusoidal a.c. supply. The circuit has a resistance of 50 Ω to current in one direction and, due to the rectifier, an infinite resistance to current in the reverse direction. Calculate:

(a) the readings on the ammeters;
(b) the form and peak factors of the current wave.

---

* A thermal ammeter is an instrument the operation of which depends upon the heating effect of a current.

Fig. 9.13 Waveforms of voltage, current and power for Example 9.4

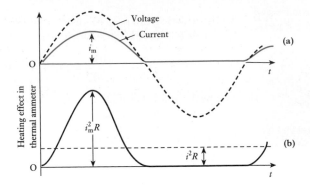

(a) Maximum value of the voltage

$$V_m = \frac{V}{0.707} = \frac{110}{0.707} = 155.5 \text{ V}$$

therefore maximum value of the current

$$I_m = \frac{V_m}{R} = \frac{155.5}{50} = 3.11 \text{ A}$$

During the positive half-cycle the current is proportional to the voltage and is therefore sinusoidal, as shown in Fig. 9.13(a); therefore average value of current over the positive half-cycle

$$I_{av} = 0.637 I_m = 0.637 \times 3.11 = 1.98 \text{ A}$$

During the negative half-cycle, the current is zero. Owing, however, to the inertia of the moving system, the moving-coil ammeter reads the average value of the current over the *whole* cycle, therefore reading on moving-coil ammeter is

$$\frac{1.98}{2} = 0.99 \text{ A}$$

The variation of the heating effect in the thermal ammeter is shown in Fig. 9.13(b), the maximum power being $I_m^2 R$, where R is the resistance of the instrument.

From expression [9.13] it is seen that the average heating effect over the positive half-cycle is $\frac{1}{2} I_m^2 R$; and since no heat is generated during the second half-cycle, it follows that the average heating effect over a complete cycle is $\frac{1}{4} I_m^2 R$.

If I is the direct current which would produce the same heating effect

$$I^2 R = \frac{1}{4} I_m^2 R$$

$$\therefore \qquad I = \frac{1}{2} I_m = \frac{3.11}{2} = 1.555 \text{ A}$$

i.e. reading on thermal ammeter = **1.56 A**.

A mistake that can very easily be made is to calculate the r.m.s. value of the current over the positive half-cycle as 0.707×3.11, namely 2.2 A, and then say that the reading on the thermal ammeter is half this value, namely 1.1 A. The importance of working out such a problem from first principles should now be evident.

(b) From equation [9.17], form factor is

$$k_f = \frac{I}{I_{av}} = \frac{1.555}{0.99} = 1.57$$

and from equation [9.18], peak factor is

$$k_p = \frac{I_m}{I} = \frac{3.11}{1.555} = 2.0$$

9.7	Average and r.m.s. values of non-sinusoidal currents and voltages

Having demonstrated the determination of average and r.m.s. values for sinusoidal currents and voltages, it is a relatively short step to consider non-sinusoidal quantities. This can easily be done by considering further examples.

Example 9.5

A current has the following steady values in amperes for equal intervals of time changing instantaneously from one value to the next (Fig. 9.14):

$$0, 10, 20, 30, 20, 10, 0, -10, -20, -30, -20, -10, 0, \text{etc.}$$

Calculate the r.m.s. value of the current and its form factor.

Because of the symmetry of the waveform, it is only necessary to calculate the values over the first half-cycle.

Fig. 9.14 Part of Example 9.5

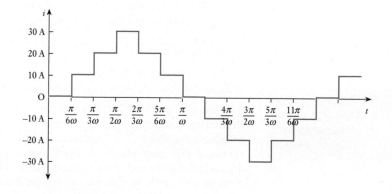

$$I_{av} = \frac{\text{area under curve}}{\text{length of base}}$$

$$= \frac{0\left(\frac{\pi}{6\omega} - 0\right) + 10\left(\frac{2\pi}{6\omega} - \frac{\pi}{6\omega}\right) + 20\left(\frac{3\pi}{6\omega} - \frac{2\pi}{6\omega}\right) + 30\left(\frac{4\pi}{6\omega} - \frac{3\pi}{6\omega}\right) + 20\left(\frac{5\pi}{6\omega} - \frac{4\pi}{6\omega}\right) + 10\left(\frac{6\pi}{6\omega} - \frac{5\pi}{6\omega}\right)}{\frac{\pi}{\omega} - 0}$$

$$= 15.0\,\text{A}$$

$$I^2 = \frac{0^2\left(\frac{\pi}{6\omega} - 0\right) + 10^2\left(\frac{2\pi}{6\omega} - \frac{\pi}{6\omega}\right) + 20^2\left(\frac{3\pi}{6\omega} - \frac{2\pi}{6\omega}\right) + 30^2\left(\frac{4\pi}{6\omega} - \frac{3\pi}{6\omega}\right) + 20^2\left(\frac{5\pi}{6\omega} - \frac{4\pi}{6\omega}\right) + 10^2\left(\frac{6\pi}{6\omega} - \frac{5\pi}{6\omega}\right)}{\frac{\pi}{\omega} - 0}$$

$$= 316$$

$$I = \sqrt{316} = 17.8\,\text{A}$$

$$k_f = \frac{I}{I_{av}} = \frac{17.8}{15.0} = 1.19$$

Example 9.6 Calculate the form factor for each of the waveforms in Fig. 9.15.

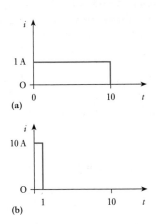

(a)

(b)

Fig. 9.15 Part of Example 9.6

For Fig. 9.15(a):

$$I_{av} = \frac{1(10 - 0)}{10 - 0} = 1.0\,\text{A}$$

$$I = \left(\frac{1^2(10 - 0)}{10 - 0}\right)^{\frac{1}{2}} = 1.0\,\text{A}$$

$$k_f = \frac{I}{I_{av}} = \frac{1.0}{1.0}$$

$$= 1.0$$

For Fig. 9.15(b):

$$I_{av} = \frac{10(1 - 0) + 0(10 - 1)}{10 - 0} = 1.0\,\text{A}$$

$$I = \left(\frac{10^2(1 - 0) + 0^2(10 - 1)}{10 - 0}\right)^{\frac{1}{2}} = 3.16\,\text{A}$$

$$k_f = \frac{I}{I_{av}} = \frac{3.16}{1.0}$$

$$= 3.16$$

It will be noted that the first waveform is that of direct current in which the r.m.s. current and the mean current have the same value. It is for this reason that the r.m.s. value of an alternating current may be equated to the mean value of a direct current.

9.8 Representation of an alternating quantity by a phasor

Suppose OA in Fig. 9.16(a) to represent to scale the maximum value of an alternating quantity, say, current, i.e. OA = I_m. Also, suppose OA to rotate anticlockwise about O at a uniform angular velocity. This is purely a conventional direction which has been universally adopted. An arrowhead is drawn at the outer end of the phasor, partly to indicate which end is assumed to move and partly to indicate the precise length of the phasor when two or more phasors happen to coincide.

Figure 9.16(a) shows OA when it has rotated through an angle θ from the position occupied when the current was passing through its zero value. If

Fig. 9.16 Phasor representation of an alternating quantity

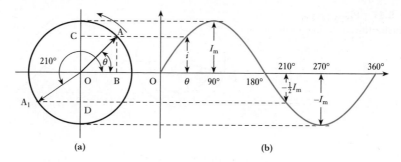

(a) (b)

AB and AC are drawn perpendicular to the horizontal and vertical axes respectively:

$$OC = AB = OA \sin \theta$$

$$= I_m \sin \theta$$

$$= i, \text{ namely the value of the current at that instant}$$

Hence the projection of OA on the vertical axis represents to scale the instantaneous value of the current. Thus when $\theta = 90°$, the projection is OA itself; when $\theta = 180°$, the projection is zero and corresponds to the current passing through zero from a positive to a negative value; when $\theta = 210°$, the phasor is in position OA_1, and the projection $= OD = \frac{1}{2}OA_1 = -\frac{1}{2}I_m$; and when $\theta = 360°$, the projection is again zero and corresponds to the current passing through zero from a negative to a positive value. It follows that OA rotates through one revolution or 2π radians in one cycle of the current wave.

If f is the frequency in hertz, then OA rotates through f revolutions of $2\pi f$ radians in 1 s. Hence the angular velocity of OA is $2\pi f$ radians per second and is denoted by the symbol ω (omega), i.e.

$$\omega = 2\pi f \quad \text{radians per second} \tag{9.19}$$

If the time taken by OA in Fig. 9.16 to rotate through an angle θ radians is t seconds, then

$$\theta = \text{angular velocity} \times \text{time}$$

$$= \omega t = 2\pi f t \text{ radians}$$

We can therefore express the instantaneous value of the current thus:

$$i = I_m \sin \theta = I_m \sin \omega t$$

$$\therefore \quad i = I_m \sin 2\pi f t \tag{9.20}$$

Let us next consider how two quantities such as voltage and current can be represented by a phasor diagram. Figure 9.17(b) shows the voltage leading the current by an angle ϕ. In Fig. 9.17(a), OA represents the maximum value of the current and OB that of the voltage. The angle between OA and OB must be the same angle ϕ as in Fig. 9.17(b). Consequently when OA is along the horizontal axis, the current at that instant is zero and the value of the

Fig. 9.17 Phasor representation of quantitites differing in phase

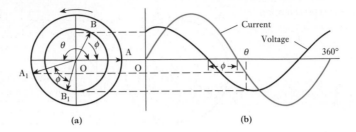

(a)　　　　　　　　(b)

voltage is represented by the projection of OB on the vertical axis. These values correspond to instant O in Fig. 9.17(b).

After the phasors have rotated through an angle θ, they occupy positions OA_1 and OB_1 respectively, with OB_1 still leading OA_1 by the same angle ϕ; and the instantaneous values of the current and voltage are again given by the projections of OA_1 and OB_1 on the vertical axis, as shown by the horizontal dotted lines.

If the instantaneous value of the current is represented by

$$i = I_m \sin \theta$$

then the instantaneous value of the voltage is represented by

$$v = V_m \sin(\theta + \phi)$$

where　　$I_m = OA$　　and　　$V_m = OB$ in Fig. 9.17(a).

The current in Fig. 9.17 is said to *lag* the voltage by an angle ϕ which is the *phase difference* between the two phasors. The phase difference remains constant irrespective of the phasor positions. When one sine wave passes through the zero following another, it is said to lag. Thus in Fig. 9.17, the current lags the voltage.

9.9　Addition and subtraction of sinusoidal alternating quantities

Suppose OA and OB in Fig. 9.18 to be phasors representing to scale the maximum values of, say, two alternating voltages having the same frequency but differing in phase by an angle ϕ. Complete the parallelogram OACB and draw the diagonal OC. Project OA, OB and OC on to the vertical axis. Then for the positions shown in Fig. 9.18:

Instantaneous value of OA = OD

Instantaneous value of OB = OE

and　　Instantaneous value of OC = OF

Since AC is parallel and equal to OB, DF = OE,

∴　　　OF = OD + DF = OD + OE

i.e. the instantaneous value of OC equals the sum of the instantaneous values of OA and OB. Hence OC represents the maximum value of the resultant voltage to the scale that OA and OB represent the maximum values of the separate voltages. Therefore OC is termed the *phasor sum* of OA and OB; and it is evident that OC is less than the arithmetic sum of OA and OB except when the latter are in phase with each other. This is the reason why it is seldom correct in a.c. work to add voltages or currents together arithmetically.

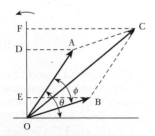

Fig. 9.18　Addition of phasors

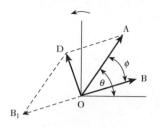

Fig. 9.19 Subtraction of phasors

If voltage OB is to be subtracted from OA, then OB is produced backwards so that OB_1 is equal and opposite to OB (Fig. 9.19). The diagonal OD of the parallelogram drawn on OA and OB_1 represents the *phasor differences* of OA and OB.

For simplicity, OA can be represented by **A** and OB as **B**, bold letters being used to indicate the appropriate phasors. It follows that

$$\mathbf{C} = \mathbf{A} + \mathbf{B} \quad \text{and} \quad \mathbf{D} = \mathbf{A} - \mathbf{B}$$

Example 9.7

The instantaneous values of two alternating voltages are represented respectively by $v_1 = 60 \sin \theta$ volts and $v_2 = 40 \sin(\theta - \pi/3)$ volts. Derive an expression for the instantaneous value of:

(a) the sum;
(b) the difference of these voltages.

(a) It is usual to draw the phasors in the position corresponding to $\theta = 0$,* i.e. OA in Fig. 9.20 is drawn to scale along the x axis to represent 60 V, and OB is drawn $\pi/3$ radians or 60° behind OA to represent 40 V. The diagonal OC of the parallelogram drawn on OA and OB represents the phasor sum of OA and OB. By measurement, OC = 87 V and angle ϕ between OC and the x-axis is 23.5°, namely 0.41 rad; hence:

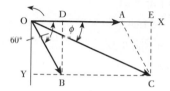

Fig. 9.20 Addition of phasors for Example 9.7

instantaneous sum of the two voltages = $87 \sin(\theta - 23.5°)$ V

Alternatively, this expression can be found thus:

Horizontal component of OA = 60 V

Horizontal component of OB = OD = 40 cos 60° = 20 V

∴ Resultant horizontal component = OA + OD = 60 + 20

= 80 V = OE in Fig. 9.20

Vertical component of OA = 0

Vertical component of OB = BD = −40 sin 60°

= −34.64 V

∴ Resultant vertical component = −34.64 V = CE

The minus sign merely indicates that the resultant vertical component is *below* the horizontal axis and that the resultant voltage must therefore lag relative to the reference phasor OA. Hence maximum value of resultant voltage is

$$OC = \sqrt{\{(80)^2 + (-34.64)^2\}}$$

$$= 87.2 \text{ V}$$

---

* The idea of a phasor rotating continuously serves to establish its physical significance, but its application in circuit analysis is simplified by *fixing* the phasor in position corresponding to $t = 0$, as in Fig. 9.20, thereby eliminating the time function. Such a phasor represents the magnitude of the sinusoidal quantity and its phase relative to a reference quantity, e.g. in Fig. 9.20 phasor OB lags the reference phasor OA by 60°.

If ϕ is the phase difference between OC and OA

$$\tan \phi = EC/OE = -\frac{34.64}{80} = -0.433$$

$\therefore$ $\qquad\qquad \phi = -23.4° = -0.41$ rad

and instantaneous value of resultant voltage is

87.2 sin$(\theta - 23.5°)$ V

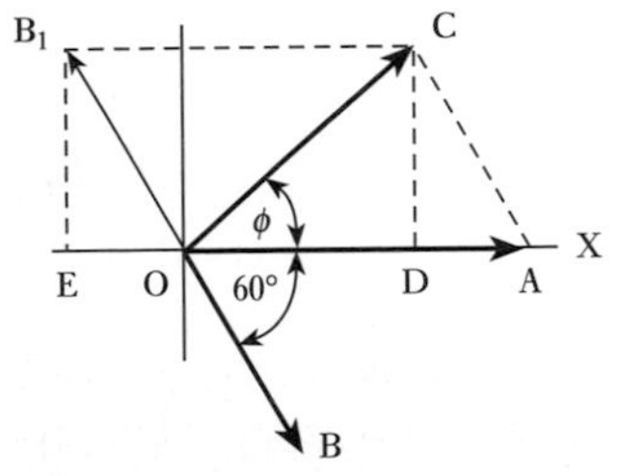

(b) The construction for subtracting OB from OA is shown in Fig. 9.21. By measurement, OC = 53 V and $\phi = 41° = 0.715$ rad. Therefore instantaneous difference of the two voltages is

53 sin$(\theta + 40.9°)$ V

Alternatively, resultant horizontal component is

OA – OE = 60 – 20 = 40 V = OD in Fig. 9.21

and   Resultant vertical component = B_1E = 34.64 V

$\qquad\qquad\qquad\qquad\qquad$ = DC in Fig. 9.21

therefore maximum value of resultant voltage is

$$OC = \sqrt{\{(40)^2 + (34.64)^2\}}$$

$$= 52.9 \text{ V}$$

and $\tan \phi = DC/OD = \dfrac{34.64}{40}$

$$= 0.866$$

$\therefore$ $\qquad\qquad \phi = 40.9° = 0.714$ rad

and instantaneous value of resultant voltage is

52.9 sin$(\theta + 40.9°)$ V

Fig. 9.21   Subtraction of phasors for Example 9.7

| 9.10 | Phasor diagrams drawn with r.m.s. values instead of maximum values |

It is important to note that when alternating voltages and currents are represented by phasors it is assumed that their waveforms are sinusoidal. It has already been shown that for sine waves the r.m.s. or effective value is 0.707 times the maximum value. Furthermore, ammeters and voltmeters are almost invariably calibrated to read the r.m.s. values. Consequently it is much more convenient to make the length of the phasors represent r.m.s. rather than maximum values. If the phasors of Fig. 9.21, for instance, were drawn to represent to scale the r.m.s. instead of the maximum values of the voltages, the shape of the diagram would remain unaltered and the phase relationships between the various quantities would remain unaffected. Hence in all phasor diagrams from now onwards, the lengths of the phasors will, for convenience, represent the r.m.s. values. This is the usual practice.

9.11   **Alternating system frequencies in practice**

We have discussed alternating voltages, currents and frequencies at some length. Before progressing to the analysis of a.c. circuits, it would be appropriate to consider what values we are likely to meet in practice.

Most electrical supplies operate at 50 or 60 Hz, with domestic supplies at 110 V up to 230 V. However, the power is distributed at higher voltages such as 11 000 V and transmitted at such voltages as 275 kV. Currents can be anything up to a few thousand amperes.

The sounds we hear depend on frequency. We can produce sound by using electrical signals between 15 Hz and 20 kHz, although not many of us can hear the upper limit. As the frequencies increase, we find signals which can be used to transmit radio, television and other communications information. In particular most of us are familiar with identifying radio stations by a frequency between 88 and 108 MHz. Frequencies above and below this range are used for television signals.

Frequencies above 300 MHz are known as microwave frequencies. This range can rise up to 300 GHz, thus we can experience remarkably high values of frequency in practice. However, in most systems at high frequencies the voltages and currents are normally very small, e.g. millivolts and microamperes.

The ranges of frequency are indicated in Fig. 9.22.

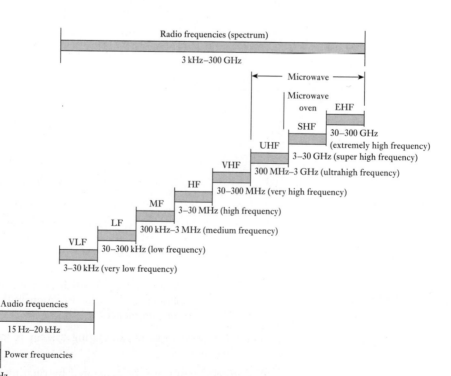

Fig. 9.22   Frequency ranges

Summary of important formulae

Instantaneous value of e.m.f. generated in a coil rotating in a uniform magnetic field is

$$e = E_m \sin \theta$$
$$= 2\pi BANn \sin \theta \text{ volts} \qquad [9.3]$$

$$f = \frac{1}{T} \qquad [9.4]$$

$$f = np \qquad [9.5]$$

For n equidistant mid-ordinates over half a cycle

$$\text{Average value} = \frac{i_1 + i_2 + \ldots + i_n}{n} \qquad [9.6]$$

and r.m.s. or effective value is

$$\sqrt{\left(\frac{i_1^2 + i_2^2 + \ldots + i_n^2}{n}\right)} \qquad [9.8]$$

For sinusoidal waves

$$\text{Average value} = 0.637 \times \text{maximum value}$$
$$I_{av} = 0.637 I_m \qquad [9.12]$$

r.m.s. or effective value is $0.707 \times$ maximum value

$$I = 0.707 I_m$$
$$= \frac{1}{\sqrt{2}} I_m \qquad [9.14]$$

$$\text{Form factor} = \frac{\text{r.m.s. value}}{\text{average value}} \qquad [9.10]$$
$$k_f = 1.11 \text{ for a sine wave}$$

$$\text{Peak or crest factor} = \frac{\text{peak or maximum value}}{\text{r.m.s. value}} \qquad [9.11]$$
$$k_p = 1.414 \text{ for a sine wave}$$

Terms and concepts

An **alternating** system is one in which the voltages and currents vary in a repetitive manner. A cycle of variation is the sequence of change before repetition commences.

The most basic form of alternating system is based on a **sinusoidal** variation.

A sinusoidal e.m.f. can be generated by rotating a rectangular coil in a uniform magnetic field although in practical terms this would be a most inefficient method.

The time taken to complete a cycle is the **period**.

Terms and concepts continued	

The **frequency** is the number of cycles completed in a second.

The **average value** of an alternating waveform has to be taken over half a cycle. The application of the average value is somewhat limited.

The **root-mean-square value** (r.m.s.) of an alternating waveform can be taken over half a cycle or over a full cycle. It is the one most generally used in electrical alternating systems.

A **phasor** is a line drawn to represent a sinusoidal alternating quantity. It is drawn to scale and its angle relative to the horizontal represents its phase shift in time.

Phasors can be added and subtracted so long as they represent like quantities.

Phasor diagrams can be used to represent r.m.s. quantities in which case they are frozen in time.

In practice electrical frequencies can vary from about 15 Hz to 300 GHz depending on the application.

Exercises 9

1. A coil is wound with 300 turns on a square former having sides 50 mm in length. Calculate the maximum value of the e.m.f. generated in the coil when it is rotated at 2000 r/min in a uniform magnetic field of density 0.8 T. What is the frequency of this e.m.f.?

2. Explain what is meant by the terms *waveform*, *frequency* and *average value*.

    A square coil of side 10 cm, having 100 turns, is rotated at 1200 r/min about an axis through the centre and parallel with two sides in a uniform magnetic field of density 0.4 T. Calculate: (a) the frequency; (b) the root-mean-square value of the induced e.m.f.; (c) the instantaneous value of the induced e.m.f. when the coil is at a position 40° after passing its maximum induced voltage.

3. A rectangular coil, measuring 30 cm by 20 cm and having 40 turns, is rotated about an axis coinciding with one of its longer sides at a speed of 1500 r/min in a uniform magnetic field of flux density 0.075 T. Find, from first principles, an expression for the instantaneous e.m.f. induced in the coil, if the flux is at right angles to the axis of rotation. Evaluate this e.m.f. at an instant 0.002 s after the plane of the coil has been perpendicular to the field.

4. The following ordinates were taken during a half-cycle of a symmetrical alternating-current wave, the current varying in a linear manner between successive points:

Phase angle, in degrees	0	15	30	45	60	75	90
Current, in amperes	0	3.6	8.4	14.0	19.4	22.5	25.0

Phase angle, in degrees	105	120	135	150	165	180
Current, in amperes	25.2	23.0	15.6	9.4	4.2	0

Determine: (a) the mean value; (b) the r.m.s. value; (c) the form factor.

5. Explain the significance of the root-mean-square value of an alternating current or voltage waveform. Define the form factor of such a waveform.

    Calculate from first principles the r.m.s. value and form factor of an alternating voltage having the following values over half a cycle, both half-cycles being symmetrical about the zero axis:

Time (ms)	0	1	2	3	4
Voltage (V)	0	100	100	100	0

These voltage values are joined by straight lines,

6. A triangular voltage wave has the following values over one half-cycle, both half-cycles being symmetrical about the zero axis:

Time (ms)	0	10	20	30	40	50	60	70	80	90	100
Voltage (V)	0	2	4	6	8	10	8	6	4	2	0

Exercises 9 continued

Plot a half-cycle of the waveform and hence determine: (a) the average value; (b) the r.m.s. value; (c) the form factor.

7. Describe, and explain the action of, an ammeter suitable for measuring the r.m.s. value of a current.

    An alternating current has a periodic time $2T$. The current for a time one-third of T is 50 A; for a time one-sixth of T, it is 20 A; and zero for a time equal to one-half of T. Calculate the r.m.s. and average values of this current.

8. A triangular voltage wave has a periodic time of $\frac{3}{100}$ s . For the first $\frac{2}{100}$ s of each cycle it increases uniformly at the rate of 1000 V/s, while for the last $\frac{1}{100}$ s it falls away uniformly to zero. Find, graphically or otherwise: (a) its average value; (b) its r.m.s. value; (c) its form factor.

9. Define the root-mean-square value of an alternating current. Explain why this value is more generally employed in a.c. measurements than either the average or the peak value. Under what circumstances would it be necessary to know (a) the average and (b) the peak value of an alternating current or voltage?

    Calculate the ratio of the peak values of two alternating currents which have the same r.m.s. values, when the waveform of one is sinusoidal and that of the other triangular. What effect would lack of symmetry of the triangular wave about its peak value have upon this ratio?

10. A voltage, $100 \sin 314t$ volts, is maintained across a circuit consisting of a half-wave rectifier in series with a 50 Ω resistor. The resistance of the rectifier may be assumed to be negligible in the forward direction and infinity in the reverse direction. Calculate the average and the r.m.s. values of the current.

11. State what is meant by the root-mean-square value of an alternating current and explain why the r.m.s. value is usually more important than either the maximum or the mean value of the current.

    A moving-coil ammeter and a moving-iron ammeter are connected in series with a rectifier across a 110 V (r.m.s.) a.c. supply. The total resistance of the circuit in the conducting direction is 60 Ω and that in the reverse direction may be taken as infinity. Assuming the waveform of the supply voltage to be sinusoidal, calculate from first principles the reading on each ammeter.

12. If the waveform of a voltage has a form factor of 1.15 and a peak factor of 1.5, and if the peak value is 4.5 kV, calculate the average and the r.m.s. values of the voltage.

13. An alternating current was measured by a d.c. milliammeter in conjunction with a full-wave rectifier. The reading on the milliammeter was 7.0 mA. Assuming the waveform of the alternating current to be sinusoidal, calculate: (a) the r.m.s. value; and (b) the maximum value of the alternating current.

14. An alternating current, when passed through a resistor immersed in water for 5 min, just raised the temperature of the water to boiling point. When a direct current of 4 A was passed through the same resistor under identical conditions, it took 8 min to boil the water. Find the r.m.s. value of the alternating current. Neglect factors other than heat given to the water. If a rectifier type of ammeter connected in series with the resistor read 5.2 A when the alternating current was flowing, find the form factor of the alternating current.

15. Explain what is meant by the *r.m.s. value* of an alternating current.

    In a certain circuit supplied from 50 Hz mains, the potential difference has a maximum value of 500 V and the current has a maximum value of 10 A. At the instant $t = 0$, the instantaneous values of the p.d. and the current are 400 V and 4 A respectively, both increasing positively. Assuming sinusoidal variation, state trigonometrical expressions for the instantaneous values of the p.d. and the current at time t. Calculate the instantaneous values at the instant $t = 0.015$ s and find the angle of phase difference between the p.d. and the current. Sketch the phasor diagram.

16. Explain with the aid of a sketch how the r.m.s. value of an alternating current is obtained.

    An alternating current i is represented by $i = 10 \sin 942t$ amperes. Determine: (a) the frequency; (b) the period; (c) the time taken from $t = 0$ for the current to reach a value of 6 A for a first and second time; (d) the energy dissipated when the current flows through a 20 Ω resistor for 30 min.

17. (a) Explain the term *r.m.s. value* as applied to an alternating current.

    (b) An alternating current flowing through a circuit has a maximum value of 70 A, and lags the applied voltage by 60°. The maximum value of the voltage is 100 V, and both current and voltage waveforms are sinusoidal. Plot the current and voltage waveforms in their correct relationship for the positive half of the voltage. What is the value of the current when the voltage is at a positive peak?

18. Two sinusoidal e.m.f.s of peak values 50 V and 20 V respectively but differing in phase by 30° are induced

in the same circuit. Draw the phasor diagram and find the peak and r.m.s. values of the resultant e.m.f.

19. Two impedances are connected in parallel to the supply, the first takes a current of 40 A at a lagging phase angle of 30°, and the second a current of 30 A at a leading phase angle of 45°. Draw a phasor diagram to scale to represent the supply voltage and these currents. From this diagram, by construction, determine the total current taken from the supply and its phase angle.

20. Two circuits connected in parallel take alternating currents which can be expressed trigonometrically as $i_1 = 13 \sin 314t$ amperes and $i_2 = 12 \sin(314t + \pi/4)$ amperes. Sketch the waveforms of these currents to illustrate maximum values and phase relationships.

    By means of a phasor diagram drawn to scale, determine the resultant of these currents, and express it in trigonometric form. Give also the r.m.s. value and the frequency of the resultant current.

21. The voltage drops across two components, when connected in series across an a.c. supply, are: $v_1 = 180 \sin 314t$ volts and $v_2 = 120 \sin(314t + \pi/3)$ volts respectively. Determine with the aid of a phasor diagram: (a) the voltage of the supply in trigonometric form; (b) the r.m.s. voltage of the supply; (c) the frequency of the supply.

22. Three e.m.f.s, $e_A = 50 \sin \omega t$, $e_B = 80 \sin(\omega t - \pi/6)$ and $e_C = 60 \cos \omega t$ volts, are induced in three coils connected in series so as to give the phasor sum of the three e.m.f.s. Calculate the maximum value of

the resultant e.m.f. and its phase relative to e.m.f. e_A. Check the results by means of a phasor diagram drawn to scale. If the connections to coil B were reversed, what would be the maximum value of the resultant e.m.f. and its phase relative to e_A?

23. Find graphically or otherwise the resultant of the following four voltages:
    $e_1 = 25 \sin \omega t$;
    $e_2 = 30 \sin(\omega t + \pi/6)$;
    $e_3 = 30 \cos \omega t$;
    $e_4 = 20 \sin(\omega t - \pi/4)$.
    Express the answer in a similar form.

24. Four e.m.f.s, $e_1 = 100 \sin \omega t$, $e_2 = 80 \sin(\omega t - \pi/6)$, $e_3 = 120 \sin(\omega t + \pi/4)$ and $e_4 = 100 \sin(\omega t - 2\pi/3)$, are induced in four coils connected in series so that the sum of the four e.m.f.s is obtained. Find graphically or by calculation the resultant e.m.f. and its phase difference with (a) e_1 and (b) e_2.

    If the connections to the coil in which the e.m.f. e_2 is induced are reversed, find the new resultant e.m.f.

25. The currents in three circuits connected in parallel to a voltage source are: (a) 4 A in phase with the applied voltage; (b) 6 A lagging the applied voltage by 30°; (c) 2 A leading the applied voltage by 45°. Represent these currents to scale on a phasor diagram, showing their correct relative phase displacement with each other. Determine, graphically or otherwise, the total current taken from the source, and its phase angle with respect to the supply voltage.

Chapter ten

Single-phase Series Circuits

Contents

Now that we are familiar with alternating currents and voltages, we can apply them in turn to each of the three basic components of a circuit, i.e. to a resistor, to an inductor and to a capacitor. We shall find that each responds in a completely different manner with the result that the current and voltage do not rise and fall at the same time unless the circuit only contains resistance. For inductors and capacitors, the relationship between voltage and current is termed the reactance and we find that in practice most circuits contain both resistance and reactance. We shall therefore look at circuits in which both resistance and reactance appear in series.

The observations that we make in this chapter form the basis of the manner in which we talk about a.c. systems. For this reason, its effects range far beyond the mere analysis of series a.c. circuits.

| **10.1** | **Basic a.c. circuits** |

In Chapter 9, we were introduced to a variety of waveforms which apply to alternating currents and voltages. In order to make our approach as simple as possible, we will limit the content of this chapter to circuits which contain a single generator producing a pure sinusoidal voltage. As previously noted, this is a reasonably good approximation to the electricity supply we meet at home. Such circuits are termed single-phase circuits.

| **10.2** | **Alternating current in a resistive circuit** |

Consider a circuit having a resistance R ohms connected across the terminals of an a.c. generator G, as in Fig. 10.1, and suppose the alternating voltage to be represented by the sine wave of Fig. 10.2. If the value of the voltage at any instant B is v volts, the value of the current at that instant is given by

$$i = \frac{v}{R} \text{ amperes}$$

When the voltage is zero, the current is also zero; and since the current is proportional to the voltage, the waveform of the current is exactly the same as that of the voltage. Also the two quantities are *in phase* with each other; that is, they pass through their zero values at the same instant and attain their maximum values in a given direction at the same instant. Hence the current wave is as shown in colour in Fig. 10.2.

If V_m and I_m are the maximum values of the voltage and current respectively, it follows that

$$I_m = \frac{V_m}{R} \qquad\qquad [10.1]$$

But the r.m.s. value of a sine wave is 0.707 times the maximum value, so that

$$\text{r.m.s. value of voltage} = V = 0.707 V_m$$

and   r.m.s. value of current $= I = 0.707 I_m$

Substituting for I_m and V_m in equation [10.1] we have

$$\frac{I}{0.707} = \frac{V}{0.707R}$$

$$I = \frac{V}{R} \qquad\qquad [10.2]$$

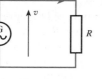

Fig. 10.1 Circuit with resistance only

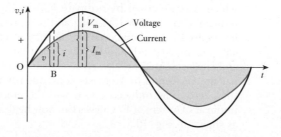

Fig. 10.2 Voltage and current waveforms for a resistive circuit

Hence Ohm's law can be applied without any modification to an a.c. circuit possessing resistance only.

If the instantaneous value of the applied voltage is represented by

$$v = V_m \sin \omega t$$

then instantaneous value of current in a resistive circuit is

$$i = \frac{V_m}{R} \sin \omega t \qquad\qquad [10.3]$$

Fig. 10.3 Phasor diagram for a resistive circuit

The phasors representing the voltage and current in a resistive circuit are shown in Fig. 10.3. The two phasors are actually coincident, but are drawn slightly apart so that the identity of each may be clearly recognized. As mentioned on p. 215, it is usual to draw the phasors in the position corresponding to $\omega t = 0$. Hence the phasors representing the voltage and current of expression [10.3] are drawn along the x-axis.

Finally let us briefly return to Fig. 10.1. The symbol used to represent the source generator was circular. Such a symbol indicates that the generator was a rotating machine but this only arises in power situations. In electronics situations, the a.c. source is static and therefore it is better to use the general symbol shown in Fig. 10.4, i.e. a square, which represents any form of source. The sinusoid indicates that it is an a.c. source and the 1 is optional. Normally it would be included in power situations and left out in electronics and communications applications.

10.3   Alternating current in an inductive circuit

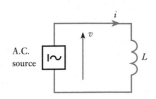

Fig. 10.4 Circuit with inductance only

Let us consider the effect of a sinusoidal current flowing through a coil having an inductance of L henrys and a negligible resistance, as in Fig. 10.4. For instance, let us consider what is happening during the first quarter-cycle of Fig. 10.5. This quarter-cycle has been divided into three equal intervals, OA, AC and CF seconds. During interval OA, the current increases from zero to AB; hence the average rate of change of current is AB/OA amperes per second, and is represented by ordinate JK drawn midway between O and A. From expression [8.2], the e.m.f., in volts, induced in a coil is

$$L \times \text{rate of change of current in amperes per second}$$

consequently, the average value of the induced e.m.f. during interval OA is $L \times$ AB/OA, namely $L \times$ JK volts, and is represented by ordinate JQ in Fig. 10.5.

Similarly, during interval AC, the current increases from AB to CE, so that the average rate of change of current is DE/AC amperes per second, which is represented by ordinate LM in Fig. 10.5; and the corresponding induced e.m.f. is $L \times$ LM volts and is represented by LR. During the third interval CF, the average rate of change of current is GH/CF, namely NP amperes per second; and the corresponding induced e.m.f. is $L \times$ NP volts and is represented by NS. At instant F, the current has ceased growing but has not yet begun to decrease; consequently the rate of change of current is then zero. The induced e.m.f. will therefore have decreased from a maximum at O to zero at F. Curves can now be drawn through the derived points, as shown in Fig. 10.5.

Fig. 10.5 Waveforms of
current, rate of change of current
and induced e.m.f.

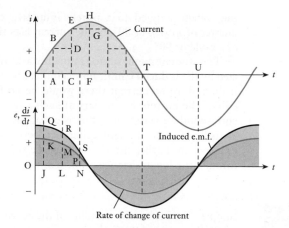

During the second quarter-cycle, the current decreases, so that the rate of change of current is negative and the induced e.m.f. becomes positive, tending to prevent the current decreasing. Since the sine wave of current is symmetrical about ordinate FH, the curves representing the rate of change of current and the e.m.f. induced in the coil will be symmetrical with those derived for the first quarter-cycle. Since the rate of change of current at any instant is proportional to the slope of the current wave at that instant, it is evident that the value of the induced e.m.f. increases from zero at F to a maximum at T and then decreases to zero at U in Fig. 10.5.

By using shorter intervals, for example by taking ordinates at intervals of 10° and noting the corresponding values of the ordinates with the aid of a calculator with trigonometric functions, it is possible to derive fairly accurately the shapes of the curves representing the rate of change of current and the induced e.m.f.

From Fig. 10.5 it will be seen that the induced e.m.f. attains its maximum positive value a quarter of a cycle before the current has done the same thing – in fact, it goes through all its variations a quarter of a cycle before the current has gone through similar variations. Hence the induced e.m.f. is said to lead the current by a quarter of a cycle or the current is said to lag the induced e.m.f. by a quarter of a cycle.

Since the resistance of the coil is assumed negligible, we can regard the whole of the applied voltage as being the induced e.m.f. Hence the curve of applied voltage in Fig. 10.6 can be drawn the same as that of the induced e.m.f.; and since the latter is sinusoidal, the wave of applied voltage must also be a sine curve.

From Fig. 10.6 it is seen that the applied voltage attains its maximum positive value a quarter of a cycle earlier than the current; in other words,

Fig. 10.6 Voltage and current
waveforms for a purely inductive
circuit

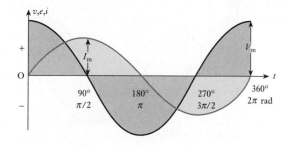

the voltage applied to a purely inductive circuit leads the current by a quarter of a cycle or 90°, or the current lags the applied voltage by a quarter of a cycle or 90°.

The student might quite reasonably ask: If the applied voltage is neutralized by the induced e.m.f., how can there be any current? The answer is that if there were no current there would be no flux, and therefore no induced e.m.f. The current has to vary at such a rate that the e.m.f. induced by the corresponding variation of flux is equal and opposite to the applied voltage. Actually there is a slight difference between the applied voltage and the induced e.m.f., this difference being the voltage required to send the current through the low resistance of the coil.

10.4	Current and voltage in an inductive circuit

Suppose the instantaneous value of the current through a coil having inductance L henrys and negligible resistance to be represented by

$$i = I_m \sin \omega t = I_m \sin 2\pi f t \qquad [10.4]$$

where t is the time, in seconds, after the current has passed through zero from negative to positive values, as shown in Fig. 10.7.

Suppose the current to increase by $\mathrm{d}i$ amperes in $\mathrm{d}t$ seconds, then instantaneous value of induced e.m.f. is

$$e = L \cdot \frac{\mathrm{d}i}{\mathrm{d}t}$$

$$= LI_m \frac{\mathrm{d}}{\mathrm{d}t}(\sin 2\pi f t)$$

$$= 2\pi f L I_m \cos 2\pi f t$$

$$e = 2\pi f L I_m \sin\left(2\pi f t + \frac{\pi}{2}\right) \qquad [10.5]$$

Since f represents the number of cycles per second, the duration of 1 cycle $= 1/f$ seconds. Consequently when

$$t = 0, \quad \cos 2\pi f t = 1$$

and　　　Induced e.m.f. $= 2\pi f L I_m$

When　　　　　　　　$t = 1/(2f), \quad \cos 2\pi f t = \cos \pi = -1$

and　　　Induced e.m.f. $= -2\pi f L I_m$

Fig. 10.7　Voltage and current waveforms for a purely inductive circuit

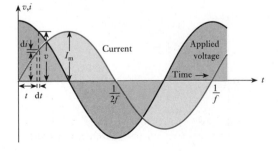

Hence the induced e.m.f. is represented by the curve in Fig. 10.7, leading the current by a quarter of a cycle.

Since the resistance of the circuit is assumed negligible, the whole of the applied voltage is equal to the induced e.m.f., therefore instantaneous value of applied voltage is

$$v = e$$

$$= 2\pi f L I_m \cos 2\pi f t$$

$$v = 2\pi f L I_m \sin(2\pi f t + \pi/2) \qquad\qquad [10.6]$$

Comparison of expressions [10.4] and [10.6] shows that the applied voltage leads the current by a quarter of a cycle. Also, from expression [10.6], it follows that the maximum value V_m of the applied voltage is $2\pi f L I_m$, i.e.

$$V_m = 2\pi f L I_m \quad \text{so that} \quad \frac{V_m}{I_m} = 2\pi f L$$

If I and V are the r.m.s. values, then

$$\frac{V}{I} = \frac{0.707 V_m}{0.707 I_m} = 2\pi f L$$

$$= \textit{inductive reactance}$$

Inductive reactance   Symbol: X_L Unit: ohm (Ω)

The inductive reactance is expressed in ohms and is represented by the symbol X_L. Hence

$$I = \frac{V}{2\pi f L} = \frac{V}{X_L} \qquad\qquad [10.7]$$

where $X_L = 2\pi f L$

The inductive reactance is proportional to the frequency and the current produced by a given voltage is inversely proportional to the frequency, as shown in Fig. 10.8.

The phasor diagram for a purely inductive circuit is given in Fig. 10.9, where E represents the r.m.s. value of the e.m.f. induced in the circuit, and V, equal to E, represents the r.m.s. value of the applied voltage.

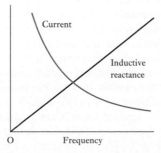

Fig. 10.8   Variation of reactance and current with frequency for a purely inductive circuit

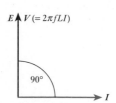

Fig. 10.9   Phasor diagram for a purely inductive circuit

10.5 Mechanical analogy of an inductive circuit

One of the most puzzling things to a student commencing the study of alternating currents is the behaviour of a current in an inductive circuit. For instance, why should the current in Fig. 10.6 be at its maximum value when there is no applied voltage? Why should there be no current when the applied voltage is at its maximum? Why should it be possible to have a voltage applied in one direction and a current flowing in the reverse direction, as is the case during the second and fourth quarter-cycles in Fig. 10.6?

It might therefore be found helpful to consider a simple mechanical analogy – the simpler the better. In mechanics, the *inertia* of a body opposes any change in the *speed* of that body. The effect of inertia is therefore analogous to that of *inductance* in opposing any change in the *current*.

Suppose we take a heavy metal cylinder C (Fig. 10.10), and roll it backwards and forwards on a horizontal surface between two extreme positions A and B. Let us consider the forces and the speed while C is being rolled from A to B. At first the speed is zero, but the force applied to the body is at its maximum, causing C to accelerate towards the right. This applied force is reduced – as indicated by the length of the arrows in Fig. 10.10 – until it is zero when C is midway between A and B; C ceases to accelerate and will therefore have attained its maximum speed from left to right.

Fig. 10.10 Mechanical analogy of a purely inductive circuit

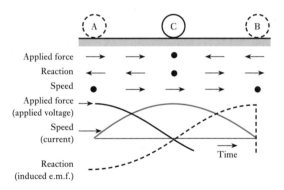

Immediately after C has passed the mid-point, the direction of the applied force is reversed and increased until the body is brought to rest at B and then begins its return movement.

The reaction of C, on the other hand, is equal and opposite to the applied force and corresponds to the e.m.f. induced in the inductive circuit.

From an inspection of the arrows in Fig. 10.10 it is seen that the speed in a given direction is a maximum of a quarter of a complete oscillation after the applied force has been a maximum in the same direction, but a quarter of an oscillation before the reaction reaches its maximum in that direction. This is analogous to the current in a purely inductive circuit lagging the applied voltage by a quarter of a cycle. Also it is evident that when the speed is a maximum the applied force is zero, and that when the applied force is a maximum the speed is zero; and that during the second half of the movement indicated in Fig. 10.10, the direction of motion is opposite to that of the applied force. These relationships correspond exactly to those found for a purely inductive circuit.

| **10.6** | **Resistance and inductance in series** |

Having considered the effects of resistance and inductance separately in a circuit, it is now necessary to consider their combined effects. This can be most simply achieved by connecting the resistance and inductance in series, as shown in Fig. 10.11(a).

Fig. 10.11
Resistance and inductance in series:
(a) circuit diagram;
(b) phasor diagram;
(c) instantaneous phasor diagram;
(d) wave diagram

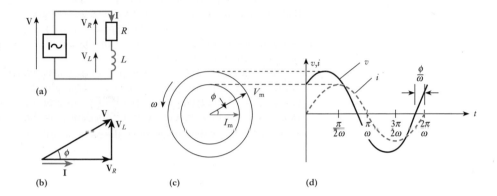

The phasor diagram results from an application of Kirchhoff's second law. For convenience, the current is taken as reference since it is common to all the elements of a series circuit. The circuit voltage may then be derived from the following relations:

$$\mathbf{V}_R = \mathbf{I}R, \text{ where } \mathbf{V}_R \text{ is in phase with } \mathbf{I}$$

$$\mathbf{V}_L = \mathbf{I}X_L, \text{ where } \mathbf{V}_L \text{ leads } \mathbf{I} \text{ by } 90°$$

$$\boxed{\mathbf{V} = \mathbf{V}_R + \mathbf{V}_L} \quad \text{(phasor sum)} \tag{10.8}$$

It will be remembered that bold symbols represent phasor quantities.

In the phasor diagram, shown in Fig. 10.11(b), the total voltage is thus obtained from relation [10.8], which is a complexor summation. The arithmetical sum of V_R and V_L is incorrect, giving too large a value for the total voltage V.

The angle of phase difference between **V** and **I** is termed the phase angle and is represented by ϕ. Also

$$V = (V_R^2 + V_L^2)^{\frac{1}{2}}$$
$$= (I^2R^2 + I^2X_L^2)^{\frac{1}{2}}$$
$$= I(R^2 + X_L^2)^{\frac{1}{2}}$$

Hence $\boxed{V = IZ}$ volts $\tag{10.9}$

where $Z = (R^2 + X_L^2)^{\frac{1}{2}}$

or $\boxed{Z = (R^2 + \omega^2 L^2)^{\frac{1}{2}} \text{ ohms}}$ $\tag{10.10}$

Here Z is termed the impedance of the circuit. Relation [10.9] will be seen to be a development of the relation $V = IR$ used in d.c. circuit analysis. However, for any given frequency, the impedance is constant and hence Ohm's law also applies to a.c. circuit analysis.

Impedance   Symbol: Z Unit: ohm (Ω)

The instantaneous phasor diagram, and the resulting wave diagram, show that the current lags the voltage by a phase angle greater than 0° but less than 90°. The phase angle between voltage and current is determined by the ratio of resistance to inductive reactance in the circuit. The greater the value of this ratio, the less will be the angle ϕ.

This statement can be developed by again considering the phasor diagram. Each side of the summation triangle has the same factor I. Consequently the triangle can be drawn to some other scale using only the values of resistance, reactance and impedance, as shown in Fig. 10.12. Such a triangle is termed an impedance triangle.

Fig. 10.12   Voltage and impedance triangles. (a) Voltage diagram; (b) impedance diagram

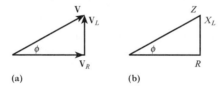

(a)   (b)

Just as in Fig. 10.11(b), the triangle is again right-angled. This compares with relation [10.10]. By the geometry of the diagram:

$$\phi = \tan^{-1}\frac{V_L}{V_R} = \tan^{-1}\frac{IX_L}{IR}$$

$$\phi = \tan^{-1}\frac{X_L}{R} \qquad\qquad [10.11]$$

To emphasize that the current lags the voltage, it is usual to give either the resulting angle as a negative value or else to use the word 'lag' after the angle. This is illustrated in Example 10.1.

The phase angle may also be derived as follows:

$$\phi = \cos^{-1}\frac{V_R}{V} = \cos^{-1}\frac{R}{Z} \qquad\qquad [10.12]$$

hence $\phi = \cos^{-1}\dfrac{R}{(R^2 + \omega^2 L^2)^{\frac{1}{2}}}$

Example 10.1

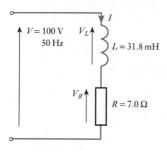

Fig. 10.13 Circuit diagram for Example 10.1

A resistance of $7.0\ \Omega$ is connected in series with a pure inductance of 31.8 mH and the circuit is connected to a 100 V, 50 Hz, sinusoidal supply (Fig. 10.13). Calculate:

(a) the circuit current;
(b) the phase angle.

$$X_L = 2\pi fL = 2\pi 50 \times 31.8 \times 10^{-3} = 10.0\ \Omega$$

$$Z = (R^2 + X_L^2)^{\frac{1}{2}} = (7.0^2 + 10.0^2)^{\frac{1}{2}} = 12.2\ \Omega$$

$$I = \frac{V}{Z} = \frac{100}{12.2} = 8.2\ \text{A}$$

$$\phi = \tan^{-1}\frac{X_L}{R} = \tan^{-1}\frac{10.0}{7.0} = 55.1°\ \text{lag or} -55.1°$$

Example 10.2

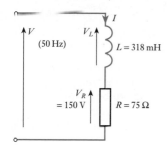

Fig. 10.14 Circuit diagram for Example 10.2

A pure inductance of 318 mH is connected in series with a pure resistance of $75\ \Omega$. The circuit is supplied from a 50 Hz sinusoidal source and the voltage across the $75\ \Omega$ resistor is found to be 150 V (Fig. 10.14). Calculate the supply voltage.

$$V_R = 150\ \text{V}$$

$$I = \frac{V}{R} = \frac{150}{75} = 2\ \text{A}$$

$$X_L = 2\pi fL = 2\pi 50 \times 318 \times 10^{-3} = 100\ \Omega$$

$$V_L = IX_L = 2 \times 100 = 200\ \text{V}$$

$$V = (V_R^2 + V_L^2)^{\frac{1}{2}} = (150^2 + 200^2)^{\frac{1}{2}} = 250\ \text{V}$$

Alternatively

$$Z = (R^2 + X_L^2)^{\frac{1}{2}} = (75^2 + 100^2)^{\frac{1}{2}} = 125\ \Omega$$

$$V = IZ = 2 \times 125 = 250\ \text{V}$$

Example 10.3

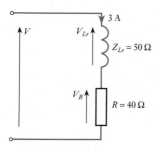

Fig. 10.15 Circuit diagram for Example 10.3

A coil, having both resistance and inductance, has a total effective impedance of $50\ \Omega$ and the phase angle of the current through it with respect to the voltage across it is 45° lag. The coil is connected in series with a $40\ \Omega$ resistor across a sinusoidal supply (Fig. 10.15). The circuit current is 3.0 A; by constructing a phasor diagram, estimate the supply voltage and the circuit phase angle.

$$V_R = IR = 3 \times 40 = 120\ \text{V}$$

$$V_{Lr} = IZ_{Lr} = 3 \times 50 = 150\ \text{V}$$

The use of subscript notation should be noted in the previous line. It would have been incorrect to write that $V_{Lr} = IZ$, since Z is used to represent the total circuit impedance. In more complex problems, numbers can be used, i.e. Z_1, Z_2, Z_3, etc. In this example, such a procedure would be tedious.

The phasor diagram (Fig. 10.16) is constructed by drawing the phasor $\mathbf{V}_R$ to some appropriate scale. The direction of this phasor will coincide with that of the current I. Since the voltage across the coil will lead the current by

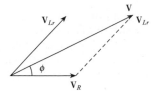

Fig. 10.16 Phasor diagram for Example 10.3

$45°$, phasor $\mathbf{V}_{Lr}$ can also be drawn. Complexor summation of the two voltages gives an estimate of the total voltage.

From the diagram

$$V = 250 \text{ V}$$

$$\phi = 25° \text{ lag}$$

We could have calculated the solution as follows:

$$V^2 = V_R^2 + V_{Lr}^2 + 2V_R V_{Lr} \cos \phi_{Lr}$$

$$= 120^2 + 150^2 + 2 \cdot 120 \cdot 150 \cdot 0.707$$

$$= 62\ 500$$

$$\therefore \qquad V = 250 \text{ V}$$

$$\cos \phi = \frac{V_R + V_{Lr} \cos \phi_{Lr}}{V} = \frac{120 + (150 \times 0.707)}{250}$$

$$= 0.904$$

$$\therefore \qquad \phi = 25° \text{ lag}$$

10.7	Alternating current in a capacitive circuit

Figure 10.17 shows a capacitor C connected in series with an ammeter A across the terminals of an a.c. source; and the alternating voltage applied to C is represented in Fig. 10.18. Suppose this voltage to be positive when it makes plate D positive relative to plate E.

If the capacitance is C farads, then from expression [5.18], the charging current i is given by

$$i = C \times \text{rate of change of p.d.}$$

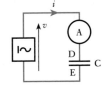

Fig. 10.17 Circuit with capacitance only

In Fig. 10.18, the p.d. is increasing positively at the maximum rate at instant zero; consequently the charging current is also at its maximum positive value at that instant. A quarter of a cycle later, the applied voltage has reached its maximum value V_m, and for a very brief interval of time the p.d. is neither increasing nor decreasing, so that there is no current. During the next quarter of a cycle, the applied voltage is decreasing. Consequently the capacitor discharges, the discharge current being in the negative direction.

When the voltage is passing through zero, the slope of the voltage curve is at its maximum, i.e. the p.d. is varying at the maximum rate; consequently the current is also a maximum at that instant.

Fig. 10.18 Voltage and current waveforms for a purely capacitive circuit

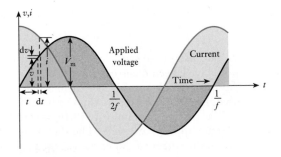

10.8 Current and voltage in a capacitive circuit

In this case we start with the voltage wave, whereas with inductance we started with the current wave. The reason for this is that in the case of inductance, we derive the induced e.m.f. by differentiating the current expression; whereas with capacitance, we derive the current by differentiating the voltage expression.

Suppose that the instantaneous value of the voltage applied to a capacitor having capacitance C farads is represented by

$$v = V_m \sin \omega t = V_m \sin 2\pi ft \qquad [10.13]$$

If the applied voltage increases by dv volts in dt seconds (Fig. 10.18) then, from equation [5.18], instantaneous value of current is

$$i = C\frac{dv}{dt}$$

$$= C\frac{d}{dt}(V_m \sin 2\pi ft)$$

$$= 2\pi fCV_m \cos 2\pi ft$$

$$i = 2\pi fCV_m \sin\left(2\pi ft + \frac{\pi}{2}\right) \qquad [10.14]$$

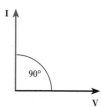

Fig. 10.19 Phasor diagram for a purely capacitive circuit

Comparison of expressions [10.13] and [10.14] shows that the current leads the applied voltage by a quarter of a cycle, and the current and voltage can be represented by phasors as in Fig. 10.19.

From expression [10.14] it follows that the maximum value I_m of the current is $2\pi fCV_m$,

$$\frac{V_m}{I_m} = \frac{1}{2\pi fC}$$

Hence, if I and V are the r.m.s. values

$$\frac{V}{I} = \frac{1}{2\pi fC} = capacitive\ reactance \qquad [10.15]$$

The capacitive reactance is expressed in ohms and is represented by the symbol X_C. Hence

$$I = 2\pi fCV = \frac{V}{X_C}$$

$$\therefore \qquad X_C = \frac{1}{2\pi fC} \qquad [10.16]$$

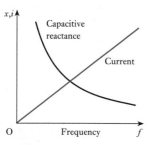

Fig. 10.20 Variation of reactance and current with frequency for a purely capacitive circuit

The capacitive reactance is inversely proportional to the frequency, and the current produced by a given voltage is proportional to the frequency, as shown in Fig. 10.20.

Capacitive reactance   Symbol: X_C Unit: ohm (Ω)

Example 10.4

A 30 μF capacitor is connected across a 400 V, 50 Hz supply. Calculate:

  (a) the reactance of the capacitor;

  (b) the current.

  (a) From expression [10.16]:

$$\text{reactance } X_C = \frac{1}{2 \times 3.14 \times 50 \times 30 \times 10^{-6}} = \mathbf{106.2 \ \Omega}$$

  (b) From expression [10.15]:

$$\text{Current} = \frac{400}{106.2} = \mathbf{3.77 \ A}$$

10.9   Analogies of a capacitance in an a.c. circuit

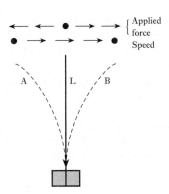

Fig. 10.21 Mechanical analogy of a capacitive circuit

If the piston P in Fig. 5.5 is moved backwards and forwards, the to-and-fro movement of the water causes the diaphragm to be distended in alternate directions. This hydraulic analogy, when applied to capacitance in an a.c. circuit, becomes rather complicated owing to the inertia of the water and of the piston, and as we do not want to take the effect of inertia into account at this stage, it is more convenient to consider a very light flexible strip L (Fig. 10.21), such as a metre rule, having one end rigidly clamped. Let us apply an alternating force comparatively slowly by hand so as to oscillate L between positions A and B.

When L is in position A, the applied force is at its maximum towards the *left*. As the force is reduced, L moves towards the *right*. Immediately L has passed the centre position, the applied force has to be increased towards the right, while the speed in this direction is decreasing. These variations are indicated by the lengths of the arrows in Fig. 10.21. From the latter it is seen that the speed towards the right is a maximum a quarter of a cycle before the applied force is a maximum in the same direction. The speed is therefore the analogue of the alternating current, and the applied force is that of the applied voltage. Hence capacitance in an electrical circuit is analogous to elasticity in mechanics, whereas inductance is analogous to inertia (section 10.5).

10.10   Resistance and capacitance in series

The effect of connecting resistance and capacitance in series is illustrated in Fig. 10.22. The current is again taken as reference.

    The circuit voltage is derived from the following relations:

$$\mathbf{V}_R = \mathbf{I}R, \text{ where } \mathbf{V}_R \text{ is in phase with } \mathbf{I}$$

$$\mathbf{V}_C = \mathbf{I}X_C, \text{ where } \mathbf{V}_C \text{ lags } \mathbf{I} \text{ by } 90°$$

$$\mathbf{V} = \mathbf{V}_R + \mathbf{V}_C \quad \text{(phasor sum)}$$

$$\text{Also} \quad V = (V_R^2 + V_C^2)^{\frac{1}{2}}$$

$$= (I^2R^2 + I^2X_C^2)^{\frac{1}{2}}$$

$$= I(R^2 + X_C^2)^{\frac{1}{2}}$$

$$\text{Hence} \quad V = IZ$$

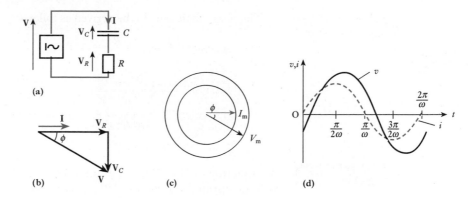

Fig. 10.22 Resistance and capacitance in series: (a) circuit diagram; (b) phasor diagram; (c) instantaneous phasor diagram; (d) wave diagram

where $Z = (R^2 + X_C^2)^{\frac{1}{2}}$ [10.17]

and $Z = \left(R^2 + \dfrac{1}{\omega^2 C^2} \right)^{\frac{1}{2}}$

Again Z is the impedance of the circuit. For any given frequency, the impedance remains constant and is thus the constant used in Ohm's law, i.e. the impedance is the ratio of the voltage across the circuit to the current flowing through it, other conditions remaining unchanged.

The instantaneous phasor diagram, and the resulting wave diagram, show that the current leads the applied voltage by a phase angle greater than 0° but less than 90°. The phase angle between voltage and current is determined by the ratio of resistance to capacitive reactance in the circuit. The greater the value of this ratio, the less will be the angle ϕ. This can be illustrated by drawing the impedance triangle for the circuit, as shown in Fig. 10.23.

Fig. 10.23 Voltage and impedance diagrams. (a) Voltage diagram; (b) impedance diagram

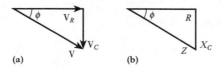

By the geometry of the diagram:

$$\phi = \tan^{-1} \frac{V_C}{V_R} = \tan^{-1} \frac{IX_C}{IR}$$

$$\phi = \tan^{-1} \frac{X_C}{R}$$ [10.18]

To emphasize that the current leads the voltage, it is usual either to give the resulting angle as a positive value or else to use the word 'lead' after the angle. This is illustrated in Example 10.5.

The phase angle can also be derived as follows:

$$\phi = \cos^{-1}\frac{V_R}{V} = \cos^{-1}\frac{R}{Z}$$

$$\phi = \cos^{-1}\frac{R}{\left(R^2 + \dfrac{1}{\omega^2 C^2}\right)^{\frac{1}{2}}} \qquad [10.19]$$

Example 10.5

A capacitor of 8.0 μF takes a current of 1.0 A when the alternating voltage applied across it is 230 V. Calculate:

(a) the frequency of the applied voltage;
(b) the resistance to be connected in series with the capacitor to reduce the current in the circuit to 0.5 A at the same frequency;
(c) the phase angle of the resulting circuit.

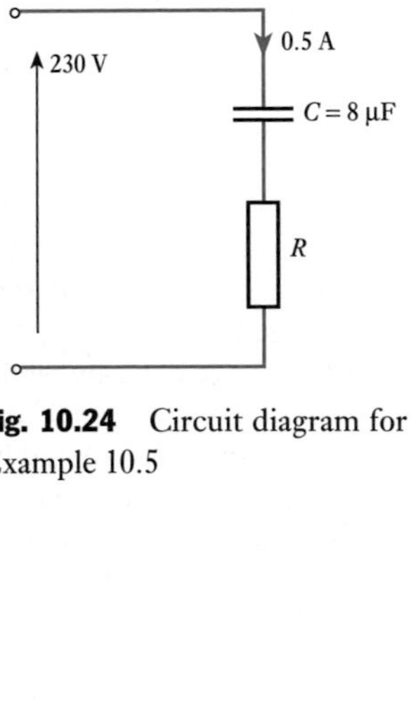

(a) $X_C = \dfrac{V}{I} = \dfrac{230}{1.0} = 230\ \Omega$

$\qquad\qquad = \dfrac{1}{2\pi f C}$

$\therefore\qquad f = \dfrac{1}{2\pi C X_C} = \dfrac{1}{2\pi \times 8 \times 10^{-6} \times 230} = 86.5\ \text{Hz}$

(b) When a resistance is connected in series with the capacitor, the circuit is now as given in Fig. 10.24.

$$Z = \frac{V}{I} = \frac{230}{0.5} = 460\ \Omega$$

$$= (R^2 + X_C^2)^{\frac{1}{2}}$$

but $X_C = 230\ \Omega$

hence $R = 398\ \Omega$

(c) $\phi = \cos^{-1}\dfrac{R}{Z} = \cos^{-1}\dfrac{398}{460} = +30°\ \text{or}\ 30°\ \textbf{lead}$

Fig. 10.24 Circuit diagram for Example 10.5

Circuit labels: 230 V, 0.5 A, $C = 8\ \mu$F, R

10.11   **Alternating current in an RLC circuit**

We have already considered resistive, inductive and capacitive circuits separately. However, we know that a practical inductor possesses inductance and resistance effectively in series. It follows that our analysis of R and L in series is equivalent to the analysis of a circuit including a practical inductor.

We can now consider the general case of R, L and C in series. This combines the instances of R and L in series with that of R and C in series. However, by producing the general case, we can adapt the results to the other two cases by merely omitting the capacitive or the inductive reactance from the expressions derived for the general case.

Before we start the general analysis, let us remind ourselves about the drawing of the phasor diagrams. Sometimes it is hard to know where to start, but

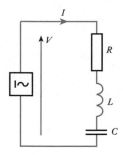

Fig. 10.25 Circuit with R, L and C in series

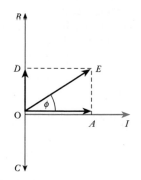

Fig. 10.26 Phasor diagram for Fig. 10.25

the rule is simple – start with the quantity that is common to the components of the circuit. We are dealing with a series circuit, therefore the current is the common quantity so that in Fig. 10.26 the current phasor is the first to be drawn. Later we will come to parallel circuits in which case the voltage is the common quantity, hence the voltage phasor is the first to be drawn.

Figure 10.25 shows a circuit having resistance R ohms, inductance L henrys and capacitance C farads in series, connected across an a.c. supply of V volts (r.m.s.) at a frequency of f hertz. Let I be the r.m.s. value of the current in amperes.

From section 10.2, the p.d. across R is RI volts in phase with the current and is represented by phasor OA in phase with OI in Fig. 10.26. From section 10.4, the p.d. across L is $2\pi fLI$, and is represented by phasor OB, leading the current by 90°; and from section 10.7, the p.d. across C is $I/(2\pi fC)$ and is represented by phasor OC lagging the current by 90°.

Since OB and OC are in direct opposition, their resultant is OD = OB − OC, OB being assumed greater than OC in Fig. 10.26; and the supply voltage is the phasor sum of OA and OD, namely OE. From Fig. 10.26,

$$OE^2 = OA^2 + OD^2 = OA^2 + (OB - OC)^2$$

$$\therefore \quad V^2 = (RI)^2 + \left(2\pi fLI - \frac{I}{2\pi fC}\right)^2$$

so that

$$I = \frac{V}{\sqrt{\left\{R^2 + \left(2\pi fL - \frac{1}{2\pi fC}\right)^2\right\}}} = \frac{V}{Z} \qquad [10.20]$$

where $Z = impedance$ of circuit in ohms

$$Z = \frac{V}{I} = \sqrt{\left\{R^2 + \left(2\pi fL - \frac{1}{2\pi fC}\right)^2\right\}} \qquad [10.21]$$

From this expression it is seen that

$$\text{Resultant reactance} = 2\pi fL - \frac{1}{2\pi fC}$$

$$= \text{inductive reactance} - \text{capacitive reactance}$$

If ϕ is the phase difference between the current and the supply voltage

$$\tan \phi = \frac{AE}{OA} = \frac{OD}{OA} = \frac{OB - OC}{OA} = \frac{2\pi fLI - I/(2\pi fC)}{RI}$$

$$= \frac{\text{inductive reactance} - \text{capacitive reactance}}{\text{resistance}}$$

$$\therefore \quad \tan \phi = \frac{X_L - X_C}{R} \qquad [10.22]$$

$$\cos \phi = \frac{OA}{OE} = \frac{RI}{ZI} = \frac{\text{resistance}}{\text{impedance}}$$

∴ $$\cos \phi = \frac{R}{Z}$$ [10.23]

and

$$\sin \phi = \frac{AE}{OE} = \frac{\text{resultant reactance}}{\text{impedance}}$$

∴ $$\sin \phi = \frac{X}{Z}$$ [10.24]

If the inductive reactance is greater than the capacitive reactance, tan ϕ is positive and the current lags the supply voltage by an angle ϕ; if less, tan ϕ is negative, signifying that the current leads the supply voltage by an angle ϕ. Note the case where $X_L = X_C$, and $I = V/R$. The current is in phase with the voltage. This condition is termed series resonance and is discussed in Chapter 14.

Example 10.6

A coil having a resistance of 12 Ω and an inductance of 0.1 H is connected across a 100 V, 50 Hz supply. Calculate:

(a) the reactance and the impedance of the coil;
(b) the current;
(c) the phase difference between the current and the applied voltage.

When solving problems of this kind, students should first of all draw a circuit diagram (Fig. 10.27) and insert all the known quantities. They should then proceed with the phasor diagram (Fig. 10.28). It is not essential to draw the phasor diagram to exact scale, but it is helpful to draw it approximately correctly since it is then easy to make a rough check of the calculated values.

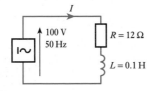

Fig. 10.27 Circuit diagram for Example 10.6

(a)   Reactance $= X_L = 2\pi f L$

$$= 2\pi \times 50 \times 0.1 = \mathbf{31.4 \ \Omega}$$

Impedance $= Z = \sqrt{(R^2 + X_L^2)}$

$$= \sqrt{(12^2 + 31.4^2)} = \mathbf{33.6 \ \Omega}$$

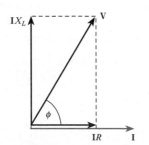

Fig. 10.28 Phasor diagram for Example 10.6

(b)   Current $= I = \dfrac{V}{Z} = \dfrac{100}{33.6} = \mathbf{2.97 \ A}$

(c) $\tan \phi = \dfrac{X}{R} = \dfrac{31.4}{12} = 2.617$

∴ $\phi = \mathbf{69°}$

Example 10.7

A metal-filament lamp, rated at 750 W, 100 V, is to be connected in series with a capacitor across a 230 V, 60 Hz supply. Calculate:

(a) the capacitance required;
(b) the phase angle between the current and the supply voltage.

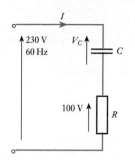

Fig. 10.29 Circuit diagram for Example 10.7

(a) The circuit is given in Fig. 10.29, where R represents the lamp. In the phasor diagram of Fig. 10.30, the voltage V_R across R is in phase with the current I, while the voltage V_C across C lags I by 90°. The resultant voltage V is the phasor sum of V_R and V_C, and from the diagram:

$$V^2 = V_R^2 + V_C^2$$

$$\therefore \qquad (230)^2 = (100)^2 + V_C^2$$

$$\therefore \qquad V_C = 270 \text{ V}$$

$$\text{Rated current of lamp} = \frac{750 \text{ W}}{100 \text{ V}} = 7.5 \text{ A}$$

From equation [10.15]

$$7.5 = 2 \times 3.14 \times 60 \times C \times 207$$

$$C = 96 \times 10^{-6} \text{ F} = \textbf{96 } \mu\textbf{F}$$

(b) If ϕ is the phase angle between the current and the supply voltage

$$\cos \phi = \frac{V_R}{V} \quad \text{(from Fig. 10.30)}$$

$$= \frac{100}{230} = 0.435$$

$$\phi = \textbf{64°12}'$$

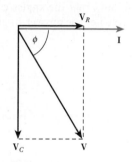

Fig. 10.30 Phasor diagram for Example 10.7

Example 10.8

A circuit having a resistance of 12 Ω, an inductance of 0.15 H and a capacitance of 100 μF in series, is connected across a 100 V, 50 Hz supply. Calculate:

(a) the impedance;
(b) the current;
(c) the voltages across R, L and C;
(d) the phase difference between the current and the supply voltage.

The circuit diagram is the same as that of Fig. 10.25.

(a) From equation [10.21],

$$Z = \sqrt{\left\{(12)^2 + \left(2 \times 3.14 \times 50 \times 0.15 - \frac{10^6}{2 \times 3.14 \times 50 \times 100}\right)^2\right\}}$$

$$= \sqrt{\{144 + (47.1 - 31.85)^2\}} = \textbf{19.4 } \Omega$$

(b) Current $= \dfrac{V}{Z} = \dfrac{100}{19.4} = \textbf{5.15 A}$

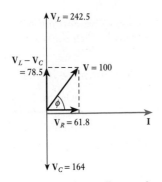

Fig. 10.31 Phasor diagram for Example 10.8

(c)   Voltage across $R = V_R = 12 \times 5.15 = 61.8$ V

Voltage across $L = V_L = 47.1 \times 5.15 = 242.5$ V

and   Voltage across $C = V_C = 31.85 \times 5.15 = 164.0$ V

These voltages and current are represented by the respective phasors in Fig. 10.31.

(d) Phase difference between current and supply voltage is

$$\phi = \cos^{-1}\frac{V_R}{V} = \cos^{-1}\frac{61.8}{100} = 51°50'$$

Or, alternatively, from equation [10.22],

$$\phi = \tan^{-1}\frac{47.1 - 31.85}{12} = \tan^{-1}1.271 = 51°48'$$

Note: the determined values for ϕ are slightly different and this is because we have inferred too great an accuracy to the angles. Given that the input information is only accurate to two decimal places, it follows that the angles can only be given to about one decimal point of a degree, i.e. the answer might better be given as 51.8°.

Summary of important formulae

For a purely resistive circuit

$$V = IR \tag{10.2}$$

For a purely inductive circuit

$$V = IX_L = 2\pi fLI = \omega LI \tag{10.7}$$

$$v = 2\pi fLI_m \sin(2\pi ft + \pi/2) \tag{10.6}$$

For a purely capacitive circuit

$$V = IX_C = I/2\pi fC = I/\omega C \tag{10.16}$$

$$i = 2\pi fCV_m \sin(2\pi ft + \pi/2) \tag{10.14}$$

For R and L in series

$$V = IZ \tag{10.9}$$

$$Z = (R^2 + \omega^2 L^2)^{\frac{1}{2}} = (R^2 + X_L^2)^{\frac{1}{2}} \tag{10.10}$$

For R and C in series

$$Z = \left(R^2 + \frac{1}{\omega^2 C^2}\right)^{\frac{1}{2}} = (R^2 + X_C^2)^{\frac{1}{2}} \tag{10.17}$$

For R, L and C in series

$$Z = \left\{R^2 + \left(\omega L - \frac{1}{\omega C}\right)^2\right\}^{\frac{1}{2}} = \{R^2 + (X_L - X_C)^2\}^{\frac{1}{2}} \tag{10.21}$$

Terms and concepts

If a circuit is purely resistive, the current is in phase with the voltage. If it is purely inductive, the current lags the voltage by 90°. If the circuit is purely capacitive, the current leads the voltage by 90°.

If a circuit contains both resistance and inductance, the current lags the voltage by an angle less than 90° but the angle is greater than 0°.

If a circuit contains both resistance and capacitance, the current leads the voltage by an angle less than 90° but the angle is greater than 0°.

If a circuit contains resistance, inductance and capacitance, the current may lead, lag or be in phase with the voltage depending on the relative values of the inductive and capacitive reactances.

The **reactance** of an inductor rises with frequency.

The **reactance** of a capacitor falls with frequency.

Exercises 10

1. A closed–circuit, 500-turn coil, of resistance 100 Ω and negligible inductance, is wound on a square frame of 40 cm side. The frame is pivoted at the mid-points of two opposite sides and is rotated at 250 r/min in a uniform magnetic field of 60 mT. The field direction is at right angles to the axis of rotation. For the instant that the e.m.f. is maximum: (a) draw a diagram of the coil, and indicate the direction of rotation, and of the current flow, the magnetic flux, the e.m.f. and the force exerted by the magnetic field on the conductors; (b) calculate the e.m.f., the current and the torque, and hence verify that the mechanical power supply balanced the electric power produced.

2. An alternating p.d. of 100 V (r.m.s.), at 50 Hz, is maintained across a 20 Ω non-reactive resistor. Plot to scale the waveforms of p.d. and current over one cycle. Deduce the curve of power and state its mean value.

3. An inductor having a reactance of 10 Ω and negligible resistance is connected to a 100 V (r.m.s.) supply. Draw to scale, for one half-cycle, curves of voltage and current, and deduce and plot the power curve. What is the mean power over the half-cycle?

4. A coil having an inductance of 0.2 H and negligible resistance is connected across a 100 V a.c. supply. Calculate the current when the frequency is: (a) 30 Hz; and (b) 500 Hz.

5. A coil of inductance 0.1 H and negligible resistance is connected in series with a 25 Ω resistor. The circuit is energized from a 230 V, 50 Hz source. Calculate: (a) the current in the circuit; (b) the p.d. across the coil; (c) the p.d. across the resistor; (d) the phase angle of the circuit. Draw to scale a phasor diagram representing the current and the component voltages.

6. A coil connected to a 230 V, 50 Hz sinusoidal supply takes a current of 10 A at a phase angle of 30°. Calculate the resistance and inductance of, and the power taken by, the coil. Draw, for one half-cycle, curves of voltage and current, and deduce and plot the power curve. Comment on the power curve.

7. A 15 Ω non-reactive resistor is connected in series with a coil of inductance 0.08 H and negligible resistance. The combined circuit is connected to a 240 V, 50 Hz supply. Calculate: (a) the reactance of the coil; (b) the impedance of the circuit; (c) the current in the circuit; (d) the power factor of the circuit; (e) the active power absorbed by the circuit.

8. The potential difference measured across a coil is 20 V when a direct current of 2 A is passed through it. With an alternating current of 2 A at 40 Hz, the p.d. across the coil is 140 V. If the coil is connected to a 230 V, 50 Hz supply, calculate: (a) the current; (b) the active power; (c) the power factor.

9. A non-inductive load takes a current of 15 A at 125 V. An inductor is then connected in series in order that the same current shall be supplied from 240 V, 50 Hz mains. Ignore the resistance of the inductor and calculate: (a) the inductance of the inductor; (b) the impedance of the circuit; (c) the phase difference between the current and the applied voltage. Assume the waveform to be sinusoidal.

10. A series a.c. circuit, ABCD, consists of a resistor AB, an inductor BC, of resistance R and inductance L, and a resistor CD. When a current of 6.5 A flows through the circuit, the voltage drops across various points are: $V_{AB} = 65$ V; $V_{BC} = 124$ V; $V_{AC} = 149$ V. The supply voltage is 220 V at 50 Hz. Draw a phasor diagram

to scale showing all the resistive and reactive volt drops and, from the diagram, determine: (a) the volt drop V_{BD} and the phase angle between it and the current; (b) the resistance and inductance of the inductor.

11. A coil of 0.5 H inductance and negligible resistance and a 200 Ω resistor are connected in series to a 50 Hz supply. Calculate the circuit impedance.

 An inductor in a radio receiver has to have a reactance of 11 kΩ at a frequency of 1.5 MHz. Calculate the inductance (in millihenrys).

12. A coil takes a current of 10.0 A and dissipates 1410 W when connected to a 230 V, 50 Hz sinusoidal supply. When another coil is connected in parallel with it, the total current taken from the supply is 20.0 A at a power factor of 0.866. Determine the current and the overall power factor when the coils are connected in series across the same supply.

13. When a steel-cored reactor and a non-reactive resistor are connected in series to a 150 V a.c. supply, a current of 3.75 A flows in the circuit. The potential differences across the reactor and across the resistor are then observed to be 120 V and 60 V respectively. If the d.c. resistance of the reactor is 4.5 Ω, determine the core loss in the reactor and calculate its equivalent series resistance.

14. A single-phase network consists of three parallel branches, the currents in the respective branches being represented by: $i_1 = 20 \sin 314t$ amperes; $i_2 = 30 \sin(314t - \pi/4)$ amperes; and $i_3 = 18 \sin(314t + \pi/2)$ amperes. (a) Using a scale of 1 cm = 5 A, draw a phasor diagram and find the total maximum value of current taken from the supply and the overall phase angle. (b) Express the total current in a form similar to that of the branch currents. (c) If the supply voltage is represented by $200 \sin 314t$ volts, find the impedance, resistance and reactance of the network.

15. A non-inductive resistor is connected in series with a coil across a 230 V, 50 Hz supply. The current is 1.8 A and the potential differences across the resistor and the coil are 80 V and 170 V respectively. Calculate the inductance and the resistance of the coil, and the phase difference between the current and the supply voltage. Also draw the phasor diagram representing the current and the voltages.

16. An inductive circuit, in parallel with a non-inductive resistor of 20 Ω, is connected across a 50 Hz supply. The currents through the inductive circuit and the non-inductive resistor are 4.3 A and 2.7 A respectively, and the current taken from the supply is 5.8 A. Find: (a) the active power absorbed by the inductive branch; (b) its inductance; and (c) the power factor of the combined network. Sketch the phasor diagram.

17. A coil having a resistance of 15 Ω and an inductance of 0.2 H is connected in series with another coil having a resistance of 25 Ω and an inductance of 0.04 H to a 230 V, 50 Hz supply. Draw to scale the complete phasor diagram for the circuit and determine: (a) the voltage across each coil; (b) the active power dissipated in each coil; (c) the power factor of the circuit as a whole.

18. Two identical coils, each of 25 Ω resistance, are mounted coaxially a short distance apart. When one coil is supplied at 100 V, 50 Hz, the current taken is 2.1 A and the e.m.f. induced in the other coil on open-circuit is 54 V. Calculate the self-inductance of each coil and the mutual inductance between them. What current will be taken if a p.d. of 100 V, 50 Hz is supplied across the two coils in series?

19. Two similar coils have a coupling coefficient of 0.6. Each coil has a resistance of 8 Ω and a self-inductance of 2 mH. Calculate the current and the power factor of the circuit when the coils are connected in series (a) cumulatively and (b) differentially, across a 10 V, 5 kHz supply.

20. A two-wire cable, 8 km long, has a capacitance of 0.3 μF/km. If the cable is connected to a 11 kV, 60 Hz supply, calculate the value of the charging current. The resistance and inductance of the conductors may be neglected.

21. Draw, to scale, phasors representing the following voltages, taking e_1 as the reference phasor:
 $e_1 = 80 \sin \omega t$ volts; $e_2 = 60 \cos \omega t$ volts;
 $e_3 = 100 \sin(\omega t - \pi/3)$ volts.
 By phasor addition, find the sum of these three voltages and express it in the form of $E_m \sin(\omega t \pm \phi)$.

 When this resultant voltage is applied to a circuit consisting of a 10 Ω resistor and a capacitor of 17.3 Ω reactance connected in series find an expression for the instantaneous value of the current flowing, expressed in the same form.

Chapter eleven   Single-phase Parallel Networks

Contents

As sure as night follows day, we can be sure that if we have series-connected impedances then there will be parallel-connected impedances. We have discovered how to determine the current in an impedance so, if two are connected in parallel, then we can find the current in each branch and add the two currents – it is as simple as that!

Well, maybe it is not just that simple but we can make things easier still by introducing polar notation. This allows us to operate with impedances in much the same manner as we did with resistors except that we have to allow for the phase differences.

11.1	Basic a.c. parallel circuits

It is most common to think of circuits and networks being supplied from a voltage source. A circuit comprises a single load being supplied from the voltage source. However, there is no reason not to supply a second load from the same supply, in which case the loads are in parallel.

However, while it is relatively simple to consider parallel circuits supplied from a d.c. source, we have to allow for the phase difference between the currents in the parallel branches. It is therefore necessary to develop our analysis of a.c. parallel circuits.

11.2	Simple parallel circuits

There are two arrangements of simple parallel circuits which require analysis; these are resistance in parallel with inductance and resistance in parallel with capacitance.

When analysing a parallel circuit, it should be remembered that it consists of two or more series circuits connected in parallel. Therefore each branch of the circuit can be analysed separately as a series circuit and then the effect of the separate branches can be combined by applying Kirchhoff's first law, i.e. the currents of the branches can be added complexorially: that is, by phasor diagram.

The circuit for resistance and inductance in parallel is shown in Fig. 11.1(a). In the resistive branch, the current is given by

$$I_R = \frac{V}{R}, \text{ where } I_R \text{ and } V \text{ are in phase}$$

Fig. 11.1 Resistance and inductance in parallel. (a) Circuit diagram; (b) phasor diagram

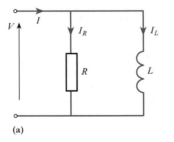
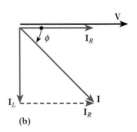

(a)   (b)

In the inductive branch, the current is given by

$$I_L = \frac{V}{X_L}, \text{ where } \mathbf{I}_L \text{ lags } \mathbf{V} \text{ by } 90°$$

The resulting phasor diagram is shown in Fig. 11.1(b). The voltage which is common to both branches is taken as reference. Since parallel circuits are more common, this is one reason that it is usual to take the voltage as reference in circuit analysis. The total supply current I is obtained by adding the branch currents complexorially, i.e.

$$\mathbf{I} = \mathbf{I}_R + \mathbf{I}_L \quad \text{(phasor sum)} \tag{11.1}$$

From the complexor diagram:

$$I = (I_R^2 + I_L^2)^{\frac{1}{2}}$$

$$= \left\{ \left(\frac{V}{R} \right)^2 + \left(\frac{V}{X_L} \right)^2 \right\}^{\frac{1}{2}}$$

$$= V \left(\frac{1}{R^2} + \frac{1}{X_L^2} \right)^{\frac{1}{2}}$$

$$\frac{V}{I} = Z = \frac{1}{\left(\dfrac{1}{R^2} + \dfrac{1}{X_L^2} \right)^{\frac{1}{2}}} \qquad\qquad [11.2]$$

It can be seen from the phasor diagram that the phase angle ϕ is a lagging angle.

$$\phi = \tan^{-1} \frac{I_L}{I_R} = \tan^{-1} \frac{R}{X_L} = \tan^{-1} \frac{R}{\omega L} \qquad\qquad [11.3]$$

Also $\qquad \phi = \cos^{-1} \dfrac{I_R}{I}$

$$\therefore \qquad \phi = \cos^{-1} \frac{Z}{R} \qquad\qquad [11.4]$$

In the case of resistance and capacitance connected in parallel, as shown in Fig. 11.2(a), the current in the resistive branch is again given by

$$I_R = \frac{V}{R}, \text{ where } \mathbf{I}_R \text{ and } \mathbf{V} \text{ are in phase}$$

In the capacitive branch, the current is given by

$$I_C = \frac{V}{X_C}, \text{ where } \mathbf{I}_C \text{ leads } \mathbf{V} \text{ by } 90°$$

The phasor diagram is constructed in the usual manner based on the relation

$$\mathbf{I} = \mathbf{I}_R + \mathbf{I}_C \quad \text{(phasor sum)} \qquad\qquad [11.5]$$

Fig. 11.2 Resistance and capacitance in parallel. (a) Circuit diagram; (b) phasor diagram

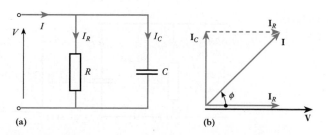

(a)   (b)

From the phasor diagram:

$$I = (I_R^2 + I_C^2)^{\frac{1}{2}}$$

$$= \left\{ \left(\frac{V}{R}\right)^2 + \left(\frac{V}{X_C}\right)^2 \right\}^{\frac{1}{2}}$$

$$= V\left(\frac{1}{R^2} + \frac{1}{X_C^2}\right)^{\frac{1}{2}}$$

$$\frac{V}{I} = Z = \frac{1}{\left(\dfrac{1}{R^2} + \dfrac{1}{X_C^2}\right)^{\frac{1}{2}}} \tag{11.6}$$

It can be seen from the phasor diagram that the phase angle ϕ is a leading angle. It follows that parallel circuits behave in a similar fashion to series circuits in that the combination of resistance with inductance produces a lagging circuit while the combination of resistance with capacitance gives rise to a leading circuit.

$$\phi = \tan^{-1}\frac{I_C}{I_R} = \tan^{-1}\frac{R}{X_C} = \tan^{-1} R\omega C \tag{11.7}$$

Also $$\phi = \cos^{-1}\frac{I_R}{I}$$

$\therefore$ $$\phi = \cos^{-1}\frac{Z}{R} \tag{11.8}$$

Example 11.1   A circuit consists of a 115 Ω resistor in parallel with a 41.5 μF capacitor and is connected to a 230 V, 50 Hz supply (Fig. 11.3). Calculate:

(a) the branch currents and the supply current;
(b) the circuit phase angle;
(c) the circuit impedance.

Fig. 11.3   Circuit and phasor diagrams for Example 11.1

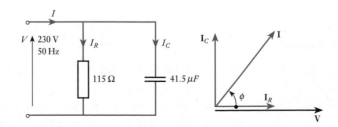

$$I_R = \frac{V}{R} = \frac{230}{115} = \mathbf{2.0\ A}$$

$$X_C = \frac{1}{2\pi f C} = \frac{1}{2\pi 50 \times 41.5 \times 10^{-6}} = \mathbf{76.7\ \Omega}$$

$$I_C = \frac{V}{X_C} = \frac{230}{76.7} = \mathbf{3.0\ A}$$

$$I = (I_R^2 + I_C^2)^{\frac{1}{2}} = (2.0^2 + 3.0^2)^{\frac{1}{2}} = \mathbf{3.6\ A}$$

$$\phi = \cos^{-1}\frac{I_R}{I} = \cos^{-1}\frac{2.0}{3.6} = \mathbf{56.3° \ lead}$$

$$Z = \frac{V}{I} = \frac{230}{3.6} = \mathbf{63.9\ \Omega}$$

Example 11.2   Three branches, possessing a resistance of 50 Ω, an inductance of 0.15 H and a capacitance of 100 μF respectively, are connected in parallel across a 100 V, 50 Hz supply. Calculate:

(a) the current in each branch;
(b) the supply current;
(c) the phase angle between the supply current and the supply voltage.

Fig. 11.4   Circuit diagram for Example 11.2

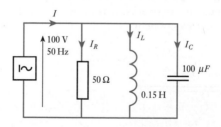

(a) The circuit diagram is given in Fig. 11.4, where I_R, I_L and I_C represent the currents through the resistance, inductance and capacitance respectively.

$$I_R = \frac{100}{50} = \mathbf{2.0\ A}$$

$$I_L = \frac{100}{2 \times 3.14 \times 50 \times 0.15} = \mathbf{2.12\ A}$$

and $I_C = 2 \times 3.14 \times 50 \times 100 \times 10^{-6} \times 100 = \mathbf{3.14\ A}$

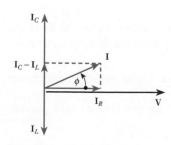

Fig. 11.5   Phasor diagram for Example 11.2

In the case of parallel branches, the first phasor (Fig. 11.5) to be drawn is that representing the quantity that is common to those circuits, namely the voltage. Then $\mathbf{I}_R$ is drawn in phase with $\mathbf{V}$, $\mathbf{I}_L$ lagging 90° and $\mathbf{I}_C$ leading 90°.

(b) The capacitor and inductor branch currents are in antiphase, hence the resultant of I_C and I_L is

$$I_C - I_L = 3.14 - 2.12$$
$$= 1.02 \text{ A, leading by } 90°$$

The current I taken from the supply is the resultant of I_R and $(I_C - I_L)$, and from Fig. 11.5:

$$I^2 = I_R^2 + (I_C - I_L)^2 = 2^2 + (1.015)^2 = 5.03$$

∴ $I = 2.24 \text{ A}$

(c) From Fig. 11.5:

$$\cos \phi = \frac{I_R}{I} = \frac{2}{2.24} = 0.893$$

$$\phi = 26°45'$$

Since I_C is greater than I_L, the supply current leads the supply voltage by **26°45'**.

11.3 Parallel impedance circuits

The analysis of impedances in parallel is similar to that of section 11.2 in that the voltage is taken as reference and the branch currents are calculated with respect to the voltage. However, the summation of the branch currents is now made more difficult since they do not necessarily remain either in phase or quadrature with one another. Thus before it is possible to analyse parallel impedance networks, it is necessary to introduce a new analytical device – current components.

Fig. 11.6 Components of a current. (a) Lagging power factor; (b) leading power factor

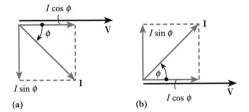

Consider Fig. 11.6 in which the current **I** is shown to lag (or lead) the voltage **V** by a phase angle ϕ. This current may be made up by two components at right angles to one another:

1. $I \cos \phi$, which is in phase with the voltage and is termed the *active* or *power* component.
2. $I \sin \phi$, which is in quadrature with the voltage and is termed the *quadrature* or *reactive* component.

By the geometry of the diagram:

$$I^2 = (I \cos \phi)^2 + (I \sin \phi)^2$$

Fig. 11.7 Addition of current phasors

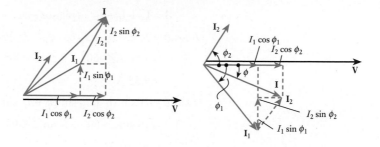

The reactive component will either lag (or lead) the voltage by 90° depending on whether the current **I** lags (or leads) the voltage **V**.

Consider the addition of the currents $\mathbf{I}_1$ and $\mathbf{I}_2$ as shown in Fig. 11.7, i.e.

$$\mathbf{I} = \mathbf{I}_1 + \mathbf{I}_2 \quad \text{(phasor sum)}$$

The value of **I** can be achieved by drawing a phasor diagram to scale, but this is not generally practicable. It can, however, be calculated if the currents are resolved into components, then

$$I \cos \phi = I_1 \cos \phi_1 + I_2 \cos \phi_2$$

$$I \sin \phi = I_1 \sin \phi_1 + I_2 \sin \phi_2$$

But

$$I^2 = (I \cos \phi)^2 + (I \sin \phi)^2$$

Hence

$$I^2 = (I_1 \cos \phi_1 + I_2 \cos \phi_2)^2 + (I_1 \sin \phi_1 + I_2 \sin \phi_2)^2 \qquad [11.9]$$

Also

$$\phi = \tan^{-1} \frac{I_1 \sin \phi_1 + I_2 \sin \phi_2}{I_1 \cos \phi_1 + I_2 \cos \phi_2}$$

$$\phi = \cos^{-1} \frac{I_1 \cos \phi_1 + I_2 \cos \phi_2}{I} \qquad [11.10]$$

Example 11.3

A parallel network consists of branches A, B and C. If $I_A = 10\angle{-60°}$ A, $I_B = 5\angle{-30°}$ A and $I_C = 10\angle{90°}$ A, all phase angles, being relative to the supply voltage, determine the total supply current.

With reference to the circuit and phasor diagrams in Fig. 11.8

Fig. 11.8 Circuit and phasor diagrams for Example 11.3

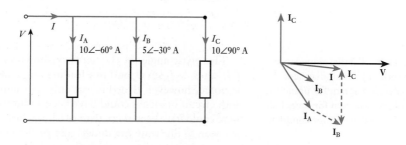

$$\mathbf{I} = \mathbf{I}_A + \mathbf{I}_B + \mathbf{I}_C \quad \text{(phasor sum)}$$

$$I \cos \phi = I_A \cos \phi_A + I_B \cos \phi_B + I_C \cos \phi_C$$

$$= 10 \cos\angle{-60°} + 5 \cos\angle{-30°} + 10 \cos\angle{90°}$$

$$= 9.33 \text{ A}$$

$$I \sin \phi = I_A \sin \phi_A + I_B \sin \phi_B + I_C \sin \phi_C$$

$$= 10 \sin\angle{-60°} + 5 \sin\angle{-30°} + 10 \sin\angle{90°}$$

$$= -1.16 \text{ A}$$

The negative sign indicates that the reactive current component is lagging, so the overall power factor will also be lagging.

$$I = ((I \cos \phi)^2 + (I \sin \phi)^2)^{\frac{1}{2}}$$

$$= (9.33^2 + 1.16^2)^{\frac{1}{2}} = 9.4 \text{ A}$$

$$\phi = \tan^{-1} \frac{I \sin \phi}{I \cos \phi} = \tan^{-1} \frac{1.16}{9.33} = 7.1° \text{ lag}$$

$$\mathbf{I} = \mathbf{9.4\angle{-7.1°} \ A}$$

Consider the circuit shown in Fig. 11.9 in which two series circuits are connected in parallel. To analyse the arrangement, the phasor diagrams for each branch have been drawn as shown in Figs 11.9(b) and 11.9(c). In each branch the current has been taken as reference; however, when the branches are in parallel, it is easier to take the supply voltage as reference, hence Figs 11.9(b) and 11.9(c) have been separately rotated and then superimposed on one another to give Fig. 11.9(d). The current phasors may then be added to give the total current in correct phase relation to the voltage. The analysis of the diagram is carried out in the manner noted above.

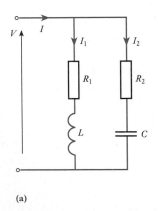

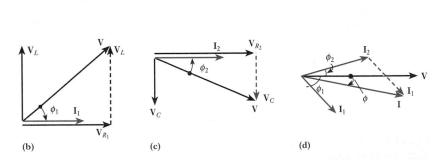

(a)   (b)   (c)   (d)

Fig. 11.9 Parallel-impedance network: (a) circuit diagram; (b) phasor diagram for branch 1; (c) phasor diagram for branch 2; (d) phasor diagram for complete circuit

The phase angle for the network shown in Fig. 11.9(a) is a lagging angle if $I_1 \sin \phi_1 > I_2 \sin \phi_2$ and is a leading angle if $I_1 \sin \phi_1 < I_2 \sin \phi_2$. It should be noted, however, that this was only an example of the method of analysis. Both circuit branches could have been inductive or capacitive. Alternatively, there could have been more than two branches. The main concern of this study has been to illustrate the underlying principles of the method of analysis.

| Example 11.4 | A coil of resistance 50 Ω and inductance 0.318 H is connected in parallel with a circuit comprising a 75 Ω resistor in series with a 159 μF capacitor. The resulting circuit is connected to a 230 V, 50 Hz a.c. supply (Fig. 11.10). Calculate: |

    (a) the supply current;
    (b) the circuit impedance, resistance and reactance.

Fig. 11.10 Circuit and phasor diagrams for Example 11.4

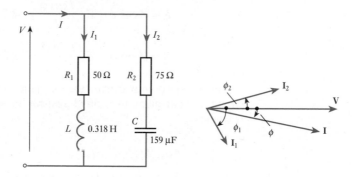

(a) $X_L = 2\pi f L = 2\pi 50 \times 0.318 = 100 \ \Omega$

$$Z_1 = (R_1^2 + X_L^2)^{\frac{1}{2}} = (50^2 + 100^2)^{\frac{1}{2}} = 112 \ \Omega$$

$$I_1 = \frac{V}{Z_1} = \frac{230}{112} = 2.05 \ \text{A}$$

$$\phi_1 = \cos^{-1}\frac{R_1}{Z_1} = \cos^{-1}\frac{50}{112} = \cos^{-1}0.447 = 63.5° \ \text{lag}$$

$$\mathbf{I_1} = 2.05\angle{-63.5°} \ \text{A}$$

$$X_C = \frac{1}{2\pi f C} = \frac{1}{2\pi 50 \times 159 \times 10^{-6}} = 20 \ \Omega$$

$$Z_2 = (R_2^2 + X_C^2)^{\frac{1}{2}} = (75^2 + 20^2)^{\frac{1}{2}} = 77.7 \ \Omega$$

$$I_2 = \frac{V}{Z_2} = \frac{230}{77.7} = 2.96 \ \text{A}$$

$$\phi_2 = \tan^{-1}\frac{X_C}{R_2} = \tan^{-1}\frac{20}{77.7} = \tan^{-1}0.267 = 15° \ \text{lead}$$

In this last equation the solution incorporating the use of the tangent is used because ϕ_2 is relatively small.

$$\mathbf{I_2} = 2.96\angle{15°} \ \text{A}$$

$$\mathbf{I} = \mathbf{I_1} + \mathbf{I_2} \quad \text{(phasor sum)}$$

$$I \cos \phi = I_1 \cos \phi_1 + I_2 \cos \phi_2$$

$$= 2.05 \cos\angle{-63.5°} + 2.96 \cos\angle{15°} = 3.77 \ \text{A}$$

$$I \sin \phi = I_1 \sin \phi_1 + I_2 \sin \phi_2$$
$$= 2.05 \sin\angle{-63.5°} + 2.96 \sin\angle{15°} = -1.07 \text{ A}$$
$$I = ((I \cos \phi)^2 + (I \sin \phi)^2)^{\frac{1}{2}} = (3.77^2 + 1.07^2)^{\frac{1}{2}}$$
$$= \mathbf{3.9 \text{ A}}$$

(b)
$$Z = \frac{V}{I} = \frac{230}{3.92} = 58.7 \ \Omega$$

$$R = Z \cos \phi = Z \cdot \frac{I \cos \phi}{I} = 58.7 \times \frac{3.77}{3.92} = \mathbf{56 \ \Omega}$$

$$X = Z \sin \phi = Z \cdot \frac{I \sin \phi}{I} = 58.7 \times \frac{1.07}{3.92} = \mathbf{16 \ \Omega}$$

Since $I \sin \phi$ is negative, the reactance must be inductive. Thus the circuit is equivalent to a 56 Ω resistor in series with a 16 Ω inductive reactance.

11.4   Polar impedances

In Example 11.4, the impedance was derived from the current and voltage. However, it may be questioned why could not the parallel impedances have been handled in a similar manner to parallel resistors? Consider then three impedances connected in parallel as shown in Fig. 11.11.

Fig. 11.11   Polar impedances in parallel

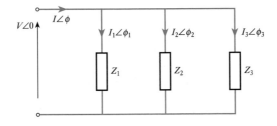

In the first branch

$$\mathbf{V = I_1 Z_1}$$

However, if consideration is given to the phase angles of **V** and **I** then to maintain balance, an impedance must also act like a complexor and have a phase angle, i.e.

$$V\angle 0 = I_1\angle\phi_1 \cdot Z_1\angle{-\phi_1}$$

The impedance phase angle is the conjugate of the circuit phase angle. This compares with the impedance triangles previously shown (apart from the reversal of the 'polarity').

In complexor notation:

$$\mathbf{I = I_1 + I_2 + I_3} \quad \text{(phasor sum)}$$

In polar notation:

$$I\angle\phi = I_1\angle\phi_1 + I_2\angle\phi_2 + I_3\angle\phi_3$$

$$\therefore \quad \frac{V\angle 0}{Z\angle-\phi} = \frac{V\angle 0}{Z_1\angle-\phi_1} + \frac{V\angle 0}{Z_2\angle-\phi_2} + \frac{V\angle 0}{Z_3\angle-\phi_3}$$

$$\therefore \quad \frac{1}{Z\angle-\phi} = \frac{1}{Z_1\angle-\phi_1} + \frac{1}{Z_2\angle-\phi_2} + \frac{1}{Z_3\angle-\phi_3} \qquad [11.11]$$

This relation compares with that for parallel resistors, but it has the complication of having to consider the phase angles. Because of this, it is not considered, at this introductory stage, prudent to use the polar approach to the analysis of parallel impedances; the method used in Example 11.4 is more suitable and less prone to error.

It is most important that the impedance phase angles are not ignored. If we were to use the magnitudes only of the impedances, we would have the following statement **which is completely wrong!**

$$\frac{1}{Z} = \frac{1}{Z_1} + \frac{1}{Z_2} + \frac{1}{Z_3}$$

Always remember to use the phase angles!

A similar situation occurs when impedances are connected in series. Consider the case shown in Fig. 11.12.

Fig. 11.12   Polar impedances in series

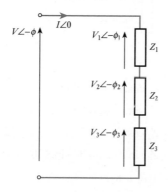

In complexor notation:

$$\mathbf{V} = \mathbf{V}_1 + \mathbf{V}_2 + \mathbf{V}_3 \quad \text{(phasor sum)}$$

In polar notation:

$$V\angle-\phi = V_1\angle-\phi_1 + V_2\angle-\phi_2 + V_3\angle-\phi_3$$

$$I\angle 0 \cdot Z\angle-\phi = I\angle 0 \cdot Z_1\angle-\phi_1 + I\angle 0 \cdot Z_2\angle-\phi_2 + I\angle 0 \cdot Z_3\angle-\phi_3$$

$$Z\angle-\phi = Z_1\angle-\phi_1 + Z_2\angle-\phi_2 + Z_3\angle-\phi_3 \qquad [11.12]$$

However, it has been shown in section 10.11 that in a series circuit

$$Z \cos\phi = Z_1 \cos\phi_1 + Z_2 \cos\phi_2 + Z_3 \cos\phi_3$$

hence

$$R = R_1 + R_2 + R_3 \qquad [11.13]$$

Similarly

$$X = X_1 + X_2 + X_3 \qquad [11.14]$$

As previously, it would have been incorrect to state that

$$Z = Z_1 + Z_2 + Z_3$$

It may therefore be concluded that, while it is practical to deal with impedances in series using polar notation, it is not practical to deal with impedances in parallel in this manner. Parallel network calculations are better approached on the basis of analysing the branch currents.

Example 11.5   Two impedances of $20\angle-45° \ \Omega$ and $30\angle30° \ \Omega$ are connected in series across a certain supply and the resulting current is found to be 10 A (Fig. 11.13). If the supply voltage remains unchanged, calculate the supply current when the impedances are connected in parallel.

Fig. 11.13 Circuit diagram for Example 11.5. Impedances connected in series

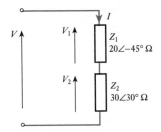

$$R_1 = Z_1 \cos\angle-\phi_1 = 20 \cos\angle-45° = 14.1 \ \Omega$$

$$X_1 = Z_1 \sin\angle-\phi_1 = 20 \sin\angle-45° = -14.1 \ \Omega, \text{ i.e. capacitive}$$

$$R_2 = Z_2 \cos\angle-\phi_2 = 30 \cos\angle30° = 26.0 \ \Omega$$

$$X_2 = Z_2 \sin\angle-\phi_2 = 30 \sin\angle30° = 15.0 \ \Omega, \text{ i.e. inductive}$$

$$Z = \{(R_1 + R_2)^2 + (X_1 + X_2)^2\}^{\frac{1}{2}}$$

$$= \{(14.1 + 26.0)^2 + (-14.1 + 15.0)^2\}^{\frac{1}{2}}$$

$$= 40.1 \ \Omega$$

$$V = IZ = 10 \times 40.1 = 401 \ \text{V}$$

We now connect the impedances in parallel as shown in Fig. 11.14 and calculate the current in each branch.

Fig. 11.14 Circuit diagram for Example 11.5. Impedance connected in parallel

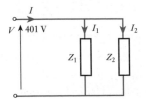

$$I_1 = \frac{V}{Z_1} = \frac{401}{20} = 20.1 \text{ A}$$

$$\phi_1 = 45°$$

$$I_1 \cos \phi_1 = 20.1 \times 0.707 = 14.2 \text{ A}$$

$$I_1 \sin \phi_1 = 20.1 \times 0.707 = 14.2 \text{ A}$$

$$I_2 = \frac{V}{Z_2} = \frac{401}{30} = 13.4 \text{ A}$$

$$\phi_2 = -30°$$

$$I_2 \cos \phi_2 = 13.4 \times 0.866 = 11.6 \text{ A}$$

$$I_2 \sin \phi_2 = 13.4 \times (-0.50) = -6.7 \text{ A}$$

For total current

$$I \cos \phi = I_1 \cos \phi_1 + I_2 \cos \phi_2$$

$$= 14.2 + 11.6$$

$$= 25.8 \text{ A}$$

$$I \sin \phi = I_1 \sin \phi_1 + I_2 \sin \phi_2$$

$$= 14.2 - 6.7 = 7.5 \text{ A}$$

$$I = ((I \cos \phi)^2 + (I \sin \phi)^2)^{\frac{1}{2}}$$

$$= (25.8^2 + 7.5^2)^{\frac{1}{2}}$$

$$= \mathbf{26.9 \text{ A}}$$

11.5 Polar admittances

An alternative approach to parallel a.c. circuits using polar notation can be made through admittance instead of impedance. The admittance is the inverse of the impedance in the same way that the conductance is the inverse of the resistance. The admittance Y is measured in siemens (abbreviated to S).

Thus in any branch of a parallel network

$$\frac{V}{Z} = I = VY \qquad\qquad [11.15]$$

Admittance   Symbol: Y Unit: siemens (S)

When the phase angles are included in this relation, it becomes

$$I\angle\phi = V\angle 0 \cdot Y\angle\phi = \frac{V\angle 0}{Z\angle-\phi}$$

$$Y\angle\phi = \frac{1}{Z\angle-\phi}$$

The resulting change in sign of the phase angle should be noted when the inversion takes place. Hence from relation [11.11]:

$$Y\angle\phi = Y_1\angle\phi_1 + Y_2\angle\phi_2 + Y_3\angle\phi_3 \qquad [11.16]$$

Hence $Y\cos\phi = Y_1\cos\phi_1 + Y_2\cos\phi_2 + Y_3\cos\phi_3$

$$G = G_1 + G_2 + G_3$$

Here G is the conductance of the circuit as in the d.c. circuit analysis. This must be the case since the current and voltage are in phase; this corresponds to the resistance of a circuit. Also

$$Y\sin\phi = Y_1\sin\phi_1 + Y_2\sin\phi_2 + Y_3\sin\phi_3$$

$$B = B_1 + B_2 + B_3$$

where B is termed the susceptance of the circuit and is the reactive component of the admittance.

Susceptance   Symbol: B Unit: siemens (S)

Figure 11.15(a) and (b) respectively show the impedance and admittance triangles.

Fig. 11.15 (a) Impedance triangle and (b) admittance triangle for capacitive circuit

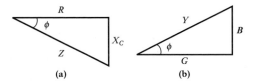

For the conductance

$$G = Y\cos\phi = Y \cdot \frac{R}{Z} = \frac{1}{Z} \cdot \frac{R}{Z}$$

∴ $$G = \frac{R}{Z^2} \qquad [11.17]$$

For the susceptance

$$B = Y\sin\phi = Y \cdot -\frac{X}{Z} = \frac{1}{Z} \cdot -\frac{X}{Z}$$

∴ $$B = -\frac{X}{Z^2} \qquad [11.18]$$

The negative sign in this expression is due to the change of sign of the phase angle noted above. Except in a purely resistive circuit, it must be remembered that $G \ne 1/R$.

Example 11.6 Three impedances $10\angle-30°\ \Omega$, $20\angle60°\ \Omega$ and $40\angle0°\ \Omega$ are connected in parallel (Fig. 11.16). Calculate their equivalent impedance.

$$Y_1\angle\phi_1 = \frac{1}{Z_1\angle-\phi_1} = \frac{1}{10\angle-30°} = 0.1\angle30°\ \text{S}$$

Fig. 11.16 Circuit diagram for Example 11.6

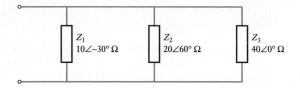

Similarly

$$Y_2 \angle \phi_2 = 0.05 \angle -60° \text{ S}$$

$$Y_3 \angle \phi_3 = 0.025 \angle 0° \text{ S}$$

$$G = G_1 + G_2 + G_3$$

$$= 0.1 \cos\angle 30° + 0.05 \cos\angle -60° + 0.025 \cos\angle 0°$$

$$= 0.087 + 0.025 + 0.025 = 0.137 \text{ S}$$

$$B = B_1 + B_2 + B_3$$

$$= 0.1 \sin\angle 30° + 0.05 \sin\angle -60° + 0.025 \sin\angle 0°$$

$$= 0.05 - 0.043 + 0.0 = 0.007 \text{ S}$$

$$Y = (G^2 + B^2)^{\frac{1}{2}} = (0.137^2 + 0.007^2)^{\frac{1}{2}}$$

$$= 0.137 \text{ S}$$

$$\phi = \tan^{-1} \frac{B}{G} = \tan^{-1} \frac{0.007}{0.137} = 3°$$

$$Z \angle -\phi = \frac{1}{Y \angle \phi} = \frac{1}{0.137 \angle 3°} = 7.32 \angle -3° \text{ } \Omega$$

Summary of important formulae

For R and L in parallel

$$\mathbf{I} = \mathbf{I}_R + \mathbf{I}_L \quad \text{(phasor sum)} \tag{11.1}$$

For R and C in parallel

$$\mathbf{I} = \mathbf{I}_R + \mathbf{I}_C \quad \text{(phasor sum)} \tag{11.5}$$

Terms and concepts

Parallel networks are simply solved by treating each branch as a simple series circuit and then adding the branch currents.

Alternatively we can manipulate branch impedances provided they are expressed in polar form.

The **admittance** is the inverse of the impedance. The in-phase component of the admittance is the **conductance** and the quadrate component is the **susceptance**.

Exercises 11

1. In order to use three 110 V, 60 W lamps on a 230 V, 50 Hz supply, they are connected in parallel and a capacitor is connected in series with the group. Find: (a) the capacitance required to give the correct voltage across the lamps; (b) the power factor of the network. If one of the lamps is removed, to what value will the voltage across the remaining two rise, assuming that their resistances remain unchanged?

2. A 130 Ω resistor and a 30 μF capacitor are connected in parallel across a 200 V, 50 Hz supply. Calculate: (a) the current in each branch; (b) the resultant current; (c) the phase difference between the resultant current and the applied voltage; (d) the active power; and (e) the power factor. Sketch the phasor diagram.

3. A resistor and a capacitor are connected in series across a 150 V a.c. supply. When the frequency is 40 Hz the current is 5 A, and when the frequency is 50 Hz the current is 6 A. Find the resistance and capacitance of the resistor and capacitor respectively. If they are now connected in parallel across the 150 V supply, find the total current and its power factor when the frequency is 50 Hz.

4. A series circuit consists of a non-inductive resistor of 10 Ω, an inductor having a reactance of 50 Ω and a capacitor having a reactance of 30 Ω. It is connected to a 230 V a.c. supply. Calculate: (a) the current; (b) the voltage across each component. Draw to scale a phasor diagram showing the supply voltage and current and the voltage across each component.

5. A coil having a resistance of 20 Ω and an inductance of 0.15 H is connected in series with a 100 μF capacitor across a 230 V, 50 Hz supply. Calculate: (a) the active and reactive components of the current; (b) the voltage across the coil. Sketch the phasor diagram.

6. A p.d. of 100 V at 50 Hz is maintained across a series circuit having the following characteristics: $R = 10$ Ω, $L = 100/\pi$ mH, $C = 500/\pi$ μF. Draw the phasor diagram and calculate: (a) the current; (b) the active and reactive components of the current.

7. A network consists of three branches in parallel. Branch A is a 10 Ω resistor, branch B is a coil of resistance 4 Ω and inductance 0.02 H, and branch C is an 8 Ω resistor in series with a 200 μF capacitor.

The combination is connected to a 100 V, 50 Hz supply. Find the various branch currents and then, by resolving into in-phase and quadrature components, determine the total current taken from the supply. A phasor diagram showing the relative positions of the various circuit quantities should accompany your solution. It need not be drawn to scale.

8. A coil, having a resistance of 20 Ω and an inductance of 0.0382 H, is connected in parallel with a circuit consisting of a 150 μF capacitor in series with a 10 Ω resistor. The arrangement is connected to a 230 V, 50 Hz supply. Determine the current in each branch and, sketching a phasor diagram, the total supply current.

9. A 31.8 μF capacitor, a 127.5 mH inductor of resistance 30 Ω and a 100 Ω resistor are all connected in parallel to a 200 V, 50 Hz supply. Calculate the current in each branch. Draw a phasor diagram to scale to show these currents. Find the total current and its phase angle by drawing or otherwise.

10. A 200 V, 50 Hz sinusoidal supply is connected to a parallel network comprising three branches A, B and C, as follows: A, a coil of resistance 3 Ω and inductive reactance 4 Ω; B, a series circuit of resistance 4 Ω and capacitive reactance 3 Ω; C, a capacitor. Given that the power factor of the combined circuit is unity, find: (a) the capacitance of the capacitor in microfarads; (b) the current taken from the supply.

11. Two circuits, A and B, are connected in parallel to a 115 V, 50 Hz supply. The total current taken by the combination is 10 A at unity power factor. Circuit A consists of a 10 Ω resistor and a 200 μF capacitor connected in series; circuit B consists of a resistor and an inductive reactor in series. Determine the following data for circuit B: (a) the current; (b) the impedance; (c) the resistance; (d) the reactance.

12. A parallel network consists of two branches A and B. Branch A has a resistance of 10 Ω and an inductance of 0.1 H in series. Branch B has a resistance of 20 Ω and a capacitance of 100 μF in series. The network is connected to a single-phase supply of 230 V at 50 Hz. Calculate the magnitude and phase angle of the current taken from the supply. Verify your answers by measurement from a phasor diagram drawn to scale.

Chapter twelve   Power in AC Circuits

Objectives

When you have studied this chapter, you should

- have an understanding of active and reactive powers

- be capable of analysing the power in an a.c. circuit containing one component

- be aware that active power is dissipated

- be aware that reactive power is not dissipated

- have an understanding of the powers associated with an a.c. series circuit

- be familiar with power factor

- be capable of analysing the power factor of an a.c. circuit

- recognize the importance of power factor in practice

- be familiar with a technique of measuring the power in a single-phase circuit

Contents

AC circuits deliver power to resistive and reactive loads. We find that in the case of resistive loads, the energy is dissipated in the same way as a direct current dissipates energy in a resistor. However, we find a completely different situation with reactive loads – here the energy is first delivered to the load and then it is returned to the source and then it is returned to the load and so on. It is like watching an unending rally in tennis as the ball of energy flies to and fro.

The power which gives rise to energy dissipation is the active power. The power describing the rate of energy moving in and out of reactances is reactive power and is an essential part of the energy transfer system.

We find therefore that we have to mix active and reactive powers and this leads us to talk about power factors. Most of us eventually meet with power factors when it comes to paying commercial electricity bills so it is a good idea to know for what we are paying.

12.1 The impossible power

When alternating current systems were first introduced, learned scientists claimed that it was impossible to deliver energy by such a means. Their argument was that power transfer would take place during the first half of the cycle – and then it would transfer back during the second half.

Curiously there was some truth in what they claimed, but they had overlooked the basic relationship $p = i^2R$. The square of the current means that the power is positive no matter whether the current has a positive or a negative value. But it is only the resistive element that dissipates energy from the circuit. Inductors and capacitors do not dissipate energy which supports the theory of the impossible power.

Let us therefore examine in more detail the energy transfer process which takes place first in resistive circuits and then in reactive circuits.

12.2 Power in a resistive circuit

In section 9.6 it was explained that when an alternating current flows through a resistor of R ohms, the average heating effect over a complete cycle is I^2R watts, where I is the r.m.s. value of the current in amperes.

If V volts is the r.m.s. value of the applied voltage, then for a non-reactive circuit having constant resistance R ohms, $V = IR$.

The waveform diagrams for resistance are shown in Fig. 12.1. To the current and voltage waves, there have been added the waves of the product vi. Since the instantaneous values of vi represent the instantaneous power p, it follows that these waves are the power waves. Because the power is continually fluctuating, the power in an a.c. circuit is taken to be the average value of the wave.

Fig. 12.1 Waveform diagrams for a resistive circuit

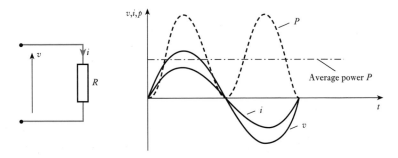

In the case of the pure resistance, the average power can be most easily obtained from the definition of the r.m.s. current in the circuit, i.e.

$$P = I^2R \qquad [12.1]$$

This relation can also be expressed as

$$P = VI \qquad [12.2]$$

Hence the power in a non-reactive circuit is given by the product of the ammeter and voltmeter readings, exactly as in a d.c. circuit.

The power associated with energy transfer from the electrical system to another system such as heat, light or mechanical drives is termed active power, thus the average given by I^2R is the active power of the arrangement.

Alternatively, the average power can be derived from a formal analysis of the power waveform.

$$P = \frac{\omega}{2\pi} \int_0^{\frac{2\pi}{\omega}} (V_m \sin \omega t \cdot I_m \sin \omega t) \, dt$$

$$= V_m I_m \frac{\omega}{2\pi} \int_0^{\frac{2\pi}{\omega}} (\sin^2 \omega t) \, dt$$

$$= V_m I_m \frac{\omega}{2\pi} \int_0^{\frac{2\pi}{\omega}} \left(\frac{1 - \cos 2\omega t}{2} \right) dt$$

From this relation it can be seen that the wave has a frequency double that of the component voltage and current waves. This can be seen in Fig. 12.1; however, it also confirms that the wave is sinusoidal although it has been displaced from the horizontal axis.

$$P = V_m I_m \frac{\omega}{2\pi} \left[\frac{t}{2} - \frac{\sin 2\omega t}{4\omega} \right]_0^{\frac{2\pi}{\omega}}$$

$$= V_m I_m \frac{\omega}{2\pi} \cdot \frac{2\pi}{2\omega}$$

$$P = \frac{V_m I_m}{2} \qquad\qquad [12.3]$$

$$P = VI$$

12.3 Power in a purely inductive circuit

Consider a coil wound with such thick wire that the resistance is negligible in comparison with the inductive reactance X_L ohms. If such a coil is connected across a supply voltage V, the current is given by $I = V/X_L$ amperes. Since the resistance is very small, the heating effect and therefore the active power are also very small, even though the voltage and the current are large. Such a curious conclusion – so different from anything we have experienced in d.c. circuits – requires fuller explanation if its significance is to be properly understood. Let us therefore consider Fig. 12.2, which shows the applied voltage and the current for a purely inductive circuit, the current lagging the voltage by a quarter of a cycle.

The power at any instant is given by the product of the voltage and the current at that instant; thus at instant L, the applied voltage is LN volts and the current is LM amperes, so that the power at that instant is LN × LM watts and is represented to scale by LP.

Fig. 12.2 Power curve for a purely inductive circuit

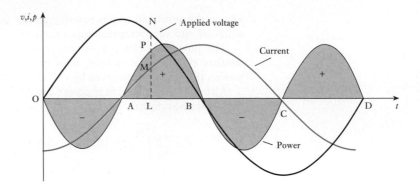

By repeating this calculation at various instants we can deduce the curve representing the variation of power over one cycle. It is seen that during interval OA the applied voltage is positive, but the current is negative, so that the power is negative; and that during interval AB, both the current and the voltage are positive, so that the power is positive.

The power curve is found to be symmetrical about the horizontal axis OD. Consequently the shaded areas marked '−' are exactly equal to those marked '+', so that the mean value of the power over the complete cycle OD is zero.

It is necessary, however, to consider the significance of the positive and negative areas if we are to understand what is really taking place. So let us consider an a.c. generator P (Fig. 12.3) connected to a coil Q whose resistance is negligible, and let us assume that the voltage and current are represented by the graphs in Fig. 12.2. At instant A, there is no current and therefore no magnetic field through and around Q. During interval AB, the growth of the current is accompanied by a growth of flux as shown by the dotted lines in Fig. 12.3. But the existence of a magnetic field involves some kind of a strain in the space occupied by the field and the storing up of energy in that field, as already dealt with in section 8.10. The current, and therefore the magnetic energy associated with it, reach their maximum values at instant B; and, since the loss in the coil is assumed negligible, it follows that at that instant the whole of the energy supplied to the coil during interval AB, and represented by the shaded area marked '+', is stored up in the magnetic field.

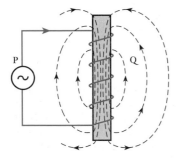

Fig. 12.3 Magnetic field of an inductive circuit

During the interval BC the current and its magnetic field are decreasing; and the e.m.f. induced by the collapse of the magnetic flux is in the same direction as the current. But any circuit in which the current and the induced or generated e.m.f. are in the same direction acts as a source of electrical energy (see section 8.2). Consequently the coil is now acting as a generator transforming the energy of its magnetic field into electrical energy, the latter being sent to generator P to drive it as a motor. The energy thus returned is represented by the shaded area marked '−' in Fig. 12.2; and since the positive and negative areas are equal, it follows that during alternate quarter-cycles electrical energy is being sent from the generator to the coil, and during the other quarter-cycles the same amount of energy is sent back from the coil to the generator. Consequently the net energy absorbed by the coil during a complete cycle is zero; in other words, the average power over a complete cycle is zero.

12.4 Power in a purely capacitive circuit

In this case, the current leads the applied voltage by a quarter of a cycle, as shown in Fig. 12.4; and by multiplying the corresponding instantaneous values of the voltage and current, we can derive the curve representing the variation of power. During interval OA, the voltage and current are both positive so that the power is positive, i.e. power is being supplied from the generator to the capacitor, and the shaded area enclosed by the power curve during interval OA represents the value of the electrostatic energy stored in the capacitor at instant A.

Fig. 12.4 Power curve for a purely capacitive circuit

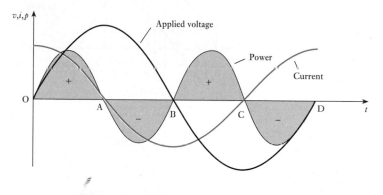

During interval AB, the p.d. across the capacitor decreases from its maximum value to zero and the whole of the energy stored in the capacitor at instant A is returned to the generator; consequently the net energy absorbed during the half-cycle OB is zero. Similarly, the energy absorbed by the capacitor during interval BC is returned to the generator during interval CD. Hence the average power over a complete cycle is zero.

That the inductive and capacitive circuits do not dissipate power can be proved by an analysis of the power wave. Consider the case of the capacitor.

$$P = \frac{\omega}{2\pi} \int_0^{\frac{2\pi}{\omega}} (V_\mathrm{m} \sin \omega t \cdot I_\mathrm{m} \cos \omega t)\,\mathrm{d}t$$

$$= V_\mathrm{m} I_\mathrm{m} \frac{\omega}{2\pi} \int_0^{\frac{2\pi}{\omega}} (\sin \omega t \cdot \cos \omega t)\,\mathrm{d}t$$

$$= V_\mathrm{m} I_\mathrm{m} \frac{\omega}{2\pi} \int_0^{\frac{2\pi}{\omega}} \frac{\sin 2\omega t}{2} \cdot \mathrm{d}t$$

$$= V_\mathrm{m} I_\mathrm{m} \frac{\omega}{2\pi} \left[-\frac{\cos 2\omega t}{4\omega} \right]_0^{\frac{2\pi}{\omega}}$$

$$P = 0 \hspace{5cm} [12.4]$$

12.5

Power in a circuit with resistance and reactance

Let us consider the general case of the current differing in phase from the applied voltage; thus in Fig. 12.5(a), the current is shown lagging the voltage by an angle ϕ.

Let instantaneous value of voltage be

$$v = V_m \sin \omega t$$

then instantaneous value of current is

$$i = I_m \sin(\omega t - \phi)$$

Fig. 12.5 Voltage, current and power curves

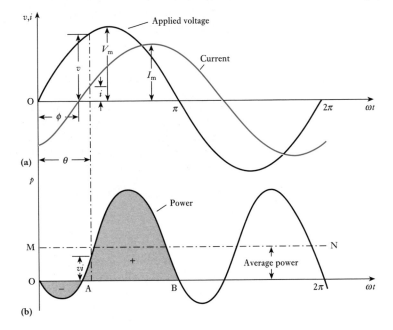

At any instant, the value of the power is given by the product of the voltage and the current at that instant, i.e. instantaneous value of power $= vi$ watts.

By multiplying the corresponding instantaneous values of voltage and current, the curve representing the variation of power in Fig. 12.5(b) can be derived, i.e. instantaneous power is

$$vi = V_m \sin \omega t \cdot I_m \sin(\omega t - \phi)$$
$$= \tfrac{1}{2}V_m I_m\{\cos \phi - \cos(2\omega t - \phi)\}$$
$$= \tfrac{1}{2}V_m I_m \cos \phi - \tfrac{1}{2}V_m I_m \cos(2\omega t - \phi)$$

From this expression, it is seen that the instantaneous value of the power consists of two components:

1. $\tfrac{1}{2}V_m I_m \cos \phi$, which contains no reference to ωt and therefore remains constant in value.
2. $\tfrac{1}{2}V_m I_m \cos(2\omega t - \phi)$, the term $2\omega t$ indicating that it varies at twice the supply frequency; thus in Fig. 12.5(b) it is seen that the power undergoes two cycles of variation for one cycle of the voltage wave. Furthermore, since the average value of a cosine curve over a *complete* cycle is zero, it

follows that this component does not contribute anything towards the *average* value of the power taken from the generator.

Hence, average power over one cycle is

$$\tfrac{1}{2}V_{m}I_{m}\cos\phi = \frac{V_{m}}{\sqrt{2}}\cdot\frac{I_{m}}{\sqrt{2}}\cdot\cos\phi$$

∴ $$P = VI\cos\phi$$ [12.5]

where V and I are the r.m.s. values of the voltage and current respectively. In Fig. 12.5(b), the average power is represented by the height above the horizontal axis of the dotted line MN drawn midway between the positive and negative peaks of the power curve.

It will be noticed that during interval OA in Fig. 12.5(b), the power is negative, and the shaded negative area represents energy returned from the circuit to the generator. The shaded positive area during interval AB represents energy supplied from the generator to the circuit, and the difference between the two areas represents the net energy absorbed by the circuit during interval OB. The larger the phase difference between the voltage and current, the smaller is the difference between the positive and negative areas and the smaller, therefore, is the average power over the complete cycle.

The average power over the complete cycle is the active power, which is measured in watts.

The product of the voltage and the current in an a.c. circuit is termed the apparent power.

Apparent Power   Symbol: S Unit: volt ampere (VA)

$$S = VI$$ [12.6]

∴ $$P = VI\cos\phi$$

and $$P = S\cos\phi$$ [12.7]

Example 12.1

A coil having a resistance of 6 Ω and an inductance of 0.03 H is connected across a 50 V, 60 Hz supply. Calculate:

(a) the current;
(b) the phase angle between the current and the applied voltage;
(c) the apparent power;
(d) the active power.

(a) The phasor diagram for such a circuit is given in Fig. 12.6.

$$\text{Reactance of circuit} = 2\pi fL = 2 \times 3.14 \times 60 \times 0.03$$

$$= 11.31\ \Omega$$

From equation [10.10]

$$\text{Impedance} = \sqrt{\{6^{2} + (11.31)^{2}\}} = 12.8\ \Omega$$

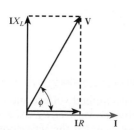

Fig. 12.6 Phasor diagram for Example 12.1

and $$\text{Current} = \frac{50}{12.8} = 3.9\ \text{A}$$

(b) From equation [10.11]

$$\tan \phi = \frac{X}{R} = \frac{11.31}{6} = 1.885$$

$$\phi = 62°3'$$

(c) Apparent power $S = 50 \times 3.91 = \mathbf{196\ VA}$

(d) 　　Active power = apparent power $\times \cos \phi$

$$= 195.5 \times 0.469 = \mathbf{92\ W}$$

alternatively:

$$\text{Active power} = I^2R = (3.91)^2 \times 6 = \mathbf{92\ W}$$

| Example 12.2 |

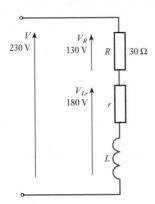

Fig. 12.7　Circuit diagram for Example 12.2

An inductor coil is connected in series with a pure resistor of 30 Ω across a 230 V, 50 Hz supply. The voltage measured across the coil is 180 V and the voltage measured across the resistor is 130 V (Fig. 12.7). Calculate the power dissipated in the coil.

The complexor diagram is constructed by first drawing the complexor I (Fig. 12.8). The resistor voltage complexor V_R is then drawn in phase with I. Since neither the coil phase angle nor the circuit phase angle are known, it is necessary to derive the remainder of the diagram by construction to scale. Circles of radius V and V_{Lr} are drawn radiating from the appropriate ends of V_R. The point of intersection of the circles satisfies the relation

$$\mathbf{V} = \mathbf{V}_R + \mathbf{V}_{Lr} \quad \text{(phasor sum)}$$

By the geometry of the diagram:

$$V^2 = V_R^2 + V_{Lr}^2 + 2V_R V_{Lr} \cos \phi_{Lr}$$

$$230^2 = 130^2 + 180^2 + 2 \times 130 \times 180 \times \cos \phi_{Lr}$$

$$\cos \phi_{Lr} = 0.077 \text{ lag}$$

$$I = \frac{V_R}{R} = \frac{130}{30} = 4.33 \text{ A}$$

$$P_r = V_{Lr} I \cos \phi_{Lr} = 180 \times 4.33 \times 0.077 = \mathbf{60\ W}$$

Alternatively:

$$Z_{Lr} = \frac{V_{Lr}}{I} = \frac{180}{4.33} = 41.5 \ \Omega$$

$$r = Z_{Lr} \cos \phi_{Lr} = 41.5 \times 0.077 = 3.20 \ \Omega$$

$$P_r = I^2 r = 4.33^2 \times 3.20 = \mathbf{60\ W}$$

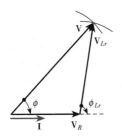

Fig. 12.8　Phasor diagram for Example 12.2

| 12.6 | Power factor |

In a.c. work, the product of the r.m.s. values of the applied voltage and current is VI. It has already been shown that the active power $P = VI \cos \phi$ and the value of $\cos \phi$ has to lie between 0 and 1. It follows that the active power P can be either equal to or less than the product VI, which is termed the apparent power and is measured in voltamperes (VA).

The ratio of the active power P to the apparent power S is termed the power factor, i.e.

$$\frac{\text{Active power } P \text{ in watts}}{\text{Apparent power } S \text{ in voltamperes}} = \text{power factor}$$

∴
$$\cos \phi = \frac{P}{S} = \frac{P}{VI} \qquad [12.8]$$

or | Active power P = apparent power S × power factor | [12.9]

Comparison of expressions [12.5] and [12.8] shows that for *sinusoidal* voltage and current:

power factor = $\cos \phi$

From the general phasor diagram of Fig. 10.26 for a *series* circuit, it follows that

$$\cos \phi = \frac{IR}{V} = \frac{IR}{IZ} = \frac{\text{resistance}}{\text{impedance}}$$

∴
$$\cos \phi = \frac{R}{Z} \qquad [12.10]$$

It has become the practice to say that the power factor is *lagging* when the *current lags the supply voltage*, and *leading* when the *current leads the supply voltage*. This means that the supply voltage is regarded as the reference quantity.

Example 12.3

An inductor coil is connected to a supply of 230 V at 50 Hz and takes a current of 5.0 A. The coil dissipates 750 W (Fig. 12.9). Calculate:

(a) the resistance and the inductance of the coil;
(b) the power factor of the coil.

In this example, the symbol r will be used to denote the resistance of the coil instead of R. This is done to draw attention to the fact that the resistance is not a separate component of the circuit but is an integral part of the inductor coil. This device was also used in Example 12.2.

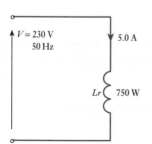

Fig. 12.9 Circuit diagram for Example 12.3

(a)
$$Z = \frac{V}{I} = \frac{230}{5} = 46 \ \Omega$$

$$r = \frac{P}{I^2} = \frac{750}{5^2} = 30 \ \Omega$$

$$X_L = (Z^2 - r^2)^{\frac{1}{2}} = (46^2 - 30^2)^{\frac{1}{2}} = 34.87 \ \Omega$$

$$L = \frac{X_L}{2\pi f} = \frac{34.87}{2\pi 50} = \frac{34.87}{314} = 0.111 \text{ H} = \textbf{111 mH}$$

(b)   Power factor = $\cos \phi = \frac{P}{S} = \frac{R}{VI} = \frac{750}{230 \times 5} = \textbf{0.65 lag}$

12.7 Active and reactive currents

If a current I lags the applied voltage V by an angle ϕ, as in Fig. 12.10, it can be resolved into two components, OA in phase with the voltage and OB lagging by 90°.

If the phasor diagram in Fig. 12.10 refers to a circuit possessing resistance and inductance in series, OA and OB must not be labelled I_R and I_L respectively. Such terms should only be applied to branch currents as would be the case if R and L were in parallel. This error of applying parallel terms to series circuits is easily made by beginners – **you have been warned!**

Since

$$\text{power} = IV \cos \phi = V \times \text{OI} \cos \phi = V \times \text{OA} \text{ watts}$$

therefore OA is termed the *active* component of the current, i.e.

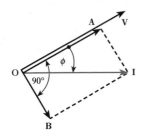

Fig. 12.10 Active and reactive components of current

$$\boxed{\text{Active component of current} = I \cos \phi} \qquad [12.11]$$

Power due to component OB is

$$V \times \text{OB} \cos 90° = 0,$$

so that OB is termed the *reactive* component of the current, i.e.

$$\boxed{\text{Reactive component of current} = I \sin \phi} \qquad [12.12]$$

and   Reactive power Q in vars $= VI \sin \phi$

The term 'var' is short for voltampere reactive.

$$\therefore \qquad \boxed{Q = VI \sin \phi} \qquad [12.13]$$

also $P^2 + Q^2 = (VI \cos \phi)^2 + (VI \sin \phi)^2$

$$= (VI)^2(\cos^2 \phi + \sin^2 \phi) = (VI)^2 = S^2$$

$$\boxed{S^2 = P^2 + Q^2} \qquad [12.14]$$

Example 12.4

A single-phase motor operating off a 400 V, 50 Hz supply is developing 10 kW with an efficiency of 84 per cent and a power factor (p.f.) of 0.7 lagging. Calculate:

(a) the input apparent power;
(b) the active and reactive components of the current;
(c) the reactive power (in kilovars).

$$(a) \quad \text{Efficiency} = \frac{\text{output power in watts}}{\text{input power in watts}}$$

$$= \frac{\text{output power in watts}}{IV \times \text{p.f.}}$$

$$\therefore \qquad 0.84 = \frac{10 \times 1000}{IV \times 0.7}$$

so that $IV = 17\,000$ VA

$$\therefore \qquad \text{input} = \mathbf{17.0 \ kVA}$$

(b)   Current taken by motor $= \dfrac{\text{input volt amperes}}{\text{voltage}}$

$$= \frac{17\ 000}{400} = 42.5\ \text{A}$$

therefore active component of current is

$$42.5 \times 0.7 = 29.75\ \text{A}$$

Since

$$\sin \phi = \surd(1 - \cos^2 \phi) = \surd\{1 - (0.7)^2\}$$

$$= 0.714$$

therefore reactive component of current is

$$42.5 \times 0.714 = \mathbf{30.4\ A}$$

(c)   Reactive power $= 400 \times \dfrac{30.35}{1000}$

$$= \mathbf{12.1\ kvar}$$

Example 12.5

Calculate the capacitance required in parallel with the motor of Example 12.4 to raise the supply power factor to 0.9 lagging.

The circuit and phasor diagrams are given in Figs 12.11 and 12.12 respectively, M being the motor taking a current I_M of 42.5 A.

Current I_C taken by the capacitor must be such that when combined with I_M, the resultant current I lags the voltage by an angle ϕ, where $\cos \phi = 0.9$. From Fig. 12.12

$$\text{active component of } I_M = I_M \cos \phi_M$$

$$= 42.5 \times 0.7$$

$$= 29.75\ \text{A}$$

and active component of I is

$$I \cos \phi = I \times 0.9$$

These components are represented by OA in Fig. 12.12.

$$\therefore \qquad I = \frac{29.75}{0.9} = 33.06\ \text{A}$$

Reactive component of $I_M = I_M \sin \phi_M$

$$= 30.35\ \text{A (from Example 12.4)}$$

and reactive component of I is

$$I \sin \phi = 33.06\surd\{1 - (0.9)^2\}$$

$$= 33.06 \times 0.436$$

$$= 14.4\ \text{A}$$

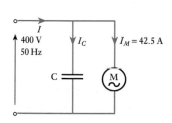

Fig. 12.11   Circuit diagram for Example 12.5

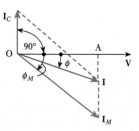

Fig. 12.12   Phasor diagram for Fig. 12.11

From Fig. 12.12 it will be seen that

$$I_C = \text{reactive component of } I_M - \text{reactive component of } I$$
$$= 30.35 - 14.4$$
$$= 15.95 \text{ A}$$

But $I_C = 2\pi f C V$

∴ $15.95 = 2 \times 3.14 \times 50 \times C \times 400$

and $C = 127 \times 10^{-6} \text{F} = \textbf{127} \, \boldsymbol{\mu}\textbf{F}$

From Example 12.5 it will be seen that the effect of connecting a **127** μF capacitor in parallel with the motor is to reduce the current taken from the supply from 42.5 to 33.1 A, without altering either the current or the power taken by the motor. This enables an economy to be effected in the size of the generating plant and in the cross-sectional area of conductor in the supply cable.

Example 12.6

An a.c. generator is supplying a load of 300 kW at a power factor of 0.6 lagging. If the power factor is raised to unity, how much more power (in kilowatts) can the generator supply for the same kilovolt-ampere loading?

Since the power in kW is number of kilovoltamperes × power factor, therefore number of kilovoltamperes is

$$\frac{300}{0.6} = 500 \text{ kVA}$$

When the power factor is raised to unity:

Number of kilowatts = number of kilovoltamperes = 500 kW

Hence increased power supplied by generator is

$$500 - 300 = \textbf{200 kW}$$

12.8   The practical importance of power factor

If an a.c. generator is rated to give, say, 2000 A at a voltage of 400 V, it means that these are the highest current and voltage values the machine can give without the temperature exceeding a safe value. Consequently the rating of the generator is given as $400 \times 2000/1000 = 800$ kVA. The phase difference between the voltage and the current depends upon the nature of the load and not upon the generator. Thus if the power factor of the load is unity, the 800 kVA are also 800 kW, and the engine driving the generator has to be capable of developing this power together with the losses in the generator. But if the power factor of the load is, say, 0.5, the power is only 400 kW, so that the engine is developing only about one-half of the power of which it is capable, though the generator is supplying its rated output of 800 kVA.

Similarly, the conductors connecting the generator to the load have to be capable of carrying 2000 A without excessive temperature rise. Consequently they can transmit 800 kW if the power factor is unity, but only 400 kW at 0.5 power factor for the same rise of temperature.

It is therefore evident that the higher the power factor of the load, the greater is the *active power* that can be generated by a given generator and transmitted by a given conductor.

The matter may be put another way by saying that, for a *given power*, the lower the power factor, the larger must be the size of the source to generate that power and the greater must be the cross-sectional area of the conductor to transmit it; in other words, the greater is the cost of generation and transmission of the electrical energy. This is the reason why supply authorities do all they can to improve the power factor of their loads, either by the installation of capacitors or special machines or by the use of tariffs which encourage consumers to do so.

Electronics engineers generally have little interest in power factor except when paying for their power supplies. Electronic circuits for the most part deal with such small levels of power that the additional heating effects due to the current not being in phase with the voltage are negligible.

12.9	**Measurement of power in a single-phase circuit**

Since the product of the voltage and current in an a.c. circuit must be multiplied by the power factor to give the active power in watts, the most convenient method of measuring the power is to use a wattmeter.

Summary of important formulae

For a general circuit

$$\text{Active power } P = VI \cos \phi \quad \text{(watts)} \qquad [12.5]$$

$$\text{Reactive power } Q = VI \sin \phi \quad \text{(vars)} \qquad [12.13]$$

$$\text{Apparent power } S = VI \qquad \text{(voltamperes)}$$

$$\text{Power factor (p.f.) } \cos \phi = P/S \qquad [12.8]$$

$$S^2 = P^2 + Q^2 \qquad [12.14]$$

Terms and concepts

The **active power**, sometimes also referred to as the **real power**, is the rate of energy conversion or dissipation taken as an average over one or more complete cycles.

The **reactive power** is the peak rate of energy storage in the reactive elements of a circuit. The average rate of energy storage is zero, the energy continually flowing into and back out from the reactive components. The reactive power is sometimes referred to as the imaginary power – a curious way of saying that it is not real.

The **apparent power** is the product of the r.m.s. voltage and current and is related to the active power by the **power factor**. The apparent power is a useful means of rating certain equipment, bearing in mind that conductor heat losses occur whether or not the current is in phase with the voltage.

Exercises 12

1. A single-phase motor takes 8.3 A at a power factor of 0.866 lagging when connected to a 230 V, 50 Hz supply. Two similar capacitors are connected in parallel with each other to form a capacitance bank. This capacitance bank is now connected in parallel with the motor to raise the power factor to unity. Determine the capacitance of each capacitor.

2. (a) A single-phase load of 5 kW operates at a power factor of 0.6 lagging. It is proposed to improve this power factor to 0.95 lagging by connecting a capacitor across the load. Calculate the kVA rating of the capacitor.

   (b) Give reasons why it is to consumers' economic advantage to improve their power factor with respect to the supply, and explain the fact that the improvement is rarely made to unity in practice.

3. A 25 kVA single-phase motor has a power factor of 0.8 lag. A 10 kVA capacitor is connected for power-factor correction. Calculate the input apparent power in kVA taken from the mains and its power factor when the motor is (a) on half load; (b) on full load. Sketch a phasor diagram for each case.

4. A single-phase motor takes 50 A at a power factor of 0.6 lagging from a 230 V, 50 Hz supply. What value of capacitance must a shunting capacitor have to raise the overall power factor to 0.9 lagging? How does the installation of the capacitor affect the line and motor currents?

5. A 230 V, single-phase supply feeds the following loads: (a) incandescent lamps taking a current of 8 A at unity power factor; (b) fluorescent lamps taking a current of 5 A at 0.8 leading power factor; (c) a motor taking a current of 7 A at 0.75 lagging power factor. Sketch the phasor diagram and determine the total current, active power and reactive power taken from the supply and the overall power factor.

6. The load taken from an a.c. supply consists of: (a) a heating load of 15 kW; (b) a motor load of 40 kVA at 0.6 power factor lagging; (c) a load of 20 kW at 0.8 power factor lagging. Calculate the total load from the supply (in kW and kVA) and its power factor. What would be the kvar rating of a capacitor to bring the power factor to unity and how would the capacitor be connected?

7. A cable is required to supply a welding set taking a current of 225 A at 110 V alternating current, the average power factor being 0.5 lagging. An available cable has a rating of 175 A and it is decided to use this cable by installing a capacitor across the terminals of the welding set. Find: (a) the required capacitor current and reactive power to limit the cable current to 175 A; (b) the overall power factor with the capacitor in circuit.

Chapter thirteen

Complex Notation

Objectives

When you have studied this chapter, you should

- have an understanding of the j operator

- be capable of using the j operator to add and subtract complex quantities

- be capable of expressing voltages, currents and impedances in complex notation

- be capable of applying complex notation to the analysis of series and parallel circuits

- have an understanding of expressing admittances in complex notation

- be capable of analysing the power in a circuit using complex notation

Contents

In the preceding chapters on a.c. circuit theory, problems have been solved exclusively by the use of phasors. As circuit problems become more involved, the techniques of complex algebra are used to *complement* those of phasors to simplify the solution process.

Complex algebra is based upon the fact that phasors, having a magnitude and phase angle, can be resolved into two components at right angles to each other. Modern calculators can rapidly convert between these rectangular or complex quantities and polar (phasor) quantities and this facility should be mastered.

The rectangular notation is particularly useful for the addition and subtraction of complex quantities like impedances, currents, voltages and power. The phasor notation, in turn, simplifies the multiplication and division of these complex quantities.

13.1 **The j operator**

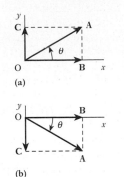

(a)

(b)

Fig. 13.1 Resolution of phasors

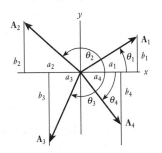

Fig. 13.2 Resolution of phasors

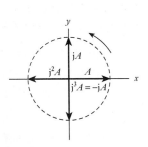

Fig. 13.3 Significance of the j operator

In Chapters 10–12, problems on a.c. circuits were solved with the aid of phasor diagrams. So long as the circuits are fairly simple, this method is satisfactory; but with a more involved circuit, such as that of Fig. 13.17, the calculation can be simplified by using complex algebra. This system enables equations representing alternating voltages and currents and their phase relationships to be expressed in simple algebraic form. It is based upon the idea that a phasor can be resolved into two components at right angles to each other. For instance, in Fig. 13.1(a), phasor **OA** can be resolved into components OB along the x-axis and OC along the y-axis, where OB = OA cos θ and OC = OA sin θ. It would obviously be incorrect to state that OA = OB + OC, since OA is actually $\sqrt{(OB^2 + OC^2)}$; but by introducing a symbol j to denote that OC is the component along the y-axis, we can represent the phasor thus:

$$\mathbf{OA} = \text{OB} + \text{jOC} = \text{OA}(\cos\theta + \text{j}\sin\theta)$$

It will be recalled that a symbol in bold type represents a phasor or complex quantity, whereas a symbol in normal type represents a magnitude.

The phasor **OA** may alternatively be expressed thus:

$$\mathbf{OA} = \text{OA}\angle\theta$$

If OA is occupying the position shown in Fig. 13.1(b), the vertical component is negative, so that

$$\mathbf{OA} = \text{OB} - \text{jOC} = \text{OA}\angle-\theta$$

Figure 13.2 represents four phasors occupying different quadrants. These phasors can be represented thus:

$$\mathbf{A}_1 = a_1 + \text{j}b_1 = A_1\angle\theta_1 \qquad\qquad [13.1]$$

where

$$A_1 = \sqrt{(a_1^2 + b_1^2)} \quad \text{and} \quad \tan\theta_1 = b_1/a_1$$

Similarly

$$\mathbf{A}_2 = -a_2 + \text{j}b_2 = A_2\angle\theta_2$$

$$\mathbf{A}_3 = -a_3 - \text{j}b_3 = A_3\angle-\theta_3$$

and $\qquad \mathbf{A}_4 = a_4 - \text{j}b_4 = A_4\angle-\theta_4$

The symbol j, when applied to a phasor, alters its direction by 90° in an anticlockwise direction, without altering its length, and is consequently referred to as an *operator*. For example, if we start with a phasor A in phase with the x-axis, as in Fig. 13.3, then jA represents a phasor of the same length upwards along the y-axis. If we apply the operator j to jA, we turn the phasor anticlockwise through another 90°, thus giving jj$A$ or j^2A in Fig. 13.3. The symbol j^2 signifies that we have applied the operator j twice in succession, thereby rotating the phasor through 180°. This reversal of the phasor is equivalent to multiplying by −1, i.e. j$^2A = -A$, so that j^2 can be regarded as being numerically equal to −1 and j = $\sqrt{(-1)}$.

When mathematicians were first confronted by an expression such as $A = 3 + \text{j}4$, they thought of j4 as being an imaginary number rather than a real number; and it was Argand, in 1806, who first suggested that an expression of this form could be represented graphically by plotting the 3 units of

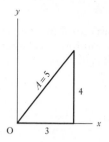

Fig. 13.4 Representation of $A = 3 + j4$

real number in the above expression along the x-axis and the 4 units of *imaginary* or j number along the y-axis, as in Fig. 13.4. This type of number, combining real and imaginary numbers, is termed a *complex number*.

The term *imaginary*, though still applied to numbers containing j, such as j4 in the above expression, has long since lost its meaning of unreality, and the terms *real* and *imaginary* have become established as technical terms, like *positive* and *negative*. For instance, if a current represented by $I = 3 + j4$ amperes were passed through a resistor of, say, 10 Ω, the power due to the 'imaginary' component of the current would be $4^2 \times 10 = 160$ W, in exactly the same way as that due to the 'real' component would be $3^2 \times 10 = 90$ W. From Fig. 13.4 it is seen that the actual current would be 5 A. When this current flows through a 10 Ω resistor, the power is $5^2 \times 10 = 250$ W, namely the sum of the powers due to the real and imaginary components of the current.

Since the real component of a complex number is drawn along the reference axis, namely the x-axis, and the imaginary component is drawn at right angles to that axis, these components are sometimes referred to as the *in-phase* and *quadrature* components respectively.

We can now summarize the various ways of representing a complex number algebraically:

$$A = a + jb \text{ (rectangular or Cartesian notation)}$$

$$= A(\cos\theta + j\sin\theta) \text{ (trigonometric notation)}$$

$$= A\angle\theta \text{ (polar notation)}.$$

13.2 Addition and subtraction of phasors

Suppose A_1 and A_2 in Fig. 13.5 to be two phasors to be added together. From this phasor diagram, it is evident that

$$A_1 = a_1 + jb_1 \quad \text{and} \quad A_2 = a_2 + jb_2$$

It was shown in section 9.9 that the resultant of A_1 and A_2 is given by A, the diagonal of the parallelogram drawn on A_1 and A_2. If a and b are the real and imaginary components respectively of A, then $A = a + jb$.

But it is evident from Fig. 13.5 that

$$a = a_1 + a_2 \quad \text{and} \quad b = b_1 + b_2$$

$$A = a_1 + a_2 + j(b_1 + b_2)$$

$$= (a_1 + jb_1) + (a_2 + jb_2)$$

$$= A_1 + A_2$$

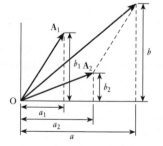

Fig. 13.5 Addition of phasors

Figure 13.6 shows the construction for subtracting phasor A_2 from phasor A_1. If B is the phasor difference of these quantities and if a_1 is assumed less than a_2, then the real component of B is negative.

$$\therefore \qquad B = -a + jb$$

$$= a_1 - a_2 + j(b_1 - b_2)$$

$$= (a_1 + jb_1) - (a_2 + jb_2)$$

$$= A_1 - A_2$$

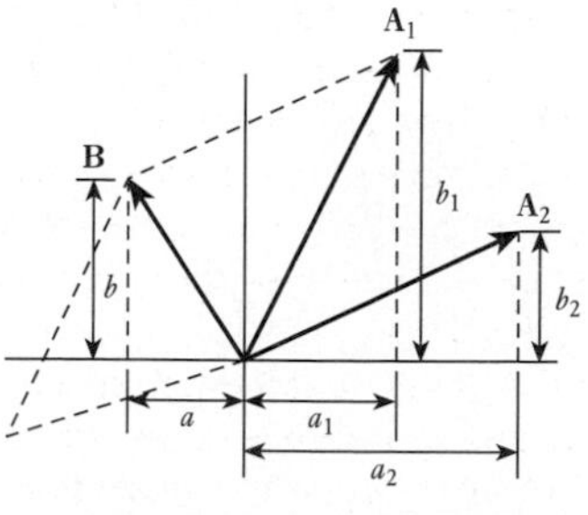

Fig. 13.6 Subtraction of phasors

13.3 Voltage, current and impedance

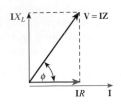

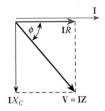

Fig. 13.7 Phasor diagram for R and L in series

Let us first consider a simple circuit possessing resistance R in *series* with an inductive reactance X_L. The phasor diagram is given in Fig. 13.7, where the current phasor is taken as the reference quantity and is therefore drawn along the x-axis, i.e.

$$\mathbf{I} = I + j0 = I\angle 0$$

From Fig. 13.7 it is evident that

$$\mathbf{V} = \mathbf{I}R + j\mathbf{I}X_L = \mathbf{I}(R + jX_L) = \mathbf{IZ} = IZ\angle\phi$$

where $\boxed{\mathbf{Z} = R + jX_L = Z\angle\phi}$ [13.2]

and $\tan\phi = \dfrac{X_L}{R}$

In the expression $\mathbf{V} = \mathbf{IZ}$, $\mathbf{V}$ and $\mathbf{I}$ differ from $\mathbf{Z}$ in that they are associated with time-varying quantities, whereas $\mathbf{Z}$ is a complex number independent of time and is therefore not a phasor in the sense that $\mathbf{V}$ and $\mathbf{I}$ are phasors. It has, however, become the practice to refer to all complex numbers used in a.c. circuit calculations as phasors.

Figure 13.8 gives the phasor diagram for a circuit having a resistance R in series with a capacitive reactance X_C. From this diagram:

$$\mathbf{V} = \mathbf{I}R - j\mathbf{I}X_C = \mathbf{I}(R - jX_C) = \mathbf{IZ} = IZ\angle -\phi$$

where $\boxed{\mathbf{Z} = R - jX_C = Z\angle -\phi}$ [13.3]

Fig. 13.8 Phasor diagram for R and C in series

and $\tan\phi = -\dfrac{X_C}{R}$

Hence for the general case of a circuit having R, X_L and X_C in series,

$$\mathbf{Z} = R + j(X_L - X_C) \quad \text{and} \quad \tan\phi = \frac{X_L - X_C}{R}$$

Example 13.1

Express in rectangular and polar notations, the impedance of each of the following circuits at a frequency of 50 Hz:

(a) a resistance of 20 Ω in series with an inductance of 0.1 H;
(b) a resistance of 50 Ω in series with a capacitance of 40 μF;
(c) circuits (a) and (b) in series.

If the terminal voltage is 230 V at 50 Hz, calculate the value of the current in each case and the phase of each current relative to the applied voltage.

(a)   For 50 Hz

$$\omega = 2\pi \times 50 = 314 \text{ rad/s}$$

$\therefore$ $\mathbf{Z} = 20 + j314 \times 0.1 = 20 + j31.4 \ \Omega$

Hence $Z = \sqrt{\{(20)^2 + (31.4)^2\}} = 37.2 \ \Omega$

and $I = \dfrac{230}{37.2} = 6.18 \text{ A}$

If ϕ is the phase difference between the applied voltage and the current

$$\tan \phi = \frac{31.4}{20} = 1.57$$

$\therefore$ $\phi = 57°30'$, current lagging

The impedance can also be expressed:

$$\mathbf{Z} = 37.2\angle 57°30' \ \Omega$$

This form is more convenient than that involving the j term when it is required to find the product or the quotient of two complex numbers; thus,

$$A\angle\alpha \times B\angle\beta = AB\angle(\alpha + \beta)$$

and $\quad A\angle\alpha/B\angle\beta = (A/B)\angle(\alpha - \beta)$

If the applied voltage is taken as the reference quantity, then

$$\mathbf{V} = 230\angle 0° \text{ volts}$$

$\therefore$ $$\mathbf{I} = \frac{230\angle 0°}{37.2\angle 57°30'} = 6.18\angle -57°30' \ \mathbf{A}$$

(b) $\quad \mathbf{Z} = 50 - j\dfrac{10^6}{314 \times 40} = 50 - j79.6\,\Omega$

$\therefore$ $\quad Z = \sqrt{\{(50)^2 + (79.6)^2\}} = 94 \ \Omega$

and $\quad I = \dfrac{230}{94} = 2.447 \text{ A}$

$$\tan \phi = -\frac{79.6}{50} = -1.592$$

$\therefore$ $\phi = 57°52'$, current leading

The impedance can also be expressed thus:

$$\mathbf{Z} = 94\angle -57°52' \ \Omega$$

$\therefore$ $$\mathbf{I} = \frac{230\angle 0°}{94\angle -57°62'} = 2.45\angle 57°52' \ \mathbf{A}$$

(c) $\quad \mathbf{Z} = 20 + j31.4 + 50 - j79.6 = 70 - j48.2 \ \Omega$

$\therefore$ $\quad Z = \sqrt{\{(70)^2 + (48.2)^2\}} = 85 \ \Omega$

and $\quad I = \dfrac{230}{85} = 2.706 \text{ A}$

$$\tan \phi = -\frac{48.2}{70} = -0.689$$

$\therefore$ $\phi = 34°34'$, current leading

The impedance can also be expressed as

$$\mathbf{Z} = 85\angle -34°34' \ \Omega$$

so that $\quad \mathbf{I} = \dfrac{230\angle 0°}{85\angle -34°34'} = 2.70\angle 34°34' \ \mathbf{A}$

| Example 13.2 | Calculate the resistance and the inductance or capacitance in series for each of the following impedances: |

(a) $10 + j15 \ \Omega$;
(b) $-j80 \ \Omega$;
(c) $50\angle30° \ \Omega$;
(d) $120\angle-60° \ \Omega$.

Assume the frequency to be 50 Hz.

(a) For $\mathbf{Z} = 10 + j15 \ \Omega$

$$\text{Resistance} = 10 \ \Omega$$

and   Inductive reactance $= 15 \ \Omega$

$$\text{Inductance} = \frac{15}{314} = 0.0478 \ \text{H} = 47.8 \ \text{mH}$$

(b) For $\mathbf{Z} = -j80 \ \Omega$

$$\text{Resistance} = 0$$

and   Capacitive reactance $= 80 \ \Omega$

$$\text{Capacitance} = \frac{1}{314 \times 80} \ \text{F} = 39.8 \ \mu\text{F}$$

(c) Figure 13.9(a) is an impedance triangle representing $50\angle30° \ \Omega$. From this diagram, it follows that the reactance is inductive and that $R = Z \cos \phi$ and $X_L = Z \sin \phi$,

$$\therefore \qquad \mathbf{Z} = R + jX_L$$
$$= Z(\cos \phi + j \sin \phi)$$
$$= 50(\cos 30° + j \sin 30°)$$
$$= 43.3 + j25 \ \Omega$$

Hence   Resistance $= 43.3 \ \Omega$

and   Inductive reactance $= 25 \ \Omega$

so that   Inductance $= \dfrac{25}{314} = 0.0796 \ \text{H} = 79.6 \ \text{mH}$

(d) Figure 13.9(b) is an impedance triangle representing $120\angle-60° \ \Omega$. It will be seen that the reactance is capacitive, so that

$$\mathbf{Z} = R - jX_C$$
$$= Z(\cos \phi - j \sin \phi)$$
$$= 120(\cos 60° - j \sin 60°)$$
$$= 60 - j103.9 \ \Omega$$

Hence   Resistance $= 60 \ \Omega$

and   Capacitive reactance $= 103.9 \ \Omega$

$$\therefore \qquad\qquad \text{Capacitance} = \frac{10^6}{314 \times 103.9} = 30.7 \ \mu\text{F}$$

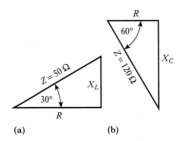

Fig. 13.9 Impedance triangles for Example 13.2(c) and (d)

13.4 Admittance, conductance and susceptance

When resistors having resistances R_1, R_2, etc. are in parallel, the equivalent resistance R is given by

$$\frac{1}{R} = \frac{1}{R_1} + \frac{1}{R_2} + \ldots$$

In d.c. work the reciprocal of the resistance is known as *conductance*. It is represented by symbol G and the unit of conductance is the *siemens*. Hence, if circuits having conductances G_1, G_2, etc. are in parallel, the total conductance G is given by

$$G = G_1 + G_2 + \ldots$$

In a.c. work the conductance is the reciprocal of the resistance *only when the circuit possesses no reactance*. This matter is dealt with more fully in section 13.5.

If circuits having impedances Z_1, Z_2, etc. are connected in parallel across a supply voltage V, then

$$I_1 = \frac{V}{Z_1} \quad I_2 = \frac{V}{Z_2}, \quad \text{etc.}$$

If Z is the equivalent impedance of Z_1, Z_2, etc. in parallel and if I is the resultant current, then, using complex notation, we have

$$\mathbf{I} = \mathbf{I}_1 + \mathbf{I}_2 + \ldots$$

$$\therefore \qquad \frac{\mathbf{V}}{\mathbf{Z}} = \frac{\mathbf{V}}{\mathbf{Z}_1} + \frac{\mathbf{V}}{\mathbf{Z}_2} + \ldots$$

so that

$$\frac{1}{\mathbf{Z}} = \frac{1}{\mathbf{Z}_1} + \frac{1}{\mathbf{Z}_2} + \ldots \qquad [13.4]$$

The reciprocal of impedance is termed *admittance* and is represented by the symbol Y, the unit being again the *siemens* (abbreviation, S). Hence, we may write expression [13.4] thus:

$$\mathbf{Y} = \mathbf{Y}_1 + \mathbf{Y}_2 + \ldots \qquad [13.5]$$

It has already been shown that impedance can be resolved into a real component R and an imaginary component X, as in Fig. 13.10(a). Similarly, an admittance may be resolved into a real component termed *conductance* and an imaginary component termed *susceptance*, represented by symbols G and B respectively as in Fig. 13.10(b), i.e.

$$\mathbf{Y} = G - jB \quad \text{and} \quad \tan\phi = -B/G$$

The significance of these terms will be more obvious when we consider their application to actual circuits.

13.5 RL series circuit admittance

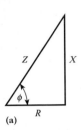

Fig. 13.10 (a) Impedance and (b) admittance triangles

The phasor diagram for this circuit has already been given in Fig. 13.7. From the latter it will be seen that the resultant voltage can be represented thus:

$$\mathbf{V} = \mathbf{I}R + j\mathbf{I}X_L$$

$$\mathbf{Z} = \frac{\mathbf{V}}{\mathbf{I}} = R + jX_L$$

The method of transferring the j term from the denominator to the numerator is known as 'rationalizing'; thus

$$\frac{1}{a + jb} = \frac{a - jb}{(a + jb)(a - jb)} = \frac{a - jb}{a^2 + b^2} \qquad [13.6]$$

If Y is the admittance of the circuit, then

$$\mathbf{Y} = \frac{1}{\mathbf{Z}} = \frac{1}{R + jX_L} = \frac{R - jX_L}{R^2 + X_L^2}$$

$$= \frac{R}{R^2 + X_L^2} - \frac{jX_L}{R^2 + X_L^2} = G - jB_L$$

∴ $$\mathbf{Y} = G - jB_L \qquad [13.7]$$

where $$G = \text{conductance} = \frac{R}{R^2 + X_L^2} = \frac{R}{Z^2} \qquad [13.8]$$

and $B_L =$ inductive susceptance

$$B_L = \frac{X_L}{R^2 + X_L^2} = \frac{X_L}{Z^2} \qquad [13.9]$$

Note that though the inductive reactance ($+jX_L$) is positive, the inductive susceptance is negative ($-jB_L$) as shown in Fig. 13.10. Note also that the impedance and admittance triangles are similar.

From equation [13.8] it is evident that if the circuit has no reactance, i.e. if $X_L = 0$, then the conductance is $1/R$, namely the reciprocal of the resistance. Similarly, from equation [13.9] it follows that if the circuit has no resistance, i.e. if $R = 0$, the susceptance is $1/X_L$, namely the reciprocal of the reactance. In general, we define the *conductance* of a *series* circuit as the ratio of the resistance to the square of the impedance and the *susceptance* as the ratio of the reactance to the square of the impedance.

13.6 RC series circuit admittance

Figure 13.8 gives the phasor diagram for this circuit. From this diagram it follows that

$$\mathbf{V} = \mathbf{I}R - j\mathbf{I}X_C$$

∴ $$\mathbf{Z} = \frac{\mathbf{V}}{\mathbf{I}} = R - jX_C$$

and $$\mathbf{Y} = \frac{1}{R - jX_C} = \frac{R + jX_C}{R^2 + X_C^2} = \frac{R}{R^2 + X_C^2} + \frac{jX_C}{R^2 + X_C^2} = G + jB_C$$

$$\therefore \qquad \mathbf{Y} = G + \mathrm{j}B_C \qquad\qquad [13.10]$$

Note that though the capacitive reactance $(-\mathrm{j}X_C)$ is negative, the capacitive susceptance is positive $(+\mathrm{j}B_C)$.

$$B_C = \frac{X_C}{R^2 + X_C^2} = \frac{X_C}{Z^2} \qquad\qquad [13.11]$$

It has been seen that in the complex expression for an *inductive* circuit, the *impedance* has a *positive* sign in front of the imaginary component, whereas the imaginary component of the *admittance* is preceded by a negative sign. On the other hand, for a *capacitive* circuit, the imaginary component of the *impedance* has a *negative* sign and that of the *admittance* has a *positive* sign.

Remember that inverting a complex number changes the sign of the j term! Thus an impedance with a negative (capacitive) reactance component results in an admittance with a positive (capacitive) susceptance component. And an impedance with a positive (inductive) reactance component results in an admittance with a negative (inductive) susceptance component.

13.7 Parallel admittance

(a) Inductive reactance

From the circuit and phasor diagrams of Figs 13.11 and 13.12 respectively, it follows that

$$\mathbf{I} = \mathbf{I}_R + \mathbf{I}_L = \frac{\mathbf{V}}{R} - \frac{\mathrm{j}\mathbf{V}}{X_L}$$

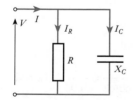

Fig. 13.11   *R* and *L* in parallel

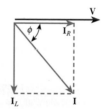

Fig. 13.12   Phasor diagram for Fig. 13.11

$$\therefore \qquad \mathbf{Y} = \frac{\mathbf{I}}{\mathbf{V}} = \frac{1}{R} - \frac{\mathrm{j}}{X_L} = G - \mathrm{j}B_L \qquad\qquad [13.12]$$

(b) Capacitive reactance

From Figs 13.13 and 13.14 it follows that

$$\mathbf{I} = \mathbf{I}_R + \mathbf{I}_C = \frac{\mathbf{V}}{R} + \frac{\mathrm{j}\mathbf{V}}{X_C}$$

Fig. 13.13   *R* and *C* in parallel

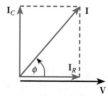

Fig. 13.14   Phasor diagram for Fig. 13.13

$$\mathbf{Y} = \frac{\mathbf{I}}{\mathbf{V}} = \frac{1}{R} + \frac{\mathrm{j}}{X_C} = G + \mathrm{j}B_C \qquad\qquad [13.13]$$

From expressions [13.12] and [13.13], it will be seen that if the admittance of a circuit is $(0.2 - j0.1)$ S, such a network can be represented as a resistance of 5 Ω in *parallel* with an inductive reactance of 10 Ω; whereas if the impedance of a circuit is $(5 + j10)$ Ω, such a network can be represented as a resistance of 5 Ω in *series* with an inductive reactance of 10 Ω.

Example 13.3 Express in rectangular notation the admittance of circuits having the following impedances:

   (a) $(4 + j6)$ Ω;
   (b) $20\angle{-30°}$ Ω.

   (a) $\mathbf{Z} = 4 + j6$ Ω

   ∴ $\mathbf{Y} = \dfrac{1}{4 + j6} = \dfrac{4 - j6}{16 + 36} = 0.077 - j0.115$ S

   (b) $\mathbf{Z} = 20\angle{-30°} = 20(\cos 30° - j \sin 30°)$

            $= 20(0.866 - j0.5) = 17.32 - j10$ Ω

   ∴ $\mathbf{Y} = \dfrac{1}{17.32 - j10} = \dfrac{17.32 + j10}{400}$

            $= \mathbf{0.043 + j0.025}$ S

Alternatively:

            $\mathbf{Y} = \dfrac{1}{20\angle{-30°}} = 0.05\angle{30°}$

            $= \mathbf{0.043 + j0.025}$ S

Example 13.4 The admittance of a circuit is $(0.05 - j0.08)$ S. Find the values of the resistance and the inductive reactance of the circuit if they are

   (a) in parallel;
   (b) in series.

   (a) The conductance of the circuit is 0.05 S and its inductive susceptance is 0.08 S. From equation [13.12] it follows that if the circuit consists of a resistance in parallel with an inductive reactance, then

$$\text{Resistance} = \frac{1}{\text{conductance}} = \frac{1}{0.05} = 20 \text{ Ω}$$

and $$\text{Inductive reactance} = \frac{1}{\text{inductive susceptance}} = \frac{1}{0.08} = 12.5 \text{ Ω}$$

   (b) Since

        $\mathbf{Y} = 0.05 - j0.08$ S

   ∴ $\mathbf{Z} = \dfrac{1}{0.05 - j0.08} = \dfrac{0.05 + j0.08}{0.0089} = 5.62 + j8.99$ Ω

Fig. 13.15 Circuit diagrams for Example 13.4

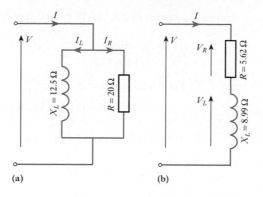

(a) (b)

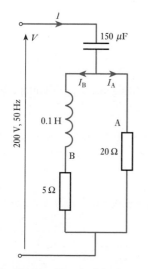

(a) (b)

Fig. 13.16 Phasor diagrams for Example 13.4

Hence if the circuit consists of a resistance in series with an inductance, the resistance is 5.62 Ω and the inductive reactance is 8.99 Ω. The two circuit diagrams are shown in Figs 13.15(a) and (b), and their phasor diagrams are given in Figs 13.16(a) and (b) respectively. The two circuits are equivalent in that they take the same current I for a given supply voltage V, and the phase difference ϕ between the supply voltage and current is the same in the two cases.

Example 13.5

A network is arranged as indicated in Fig. 13.17, the values being as shown. Calculate the value of the current in each branch and its phase relative to the supply voltage. Draw the complete phasor diagram.

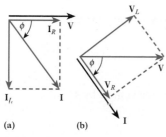

Fig. 13.17 Circuit diagram for Example 13.5

$$\mathbf{Z}_A = 20 + j0 \ \Omega$$

$$\therefore \qquad \mathbf{Y}_A = 0.05 + j0 \ \text{S}$$

$$\mathbf{Z}_B = 5 + j314 \times 0.1 = 5 + j31.4 \ \Omega$$

$$\therefore \qquad \mathbf{Y}_B = \frac{1}{5 + j31.4} = \frac{5 - j31.4}{1010} = 0.004\ 95 - j0.0311 \ \text{S}$$

If $\mathbf{Y}_{AB}$ is the combined admittance of circuits A and B

$$\mathbf{Y}_{AB} = 0.05 + 0.004\ 95 - j0.0311$$

$$= 0.054\ 95 - j0.0311 \ \text{S}$$

If $\mathbf{Z}_{AB}$ is the equivalent impedance of circuits A and B

$$\mathbf{Z}_{AB} = \frac{1}{0.054\ 95 - j0.0311}$$

$$= \frac{0.054\ 95 + j0.0311}{0.003\ 987}$$

$$= 13.78 + j7.8 \ \Omega$$

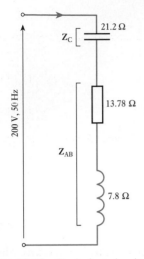

Fig. 13.18   Equivalent circuit of Fig. 13.17

Hence the circuit of Fig. 13.17 can be replaced by that shown in Fig. 13.18.

$$Z_C = -j\frac{10^6}{314 \times 150} = -j21.2 \ \Omega$$

$\therefore$ Total impedance $= \mathbf{Z}$

$$= 13.78 + j7.8 - j21.2 = 13.78 - j13.4$$

$$= \sqrt{\{(13.78)^2 + (13.4)^2\}} \angle\tan^{-1} -\frac{13.4}{13.78}$$

$$= 19.22\angle-44°12' \ \Omega$$

If the supply voltage is $\mathbf{V} = 200\angle0°$ V, therefore supply current is

$$\mathbf{I} = \frac{200\angle0°}{19.22\angle-44°12'} = 10.4\angle44°12' \ A$$

i.e. the supply current is 10.4 A leading the *supply* voltage by 44°12′.
    The p.d. across circuit AB $= \mathbf{V}_{AB} = \mathbf{I}\mathbf{Z}_{AB}$.

But $\mathbf{Z}_{AB} = 13.78 + j7.8$

$$= \sqrt{\{(13.78)^2 + (7.8)^2\}} \angle\tan^{-1}\frac{7.8}{13.78}$$

$$= 15.85\angle29°30' \ \Omega$$

$\therefore$ $\mathbf{V}_{AB} = 10.4\angle44°12' \times 15.85\angle29°30'$

$$= 164.8\angle73°42' \ V$$

Since $\mathbf{Z}_A = 20 + j0 = 20\angle0° \ \Omega$

$\therefore$ $\mathbf{I}_A = \dfrac{164.8\angle73°42'}{20\angle0°} = 8.24\angle73°42' \ A$

i.e. the current through branch A is 8.24 A leading the *supply* voltage by 73°42′. Similarly

$$\mathbf{Z}_B = 5 + j31.4 = 31.8\angle80°58' \ \Omega$$

$\therefore$ $\mathbf{I}_B = \dfrac{164.8\angle73°42'}{31.8\angle80°58'} = 5.18\angle-7°16' \ A$

i.e. the current through branch B is 5.18 A lagging the *supply* voltage by 7°16′. Impedance of C is

$$\mathbf{Z}_C = -j21.2 = 21.2\angle-90° \ \Omega$$

therefore p.d. across C is

$$\mathbf{V}_C = \mathbf{I}\mathbf{Z}_C = 10.4\angle44°12' \times 21.2\angle-90° = 220\angle-45°48' \ V$$

The various voltages and currents of this example are represented by the respective phasors in Fig. 13.19.

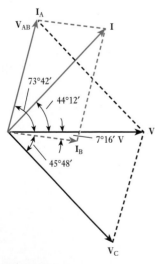

Fig. 13.19   Phasor diagram for Example 13.5

13.8 **Calculation of power, using complex notation**

Suppose the alternating voltage across and the current in a circuit to be represented respectively by

$$\mathbf{V} = V\angle\alpha = V(\cos\alpha + \text{j}\sin\alpha) = a + \text{j}b \qquad [13.14]$$

and

$$\mathbf{I} = I\angle\beta = I(\cos\beta + \text{j}\sin\beta) = c + \text{j}d \qquad [13.15]$$

Since the phase difference between the voltage and current is $(\alpha - \beta)$, active power is

$$VI\cos(\alpha - \beta)$$
$$= VI(\cos\alpha \cdot \cos\beta + \sin\alpha \cdot \sin\beta)$$
$$= ac + bd \qquad [13.16]$$

i.e. the power is given by the sum of the products of the real components and of the imaginary components. Reactive power is

$$VI\sin(\alpha - \beta)$$
$$= VI(\sin\alpha \cdot \cos\beta - \cos\alpha \cdot \sin\beta)$$
$$= bc - ad \qquad [13.17]$$

If we had proceeded by multiplying equation [13.14] by equation [13.15], the result would have been

$$(ac - bd) + \text{j}(bc + ad)$$

The terms within the brackets represent neither the active power nor the reactive power. The correct expressions for these quantities are derived by multiplying the voltage by the *conjugate* of the current, the conjugate of a complex number being a quantity that differs only in the sign of the imaginary component; thus the conjugate of $c + \text{j}d$ is $c - \text{j}d$. Hence

$$(a + \text{j}b)(c - \text{j}d) = (ac + bd) + \text{j}(bc - ad)$$
$$= (\text{active power}) + \text{j}(\text{reactive power})$$
$$S = P + \text{j}Q \qquad [13.18]$$

Example 13.6

The p.d. across and the current in a circuit are represented by $(100 + \text{j}200)$ V and $(10 + \text{j}5)$ A respectively. Calculate the active power and the reactive power.

From equation [13.16], active power is

$$(100 \times 10) + (200 \times 5) = 2000 \text{ W}$$

From equation [13.17], reactive power is

$$(200 \times 10) - (100 \times 5) = 1500 \text{ var}$$

Alternatively:

$$100 + \text{j}200 = 223.6\angle63°26' \text{ V}$$

and

$$10 + \text{j}5 = 11.18\angle26°34' \text{ A}$$

therefore phase difference between voltage and current is

$$63°26' - 26°34' = 36°52'$$

Hence active power is

$$223.6 \times 11.18 \cos 36°52' = \textbf{2000 W}$$

and reactive power is

$$223.6 \times 11.18 \sin 36°52' = \textbf{1500 var}$$

Summary of important formulae

$$\mathbf{A} = a + jb = A(\cos \theta + j \sin \theta) = A\angle\theta$$

where

$$A = \sqrt{(a^2 + b^2)} \quad \text{and} \quad \theta = \tan^{-1}\frac{b}{a}$$

$$A\angle\alpha \times B\angle\beta = AB\angle(\alpha + \beta)$$

$$\frac{A\angle\alpha}{B\angle\beta} = \frac{A}{B}\angle(\alpha - \beta)$$

$$\frac{1}{a + jb} = \frac{a - jb}{a^2 + b^2} \qquad [13.6]$$

For a circuit having R and L in *series*

$$\mathbf{Z} = R + jX_L = Z\angle\phi \qquad [13.2]$$

$$\text{Admittance} = \mathbf{Y} = \frac{1}{\mathbf{Z}} = \frac{R}{Z^2} - \frac{jX_L}{Z^2} = G - jB_L$$

$$= Y\angle{-\phi} \qquad [13.7]$$

For a circuit having R and C in *series*

$$\mathbf{Z} = R - jX_C = Z\angle{-\phi} \qquad [13.3]$$

and

$$\mathbf{Y} = \frac{R}{Z^2} + \frac{jX_C}{Z^2} = G + jB_C = Y\angle\phi \qquad [13.10]$$

Conductance is

$$G = \frac{R}{Z^2}$$

and is $1/R$ only when $X = 0$. Susceptance is

$$B = \frac{X}{Z^2}$$

and is $1/X$ only when $R = 0$.

Summary of important formulae continued

For impedances

$$\mathbf{Z}_1 = R_1 + jX_1 \quad \text{and} \quad \mathbf{Z}_2 = R_2 + jX_2$$

in *series*, total impedance is

$$\mathbf{Z} = \mathbf{Z}_1 + \mathbf{Z}_2 = (R_1 + R_2) + j(X_1 + X_2) = Z\angle\phi$$

where $Z = \surd\{(R_1 + R_2)^2 + (X_1 + X_2)^2\}$

and $\phi = \tan^{-1}\dfrac{X_1 + X_2}{R_1 + R_2}$

For a circuit having R and L in *parallel*

$$\mathbf{Y} = \frac{1}{R} - \frac{j}{X_L} = G - jB_L = Y\angle-\phi \qquad [13.12]$$

For a circuit having R and C in *parallel*

$$\mathbf{Y} = \frac{1}{R} + \frac{j}{X_C} = G + jB_C = Y\angle\phi \qquad [13.13]$$

For admittances

$$\mathbf{Y}_1 = G_1 + jB_1 \quad \text{and} \quad \mathbf{Y}_2 = G_2 + jB_2$$

in *parallel*, total admittance is

$$\mathbf{Y} = \mathbf{Y}_1 + \mathbf{Y}_2 = (G_1 + G_2) + j(B_1 + B_2)$$

$$= Y\angle\phi$$

where $Y = \surd\{(G_1 + G_2)^2 + (B_1 + B_2)^2\}$

and $\phi = \tan^{-1}\dfrac{B_1 + B_2}{G_1 + G_2}$

If $\mathbf{V} = a + jb$

and $\mathbf{I} = c + jd$

active power $= ac + bd$ [13.16]

and reactive power $= bc - ad$ [13.17]

$\mathbf{S} = P + jQ$ [13.18]

Terms and concepts

A **complex number** is one which represents the horizontal and vertical components of a polar number separately. The horizontal component is the **real component** (hence real power in Chapter 12) and the vertical component is the **imaginary component**.

Voltages, currents and impedances can all be represented by complex numbers.

However, care should be taken that complex voltages and complex currents contain time information whereas complex impedances are merely independent operators.

<table>
<tr><td>**Terms and concepts continued**</td><td>Complex notation is especially useful when dealing with parallel networks since it simplifies both the addition (and subtraction) of the branch currents and also the manipulation of the impedance which is difficult if expressed in polar notation.

Power can be expressed in complex form but if we wish to obtain the power from a voltage and current, we need to use the conjugate of the current: this removes the time information which otherwise distorts the solution.</td></tr>
</table>

Exercises 13

1. Express in rectangular and polar notations the phasors for the following quantities: (a) $i = 10 \sin \omega t$; (b) $i = 5 \sin(\omega t - \pi/3)$; (c) $v = 40 \sin(\omega t + \pi/6)$.

   Draw a phasor diagram representing the above voltage and currents.

2. With the aid of a simple diagram, explain the j-notation method of phasor quantities.

   Four single-phase generators whose e.m.f.s can be represented by: $e_1 = 20 \sin \omega t$; $e_2 = 40 \sin(\omega t + \pi/2)$; $e_3 = 30 \sin(\omega t - \pi/6)$; $e_4 = 10 \sin(\omega t - \pi/3)$; are connected in series so that their resultant e.m.f. is given by $e = e_1 + e_2 + e_3 + e_4$. Express each e.m.f. and the resultant in the form $a \pm jb$. Hence find the maximum value of e and its phase angle relative to e_1.

3. Express each of the following phasors in polar notation and draw the phasor diagram: (a) $10 + j5$; (b) $3 - j8$.

4. Express each of the following phasors in rectangular notation and draw the phasor diagram: (a) $20\angle 60°$; (b) $40\angle -45°$.

5. Add the two phasors of Q. 3 and express the result in: (a) rectangular notation; (b) polar notation. Check the values by drawing a phasor diagram to scale.

6. Subtract the second phasor of Q. 3 from the first phasor and express the result in: (a) rectangular notation; (b) polar notation. Check the values by means of a phasor diagram drawn to scale.

7. Add the two phasors of Q. 4 and express the result in: (a) rectangular notation; (b) polar notation. Check the values by means of a phasor diagram drawn to scale.

8. Subtract the second phasor of Q. 4 from the first phasor and express the result in: (a) rectangular notation; (b) polar notation. Check the values by a phasor diagram drawn to scale.

9. Calculate the resistance and inductance or capacitance in *series* for each of the following impedances, assuming the frequency to be 50 Hz: (a) $50 + j30 \, \Omega$; (b) $30 - j50 \, \Omega$; (c) $100\angle 40° \, \Omega$; (d) $40\angle 60° \, \Omega$.

10. Derive expressions, in rectangular and polar notations, for the admittances of the following impedances:

(a) $10 + j15 \, \Omega$; (b) $20 - j10 \, \Omega$; (c) $50\angle 20° \, \Omega$; (d) $10\angle -70° \, \Omega$.

11. Derive expressions in rectangular and polar notations, for the impedances of the following admittances; (a) $0.2 + j0.5$ siemens; (b) $0.08\angle -30°$ siemens.

12. Calculate the resistance and inductance or capacitance in *parallel* for each of the following admittances, assuming the frequency to be 50 Hz: (a) $0.25 + j0.06$ S; (b) $0.05 - j0.1$ S; (c) $0.8\angle 30°$ S; (d) $0.5\angle -50°$ S.

13. A voltage, $v = 150 \sin(314t + 30°)$ volts, is maintained across a coil having a resistance of $20 \, \Omega$ and an inductance of 0.1 H. Derive expressions for the r.m.s. values of the voltage and current phasors in: (a) rectangular notation; (b) polar notation. Draw the phasor diagram.

14. A voltage, $v = 150 \sin(314t + 30°)$ volts, is maintained across a circuit consisting of a $20 \, \Omega$ non-reactive resistor in series with a loss-free $100 \, \mu F$ capacitor. Derive an expression for the r.m.s. value of the current phasor in: (a) rectangular notation; (b) polar notation. Draw the phasor diagram.

15. Calculate the values of resistance and reactance which, when in parallel, are equivalent to a coil having a resistance of $20 \, \Omega$ and a reactance of $10 \, \Omega$.

16. The impedance of two parallel branches can be represented by $(24 + j18) \, \Omega$ and $(12 - j22) \, \Omega$ respectively. If the supply frequency is 50 Hz, find the resistance and inductance or capacitance of each circuit. Also, derive a symbolic expression in polar form for the admittance of the combined circuits, and thence find the phase angle between the applied voltage and the resultant current.

17. A coil of resistance $25 \, \Omega$ and inductance 0.044 H is connected in parallel with a branch made up of a $50 \, \mu F$ capacitor in series with a $40 \, \Omega$ resistor, and the whole is connected to a 200 V, 50 Hz supply. Calculate, using symbolic notation, the total current taken from the supply and its phase angle, and draw the complete phasor diagram.

18. The current in a circuit is given by 4.5 + j12 A when the applied voltage is 100 + j150 V. Determine: (a) the complex expression for the impedance, stating whether it is inductive or capacitive; (b) the active power; (c) the phase angle between voltage and current.

19. Explain how alternating quantities can be represented by complex numbers.

    If the potential difference across a circuit is represented by 40 + j25 V, and the circuit consists of a coil having a resistance of 20 Ω and an inductance of 0.06 H and the frequency is 79.5 Hz, find the complex number representing the current in amperes.

20. The impedances of two parallel branches can be represented by (20 + j15) Ω and (10 − j60) Ω respectively. If the supply frequency is 50 Hz, find the resistance and the inductance or capacitance of each branch. Also, derive a complex expression for the admittance of the combined network, and thence find the phase angle between the applied voltage and the resultant current. State whether this current is leading or lagging relatively to the voltage.

21. An alternating e.m.f. of 100 V is induced in a coil of impedance 10 + j25 Ω. To the terminals of this coil there is joined a circuit consisting of two parallel impedances, one of 30 − j20 Ω and the other of 50 + j0 Ω. Calculate the current in the coil in magnitude and phase with respect to the induced voltage.

22. A circuit consists of a 30 Ω non-reactive resistor in series with a coil having an inductance of 0.1 H and a resistance of 10 Ω. A 60 μF loss-free capacitor is connected in parallel with the *coil*. The network is connected across a 200 V, 50 Hz supply. Calculate the value of the current in each branch and its phase relative to the supply voltage.

23. An impedance of 2 + j6 Ω is connected in series with two impedances of 10 + j4 Ω and 12 − j8 Ω, which are in parallel. Calculate the magnitude and power factor of the main current when the combined circuit is supplied at 200 V.

24. The p.d. across and the current in a given circuit are represented by (200 + j30) V and (5 − j2) A respectively. Calculate the active power and the reactive power. State whether the reactive power is leading or lagging.

25. Had the current in Q. 24 been represented by (5 + j2) A, what would have been the active power and the reactive power? Again state whether the reactive power is leading or lagging.

26. A p.d. of 200∠30° V is applied to two branches connected in parallel. The currents in the respective branches are 20∠60° A and 40∠30° A. Find the apparent power (in kVA) and the active power (in kW) in each branch and in the main network. Express the current in the main network in the form $A + jB$.

Chapter fourteen

Resonance in AC Circuits

Objectives

When you have studied this chapter, you should

- have an understanding of the response of L and C to frequency variation
- be capable of analysing the response of an RLC series circuit to frequency variation
- be aware that resonance involves the oscillation of energy between the inductive and capacitive components
- have an understanding of resonance in a simple parallel network
- be capable of analysing the resonant condition of simple series and parallel circuits
- understand the term Q factor for both series and parallel resonant circuits

Contents

Having become familiar with the a.c. analysis of series circuits having resistance, inductance and capacitance, it is now possible to question what happens when the inductive and capacitive reactances of an a.c. circuit are equal in magnitude. This chapter will discuss this condition for both series and parallel a.c. circuits. It will lead to an understanding of an important circuit condition known as *resonance*, which may be achieved either by varying the frequency or by varying one of the components, usually the capacitance, until the inductive and capacitive reactances are equal. Resonance makes the circuit respond in a markedly different manner at the resonant frequency than at other frequencies, a behaviour which is the basis of the selection process used in many communications systems. The tuning of a radio is a common example. Series resonance not only enables a signal of given frequency to be considerably magnified but also separates it from signals of other frequencies.

Parallel resonant circuits will also be analysed. This type of circuit, when used in communication applications, is referred to as a *rejector*, since its impedance is a maximum and the resultant current a minimum at resonance. This is in contrast to the series resonant circuit in which the impedance is a minimum and the resultant current a maximum at the resonant frequency.

14.1 Introduction

When introducing a.c. circuits, a supply was defined by its voltage and frequency. For many applications, they are constant; for example, the source of supply to our homes. It is also true for many data and control circuit applications. However, many communications systems involve circuits in which either the supply voltage (in such applications it is usually called a signal) operates with a varying frequency, or a number of signals operate together, each with its own frequency. An understanding of communications systems including radio, television and telephones, as well as machine control systems, requires a knowledge of how circuits are affected by a variation of the frequency. In particular, the condition known as resonance will be investigated, which will be introduced by analysing the effect of frequency variation on the capacitive and inductive reactances of a series RLC circuit.

14.2 Frequency variation in a series *RLC* circuit

Let us consider the series RLC circuit in Fig. 14.1.

We have seen in section 10.11 that the impedance Z of this circuit is given by

$$Z = \sqrt{\left\{ R^2 + \left(\omega L - \frac{1}{\omega C} \right)^2 \right\}}$$

[14.1]

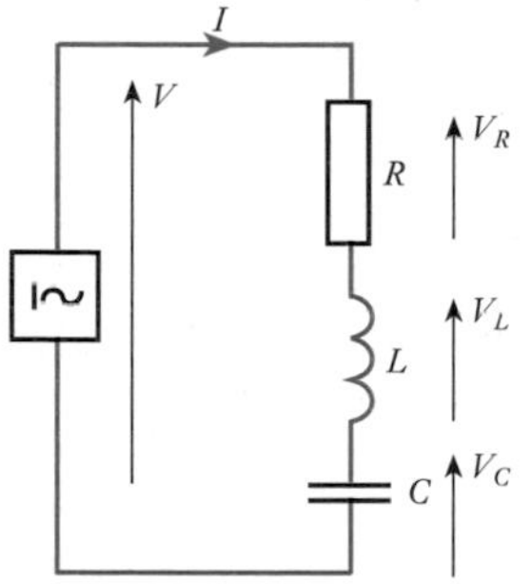

The value of the reactance X of the circuit $\omega L - 1/(\omega C)$ (i.e. inductive reactance − capacitive reactance) will depend on frequency.

For the inductive reactance:

$$|X_L| = \omega L = 2\pi f L$$

which will increase with frequency.

For the capacitive reactance:

$$|X_C| = \frac{1}{\omega C} = \frac{1}{2\pi f C}$$

which is largest at low frequencies.

Fig. 14.1 Circuit with R, L and C in series

Fig. 14.2 Inductive reactance increases linearly with frequency

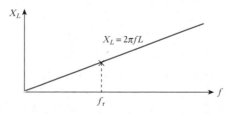

Fig. 14.3 Capacitive reactance decreases with frequency

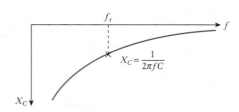

By comparing Figs 14.2 and 14.3, it can be seen that:

- at frequency f_r, $|X_L| = |X_C|$ so the impedance Z, from equation [14.1], is purely resistive;
- below f_r, $|X_L| < |X_C|$ so the circuit is capacitive;
- above f_r, $|X_L| > |X_C|$ so the circuit is inductive.

The overall variation of the impedance $|Z|$ can be seen in Fig. 14.4. This figure shows that, for frequency f_r, the inductive reactance AB and the capacitive reactance AC are equal in magnitude so that the resultant reactance is zero. Consequently, the impedance is then only the resistance AD of the circuit. Furthermore, as the frequency is reduced below f_r or increased above f_r, the impedance increases and therefore the current decreases. The actual shapes and relative magnitudes of these curves depend on the actual values of R, L and C in the series resonant circuit.

Also, it will be seen from Fig. 14.5 that when the frequency is f_r, the voltages across L and C are equal (but opposite in phase so they cancel) and each is much greater than the supply voltage. Such a condition is referred to

Fig. 14.4 Variation of reactance and impedance with frequency

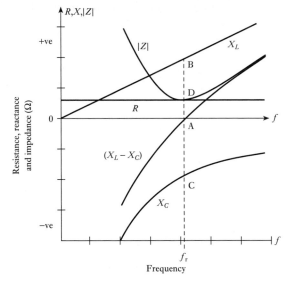

Fig. 14.5 Effect of frequency variation on voltages across R, L and C

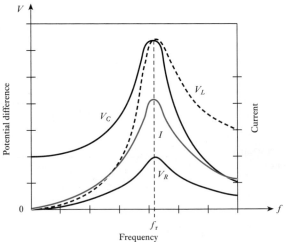

as *resonance*, an effect that is extremely important in communications, e.g. radio, partly because it provides a simple method of increasing the sensitivity of a receiver and partly because it provides *selectivity*, i.e. it enables a signal of given frequency to be considerably magnified so that it can be separated from signals of other frequencies.

The phase of the circuit impedance (from equation [10.22]) is given by

$$\phi = \tan^{-1}\frac{(X_L - X_C)}{R}$$

[14.2]

and will also depend on frequency as the values of X_L and X_C change. Figure 14.6 shows phasor diagrams of impedance below f_r, at f_r and above f_r.

Hence it can be seen that:

- below resonance $X_L < X_C$, ϕ is negative, the circuit is capacitive;
- at resonance (f_r) $X_L = X_C$, ϕ is zero, the circuit is purely resistive;
- above resonance $X_L > X_C$, ϕ is positive, the circuit is inductive.

Figure 14.7 shows the overall variation of phase as the frequency is increased.

Fig. 14.6 Impedance diagrams (a) below, (b) at and (c) above the resonant frequency

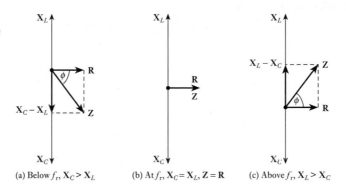

(a) Below f_r, $X_C > X_L$ (b) At f_r, $X_C = X_L$, $Z = R$ (c) Above f_r, $X_L > X_C$

Fig. 14.7 Variation of magnitude $|I|$ and phase ϕ of current with frequency in a series *RLC* circuit

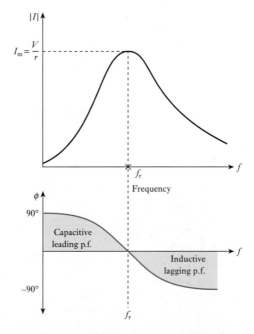

14.3 The resonant frequency of a series *RLC* circuit

At the frequency f_r, $|X_L| = |X_C|$:

$$2\pi f_r L = \frac{1}{2\pi f_r C}$$

so

$$f_r = \frac{1}{2\pi\sqrt{(LC)}} \qquad [14.3]$$

At this frequency f_r, known as the resonant frequency, $Z = R$ and $I = V/R$. The angular frequency ω_r, at resonance, is

$$\omega_r = \frac{1}{\sqrt{(LC)}}$$

14.4 The current in a series *RLC* circuit

Since

$$I = \frac{V}{Z\angle\phi} = \frac{V\angle{-\phi}}{Z}$$

from equations [14.1] and [14.2]:

$$I = \frac{V}{\sqrt{\left\{ R^2 + \left(\omega L - \dfrac{1}{\omega C} \right)^2 \right\}}} \angle -\tan^{-1}\frac{(\omega L - 1/(\omega C))}{R} \qquad [14.4]$$

The variation of the magnitude and phase of the current with frequency is shown in Fig. 14.7. I is a maximum when $\omega L = 1/(\omega C)$, when the circuit is resistive ($\phi = 0$). Hence

$$I_m = \frac{V}{R}$$

14.5 Voltages in a series *RLC* circuit

The voltages, shown in Fig. 14.1, across the inductor (V_L) and the capacitor (V_C) are 180° out of phase with each other. They are both 90° out of phase with the voltage across the resistor. The current I and V_R are always in phase. Figure 14.8 shows the phasor diagram of the voltages in the series *RLC* circuit below and above the resonant frequency and at the resonant frequency f_r.

Fig. 14.8 Phasor diagram of series *RLC* circuit.
(a) Capacitive, **I** leads **V**. Below resonant frequency f_r.
(b) Resistive **V** and **I** in phase. At resonant frequency f_r.
(c) Inductive, **I** lags **V**. Above resonant frequency f_r

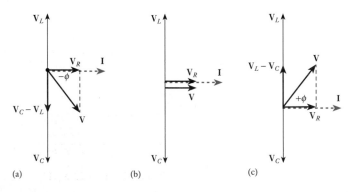

(a)   (b)   (c)

14.6   Quality factor Q

At resonance, the voltages across L and C can be very much greater than the applied voltage:

$$V_C = IX_C \angle{-90°}$$

Substituting for I (refer to equation [14.4], and neglecting phase),

$$|V_C| = \frac{V \cdot X_C}{\sqrt{\{R^2 + (X_L - X_C)^2\}}} \qquad [14.5]$$

At resonance $X_L = X_C$, $\omega = \omega_r$ and $X_C = 1/(\omega_r C)$. So

$$V_C = \frac{V}{R} X_C$$

$$V_C = \frac{V}{\omega_r CR} = QV \qquad [14.6]$$

where $\quad Q = \dfrac{1}{\omega_r CR}$

Q is termed the Q factor or voltage magnification, because V_C equals Q multiplied by the source voltage V:

$$Q = \frac{1}{\omega_r CR} = \frac{\omega_r L}{R} \qquad [14.7]$$

Also, since $\omega_r = 1/\sqrt{(LC)}$

$$Q = \frac{1}{R}\sqrt{\frac{L}{C}} \qquad [14.8]$$

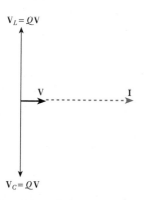

Fig. 14.9 Voltage magnification Q in series resonant circuit

In a series RLC circuit, values of V_L and V_C can actually be very large at resonance and can lead to component damage if not recognized and subject to careful design. Figure 14.9 illustrates the voltage magnification in a series resonant circuit.

Example 14.1

A circuit, having a resistance of $4.0\,\Omega$ and inductance of $0.50\,H$ and a variable capacitance in series, is connected across a $100\,V$, $50\,Hz$ supply. Calculate:

    (a) the capacitance to give resonance;
    (b) the voltages across the inductance and the capacitance;
    (c) the Q factor of the circuit.

(a) For resonance:

$$2\pi f_r L = 1/(2\pi f_r C)$$

$$\therefore \qquad C = \frac{1}{(2 \times 3.14 \times 50)^2 \times 0.5}$$

$$= 20.3 \times 10^{-6}\,F = \mathbf{20.3\ \mu F}$$

(b) At resonance:

$$I = \frac{V}{R} = \frac{100}{4} = 25 \text{ A}$$

$\therefore$ p.d. across inductance $= V_L \times 2 \times \pi \times 50 \times 0.5 \times 25$

$$= 3930 \text{ V}$$

and   p.d. across capacitor $= V_C = \mathbf{3930 \text{ V}}$

Or alternatively:

$$V_C = IX_C$$

$\therefore$ $$V_C = \frac{25 \times 10^6}{2 \times 3.14 \times 50 \times 20.3} = \mathbf{3930 \text{ V}}$$

(c) From equation [14.7]

$$Q \text{ factor} = \frac{2 \times 3.14 \times 50 \times 0.5}{4} = \mathbf{39.3}$$

The voltages and current are represented by the respective phasors, but not to scale, in Fig. 14.9. Figure 14.10 shows how the current taken by this circuit varies with frequency, the applied voltage being assumed constant at 100 V.

Fig. 14.10 Variation of current with frequency for circuit of Example 14.1

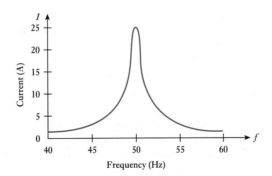

Example 14.2

A coil of resistance 5.0 Ω and inductance 1.0 mH is connected in series with a 0.20 μF capacitor. The circuit is connected to a 2.0 V, variable frequency supply (Fig. 14.11). Calculate the frequency at which resonance occurs, the voltages across the coil and the capacitor at this frequency and the Q factor of the circuit.

Now

$$f_r = \frac{1}{2\pi\sqrt{(LC)}}$$

$$= \frac{1}{2\pi\sqrt{(1 \times 10^{-3} \times 0.2 \times 10^{-6})}}$$

$$= \mathbf{11.25 \text{ kHz}}$$

Fig. 14.11 Circuit and phasor diagrams for Example 14.2

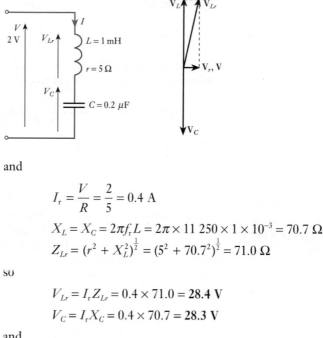

and

$$I_r = \frac{V}{R} = \frac{2}{5} = 0.4 \text{ A}$$

$$X_L = X_C = 2\pi f_r L = 2\pi \times 11\ 250 \times 1 \times 10^{-3} = 70.7\ \Omega$$

$$Z_{Lr} = (r^2 + X_L^2)^{\frac{1}{2}} = (5^2 + 70.7^2)^{\frac{1}{2}} = 71.0\ \Omega$$

so

$$V_{Lr} = I_r Z_{Lr} = 0.4 \times 71.0 = \textbf{28.4 V}$$

$$V_C = I_r X_C = 0.4 \times 70.7 = \textbf{28.3 V}$$

and

$$Q \text{ factor} = \frac{\omega_r L}{r} = \frac{70.7}{5} = \textbf{14.1}$$

In Example 14.1 the voltages across the inductance and the capacitance at resonance are each nearly 40 times the supply voltage while the voltage magnification in Example 14.2 is about 14.

14.7 Oscillation of energy at resonance

The capacitive and inductive reactances store energy that oscillates between them, the energy being at one moment stored as electrostatic energy in the capacitor, and a quarter of a cycle later as magnetic energy in the inductor. At the resonant frequency, when the capacitive and inductive reactances are equal, they transfer equal energy, and the circuit appears resistive. The maximum magnetic energy stored in L at any instant is $\frac{1}{2}LI_m^2$ joules, where I_m is the maximum value of current in the inductor, and the maximum electrostatic energy in C at instant B is $\frac{1}{2}CV_m^2$ joules, where V_m represents the maximum value of the voltage across the capacitor. However, energy is dissipated as I^2R losses in the resistance of the circuit as the energy is passed backwards and forwards between L and C.

This leads to a more general definition of Q factor. It is defined as the ratio of the reactive power, of either the capacitor or the inductor to the average power of the resistor at resonance:

$$Q = \frac{\textbf{reactive power}}{\textbf{average power}} \qquad [14.9]$$

The lower the value of R, the lower is the power dissipated in the resistor. The value of Q is in turn higher and the more defined is the resonance peak.

For inductive reactance X_L at resonance:

$$Q = \frac{\text{reactive power}}{\text{average power}} = \frac{I^2 X_L}{I^2 R} = \frac{X_L}{R} = \frac{\omega_r L}{R}$$

as derived in equation [14.7].

For capacitive reactance X_C at resonance:

$$Q = \frac{\text{reactive power}}{\text{average power}} = \frac{I^2 X_C}{I^2 R} = \frac{X_C}{R} = \frac{1}{\omega_r C R}$$

as derived in equation [14.7].

14.8   Mechanical analogy of a resonant circuit

It was pointed out in section 10.5 that inertia in mechanics is analogous to inductance in the electrical circuit, and in section 10.9 that elasticity is analogous to capacitance. A very simple mechanical analogy of an electrical circuit possessing inductance, capacitance and a very small resistance can therefore be obtained by attaching a mass W (Fig. 14.12) to the lower end of a helical spring S, the upper end of which is rigidly supported. If W is pulled down a short distance and then released, it will oscillate up and down with gradually decreasing amplitude. By varying the mass of W and the length of S it can be shown that the greater the mass and the more flexible the spring, the lower is the natural frequency of oscillation of the system.

If W is set into a slight oscillation and then given a small downward tap each time it is moving downwards, the oscillations may be made to grow to a large amplitude. In other words, when the frequency of the applied force is the same as the natural frequency of oscillation, a small force can build up large oscillations, the work done by the applied force being that required to supply the losses involved in the transference of energy backwards and forwards between the kinetic and potential forms of energy.

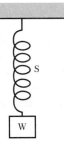

Fig. 14.12 Mechanical analogy of a resonant circuit

Examples of mechanical resonance are very common; for instance, the rattling of a loose member of a vehicle at a particular speed or of a loudspeaker diaphragm when reproducing a sound of a certain pitch, and the oscillations of the pendulum of a clock and of the balance wheel of a watch due to the small impulse given regularly through the escapement mechanism from the mainspring.

14.9   Series resonance using complex notation

Consider the circuit diagram of Fig. 14.13 in which the inductive and capacitive reactances are presented in their complex form.

From Kirchhoff's voltage law:

$$\mathbf{V} = \mathbf{V}_R + \mathbf{V}_L + \mathbf{V}_C \text{ (phasor sum)}$$

$$= IR + Ij\omega L + \frac{I}{j\omega C}$$

$$= I\left\{ R + j\left(\omega L - \frac{1}{\omega C} \right) \right\}$$

Fig. 14.13   Series resonant RLC circuit using j operator

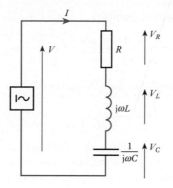

$\therefore$ $$I = \frac{V}{R + j\left(\omega L - \dfrac{1}{\omega C}\right)} = \frac{V}{Z}$$ [14.10]

At resonance, we have already seen that $\omega L = 1/(\omega C)$ and $Z = R$.

14.10   Bandwidth

The bandwidth of a circuit is defined as the frequency range between the half-power points when $I = I_{max}/\sqrt{2}$. This is illustrated in Fig. 14.14.

The bandwidth, BW, equals $\omega_2 - \omega_1$, where the frequencies ω_2 and ω_1 are referred to as half-power points or frequencies. They are also referred to as cut-off frequencies. The term half-power frequency can be justified by consideration of the conditions for maximum and half-power for the series RLC circuit.

At maximum power, when $\omega = \omega_r$,

$$I_{max} = \frac{V}{R}$$

$\therefore$ $$P_{max} = I_{max}^2 R = \frac{V^2}{R}$$

Fig. 14.14   The resonance peak, bandwidth and half-power frequencies

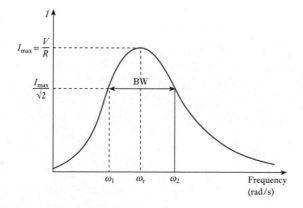

At the half-power points

$$P_1 = P_2 = \frac{I_{max}^2 \cdot R}{2} = \left(\frac{I_{max}}{\sqrt{2}}\right)^2 \cdot R$$

Thus, the condition for half-power is given when

$$|I| = \frac{I_{max}}{\sqrt{2}} = \frac{V}{R\sqrt{2}}$$

The vertical lines either side of $|I|$ indicate that only the magnitude of the current is under consideration – but the phase angle will not be neglected. To obtain the condition for half-power in the circuit, when $|Z| = R\sqrt{2}$, refer to the impedance diagrams of Fig. 14.6. It can be deduced that this occurs both above and below the resonant frequency when, including the phase angle,

$$Z = R\sqrt{2}\angle\pm45°$$

or, to use the complex form,

$$Z = R(1 \pm j1)$$

Thus for half-power:

$$I = \frac{V}{R(1 \pm j1)} \quad \text{and} \quad Z = R(1 \pm j1) \qquad [14.11]$$

Note that, at the half-power points, the phase angle of the current is 45°. Below the resonant frequency, at ω_1, the circuit is capacitive and $Z = R(1 - j1)$. Above the resonant frequency, at ω_2, the circuit is inductive and $Z = R(1 + j1)$. If we consider, from equation [14.10],

$$Z = R + j\left(\omega L - \frac{1}{\omega C}\right)$$

$$= R\left\{1 + j\left(\frac{\omega L}{R} - \frac{1}{\omega C R}\right)\right\}$$

at the half-power points, from equation [14.11],

$$\frac{\omega L}{R} - \frac{1}{\omega C R} = \pm 1$$

Since, from equation [14.7],

$$Q = \frac{1}{\omega_r C R} = \frac{\omega_r L}{R}$$

then, at the half-power points,

$$Q\left(\frac{\omega}{\omega_r} - \frac{\omega_r}{\omega}\right) = \pm 1$$

For ω_2: $\quad Q\left(\frac{\omega_2}{\omega_r} - \frac{\omega_r}{\omega_2}\right) = 1 \qquad [14.12]$

For ω_1: $\boxed{Q\left(\dfrac{\omega_1}{\omega_r} - \dfrac{\omega_r}{\omega_1}\right) = -1}$ [14.13]

It can be deduced directly from equations [14.12] and [14.13] that

$$\omega_2 = \frac{\omega_r}{2Q} + \omega_r\sqrt{1 + -\frac{1}{4Q^2}}$$

and $$\omega_1 = \frac{-\omega_r}{2Q} + \omega_r\sqrt{1 + \frac{1}{4Q^2}}$$

Hence the bandwidth

$$BW = \omega_2 - \omega_1 = \frac{\omega_r}{Q}$$

i.e. $\boxed{\textbf{Bandwidth} = \dfrac{\textbf{resonant frequency}}{Q \textbf{ factor}}}$ [14.14]

Note also that $\omega_1\omega_2 = \omega_r^2$.

14.11 Selectivity

It can be deduced from equation [14.14] that the shape of the resonance curve depends on the Q factor. The bandwidth, the range of frequencies for which the power is greater than half-power, is narrower, the higher Q is.

A circuit is said to be selective if the response has a sharp peak and narrow bandwidth and is achieved with a high Q factor. Q is therefore a measure of selectivity. It should be noted that in practice the curve of $|I|$ against ω is not symmetrical about the resonant frequency. This can be observed in Fig. 14.15. In fact, the curve is the inverse of the impedance Z (see Fig. 14.4), but the larger the value of Q, the more symmetrical the curve appears about the resonant frequency. It can be assumed here that the resonant frequency lies at the midpoint of the bandwidth.

In order to obtain higher selectivity, Q must be large.

Fig. 14.15 The effect of Q on I_{max} and on the bandwidth (BW)

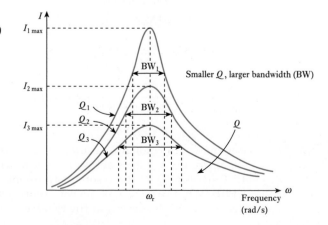

Since

$$Q = \frac{\omega_r L}{R} = \frac{1}{\omega_r CR}$$

thus, for high selectivity, R must be small. This means that the total series resistance of the circuit including the source resistance must be small. Therefore a series tuned circuit must be driven by a voltage source having a low internal resistance if it is to exhibit a resonance peak and be selective. Examples 14.3 and 14.4 illustrate the design of series resonant circuits to be selective of a small range of frequencies, the bandwidth. Such circuits are the basis of circuits known as *bandpass* or *passband filters*, which are explored further in Chapter 18.

Example 14.3 The bandwidth of a series resonant circuit is 500 Hz. If the resonant frequency is 6000 Hz, what is the value of Q? If $R = 10\ \Omega$, what is the value of the inductive reactance at resonance? Calculate the inductance and capacitance of the circuit.

From equation [14.14]:

$$\text{Bandwidth (BW)} = \frac{\text{resonant frequency}}{Q \text{ factor}}$$

Hence

$$Q = \frac{f_r}{\text{BW}} = \frac{6000}{500} = 12$$

From equation [14.7]:

$$Q = \frac{X_L}{R}$$

$$\therefore \quad X_L = QR = 12 \times 10 = 120\ \Omega$$

$$X_L = 2\pi f_r L$$

$$\therefore \quad L = \frac{X_L}{2\pi f_r} = \frac{120}{2\pi 6000} = 3.18\ \text{mH}$$

$$|X_L| = |X_C| = 120\ \Omega$$

$$\therefore \quad X_C = \frac{1}{2\pi f_r C} = 120\ \Omega$$

$$C = \frac{1}{2\pi \times 6000 \times 120} = 0.22\ \mu\text{F}$$

Example 14.4 For the series resonant circuit in Fig. 14.16, calculate I at resonance. What are the voltages across the three series components, R, L and C? Calculate Q. If the resonant frequency is 6000 Hz find the bandwidth. What are the half-power frequencies and what is the power dissipated in the circuit at the two frequencies?

Fig. 14.16 Circuit for Example 14.4

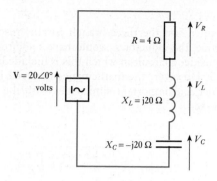

At the resonant frequency, $Z = R$.

$$\therefore \qquad I_{max} = \frac{20\angle 0°}{4} = \mathbf{5\angle 0°}$$

$$V_R = I_{max}R = 5\angle 0° \times 4\ \Omega$$

$$V_R = \mathbf{20\angle 0°\ V}$$

$$V_L = I_{max}X_L = 5\angle 0.20°\angle 90°$$

$$V_L = \mathbf{100\angle 90°\ V}$$

$$V_C = I_{max}X_C = 5\angle 0.20°\angle -90°$$

$$V_C = \mathbf{100\angle -90°\ V}$$

$$Q = \frac{X_L}{R} = \frac{20}{4} = 5$$

From equation [14.14]:

$$\text{Bandwidth} = \frac{\text{resonant frequency}}{Q\text{ factor}}$$

$$\therefore \qquad \text{BW} = \frac{6000}{5} = \mathbf{1200\ Hz}$$

The half-power frequencies are at frequencies 600 Hz above and below the resonant frequency of 6000 Hz:

$$\therefore \qquad f_2 = 6000 + 600 = \mathbf{6600\ Hz}$$

$$f_1 = 6000 - 600 = \mathbf{5400\ Hz}$$

$$P_{max} = I_{max}^2 R$$

$$= 5^2 \times 4\ \Omega$$

$$= \mathbf{100\ W}$$

At the half-power frequencies f_1 and f_2:

$$P_{\text{half-power}} = \frac{1}{2}P_{max} = \mathbf{50\ W}$$

14.12 Parallel resonance

Consider first the three-branch parallel resonant circuit in Fig. 14.17. In practice, the inductor L would have some resistance, so it is more practical to consider a circuit in which this is included, as will be discussed in section 14.15. However, the mathematics involved in the analysis of Fig. 14.17 is simpler and important similarities with the series resonant circuit can be discussed.

Fig. 14.17 The three-branch parallel resonant circuit

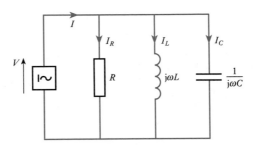

In Fig. 14.17, $V = IZ$, where Z is the net impedance of the three parallel branches. In parallel circuits of this nature, it is simpler to consider the total admittance Y of the three branches. Thus

$$V = IZ = \frac{I}{Y}$$

where $Y = G + \dfrac{1}{j\omega L} + j\omega C$

$$= G - \frac{j}{\omega L} + j\omega C$$

$\therefore$ $Y = G + j\left(\omega C - \dfrac{1}{\omega L}\right)$

so $$V = \frac{I}{G + j\left(\omega C - \dfrac{1}{\omega L}\right)}$$ [14.15]

Hence by comparison with the series resonant circuit equation [14.10], we have dual equations:

Series	Parallel
$I = \dfrac{V}{R + j\left(\omega L - \dfrac{1}{\omega C}\right)} = \dfrac{V}{Z}$	$V = \dfrac{I}{G + j\left(\omega C - \dfrac{1}{\omega L}\right)} = \dfrac{I}{Y}$

Therefore, the same results apply, except that V is used instead of I, Y instead of Z and G instead of R. Thus for both a parallel and a series resonant circuit, resonance occurs when

$$\omega C = \frac{1}{\omega L}$$

Therefore: $\omega_r = \dfrac{1}{\sqrt{(LC)}}$

At the resonant frequency, $Y = G$, the conductance of the parallel resistance, and $I = VG$.

14.13 Current magnification

At resonance, $|X_L| = |X_C|$. Note that the vertical lines either side of these symbols indicate that the magnitudes of the inductive and capacitive reactances are equal. (The phases are opposite.) From Fig. 14.17, at the resonant frequency ω_r,

$$I_L = \frac{V}{j\omega_r L}$$

Since, at resonance, $V = I/G$

$$I_L = \frac{I}{j\omega_r LG}$$

$\therefore \qquad |I_L| = \dfrac{I}{\omega_r LG} = QI$

where Q is the current magnification

$\therefore \qquad Q = \dfrac{1}{\omega_r LG} = \dfrac{\omega_r C}{G} = \dfrac{B}{G} = \dfrac{R}{X}$ \hfill [14.16]

where B is the inductive or capacitive susceptance and X is the inductive or capacitive reactance.

By substituting $\omega_r = 1/\sqrt{(LC)}$ into equation [14.16],

$$Q = \frac{1}{G}\sqrt{\frac{C}{L}} = R\sqrt{\frac{C}{L}}$$

From equation [14.9], at resonance,

$$Q = \frac{\text{reactive power}}{\text{average power}}$$

For the parallel circuit:

$$Q = \frac{V^2/X}{V^2/R} = \frac{R}{X} = \frac{1}{\omega_r LG}$$

This is as expected from equation [14.16], which confirms that Q has the same inherent definition for both parallel and series circuits. It may appear, at first glance, that the expressions for Q for a series and a parallel resonant circuit are quite different. It will be shown, in section 14.15, that they are the same. Meanwhile, it should be remembered that R and X in equation [14.16] are parallel circuit components – unlike R and X in the series circuit.

14.14　　　**Parallel and series equivalents**

It has already been mentioned that, in practice, the inductor L would have some resistance. It is more practical therefore to consider a circuit in which this is included, such as, for example, that of Fig. 14.18.

Fig. 14.18 The two-branch parallel resonant circuit

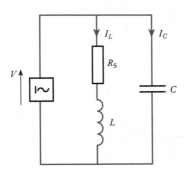

However, if the inductive branch of the circuit of Fig. 14.18, containing series inductance and resistance, can be represented mathematically as two parallel components of inductance and resistance, the analysis of the two-branch circuit will revert to that carried out in section 14.12 for a three-branch circuit. Consider, therefore, the circuits of Fig. 14.19.

Fig. 14.19 Series and parallel equivalent circuits

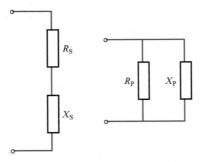

For equivalence:

$$R_S + jX_S = \frac{R_P \cdot jX_P}{R_P + jX_P}$$

$$= \frac{(R_P - jX_P)R_P \cdot jX_P}{R_P^2 + X_P^2}$$

$$= \frac{R_P \cdot X_P^2}{R_P^2 + X_P^2} + j\frac{R_P^2 \cdot X_P}{R_P^2 + X_P^2}$$

$$= \frac{R_P}{1 + \dfrac{R_P^2}{X_P^2}} + j\frac{X_P}{1 + \dfrac{X_P^2}{R_P^2}}$$

Since, from equation [14.16], in a parallel circuit $Q = R_P/X_P$

$$R_S + jX_S = \frac{R_P}{1 + Q^2} + \frac{jX_P}{1 + \dfrac{1}{Q^2}}$$

Equating real and imaginary terms,

$$R_S = \frac{R_P}{1+Q^2} \approx \frac{R_P}{Q^2} \quad \text{when } Q \gg 1$$

$$X_S = \frac{X_P}{1+Q^2} \approx X_P \quad \text{when } Q \gg 1$$

Similarly

$$R_P = (1+Q^2)R_S \approx Q^2 R_S \quad \text{when } Q \gg 1$$

and

$$X_P = (1+Q^2)X_S \approx X_S \quad \text{when } Q \gg 1$$

It can be assumed that, in resonant circuits typically designed to have high Q factors, such an approximation is justified practically. These equivalents can now be used to analyse the two-branch parallel resonant circuit.

<table>
<tr><td>**14.15**</td><td>**The two-branch parallel resonant circuit**</td></tr>
</table>

As has already been mentioned, it is more practical to consider parallel resonance in the two-branch parallel circuit of Fig. 14.18. However, using the method described in section 14.14, the series L and R of the coil can be converted into their parallel equivalent circuit after which the three-branch circuit analysis of section 14.12 applies. For series and parallel equivalence in Fig. 14.20, using the equivalents derived in section 14.14 for high values of Q (>10):

$$R_P = Q^2 R_S \quad \text{and} \quad L_P = L_S$$

Fig. 14.20 Series and parallel equivalent circuits

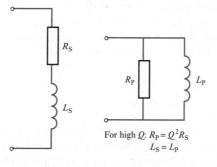

For high Q: $R_P = Q^2 R_S$
$L_S = L_P$

Figure 14.18 can be redrawn as shown in Fig. 14.21.

Fig. 14.21 (a) Two-branch and (b) equivalent three-branch parallel resonant circuits

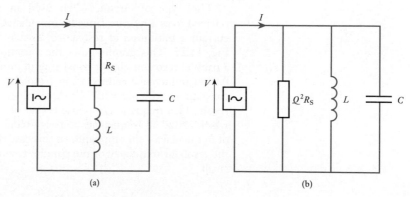

(a)

(b)

The expressions developed in section 14.12 now apply, i.e. equation [14.15] and $\omega_r = 1/\sqrt{(LC)}$. It should be noted again that, as explained in section 14.13, Q is the same for both parallel and series resonant circuits:

<table>
<tr><td>Series resonant circuit</td><td>Parallel resonant circuit</td></tr>
</table>

$$Q_S = \frac{\omega_r L}{R_S} \qquad\qquad Q_P = \frac{R_P}{\omega_r L}$$

$$\omega_r L = Q_S R_S \qquad\qquad \omega_r L = \frac{R_P}{Q_P}$$

Since $\quad Q_S = Q_P$

$$\omega_r L = Q_S R_S = \frac{R_P}{Q_P}$$

and $R_P = Q^2 R_S$ as expected.

Since

$$\omega_1^2 = \frac{1}{LC} \quad \text{and} \quad Q = \frac{\omega_r L}{R}$$

$$R_P = Q^2 R_S = \left(\frac{\omega_r L}{R_S}\right)^2 R_S = \frac{L}{CR_S}$$

This quantity is particularly significant because it is the impedance of the parallel network at resonance and is equivalent to a resistor of $L/(CR_S)$ ohms. It is known as the *dynamic resistance* and also, though it is wholly resistive, as the *dynamic impedance*:

$$Z_r = \frac{L}{CR_S} \tag{14.17}$$

It is clear that the lower the resistance of the coil, the higher is the dynamic impedance of the parallel circuit. For high-Q coils, R_S is small, which makes Z_r large as required for good selectivity. Example 14.5 illustrates a calculation of Z_r.

This type of circuit, when used in communication applications, is referred to as a *rejector*, since its impedance is a maximum and the resultant current a minimum at resonance, as may be observed from the graph of Fig. 14.22. This graph shows the variation of current with frequency in a parallel resonant circuit consisting of an inductor of 4.0 Ω resistance and 0.50 H inductance connected in parallel with a 20.3 μF capacitor across a constant a.c. voltage of 100 V. Equation [14.15] might be used to verify this result. This frequency response is that of a stopband filter, see Chapter 18, which allows all frequencies to pass except those lying within a small range of frequencies – the stopband. In this case the reject or stopband lies around the resonant frequency of the parallel resonant circuit, where the current is small.

Fig. 14.22 Resonance curve for a rejector

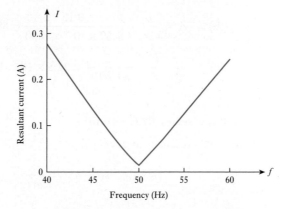

In conclusion, it should be noted that a more exact expression for the resonant frequency in a two-branch parallel resonant circuit is given by:

$$f_r = \frac{1}{2\pi} \sqrt{\left(\frac{1}{LC} - \frac{R_S^2}{L^2} \right)}$$

If R_S is very small compared with $\omega_r L$, as in communications circuits,

$$f_r = \frac{1}{2\pi\sqrt{(LC)}}$$

which is the same as the resonant frequency of a series circuit. This is an acceptable approximation for the resonant frequency of the two-branch circuit for the purpose of understanding the principles of parallel resonance.

Example 14.5

A coil of 1 kΩ resistance and 0.15 H inductance is connected in parallel with a variable capacitor across a 2.0 V, 10 kHz a.c. supply as shown in Fig. 14.23. Calculate:

(a) the capacitance of the capacitor when the supply current is a minimum;
(b) the effective impedance Z_r of the network at resonance;
(c) the supply current.

The network supply current is a minimum when the network is in resonance.

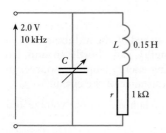

Fig. 14.23 Circuit diagram for Example 14.5

(a) $$f_r = \frac{1}{2\pi\sqrt{(LC)}}$$

$$f_r^2 = \frac{1}{4\pi^2 LC}$$

∴ $$C = \frac{1}{4\pi^2 L f_r^2}$$

$$= \frac{1}{4\pi^2 \times 0.15^2 \times 10^8}$$

$$= 0.169 \times 10^8 = 1.69 \text{ nF}$$

(b) $Z_r = \dfrac{L}{CR_S} = \dfrac{0.15}{1.69 \times 10^{-9} \times 1000} = 89 \text{ k}\Omega$

(c) $I_S = \dfrac{V}{Z_r} = \dfrac{2}{89 \times 10^3} = 22.5 \times 10^{-6} \text{ A}$

Summary of important formulae

For a series RLC circuit:

At the resonant frequency f_r

$$|X_L| = |X_C| \quad \text{so} \quad \omega_r L = \frac{1}{\omega_r C}$$

$$f_r = \frac{1}{2\pi \sqrt{(LC)}} \tag{14.3}$$

Q factor: $Q = \dfrac{1}{\omega_r CR} = \dfrac{\omega_r L}{R}$ [14.7]

$$Q = \frac{\text{reactive power}}{\text{average power}} \tag{14.9}$$

$$\text{Bandwidth (BW)} = \frac{\text{resonant frequency}}{Q \text{ factor}} \tag{14.14}$$

For a parallel RLC network:

Q factor: $Q = \dfrac{1}{\omega_r LG} = \dfrac{\omega_r C}{G} = \dfrac{B}{G} = \dfrac{R}{X}$ [14.16]

Resonant frequency: $f_r = \dfrac{1}{2\pi} \sqrt{\left(\dfrac{1}{LC} - \dfrac{R_S^2}{L^2} \right)}$

If R_S is very small compared with $\omega_r L$, as in communications circuits,

$$f_r = \frac{1}{2\pi \sqrt{(LC)}}$$

which is the same as the resonant frequency of a series circuit.

Dynamic impedance: $Z_r = \dfrac{L}{CR_S}$ [14.17]

Terms and concepts

Resonance occurs when the peak energies stored by the inductor and the capacitor are equal and hence this energy can shuttle to and fro between these components without taking energy from the source. In practice the transfer incurs loss in the resistance of the circuit.

In an **RLC** series circuit, resonance occurs when the supply voltage and current are in phase.

At the resonant frequency of a series resonant circuit, the impedance is a minimum and equal to the circuit resistance. The magnitudes of the inductive and capacitive reactance are equal and therefore cancel each other.

Terms and concepts continued	The voltages which appear across the reactive components can be many times greater than that of the supply. The factor of magnification, the voltage magnification in the series circuit, is called the **Q factor**.

An **RLC** series circuit accepts maximum current from the source at resonance and for that reason is called an **acceptor circuit**.

In a parallel **RLC** network, the expression for the resonant frequency can be regarded as being the same as the expression for the equivalent series circuit.

The lowest current from the source occurs at the resonant frequency of a parallel circuit hence it is called a **rejector circuit**.

At resonance, the current in the branches of the parallel circuit can be many times greater than the supply current. The factor of magnification, the current magnification in the parallel circuit, is again called the **Q factor**.

At the resonant frequency of a resonant parallel network, the impedance is wholly resistive. The value of this impedance is known as the **dynamic resistance** and also, though the quantity is resistive, as the **dynamic impedance**.

Exercises 14

1. A series circuit comprises an inductor, of resistance 10 Ω and inductance 159 μH, and a variable capacitor connected to a 50 mV sinusoidal supply of frequency 1 MHz. What value of capacitance will result in resonant conditions and what will then be the current? For what values of capacitance will the current at this frequency be reduced to 10 per cent of its value at resonance?

2. A circuit consists of a 10 Ω resistor, a 30 mH inductor and a 1 μF capacitor, and is supplied from a 10 V variable-frequency source. Find the frequency for which the voltage developed across the capacitor is a maximum and calculate the magnitude of this voltage.

3. Calculate the voltage magnification created in a resonant circuit connected to a 230 V a.c. supply consisting of an inductor having inductance 0.1 H and resistance 2 Ω in series with a 100 μF capacitor. Explain the effects of increasing the above resistance value.

4. A series circuit consists of a 0.5 μF capacitor, a coil of inductance 0.32 H and resistance 40 Ω and a 20 Ω non-inductive resistor. Calculate the value of the resonant frequency of the circuit. When the circuit is connected to a 30 V a.c. supply at this resonant frequency, determine: (a) the p.d. across each of the three components; (b) the current flowing in the circuit; (c) the active power absorbed by the circuit.

5. An e.m.f. whose instantaneous value at time t is given by 283 sin(314t + π/4) volts is applied to an inductive circuit and the current in the circuit is 5.66 sin(314t − π/6) amperes. Determine: (a) the frequency of the e.m.f.; (b) the resistance and inductance of the circuit; (c) the active power absorbed. If series capacitance is added so as to bring the circuit into resonance at this frequency and the above e.m.f. is applied to the resonant circuit, find the corresponding expression for the instantaneous value of the current. Sketch a phasor diagram for this condition.

   Explain why it is possible to have a much higher voltage across a capacitor than the supply voltage in a series circuit.

6. A coil, of resistance R and inductance L, is connected in series with a capacitor C across a variable-frequency source. The voltage is maintained constant at 300 mV and the frequency is varied until a maximum current of 5 mA flows through the circuit at 6 kHz. If, under these conditions, the Q factor of the circuit is 105, calculate: (a) the voltage across the capacitor; (b) the values of R, L and C.

7. A constant voltage at a frequency of 1 MHz is maintained across a circuit consisting of an inductor in series with a variable capacitor. When the capacitor is set to 300 pF, the current has its maximum value.

Exercises 14 continued

When the capacitance is reduced to 284 pF, the current is 0.707 of its maximum value. Find (a) the inductance and the resistance of the inductor and (b) the Q factor of the inductor at 1 MHz. Sketch the phasor diagram for each condition.

8. A coil of resistance 12 Ω and inductance 0.12 H is connected in parallel with a 60 μF capacitor to a 100 V variable-frequency supply. Calculate the frequency at which the circuit will behave as a non-reactive resistor, and also the value of the dynamic impedance. Draw for this condition the complete phasor diagram.

9. Calculate, from first principles, the impedance at resonance of a circuit consisting of a coil of inductance 0.5 mH and effective resistance 20 Ω in parallel with a 0.0002 μF capacitor.

10. A coil has resistance of 400 Ω and inductance of 318 μH. Find the capacitance of a capacitor which, when connected in parallel with the coil, will produce resonance with a supply frequency of 1 MHz. If a second capacitor of capacitance 23.5 pF is connected in parallel with the first capacitor, find the frequency at which resonance will occur.

Chapter fifteen

Network Theorems Applied to AC Networks

Objectives

When you have studied this chapter, you should

- be familiar with the relevance of Kirchhoff's laws to the analysis of a.c. networks
- be capable of analysing a.c. networks by the application of Kirchhoff's laws
- be capable of analysing a.c. networks by the application of Nodal analysis
- be capable of analysing a.c. networks by the application of Thévenin's theorem
- be capable of analysing a.c. networks by the application of Norton's theorem
- be capable of transforming a star-connected impedance load into a delta-connected impedance load and vice versa
- be familiar with the condition required for maximum power to be developed in an impedance load
- have an understanding of the significance the maximum power condition has in practice

Contents

Chapter 4 introduced the main methods of network analysis and applied them to the solution of circuits having resistance and direct voltages. In this chapter, the same methods will be applied to the solution of circuits having alternating voltages and impedances, instead of just resistances, due to the resistance, inductance and capacitance of a.c. networks. The methods are, in fact, readily applicable using the techniques of complex numbers developed in Chapter 13. Kirchhoff's laws, Nodal analysis, the Superposition theorem, Thévenin's and Norton's theorems, star/delta transformation and the maximum power transfer theorem will all be discussed and their application to a.c. networks illustrated.

15.1   One stage further

In Chapter 4, we were introduced to a number of techniques by which we could analyse d.c. circuit and network performance. Since then we have been introduced to a.c. circuits and networks. Each introduction involved a degree of complexity so far as handling the mathematical terms. Could it be that things become too difficult to handle when we apply a.c. signals to complicated networks?

Fortunately we have been introduced to complex notation. Provided we apply it to the a.c. quantities, we can readily apply the network theorems in almost the same ways which we used in the d.c. circuits and networks. As with the theorems for d.c. circuits and networks, the easiest way to demonstrate the analysis of a.c. circuits and networks is to look at a number of examples.

15.2   Kirchhoff's laws and network solution

Let us look at a simple application of Kirchhoff's laws to an a.c. network.

Example 15.1

For the network shown in Fig. 15.1, determine the supply current and the source e.m.f.

Fig. 15.1 Circuit diagram for Example 15.1

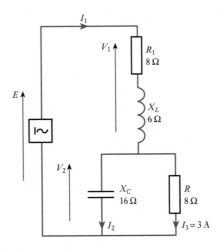

Let $\quad \mathbf{I}_3 = (3 + \mathrm{j}0)\ \mathrm{A}$

$\therefore \qquad \mathbf{V}_2 = \mathbf{I}_3 R = (3 + \mathrm{j}0)(8 + \mathrm{j}0) = (24 + \mathrm{j}0)\ \mathrm{V}$

$$\mathbf{I}_2 = \frac{V_2}{X_C} = \frac{(24 + \mathrm{j}0)}{(0 - \mathrm{j}16)} = (0 + \mathrm{j}1.5)\ \mathrm{A}$$

$$\mathbf{I}_1 = \mathbf{I}_2 + \mathbf{I}_3 = (0 + \mathrm{j}1.5) + (3 + \mathrm{j}0) = (3 + \mathrm{j}1.5)\ \mathrm{A}$$

$\therefore \qquad I_1 = (3^2 + 1.5^2)^{\frac{1}{2}} = \mathbf{3.35\ A}$

$$\mathbf{V}_1 = \mathbf{I}_1(R + \mathrm{j}X_L) = (3 + \mathrm{j}1.5)(8 + \mathrm{j}6) = (15 + \mathrm{j}30)\ \mathrm{V}$$

$$\mathbf{E} = \mathbf{V}_1 + \mathbf{V}_2 = (15 + \mathrm{j}30) + (24 + \mathrm{j}0) = (39 + \mathrm{j}30)\ \mathrm{V}$$

$\therefore \qquad E = (39^2 + 30^2)^{\frac{1}{2}} = \mathbf{49.2\ V}$

This is not the only form of solution which could have been given. For instance, we could have applied the current-sharing rule to the current in the 8 Ω resistor to obtain the supply current directly. Or we could have determined the effective impedance of the network and hence obtained the source e.m.f. by obtaining the product of supply current and effective impedance. It is always preferable to seek the simplest solution but any solution which appeals is valid.

Let us look at other examples illustrating the direct and most simple approach.

Example 15.2   Given the network shown in Fig. 15.2, determine I_1, E, I_2 and I.

Fig. 15.2   Circuit diagram for Example 15.2

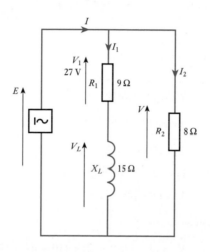

$$I_1 = \frac{V_1}{R_1} = \frac{(27 + j0)}{9 + j0} = (3 + j0)\ A$$

$$\therefore \qquad I_1 = (3^2 + 0^2)^{\frac{1}{2}} = 3\ A$$

$$V_L = IX_L = (3 + j0)(0 + j15) = (0 + j45)\ V$$

$$E = V_1 + V_L = (27 + j0) + (0 + j45)$$

$$= (27 + j45)\ V$$

$$\therefore \qquad E = (27^2 + 45^2)^{\frac{1}{2}} = \mathbf{52.5\ V}$$

$$I_2 = \frac{E}{R_2} = \frac{(27 + j45)}{8 + j0} = (3.38 + j5.63)\ A$$

$$\therefore \qquad I_2 = (3.38^2 + 5.63^2)^{\frac{1}{2}} = 6.56\ A$$

$$I = I_1 + I_2 = (3 + j0) + (3.38 + j5.64)$$

$$= (6.38 + j5.63)\ A$$

$$\therefore \qquad I = (6.38^2 + 5.63^2)^{\frac{1}{2}} = \mathbf{8.5\ A}$$

Example 15.3   For the network shown in Fig. 15.3, the power dissipated in R is 20 W. Determine I and V.

Fig. 15.3 Circuit diagram for Example 15.3

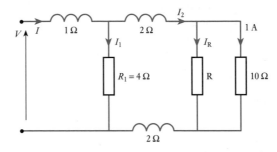

Let current in 10 Ω resistor be $(1 + j0)$ A. Therefore voltage across resistor is

$$\mathbf{V}_{10} = (1 + j0)10 = (10 + j0) \text{ V}$$

For resistor R:

$$\mathbf{P} = (20 + j0) \text{ W} = (10 + j0)\mathbf{I}_{R}$$

$$\therefore \qquad \mathbf{I}_{R} = (2 - j0) \text{ A} = (2 + j0) \text{ A}$$

$$\mathbf{I}_{2} = (1 + j0) + (2 + j0) = (3 + j0) \text{ A}$$

Voltage across R_1 is given by

$$\mathbf{V}_{1} = (3 + j0)(0 + j2) + (10 + j0) + (3 + j0)(0 + j2)$$
$$= (10 + j12) \text{ V}$$

$$\mathbf{I}_{1} = \frac{(10 + j12)}{(4 + j0)} = (2.5 + j3) \text{ A}$$

$$\mathbf{I} = \mathbf{I}_{1} + \mathbf{I}_{2} = (2.5 + j3) + (3 + j0) = (5.5 + j3) \text{ A}$$

$$\therefore \qquad I = (5.5^2 + 15^2)^{\frac{1}{2}} = \mathbf{6.3 \, A}$$

Voltage across 1 Ω resistor is given by

$$(5.5 + j3)(1 + j0) = (5.5 + j3) \text{ V}$$

$$\therefore \qquad \mathbf{V} = (5.5 + j3) + (10 + j12) = (15.5 + j15) \text{ V}$$

$$\therefore \qquad V = (15.5^2 + 15^2)^{\frac{1}{2}} = \mathbf{21.6 \, V}$$

Example 15.3 shows that quite a complicated network can be analysed by a direct approach using a simple application of Kirchhoff's laws. It is not always possible to proceed in this way. For instance, let us look again at the network used in Example 15.1, but pose a slightly different situation.

| Example 15.4 | For the network shown in Fig. 15.4, determine the supply current I_1 and the branch current I_3. |

Fig. 15.4 Circuit diagram for Example 15.4

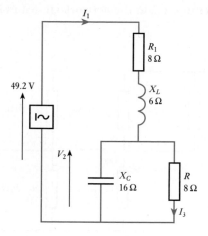

Essentially this network consists of three parts in series, but one of them comprises X_C and R in parallel. These can be replaced by an equivalent impedance thus

$$\mathbf{Z_e} = \frac{(R)(-jX_C)}{R - jX_C} = \frac{-j128}{8 - j16} = \frac{(-j128)(8 + j16)}{(8^2 + 16^2)}$$

$$= (6.4 - j3.2)\ \Omega$$

Replace R and X_C by Z_e as shown in Fig. 15.5.

Now the network has been reduced to a simple series circuit, the total effective impedance is

$$\mathbf{Z_1} = 8 + j6 + (6.4 - j3.2) = (14.4 + j2.8)\ \Omega$$

$$\therefore \quad Z_1 = (14.4^2 + 2.8^2)^{\frac{1}{2}} = 14.7\ \Omega$$

$$\therefore \quad I_1 = \frac{49.2}{14.7} = 3.35\ \text{A}$$

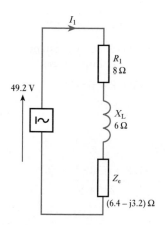

Fig. 15.5 Circuit diagram for Example 15.4

$$I_3 = \left| \frac{-jX_C}{R - jX_C} \right| \cdot I_1 = \left| \frac{-j16}{8 - j16} \right| \cdot 3.35 = \frac{16}{(8^2 + 16^2)^{\frac{1}{2}}} \cdot 3.35$$

$$= 3\ \text{A}$$

In this example, the voltage and current information come from the source and not from the load. It is for this reason that the calculation has to be based on an analysis of the impedances. The calculation was therefore based on network reduction, i.e. by replacing two components by one equivalent impedance.

It is worth noting how we slipped out of using complex notation whenever it was not needed. Had we determined first I_1 and then I_3 in complex form, we would have obtained the same outcomes, but we would have undertaken unnecessary calculations.

Let us now look at a network in which network reduction transforms a complicated problem into a reasonably simple one.

Example 15.5   Determine V_{AB} in the network shown in Fig. 15.6.

Fig. 15.6   Circuit diagram for Example 15.5

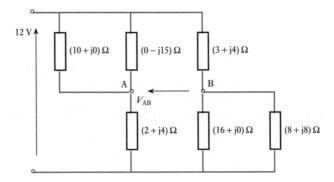

This is quite a complicated network. However, there are two instances of parallel impedances. For the $(10 + j0)$ Ω and $(0 - j15)$ Ω impedances

$$\mathbf{Z} = \frac{(10 + j0) \cdot (0 - j15)}{10 + j0 + 0 - j15} = \frac{-j150}{10 - j15} = \frac{-j150(10 + j15)}{10^2 + 15^2}$$

$$= (6.92 - j4.62) \; \Omega$$

For the $(16 + j0)$ Ω and $(8 + j8)$ Ω in parallel

$$\mathbf{Z} = \frac{(16 + j0) \cdot (8 + j8)}{16 + j0 + 8 + j8} = \frac{128 + j128}{24 + j8} = \frac{(128 + j128)(24 - j8)}{24^2 + 8^2}$$

$$= (6.4 - j3.2) \; \Omega$$

We can now redraw the network as shown in Fig. 15.7.

Fig. 15.7   Circuit diagram for Example 15.5

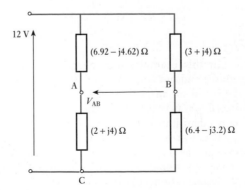

$$V_{AC} = \frac{2 + j4}{(6.92 - j4.62) + (2 + j4)} \cdot (12 + j0) = \frac{24 + j48}{8.92 - j0.62}$$

$$= (2.62 + j5.54)\ V$$

$$V_{BC} = \frac{6.4 - j3.2}{(3 + j4) + (6.4 - j3.2)} \cdot (12 + j0) = \frac{76.8 - j38.4}{32.0 - j16.0}$$

$$= (2.4 + j0)\ V$$

$$V_{AB} = V_{AC} - V_{BC} = 2.62 + j5.54 - 2.4 - j0 = (0.22 + j5.54)\ V$$

$$\therefore \qquad V_{AB} = (0.22^2 + 5.55^2)^{\frac{1}{2}} = \mathbf{5.55\ V}$$

Once again we have observed two methods of circuit analysis. We have already been confronted with this difficult choice; we met it in section 4.2. The decision as to which choice we should take remains the same – if the information provided relates to the component voltages or currents then take the first method of approach, i.e. applying the simplest forms of Kirchhoff's laws to the components. However, if the information depends on the supply voltage or current then we should try network reduction techniques.

We were shown in section 4.2 that resistors were not always either in series or in parallel. Impedances are no different and Example 15.6 demonstrates a common situation – an a.c. bridge.

Example 15.6

For the network shown in Fig. 15.8, calculate the current in the $(3 + j4)\ \Omega$ impedance.

In this network, the impedances are neither in series nor in parallel and hence a more difficult method of analysis must be employed. Let the currents be as indicated in Fig. 15.9 and hence the volt drops can be determined. These volt drops can then be related by Kirchhoff's second law.

Fig. 15.8 Circuit diagram for Example 15.6

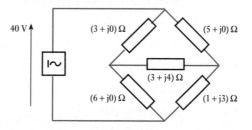

Fig. 15.9 Circuit diagram for Example 15.6

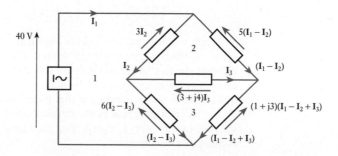

In loop 1:

$$40 + j0 = 3\mathbf{I}_2 + 6(\mathbf{I}_2 - \mathbf{I}_3)$$

$$\therefore \qquad 40 = 9\mathbf{I}_2 - 6\mathbf{I}_3$$

$$\therefore \qquad \mathbf{I}_2 = \frac{40 + 6\mathbf{I}_3}{9} \qquad\qquad [1]$$

In loop 2:

$$0 = 5(\mathbf{I}_1 - \mathbf{I}_2) - (3 + j4)\mathbf{I}_3 - 3\mathbf{I}_2$$

$$0 = 5\mathbf{I}_1 - 8\mathbf{I}_2 - (3 + j4)\mathbf{I}_3 \qquad\qquad [2]$$

In loop 3:

$$0 = (3 + j4)\mathbf{I}_3 + (1 + j3)(\mathbf{I}_1 - \mathbf{I}_2 + \mathbf{I}_3) - 6(\mathbf{I}_2 - \mathbf{I}_3)$$

$$\therefore \qquad 0 = (1 + j3)\mathbf{I}_1 - (7 + j3)\mathbf{I}_2 + (10 + j7)\mathbf{I}_3 \qquad\qquad [3]$$

Substituting equation [1] in [2]

$$0 = 5\mathbf{I}_1 - \frac{8(40 + 6\mathbf{I}_3)}{9} - (3 + j4)\mathbf{I}_3$$

$$\therefore \qquad 0 = 45\mathbf{I}_1 - 320 - 48\mathbf{I}_3 - (27 + j36)\mathbf{I}_3$$

and $\qquad 320 = 45\mathbf{I}_1 - (75 + j36)\mathbf{I}_3 \qquad\qquad [4]$

Substituting equation [1] in [3]:

$$0 = (1 + j3)\mathbf{I}_1 - \frac{(7 + j3)(40 + 6\mathbf{I}_3)}{9} + (10 + j7)\mathbf{I}_3$$

$$= (9 + j27)\mathbf{I}_1 - (280 + j120) + (48 + j45)\mathbf{I}_3$$

$$\therefore \qquad (280 + j120) = (9 + j27)\mathbf{I}_1 + (48 + j45)\mathbf{I}_3 \qquad\qquad [5]$$

Multiply equation [4] by $(9 + j27)$

$$320(9 + j27) = 45(9 + j27)\mathbf{I}_1 - (75 + j36)(9 + j27)\mathbf{I}_3 \qquad\qquad [6]$$

Multiply equation [5] by 45

$$45(280 + j120) = 45(9 + j27)\mathbf{I}_1 + 45(48 + j45)\mathbf{I}_3 \qquad\qquad [7]$$

Taking equation [7] from equation [6]

$$9720 - j3240 = (3807 + 4509)\mathbf{I}_3$$

$$\therefore \qquad \mathbf{I}_3 = \frac{9720 - j3240}{3807 + j4509} = (0.65 - j1.62) \text{ A}$$

$$\therefore \qquad I_3 = 1.75 \text{ A}$$

When we analysed networks of the form considered in Example 15.6 when supplied from a direct voltage source, the resulting calculation was quite difficult. In fact in section 4.2, we noted that we had to proceed with great caution as mistakes can easily be made. Having completed Example 15.6, we quickly came to the conclusion that there must be better methods of analysis.

Let us then progress to the application of the variety of theorems which we met in Chapter 4, except that this time we will consider them in the context of a.c. networks.

15.3 Nodal analysis (Node Voltage method)

We have already discussed the method in Chapter 4 in relation to d.c. circuit analysis. Now, Nodal analysis is applied to an a.c. circuit in which we consider impedances instead of resistances.

Example 15.7

Calculate the output voltage V_o in the circuit of Fig. 15.10 using nodal analysis.

Fig. 15.10 Circuit diagram for Example 15.7

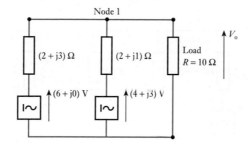

Apply Kirchhoff's current law ($\Sigma I = 0$) at node 1:

$$\frac{V_o - (6 + j0)}{2 + j3} + \frac{V_o - (4 + j3)}{2 + j1} + \frac{V_o}{10} = 0$$

Multiply by $(2 + j1)(2 + j3)$:

$$(2 + j1)(V_o - 6) + (2 + j3)V_o - (2 + j3)(4 + j3) + \frac{(2 + j1)(2 + j3)V_o}{10} = 0$$

$$V_o[2 + j1] + V_o[2 + j3] + \frac{V_o}{10}[4 - 3 + j8] - 12 - j6 - [-1 + j18] = 0$$

$$4.1V_o + j4.8V_o = 12 + j6 - 1 + j18$$

$$V_o = \frac{11 + j24}{4.1 + j4.8} = \frac{26.4\angle65.4°}{6.31\angle49.5°}$$

$$V_o = 4.18\angle15.9°$$

15.4 Superposition theorem

We already know the concept of this theorem from section 4.5 so let us apply it to an example.

Example 15.8

By means of the Superposition theorem, determine the currents in the network shown in Fig. 15.11(a).

Because there are two sources of e.m.f. in the network, two separate networks need to be considered, each having one source of e.m.f. Figure 15.11(b) shows the network with the $(6 + j8)$ V source replaced by a short-circuit, there being no internal impedance indicated. Also, Fig. 15.11(c) shows the network with the $(10 + j0)$ V source similarly replaced.

Fig. 15.11　Circuit diagrams for
Example 15.8

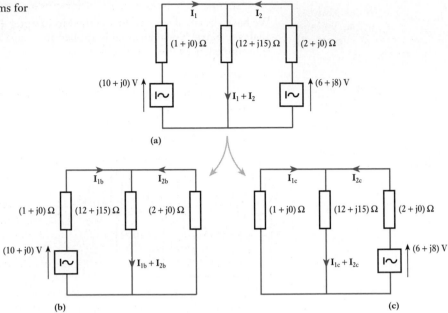

(a)

(b)　　　　　　　　　　　　　　　　　　　(c)

For the Fig. 15.11(b) arrangement, the total impedance is

$$\mathbf{Z} = 1 + j0 + \frac{(12 + j15)(2 + j0)}{12 + j15 + 2 + j0} = 1 + \frac{24 + j30}{14 + j15}$$

$$= (2.87 + j0.14) \ \Omega$$

$$\therefore \qquad \mathbf{I}_{1b} = \frac{(10 + j0)}{2.87 + j0.14} = \frac{10(2.87 - j0.14)}{2.87^2 + 0.14^2} = (3.48 - j0.17) \ \text{A}$$

$$\mathbf{I}_{2b} = -(3.48 - j0.17)\frac{(12 + j15)}{(12 + j15) + (2 + j0)} = -(3.24 + j0.09) \ \text{A}$$

$$\therefore \qquad \mathbf{I}_{1b} + \mathbf{I}_{2b} = (3.48 - j0.17) - (3.24 + j0.09) = (0.24 - j0.26) \ \text{A}$$

For the Fig. 15.11(c) arrangement, the total impedance is

$$\mathbf{Z} = (2 + j0) + \frac{(12 + j15)(1 + j0)}{(12 + j15 + 1 + j0)} = 2 + \frac{12 + j15}{13 + j15}$$

$$= (2.97 + j0.04) \ \Omega$$

$$\mathbf{I}_{2c} = \frac{(6 + j8)}{(2.97 + j0.04)} = \frac{(6 + j8)(2.97 - j0.04)}{(8.823)}$$

$$= (2.06 + j2.67) \ \text{A}$$

$$\mathbf{I}_{1c} = -(2.06 + j2.67)\frac{(12 + j15)}{(12 + j15) + (1 + j0)}$$

$$= -(1.83 + j2.59) \ \text{A}$$

$$\therefore \qquad \mathbf{I}_{1c} + \mathbf{I}_{2c} = (0.23 + j0.08) \ \text{A}$$

Putting the two arrangements back together, we obtain:

$$\mathbf{I}_1 = \mathbf{I}_{1b} + \mathbf{I}_{1c} = (3.48 - j0.17) - (1.83 + j2.59)$$

$$= (1.65 - j2.76)\ \text{A}$$

$$\therefore \qquad I_1 = \mathbf{3.22\ A}$$

$$\mathbf{I}_2 = \mathbf{I}_{2b} + \mathbf{I}_{2c} = (-3.24 - j0.09) + (2.06 + j2.67)$$

$$= (-1.18 + j2.58)\ \text{A}$$

$$\therefore \qquad I_2 = \mathbf{2.84\ A}$$

$$\mathbf{I}_1 + \mathbf{I}_2 = (1.65 - j2.76) + (-1.18 + j2.58) = (0.47 - j0.18)\ \text{A} = \mathbf{I}_3$$

$$\therefore \qquad I_3 = \mathbf{0.50\ A}$$

In this problem, it will be noted that the currents flowing in the sources are relatively large, especially when it is noted that both source e.m.f.s are 10 V. The reason is that the two voltages, although equal in magnitude, are not in phase opposition.

When applying the Superposition theorem, as in Example 15.8, it is assumed that the operational frequency of both sources is the same. If the frequencies were different, we could still determine the component currents, but they could not be added together. However, sources with different frequencies take us into more advanced techniques which we need only to consider in circuits such as communications circuits.

15.5 Thévenin's theorem

As with the concept of the Superposition theorem, the principles of application are the same except that we consider impedances instead of resistances. Thus the theorem can be stated as follows:

> Any two-terminal a.c. network can be replaced by an equivalent circuit consisting of a voltage source equal to the open circuit voltage at the terminals and a series impedance equal to the internal impedance as seen from the terminals.

Many networks from those involved with the highest to the lowest power levels can be represented in this way. The principle of operation needs to be demonstrated by a number of situations.

Example 15.9 Determine the Thévenin equivalent circuit for the network supplying the load shown in Fig. 15.12.

Fig. 15.12 Circuit diagram for Example 15.9

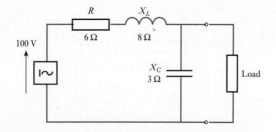

Let

$$\mathbf{V} = (100 + \text{j}0) \text{ V}$$

To determine the open-circuit voltage, let us consider the network shown in Fig. 15.13.

Fig. 15.13 Circuit diagram for Example 15.9

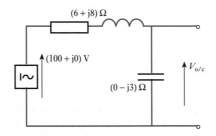

$$\frac{\mathbf{V}_{\text{o/c}}}{\mathbf{V}} = \frac{0 - \text{j}3}{(6 + \text{j}8) + (0 - \text{j}3)} = \frac{(-\text{j}3)(6 - \text{j}5)}{61} = (-0.246 - \text{j}0.295)$$

$$\therefore \qquad \mathbf{V}_{\text{o/c}} = (-24.6 - \text{j}29.5) \text{ V}$$

To determine the internal impedance, short-circuit the source and consider the circuit shown in Fig. 15.14.

$$\mathbf{Z}_{\text{in}} = \frac{(0 - \text{j}3) \cdot (6 + \text{j}8)}{(0 - \text{j}3) + (6 + \text{j}8)} = \frac{24 - \text{j}18}{6 + \text{j}5} = \frac{(24 - \text{j}18) \cdot (6 - \text{j}5)}{6^2 + 5^2}$$

$$= (0.89 - \text{j}3.74) \text{ }\Omega$$

Fig. 15.14 Circuit diagram for Example 15.9

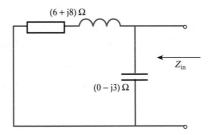

The equivalent circuit is therefore that shown in Fig. 15.15.

Fig. 15.15 Circuit diagram for Example 15.9

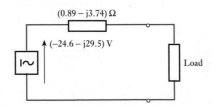

Example 15.10	For the circuit shown in Fig. 15.16, A and B represent the two terminals of a network. Calculate the load current. Note that this problem was solved earlier using nodal analysis. In Example 15.7, V_{AB} was calculated.

Fig. 15.16 Circuit diagram for Example 15.10

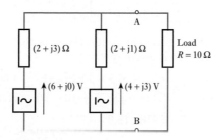

Fig. 15.17 Circuit diagram for Example 15.10

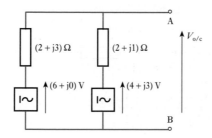

The open-circuit voltage can be obtained from the circuit shown in Fig. 15.17. The circuit will have a circulating current as follows:

$$I = \frac{(6 + j0) - (4 + j3)}{(2 + j3) + (2 + j1)} = \frac{2 - j3}{4 + j4}$$

$$= \frac{(2 - j3)(4 - j4)}{32} = \frac{-4 - j20}{32}$$

$$= -0.125 - j0.625$$

$$\therefore \quad V_{o/c} = [6 + j0] + (0.125 + j0.625)(2 + j3)$$

$$= 6 - 1.875 + 0.25 + j1.25 + j0.375$$

$$= 4.375 + j1.625 \text{ V}$$

or $\quad V_{o/c} = 4.66\angle 20.4°$

To obtain the internal impedance, the sources are short-circuited and the circuit shown in Fig. 15.18 is used to determine the value.

$$Z_{in} = \frac{(2 + j3) \cdot (2 + j1)}{(2 + j3) + (2 + j1)} = \frac{1 + j8}{4 + j4} = (1.125 + j0.875) \; \Omega$$

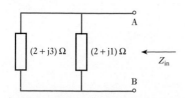

Fig. 15.18 Circuit diagram for Example 15.10

Fig. 15.19   Circuit diagram for Example 15.10

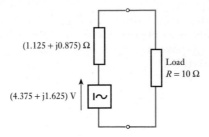

The total equivalent circuit becomes that shown in Fig. 15.19. Hence the load current is given by

$$I = \frac{4.375 + j1.625}{1.125 + j0.875 + (10 + j0)} = \frac{4.66 \angle 20.4°}{11.16 \angle 4.5°}$$

$$I = 0.418 \angle 15.9°$$

Compare this method of solution with that of Example 15.7.

Example 15.11   Determine the Thévenin equivalent circuit for the network shown in Fig. 15.20.

Fig. 15.20   Circuit diagram for Example 15.11

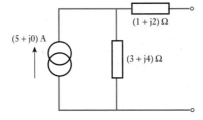

The open-circuit voltage is given by

$$V_{o/c} = (5 + j0) \cdot (3 + j4) = (15 + j20) \text{ V}$$

The internal impedance is given by

$$Z_{in} = (1 + j2) + (3 + j4) = (4 + j6) \text{ Ω}$$

The equivalent circuit is therefore that shown in Fig. 15.21.

Fig. 15.21   Circuit diagram for Example 15.11

Example 15.12   For the bridge network shown in Fig. 15.22, determine the current in the 10 Ω resistor.

Fig. 15.22   Circuit diagram for Example 15.12

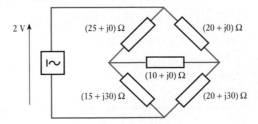

Fig. 15.23 Circuit diagram for Example 15.12

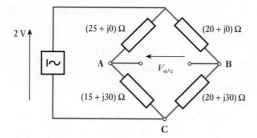

To determine the equivalent circuit as seen from the 10 Ω resistor consider first the open-circuit as illustrated in Fig. 15.23. Let

$$\mathbf{V} = (2 + j0) \text{ V}$$

$$\mathbf{V}_{AB} = \mathbf{V}_{AC} - \mathbf{V}_{BC}$$

$$= \left[\frac{15 + j30}{(15 + j30) + (25 + j0)} \cdot (2 + j0) \right] - \left[\frac{20 + j30}{(20 + j30) + (20 + j0)} \cdot (2 + j0) \right]$$

$$= \frac{30 + j60}{40 + j30} - \frac{40 + j60}{40 + j30} = \frac{-10 + j0}{40 + j30} = (-0.16 + j0.12) \text{ V}$$

Obtaining the internal impedance requires the source to be short-circuited as shown in Fig. 15.24.

Fig. 15.24 Circuit diagram for Example 15.12

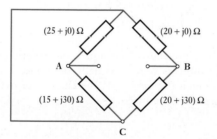

Sometimes it is difficult to see the way in which to reduce the network in Fig. 15.24. It may help to redraw it as shown in Fig. 15.25. Careful comparison of Figs 15.24 and 15.25 will show that they are effectively the same.

Fig. 15.25 Alternative circuit diagram to Fig. 15.24

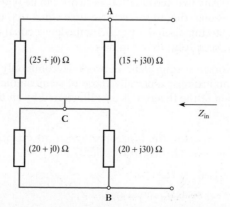

$$Z_{in} = \frac{(15 + j30)(25 + j0)}{(15 + j30) + (25 + j0)} + \frac{(20 + j30)(20 + j0)}{(20 + j30) + (20 + j0)}$$

$$= \frac{375 + j750}{40 + j30} + \frac{400 + j600}{40 + j30} = \frac{775 + j1350}{40 + j30}$$

$$= (28.6 + j12.3) \ \Omega$$

The Thévenin equivalent circuit becomes that shown in Fig. 15.26.

Fig. 15.26 Circuit diagram for Example 15.12

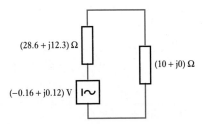

$(28.6 + j12.3) \Omega$

$(10 + j0) \Omega$

$(-0.16 + j0.12) \text{ V}$

Current in 10 Ω resistor is given by

$$I = \frac{-0.16 + j0.12}{(28.6 + j12.3) + (10 + j0)} = \frac{-0.16 + j0.12}{38.6 + j12.3}$$

$\therefore \qquad I = 6.17 \text{ mA}$

It is interesting to compare this solution with that of Example 15.6 since the problems were essentially the same. The length of the second solution was much the same, but the chances of making an analytical mistake were considerably reduced.

15.6   Norton's theorem

As with the concept of Thévenin's theorem, the principles of application are the same except that again we consider impedances instead of resistances. Thus the theorem can be stated as follows:

Any two-terminal a.c. network can be replaced by an equivalent circuit consisting of a current source equal to the short-circuit current at the terminals and a parallel impedance equal to the internal impedance as seen from the terminals.

Norton's equivalent network is commonly associated with light current applications, especially those of semiconductor devices. Again the application of the theorem is best illustrated by an example.

Example 15.13 Determine the Norton equivalent circuit for the network supplying the load shown in Fig. 15.27.

Let

$$V = (100 + j0) \text{ V}$$

Fig. 15.27 Circuit diagram for Example 15.13

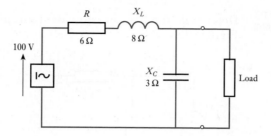

To determine the short-circuit current, let us consider the network shown in Fig. 15.28.

Fig. 15.28 Circuit diagram for Example 15.13

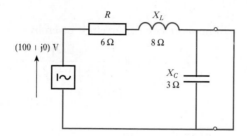

$$\mathbf{I}_{s/c} = \frac{100 + j0}{6 + j8} = \frac{100(6 - j8)}{100} = (6 - j8)\ \text{A}$$

To determine the internal impedance, short-circuit the source and consider the network shown in Fig. 15.29.

Fig. 15.29 Circuit diagram for Example 15.13

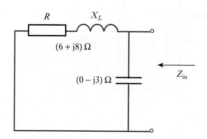

$$\mathbf{Z}_{in} = \frac{(0 - j3) \cdot (6 + j8)}{(0 - j3) + (6 + j8)} = \frac{24 - j18}{6 + j5} = \frac{(24 - j18) \cdot (6 - j5)}{6^2 + 5^2}$$
$$= (0.89 - j3.74)\ \Omega$$

The equivalent network is therefore that shown in Fig. 15.30.

Fig. 15.30 Circuit diagram for Example 15.13

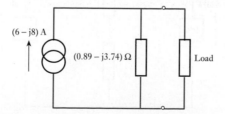

Example 15.14   Determine the Norton equivalent circuit for the network shown in Fig. 15.31.

Fig. 15.31   Circuit diagram for Example 15.14

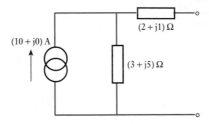

The short-circuit current is given by

$$\mathbf{I}_{s/c} = (10 + j0) \cdot \frac{3 + j5}{(3 + j5) + (2 + j1)} = \frac{30 + j50}{5 + j6}$$

$$= \frac{(30 + j50)(5 - j6)}{61} = (7.38 - j0.82) \text{ A}$$

The internal impedance is given by

$$\mathbf{Z}_{in} = (3 + j5) + (2 + j1) = (5 + j6) \ \Omega$$

The equivalent circuit therefore is that shown in Fig. 15.32.

Fig. 15.32   Circuit diagram for Example 15.14

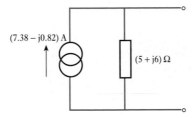

Example 15.15   Calculate the potential difference across the 2 Ω resistor in the network shown in Fig. 15.33.

Fig. 15.33   Circuit diagram for Example 15.15

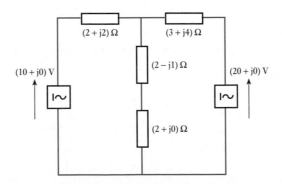

Fig. 15.34 Circuit diagram for
Example 15.15

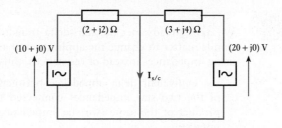

Short-circuit the branch containing the 2 Ω resistor as shown in Fig. 15.34.

$$I_{s/c} = \frac{10 + j0}{2 + j2} + \frac{20 + j0}{3 + j4} = \frac{20 - j20}{2^2 + 2^2} + \frac{60 - j80}{3^2 + 4^2}$$

$$= 2.5 - j2.5 + 2.4 - j3.2 = (4.9 - j5.7) \text{ A}$$

To obtain the internal impedance, replace the sources with short-circuits and look into the network from the ends of the branch.

$$Z_{in} = \frac{(2 + j2) \cdot (3 + j4)}{(2 + j2) + (3 + j4)} = \frac{-2 + j14}{5 + j6} = \frac{(-2 + j14)(5 - j6)}{61}$$

$$= (1.21 + j1.34) \text{ Ω}$$

The circuit therefore reduces to that given in Fig. 15.35.

Fig. 15.35 Circuit diagram for
Example 15.15

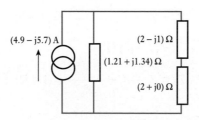

For the 2 Ω resistor:

$$I = (4.9 - j5.7) \frac{(1.21 + j1.34)}{(1.21 + j1.34) + [(2 + j0) + (2 - j1)]}$$

$$= (4.9 - j5.7) \frac{(1.21 + j1.34)}{(5.21 + j0.34)} = 7.52\angle -49.3° \cdot \frac{1.81\angle 47.9°}{5.22\angle 3.7°}$$

$$= 2.61\angle -5.1° \text{ A}$$

Therefore for the 2 Ω resistor:

$$V = 2.61\angle -5.1° \times 2 = 5.22\angle -5.1° \text{ V} \quad \text{or} \quad (5.20 - j0.46) \text{ V}$$

Hence

$$V = \mathbf{5.22 \text{ V}}$$

15.7 Star–delta transformation

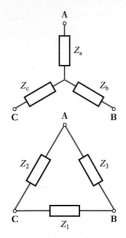

Fig. 15.36 Star–delta transformation

We have already met the star–delta transformation in section 4.10. It is a simple matter to change the application to a.c. networks; this is effected by using impedances instead of resistances, thus:

> the equivalent delta impedance between two terminals is the sum of the two star impedances connected to those terminals plus the product of the same two star impedances divided by the third star impedance.

Let us apply this transformation to the arrangements shown in Fig. 15.36.

$$Z_1 = Z_b + Z_c + \frac{Z_b Z_c}{Z_a} \qquad [15.1]$$

$$Z_2 = Z_c + Z_a + \frac{Z_c Z_a}{Z_b} \qquad [15.2]$$

$$Z_3 = Z_a + Z_b + \frac{Z_a Z_b}{Z_c} \qquad [15.3]$$

Example 15.16

Determine the delta equivalent circuit for the star circuit shown in Fig. 15.37.

Fig. 15.37 Circuit diagram for Example 15.16

$$\mathbf{Z}_1 = Z_b + Z_c + \frac{Z_b \cdot Z_c}{Z_a} = (3 + j1) + (1 + j2) + \frac{(3 + j1) \cdot (1 + j2)}{(1 + j2)}$$

$$= \mathbf{(7 + j4)} \ \Omega$$

$$\mathbf{Z}_2 = Z_c + Z_a + \frac{Z_c \cdot Z_a}{Z_b} = (1 + j2) + (1 + j2) + \frac{(1 + j2) \cdot (1 + j2)}{3 + j1}$$

$$= 2 + j4 + \frac{-3 + j4}{3 + j1} = 2 + j4 + \frac{(-3 + j4) \cdot (3 - j1)}{3^2 + 1^2}$$

$$= \mathbf{(1.5 + j5.5)} \ \Omega$$

$$\mathbf{Z}_3 = Z_a + Z_b + \frac{Z_a \cdot Z_b}{Z_c} = (1 + j2) + (3 + j1) + \frac{(1 + j2) \cdot (3 + j1)}{(1 + j2)}$$

$$= \mathbf{(7 + j4)} \ \Omega$$

The delta equivalent circuit is shown in Fig. 15.38.

Fig. 15.38 Delta circuit diagram for Example 15.16

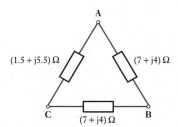

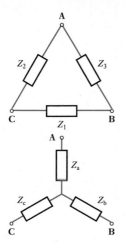

Fig. 15.39 Delta–star transformation

15.8 Delta–star transformation

We have already met the delta–star transformation in section 4.9. It is a simple matter to change the application to d.c. networks; this is effected by using impedances instead of resistances, thus:

> the equivalent star impedance connected to a given terminal is equal to the product of the two delta impedances connected to the same terminal divided by the sum of the delta impedances.

Again let us apply this transformation to the arrangements shown in Fig. 15.39.

$$Z_a = \frac{Z_2 Z_3}{Z_1 + Z_2 + Z_3} \qquad [15.4]$$

$$Z_b = \frac{Z_3 Z_1}{Z_1 + Z_2 + Z_3} \qquad [15.5]$$

$$Z_c = \frac{Z_1 Z_2}{Z_1 + Z_2 + Z_3} \qquad [15.6]$$

Not all delta–star transformations arise from diagrams in which the delta connection is readily recognized. Example 15.17 illustrates an instance in which we have to find the delta connection.

Example 15.17 Determine the current in the load shown in Fig. 15.40.

Fig. 15.40 Circuit diagram for Example 15.17

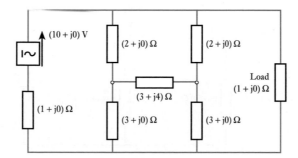

In this example, the solution would be simple if it were not for the $(1 + j0)\ \Omega$ resistance in series with the source. The alternative which is most attractive is the application of Thévenin's theorem, but how can we handle the complicated network? Here we need to observe the two $(2 + j0)\ \Omega$ impedances and the $(3 + j4)\ \Omega$ impedance; these are connected in delta. Let us transform them into the equivalent star.

$$\mathbf{Z}_a = \frac{\mathbf{Z}_2 \mathbf{Z}_3}{\mathbf{Z}_1 + \mathbf{Z}_2 + \mathbf{Z}_3} = \frac{(2 + j0) + (2 + j0)}{(2 + j0) + (2 - j0) + (3 + j4)}$$

$$= \frac{4 + j0}{7 + j4} = \frac{28 - j16}{65} = (0.43 - j0.25)\ \Omega$$

$$\mathbf{Z}_b = \mathbf{Z}_c = \frac{(2 + j0) + (2 + j4)}{(2 + j0) + (2 + j0) + (3 + j4)} = \frac{5 + j4}{7 + j4} = \frac{51 - j8}{65}$$

$$= (0.78 - j0.12)\ \Omega$$

The network therefore can be redrawn as shown in Fig. 15.41.

Fig. 15.41   Circuit diagram for Example 15.17

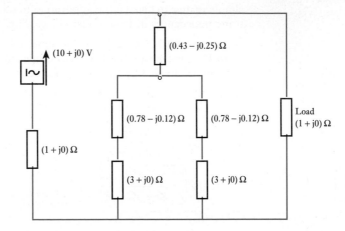

The series impedances can be added to give $(3.78 - j0.12)$ Ω in each of the parallel branches. There being two identical branches in parallel, they can be replaced by an equivalent impedance of $(1.89 - j0.06)$ Ω as shown in Fig. 15.42.

Fig. 15.42   Circuit diagram for Example 15.17

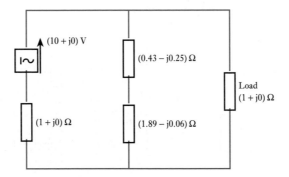

To solve the problem, we can now determine the Thévenin equivalent circuit supplying the load.

The series impedances can be summed to give

$$\mathbf{Z} = (2.32 - j0.31) \ \Omega$$

$$\mathbf{Z}_{in} = \frac{(2.32 - j0.31) \cdot (1 + j0)}{(2.32 - j0.31) + (1 + j0)} = \frac{2.32 - j0.31}{3.32 - j0.31} = (0.70 - j0.03) \ \Omega$$

$$\mathbf{V}_{o/c} = (10 + j0)$$

The total network therefore reduces to that shown in Fig. 15.43.

$$\mathbf{I} = \frac{10 + j0}{(0.70 - j0.03) + (1 + j0)} = \frac{10 + j0}{1.70 + j0.03}$$

$$\therefore \qquad I = 5.88 \ \text{A}$$

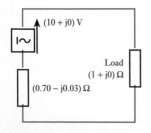

Fig. 15.43   Equivalent circuit diagram for Example 15.18

15.9 Maximum power transfer

In section 4.11, we found that maximum power transfer occurs in a d.c. circuit when the load resistance is equal to the internal source resistance. In an a.c. circuit, it is tempting to expect it to be sufficient to replace resistance by impedance, i.e. that it would be sufficient for the load impedance to be equal to the internal source impedance.

However, in section 4.11, we observed that maximum current and hence maximum power was developed when the current was in phase with the supply voltage. Applying this observation to the circuit in Fig. 15.44, the current will be in phase with the voltage provided the total reactance is zero. This will be achieved if the reactance of the load is equal in magnitude to that of the source but opposite in nature. Thus if the source reactance is $+jX$ then the load impedance should be $-jX$.

Fig. 15.44 Maximum power transfer

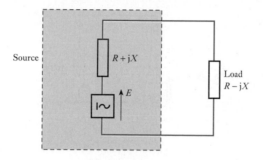

Taking into account the source resistance, maximum power transfer occurs when the internal source impedance is $R + jX$ and the load impedance is $R - jX$, i.e. one is the conjugate impedance of the other.

If the phase angle of the load cannot be varied, the maximum power occurs when the magnitude of the load impedance is equal to the magnitude of the source impedance.

Example 15.18

For the network shown in Fig. 15.45, determine the impedance of the load which will dissipate maximum power, and determine the maximum power.

Fig. 15.45 Circuit diagram for Example 15.18

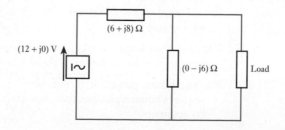

The equivalent internal impedance is given by

$$Z_{in} = \frac{(6 + j8) \cdot (0 - j6)}{(6 + j8) + (0 - j6)} = \frac{48 - j36}{6 + j2} = \frac{(24 - j18)(3 - j1)}{3^2 + 1^2}$$

$$= \frac{54 - j78}{10} = (5.4 - j7.8)\,\Omega$$

$$V_{o/c} = (12 + j0) \cdot \frac{(0 - j6)}{(6 + j8) + (0 - j6)} = \frac{0 - j72}{(6 + j2)} = \frac{-j36(3 - j1)}{10}$$

$$= \frac{-36 - j108}{10} = (-3.6 - j10.8)\,V$$

The network can therefore be represented as shown in Fig. 15.46.

Fig. 15.46 Circuit diagram for Example 15.18

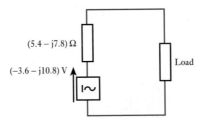

(5.4 − j7.8) Ω

(−3.6 − j10.8) V

Load

For maximum power, the load should have conjugate impedance to that of the source internal impedance

$$\therefore \qquad Z_{LOAD} = (5.4 + j7.8)\,\Omega$$

Therefore total impedance of circuit is given by

$$Z = (5.4 - j7.8) + (5.4 + j7.8) = (10.8 + j0)\,\Omega$$

$$\therefore \qquad I = \frac{(-3.6 - j10.8)}{(10.8 + j0)} = (-0.33 - j1.0)\,A$$

$$\therefore \qquad I = 1.05\,A$$

$$\therefore \qquad P_{LOAD} = I^2 R = 1.05^2 \times 5.4$$

$$= \mathbf{6.0\,W}$$

Terms and concepts

By the application of complex notation, any of the network theorems which were used in d.c. networks can also be applied to a.c. networks.

This has been largely demonstrated in this chapter because the examples were merely a.c. versions of those which appeared in Chapter 4.

For maximum power, it is necessary to match the impedances which requires equal resistance components but equal and opposite reactance components.

Exercises 15

Use the Superposition theorem to solve Exercises 1 to 5.

1. For the network shown in Fig. A, determine the current in X_L. Use both the Superposition theorem and Nodal analysis.

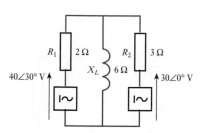

Fig. A

2. For the network shown in Fig. B, determine the current in X_L.

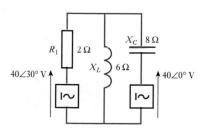

Fig. B

3. For the network shown in Fig. C, determine the current I.

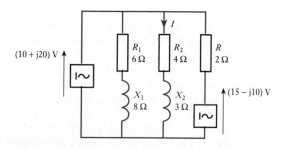

Fig. C

4. Determine the voltage V for the network shown in Fig. D.

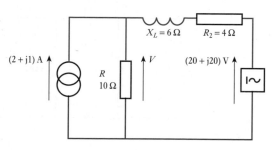

Fig. D

5. For the network shown in Fig. E, determine the voltage across and the current in X_L.

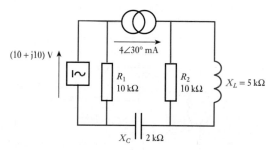

Fig. E

Use Thévenin's theorem to solve Exercises 6 to 10.

6. Determine the Thévenin equivalent circuit for the network shown in Fig. F.

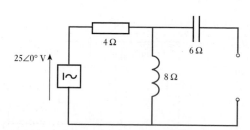

Fig. F

Exercises 15 continued

7. Determine the Thévenin equivalent circuit for the network shown in Fig. G.

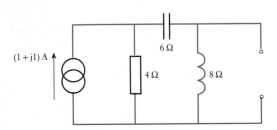

Fig. G

8. For the network shown in Fig. D, determine the voltage V.

9. For the network shown in Fig. C, determine the current I.

10. For the network shown in Fig. H, determine the current I.

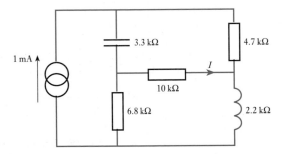

Fig. H

Use Norton's theorem to solve Exercises 11 to 15.

11. For the network shown in Fig. I, determine the Norton equivalent network.

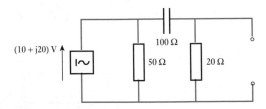

Fig. I

12. For the network shown in Fig. G, determine the Norton equivalent network.

13. For the network shown in Fig. E, determine the current in X_L.

14. For the network shown in Fig. J, determine the current in R_L. Use both Norton's theorem and nodal analysis.

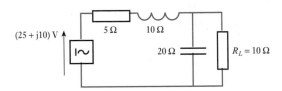

Fig. J

15. For the network shown in Fig. K, determine the current in the 60 Ω resistor.

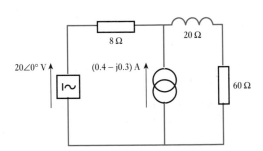

Fig. K

Use the maximum power theorem to solve Exercises 16 to 20.

16. For the network shown in Fig. F, determine the load resistance which will dissipate maximum power when connected to the output terminals.

17. For the network shown in Fig. G, determine the resistive and reactive components of a load impedance which will dissipate maximum power when connected to the output terminals.

18. For the network shown in Fig. J, R_L is replaced by an impedance. Determine the maximum power which it can dissipate.

Exercises 15 continued

19. For the network shown in Fig. K, the 60 Ω resistor is replaced by a variable resistor. Determine the maximum power it can dissipate and its resistance.

20. For the network shown in Fig. L, determine the maximum power which can be dissipated by Z.

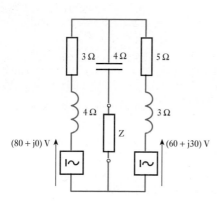

Fig. L

Use the star–delta and delta–star transforms for Exercises 21 to 24.

21. Determine the delta equivalent network for that shown in Fig. M.

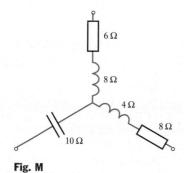

Fig. M

22. Determine the star equivalent network for that shown in Fig. N.

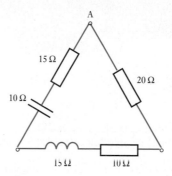

Fig. N

23. Determine the input impedance to the network shown in Fig. O.

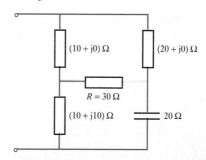

Fig. O

24. For the network shown in Fig. P, determine the current I.

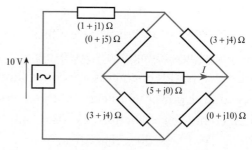

Fig. P

Chapter sixteen — Micropower or Megapower

So far we have studied a wide variety of electrical principles with only a few references to their practical applications. As we come to the end of Section One, it is time to consider the factors which influence the relationships between principles and practice.

In particular, we have to consider the reasons that some engineers describe themselves as power engineers whereas others call themselves electronics engineers. Depending on that choice, we can select and interpret those general topics which appear most significant to our further studies.

Although our interests have a bias one way or another, it is worth observing that more and more power engineers are using electronic devices in their work yet equally more and more electronics engineers are controlling power systems. The division between the megapower power engineer and the micropower electronics engineer is becoming less distinct and few can afford to study only power or only electronics. So are you mega or micro? And how does that affect your approach to further study?

16.1   Introduction

We have now reached the end of Section One of this book, *Electrical Principles*, this section having been dedicated to the introduction of basic electrical principles. However, we have now developed a sufficient understanding of engineering to let us consider further the roles of the electronics engineer and the power engineer to whom reference was made in section 2.1.

Both types of engineers use the same principles which have appeared in this section of the book, and both deal with electricity. So, in what ways are such similar individuals different? In other words, how do we separate the sheep from the goats?

We could quite simply say that the difference lies in the applications to which they apply themselves. For many this division might work, but for others there is the apparent involvement of electronic control equipment in power systems operation and vice versa. Instead there are four underlying factors which we now consider.

16.2   System efficiency

In section 2.10, we noted that carbon resistors with a nominal resistance could actually have a resistance variation 5 per cent, 10 per cent or more above or below the nominal value. Here we have to ask the question – does this matter? Well it does if we want to obtain a particular performance from a circuit. This can be achieved by adding a variable resistor into the circuit so that the total resistance can be adjusted until we get exactly the circuit response being sought.

By introducing the variable resistor, we also introduce a device which is producing an I^2R loss. The consequence is that, in terms of energy efficiency, we have made a system which is not at all efficient. Let us consider a simple circuit in Example 16.1.

Example 16.1

A resistor of nominal resistance 10 kΩ has an actual resistance of 11 kΩ. It is connected into the circuit shown in Fig. 16.1 and it is required that the voltage across the resistor should be 1.8 V. Determine the power dissipation for the variable resistor and the efficiency of the circuit.

Fig. 16.1 Circuit for Example 16.1

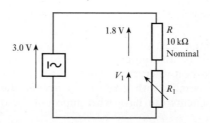

$$V_1 = 3.0 - 1.8 = 1.2 \text{ V}$$

$$\frac{V}{V_1} = \frac{R}{R_1}$$

$$R_1 = 7.33 \text{ k}\Omega$$

$$P_1 = \frac{V_1^2}{R_1} = \frac{1.2^2}{7.33 \times 10^{-6}} = 0.196 \times 10^{-6} \text{ W} = \mathbf{196 \ \mu W}$$

$$P = \frac{V^2}{R} = 0.295 \times 10^{-6} \text{ W} = \mathbf{295 \ \mu W}$$

$$\eta = \frac{295}{196 + 296} = \mathbf{0.60 \text{ or } 60 \text{ per cent}}$$

Looking at the solution, our first reaction must be that 60 per cent efficiency is very poor. But then we should consider the rate of waste which is only about 200 μW. This is so small that it is negligible. This is particularly the case if the output power is being derived from a relatively large source so that the rate of waste energy is insignificant.

Let us now consider a very different situation. The transmission of electrical power across the land depends on overhead lines operating at extremely high voltages. When we come to the end of a line, a transformer changes the voltage to a lower one at which the power can be distributed. Such a transformer could be rated to handle 1000 MW, sufficient to supply a large town. Typically a transformer has an efficiency in excess of 99 per cent which sounds very good – but is it that good?

Ninety-nine per cent efficient means that there is a 1 per cent loss. One per cent of 1000 MW is 10 MW – the equivalent of 10 000 electric heaters all standing in the open wasting energy! Also 10 MW could be sufficient to supply a village of about 1000 houses.

Even a small increase in efficiency can be significant. For instance, if the efficiency was increased to 99.05 per cent, then the reduction in the rate of waste would be 500 kW, itself quite a large power.

Therefore a factor which an engineer might have to consider is whether or not the efficiency is significant. Generally electronics engineers can almost ignore energy efficiency, whereas power engineers are greatly concerned with energy efficiency. However, this categorization is changing with the growth of mobile electronics, which have finite power available. This provides our first criterion for dividing the two types of engineer – yet the analysis of efficiency is derived from the electrical principles which are shared by both. It is a simple matter of your viewpoint.

16.3   Effects of waste heat

In section 2.9, we met the reality that there is no such thing as the perfect conductor. Instead every conductor has some resistance and therefore dissipates heat when current flows through it.

In circuits with very small currents, the conductors joining the circuit components can have what appear to be quite large values of resistance.

Let us again consider Example 16.1 and assume an artificially high value of resistance for the conductors. If we assume 11 Ω, we can readily observe that the rate of energy dissipation by the conductor will be 295 pW – which is so small that we can neglect it. Certainly the conductor will not heat significantly as a consequence of this heating effect – but because of their small size, the resistor components could become quite warm; hence the need to ensure that they are correctly selected as described in section 2.10.

Going to the opposite end of the spectrum, power engineers deal with very large currents, say 10 kA and more. Thus a cable of perhaps 10 μΩ resistance would dissipate waste energy at a rate of 1 kW. Depending on the cable dimensions, this could be sufficient to raise the temperature of the conductor. What matters is whether or not this temperature rise would be sufficient to damage the insulant around the conductor. The limitation on the current will depend on the temperature which results and the ability of the insulator to resist damage as a consequence of operating at a higher temperature.

Electrical machines involve windings rather like those which we met in Chapters 7 and 8. Because the conductor is wound in a coil to form a winding, it is more difficult to remove the waste energy, i.e. the heat loss, and therefore quite high temperatures can be built up in the middle of the winding. This is especially limiting to the currents which can be sustained without damage to the insulant. Sooner or later, most of us will meet with a motor which is smoking because it is too hot, say by working our DIY drill too hard.

Therefore, a second factor which an engineer might have to consider is whether or not the temperature rise as a consequence of waste energy is significant. Electronics engineers and power engineers are greatly concerned with temperature rise, but for different reasons. This degree of concern therefore provides our second criterion for dividing the types of engineer – yet the cause of energy waste is the I^2R loss derived from a common electrical principle. Its relevance is again simply a matter of your viewpoint.

Later in section 20.6, we will meet the semiconductor junction. This is particularly susceptible to the effects of temperature, so if it is operated in too high an ambient temperature, the junction breaks down and has to be replaced. It is for this reason that we are told not to keep electronic cameras and radios out in the heat of the sun. Therefore, although we are usually concerned about temperature rise due to waste energy coming from the I^2R loss, there are other causes which we need to bear in mind.

16.4 The length of the circuit

How long is long? If we were to ask this question, we would find that we might get a surprising variety of answers. This is especially the case with electrical engineers.

Generally, we tend to think of length in terms which apply to ourselves. You might very well think of length as being a measure of travel, so that a long journey might be 200 km or 100 miles. Electrical energy has to travel along transmission lines and we find that power engineers would consider anything in excess of 200 km or 100 miles to be long.

No conductor system is without loss, so when we use smaller wires then the length which might be considered long is reduced. Thus 200 m might be long when supplying a large motor in a factory and 40 m might be long when wiring a house.

When we look at electronic systems, and in particular when we look at data control and communication systems, we find that our understanding of length is completely changed. We have already observed in section 7.3 that the absolute permeability and permittivity are related to the speed of light. Taken a stage further, we can observe that electrical signals move through space with the same speed.

In control and communication systems, we find that one limitation to their speeds of operation is limited by the time taken for a signal to move from one component to the next. Thus if the distance between components is initially 2 mm and is reduced to 1 mm then the speed of operation could be doubled. Techniques of component manufacture have been developed such that about a million circuit items can be mounted on an area of about 1 mm^2. In such instances, as are found in microprocessor chips, a long distance seems to be 1 μm.

Thus we find that what might be considered to be long varies dramatically according to the context in which the question arises.

16.5   Accuracy

Let us consider an electronic application with which we are all familiar – the television set. Every second a beam of light crosses the 625 lines which go to make up the picture; this happens not just once but 25 times every second. If the beam of light started at the wrong time or if it moved with the wrong speed, the picture would break up or roll over. The result is that the electronic circuits which control the production of our television picture require to be of an exceptionally high accuracy.

Yet we have already observed that electronic circuit components have manufacturing tolerances which are anything but accurate. Although we have discussed resistors having tolerances up to and in excess of 20 per cent, other components such as transistors can vary by as much as 50 per cent. Therefore we find that highly accurate outcomes are produced by highly inaccurate components. Such is the world of electronics.

By now we can anticipate that the power systems are bound to be different. And our anticipation is fulfilled. But perhaps we should consider the reasons why this should be so. Many power systems are used to deliver electrical energy, say, to a street of houses. We cannot readily predict when No. 3 is going to switch on a cooker, or No. 5 is switching off its hot water tank, or No. 14 switches on its lights. It does not matter what happens – we still expect a steady supply voltage which will deliver the power appropriate to the equipment used.

Similarly an electric motor might have to vary its output according to the load applied. For instance the motor driving an electric train will work harder when the train has passengers than when it is empty. And again will one train be starting or will two or three all be demanding a supply of energy at the one time?

So the total demand in power systems is a lottery – yet we expect the total system to supply the right amount of energy no matter what is happening. This requires great accuracy in the supply system to supply a most inaccurate load.

16.6 The way ahead

We are now ready to apply the electrical principles to which we have been introduced in Section One. Section Two, *Electronic Egineering*, will deal with electronics systems while Section Three, *Power Engineering*, will deal with power systems. In other words, we have one section for micropower and another section for megapower, both with the same underlying principles.

In Section Two, the electronic systems have to be considered in the following contexts:

1. The power levels are very low and therefore energy efficiency may be of less importance, except for portable devices.
2. Even with poor efficiency, the waste heat is rarely significant, except where heating adversely affects components.
3. The reduction of distance between circuit components has led to miniaturization.
4. The outcomes of electronic circuits generally seek exceptionally high levels of accuracy by means of components with poor tolerances.

In Section Three, the power systems have to be considered in the following contexts:

1. The power levels are high and therefore energy efficiency is most important.
2. Even with high efficiency, the waste heat is significant and limiting to applications.
3. The distance to a load is only occasionally significant but generally does not feature.
4. The outcomes of power systems can be quite variable, yet are achieved using components with high tolerances.

It is easy to overlook these when we start to develop circuits in the following sections. In particular we will be undertaking circuit analyses as though the component values were absolute. Instead we should remember that the examples and calculations are only intended to illustrate the responses we might expect rather than the definitive outcomes. So let us proceed to find out what happens when we apply our electrical principles to a variety of applications.

Terms and concepts

We have observed that there are four underlying factors which affect our judgement as engineers.

The **system efficiency** in terms of energy loss is most significant when considering large amounts of energy. Generally we can afford to lose 1 per cent of 1 W but 1 per cent of 1 MW is costly.

The **waste heat** gives rise to the rating of equipment, so we consider the rate of loss and how quickly we can dispose of it.

We have to consider the size of the system. It makes a great deal of difference whether the load is beside the source or a long way distant – and how do we judge what is a long way distant?

The final factor is our desire for accuracy leading us to realize how seldom we are sure of what we require by way of accuracy.

The ways in which we respond to these factors vary greatly depending on the applications – and it is now time to proceed from developing general electrical principles to their applications first in electronic systems and second in power systems.

Section two   Electronic Engineering

Chapter seventeen

Electronic Systems

Objectives

When you have studied this chapter, you should

- be aware of the concept of systems
- understand the idea that systems are constructed from subsystems
- be familiar with the basic concepts of an amplifier and attenuator
- be capable of working out the values of gain for a system
- be able to illustrate systems using block diagrams

Contents

Everyone is familiar with electronic devices, whether it is the radio, television, mobile phones or computers. However, most people have no knowledge of what goes on inside these devices and that is the way they like it. Fortunately for them there are others who are more curious. Electronic devices generally have hundreds or even thousands of components. To be able to understand their functions, they must be split into groups and placed in blocks. These blocks may then be related by means of systems.

In this chapter, systems, and the relationships between blocks which make them up, will be introduced. In particular, the amplifier, an important block of many systems, will be presented. The attenuator will follow this. By the end of the chapter, you will be able to consider electronic systems and relate amplifiers by virtue of their voltage and current gain.

17.1 Introduction to systems

The subject of electronics is one that is difficult to define since it refers to an extremely wide range of electrical technology.

Electronics could deal with the detailed circuitry, but at this stage it is better to stand back and look at systems which operate electronically. To do this, perhaps we should first be clear about what is meant by a system.

The term *system* can be applied to quite a range of situations. For instance, the process of communicating by letter is a system. The process can be split up into a number of parts starting with the posting of the letter and finishing with its delivery. In between, the letter will have been transported to a sorting station where it is selected for a further stage of transport. Therefore the postal system consists of a well-defined series of procedures.

In much the same way, a railway system consists of a variety of well-defined routes and stations; a telephone system consists of a variety of communication links including cables and satellites; an educational system takes pupils through primary and secondary education into tertiary education; and so on.

Any system normally has to be complete if it is to function. If one of the transport arrangements for the postal service were to break down, then the letter could not get from the point of sending to the point of delivery. Similarly, if a bridge on the railway were damaged then the train and its passengers could not travel from the point of departure to the point of arrival.

Such problems can be solved by introducing a temporary alternative. For instance, the postal service could obtain an alternative vehicle to replace that which had broken down, while the railway might hire buses to ferry the passengers past the damaged bridge.

Large systems usually can be broken down into smaller systems which are called *subsystems*. For instance, the postal service had a series of subsystems most of which were concerned with either transporting the letter from one sorting station to another or the sorting stations themselves. Such an arrangement can be shown diagrammatically as in Fig. 17.1.

Fig. 17.1 A postal system

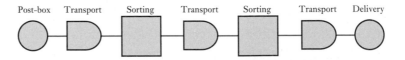

Post-box   Transport   Sorting   Transport   Sorting   Transport   Delivery

Systems have inputs and outputs such as the post-box where the system started and the letter-box into which the letter was delivered at the end of the system. In simple cases, a system has a well-defined boundary with a single input and a single output. A typical simple system is exemplified by a ski–lift. Such a system has only one route with no alternatives along the way. Its operation is all or nothing.

Many electronic systems fall into the category of being a simple system and can be represented as shown in Fig. 17.2.

Fig. 17.2 A simple system

17.2 Electronic systems

In practice many electronic systems are complicated; fortunately these complicated systems can be divided into smaller and simpler systems similar to the form of that shown in Fig. 17.2. By breaking an electronic system into these component subsystems, we can more easily understand the function of the entire system. Most significantly, we do not need to know precisely what is happening within each of the subsystems. Rather our concern is to know the action of the subsystem or, to put it another way, what is the relationship between the output and the input?

A typical electronic subsystem consists of an amplifier in which the output is several times larger than the input, i.e. its function is to magnify. If we were to input a voltage of 10 mV and produce a resulting output of 100 mV then the voltage has been magnified 10 times. Such an amplifier could then be described as amplifying by a factor of 10.

The fact that we have no idea how this was achieved may seem peculiar, yet it reflects the manufacturing techniques now used in electronics. It is quite usual to have what looks like a single component which acts as an amplifier. In fact it contains hundreds if not thousands of components in a sealed unit. If one component fails, the complete unit is useless and has to be replaced, but such is the cheapness of the unit that this is preferable to the cost of repairing the unit. In fact, the components are so small that repair is effectively impossible even if we wished to try.

The approach of using subsystems to understand an electronic system is often referred to as 'black box' approach. This might be thought to refer to the units which usually are encapsulated in a black material, but it is really emphasizing the fact that we cannot see what makes up the contents of the unit.

There are many types of electronic subsystems, but there are two with which we wish to be initially familiar: these are the amplifier and the attenuator.

17.3 Basic amplifiers

We have already met the most simple form of amplifier when we considered the device which produced an output voltage of 100 mV when an input voltage of 10 mV was applied to it. However, no mention was made of the associated current which might readily have the same output and input values. If this were the case, then the output power would be 10 times that of the input.

This raises the question, from where has the additional power come? In practice the output power comes from a separate power supply – the amount made available depends on the input power. The output power may include the input power, but it could just as readily be completely derived from the separate power source.

The action is an example of gain. That is to say, a small input signal power controls a larger output signal power. Certain devices not normally referred to as amplifiers do nevertheless come into this category by definition. In a relay, for example, the power required by the coil to close the contacts can be considerably less than that involved in the circuit switched by the contacts. Another example is the separately excited d.c. generator being driven at constant speed. Here the power being fed to a load connected

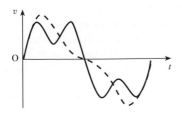

Fig. 17.3 Signal waveforms

across the output terminals can be controlled by a relatively small power fed to its field winding.

In the basic amplifier, there is a further characteristic that it must exhibit. The waveform of the input voltage or current, called a signal, must be maintained to a fairly high degree of accuracy in the output signal. This can be illustrated by comparing two signals representing the same note played by two different musical instruments. They are both of the same fundamental frequency representing the pitch, but the waveforms differ, representing different tones. This is illustrated in Fig. 17.3.

Fig. 17.4 Amplifier block diagram

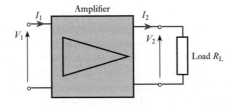

Figure 17.4 shows an amplifier with a resistive load R_L connected across the output terminals. The basic parameters of the amplifier are as follows:

$$\text{Voltage gain } (G_v) = \frac{\text{output signal voltage}}{\text{input signal voltage}}$$

$$G_v = \frac{V_2}{V_1} \qquad [17.1]$$

$$\text{Current gain } (G_i) = \frac{\text{output signal current}}{\text{input signal current}}$$

$$G_i = \frac{I_2}{I_1} \qquad [17.2]$$

$$\text{Power gain } (G_p) = \frac{\text{output signal power}}{\text{input signal power}}$$

$$G_p = \frac{V_2 \times I_2}{V_1 \times I_1}$$

$$G_p = G_v G_i \qquad [17.3]$$

Figure 17.5 shows typical waveforms where the signals are assumed to vary sinusoidally with time. The waveforms met in practice tend to be more complex, as illustrated in Fig. 17.3, but it can be shown that such waves are formed from a series of pure sine waves the frequencies of which are exact multiples of the basic or fundamental frequency. These sine waves are known as harmonics. The use of sine waves in the analysis and testing of amplifiers is therefore justified.

Fig. 17.5 Amplifier signal waveforms

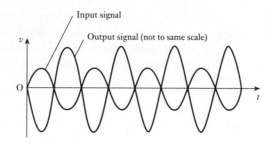

Examination of the waveforms in Fig. 17.5 shows that the output signal voltage is 180° out of phase with the input signal voltage. Such an amplifier is known as an *inverting amplifier* and the gain as defined by relation [17.1] is negative. These amplifiers are very common in practice although non-inverting types, where the output and input signals are in phase, are also met. Some amplifiers have alternative inputs for inverting and non-inverting operation. Amplifiers will be discussed in more detail in Chapter 19.

So far only the input and output signals have been considered in the operation of the amplifier. It is necessary, however, to provide a source of power from which is obtained the output signal power fed to the load, the magnitude of this signal power being controlled by the magnitude of the input signal. The source itself has to be a direct current one and could be a dry battery or a rectified supply as described in Chapter 21. The magnitude of this supply voltage depends on the type of device used in the amplifier.

Example 17.1

An amplifier has an input voltage of 8 mV and an output voltage of 480 mV. The input current in phase with the voltage is 200 μA and the power gain is 3000. Determine:

(a) the voltage gain;
(b) the output power;
(c) the output current.

(a) $\quad G_v = \dfrac{v_o}{v_i} = \dfrac{480 \times 10^{-3}}{8 \times 10^{-3}}$

$\quad\quad = 60$

$\quad P_i = v_i i_i = 480 \times 10^{-3} \times 200 \times 10^{-6}$

$\quad\quad = 96\ \mu W$

(b) $\quad P_o = G_p P_i = 3000 \times 96 \times 10^{-6}$

$\quad\quad = 288\ mW$

(c) $\quad i_o = G_i i_i = \dfrac{G_p}{G_v} \times i_i = \dfrac{3000}{60} \times 200 \times 10^{-6}$

$\quad\quad = 10\ mA$

17.4 Basic attenuators

Again let us consider a simple electronic black box system as shown in Fig. 17.6. The input voltage is applied to the input terminals which we will call the input port, and the output voltage will be observed at the output terminals which we will call the output port.

Fig. 17.6 A simple electronic system

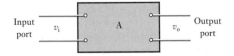

If the input is a voltage v_i and the output voltage is v_o, the relationship between the two can be expressed as

$$v_o = A v_i \qquad [17.4]$$

This is the general situation in which A is the action produced by the black box and is called the transfer function. Generally when A is greater than unity, the system is acting as an amplifier. However, when A is less than unity, the system is acting as an attenuator. Attenuators do not require a separate power supply and therefore the output power is entirely derived from the input power.

17.5 Block diagrams

The circuit block diagrams have all been shown with two input terminals and two output terminals. In practice, one of the input and one of the output terminals are connected to earth and are therefore common. This is shown in Fig. 17.7(a). It is often convenient to dispense with the earth connections and the diagram can be simplified to that shown in Fig. 17.7(b). It follows that the amplifier block diagram can be reduced from that shown in Fig. 17.4 to that shown in Fig. 17.8.

Fig. 17.7 Reduction of block diagrams

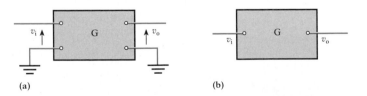

(a)   (b)

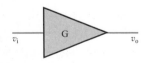

Fig. 17.8 An amplifier symbol

A block diagram is the first stage to the understanding of an electronic system. However, it does not specify how each box functions nor what is contained within each box. Also a block diagram does not indicate where the box is to be found on the physical system. Thus the block diagram is only intended to represent the operation of the complete system.

Using a block diagram, should an electronic system fail to operate correctly, it is a relatively simple procedure to locate which box or subsystem has broken down. Knowing which box or subsystem is faulty enables us to repair or replace it, generally the latter. It can also lead to consideration of change of system design to ensure that the problem does not reoccur.

17.6 Layout of block diagrams

It is worth noting that there are some general principles to the layout of block diagrams:

1. The input normally appears at the left-hand side and the output at the right-hand side, just as in ordinary circuit diagrams.
2. Power supply lines, if included, are usually drawn with the unearthed line to the top and the earthed line below. The supply lines can be positive or negative with respect to earth.
3. Some boxes are drawn in the rectangular form already shown. However, amplifiers are shown as triangles with or without a surrounding rectangle. In Chapter 27 we will be introduced to other shapes found in logic systems.
4. Arrows are sometimes added to shows the direction of power flow.

The flow of power is an awkward concept since power is actually a rate of energy flow – and a rate is not therefore a concept which can flow. In fact power flow refers to the direction of energy movement.

Summary of important formulae

For an amplifier,

voltage gain, $\quad G_v = V_o / V_i$ $\hfill$ [17.1]

current gain, $\quad G_i = I_o / I_i$ $\hfill$ [17.2]

power gain, $\quad G_p = G_v G_i$ $\hfill$ [17.3]

For an attenuator,

$$v_o = A v_i \hfill [17.4]$$

Terms and concepts

An electronic system can be made from a number of subsystems such as **amplifiers**, **attenuators** and **transducers**. It is not essential that we are aware of the detailed construction of a subsystem; rather it is essential that we can relate its output to its input.

The function of an amplifier is to provide **gain**. Usually the gain is that either the current or the voltage out is directly proportional to the current or voltage in and is greater in magnitude.

Current or voltage gains normally result in **power gain**. The additional power is tapped from a source separate from that of the input current or voltage.

Attenuators have no separate power source and therefore there cannot be a power gain. The attenuation can be thought of as **negative gain**.

Systems can be illustrated by means of **block diagrams** which should be read from left to right.

Chapter eighteen

Passive Filters

Filters have a wide application in electronics equipment: in hi-fi systems and communications equipment, for example. This might be achieved with either passive or active (using op-amps or transistors) filter circuit design, though we are only going to consider passive filter circuits in this chapter. Filters are typically designed to select or reject a band of frequencies depending on the particular application. Hence, their operation is described by the way the output of the filter circuit varies with frequency. This is termed the frequency response and is often presented as a graph of output voltage versus frequency.

Normally the frequency range considered is fairly large and it is usual to plot frequency on a logarithmic scale. Similarly, where the amplitude variations are large a logarithmic scale can be used. Alternatively, the output may be expressed in logarithmic form by using the decibel notation. Consequently, before we discuss the design of filters, we will review the mathematics of logarithms applicable to this subject.

The main part of the chapter will consider the design of simple low-pass and high-pass *RC* filters but it includes a brief treatment of passband and stopband filter circuits. Since, in a.c. circuits, the output quantity is a phasor, then both the amplitude and the phase angle will be functions of frequency, giving the amplitude and phase response of the circuit. Both amplitude and phase response will be considered in detail for both low- and high-pass filters. By the end of the chapter, you will also be able to draw Bode diagrams, which enable the frequency response of filters on a decibel scale to be estimated rapidly and accurately.

18.1   Introduction

A filter changes the amplitude of an a.c. voltage as the frequency is changed. Thus, the operation of a filter is described by the way the output of the filter circuit varies with frequency. This is termed the frequency response. This information is typically presented in a graph of output voltage versus frequency.

Filter circuits are typically designed to select or reject a band of frequencies depending on the particular application. This might be achieved with either a *passive* (using a series or parallel combination of R, L and C) or *active* (using op-amps or transistors and R, L and C) filter circuit design. Filters have a wide application in electronics equipment. In hi-fi systems, filters are used to select particular bands of frequencies for amplification or attenuation as required by the listener. They are also used to filter out *noise*, unwanted signals at frequencies resulting from equipment deficiencies or signal pick-up from the surrounding medium. In communications equipment, filters are used to select the frequency band containing wanted information whilst rejecting unwanted frequencies.

In this chapter, only passive types of filter will be considered. In fact, one type of filter has already been studied – the resonant filter.

18.2   Types of filter

If the maximum output voltage of the circuit, at any frequency, is V_m, then the frequency range for which the response of the filter is greater than $V_m/\sqrt{2}$ is termed the *passband*. The range of frequencies for which the output voltage falls to less than $V_m/\sqrt{2}$ is called the *reject* or *stopband*. The frequency at which the output voltage is $V_m/\sqrt{2}$ is called a **cut-off frequency**, or **half-power frequency**.

There are four broad categories of filter – based on their frequency response:

(a)   Low-pass filter

This filter permits the passage of low-frequency signals, whilst rejecting those at higher frequencies. The frequency range for which the response of the filter (V_{out}) lies above $V_m/\sqrt{2}$ is at low frequencies as shown in Fig. 18.1(a). f_2 is the cut-off frequency, the frequency above which the response falls below $V_m/\sqrt{2}$.

(b)   High-pass filter

This filter permits the passage of high-frequency signals, whilst rejecting those at low frequencies. The frequency range for which the response of the filter (V_{out}) lies above $V_m/\sqrt{2}$ is at high frequencies as shown in Fig. 18.1(b). f_1 is the cut-off frequency, below which the response falls below $V_m/\sqrt{2}$.

(c)   Passband filter

This filter (also called a bandpass filter) permits the passage of signals in a defined frequency range, in which the response of the filter (V_{out}) lies above

$V_m/\sqrt{2}$, whilst rejecting those signals outwith this range of frequencies. This is shown in Fig. 18.1(c). The cut-off frequencies f_1 and f_2 determine the frequency range in which the response lies above $V_m/\sqrt{2}$. Below f_1 and above f_2, the response falls below $V_m/\sqrt{2}$. This response is that of the series resonant circuit discussed in Chapter 14. The range of frequencies between f_1 and f_2 is termed the *bandwidth*.

(d)   Stopband filter

This filter prevents the passage of signals in a defined frequency range, in which the response of the filter (V_{out}) lies below $V_m/\sqrt{2}$. This is shown in

Fig. 18.1   Main filter types

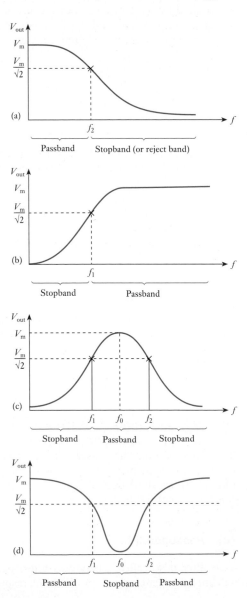

Fig. 18.1(d). The cut-off frequencies f_1 and f_2 determine the frequency range in which the response lies below $V_m/\sqrt{2}$. Below f_1 and above f_2, the response is greater than $V_m/\sqrt{2}$.

18.3 Frequency response

It is the variation of reactance of inductors and capacitors with frequency, illustrated in Fig. 18.2, that causes the output of a circuit to vary with frequency even though the input voltage is held constant.

The way in which a circuit behaves as the frequency is varied is referred to as the frequency response of the circuit. Since, in a.c. circuits, the output quantity is a phasor, then both the amplitude and the phase angle will be functions of frequency, giving the *amplitude* and *phase response* of the circuit. In many cases, the phase response is considered unimportant and the term frequency response is used to describe the amplitude response.

Normally the frequency range considered is fairly large and it is thus usual to plot frequency on a *logarithmic scale*. Similarly, where the amplitude variations are large a logarithmic scale can be used. Alternatively, the output may be expressed in logarithmic form by using the *decibel notation*.

Fig. 18.2 Variation of reactance with frequency: (a) inductive reactance; (b) capacitive reactance

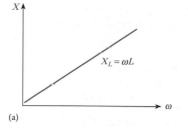
(a)

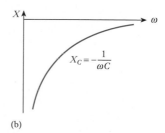
(b)

18.4 Logarithms

A knowledge of the mathematics of logarithms is required in order to understand their application to the subject of filters, frequency response using logarithmic scales and the use of decibels. This section is a brief review of the basic principles involved.

If a given value can be expressed as a power of another number:

e.g. $\quad 9 = 3^2$

then 3 is called the *base* of the power and 2 the *index*. That is,

$$9 \quad = \quad 3^2 \leftarrow \text{index (or log to base 3)}$$

$$\uparrow \qquad \uparrow$$

number   base

The index 2 can also be defined as the logarithm of the number to the base 3 and is written as

$$2 = \log_3 9$$

In general, if $X = a^x$,

$$x = \log_a X$$

Since logs (shorthand for logarithms) are the indices of powers of the given base, the rules of logs follow closely those of indices.

(a) Some important notes about logarithms

$$\log_{10}1 \quad = 0 \qquad \text{Note:} \quad 10^0 \quad = 1$$
$$\log_{10}10 \quad = 1 \qquad\qquad\qquad 10^1 \quad = 10$$
$$\log_{10}100 = 2 \qquad\qquad\qquad 10^2 \quad = 100$$
$$\log_{10}2 \quad = 0.3010 \qquad\qquad 10^{0.3010} = 2$$

(1)   The log of any number less than 1 is negative:

$$\log_{10}\frac{1}{2} = \log_{10}0.5 = -0.3010$$

$$\log_{10}\frac{1}{2} = \log_{10}0.1 = -1$$

(2)   The log quotient rule:

$$\log_{10}\frac{a}{b} = \log_{10}a - \log_{10}b$$

$$\log_{10}\frac{1}{2} = \log_{10}1 - \log_{10}2$$

$$= 0 - 0.3010$$

$$= -0.3010$$

(3)   Log product rule:

$$\log_{10}(x \times y) = \log_{10}x + \log_{10}y$$
$$\log_{10}(2 \times 3) = \log_{10}2 + \log_{10}3$$

(4) $\qquad \log_{10}y^x = x \log_{10}y$

$$\log_{10}2^3 = 3 \log_{10}2$$

(b) Antilogs

The number that corresponds to a given value of a log to a certain base is called its antilogarithm. For example:

$$8 = 2^3$$

$$\log_2 8 = 3$$

i.e. the antilog of 3 to the base of 2 is the number 8 and this is written

$$\text{antilog}_2 3 = 8 \quad (2^3)$$
$$\text{antilog}_7 2 = 49 \quad (7^2)$$

(c) Common logs

These are logs to a base of 10, as has already been discussed. This is the calculation performed on a calculator when the LOG button is used.

i.e. $100 = 10^2, \quad \log_{10}100 = 2$

When using common logs, we usually omit the base 10 and simply write

$$\log 100 = 2$$

When using logs to any other base, the base is written in

$$25 = 5^2$$

$$\log_5 25 = 2$$

Note, for example, that log 85 lies between 1 and 2 because

$$\log 10 = 1$$

$$\log 100 = 2$$

In fact log 85 = 1.9294. Here 1 is termed the *characteristic* and the decimal part is known as the *mantissa*.

For positive numbers that are less than 1:

- The characteristic is negative and one more than the number of zeros following the decimal point.
- The mantissa is still positive.

For example,

$$\log 0.0500 = -2 + 0.6990$$

This is written $\bar{2}.6990$. The bar, the negative sign above the 2, only refers to the 2. So

$$\log 0.0500 = -2 + 0.6990 = \bar{2}.6990 = -1.3010$$

Example 18.1   Show that log 90 is equal to the sum of the logs of the factors of 90.

Use a calculator to show that log 90 = 1.9542. Note that

$$90 = 3^2 \times 5 \times 2$$

Thus

$$\log 90 = 2\log 3 + \log 5 + \log 2$$
$$= (2 \times 0.4771) + 0.6990 + 0.3010$$
$$= 1.9542$$

(d) Natural logarithms

These are logs to the base e (e = 2.7183) and are written $\log_e X$ or simply ln X. Natural logarithms are obtained on a calculator by using the button marked LN. This type of logarithm should not be confused with logarithms to the base of 10 (calculator button LOG), which will be used extensively throughout this section on filters.

18.5 Log scales

The use of log scales permits a review of the response of a system for an extended range of frequencies, without a significant loss in information for any frequency range. Special graph paper is available with either one or two logarithmic scales to facilitate the plotting of frequency response information. When the paper has one log scale, the other is a linear scale; it is called *semi-log* graph paper. *Log–log* paper has both axes having logarithmic scales. In the example of semi-log paper shown in Fig. 18.3, note the linear scaling of the vertical axis and the repeating intervals of the log scale of the horizontal axis. These frequency intervals – from 1 kHz to 10 kHz, or 10 kHz to 100 kHz, for example – are known as frequency *decades*.

Fig. 18.3   An example of semi-log graph paper

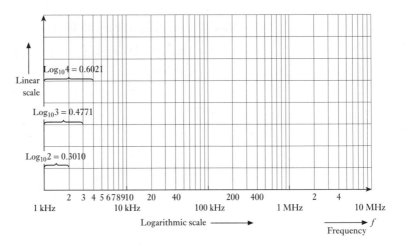

The spacing of the log scale is determined by taking the common log (to the base 10) of the number.

- The log scale does not begin at zero. It begins at 1, because $\log_{10} 1 = 0$.
- The distance on the log scale between 1 and 2 is because of the fact that $\log_{10} 2 = 0.3010$. Hence 2 appears at about 30 per cent of the full distance of a log interval (or decade).
- The distance between 1 and 3 is $\log_{10} 3 = 0.4771$. Hence 3 appears at about 48 per cent of the full distance of a log interval.
- $\log_{10} 4 = 0.6021$ (this is of course $2 \log_{10} 2$). Hence 4 appears at about 60 per cent of the full distance of a log interval.

It is clear from Fig. 18.3 that, though intervals on the log scale become smaller at the high end of each interval, a single graph can provide a detailed frequency response extending over a wide range of frequency intervals – from 1 kHz to 10 MHz as illustrated.

Example 18.2   Plot the frequency response data shown using semi-log graph paper.

Frequency (Hz)	V_{out} (V)
100	39.6
300	37.4
500	34.0
1000	25.0
3000	10.3
5000	6.3
10 000	3.2
50 000	0.64

The solution is shown in the graph of Fig. 18.4.

Fig. 18.4 Frequency response
for Example 18.2

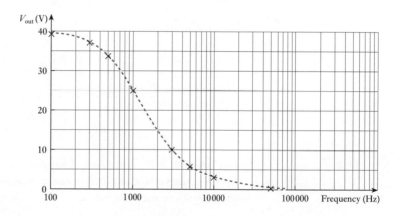

18.6   **The decibel
                     (dB)**

Consider the circuit shown in Fig. 18.5:

Fig. 18.5 Circuit to illustrate
power and voltage gain

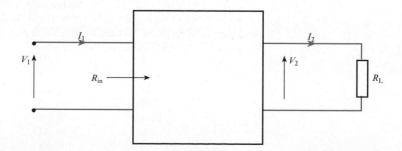

$$P_1 = \text{input power} = \frac{V_1^2}{R_{in}} = I_1^2 R_{in}$$

$$P_2 = \text{output power} = \frac{V_2^2}{R_L} = I_2^2 R_L$$

$$\text{Power gain} = \frac{P_2}{P_1} = \frac{V_2^2/R_L}{V_1^2/R_{in}}$$

The *decibel* is a logarithmic unit (to the base 10) of *power ratio* (1/10 of a *bel*). The unit is named after the Scottish engineer, Alexander Graham Bell (1847–1922).

$$1 \text{ bel} = \log_{10}\frac{P_2}{P_1} \quad \text{when} \quad \frac{P_2}{P_1} = 10$$

Note that the voltage gain is often given the symbol A; the gain in dB, the symbol G.

$$\text{Gain in dB} = G = 10 \log\frac{P_2}{P_1} = 10 \log\left(\frac{V_2}{V_1}\right)^2\frac{R_{in}}{R_L}$$

If $R_L = R_{in}$, then

$$G = 20 \log\frac{V_2}{V_1} \quad \text{or} \quad 20 \log\frac{I_2}{I_1}$$

(Note that $\log_{10}2^2 = 2 \log_{10}2$.)
   Note also that

$$G = 20 \log\frac{V_2}{V_1}$$

only when $R_L = R_{in}$. However, the condition is often relaxed and, even when $R_L \neq R_{in}$, the voltage and current gain of a circuit are defined as

$$G = 20 \log\frac{V_2}{V_1} \qquad\qquad [18.1]$$

and $G = 20 \log\dfrac{I_2}{I_1}$

Voltage gain A $\dfrac{V_2}{V_1}$	G dB $20 \log_{10}\dfrac{V_2}{V_1}$
0.707	−3 dB
1	0 dB
2	6 dB
10	20 dB
100	40 dB
1000	60 dB
10 000	80 dB
100 000	100 dB

Note therefore:

- Voltage gains are more easily plotted in dB.
- A doubling of voltage level results in a 6 dB change
- A change in voltage gain from 1 to 10 or 10 to 100 results in a 20 dB change.
- For any gain less than 1 the decibel level is negative. So for

$$\frac{V_2}{V_1} = 0.707, \ G \ dB = 20 \log 0.707 = -3 \ dB$$

The decibel is a measure of power ratio and more usually a measurement of voltage and current ratio and is dimensionless. It is standard practice in industry to plot the frequency response of filters, amplifiers and systems in total against a decibel scale.

$$\textbf{Gain (dB)} = \textbf{10 log}_{10} \ \frac{P_2}{P_1} \qquad\qquad [18.2]$$

This equation allows an extended range of interest on the same graph to be plotted, in much the same manner as the log frequency plots discussed earlier. Dealing with gains in dB of manageable value is also simpler.

For the special case of $P_2 = 2P_1$, the gain in dB is $G = 10 \log_{10} 2 = 3$ dB. Thus for a loudspeaker, a 3 dB increase in output would require twice the power. Increments of 3 dB in output level are a convenient measure because they are readily detectable to the ear.

Example 18.3

If a system has a voltage gain of 40 dB, find the applied voltage if the output voltage is 7.0 V.

Using equation [18.1]:

$$\text{Gain (dB)} = 20 \log \frac{V_2}{V_1}$$

$$40 = 20 \log \frac{7.0}{V_1}$$

$$2 = \log \frac{7.0}{V_1}$$

$$10^2 = \frac{7.0}{V_1}$$

Hence $\qquad V_1 = \textbf{70 mV}$

Example 18.4

A power level of 200 W is 3 dB above what other power level?

Using equation [18.2]:

$$\text{Gain (dB)} = 10 \log_{10} \frac{P_2}{P_1}$$

$$3 = 10 \log_{10} \frac{200}{P_1}$$

$$10^{0.3} = \frac{200}{P_1}$$

$$P_1 = \textbf{100 W}$$

18.7 **The low-pass or lag circuit**

Consider the simple RC circuit of Fig. 18.6, in which the variation of V_C with frequency is considered:

Fig. 18.6 Low-pass filter circuit

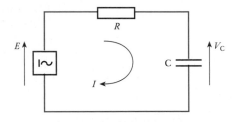

$$V_C = \frac{\dfrac{1}{j\omega C}}{R + \dfrac{1}{j\omega C}} \times E$$

$\therefore$ Gain $A = \dfrac{V_C}{E}$

$$= \frac{1}{1 + j\omega CR}$$

$$= \frac{1 - j\omega CR}{1 + (\omega CR)^2}$$

This can be expressed in polar form as:

$$\text{Gain } A = \frac{\{1 + (\omega CR)^2\}^{\frac{1}{2}}}{1 + (\omega CR)^2} \angle -\tan^{-1}\omega CR$$

i.e. $\mathbf{Gain\ } A = \dfrac{1}{\sqrt{\{1 + (\omega CR)^2\}}} \angle -\tan^{-1}\omega CR$ [18.3]

or   Gain $A = \left| \dfrac{V_C}{E} \right| \angle \theta$

where $|V_C/E|$ is the magnitude of the gain A and θ is the phase of the gain A.

By considering the solution of equation [18.3] at frequencies of $\omega = 0$, $\omega = 1/(CR)$ and $\omega = \infty$, an understanding of the frequency response of the circuit can be deduced:

At frequency $\omega = 0$: $|V_C| = E$ and $\theta = 0$

At very high frequency, $\omega \to \infty$: $|V_C| = 0$ and $\theta = -90°$

At a frequency $\omega_2 = \dfrac{1}{CR}$: $|V_C| = \dfrac{1}{\sqrt{2}}E$ and $\theta = -45°$

Fig. 18.7 Frequency response for low-pass circuit: (a) amplitude response; (b) phase response

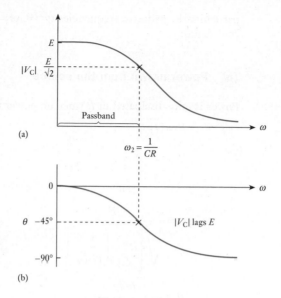

Hence with E constant, $|V_C|$ *falls* as ω increases and $\angle V_C$ *lags* further behind $\angle E$. The amplitude and phase response are illustrated in Fig. 18.7. It is clear that at low frequencies, there is a passband response. The upper end of the passband is at the frequency ω_2, variously named *the critical frequency*, *the cut-off frequency* or the *−3 dB frequency* because its output voltage is 3 dB down at this frequency from the value at $\omega = 0$. It is also called the *half-power point* or frequency, the reason for which will be discussed shortly. The stopband lies above this frequency ω_2, where the voltage V_C falls below $0.707E$ (or $E/\sqrt{2}$).

(a) Phasor diagram

The phase response indicates that V_C increasingly lags E as the frequency increases; V_C lags E by $\theta = -\tan^{-1}\omega CR$, from equation [18.3]. In particular, at the frequency $\omega_2 = 1/(CR)$, $\theta = -45°$. We can summarize the way in which V_C varies with frequency by the circle diagram of Fig. 18.8. I leads V_C by $90°$. V_R leads V_C by $90°$ and is in phase with I. As the frequency varies from 0 to ∞ the point A moves around the circumference of a semicircle with E as the diameter. The circle diagram gives the locus of V_C as ω varies. Note in

Fig. 18.8 Circle diagram showing locus of V_C as ω varies

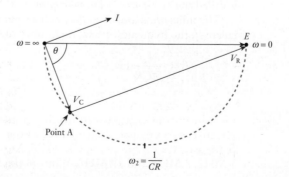

particular the values at frequencies $\omega = 0$, $\omega = 1/(CR)$ and $\omega = \infty$ previously considered.

(b) Power drawn from the source

Power is only dissipated in R since no power is consumed by the reactance.

$$P = |I|^2 R$$

$$I = \frac{E}{R + \dfrac{1}{j\omega C}}$$

$$|I| = \frac{E}{\sqrt{\left\{ R^2 + \left(\dfrac{1}{\omega C} \right)^2 \right\}}}$$

$$\therefore \qquad P = \frac{E^2 R}{R^2 + \left(\dfrac{1}{\omega C} \right)^2}$$

Values of P at frequencies $\omega = 0$, $\omega = 1/(CR)$ and $\omega = \infty$ can be calculated from this equation.

For $\omega = 0$: $P = 0$

For $\omega \to \infty$: $P = \dfrac{E^2}{R}$

Note that this is the maximum power. This result might be anticipated because at high frequencies the reactance of the capacitor becomes negligible.

For $\omega_2 = \dfrac{1}{CR}$: $P = \dfrac{E^2}{2R} = \dfrac{1}{2} P_{max}$

Thus

$$\omega_2 = \frac{1}{CR} \qquad\qquad\qquad [18.4]$$

is called the *half-power point* or *frequency*.

   The units of ω_2 are, of course, radians per second. If it is necessary to calculate f_2, the frequency in Hz, then, since $\omega_2 = 2\pi f_2$,

$$f_2 = \frac{1}{2\pi CR} \text{ Hz}$$

Example 18.5   Plot the output voltage amplitude versus log frequency for a low-pass *RC* filter having a resistor of 1 kΩ and a capacitor of 400 pF supplied from a 12 V a.c. source (Fig. 18.9). Make calculations of V_o at 100 kHz, 1 MHz, 3 MHz and 10 MHz. What is the half-power frequency?

Fig. 18.9 Circuit diagram for Example 18.5

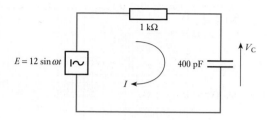

Half-power frequency is

$$f_2 = \frac{1}{2\pi CR} \text{ Hz}$$

$$= \frac{1}{2\pi \times 400 \times 10^{-12} \times 10^3}$$

$$= 398 \text{ kHz}$$

From equation [18.3]:

$$\text{Gain } A = \frac{1}{\sqrt{\{1 + (2\pi fCR)^2\}}} \angle -\tan^{-1} 2\pi fCR$$

$$= \left|\frac{V_C}{E}\right| \angle \theta$$

where E is the source voltage of 12 V.

At frequency $f = 0$: $\qquad |V_C| = 12$ V

At very high frequency, $f \to \infty$: $\quad |V_C| = 0$

At a frequency $f_2 = \dfrac{1}{2\pi CR}$: $\qquad |V_C| = \dfrac{1}{\sqrt{2}}E = 8.5$ V

At a frequency of 100 kHz:

$$|V_C| = \frac{12}{\sqrt{\{1 + (2\pi \times 100 \times 10^3 \times 400^{-12} \times 10^3)^2\}}} = 11.64 \text{ V}$$

The output voltage $|V_C|$ can be calculated in a similar way for other frequencies:

Frequency	$|V_C|$
100 kHz	11.64 V
398 kHz	8.50 V
1 MHz	4.43 V
2.5 MHz	1.89 V
10 MHz	0.48 V

The results are shown in the graph of Fig. 18.10, which shows a low-pass response. The passband extends to the half-power frequency f_2 where the output voltage $|V_C|$ has fallen to 0.707 of the input voltage.

Fig. 18.10 Frequency response for circuit of Example 18.5

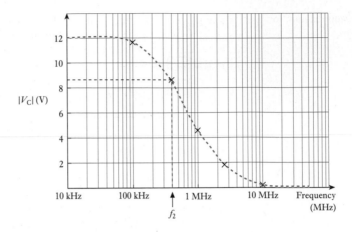

| 18.8 | **The high-pass or lead circuit** |

Consider the RC circuit of Fig. 18.11 in which, in contrast to the low-pass circuit, the variation of V_R with frequency is considered:

Fig. 18.11 High-pass filter circuit

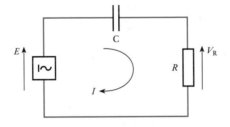

$$V_R = \frac{R}{R + \dfrac{1}{j\omega C}} \times E$$

$\therefore$ Gain $A = \dfrac{V_R}{E} = \dfrac{1}{1 + \dfrac{1}{j\omega CR}} = \dfrac{1}{1 - \dfrac{j}{\omega CR}}$

$$= \frac{1 + \dfrac{j}{\omega CR}}{1^2 + \left(\dfrac{1}{\omega CR}\right)^2}$$

which can be expressed in polar form as

$$\text{Gain } A = \frac{1}{\sqrt{\left\{1 + \left(\dfrac{1}{\omega CR}\right)^2\right\}}} \angle -\tan^{-1}\frac{1}{\omega CR} \qquad [18.5]$$

i.e.   Gain $A = \left|\dfrac{V_R}{E}\right| \angle \phi$

where $|V_R/E|$ is the magnitude of the gain A and ϕ is the phase of the gain A.

By considering the solution of equation [18.5] at frequencies of $\omega = 0$, $\omega = 1/(CR)$ and $\omega = \infty$, an understanding of the frequency response of the circuit can be deduced:

$$\text{For frequency } \omega = 0: \qquad \frac{V_R}{E} = 0\angle 90°$$

$$\text{For frequency } \omega \to \infty: \qquad \frac{V_R}{E} = 1\angle 0°$$

$$\text{For frequency } \omega_1 = \frac{1}{CR}: \qquad \frac{V_R}{E} = \frac{1}{\sqrt{2}}\angle 45°$$

This result is the same as that for the half-power frequency ω_2 for a low-pass circuit (equation [18.4]). Hence with E constant, $|V_R|$ *increases* as ω increases and $\angle V_R$ *leads* $\angle E$. The amplitude and phase response are illustrated in Fig. 18.12. It is clear that at high frequencies, there is a passband response. The lower end of the passband is at the frequency ω_1, again called the critical frequency, the *cut-off frequency* or the *−3 dB frequency* because its output voltage is 3 dB down at this frequency from the value at $\omega = \infty$. It is also called the half-power point or frequency as before. The stopband lies below the frequency ω_1, where V_R falls below $0.707E$ (i.e. $E/\sqrt{2}$).

Fig. 18.12   Frequency response for high-pass circuit: (a) amplitude response, (b) phase response

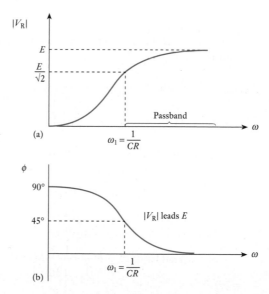

Example 18.6   Sketch the frequency response for a high-pass RC filter circuit having a resistor of 500 Ω and a capacitor of 0.2 μF (Fig. 18.13) by calculating the amplitude and phase of the gain at 1, 5, 10, 50 and 100 krad/s. What is the half-power frequency?

Fig. 18.13 Circuit diagram for
Example 18.6

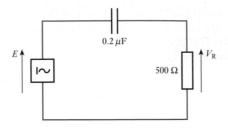

The half-power frequency is

$$\omega_1 = \frac{1}{CR} = \frac{1}{0.2 \times 10^{-6} \times 500} = \textbf{10 krad/s} \quad (1591 \text{ Hz})$$

From equation [18.5]:

$$\text{Gain } A = \left|\frac{V_R}{E}\right| \angle\phi = \frac{1}{\sqrt{\left\{1 + \left(\dfrac{1}{\omega CR}\right)^2\right\}}} \angle\tan^{-1}\frac{1}{\omega CR}$$

This equation may be rewritten, by substitution for $\omega_1 = 1/(CR)$, as

$$\text{Gain } A = \frac{1}{\sqrt{\left\{1 + \left(\dfrac{\omega_1}{\omega}\right)^2\right\}}} \angle\tan^{-1}\frac{\omega_1}{\omega}$$

Since $\omega_1 = 10$ krad/s, this simplifies further, where ω is any frequency in krad/s:

$$\text{Gain } A = \frac{1}{\sqrt{\left\{1 + \left(\dfrac{10}{\omega}\right)^2\right\}}} \angle\tan^{-1}\frac{10}{\omega}$$

Values of $\omega = 1, 5, 10, 50$ and 100 krad/s may now be inserted to provide the frequency response (amplitude and phase response) required.

ω (krad/s)	$\left\|\dfrac{V_R}{E}\right\|$ $\dfrac{1}{\sqrt{\left\{1 + \left(\dfrac{10}{\omega}\right)^2\right\}}}$	$\angle\phi$ $\angle\tan^{-1}\dfrac{10}{\omega}$
1	0.1	84
5	0.45	63
10	0.707	45
50	0.98	11
100	0.995	6

The frequency response is shown in the graph of Fig. 18.14.

Fig. 18.14 Frequency response for circuit of Example 18.6

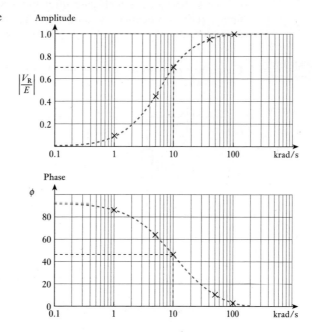

18.9 Passband (or bandpass) filter

There are a variety of techniques to produce a passband filter characteristic. We have met the resonant circuit. Another method is to use a low-pass and a high-pass filter in conjunction, connected together in series (Fig. 18.15). The components of the two circuits are selected so that the cut-off (half-power) frequency ω_1 of the high-pass filter is lower than the cut-off (half-power) frequency ω_2 of the low-pass filter.

Fig. 18.15 (a) Low-pass and high-pass filters in cascade. (b) Frequency response

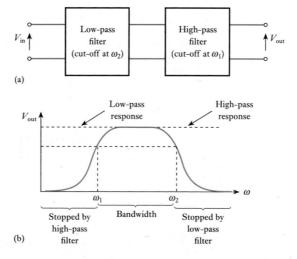

Frequencies less than ω_1 will have little effect on the output voltage because they lie in the stopband of the high-pass filter. Frequencies above ω_2 will pass through the high-pass filter but will be rejected by the low-pass filter circuit. Frequencies ω_1 and ω_2 define the bandwidth of the passband characteristic.

Example 18.7   Calculate ω_1 and ω_2 for the passband filter of Fig. 18.16.

Fig. 18.16 Circuit diagram for Example 18.7

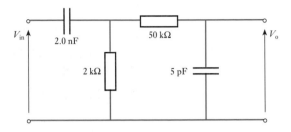

Cut-off frequency for high-pass filter:

$$\omega_1 = \frac{1}{R_1 C_1} = \frac{1}{2\ \text{k}\Omega \times 2\ \text{nF}} = \textbf{250 krad/s} \quad \textbf{(39.8 kHz)}$$

Cut-off frequency for low-pass filter:

$$\omega_2 = \frac{1}{R_2 C_2} = \frac{1}{50\ \text{k}\Omega \times 5\ \text{pF}} = \textbf{4000 krad/s} \quad \textbf{(636.6 kHz)}$$

It should be noted that $V_o \neq 0.707 V_{in}$ at these frequencies, because of the loading effect of the two filter circuits on each other. However, it is sufficient here to understand the principle of this method of producing a passband characteristic. Further analysis of such a filter is beyond the scope of this text.

A resonant circuit can also be used to produce a passband characteristic and reference should be made to the chapter on resonant circuits. Example 18.8 illustrates the principles of this method.

Example 18.8   Calculate the resonant frequency and bandwidth for the series resonant circuit of Fig. 18.17. What is the output voltage at f_r and at f_1 and f_2, the half-power frequencies? Calculate the values of f_1 and f_2.

At resonance:

$$V_o = \frac{15}{15 + 5} \times 20\ \text{V} = 15\ \text{V}_{rms}$$

Fig. 18.17 Circuit diagram for Example 18.7

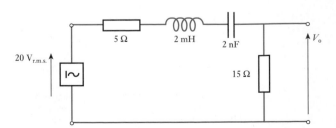

The resonant frequency:

$$f_r = \frac{1}{2\pi\sqrt{(LC)}} = \frac{1}{2\pi\sqrt{(2\times 10^{-3}\times 2\times 10^{-9})}}\text{ Hz}$$

$$= \frac{500\ 000}{2\pi} = 79\ 577\text{ Hz}$$

The output voltage at the half-power frequencies is:

$$V_o = 0.707\times 15\text{ V}_{rms} = 10.6\text{ V}_{rms}.$$

In order to calculate the bandwidth from the expression

$$\text{Bandwidth} = \frac{\text{resonant frequency}}{Q\text{ factor}}$$

Q is required from the expression

$$Q = \frac{X_L}{R}$$

at the resonant frequency where

$$X_L = 2\pi\frac{500\ 000}{2\pi}\times 2\times 10^{-3}\ \Omega = 1000\ \Omega$$

The Q factor is determined by the total resistance of the circuit, 20 Ω:

$$Q = \frac{1000}{20} = 50$$

Hence

$$\text{Bandwidth} = \frac{79\ 577}{50} = 1592\text{ Hz}$$

The half-power frequencies lie $\pm 1592/2$ Hz on either side of the resonant frequency. Thus

$$f_1 = 79\ 577 - 796 = 78\ 871\text{ Hz}$$

$$f_2 = 79\ 577 + 796 = 80\ 373\text{ Hz}$$

Figure 18.18 shows the frequency response.

Fig. 18.18 Frequency response for Example 18.8

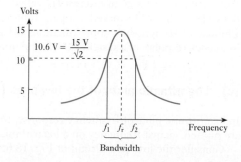

18.10 Stopband (or bandstop) filters

As in the case of the passband filter, a low–pass and a high–pass filter circuit may be used to create this characteristic. However, the two filters are used in parallel for this type of filter as shown in Fig. 18.19. Low frequencies, lower than the half-power frequency ω_2 of the low-pass filter, pass through the low-pass circuit. Frequencies above the half-power frequency ω_1 of the high-pass filter pass through the high-pass circuit. Frequencies between these two frequencies, ω_1 and ω_2, lie in the stopband of both filter circuits as shown in the frequency response curve of Fig. 18.19. Further analysis of such a filter is beyond the scope of this text.

Fig. 18.19   (a) Low-pass and high-pass filters in parallel. (b) Frequency response

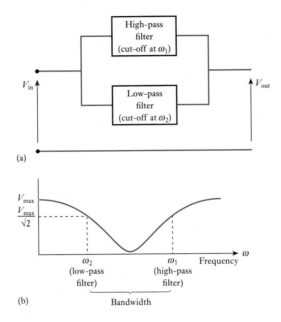

18.11 Bode plots

So far, the gain of filter circuits has been considered by direct comparison of input and output voltages over a range of frequencies. A technique for sketching the frequency response of such circuits on a decibel scale provides an excellent technique for comparing decibel levels at different frequencies.

The curves obtained of the magnitude (in dB) and phase angle of the gain over a frequency range of interest are called *Bode* plots, named after Henrik Bode (1905–81) who worked, for part of his career, at Bell Labs in the USA. The use of straight-line approximations (idealized Bode plots) enables the frequency response of a system to be found quickly and reasonably accurately.

(a)   Logarithmic plots of the low-pass (lag) circuit

In logarithmic plots of frequency response, the gain (in dB) and phase angle are plotted against frequency on a logarithmic scale.

Consider the low-pass circuit of Fig. 18.6.

From equation [18.3], the magnitude of the gain is

$$|A| = \frac{1}{\sqrt{\{1 + (\omega CR)^2\}}}$$

From equation [18.1], the gain in dB is given by the expression $G = 20 \log|A|$. Therefore

$$G = 20 \log|A| = 20 \log\{1 + (\omega CR)^2\}^{-\frac{1}{2}}$$

$$G = -10 \log\{1 + (\omega CR)^2\} \qquad\qquad [18.6]$$

(Note that $20 \log_{10} 2^{-\frac{1}{2}} = -10 \log_{10} 2$; refer to section 18.4.)

By consideration of equation [18.6] at frequencies $\omega = 0$, $\omega = 1/(CR)$ and $\omega = \infty$:

$$\text{At } \omega = 0: \qquad\qquad G = -10 \log 1 = 0 \text{ dB}$$

$$\text{At } \omega_2 = \frac{1}{CR}: \qquad\qquad G = -10 \log\{1 + 1\} = -3 \text{ dB}$$

$$\text{At } \omega \to \infty, \omega CR > 1: \quad G = -20 \log(\omega CR)$$

$$= -20 \log\left(\frac{\omega}{\omega_2}\right)$$

Since, at the half-power frequency, $\omega_2 = 1/(CR)$.

$$G = -20 \log\left(\frac{\omega}{\omega_2}\right) \text{ dB} \qquad\qquad [18.7]$$

Note that this is an approximation of equation [18.6]. The logarithmic plot (Bode plot) shown in Fig. 18.20(a) can now be explained. If a log scale of frequency is used such that $x = \log(\omega/\omega_2)$, then $G = -20x$. This gives a straight line having a slope of 20 dB/unit x. If $\log(\omega/\omega_2)$ is 1, then $\omega/\omega_2 = 10$. So each unit on the x-axis will correspond to a tenfold increase in frequency – this is known as a frequency *decade*.

To explain the graph, equation [18.7] is considered:

$$\text{When } \frac{\omega}{\omega_2} = 1, \text{ i.e. } \omega = \omega_2, G = 0 \text{ dB as } \log_{10}\frac{\omega}{\omega_2} = 0$$

Note that the exact value of G, calculated from equation [18.6], gives -3 dB at this frequency.

$$\text{When } \frac{\omega}{\omega_2} = 10, G = -20 \text{ dB and } \log_{10} 10 = 1$$

The log plot depicted in Fig. 18.20(a) is a straight line having a slope of -20 dB/decade. Note that the slope is constant and independent of the value of CR. CR only affects the half-power frequency (corner frequency or *breakpoint*).

For $\omega/\omega_2 = 2$ we have a frequency *octave* and $\log \omega/\omega_2 = 0.3$. Hence the slope may also be written as

$$G = -20 \log\left(\frac{\omega}{\omega_2}\right) = -20 \times 0.3 = -6 \text{ dB}$$

Fig. 18.20 Bode plots for low-pass filter: (a) amplitude response; (b) phase response

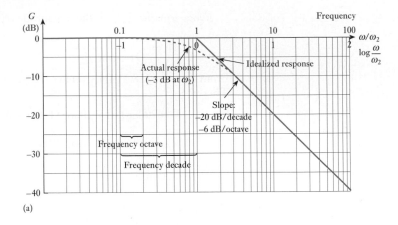

(a)

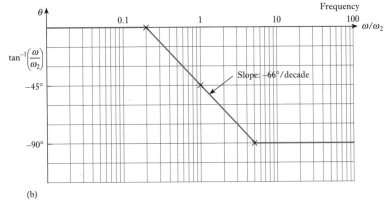

(b)

Note that **−6 dB/octave = −20 dB/decade**.

In this way, when using log plots of the frequency response, the amplitude response may be represented by two straight lines intersecting at the half-power frequency, the slope above the half-power frequency being −20 dB/decade (−6 dB/octave) as shown in Fig. 18.20(a). Similarly the phase response can be derived from equation [18.3] and a straight-line approximation drawn, as in Fig. 18.20(b). These straight-line approximations of the frequency response are referred to as *idealized Bode diagrams* depicting the magnitude and phase of the gain G for the low-pass filter circuit.

Example 18.9 Make a Bode plot of the amplitude of the gain for the low-pass filter circuit of Fig. 18.21.

The breakpoint or half-power frequency:

$$\omega_2 = \frac{1}{CR} = \frac{1}{4 \times 10^{-9} \times 1 \times 25 \times 10^3}$$

$$\omega_2 = \textbf{200 krad/s}$$

At 200 krad/s, $\omega_2 = 1/(CR)$ so actually $G = -10\log(1 + 1) = -3$ dB. The exact values of G at any frequency can be calculated from equation [18.6]. However, in the idealized (straight-line) response, the graph is reduced to

Fig. 18.21 Circuit diagram for Example 18.8

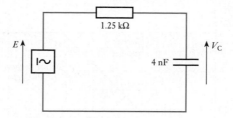

Fig. 18.22 Bode plot: amplitude response for Example 18.8

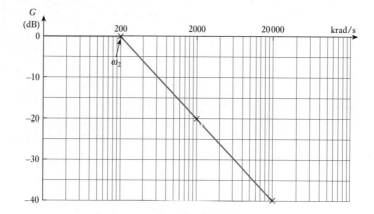

two straight lines which intersect at the breakpoint ω_2. This is shown in Fig. 18.22. Below 200 krad/s, the response is flat and equal to 0 dB. Above the breakpoint, the gain falls at −20 dB/decade. Such an approximation allows a rapid and accurate result to be determined.

(b) Logarithmic plots of the high-pass (lead) circuit

Consider the lead circuit of Fig. 18.10.

From equation [18.5], the magnitude of the gain is

$$|A| = \frac{1}{\sqrt{\left\{1 + \left(\dfrac{1}{\omega CR}\right)^2\right\}}}$$

$$\text{Gain in dB} = 20 \log|A| = 20 \log\left\{1 + \left(\frac{1}{\omega CR}\right)^2\right\}^{-\frac{1}{2}}$$

$$\therefore \qquad G = -10 \log\left\{1 + \left(\frac{1}{\omega CR}\right)^2\right\} \tag{18.8}$$

$$G = -10 \log\left\{1 + \left(\frac{\omega_1}{\omega}\right)^2\right\} \quad \text{where } \omega_1 = \frac{1}{CR} \tag{18.9}$$

Fig. 18.23 Bode plots for high-pass filter: (a) amplitude response; (b) phase response

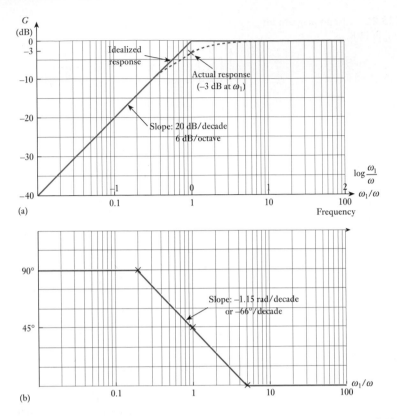

By consideration of equation [18.9] at frequencies $\omega = 0$, $\omega = 1/(CR)$ and $\omega = \infty$:

$$\text{At } \omega \to 0: \quad G = -10 \log\left(\frac{\omega_1}{\omega}\right)^2 = -20 \log\left(\frac{\omega_1}{\omega}\right)$$

$$\text{At } \omega = \omega_1: \quad G = -10 \log(1 + 1) = -3 \text{ dB}$$

$$\text{At } \omega = \infty: \quad G = 0 \text{ dB}$$

The log plot of the gain, depicted in Fig. 18.23(a), is a straight line having a slope of −20 dB/decade. Note that the slope is constant and independent of the value of CR. CR only affects the half-power frequency (corner frequency or *breakpoint*). Similarly the phase response can be derived using equation [18.5], from which

$$\angle\phi = \angle\tan^{-1}\frac{1}{\omega CR}$$

and the straight-line approximation drawn. Figure 18.23 depicts the idealized Bode diagrams for the magnitude and phase of the gain G for the high-pass filter circuit.

| Example 18.10 | Sketch the idealized (straight-line) Bode diagram of a high-pass RC filter having a 94 nF capacitor and a resistor of 169.3 Ω. Plot, using semi-log graph paper, the actual response by calculating the decibel level at f_1, the half-power frequency, and at $f_1/10$, $f_1/2$ and $2f_1$. |

Fig. 18.24 Circuit diagram for Example 18.10

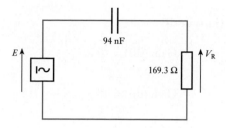

The circuit is depicted in Fig. 18.24.

$$f_1 = \frac{1}{2\pi CR} = \frac{1}{2\pi \times 94 \times 10^{-9} \times 169.3}$$

$$f_1 = \textbf{10 000 Hz}$$

From equation [18.9]:

$$G = -10 \log\left\{1 + \left(\frac{\omega_1}{\omega}\right)^2\right\} \quad \text{where } \omega_1 = \frac{1}{CR}$$

This can be rewritten as

$$G = -10 \log\left\{1 + \left(\frac{f_1}{f}\right)^2\right\} \text{ dB}$$

For $f_1/10$: $G = -10 \log\{1 + (10)^2\} = -20$ dB

For $f_1/2$: $G = -10 \log\{1 + (2)^2\} = -7$ dB

Note that for the idealized response $G = -6$ dB at this frequency.

For f_1: $G = -10 \log\{1 + (1)^2\} = -3$ dB

This frequency is, of course, the corner frequency of the idealized Bode plot at 0 dB.

For $2f_1$: $G = -10 \log\{1 + (0.5)^2\} = -1$ dB

The idealized Bode plot and the actual response are shown in the graph of Fig. 18.25.

Fig. 18.25 Bode plot: amplitude response for Example 18.10

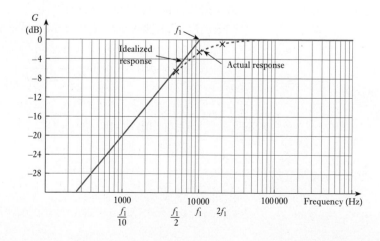

Summary of important formulae

Gain in dB

$$\text{Gain (dB)} = 20 \log_{10} \frac{V_2}{V_1} \qquad [18.1]$$

$$\text{Gain (dB)} = 10 \log_{10} \frac{P_2}{P_1} \qquad [18.2]$$

For a low-pass filter

$$\text{Gain } A = \frac{1}{\sqrt{\{1 + (\omega CR)^2\}}} \angle -\tan^{-1} \omega CR \qquad [18.3]$$

i.e. $$\text{Gain } A = \left| \frac{V_C}{E} \right| \angle \theta$$

where $|V_C/E|$ is the magnitude of the gain and θ is the phase of the gain.

$$\text{Gain (dB)} = -10 \log\{1 + (\omega CR)^2\} \qquad [18.6]$$

$$\text{Gain (dB)} = -20 \log\left(\frac{\omega}{\omega_2}\right) \text{dB} \quad \text{where } \omega_2 = \frac{1}{CR} \qquad [18.7]$$

Half-power frequency for both low- and high-pass filters

$$\omega = \frac{1}{CR} \qquad [18.4]$$

For a high-pass filter

$$\text{Gain } A = \frac{1}{\sqrt{\left\{1 + \left(\dfrac{1}{\omega CR}\right)^2\right\}}} \angle \tan^{-1} \frac{1}{\omega CR} \qquad [18.5]$$

i.e. $$\text{Gain } A = \left| \frac{V_R}{E} \right| \angle \phi$$

where $|V_R/E|$ is the magnitude of the gain and ϕ is the phase of the gain.

$$\text{Gain (dB)} = -10 \log\left\{1 + \left(\frac{1}{\omega CR}\right)^2\right\} \qquad [18.8]$$

$$\text{Gain (dB)} = -10 \log\left\{1 + \left(\frac{\omega_1}{\omega}\right)^2\right\} \quad \text{where } \omega_1 = \frac{1}{CR} \qquad [18.9]$$

Terms and concepts

Filters are typically designed to select or reject a band of frequencies depending on the particular application. Their operation is described by the way the output of the filter circuit varies with frequency. This is termed the frequency response and is often presented as a graph of output voltage versus frequency.

The frequency range considered is often large and it is usual to plot frequency on a logarithmic scale. The output (gain) of the circuit is often expressed in logarithmic form by using the decibel notation.

The **decibel (dB)** is a measure of power ratio and more usually a measurement of voltage and current ratio and is dimensionless. It is standard practice in industry to plot the frequency response of filters, amplifiers and systems in total against a decibel scale.

If the maximum output voltage of the circuit, at any frequency, is V_m, then the frequency range for which the response of the filter is greater than $V_m/\sqrt{2}$ is termed the **passband**.

The frequency at which the output voltage is $V_m/\sqrt{2}$ is called the **cut-off frequency** or **half-power frequency**.

The range of frequencies for which the output voltage falls to less than $V_m/\sqrt{2}$ is called the **reject** or **stopband**.

A **low-pass filter** permits the passage of low-frequency signals, whilst rejecting those at higher frequencies.

A **high-pass filter** permits the passage of high-frequency signals, whilst rejecting those at low frequencies.

Graphs of the magnitude (in dB) and phase angle of the gain of circuits such as filters plotted on a log frequency scale are called **Bode plots**. The use of straight-line approximations (idealized Bode plots) enables the frequency response of a system to be found quickly and accurately.

Exercises 18

1. Evaluate:

   (a) $\log_{10} 450\,000$
   (b) $\log_{10}(230)(0.9)$
   (c) $\log_{10} 0.0001$

2. Given $x = \log_{10} X$ determine X for the following values of x: (a) 3; (b) 0.04; (c) 2.1.
3. Determine $\log_{10} 48$ and show that it is equal to $\log_{10}(3 \times 2 \times 2 \times 2 \times 2)$.
4. Find the voltage gain in dB of a system when the applied signal is 3 mV and the output voltage is 1.5 V.
5. If a 3 W speaker is replaced by one with a 30 W output, what is the increase in dB?
6. An amplifier raises the voltage from 0.2 mV to 15 mV. What is the gain in dB?

7. A low-pass RC filter circuit has a resistance of 400 Ω and a capacitor of 10 μF supplied from a 12 V a.c. source. Find:

   (a) the cut-off frequency ω_2 and the magnitude of the output voltage V_C at that frequency;
   (b) the magnitude and phase of V_C with respect to E, the source voltage, at $\omega = 125$, 500 and 1000 rad/s.

8. A low-pass RC filter comprises a resistance of 159.2 Ω and a capacitor of 100 nF.

   (a) Find the cut-off frequency f_2.
   (b) Draw the frequency response of the filter circuit by calculating the magnitude and phase of the gain at frequencies of 1, 5, 10, 50 and 100 kHz.

9. A resistor of 10 kΩ and a capacitor of 2000 pF are used as both a low-pass and a high-pass filter. Sketch the

frequency response for both circuits. Calculate the magnitude and phase of the gain at frequencies of half and twice the half-power frequencies for both filter circuits.

10. Plot the frequency response for a high-pass RC filter circuit having a resistor of 400 Ω and a capacitor of 0.1 μF by calculating the amplitude and phase of the gain at 2.5, 12.5, 25, 50 and 250 krad/s.

11. Find the frequency range for which V_C is within 0.5 dB for a low-pass filter circuit having a resistor of 47 Ω and a capacitor of 5 μF.

12. A high-pass filter is constructed from a 0.1 μF capacitor and a resistor of 1 kΩ. Sketch the Bode plot (dB versus log frequency) by calculating the decibel level at f_1, the half-power frequency, and at $f_1/10$, $f_1/2$ and $2f_1$. Compare the actual response against the idealized response.

13. Sketch the Bode plot for the frequency response of a high-pass filter circuit having a resistor of 0.47 kΩ and a capacitor of 0.05 μF. Plot the actual response by calculating the decibel level at f_1, the half-power frequency, and at frequencies of $f_1/10$, $f_1/2$ and $2f_1$, and $10f_1$. Calculate and plot the phase response at these same frequencies.

Amplifier Equivalent Networks

Objectives

When you have studied this chapter, you should

- understand the idea of equivalent circuits
- be familiar with Thévenin's and Norton's theorems
- be able to use decibels to represent power ratios
- be familiar with the idea of frequency response
- have an understanding of feedback and its effects

Contents

One of the most significant units in an electronic system is the amplifier. In this chapter, two basic networks which can represent the action of an amplifier will be considered. One of them is voltage-operated, the other is current-operated. These are derived from the concepts that were introduced by Thévenin's and Norton's theorems.

The applications of these equivalent networks will be extended to allow for variation of frequency within the limits of acceptable bandwidth. By this extension it will be shown that the operation may be improved by the use of feedback.

By the end of the chapter, we shall be able to undertake exercises examining the performance of amplifiers including the effect of feedback on amplification.

19.1   Amplifier constant-voltage equivalent networks

An equivalent network is one that can replace the actual network for the purpose of analysis. The network shown in Fig. 19.1 can be used as an equivalent network for an amplifier. Remembering that the input signal only controls the flow of energy from a separate source, the network has a circuit consisting of the input resistance of the amplifier R_i, and this represents the property whereby the input circuit of the amplifier loads the source of the input signal. It follows that

Fig. 19.1 Amplifier constant-voltage equivalent network

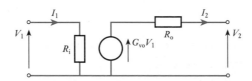

$$R_i = \frac{V_1}{I_1}$$

[19.1]

The ability of the amplifier to produce gain requires a source of controlled energy and this is represented by the voltage generator. The generator amplifies the input voltage V_1 by a gain of G_{vo} so that the voltage induced by the generator is $G_{vo}V_1$. Here G_{vo} is the voltage gain that would be obtained if the output terminals of the amplifier were open-circuited.

However, any practical voltage generator has internal resistance, a principle already considered when we were introduced to Thévenin's theorem in Chapter 4. The internal output resistance of the amplifier is R_o. The effect of R_o causes the voltage gain G_v, obtained when the amplifier is loaded, which is less than the open-circuit voltage gain G_{vo} due to the voltage drop across R_o.

In most equivalent networks it is found that the input resistance R_i has a relatively high value, while the output resistance R_o has a relatively low value. Ideally, R_i would be infinitely large, thus reducing the input current and hence the input power to zero. In practice, amplifier designs generally try to ensure that the input power is as small as possible, usually because the input signal has little power available to it. The output resistance R_o gives rise to a loss of the power produced by the equivalent generator and hence it is desirable that the loss is reduced as far as possible, which effectively requires that R_o be as small as possible.

The use of such equivalent networks is restricted to the signal quantities only. They should not be used in calculations concerned with the direct quantities associated with the amplifier. Moreover, their use assumes an exact linear relationship between input and output signals, i.e. the amplifier produces no waveform distortion.

Figure 19.2 shows an amplifier, represented by its equivalent circuit, being supplied by a source of signal E_s volts and source resistance R_s, and feeding a load resistance R_L. Expressions for the important parameters are derived as follows. Input signal voltage to amplifier is

Fig. 19.2 Amplifier with signal source and load

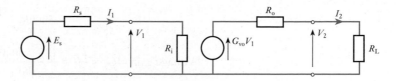

$$V_1 = \frac{R_i E_s}{R_s + R_i}$$

Output signal voltage is

$$V_2 = \frac{R_L}{R_o + R_L} G_{vo} V_1$$

∴ $$\text{Voltage gain} = G_v = \frac{V_2}{V_1} = \frac{G_{vo} R_L}{R_o + R_L}$$ [19.2]

Output signal current is

$$I_2 = \frac{G_{vo} V_1}{R_o + R_L} = \frac{G_{vo} I_1 R_i}{R_o + R_L}$$

∴ $$\text{Current gain} = G_i = \frac{I_2}{I_1} = \frac{G_{vo} R_i}{R_o + R_L}$$ [19.3]

Power gain is

$$G_p = \frac{\text{signal power into load}}{\text{signal power into amplifier}}$$

∴ $$G_p = \frac{V_2^2/R_L}{V_1^2/R_i} = \frac{I_2^2 R_L}{I_1^2 R_i} = \frac{V_2 I_2}{V_1 I_1} = G_v G_i$$ [19.4]

Example 19.1

An integrated-circuit amplifier has an open-circuit voltage gain of 5000, an input resistance of 15 kΩ and an output resistance of 25 Ω. It is supplied from a signal source of internal resistance 5.0 kΩ and it feeds a resistive load of 175 Ω. Determine the magnitude of the signal source voltage to produce an output signal voltage of 1.0 V. What value of load resistance would halve the signal voltage output for the same input?

The network is shown in Fig. 19.3.

Fig. 19.3 Part of Example 19.1

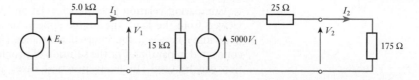

$$V_2 = \frac{175}{25 + 175} \times 5000 V_1$$

$$\text{Voltage gain} = G_v = \frac{V_2}{V_1} = \frac{175 \times 5000}{200} = 4350$$

$$\therefore \qquad V_1 = \frac{1.0}{4350} = 2.30 \times 10^{-4} \text{ V} = 230 \, \mu\text{V}$$

$$V_1 = \frac{15}{5 + 15} E_s$$

$$\therefore \qquad E_s = \frac{20 \times 230}{15} = 307 \, \mu\text{V}$$

For half the signal output the voltage gain must be halved.

$$\therefore \qquad G_v = \frac{4350}{2} = 2175 = \frac{R_L \times 5000}{25 + R_L}$$

$$\therefore \qquad 2175 \times 25 + 2175 R_1 = 5000 R_L$$

$$\therefore \qquad R_L = \frac{2175 \times 25}{2825} = 19.2 \, \Omega$$

19.2 Amplifier constant-current equivalent networks

The constant-voltage equivalent network lends itself to the analysis of amplifiers in which the signals are primarily considered in terms of voltage. However, some amplifiers, especially those using epitaxial or junction transistors, are better considered in terms of current. In order to do this, it is first of all necessary to introduce the constant-current generator. It was shown in Chapter 4 that a source of electrical energy could be represented by a source of e.m.f. in series with a resistance. This is not, however, the only form of representation. Consider such a source feeding a load resistor R_L as shown in Fig. 19.4.

From this circuit:

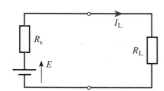

Fig. 19.4 Energy source feeding load

$$I_L = \frac{E}{R_s + R_L} = \frac{\dfrac{E}{R_s}}{\dfrac{R_s + R_L}{R_s}}$$

$$\therefore \qquad I_L = \frac{R_s}{R_s + R_L} \times I_s \qquad\qquad [19.5]$$

where $I_s = E/R_s$ is the current which would flow in a short-circuit across the output terminals of the source.

Comparing relation [19.5] with relation [3.8] it can be seen that, when viewed from the load, the source appears as a source of current (I_s) which is dividing between the internal resistance (R_s) and the load resistor (R_L) connected in parallel. For the solution of problems either form of representation can be used. In many practical cases an easier solution is obtained

Fig. 19.5 Equivalence of constant-voltage generator and constant-current generator forms of representation

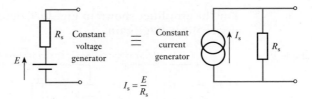

$$I_s = \frac{E}{R_s}$$

using the current form. Figure 19.5 illustrates the equivalence of the two forms.

The internal resistance of the constant-current generator must be taken as infinite, since the resistance of the complete source must be R_s as is obtained with the constant-voltage form.

The ideal constant-voltage generator would be one with zero internal resistance so that it would supply the same voltage to all loads. Conversely, the ideal constant-current generator would be one with infinite internal resistance so that it supplied the same current to all loads. These ideal conditions can be approached quite closely in practice.

Example 19.2

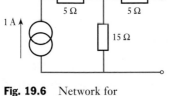

Fig. 19.6 Network for Example 19.2

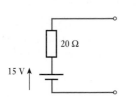

Fig. 19.7 Part of Example 19.2

Represent the network shown in Fig. 19.6 by one source of e.m.f. in series with a resistance.

Potential difference across output terminals which are open-circuited is

$$V_0 = 1 \times 15 = 15 \text{ V}$$

Resistance looking into output terminals with the current generator short-circuited is

$$5 + 15 = 20 \text{ }\Omega$$

therefore the circuit can be represented as shown in Fig. 19.7.

By using the constant-current generator principle we can create the alternative form of amplifier equivalent network which is shown in Fig. 19.8. Here the input circuit remains as before, but the output network consists of a current generator producing a current $G_{is}I_1$ where G_{is} is the current gain that would be obtained with the output terminals short-circuited.

Since the voltage and current equivalent networks are also to be equivalent to each other, it follows that

$$G_{vo}V_1 = G_{is}I_1R_o = G_{is}\frac{V_1}{R_i}R_o$$

$$\therefore \qquad G_{vo} = \frac{R_o}{R_i}G_{is} \qquad\qquad [19.6]$$

Fig. 19.8 Amplifier constant-current equivalent network

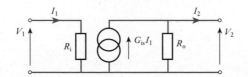

Example 19.3   For the amplifier shown in Fig. 19.9, determine the current gain and hence the voltage gain.

Fig. 19.9   Part of Example 19.3

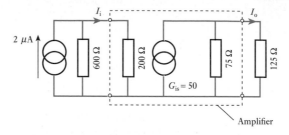

The input current I_i is given by application of relation [3.8], hence

$$I_i = \frac{600}{600 + 200} \times 2 \times 10^{-6} = 1.5 \times 10^{-6} \text{ A}$$

Hence the generator current is

$$G_{is}I_i = 50 \times 1.5 \times 10^{-6} = 75 \times 10^{-6} \text{ A}$$

and the output current I_o by application of relation [3.8] is

$$I_o = \frac{75}{75 + 125} \times 75 \times 10^{-6} = 28.1 \times 10^{-6} \text{ A}$$

$$\therefore \qquad G_i = \frac{I_o}{I_i} = \frac{28.1 \times 10^{-6}}{1.5 \times 10^{-6}} = 18.8$$

Note that the amplifier gain is the ratio of the output terminal current to the input terminal current.

$$G_v = \frac{V_o}{V_i} = \frac{28.1 \times 10^{-6} \times 125}{1.5 \times 10^{-6} \times 200} = 11.7$$

19.3   Logarithmic units

We have already looked at this in Chapter 18, but it is equally inportant for amplifiers as it was for filters to be able to use logarithmic units. It is sometimes found convenient to express the ratio of two powers P_1 and P_2 in logarithmic units known as *bels* as follows:

$$\text{Power ratio in bels} = \log \frac{P_2}{P_1} \qquad [19.7]$$

It is found that the *bel* is rather a large unit and as a result the *decibel* (one-tenth of a bel) is more common, so that

$$\text{Power ratio in decibels (dB)} = 10 \log \frac{P_2}{P_1} \qquad [19.8]$$

If the two powers are developed in the same resistance or equal resistances then

$$P_1 = \frac{V_1^2}{R} = I_1^2 R \quad \text{and} \quad P_2 = \frac{V_2^2}{R} = I_2^2 R$$

where V_1, I_1, V_2 and I_2 are the voltages across and the currents in the resistance. Therefore

$$10 \log \frac{P_2}{P_1} = 10 \log \frac{V_2^2/R}{V_1^2/R} = 10 \log \frac{V_2^2}{V_1^2}$$

$$\text{Power ratio in dB} = 20 \log \frac{V_2}{V_1} \qquad [19.9]$$

Similarly

$$\text{Power ratio in dB} = 20 \log \frac{I_2}{I_1} \qquad [19.10]$$

The relationships defined by relations [19.9] and [19.10], although expressed by ratios of voltages and currents respectively, still represent power ratios. They are, however, used extensively, although by fundamental definition erroneously, to express voltage and current ratios where common resistance values are not involved. For example, the voltage gain of an amplifier is expressed often in decibels.

Figure 19.10 shows two amplifiers connected in cascade, in which case the input of the second amplifier is the load on the first one. The overall voltage gain is given by

Fig. 19.10 Amplifiers in cascade

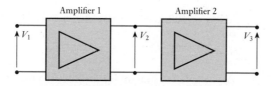

$$G_v = \frac{V_3}{V_2} \times \frac{V_2}{V_1}$$

Expressing this in decibels:

$$\text{Voltage gain in dB} = 20 \log \frac{V_3}{V_2} \times \frac{V_2}{V_1}$$

$$= 20 \log \frac{V_3}{V_2} + 20 \log \frac{V_2}{V_1}$$

Thus the overall voltage gain in decibels is equal to the sum of the voltage gains in decibels of the individual amplifiers. This is a most useful result and can be extended to any number of amplifiers in cascade. It is obviously applicable to current and power gains.

Using decibels gives a representation of one power (or voltage) with reference to another. If P_2 is greater than P_1 then P_2 is said to be $10 \log P_2/P_1$ dB *up* on P_1. For P_2 less than P_1, $\log P_2/P_1$ is negative. Since $\log P_1/P_2 = -\log P_2/P_1$ it is usual to determine the ratio greater than unity and P_2 is said to be $10 \log P_1/P_2$ *down* on P_1.

Example 19.4

The voltage gain of an amplifier when it feeds a resistive load of 1.0 kΩ is 40 dB. Determine the magnitude of the output signal voltage and the signal power in the load when the input signal is 10 mV.

$$20 \log \frac{V_2}{V_1} = 40$$

∴ $$\log \frac{V_2}{V_1} = 2.0$$

∴ $$\frac{V_2}{V_1} = 100$$

∴ $$V_2 = 100 \times 10 = 1000 \text{ mV} = \textbf{1.0 V}$$

$$P_2 = \frac{V_2^2}{R_L} = \frac{1.0^2}{1000} = \frac{1}{1000} \text{ W} = \textbf{1 mW}$$

Example 19.5

Express the power dissipated in a 15 Ω resistor in decibels relative to 1.0 mV when the voltage across the resistor is 1.5 V r.m.s.

$$P_2 = \frac{1.5^2}{15} \text{ W} = 150 \text{ mW}$$

Power level in dB relative to 1 mW is

$$10 \log \frac{150}{1} = 10 \times 2.176 = \textbf{21.76 dB}$$

Example 19.6

An amplifier has an open-circuit voltage gain of 70 dB and an output resistance of 1.5 kΩ. Determine the minimum value of load resistance so that the voltage gain is not more than 3.0 dB down on the open-circuit value. With this value of load resistance determine the magnitude of the output signal voltage when the input signal is 1.0 mV.

$$20 \log G_{\text{vo}} = 70$$

∴ $$\log G_{\text{vo}} = 3.50$$

∴ $$G_{\text{vo}} = 3160$$

$$20 \log G_{\text{v}} = 70 - 3 = 67$$

∴ $$\log G_{\text{v}} = 3.35$$

∴ $$G_{\text{v}} = 2240$$

∴ $$\frac{R_L}{R_o + R_L} 3160 = 2240$$

∴ $$3160 R_L = 2240 \times 1.5 + 2240 R_L$$

∴ $$R_L = \frac{2240 \times 1.5}{920} = \textbf{3.65 kΩ}$$

Alternatively since

$$20 \log G_{\text{vo}} - 20 \log G_{\text{v}} = 3.0$$

$$\therefore \qquad 20 \log \frac{G_{\text{vo}}}{G_{\text{v}}} = 3.0$$

$$\therefore \qquad \frac{G_{\text{vo}}}{G_{\text{v}}} = 1.41$$

$$\therefore \qquad \frac{R_{\text{L}}}{R_{\text{o}} + R_{\text{L}}} = \frac{1}{1.41}$$

$$\therefore \qquad R_{\text{L}} = \frac{1.5}{0.41} = \textbf{3.65 k}\Omega$$

$$V_2 = 2240 \times 1.0 = 2240 \text{ mV} = \textbf{2.24 V}$$

19.4 Frequency response

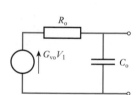

Fig. 19.11 Equivalent network for high-frequency operation

In many cases amplifier gain decreases at high frequencies and also in some cases at low frequencies. The equivalent circuits, as have been used so far, give no indication of this since they contain pure resistance only. They have therefore to be modified for the frequency ranges in which the gain decreases.

One cause of loss of gain at high frequencies is the presence of shunt capacitance across the load. This can be due to stray capacitance in the external circuit and, what is usually more important, capacitance within the amplifier itself. The effective load on the amplifier tends to zero as the frequency tends to infinity. Thus the voltage gain decreases because of the decrease in load impedance, and the current gain decreases because the shunt capacitance path drains current away from the load. Another cause of loss of gain at high frequencies is the inherent decrease of available gain in the amplifier, i.e. G_{vo} decreases with frequency. The manner in which the gain decreases due to this effect is similar to that produced by shunt capacitance, and both effects can be represented on the equivalent circuit by the connection of capacitance across the output terminals assuming that, at the frequency being considered, only one of the effects is appreciable. In practice this is often a reasonable assumption over a considerable frequency range. Figure 19.11 shows the output section of the equivalent circuit modified for high-frequency operation. Associated with the loss of gain will be a shift of phase between input and output signals from the nominal value. A more complex circuit would be required if both the effects considered above had to be taken into account simultaneously.

Loss of gain at low frequencies is due to the use of certain capacitors in the circuit. Their values are chosen such that the reactances are very small at the frequencies being used, so that little of the signal's voltage is developed across them. At low frequencies, however, appreciable signal is developed across them, resulting in a loss of signal at the load. As with the high-frequency response there is a corresponding shift of phase between input and output signals.

The *bandwidth* of an amplifier is defined, similarly to that of a filter, as the difference in frequency between the lower and upper frequencies, f_1 and f_2 respectively, at which the gain is 3.0 dB down on its maximum value.

Fig. 19.12 Frequency responses of a typical R–C coupled amplifier and a typical tuned amplifier

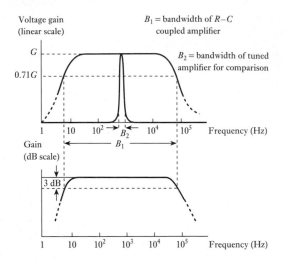

The frequency response that results in an amplifier with capacitors in its circuitry often takes the form shown in Fig. 19.12. The loss of output signal and hence the loss of gain are clearly shown both at low frequencies and at high frequencies. Note the use of a logarithmic scale for the base – this is necessary to expand the characteristic at low frequencies as otherwise it would be crushed and the form of the characteristic would be lost.

The gain may be shown either in numerical form or in logarithmic form. In the latter case the bandwidth is determined by the 3 dB points which are the half-power points. If P_m is the power associated with the maximum gain, usually termed the mid-band gain, and the corresponding output voltage is V_o, then for the half-power condition let the voltage be V_1.

$$\therefore \qquad \frac{V_1^2}{R} = \tfrac{1}{2}P_m = \frac{V_o^2}{2R}$$

$$\therefore \qquad V_1 = \frac{1}{\sqrt{2}} V_o = 0.71 V_o \qquad\qquad [19.11]$$

It follows that since the input voltage is assumed constant then the voltage gain must also have fallen to 0.71 of the mid-band gain. The frequency response bandwidth can therefore also be determined by the points at which the voltage gain has fallen to $0.71G$. Although this has been argued for voltage gain, it would equally have held for current gain.

For a d.c. amplifier the bandwidth will be simply f_2 since the gain extends down to zero frequency. The *passband* or working frequency range is that bounded by f_1 and f_2. If this lies within the range of frequencies normally audible to the ear as sound waves, e.g. 30 Hz–15 kHz, the amplifier is referred to as an *audio amplifier*. Such amplifiers are used in sound reproduction systems. The signals applied to the cathode-ray tube in television receivers require greater passbands extending from zero to several megahertz and are known as *video amplifiers*. In both audio and video amplifiers approximately constant gain over a fairly wide range of frequencies is required and they are collectively known as *broad-band amplifiers*. Other types known as *narrow-band amplifiers* are used where the bandwidth is considerably less than the

Fig. 19.13 Frequency response characteristic of a broad-band amplifier

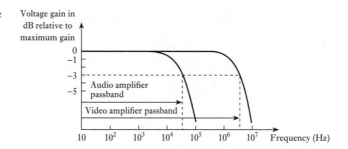

centre frequency. This type of amplifier provides selectivity between signals of different frequency. Figure 19.13 shows a typical frequency response curve for a broad-band amplifier while the curve for a narrow-band or tuned amplifier is incorporated into Fig. 19.12.

19.5 Feedback

Feedback is the process whereby a signal derived in the output section of the amplifier is fed back into the input section. In this way the amplifier can be used to provide characteristics which differ from those of the basic amplifier. The signal fed back can be either a voltage or a current, being applied in series or shunt respectively with the input signal. Moreover, the feedback signal, whether voltage or current, can be directly proportional to the output signal voltage or current. This gives rise to four basic types of feedback, i.e. series-voltage, series-current, shunt-voltage and shunt-current. The characteristics produced by these four types of feedback are similar in some respects and differ in others. Series-voltage feedback will be considered here as a representative type and Fig. 19.14 shows such a feedback amplifier.

The block marked β is that part of the network which provides the feedback voltage $V_f = \beta V_2$. In one of its simplest forms it could consist of two resistors connected across the output to form a voltage divider, the feedback voltage being the signal developed across one of the resistors. The voltage gain $G_v = A$ will be dependent on the load which the β network presents to the amplifier, although in many practical applications the values of the components used in the network are such as to present negligible loading.

From Fig. 19.14 input signal to basic amplifier is

$$V_a = V_1 + V_f = V_1 + \beta V_2$$

$$V_2 = AV_a = A(V_1 + \beta V_2)$$

$$V_2(1 - \beta G_v) = AV_1$$

$$V_2(1 - \beta A) = AV_1$$

Fig. 19.14 Series–voltage feedback amplifier

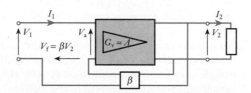

Voltage gain with feedback is

$$G_{vf} = \frac{V_2}{V_1}$$

$$G_{vf} = \frac{A}{1 - \beta A} \qquad [19.12]$$

If the magnitude of $1 - \beta A$ is greater than unity then the magnitude of G_{vf} is less than that of A and the feedback is said to be *negative* or degenerative. The simplest means of accomplishing this is to provide a phase-inverting amplifier, in which case A is negative, and for β to be a positive fraction, as would be obtained with a simple resistive voltage divider. For this particular case relation [19.12] can be written as

$$|G_{vf}| = \frac{|A|}{1 + \beta|A|}$$

where $|G_{vf}|$ and $|A|$ represent the magnitudes of the quantities.

Thus if $\beta|A| \gg 1$ then

$$G_{vf} \approx \frac{|A|}{\beta|A|}$$

i.e. $$|G_{vf}| \approx \frac{1}{\beta} \qquad [19.13]$$

Relation [19.13] illustrates that the voltage gain with negative feedback is relatively independent of the voltage gain of the basic amplifier, provided that the product $\beta|A|$ remains large compared to unity. This is one of the most important characteristics of negative series-voltage feedback amplifiers. Thus the desired voltage gain can be obtained with a high degree of stability by selection of the component values in the feedback network.

To explain further the significance of such stability, we need to be aware that A can vary, either due to ageing of the components or due to the supply voltage not remaining constant. With the feedback gain being largely dependent on β, such changes in A become insignificant. For instance, consider an amplifier which has a gain of 2000 before the introduction of feedback and let there be negative series feedback with a feedback ratio of 0.25.

$$G_{vf} = \frac{A}{1 - \beta A} \quad \text{where} \quad A = -2000$$

$$= \frac{-2000}{1 + (0.25 \times 2000)} = 3.99$$

Now let A fall to 1000, either due to ageing or change of supply voltage. Now

$$G_{vf} = \frac{-1000}{1 + (0.25 \times 1000)} = 3.98$$

Such a change in gain is almost negligible, yet the intrinsic amplifier feedback, i.e. without feedback, has fallen by 50 per cent. Thus although the overall gain is much reduced, it is now stabilized and will be able to give considerably more consistent performance.

Example 19.7

An amplifier with an open-circuit voltage gain of −1000 and an output resistance of 100 Ω feeds a resistive load of 900 Ω. Negative feedback is provided by connection of a resistive voltage divider across the output and one-fiftieth of the output voltage fed back in series with the input signal. Determine the voltage gain with feedback. What percentage change in the voltage gain with feedback would be produced by a 50 per cent change in the voltage gain of the basic amplifier due to a change in the load?

The loading effect of the feedback network can be neglected.

Fig. 19.15 Part of Example 19.7

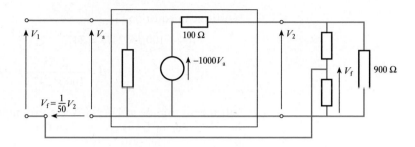

The network is shown in Fig. 19.15.

$$A_1 = G_v = \frac{900}{100 + 900}(-1000) = -900$$

Voltage gain with feedback is

$$G_{vf} = \frac{A_1}{1 - \beta A_1} = \frac{-900}{1 - (\frac{1}{50})(-900)} = \frac{-900}{1 + 18} = -47.4$$

For $\quad A_2 = -450$

$$\therefore \quad G_{vf} = \frac{-450}{1 - (\frac{1}{50})(-450)} = \frac{-450}{1 + 9} = -45.0$$

$$\Delta G_{vf} = 47.4 - 45.0 = 2.4$$

Percentage change in $G_{vf} = \dfrac{2.4}{47.4} \times 100 = \textbf{5.1 per cent}$

Example 19.8

An amplifier is required with an overall voltage gain of 100 and which does not vary by more than 1.0 per cent. If it is to use negative feedback with a basic amplifier, the voltage gain of which can vary by 20 per cent, determine the minimum voltage gain required and the feedback factor.

$$100 = \frac{A}{1 + \beta A} \qquad\qquad [1]$$

$$99 = \frac{0.8A}{1 + \beta 0.8A} \qquad\qquad [2]$$

$$\therefore \qquad 100 + 100\beta A = A \qquad\qquad\qquad [3]$$

$$99 + 79.2\beta A = 0.8A \qquad\qquad\qquad [4]$$

Equation [3] × 0.792

$$\therefore \qquad 79.2 + 79.2\beta A = 0.792A \qquad\qquad\qquad [5]$$

Equations [4] and [5]

$$19.8 = 0.008A$$

$$\therefore \qquad A = \frac{19.8}{0.008} = \mathbf{2475}$$

Substitute in equation [3]

$$100 + 100\beta \times 2475 = 2475$$

$$\therefore \qquad \beta = \frac{2375}{100 \times 2475} = \mathbf{0.009\,60}$$

If the magnitude of $1 - \beta A$ is less than unity, then from relation [19.12] the magnitude of G_{vf} is greater than that of A. The feedback is then said to be *positive* or regenerative. It is not very common, however, to use positive feedback to increase gain since a positive feedback amplifier has opposite characteristics to that of a negative feedback amplifier. Thus the stability of the voltage gain with feedback will be worse than that of the basic amplifier.

There is one case of the use of positive feedback that is of considerable practical importance. That is the case where $\beta A = 1$, which gives, from relation [19.13],

$$G_{vf} = \frac{A}{0} = \infty$$

An amplifier with infinite gain is one that can produce an output signal with no externally applied input signal. It provides its own input signal via the feedback network. Such a circuit is known as an *oscillator*, and to produce an output signal at a predetermined frequency the circuitry is arranged so that $\beta A = 1$ at that frequency only. It must be stressed that the condition $\beta A = 1$ must be satisfied both in magnitude and phase.

It is necessary to consider the condition $\beta A > 1$. This can only be a transient condition in a feedback amplifier since it means that the output signal amplitude would be increasing with time. This build-up of amplitude will eventually entail the amplifier operating in non-linear parts of its characteristics, perhaps even into the cut-off and/or saturation regions. This results in an effective decrease in A and the system settles down when $\beta A = 1$. Thus the substitution of values of βA greater than +1 in relation [19.12] has little practical significance.

In section 19.4 it was stated that loss of gain at high and low frequencies was accompanied by a change of phase shift between input and output signals. It is possible, therefore, for feedback, which is designed to be negative in the passband, to become positive at high or low frequencies and introduce the possibility of the condition $\beta A = 1$ existing. The amplifier would then oscillate at the appropriate frequency and it is said to be unstable. Much of the design work associated with nominally negative feedback amplifiers is concerned with maintaining stability against oscillation.

Example 19.9

An amplifier has a voltage gain of -1000 within the passband. At a specific frequency f_x outside the passband the voltage gain is 15 dB down on the passband value and there is zero phase between input and output signal voltages. Determine the maximum amount of negative feedback that can be used so that the feedback amplifier will be stable.

Let voltage gain at $f_x = G_{vx}$

$$\therefore \qquad 20 \log \frac{G_v}{G_{vx}} = 15$$

$$\therefore \qquad \frac{G_v}{G_{vx}} = 5.62$$

$$\therefore \qquad G_{vx} = \frac{1000}{5.62} = 178$$

Oscillation will occur if

$$\beta G_v = 1$$

$$\therefore \qquad \text{Maximum value of } \beta = \frac{1}{178} = \mathbf{0.0058}$$

19.6

Effect of feedback on input and output resistances

So far, the effect of series–voltage feedback on voltage gain has been considered. It is necessary to consider its effect on the other characteristics of the amplifier. While the introduction of series feedback will affect the magnitude of the input current I_1 for a given value of input voltage V_1 the current gain will be unaffected, i.e. a given value of I_1 will still produce the same value of I_2 as specified by relation [19.3]. It follows therefore that the power gain which is the product of the voltage and current gains will change by the same factor as the voltage gain.

The input resistance with feedback can be determined with reference to Fig. 19.14 as follows. Input resistance with feedback

$$R_{if} = \frac{V_1}{I_1}$$

$$= \frac{V_a - V_f}{I_1} = \frac{V_a - \beta V_2}{I_1} = \frac{V_a - \beta A V_a}{I_1}$$

$$= \frac{V_a(1 - \beta A)}{I_1}$$

$$R_{if} = R_i(1 - \beta A) \qquad\qquad [19.14]$$

Thus the input resistance is increased by negative series–voltage feedback and hence decreased by positive series–voltage feedback.

The output resistance with feedback can be determined from the ratio of the open-circuit output voltage $V_{2o/c}$ to the short–circuit output current $I_{2s/c}$.

$$V_{2o/c} = AV_a = G_{vo}(V_1 + V_f) = A(V_1 + \beta V_{2o/c})$$

$$V_{2o/c} = \frac{AV_1}{1 - \beta A}$$

also $\quad I_{2s/c} = \dfrac{AV_a}{R_o} = \dfrac{AV_1}{R_o}$

Note that there is no signal fed back in this case since there is no output voltage. Output resistance R_{of} with feedback is

$$R_{of} = \frac{V_{2o/c}}{I_{2s/c}}$$

$$R_{of} = \frac{R_o}{1 - \beta A} \qquad\qquad [19.15]$$

Thus the output resistance is decreased by negative series-voltage feedback and hence increased by positive series-voltage feedback.

| Example 19.10 | An amplifier has an open-circuit voltage gain of 1000, an input resistance of 2000 Ω and an output resistance of 1.0 Ω. Determine the input signal voltage required to produce an output signal current of 0.5 A in a 4.0 Ω resistor connected across the output terminals. If the amplifier is then used with negative series-voltage feedback so that one-tenth of the output signal is fed back to the input, determine the input signal voltage to supply the same output signal current. |

From relation [19.3].

$$\frac{I_2}{I_1} = \frac{AR_i}{R_o + R_L} = \frac{1000 \times 2000}{1.0 + 4.0} = 4.0 \times 10^5$$

$$I_1 = \frac{0.5}{4.0 \times 10^5} = 1.25 \times 10^{-6}\ \text{A}$$

$$V_1 = I_1 R_i = 1.25 \times 10^{-6} \times 2 \times 10^3$$

$$= 2.5\ \text{mV}$$

With feedback:

$$\frac{I_2}{I_1} = 4.0 \times 10^5 \text{ (as before)}$$

$$\therefore \qquad I_1 = 1.25 \times 10^{-6}\ \text{A}$$

$$R_{if} = R_i(1 + \beta A) = 2000\left(1 + \frac{1}{10} \times \frac{4}{1 + 4} \times 1000\right)$$

$$= 2000(1 + 80) = 162\ 000\ \Omega$$

$$\therefore \qquad V_1 = 1.25 \times 10^{-6} \times 1.62 \times 10^5$$

$$= 0.202\ \text{V}$$

$$= 202\ \text{mV}$$

19.7	Effect of feedback on bandwidth

Consider an amplifier with a frequency response characteristic of the form shown in Fig. 19.16 and operating without feedback. The bandwidth limiting frequencies are f_1 and f_2. Now apply negative feedback so that the mid-band gain falls to G_{vf}. Owing to feedback, G_{vf} is much smaller than A and the characteristic becomes flatter. It follows that the new cut-off frequencies $f_{1'}$ and $f_{2'}$ are further apart and hence the new bandwidth is increased. We may therefore conclude that the bandwidth of an amplifier is increased following the application of negative feedback.

Fig. 19.16 Effect of negative feedback on bandwidth

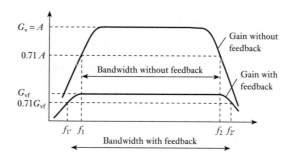

19.8	Distortion

Amplifiers produce three types of distortion:

1. *Phase distortion*. In this distortion, consider a signal made from a number of alternating voltages of differing frequencies. During the amplification, the phase shifts experienced by the different frequencies can also be different, thus the total waveform being amplified becomes distorted.

2. *Amplitude distortion*. This is also known as frequency distortion and arises when signals of differing frequencies are not all amplified to the same extent. The cause of this can be observed by reference to the frequency response characteristic shown in Fig. 19.16 in which we can observe that the gain is less at low and high frequencies.

3. *Harmonic distortion*. This arises in amplifiers which are non-linear in their response. It follows that the input signal could be a pure sinusoidally varying voltage yet the output voltage would include multiple-frequency components, e.g. a 1 kHz input signal might be accompanied by other output signals at frequencies of 2 kHz, 3 kHz and so on. These additional frequencies have been generated within the amplifier and therefore give rise to distortion. The multiple-frequency signals are termed *harmonics*, hence the name for the distortion.

Each form of distortion is present in most amplifiers to a greater or lesser extent. However, the distortions are very much dependent on the basic gain of the amplifier, hence the introduction of feedback can and does reduce distortion to a considerable extent.

Summary of important formulae

For an amplifier constant-voltage equivalent circuit,

$$G_v = \frac{G_{vo}R_L}{R_o + R_L} \qquad [19.2]$$

$$G_i = \frac{G_{vo}R_i}{R_o + R_L} \qquad [19.3]$$

For an amplifier constant-current equivalent circuit,

$$G_{vo} = \frac{R_o}{R_i} \cdot G_{is} \qquad [19.6]$$

Power ratio in decibels,

$$(dB) = 10 \log P_2/P_1 \qquad [19.8]$$

$$= 20 \log V_2/V_1 \qquad [19.9]$$

The gain of an amplifier with negative feedback,

$$G_{vf} = \frac{A}{1 - \beta A} \qquad [19.12]$$

$$R_{if} = R_i(1 - \beta A) \qquad [19.14]$$

$$R_{of} = \frac{R_o}{1 - \beta A} \qquad [19.15]$$

Terms and concepts

An amplifier can be represented by an **equivalent circuit**. The creation of this equivalent circuit can be derived from the constant-voltage equivalent circuit associated with Thévenin's theorem or from the constant-current equivalent circuit associated with Norton's theorem.

Although an equivalent circuit appears to represent a single amplifying device, it can in fact represent the cumulative effect of many devices.

The ratio of power out to power in for any unit in a system can be expressed in **decibels** (dB) which are logarithmic.

The bandwidth of an amplifier is often determined by the conditions in which the gain falls by 3 dB relative to the mid-band gain. The limiting frequencies are sometimes called the **3-dB points**.

Feedback is a circuit arrangement in which the output is partially applied to the input of an amplifier. This can have the effect of reducing the gain but the overall amplifier operation is more stable and the performance more consistent. Alternatively feedback can be used to improve the gain but this can cause instability.

Feedback can be used to improve bandwidth.

Exercises 19

1. Draw the block equivalent diagram of an amplifier and indicate on it the input and output voltages and currents. Hence produce statements of the voltage gain G_v and the power gain G_p.

2. Find the voltage gain, the current gain and the power gain of the amplifier shown in Fig. A.

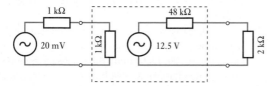

Fig. A

3. An amplification system consists of a source, an amplifier and a load. The source has an open-circuit output voltage of 25 mV and an output resistance of 2 kΩ. The amplifier has an open-circuit voltage gain of 975, an input resistance of 8 kΩ and an output resistance of 3 Ω. The load resistance is 3 Ω. Draw the equivalent circuit of this system and hence determine the voltage, current and power gains of the amplifier. If the load resistance is reduced to 2 Ω, determine the current gain of the amplifier.

4. (a) Explain the term *bandwidth* and describe the relationship between gain and bandwidth.

(b) Describe the effect on the input and output resistance of series–voltage negative feedback.

(c) An amplifier has the following parameters: input resistance, 500 kΩ; output resistance, 4.7 kΩ; short-circuit current gain, 5000. Draw the equivalent circuit for the amplifier and determine the open-circuit voltage gain.

5. (a) An amplifier has the following parameters: input resistance, 2 kΩ; output resistance, 250 Ω; open-circuit voltage gain, 1000. It is used with a load resistance of 750 Ω. Draw the Thévenin (constant-voltage generator) equivalent circuit of the amplifier and its load. For this amplifier calculate: (i) the voltage gain; (ii) the current gain.

(b) Draw the Norton (constant-current generator) equivalent circuit of the amplifier and its load.

(c) A 10 mV ideal voltage generator is applied to the amplifier input. Determine the current of the source generator for the amplifier equivalent circuit.

6. For the amplifier shown in Fig. B, determine the current gain and the voltage gain.

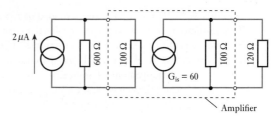

Fig. B

7. An amplifier has a voltage gain of 50 dB. Determine the output voltage when the input voltage is 2.0 mV.

8. Express in decibels the gain of an amplifier which gives an output of 10 W from an input of 0.1 W.

9. The output of a signal generator is calibrated in decibels for a resistive load of 600 Ω connected across its output terminals. Determine the terminal voltage to give: (a) 0 dB corresponding to 1 mW dissipation in the load; (b) +10 dB; (c) −10 dB.

10. An amplifier system consists of a single source, an amplifier and a loudspeaker acting as a load. The source has an open-circuit voltage output of 20 mV and an output resistance of 14 kΩ. The amplifier has an input resistance of 9 kΩ, an open-circuit voltage gain of 60 dB and an output resistance of 8 Ω. The load has a resistance of 8 Ω. Draw the equivalent circuit of this system. Calculate: (a) the current, voltage and power gains (express these gains in dB); (b) the output power if an identical loudspeaker is added in parallel with the original loudspeaker.

11. Draw and explain the gain frequency characteristic of a resistance–capacitance-coupled amplifier. Reference should be made to mid-band frequencies and to cut-off points.

An amplifier has an open-circuit voltage gain of 800, an output resistance of 20 Ω and an input resistance of 5 kΩ. It is supplied from a signal source of e.m.f. 10 mV and internal resistance 5 kΩ. If the amplifier supplies a load of 30 Ω, determine the magnitude of the output signal voltage and the power gain (expressed in decibels) of the amplifier.

12. The voltage gain of an amplifier is 62 dB when the amplifier is loaded by a 5 kΩ resistor. When the amplifier is loaded by a 10 kΩ resistor, the voltage gain is 63 dB. Determine the open-circuit voltage gain and the output resistance of that amplifier.

13. An amplifier has an input resistance of 20 kΩ and an output resistance of 15 Ω. The open-circuit gain of the amplifier is 25 dB and it has a resistive load of 135 Ω. The amplifier is supplied from an a.c. signal source of internal resistance 5 kΩ and r.m.s. amplitude 15 dB above a reference level of 1 μV. Draw the equivalent circuit of this arrangement of source, amplifier and load. Calculate: (a) the voltage at the input terminals to the amplifier; (b) the voltage across the load; (c) the power gain in dB.

14. (a) An amplifier has series-voltage feedback. With the aid of a block diagram for such an amplifier, derive the general feedback equation:

$$G = \frac{A}{1 - \beta A}$$

(b) (i) In an amplifier with a constant input a.c. signal of 0.5 V, the output falls from 25 V to 15 V when feedback is applied. Determine the fraction of the output voltage which is fed back. (ii) If, owing to ageing, the amplifier gain without feedback falls to 40, determine the new gain of the stage, assuming the same value of feedback.

15. The gain of an amplifier with feedback is 110. Given that the feedback fraction is +1.5 per cent, determine the normal gain of the amplifier.

16. When a feedback fraction of 1/60 is introduced to an amplifier, its gain changes by a factor of 2.5. Find the normal gain of the amplifier.

17. An amplifier has a gain of 120. What change in gain will occur if a −3.5 per cent feedback is introduced? If the overall gain had to be reduced to 10, what feedback would be required?

18. The voltage amplification of an amplifier is 65. If feedback fractions of 0.62 per cent, −1/50, −1/80 were introduced, express the new amplifier gain in dB.

19. The gain of an amplifier with feedback is 53, and without feedback the gain is 85. Express the feedback as a percentage and the change in gain in dB.

20. An amplifier with −4.5 per cent feedback has a gain of 12.04 dB. Determine the amplifier gain in dB without feedback.

21. An amplifier is required with a voltage gain of 100. It is to be constructed from a basic amplifier unit of voltage gain 500. Determine the necessary fraction of the output voltage that must be used as negative series voltage feedback. Hence determine the percentage change in the voltage gain of the feedback amplifier if the voltage gain of the basic amplifier: (a) decreases by 10 per cent; (b) increases by 10 per cent.

22. A series-voltage feedback amplifier has a feedback factor $\beta = 9.5 \times 10^{-4}$. If its voltage gain without feedback is 1000, calculate the voltage gain when the feedback is: (a) negative; (b) positive. What percentage increase in voltage gain without feedback would produce oscillation in the positive feedback case?

23. An amplifier has the following characteristics: open-circuit voltage gain, 75 dB; input resistance, 40 kΩ; output resistance, 1.5 kΩ. It has to be used with negative series-voltage feedback to produce an output resistance of 5.0 Ω. Determine the necessary feedback factor. With this feedback factor and a resistive load of 10 Ω determine the output signal voltage when the feedback amplifier is supplied from a signal source of 10 mV and series resistance 10 kΩ.

Chapter twenty   Semiconductor Materials

Objectives

When you have studied this chapter, you should

- be aware of the structure of atoms
- be able to describe what is happening with electrons in an electric current
- have an understanding of what a semiconductor is, in particular silicon and germanium semiconductors
- be aware of n-type and p-type materials
- have a basic knowledge of junctions and the processes at these junctions

Contents

It has been established that data may be transmitted by means of systems that include amplifiers. This concept is only useful if a device can be provided that amplifies.

In this chapter, semiconductor materials, which are fundamental to the operation of most amplifying devices, will be introduced. It is necessary to understand not only the operation of these materials, but also the effects of adding selected impurities. These impurities allow p-type and n-type materials to be produced. Then what happens when p-type meets n-type will be described. This joining of materials forms a junction diode, a device that only conducts in one direction.

By the end of the chapter, we shall be familiar with the materials from which diodes, transistors and other electronic devices are produced. We shall be familiar with diode characteristics and be prepared to develop amplifier devices such as the transistor.

20.1 Introduction

Having considered the general principles of amplifiers, we next have to consider some of the ways in which amplifier networks operate. Over the history of electronics, many devices have been used but the most common forms of device are now based on semiconductor materials. This chapter is devoted to introducing such materials. Some simple applications are introduced in Chapter 21, while the transistor, which is fundamental to most amplifiers, is described in Chapter 22.

20.2 Atomic structure

Every material is made up of one or more elements, an element being a substance composed entirely of atoms of the same kind. For instance, water is a combination of the elements hydrogen and oxygen, whereas common salt is a combination of the elements sodium and chlorine. The atoms of different elements differ in their structure, and this accounts for different elements possessing different characteristics.

Every atom consists of a relatively massive core or nucleus carrying a positive charge, around which *electrons* move in orbits at distances that are great compared with the size of the nucleus. Each electron has a mass of 9.11×10^{-31} kg and a *negative* charge, $-e$, equal to 1.602×10^{-19} C. The nucleus of every atom except that of hydrogen consists of *protons* and *neutrons*. Each proton carries a *positive* charge, e, equal in magnitude to that of an electron and its mass is 1.673×10^{-27} kg, namely 1836 times that of an electron. A neutron, on the other hand, carries no resultant charge and its mass is approximately the same as that of a proton. Under normal conditions, an atom is neutral, i.e. the total negative charge on its electrons is equal to the total positive charge on the protons.

The atom possessing the simplest structure is that of hydrogen – it consists merely of a nucleus of one proton together with a single electron which may be thought of as revolving in an orbit, of about 10^{-10} m diameter, around the proton, as in Fig. 20.1(a).

Figure 20.1(b) shows the arrangement of a helium atom. In this case the nucleus consists of two protons and two neutrons, with two electrons orbiting in what is termed the *K-shell*. The nucleus of a carbon atom has six protons and six neutrons and therefore carries a positive charge 6e. This nucleus is surrounded by six orbital electrons, each carrying a negative charge of $-e$, two electrons being in the K-shell and four in the L-shell, and their relative positions may be imagined to be as shown in Fig. 20.1(c).

Fig. 20.1 Hydrogen, helium and carbon atoms

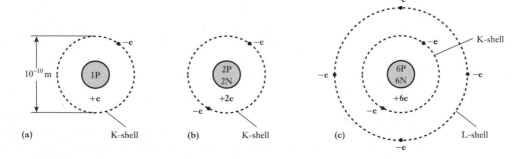

The further away an electron is from the nucleus, the smaller is the force of attraction between that electron and the positive charge on the nucleus; consequently the easier it is to detach such an electron from the atom. When atoms are packed tightly together, as in a metal, each outer electron experiences a small force of attraction towards neighbouring nuclei, with the result that such an electron is no longer bound to any individual atom, but can move at random within the metal. There electrons are termed *free* or *conduction* electrons and only a slight external influence is required to cause them to drift in a desired direction.

An atom which has lost or gained one or more free electrons is referred to as an *ion*; thus, for an atom which has lost one or more electrons, its negative charge is less than its positive charge, and such an atom is therefore termed a *positive ion*.

In semiconductor work, the materials with which we are principally concerned are silicon and germanium. These materials possess a crystalline structure, i.e. the atoms are arranged in an orderly manner. In both silicon and germanium, each atom has four electrons orbiting in the outermost shell and is therefore said to have a valency of four; or, alternatively, the atoms are said to be *tetravalent*. In the case of the silicon atom, the nucleus consists of 14 protons and 14 neutrons; and when the atom is neutral, the nucleus is surrounded by 14 electrons, four of which are *valence electrons*, one or more of which may be detached from the atom. If the four valence electrons were detached, the atom would be left with 14 units of positive charge on the protons and 10 units of negative charge on the 10 remaining electrons, thus giving an ion (i.e. an atom possessing a net positive or negative charge) carrying a net positive charge of 4**e**, where **e** represents the magnitude of the charge on an electron, namely 1.6×10^{-19} C. The neutrons possess no resultant electric charge. A tetravalent atom, isolated from other atoms, can therefore be represented as in Fig. 20.2, where the circle represents the ion carrying the net positive charge of 4**e** and the four dots represent the four valence electrons.

The cubic diamond lattice arrangement of the atoms in a perfect crystal of germanium or silicon is represented by the circles in Fig. 20.3, where atoms B, C, D and E are located at diagonally opposite corners of the six surfaces of an imaginary cube, shown dotted, and atom A is located at the centre of the cube. The length of each side of the dotted cube is about 2.7×10^{-10} m for silicon and about 2.8×10^{-10} m for germanium.

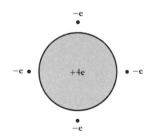

Fig. 20.2 An isolated tetravalent atom

20.3   Covalent bonds

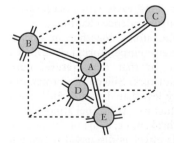

Fig. 20.3 Atomic structure of a lattice crystal

When atoms are as tightly packed as they are in a silicon or a germanium crystal, the simple arrangement of the valence electrons shown in Fig. 20.3 is no longer applicable. The four valence electrons of each atom are now shared with the adjacent four atoms: thus in Fig. 20.3, atom A shares its four valence electrons with atoms B, C, D and E. In other words, one of A's valence electrons is linked with A and B, another with A and C, etc. Similarly, one valence electron from each of atoms B, C, D and E is linked with atom A. One can imagine the arrangement to be somewhat as depicted in Fig. 20.4 where the four dots, marked $-e_A$, represent the four valence electrons of atom A, and dots $-e_B$ $-e_C$ $-e_D$ and $-e_E$ represent the valence electrons of atoms B, C, D and E respectively that are linked with atom A. The dotted lines are not intended to indicate the actual paths or the relative positions of these valence electrons, but merely that the electrons on the various dotted lines move around the two atoms enclosed by a given dotted line.

Fig. 20.4   Covalent electron bonds

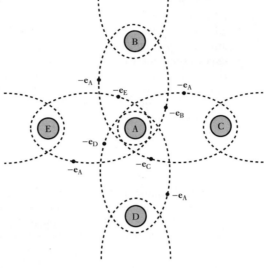

Fig. 20.5   Tetravalent atoms with covalent bonds

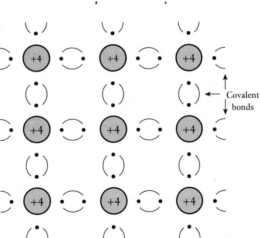

It follows that each positive ion of silicon or germanium, carrying a net charge of **4e**, has eight electrons, i.e. four *electron-pairs*, surrounding it. Each electron-pair is referred to as a *covalent bond*; and in Fig. 20.3, the covalent bonds are represented by the pairs of parallel lines between the respective atoms. An alternative two–dimensional method of representing the positive ions and the valence electrons forming the covalent bonds is shown in Fig. 20.5, where the large circles represent the ions, each with a net positive charge of **4e**, and the bracketed dots represent the valence electrons.

These covalent bonds serve to keep the atoms together in crystal formation and are so strong that at absolute zero temperature, i.e. −273 °C, there are no free electrons. Consequently, at that temperature, pure silicon and germanium behave as perfect insulators. Silicon or germanium can be regarded as 'pure' when impurities are less than 1 part in 10^{10}. Such materials are referred to as *intrinsic* semiconductors. At normal atmospheric temperature, some of the covalent bonds are broken, i.e. some of the valence electrons break away from their atoms. This effect is discussed in section 20.6, but, as a first approximation, we can assume that pure silicon and germanium are perfect insulators and that the properties utilized in semiconductor rectifiers are produced by controlled amounts of impurities introduced into pure silicon and germanium crystals.

20.4 An n-type semiconductor

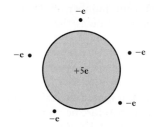

Fig. 20.6 An isolated pentavalent atom

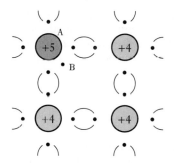

Fig. 20.7 An n-type semiconductor

Certain elements such as phosphorus, arsenic and antimony are pentavalent, i.e. each atom has five valence electrons, and an isolated pentavalent atom can be represented by an ion having a net positive charge of 5e and five valence electrons as in Fig. 20.6. When a minute trace of the order of 1 part in 10^8 of such an element is added to pure silicon or germanium, the conductivity is considerably increased. We shall now consider the reason for this effect.

When an atom of a pentavalent element such as antimony is introduced into a crystal of pure silicon, it enters into the lattice structure by replacing one of the tetravalent silicon atoms, but *only four of the five valence electrons of the antimony atom can join as covalent bonds*. Consequently the substitution of a pentavalent atom for a silicon atom provides a free electron. This state of affairs is represented in Fig. 20.7, where A is the ion of, say, an antimony atom, carrying a positive charge of 5e, with four of its valence electrons forming covalent bonds with four adjacent atoms, and B represents the unattached valence electron free to wander at random in the crystal. This random movement, however, is such that the density of these free or mobile electrons remains constant throughout the crystal and therefore there is no accumulation of free electrons in any particular region.

Since the pentavalent impurity atoms are responsible for introducing or donating free electrons into the crystal, they are termed *donors*; and a crystal doped with such impurity is referred to as an *n-type* (i.e. negative-type) semiconductor. It will be noted that each antimony ion has a *positive* charge of 5e and that the valence electrons of each antimony atom have a total *negative* charge of −5e; consequently the doped crystal is *neutral*. In other words, donors provide fixed positively charged ions and an equal number of electrons free to move about in the crystal, as represented by the circles and dots respectively in Fig. 20.8.

The greater the amount of impurity in a semiconductor, the greater is the number of free electrons per unit volume and therefore the greater is the conductivity of the semiconductor.

When there is no p.d. across the metal electrodes, S and T, attached to opposite ends of the semiconductor, the paths of the random movement of one of the free electrons may be represented by the full lines AB, BC, CD, etc. in Fig. 20.9, i.e. the electron is accelerated in direction AB until it

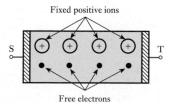

Fig. 20.8 An n-type semiconductor

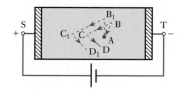

Fig. 20.9 Movement of a free electron

collides with an atom with the result that it may rebound in direction BC, etc. Different free electrons move in different directions so as to maintain the density constant; in other words, there is no resultant drift of electrons towards either S or T.

Let us next consider the effect of connecting a cell across S and T, the polarity being such that S is positive relative to T. The effect of the electric field (or potential gradient) in the semiconductor is to modify the random movement of the electron as shown by the dotted lines AB_1, B_1C_1 and C_1D_1 in Fig. 20.9, i.e. there is superimposed on the random movement a drift of the electron towards the positive electrode S, and the number of electrons entering electrode S from the semiconductor is the same as that entering the semiconductor from electrode T.

20.5 A p-type semiconductor

Materials such as indium, gallium, boron and aluminium are trivalent, i.e. each atom has only three valence electrons and may therefore be represented as an ion having a positive charge of 3e surrounded by three valence electrons. When a trace of, say, indium is added to pure silicon, the indium atoms replace the corresponding number of silicon atoms in the crystal structure, but each indium atom can provide only three valence electrons to join with the four valence electrons of adjacent silicon atoms, as shown in Fig. 20.10, where C represents the indium atom. Consequently there is an incomplete valence bond, i.e. there is a vacancy represented by the small circle D in Fig. 20.10. This vacancy is referred to as a *hole* – a term that is peculiar to semiconductors. This incomplete valent bond has the ability to attract a covalent electron from a nearby silicon atom, thereby filling the vacancy at D but creating another hole at, say, E as in Fig. 20.11(a). Similarly, the incomplete valent bond due to hole E attracts a covalent electron from another silicon atom, thus creating a new hole at, say, F as in Fig. 20.11(b).

If the semiconductor is not being subjected to an external electric field, the position of the hole moves at random from one covalent bond to another covalent bond, the speed of this random movement being about half that of free electrons, since the latter can move about with comparative ease. Different holes move in different directions so as to maintain the density of the holes (i.e. the number of holes per unit volume) uniform throughout the crystal, otherwise there would be an accumulation of positive charge in one region with a corresponding negative charge in another region.

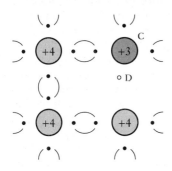

Fig. 20.10 A p-type semiconductor

Fig. 20.11 Movement of a hole in a p-type semiconductor

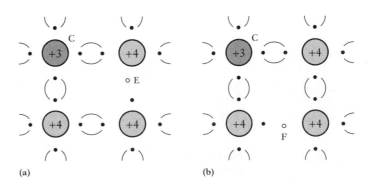

(a)   (b)

It will be seen that each of the silicon atoms associated with holes E and F in Fig. 20.11(a) and (b) respectively has a nucleus with a resultant positive charge 4e and three valence electrons having a total negative charge equal to −3e. Hence each atom associated with a hole is an ion possessing a net positive charge e, and the movement of a hole from one atom to another can be regarded as the movement of a positive charge e within the structure of the p-type semiconductor.

Silicon and germanium, doped with an impurity responsible for the formation of holes, are referred to as *p-type* (positive-type) semiconductors; and since the trivalent impurity atoms can accept electrons from adjacent silicon or germanium atoms, they are termed *acceptors*.

It will be seen from Fig. 20.11 that the trivalent impurity atom, with the *four covalent bonds complete*, has a nucleus carrying a positive charge 3e and four electrons having a total negative charge −4e. Consequently such a trivalent atom is an ion carrying a net negative charge −e. For each such ion, however, there is a hole somewhere in the crystal; in other words, the function of an acceptor is to provide *fixed* negatively charged ions and an equal number of holes as in Fig. 20.12.

Let us next consider the effect of applying a p.d. across the two opposite faces of a p-type semiconductor, as in Fig. 20.13(a), where S and T represent metal plates, S being positive relative to T. The negative ions, being locked in the crystal structure, cannot move, but the holes drift in the direction of the electric field, namely towards T. Consequently region x of the semi-conductor acquires a net negative charge and region y acquires an equal net positive charge, as in Fig. 20.13(b). These charges attract electrons from T into region y and repel electrons from region x into S, as indicated in Fig. 20.13(c). The electrons attracted from T combine with holes in region y and electrons from covalent bonds enter S, thus creating in region x new holes which move from that region towards electrode T. The rate at which holes are being neutralized near electrode T is the same as that at which they are being created near electrode S. Hence, in a p-type semiconductor, we can regard the current as being due to the drift of holes in the conventional direction, namely from the positive electrode S to the negative electrode T.

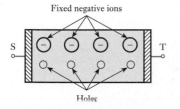

Fig. 20.12 A p-type semiconductor

Fig. 20.13 Drift of holes in a p-type semiconductor

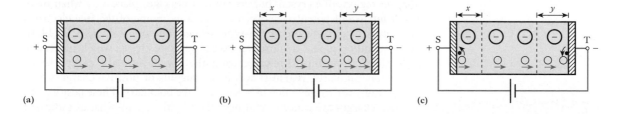

(a) (b) (c)

20.6 Junction diode

Let us now consider a crystal, one half of which is doped with p-type impurity and the other half with n-type impurity. Initially, the p-type semiconductor has mobile holes and the same number of fixed negative ions carrying exactly the same total charge as the total positive charge represented by the holes. Similarly the n-type semiconductor has mobile electrons and the same number of fixed positive ions carrying the same total charge as the total negative charge on the mobile electrons. Hence each region is initially neutral.

Fig. 20.14   Junction diode

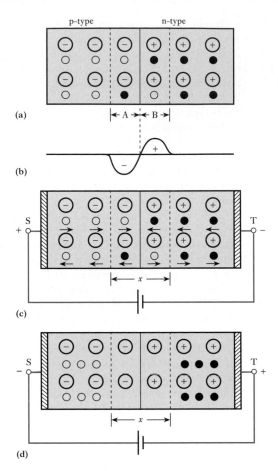

Owing to their random movements, some of the holes will diffuse across the boundary into the n-type semiconductor. It will be noted that *diffusion* takes place when there is a difference in the concentration of carriers in adjacent regions of a crystal; but *drift* of carriers takes place only when there is a difference of potential between two regions. Some of the free electrons will similarly diffuse into the p-type semiconductor, as in Fig. 20.14(a). Consequently region A acquires an excess negative charge which repels any more electrons trying to migrate from the n-type into the p-type semiconductor. Similarly, region B acquires a surplus of positive charge which prevents any further migration of holes across the boundary. These positive and negative charges are concentrated near the junction, somewhat as indicated in Fig. 20.14(b), and thus form a potential barrier between the two regions.

(a)   Forward bias

Let us next consider the effect of applying a p.d. across metal electrodes S and T, S being positive relative to T, as in Fig. 20.14(c). The direction of the electric field in the semiconductor is such as to produce a drift of holes towards the right in the p–type semiconductor and of free electrons towards the left in the n-type semiconductor. In the region of the junction, free electrons and holes combine, i.e. free electrons fill the vacancies represented by the holes. For each combination, an electron is liberated from a covalent bond in the region near positive plate S and enters that plate, thereby creating a new hole which moves through the p-type material towards the junction, as described in section 20.5. Simultaneously, an electron enters the n-region from the negative plate T and moves through the n-type semiconductor towards the junction, as described in section 20.4. The current in the diode is therefore due to hole flow in the p-region, electron flow in the n-region and a combination of the two in the vicinity of the junction.

(b) Reverse bias

When the polarity of the applied voltage is reversed, as shown in Fig. 20.14(d), the holes are attracted towards the negative electrode S and the free electrons towards the positive electrode T. This leaves a region x, known as a *depletion layer*, in which there are no holes or free electrons, i.e. there are no charge carriers in this region apart from the relatively few that are produced spontaneously by thermal agitation, as mentioned below. Consequently the junction behaves as an insulator.

In practice, there is a small current due to the fact that at room temperature, thermal agitation or vibration of atoms takes place in the crystal and some of the valence electrons acquire sufficient velocity to break away from their atoms, thereby producing *electron–hole pairs*. An electron–hole pair has a life of about 50 μs in silicon and about 100 μs in germanium. The generation and recombination of electron–hole pairs is a continuous process and is a function of the temperature. The higher the temperature, the greater is the rate at which generation and recombination of electron–hole pairs take place and therefore the lower the *intrinsic resistance* of a crystal of pure silicon or germanium.

These thermally liberated holes and free electrons are referred to as *minority carriers* because, at normal temperature, their number is very small compared with the number of *majority carriers* due to the doping of the semiconductor with donor and acceptor impurities. Hence, in a p-type semiconductor, holes form the majority carriers and electrons the minority carriers, whereas in an n-type crystal, the majority carriers are electrons and holes are the minority carriers.

When a silicon junction diode is biased in the reverse direction, the current remains nearly constant for a bias varying between about 0.1 V and the breakdown voltage. This constant value is referred to as the *saturation current* and is represented by I_s in Fig. 20.15. In practice, the reverse current increases with increase of bias, this increase being due mainly to surface leakage. In the case of a silicon junction diode in which the surface leakage is negligible, the current is given by the expression

$$i = I_s(e^{\frac{ev}{kT}} - 1)$$ [20.1]

where I_s = saturation current with negative bias,

   e = charge on electron = 1.6×10^{-19} C,

   v = p.d., in volts, across junction,

   k = Boltzmann's constant = 1.38×10^{-23} J/K

and T = thermodynamic temperature = $(273.15 + \theta)$ °C

Fig. 20.15 Static characteristic for a silicon junction diode having negligible surface leakage

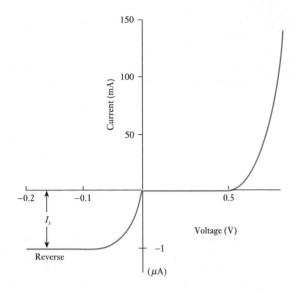

Let us assume a saturation current of, say, 10 nA; then for a temperature of 300 K (= 27 °C) we have, from expression [20.1],

$$i = 10(e^{38.6v} - 1) \text{ nanoamperes} \qquad [20.2]$$

Values of current i, calculated from expression [20.2] for various values of v, are plotted in Fig. 20.15.

20.7	Construction and static characteristics of a junction diode

Figure 20.16 shows one arrangement of a junction diode. A thin wafer or sheet W is cut from an n-type crystal, the area of the wafer being proportional to the current rating of the diode. The lower surface of the wafer is soldered to a copper plate C and a bead of indium I is placed centrally on the upper surface. The unit is then heat treated so that the indium forms a p-type alloy. A copper electrode E is soldered to the bead during the heat treatment, and the whole element is hermetically sealed in a metal or other opaque container M to protect it from light and moisture. The electrode E is insulated from the container by a bush B.

Typical voltage/current characteristics of a germanium junction diode are given in Fig. 20.17, the full lines being for a temperature of the surrounding air (i.e. ambient temperature) of 20 °C and the dotted lines for 55 °C. For a given reverse bias, the reverse current roughly doubles for every 10 °C rise of temperature. This rectifier can withstand a peak inverse voltage of about 100 V at an ambient temperature of 20 °C.

The silicon junction diode is similar in appearance to the germanium diode and typical voltage/current characteristics are given in Fig. 20.18. The properties of silicon junction diodes differ from those of germanium junction diodes in the following respects:

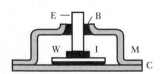

Fig. 20.16 A junction diode

Fig. 20.17 Static characteristics of a germanium junction diode

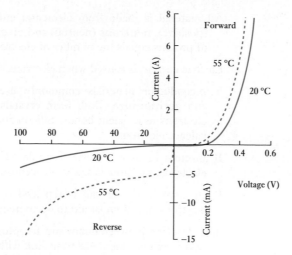

Fig. 20.18 Static characteristics of a silicon junction diode

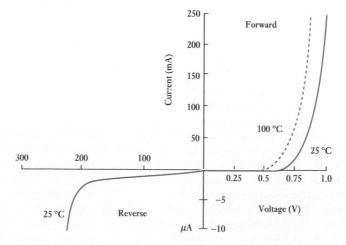

1. The forward voltage drop is roughly double that of the corresponding germanium diode.
2. The reverse current at a given temperature and voltage is approximately a hundredth of that of the corresponding germanium diode, but there is little sign of current saturation as is the case with germanium – in fact, the reverse current of a silicon diode is roughly proportional to the square root of the voltage until breakdown is approached.
3. The silicon diode can withstand a much higher reverse voltage and can operate at temperatures up to about 150–200 °C, compared with about 75–90 °C for germanium.
4. The reverse current of a silicon diode, for a given voltage, practically doubles for every 8 °C rise of temperature, compared with 10 °C for germanium.

| Terms and concepts | All material is made from elemental atoms which comprise **protons** (positive), **neutrons** (neutral) and **electrons** (negative). The number of protons equals the number of electrons in any atom. |

Electric current is caused when electrons migrate from atom to atom.

Semiconductor materials commonly used in electronics are **silicon** and **germanium**. Both form crystals with adjacent atoms sharing electrons in covalent bonds. Silicon and germanium atoms have four valence electrons.

If an atom in such a crystal is replaced by an atom with five valence electrons, the atom is said to be a **donor**.

If an atom in the crystal is replaced by an atom with three valence electrons, the atom is said to be an **acceptor**.

The **junction** between donor and acceptor materials creates a **depletion layer** because donor electrons link with the acceptor atoms.

The development of such junctions gives rise to the **diode**, a device which readily conducts in one direction but not in the other.

Exercises 20

1. Relative to a highly refined semiconductor, is a doped semiconductor: (a) more highly refined; (b) similarly refined; (c) less refined; (d) completely unrefined?

2. A piece of highly refined silicon is doped with arsenic. When connected into a circuit, is the conduction process in the semiconductor crystal due to: (a) electrons provided by the silicon; (b) holes provided by the arsenic; (c) holes provided by the silicon; (d) electrons provided by the silicon and the arsenic?

3. With reference to the semiconductor material specified in Q. 2, is the conduction process: (a) intrinsic; (b) extrinsic; (c) both but mainly extrinsic; (d) both but mainly intrinsic?

4. In the barrier layer of a p–n junction and within the p-type material, is the charge: (a) positive due to fixed atoms; (b) negative due to fixed atoms; (c) positive due to holes present; (d) negative due to electrons present?

5. A piece of doped semiconductor material is introduced into a circuit. If the temperature of the material is raised, will the circuit current: (a) increase; (b) remain the same; (c) decrease; (d) cease to flow?

6. Some highly refined semiconductor material is doped with indium. For such a material, are the majority carriers: (a) electrons from the semiconductor; (b) electrons from the indium; (c) holes from the semiconductor; (d) holes from the indium?

7. With the aid of a suitable sketch, describe the essential features of a semiconductor diode and also sketch a graph showing the forward and reverse characteristics of a typical silicon diode.

8. Explain, with reference to a semiconductor material, what is meant by: (a) intrinsic conductivity; (b) extrinsic conductivity.

9. Discuss the phenomenon of current flow in: (a) intrinsic; (b) p-type; (c) n-type semiconductors. Hence explain the rectifying action of a p–n junction.

10. A silicon diode has a reverse-bias resistance which can be considered as infinitely large. In the forward-bias direction, no current flows until the voltage across the device exceeds $600\,\text{mV}$. The junction then behaves as a constant resistance of $20\,\Omega$. From this information draw to scale the characteristic of this device.

Chapter twenty-one

Rectifiers

The diode is a device that passes current in only one direction. By appropriate use of such devices in networks, they can convert alternating current to direct current. In this chapter a variety of circuit arrangements involving one or more diodes will be investigated. The investigations will introduce certain design considerations, in particular the observation that the increased costs of using more diodes are offset by the improved network performance.

Although the rectified current is unidirectional, it will be seen that it does not have the even flow normally associated with direct current. To achieve the even flow the concept of smoothing will be introduced.

By the end of the chapter, you will be familiar with the process of rectification and be able to undertake exercises to determine the output voltages and currents when applied to resistive loads.

21.1 Rectifier circuits

Since a diode has the characteristic of having a much greater conductivity in one direction than in the other, it will produce a direct component of current when connected in series with an alternating voltage and a load. This process is known as rectification and is the main use to which diodes are put. There are numerous applications for rectification, e.g. driving a d.c. motor from a.c. mains and the production of direct-voltage supplies for electronic amplifiers.

21.2 Half-wave rectifier

While alternating currents and voltages play the leading roles in most electrical and electronic equipment, nevertheless many devices can either only operate on unidirectional currents and voltages, or at least they require such a supply as part of their mode of operation. The process of obtaining unidirectional currents and voltages from alternating currents and voltages is termed *rectification*.

The device that makes such a process possible is a diode, the ideal operating characteristic of which is given in Fig. 21.1. When the applied voltage acts in the forward direction, there is no voltage drop across the diode and a current flows unimpeded. However, when the applied voltage acts in the reverse direction, a voltage drop appears across the diode and no current flows.

Fig. 21.1 Ideal diode characteristics

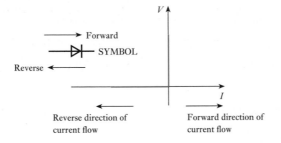

It is possible to obtain rectification by means of a single diode as indicated in Fig. 21.2. The current can only flow through the diode in one direction and thus the load current can only flow during alternate half-cycles. For this reason, the system is known as half-wave rectification. The load current, and hence the voltage drop across the load, is unidirectional and could be described as direct, although this term is more usually reserved for steady unidirectional quantities.

Fig. 21.2 Half-wave rectification

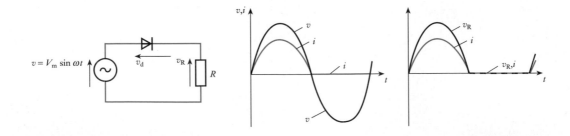

With reference to the circuit shown in Fig. 21.2:

v = instantaneous supply voltage

v_d = instantaneous voltage across the diode

v_R = instantaneous voltage across the load resistance

i = instantaneous diode current

Using Kirchhoff's second law in the closed loop:

$$v = v_d + v_R = v_d + iR$$

$$i = -\frac{1}{R}v_d + \frac{v}{R} \qquad \text{[21.1]}$$

Examination of the characteristics shown in Figs 20.17 and 20.18 indicates that v_d has a small value during the positive half-cycle when the diode conducts, but is equal to v during the negative half-cycle, there being negligible current and v_R consequently also being negligible. However, if we simplify the voltage/current characteristic to the idealized one shown in Fig. 21.1, then the value of v_d is zero and the effective resistance r_d (termed the *forward resistance*) of the diode is zero. If the supply voltage is

$$v = V_m \sin \omega t$$

then $\qquad i = \dfrac{v}{R} = \dfrac{V_m \sin \omega t}{R} = I_m \sin \omega t$ during the positive half-cycle.

Thus if the mean value of the current (neglecting the reverse current) $= I_{dc}$ then:

$$I_{dc} = \frac{1}{2\pi}\int_0^\pi I_m \sin \omega t \, d(\omega t) = \frac{I_m}{2\pi}\Big[-\cos \omega t\Big]_0^\pi$$

$$= \frac{I_m}{2\pi}[-\cos \pi + \cos 0] = \frac{I_m}{2\pi}[1 + 1]$$

$$\therefore \qquad I_{dc} = \frac{I_m}{\pi} = 0.318 I_m \qquad \text{[21.2]}$$

Similarly if the r.m.s. value of the current is I_{rms} then:

$$I_{rms} = \sqrt{\left[\frac{1}{2\pi}\int_0^\pi I_m^2 \sin^2 \omega t \, d(\omega t)\right]}$$

$$= \sqrt{\left[\frac{I_m^2}{2\pi}\int_0^\pi \tfrac{1}{2}(1 - \cos 2\omega t)\, d(\omega t)\right]}$$

$$= \sqrt{\left\{\frac{I_m^2}{2\pi} \times \frac{1}{2}\Big[\omega t + \tfrac{1}{2}\sin 2\omega t\Big]_0^\pi\right\}} = \sqrt{\left\{\frac{I_m^2}{4\pi}[\pi]\right\}}$$

$$\therefore \qquad I_{rms} = \frac{I_m}{2} = 0.5 I_m \qquad \text{[21.3]}$$

The voltage across the load during the positive half-cycles is given by

$$v_R = Ri = RI_m \sin \omega t = V_m \sin \omega t$$

Therefore, in the same way as derived for the current:

$$\text{Mean value of the load voltage} = V_{dc} = \frac{V_m}{\pi} \qquad [21.4]$$

$$\text{RMS value of the load voltage} = V_{rms} = \frac{V_m}{2} \qquad [21.5]$$

The maximum voltage, which occurs across the diode in the reverse direction, is known as the peak inverse voltage (PIV). This must be less than the breakdown voltage of the diode if it is not to conduct appreciably in the reverse direction. The peak inverse voltage for the diode in this circuit occurs when the potential of B is positive with respect to A by its maximum amount. The reverse resistance of the diode will, in the great majority of practical cases, be very much greater than the load resistance and most of the applied voltage will appear across the diode. Thus the peak inverse voltage equals approximately the peak value of the supply voltage.

Since the production of direct current from an a.c. supply is the object of the circuit, the useful power output is that produced in the load by the d.c. component of the load current. The efficiency of a rectifier circuit is defined as

$$\text{Efficiency } \eta = \frac{\text{power in the load due to d.c. component of current}}{\text{total power dissipated in the circuit}}$$

$$= \frac{I_{dc}^2 R}{I_{rms}^2 R} = \left[\frac{I_m}{\pi}\right]^2 \times \left[\frac{2}{I_m}\right]^2 = 0.405$$

This efficiency is based on $r_d = 0$. If it had a greater value, the efficiency would be reduced since the total power would be $I_{rms}^2(r_d + R)$, hence

$$\eta_m = 0.405 \qquad [21.6]$$

Waveforms for the half-wave rectifier circuit are shown in Fig. 21.3.

Fig. 21.3 Waveforms for half-wave rectifier circuit

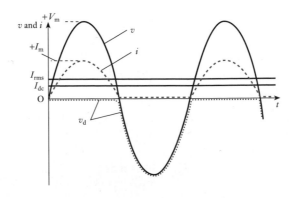

21.3
Full-wave rectifier network

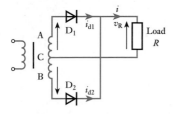

Fig. 21.4 Full-wave rectifier with resistive load

The half-wave rectifier gave rise to an output which had a unidirectional current, but the resulting current does not compare favourably with the direct current that one would expect from, say, a battery. One reason is that the half-wave rectifier is not making use of the other half of the supply waveform. It would be technically and economically advantageous if both halves were rectified and this may be achieved by a full-wave rectifier network.

The basic full-wave rectifier network is shown in Fig. 21.4. Here C is a centre tap on the secondary of the transformer, thus the e.m.f.s induced in each section of the secondary are equal, and when the potential of A is positive with respect to C, so is that of C positive with respect to B. With these polarities, diode D_1 will conduct while diode D_2 is non-conducting. When these polarities reverse, diode D_2 will conduct and D_1 will be non-conducting. In this way each diode conducts on alternate half-cycles, passing current through the load in the same direction. For

$$v_{AC} = v = V_m \sin \omega t$$

$$v_{BC} = -v = -V_m \sin \omega t$$

For identical diodes with a forward resistance $r_d = 0$ and an infinite reverse resistance, then during the period that v is positive:

$$\text{Diode 1 current} = i_{d1} = \frac{V_m \sin \omega t}{R} = I_m \sin \omega t$$

$$\text{Diode 2 current} = i_{d2} = 0$$

During the period that v is negative:

$$\text{Diode 1 current} = i_{d1} = 0$$

$$\text{Diode 2 current} = i_{d2} = \frac{-V_m \sin \omega t}{R} = -I_m \sin \omega t$$

At any instant the load current is given by

$$i = i_{d1} + i_{d2}$$

Thus in this circuit the current will repeat itself twice every cycle of the supply voltage, therefore mean value of the load current is

$$I_{dc} = \frac{1}{\pi} \int_0^\pi I_m \sin \omega t \, d(\omega t)$$

$$I_{dc} = \frac{2 I_m}{\pi} = 0.637 I_m \qquad [21.7]$$

RMS value of the load current is

$$I_{rms} = \sqrt{\left[\frac{1}{\pi} \int_0^\pi I_m^2 \sin^2 \omega t \, d(\omega t) \right]}$$

$$I_{rms} = \frac{I_m}{\sqrt{2}} = 0.707 I_m \qquad [21.8]$$

Similarly, for the load voltage, mean value of the load voltage is

$$V_{dc} = \frac{2V_m}{\pi}$$ [21.9]

RMS value of the load voltage is

$$V_{rms} = \frac{V_m}{\sqrt{2}}$$ [21.10]

For this circuit

$$\eta = \frac{I_{dc}^2 R}{I_{rms}^2 R} = \left[\frac{2I_m}{\pi}\right]^2 \times \left[\frac{\sqrt{2}}{I_m}\right]^2 = \frac{8}{\pi^2}$$

Again, since $r_d = 0$,

$$\eta = 0.81$$ [21.11]

Waveforms for the full-wave rectifier network are shown in Fig. 21.5.

Fig. 21.5 Waveforms for full-wave rectifier circuit

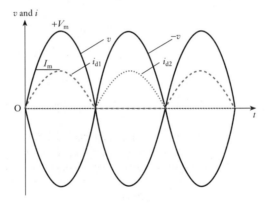

<table>
<tr><td>21.4</td><td>**Bridge rectifier network**</td></tr>
</table>

The full-wave rectifier only makes use of each half of the transformer winding for half of the time. The winding can be used all of the time, or the transformer can be omitted in certain cases by the application of the bridge rectifier network. Such a network is shown in Fig. 21.6(a).

When the potential of A is positive with respect to B, diodes D_1 and D_3 conduct and a current flows in the load. When the potential of B is positive with respect to A, diodes D_2 and D_4 conduct and the current in the load is in the same direction as before. Thus a full-wave type of output is obtained. The expressions derived for current and load voltage for the full-wave circuit will be applicable here also. Note should be taken, however, that, at any instant, two diodes are conducting: Fig. 21.6(b) shows the active circuits during each half-cycle.

The waveform diagrams shown in Fig. 21.6(c) indicate that the current i taken from the supply is purely alternating yet the load current i_R is unidirectional and therefore basically a direct current.

Fig. 21.6 Full-wave bridge rectification

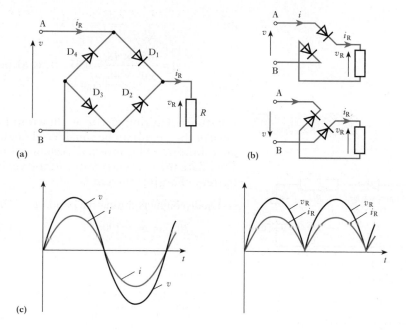

(a)

(b)

(c)

The efficiency of the network is given by

$$\eta = \frac{I_{dc}^2 R}{I_{rms}^2 R} = \left[\frac{2}{\pi} I_m\right]^2 \times \left[\frac{\sqrt{2}}{I_m}\right]^2 = \frac{8}{\pi^2} \qquad [21.12]$$

Neglecting the forward resistances of the diodes, the voltage which appears across the non-conducting diodes is equal to the supply voltage. Thus the peak inverse voltage is equal to the peak value of the supply voltage.

There are certain advantages to be obtained by using the bridge circuit as opposed to the centre-tapped secondary full-wave circuit. For a given mean load voltage, the bridge circuit uses only half the number of secondary turns, and for a diode, of given maximum peak-inverse-voltage rating, twice the mean output voltage can be obtained from it since the peak inverse voltage encountered is V_m compared to $2V_m$ for the full-wave circuit.

Example 21.1

The four diodes used in a bridge rectifier circuit have forward resistances which may be considered negligible, and infinite reverse resistances. The alternating supply voltage is 230 V r.m.s. and the resistive load is 46.0 Ω. Calculate:

(a) the mean load current;
(b) the rectifier efficiency.

(a) $\qquad I_m = \dfrac{\sqrt{2} \times 230}{46} = 7.07 \text{ A}$

$\qquad I_{dc} = \dfrac{2I_m}{\pi} = \dfrac{2 \times 7.07}{\pi} = 4.5 \text{ A}$

(b) $I_{rms} = \dfrac{I_m}{\sqrt{2}} = \dfrac{7.07}{\sqrt{2}} = 5.0 \text{ A}$

∴ $\eta = \dfrac{4.5^2 \times 48}{5.0^2 \times 48} = 0.81$ **(i.e. 81 per cent)**

Example 21.2

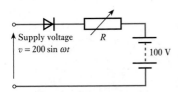

Fig. 21.7 Circuit for Example 21.2

A battery-charging circuit is shown in Fig. 21.7. The forward resistance of the diode can be considered negligible and the reverse resistance infinite. The internal resistance of the battery is negligible. Calculate the necessary value of the variable resistance R so that the battery charging current is 1.0 A.

Diode conducts during the period $v > 100$ V.

$$200 \sin \omega t = 100$$

i.e. $\sin \omega t = 0.5$

i.e. when

$$\omega t = \frac{\pi}{6} \quad \text{and} \quad \frac{5\pi}{6}$$

Therefore diode conducts when

$$\frac{\pi}{6} < \omega t < \frac{5\pi}{6}$$

During conduction

$$i = \frac{v - 100}{R} = \frac{200 \sin \omega t - 100}{R}$$

Mean value of current is

$$1.0 = \frac{1}{2\pi} \int_{\pi/6}^{5\pi/6} \frac{200 \sin \omega t - 100}{R} \, d(\omega t)$$

$$= \frac{1}{2\pi R} \Big[-200 \cos \omega t - 100 \omega t \Big]_{\pi/6}^{5\pi/6}$$

$$= \frac{1}{2\pi R} \left[-200 \cos \frac{5\pi}{6} - 100 \times \frac{5\pi}{6} + 200 \cos \frac{\pi}{6} + 100 \times \frac{\pi}{6} \right]$$

$$= \frac{1}{2\pi R} \left[-200 \left(-\frac{\sqrt{3}}{2} \right) - \frac{500\pi}{6} + 200 \left(\frac{\sqrt{3}}{2} \right) + \frac{100\pi}{6} \right]$$

$$= \frac{1}{2\pi R} \left[200\sqrt{3} - \frac{400\pi}{6} \right]$$

$$R = \frac{200\sqrt{3}}{2\pi} - \frac{400\pi}{2\pi \times 6}$$

$$R = 55.1 - 33.3 = \textbf{21.8 } \Omega$$

21.5 Smoothing

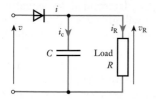

Fig. 21.8 Half-wave rectifier with capacitor input filter

The rectifier circuits so far described have produced, as required, a direct component of current in the load. There remains, however, a large alternating component. In a large number of applications it is desirable to keep this latter component small. This can be accomplished by the use of smoothing circuits, the simplest of which consists of a capacitor in parallel with the load. Figure 21.8 shows such an arrangement.

The diode conducts when the supply voltage v is more positive than the load voltage v_R. During this conduction period, if the diode forward resistance is neglected, then the load voltage is equal to the supply voltage. Therefore if:

$$v = V_m \sin \omega t$$

$$i_R = \frac{v}{R} = \frac{V_m \sin \omega t}{R}$$

$$i_C = C \frac{dv}{dt} = C\omega V_m \cos \omega t$$

$$\text{Diode current} = i = i_R + i_C = \frac{V_m}{R} \sin \omega t + \omega C V_m \cos \omega t$$

$$= \sqrt{\left[\left(\frac{V_m}{R} \right)^2 + (\omega C V_m)^2 \cdot \sin(\omega t + \phi) \right]}$$

$$\therefore \qquad i = \frac{V_m}{R} \sqrt{(1 + \omega^2 C^2 R^2)} \cdot \sin(\omega t + \phi) \qquad [21.13]$$

where $\quad \tan \phi = \dfrac{\omega C V_m}{V_m / R} = \omega C R$

If $i = 0$, i.e. diode cuts off, when $t = t_2$ then

$$\sin(\omega t_2 + \phi) = 0$$

$$\therefore \qquad \omega t_2 + \phi = \pi$$

$$\therefore \qquad \omega t_2 = \pi - \phi \qquad [21.14]$$

If $\omega C R \gg 1$ then

$$\phi \approx \frac{\pi}{2}$$

and $\quad \omega t_2 \approx \pi - \dfrac{\pi}{2} = \dfrac{\pi}{2}$

i.e. the diode ceases to conduct near the instant at which v has its positive maximum value.

While the diode is non-conducting, C will discharge through R and the load voltage will be given by

$$v_R = V e^{\frac{-(t - t_2)}{CR}}$$

where V is the capacitor voltage at the instant the diode cuts off. This will equal approximately the peak value of the supply voltage if $\omega CR \gg 1$. The diode will start to conduct again during the period that the supply voltage is positive and increasing. This instant can be determined by equating $V\,\mathrm{e}^{-(t-t_2)/CR}$ and $V_\mathrm{m} \sin \omega t$, one solution of which will give $T + t_1$, where T is the period of the supply voltage and t_1 the instant of time at which the diode starts to conduct.

In order to keep the variation in load voltage down during the period when the diode is non-conducting, a long time constant CR compared to the period of the supply voltage is required. Note, however, from relation [21.13] that the peak value of the diode current increases with C and therefore care must be taken to ensure that the maximum allowable diode peak current is not exceeded. Waveforms for the circuit are shown in Fig. 21.9.

Fig. 21.9 Waveforms for half-wave rectifier with capacitor input filter

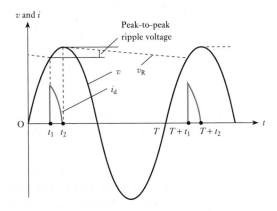

A similar analysis could be carried out for the full-wave circuit. The operation is identical to that of the half-wave circuit during the charging period, but the capacitor discharges into the load resistance for a shorter period, giving less amplitude of ripple for a given time constant during the non-conducting period. Figure 21.10 shows waveforms for a full-wave circuit with a capacitor input filter.

Fig. 21.10 Waveforms for full-wave rectifier with capacitor input filter

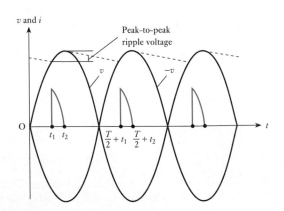

Fig. 21.11 Smoothing circuits.
(a) Series inductor filter;
(b) L–C filter; (c) π filter;
(d) resistive filter

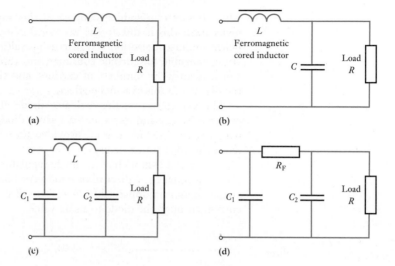

Figure 21.11 shows some other types of filter circuits used in practice for smoothing purposes. The ideal arrangement is to connect a low-valued impedance at the frequency being used and which presents a high resistance to direct current, e.g. a capacitor, in parallel with the load, and/or a high-valued impedance at the frequency being used and which presents a low resistance to direct current, e.g. an inductor, in series with the load. This has the effect of minimizing the ripple across the load while having little effect on the direct voltage developed at the input to the filter circuit. The circuit of Fig. 21.11(d) does not meet these requirements fully in so much as the resistor R will reduce the direct voltage as well as the ripple. The circuit is useful, however, in low current applications where the direct voltage drop across R can be kept small; a resistor has an economic advantage over a ferromagnetic-cored inductor.

21.6   Zener diode

If the reverse voltage across a p–n junction is gradually increased, a point is reached where the energy of the current carriers is sufficient to dislodge additional carriers. These carriers, in turn, dislodge more carriers and the junction goes into a form of avalanche breakdown characterized by a rapid increase in current as shown in Fig. 20.18. The power due to a relatively large reverse current, if maintained for an appreciable time, can easily ruin the device.

Special junction diodes, often known as *Zener diodes*, but more appropriately termed *voltage regulator diodes*, are available in which the reverse breakdown voltage is in the range of about 4 V to about 75 V, the actual voltage depending upon the type of diode.

When the voltage regulator diode is forward biased, it behaves as a normal diode. With a small reverse voltage, the current is the sum of the surface leakage current and the normal saturation current due to the thermally generated holes and electrons (section 20.6). This current is only a few microamperes; but as the reverse voltage is increased, a value is reached at which the current suddenly increases very rapidly (Fig. 20.18). As already mentioned, this is due to the increased velocity of the carriers being sufficient to cause ionization. The carriers resulting from ionization by collision are responsible for further collisions and thus produce still more carriers. Consequently the numbers of carriers, and therefore the current, increase rapidly due to this avalanche effect.

The voltage across the regulator diode after breakdown is termed the *reference voltage* and its value for a given diode remains practically constant over a wide range of current, provided the maximum permissible junction temperature is not exceeded.

Figure 21.12 shows how a voltage regulator diode (or Zener diode) can be used as a voltage stabilizer to provide a constant voltage from a source whose voltage may vary appreciably. A resistor R is necessary to limit the reverse current through the diode to a safe value.

Fig. 21.12 A voltage stabilizer

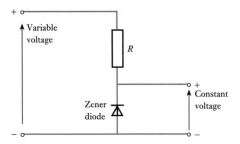

Summary of important formulae

For a half-wave rectifier,

$$I_{dc} = \frac{1}{\pi} I_m = 0.318 I_m \qquad [21.2]$$

$$I_{rms} = \tfrac{1}{2} I_{rms} = 0.5 I_m \qquad [21.3]$$

$$V_{dc} = \frac{1}{\pi} V_m = 0.318 V_m \qquad [21.4]$$

$$V_{rms} = \tfrac{1}{2} V_m = 0.5 V_m \qquad [21.5]$$

For a full-wave rectifier,

$$I_{dc} = \frac{2}{\pi} I_m = 0.637 I_m \qquad [21.7]$$

$$I_{rms} = \frac{1}{\sqrt{2}} I_m = 0.707 I_m \qquad [21.8]$$

$$V_{dc} = \frac{2}{\pi} V_m = 0.637 V_m \qquad [21.9]$$

$$V_{rms} = \frac{1}{\sqrt{2}} V_m = 0.707 V_m \qquad [21.10]$$

Terms and concepts

A **rectifier** circuit or network is generally used to convert alternating current to direct current.

A single diode can only provide **half-wave rectification**. It is not efficient and the 'direct current' can only be supplied half the time.

A **full-wave rectifier** requires at least two diodes and generally four diodes are involved. The direct current is reasonably consistent in its unidirectional flow.

Smoothing is the process of removing the worst of the output variations in the current.

A **Zener diode** is one which tends to have the same volt drop across it regardless of the current passing through it. In practice there are limits to the variation of current which it can withstand.

Exercises 21

1. If sine waves of peak values (a) 0.1 V, (b) 1 V, (c) 5 V are connected in turn in series with the device described in Q. 10 of Exercises 20 (see page 416), what is the peak current in each case, assuming that the diode is still operating within its rate limits?

2. What are the peak currents in Q. 1 if in addition a 50 Ω resistor is included in series with the sine-wave source in each case? (You might need some assistance with this one.)

3. Sketch one form of full-wave rectifier circuit together with smoothing components. If the supply frequency is 400 Hz, what is the ripple frequency?

4. A silicon diode has forward characteristics given in the following table:

V (V)	0	0.50	0.75	1.00	1.25	1.50	1.75	2.00
I (A)	0	0.1	1.5	0.3	3.3	5.0	6.7	8.5

The diode is connected in series with a 1.0 Ω resistor across an alternating voltage supply of peak value 10.0 V. Determine the peak value of the diode forward current and the value of the series resistor which would limit the peak current to 5.0 A.

5. A semiconductor diode, the forward and reverse characteristics of which can be considered ideal, is used in a half-wave rectifier circuit supplying a resistive load of 1000 Ω. If the r.m.s. value of the sinusoidal supply voltage is 230 V determine: (a) the peak diode current; (b) the mean diode current; (c) the r.m.s. diode current; (d) the power dissipated in the load.

6. Two semiconductor diodes used in a full-wave rectifier circuit have forward resistances which will be considered constant at 1.0 Ω and infinite reverse resistances. The circuit is supplied from a 300–0–300 V r.m.s. secondary winding of a transformer and the mean current in the resistive load is 10 A. Determine the resistance of the load, the maximum value of the voltage which appears across the diodes in reverse, and the efficiency of the circuit.

7. (a) Sketch network diagrams showing how diodes may be connected to the secondary winding of a suitable transformer in order to obtain unsmoothed: (i) half-wave rectification; (ii) full-wave rectification with two diodes; (iii) full-wave rectification with four diodes. For each network, show the waveform of the output voltage.

   (b) Explain the necessity of smoothing the output voltage before applying it to a transistor amplifier.

   (c) Sketch a typical network for smoothing such a supply to a transistor amplifier.

8. A full-wave rectifier circuit supplies a 2000 Ω resistive load. The characteristics of the diodes used can be considered ideal and each half of the secondary winding of the transformer develops an output voltage of 230 V. If the mean current in the load is to be limited to 100 mA by the connection of equal resistors in series with the diodes, determine the value of these resistors and the power dissipated in them.

9. Draw the circuit diagram of a full-wave bridge rectifier network supplying a resistive load.

   By reference to the circuit diagram, explain the operation of the network, including in your answer diagrams illustrating the input and output current waveforms.

   What would be the effect on the operation of the other three diodes if one of the component diodes were to become short-circuited?

Exercises 21 continued

10. Describe, with the aid of suitable diagrams, the rectifier action of a semiconductor diode. (Reference should be made to the principle of conduction in the semiconductor materials and the potential difference at the barrier layer.)

The rectifier diodes shown in Fig. A are assumed to be ideal. Calculate the peak current in each of the resistors, given that the applied voltage is sinusoidal.

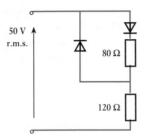

50 V r.m.s.

80 Ω

120 Ω

Fig. A

11. The four semiconductor diodes used in a bridge rectifier circuit have forward resistances which can be considered constant at 0.1 Ω and infinite reverse resistance. They supply a mean current of 10 A to a resistive load from a sinusoidally varying alternating supply of 20 V r.m.s. Determine the resistance of the load and the efficiency of the circuit.

12. A half-wave rectifier circuit is used to charge a battery of e.m.f. 12 V and negligible internal resistance. The sinusoidally varying alternating supply voltage is 24 V peak. Determine the value of resistance to be connected in series with the battery to limit the charging current to 1.0 A. What peak current would flow in the diode if the battery were reversed?

Exercises 13 and 14 should only be attempted after reading section 22.14.

13. A stabilized power supply is tested and the following results are obtained: unstabilized input voltage to the Zener circuit constant at 25 V; output voltage from the stabilizer on no-load, 16 V; output voltage from the stabilizer on full load (50 mA), 14 V; the output stabilized voltage on full load falls to 13.5 V when the unstabilized voltage is reduced to 18 V. Find: (a) the output resistance; (b) the stabilization factor.

14. A Zener diode has a characteristic which may be considered as two straight lines, one joining the points $I = 0$, $V = 0$ and $I = 0$, $V = -9$ V, and the other joining the points $I = 0$, $V = -9$ V and $I = -45$ mA, $V = -10$ V. Find: (a) its resistance after breakdown; (b) the value of the stabilization series resistor if the unstabilized voltage is 15 V and the load voltage is 9.5 V, when the load current is 30 mA; (c) the range of load current this circuit can deal with if the maximum dissipation in the Zener is 0.5 W; (d) the range of unstabilized voltage this circuit can deal with on a constant load current of 20 mA; (e) the stabilization factor.

Chapter twenty-two

Junction Transistor Amplifiers

Objectives

When you have studied this chapter, you should

- have a knowledge of the construction of a bipolar junction transistor

- be aware of the use of bipolar transistors in common-base and common-emitter circuits

- be able to interpret the static characteristics for transistor circuits

- have the skills to draw a load line for a transistor amplifier circuit

- have the ability to select circuit components

- understand the use of transistor equivalent circuits

- have an appreciation of hybrid parameters

- be able to describe how transistors operate as switches

Contents

The semiconductor materials and diodes detailed in the earlier chapters can be used in the transistor, which forms the basis of most amplifiers. There are two types of transistor, but only the bipolar junction transistor will be detailed here.

The construction of the junction transistor and its operation will be described. The ways that it can be connected into an amplifier circuit will be considered. These circuit arrangements will be analysed in order to distinguish the benefits of each.

A transistor's operation is improved when it has a stable power supply. Some simple transistor application systems will be described including a system to provide a stabilized voltage supply. Finally, the use of the transistor as a switch will be detailed, which is significant in digital systems.

By the end of this chapter, you will understand the function of the transistor amplifier. You will also be able to do exercises analysing transistor amplifier performance.

22.1 Introduction

Having become familiar with the semiconductor diode in Chapter 20, it is now possible to progress to an understanding of the transistor, which is fundamental to most amplifier arrangements, as well as many switching arrangements associated with digital systems.

There are two basic types of transistor: (1) the bipolar junction transistor; (2) the field effect transistor (FET).

Early transistor circuits depended entirely on the junction transistor and for that reason it will be considered first. However its large size has reduced its range of applications in practice.

22.2 Bipolar junction transistor

A bipolar junction transistor is a combination of two junction diodes and consists of either a thin layer of p-type semiconductor sandwiched between two n-type semiconductors, as in Fig. 22.1(a), and referred to as an n–p–n transistor, or a thin layer of an n-type semiconductor sandwiched between two p-type semiconductors, as in Fig. 22.1(b), and referred to as a p–n–p transistor. The thickness of the central layer, known as the *base*, is of the order of 25 μm (or 25×10^{-6} m).

The junction diode formed by n_1–p in Fig. 22.1(a) is biased in the forward direction by a battery B_1 so that free electrons are urged from n_1 towards p. Hence n_1 is termed an *emitter*. On the other hand, the junction diode formed by n_2–p in Fig. 22.1(a) is biased in the reverse direction by battery B_2 so that if battery B_1 were disconnected, i.e. with zero emitter current, no current would flow between n_2 and p apart from that due to the thermally generated minority carriers referred to in section 20.6. However, with B_1 connected as in Fig. 22.1(a), the electrons from emitter n_1 enter p and diffuse through the base until they come within the influence of n_2, which is connected to the positive terminal of battery B_2. Consequently the electrons which reach n_2 are collected by the metal electrode attached to n_2; hence n_2 is termed a *collector*.

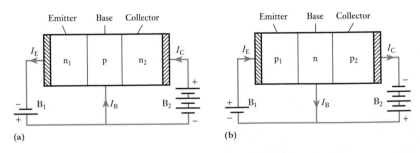

Fig. 22.1 Arrangement of a bipolar junction transistor. (a) n–p–n; (b) p–n–p

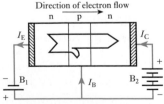

Fig. 22.2 Electron flow in an n–p–n transistor

Some of the electrons, in passing through the base, combine with holes; others reach the base terminal. The electrons which do not reach collector n_2 are responsible for the current at the base terminal, and the distribution of electron flow in an n–p–n transistor can be represented diagrammatically as in Fig. 22.2. The *conventional* directions of the current are represented by the arrows marked I_E, I_B and I_C, where I_E is the emitter current, I_B the base current and I_C the collector current.

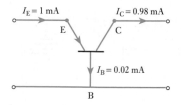

Fig. 22.3 Currents in emitter, collector and base circuits of a p–n–p transistor

By making the thickness of the base very small and the impurity concentration in the base much less than in the emitter and collector, the free electrons emerging from the emitter have little opportunity of combining with holes in the base, with the result that about 98 per cent of these electrons reach the collector.

By Kirchhoff's first law, $I_E = I_B + I_C$, so that if $I_C = 0.98I_E$, then $I_B = 0.02I_E$; thus, when $I_E = 1$ mA, $I_C = 0.98$ mA and $I_B = 0.02$ mA, as in Fig. 22.3.

In the above explanation we have dealt with the n–p–n transistor, but exactly the same explanation applies to the p–n–p transistor of Fig. 22.1(b) except that the movement of electrons is replaced by the movement of holes.

22.3 Construction of a bipolar transistor

The first step is to purify the silicon or germanium so that any impurity does not exceed about 1 part in 10^{10}. Various methods have been developed for attaining this exceptional degree of purity, and intensive research is still being carried out to develop new methods of purifying and of doping silicon and germanium.

In one form of construction of the p–n–p transistor, the purified material is grown as a single crystal, and while the material is in a molten state, an n-type impurity (e.g. antimony in the case of germanium) is added in the proportion of about 1 part in 10^8. The solidified crystal is then sawn into slices about 0.1 mm thick. Each slice is used to form the base region of a transistor; thus in the case of n-type germanium, a pellet of p-type impurity such as indium is placed on each side of the slide, the one which is to form the collector being about three times the size of that forming the emitter. One reason for the larger size of the collector bead is that the current carriers from the emitter spread outwards as they pass through the base, and the larger area of the collector enables the latter to collect these carriers more effectively. Another reason is that the larger area assists in dissipating the greater power loss at the collector–base junction. This greater loss is due to the p.d. between collector and base being greater than that between emitter and base.

The assembly is heated in a hydrogen atmosphere until the pellets melt and dissolve some of the germanium from the slice, as shown in Fig. 22.4. Leads for the emitter and collector are soldered to the surplus material in the pellets to make non-rectifying contacts, and a nickel tab is soldered to make connection to the base. The assembly is then hermetically sealed in a metal or glass container, the glass being coated with opaque paint. The transistor is thus protected from moisture and light.

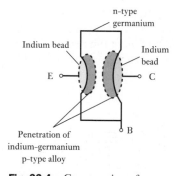

Fig. 22.4 Construction of a p–n–p germanium transistor

22.4 Common-base and common-emitter circuits

In Fig. 22.1 the transistor is shown with the base connected directly to both the emitter and collector circuits; hence the arrangement is referred to as a *common-base* circuit. The conventional way of representing this arrangement is shown in Fig. 22.5(a) and (b) for n–p–n and p–n–p transistors respectively, where E, B and C represent the emitter, base and collector terminals. The arrowhead on the line joining the emitter terminal to the base indicates the conventional direction of the current in that part of the circuit.

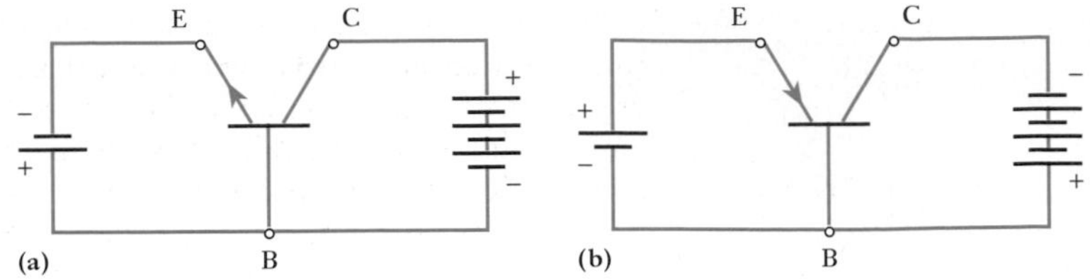

Fig. 22.5 Common-base
circuits. (a) n–p–n; (b) p–n–p

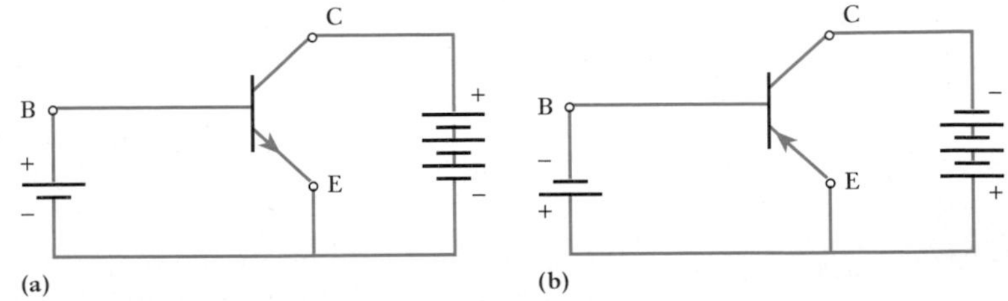

Fig. 22.6 Common-emitter
circuits. (a) n–p–n; (b) p–n–p

Figure 22.6 shows the *common-emitter* method of connecting a transistor. In these diagrams it will be seen that the emitter is connected directly to the base and collector circuits. This method is more commonly used than the common-base circuit owing to its higher input resistance and the higher current and power gains.

The common-collector circuit is used only in special cases, and will therefore not be considered here.

22.5	Static characteristics for a common-base circuit

Figure 22.7 shows an arrangement for determining the static characteristics of an n–p–n transistor used in a common-base circuit. The procedure is to maintain the value of the emitter current, indicated by A_1, at a constant value, say 1 mA, by means of the slider on R_1, and note the readings on A_2 for various values of the collector–base voltage given by voltmeter V_2. The

Fig. 22.7 Determination of
static characteristics for a
common-base n–p–n
transistor circuit

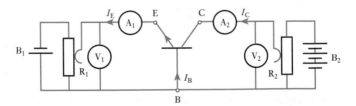

test is repeated for various values of the emitter current and the results are plotted as in Fig. 22.8.

In accordance with *BS 3363*, the current is assumed to be positive when its direction is from the external circuit towards the transistor terminal, and the voltage V_{CB} is positive when C is positive relative to B. Hence, for an n–p–n transistor, the collector current and collector–base voltage are positive but the emitter current is negative. For a p–n–p transistor, all the signs have to be reversed.

From Fig. 22.8 it will be seen that for positive values of the collector–base voltage, the collector current remains almost constant, i.e. nearly all the electrons entering the base of an n–p–n transistor are attracted to the

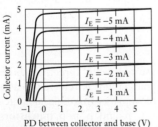

Fig. 22.8 Static characteristics
for a common-base n–p–n
transistor circuit

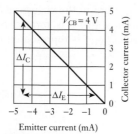

Fig. 22.9 Relationship between collector and emitter currents for a given collector–base voltage

collector. Also, for a given collector–base voltage, the collector current is practically proportional to the emitter current. This relationship is shown in Fig. 22.9 for V_{CB} equal to 4 V. The ratio of the change, ΔI_C, of the collector current to the change, ΔI_E, of the emitter current (neglecting signs), for a given collector–base voltage, is termed the *current amplification factor for a common-base circuit* and is represented by the symbol α, i.e.

$$\alpha = \frac{\Delta I_C}{\Delta I_E} \quad \text{for a given value of } V_{CB} \qquad [22.1]$$

$$= \text{slope (neglecting signs) of } \frac{I_C}{I_E} \text{ graph in Fig. 22.9}$$

22.6 Static characteristics for a common-emitter circuit

Figure 22.10 shows an arrangement for determining the static characteristics of an n–p–n transistor used in a common-emitter circuit. Again, the procedure is to maintain the *base* current, I_B, through a microammeter A_1 constant at, say, 25 μA, and to note the collector current, I_C, for various values of the collector–emitter voltage V_{CE}, the test being repeated for several values of the base current, and the results are plotted as shown in Fig. 22.11. For a given voltage between collector and emitter, e.g. for $V_{CE} = 4$ V, the relationship between the collector and base currents is practically linear as shown in Fig. 22.12.

Fig. 22.10 Determination of static characteristics for a common-emitter n–p–n transistor circuit

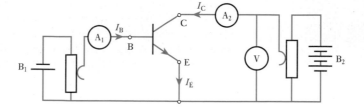

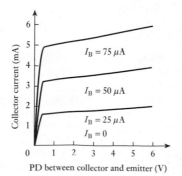

Fig. 22.11 Static characteristics for a common-emitter n–p–n transistor circuit

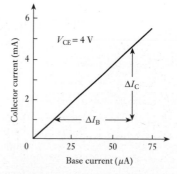

Fig. 22.12 Relationship between collector and base currents, for a given collector–emitter voltage

The ratio of the change, ΔI_C, of the collector current to the change, ΔI_B, of the base current, for a given collector–emitter voltage, is termed the *current amplification factor for a common-emitter circuit* and is represented by the symbol β, i.e.

$$\beta = \frac{\Delta I_C}{\Delta I_B} \quad \text{for a given value of } V_{CE} \tag{22.2}$$

= slope of graph in Fig. 22.12

22.7 Relationship between α and β

From Figs 22.7 and 22.10 it is seen that

$$I_E = I_C + I_B$$

$$\therefore \quad \Delta I_E = \Delta I_C + \Delta I_B$$

From expression [22.1]

$$\alpha = \frac{\Delta I_C}{\Delta I_B}$$

$$= \frac{\Delta I_C}{(\Delta I_C + \Delta I_B)}$$

$$\therefore \quad \frac{1}{\alpha} = 1 + \frac{\Delta I_B}{\Delta I_C}$$

$$= 1 + \frac{1}{\beta} = \frac{(1+\beta)}{\beta}$$

Hence

$$\alpha = \frac{\beta}{(1+\beta)} \tag{22.3}$$

and

$$\beta = \frac{\alpha}{(1-\alpha)} \tag{22.4}$$

Thus, if

$$\alpha = 0.98, \quad \beta = 0.98/0.02 = 49$$

and if

$$\alpha = 0.99, \quad \beta = 0.99/0.01 = 99$$

i.e. a small variation in α corresponds to a large variation in β. It is therefore better to determine β experimentally and calculate therefrom the corresponding value of α by means of expression [22.3].

A transistor in a common-emitter circuit has the base–emitter junction forward biased and the collector–base junction reverse biased. Under these conditions and in the absence of an input signal

$$I_E = I_B + I_C$$

If the emitter is not connected, $I_E = 0$. However, we find there is still a small collector current – this is a leakage current crossing the reverse-based collector–base junction. This leakage current is given the symbol I_{CB0} being the collector–base current with zero emitter current.

Let us now reconnect the emitter. The leakage current is still present and therefore the total collector current is given by

$$I_C = \alpha I_E + I_{CB0}$$

$$= \alpha(I_B + I_C) + I_{CB0}$$

$$I_C(1 - \alpha) = \alpha I_B + I_{CB0}$$

$$I_C = \frac{\alpha}{1 - \alpha} I_B + \frac{I_{CB0}}{1 - \alpha}$$

$$I_C = \beta I_B + \frac{I_{CB0}}{1 - \alpha}$$

If I_{CB0} is very small, the last term of this expression can be neglected and the relation reduces to that of equation [22.2]. The leakage current is not necessarily negligible. The leakage current is temperature dependent, hence an increase in temperature causes the leakage current to rise. In turn this results in an increase in collector current and a change in bias conditions. The change in bias conditions is stabilized by using a potential divider, as shown in Fig. 22.13. This holds the base voltage almost constant. It is also necessary to introduce a resistor into the emitter connection.

Fig. 22.13 Bias point stabilization

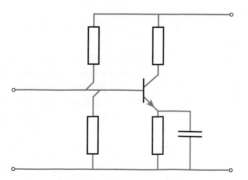

22.8 **Load line for a transistor**

Let us consider an n–p–n transistor used in a common-base circuit (Fig. 22.14), together with an a.c. source S having an internal resistance R_S, a load resistance R and bias batteries B$_1$ and B$_2$ giving the required transistor currents.

Let us assume that the voltage between the collector and base is to be 3 V when there is *no alternating voltage* applied to the emitter, and that, with an emitter current of 3 mA, the corresponding collector current is 2.9 mA. This corresponds to point D in Fig. 22.15. Also suppose the resistance R of the load to be 1000 Ω. Consequently the corresponding p.d. across R is 0.0029×1000, namely 2.9 V, and the total bias supplied by battery B$_2$ must be $3 + 2.9$, namely 5.9 V. This is represented by point P in Fig. 22.15.

Fig. 22.14   An n–p–n transistor
with load resistance R

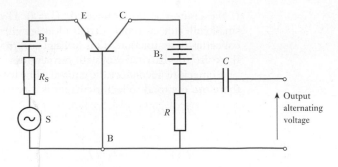

Fig. 22.15   Graphical
determination of the output
current of a transistor in a
common-base circuit

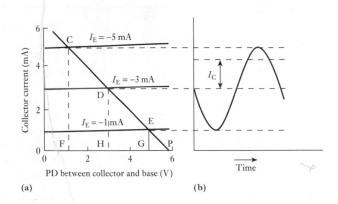

The straight line PD drawn through P and D is the *load line*, the inverse
of the slope of which is equal to the resistance of the load, e.g. HP/DH =
2.9/0.0029 = 1000 Ω. The load line is the locus of the variation of the collector current for any variation in the emitter current. Thus, if the emitter
current decreases to 1 mA, the collector current decreases to GE and the
p.d. across R decreases to GP. On the other hand, if the emitter current
increases to 5 mA, the collector current increases to FC and the p.d. across
R increases to FP. Hence, if the emitter current varies sinusoidally between
1 and 5 mA, the collector current varies as shown at (b) in Fig. 22.15. This
curve will be sinusoidal if the graphs are linear and are equally spaced over
the working range.

The function of capacitor C is to eliminate the d.c. component of the
voltage across R from the output voltage.

If I_c is the r.m.s. value of the *alternating* component of the collector
current, $I_c R$ gives the r.m.s. value of the alternating component of the
output voltage and $I_c^2 R$ gives the output power due to the alternating e.m.f.
generated in S. Note that a lower-case subscript indicates an r.m.s. value
compared with a capital for the d.c. value – the difference requires good
eyesight! Hence the larger the value of R, the greater the voltage and power
gains, the maximum value of R for a given collector supply voltage being
limited by the maximum permissible distortion of the output voltage.

A similar procedure can be used to determine the output voltage and
power for a transistor used in a common-emitter circuit.

<table>
<tr><td>**22.9**</td><td>**Transistor as an amplifier**</td></tr>
</table>

Consider again the basic action of an amplifier. The input signal shown in Fig. 22.16 controls the amount of power that the amplifier takes from the power source and converts into power in the load. We have also seen that in the transistor the collector current is controlled by the emitter or base currents. By connecting a load effectively between the collector and the common terminal, the transistor can produce gain.

Fig. 22.16 Mode of operation of an amplifier

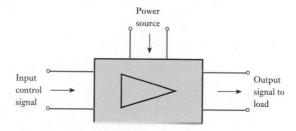

Again the input signal is generally an alternating quantity. However, the transistor requires to operate in a unidirectional mode, otherwise the negative parts of the alternating quantity would cause, say, the emitter–base junction to be reverse biased and this would prevent normal transistor action occurring. As a result, it is necessary to introduce a bias.

Figure 22.17 shows a practical transistor amplifier circuit utilizing an n–p–n transistor in the common-emitter mode. The resistor R_C is connected between the collector and the positive supply from a battery or other d.c. source. The resistor R_B is included to provide the bias current to the base of the transistor. In order to separate the direct current of the transistor arrangement from the alternating signal entering and leaving the amplifier, capacitors are included (it will be recalled that capacitors appear to pass an alternating current but do not pass a direct current). The input a.c. signal is fed into the transistor via C_1 which prevents the signal source having any effect on the steady component of the base current. Similarly, the coupling capacitor C_2 prevents the load connected across the output terminals from affecting the steady conditions of the collector. The capacitances of C_1 and C_2 are selected to ensure that their reactances are negligible at the operating frequencies and consequently they have negligible effect on the amplifier operation so far as the a.c. signal is concerned.

Fig. 22.17 Simple common-emitter transistor amplifier

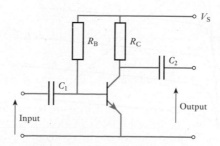

Let I_B be the steady base current, usually termed the *quiescent current*, and let V_{BE} be the quiescent base–emitter voltage; then

$$V_S = I_B R_B + V_{BE}$$

$$\therefore \qquad R_B = \frac{V_S - V_{BE}}{I_B} \qquad\qquad [22.5]$$

Often V_{BE} is significantly smaller than V_S and, approximately,

$$R_B = \frac{V_S}{I_B} \qquad\qquad [22.6]$$

Let the applied signal current to the base be given by

$$i_b = I_{b_m} \sin \omega t$$

whereby the total base current is

$$I_B + I_{b_m} \sin \omega t$$

At any instant

$$v_{ce} = V_S - i_c R_C$$

and $\qquad i_c = -\dfrac{1}{R_C} \cdot v_{ce} + \dfrac{V_S}{R_C} \qquad\qquad [22.7]$

This is the relation defining the load line of the form shown in Fig. 22.15(a) except that now it is expressed in terms that are applicable to the circuit arrangement shown in Fig. 22.17. In such a circuit arrangement, the load line applied to an appropriate set of output characteristics is shown in Fig. 22.18.

For any value of base current, i_C and v_{CE} can be taken from the intersection of the characteristic appropriate to the chosen base current and the load line.

For a linear response from an amplifier, it is necessary to operate on that part of the load line which meets the family of characteristics at regular intervals. It can be clearly seen that a value greater than A would incur a

Fig. 22.18   Load line for a bipolar transistor amplifier

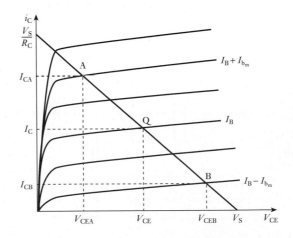

much smaller interval than the others of lesser value, all of which are reasonably equal. Here B is the lowest acceptable point of operation on the load line and Q is more or less half-way between A and B. The nearer it is to being exactly half-way then the better will be the linear amplification.

Here A, Q and B correspond to the maximum, mean and minimum values of the base current, which are $I_B + I_{b_m}$, I_B and $I_B - I_{b_m}$ respectively. The corresponding values of the collector current are I_{CA}, I_C and I_{CB} respectively, and the collector–emitter voltages are V_{CEA}, V_{CE} and V_{CEB} respectively.

The current gain

$$G_i = \frac{\Delta I_o}{\Delta I_i} = \frac{I_{CA} - I_{CB}}{2I_{b_m}}$$

[22.8]

The voltage gain

$$G_v = \frac{\Delta V_o}{\Delta V_i}$$

[22.9]

and the power gain

$$G_p = \frac{P_o}{P_i}$$

[22.10]

where P_o is the signal power in the load and P_i is the signal power into the transistor.

In practice, the voltage gain can only be determined accurately if we know the change in input voltage that has brought about the change in base current. However, little error is introduced if the collector–emitter voltage is assumed constant when calculating the base–emitter voltage producing the base current. This simplification is often taken a stage further by assuming a linear input characteristic, which leads to an equivalent circuit similar to that considered in Fig. 19.8.

Care should be taken in interpreting relation [22.10]. The powers P_i and P_o are those related to the signal frequency and they exclude the powers associated with the direct conditions, i.e. the quiescent conditions. For instance, the signal power in the load is the product of the r.m.s. signal load voltage and the r.m.s. signal load current.

It follows that

$$G_p = G_v G_i$$

[22.11]

Using the load line shown in Fig. 22.18, then the characteristic relating i_c to i_b can be derived, as shown in Fig. 22.19. This characteristic is known as the dynamic characteristic for the given values of R_C and V_S. Provided that this characteristic is linear over the operating range then we can be certain that the waveform of i_c will be identical to that of i_b. This does not imply that there will be no distortion within a transistor arrangement, because the input characteristics of the transistor are not themselves linear. It follows that the signal base current is not an exact replica of the applied signal, but the consequent distortion is generally quite small provided that the source resistance is considerably larger than the input resistance. A linear dynamic

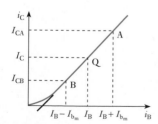

Fig. 22.19 Dynamic characteristic

characteristic is usually a good indicator that the transistor amplifier will cause little distortion.

In considering the circuit shown in Fig. 22.17 we took R_C to be the effective load, there being no other load connected across the output terminals of the amplifier. If a resistor R_L was connected across the output terminals, then the total load presented to the transistor amplifier would effectively be R_C and R_L in parallel (assuming the reactance of C_2 to be negligible). The effective load is therefore

$$R_P = \frac{R_C \cdot R_L}{R_C + R_L}$$

and the amplifier performance can be determined by drawing a load line for R_P and passing through the quiescent point on the characteristics. Note that the quiescent point Q is the same for both the load line associated with R_C and the load line associated with R_P because in both cases it refers to the zero-signal condition. The load line associated with R_C is termed the *d.c. load line*, while that associated with R_P is the *a.c. load line*.

The use of the a.c. load line is better illustrated by means of the following example.

Example 22.1

A bipolar transistor amplifier stage is shown in Fig. 22.20 and the transistor has characteristics, shown in Fig. 22.21, which may be considered linear over the working range. A 2.2 kΩ resistive load is connected across the output terminals and a signal source of sinusoidal e.m.f. 0.6 V peak and internal resistance 10 kΩ is connected to the input terminals. The input resistance of the transistor is effectively constant at 2.7 kΩ. Determine the current, voltage and power gains of the stage. The reactances of the coupling capacitors C_1 and C_2 may be considered negligible.

First determine the extremities of the d.c. load line. If

$$i_c = 0 \quad \text{then} \quad V_s = 12 \text{ V} = v_{ce}$$

If $\quad v_{ce} = 0 \quad$ then $\quad i_c = \frac{V_s}{R_c} = \frac{12}{1.8 \times 10^3} \equiv 6.7 \text{ mA}$

For the a.c. load line,

$$R_p = \frac{1.8 \times 2.2}{1.8 + 2.2} = 1.0 \text{ k}\Omega$$

$$= \frac{\Delta v}{\Delta I}$$

Fig. 22.20 Bipolar transistor amplifier stage for Example 22.1

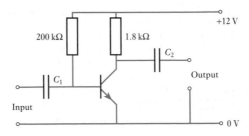

Fig. 22.21 Bipolar transistor characteristics for Example 22.1

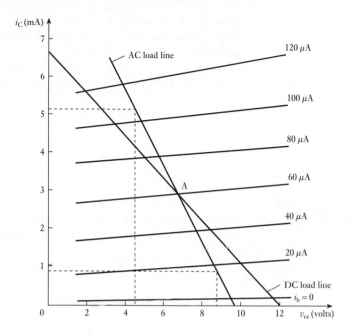

Hence the slope of the a.c. load line is

$$-\frac{1000}{1} \equiv -1.0 \text{ mA/V}$$

and the quiescent base current

$$I_{B_o} = \frac{12}{200 \times 10^3} \equiv 60 \ \mu\text{A}$$

The a.c. load line is therefore drawn with a slope of -1.0 mA/V through the quiescent point Q, which is given by the intersection of the 60 μA characteristic and the d.c. load line.

The input circuit consists of the base–emitter junction in parallel with the bias resistor, i.e. the signal passes through both in parallel. However, the transistor input resistance is 2.7 kΩ hence the shunting effect of 200 kΩ is negligible, and effectively the entire input circuit can be represented by Fig. 22.22.

The peak signal base current is

$$\frac{0.6}{(10 + 2.7) \times 10^3} \equiv 47 \ \mu\text{A}$$

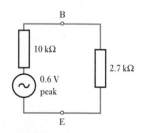

Fig. 22.22 Equivalent input circuit for Example 22.1

and hence the maximum signal base current is

$$60 + 47 = 107 \ \mu\text{A}$$

and the minimum signal base current is

$$60 - 47 = 13 \ \mu\text{A}$$

From the a.c. load line in Fig. 22.21,

$$\Delta i_c = 5.1 - 0.9 = 4.22 \text{ mA}$$

This change in collector current is shared between the $1.8\ k\Omega$ collector resistor and the load resistor of $2.2\ k\Omega$. Hence the change in output current is

$$\Delta i_o = 4.2 \times \frac{1.8 \times 10^3}{(1.8 + 2.2) \times 10^3} = 1.9\ \text{mA}$$

The change in input current is

$$\Delta i_i = 107 - 13 = 94\ \mu\text{A}$$

Therefore the current gain for the amplifier stage is

$$G_i = \frac{\Delta i_o}{\Delta i_i} = \frac{1.9 \times 10^{-3}}{94 \times 10^{-6}} = \mathbf{20}$$

Note that this is the current gain of the amplifier and should not be confused with the current gain of the transistor, which is given by

$$G_i = \frac{\Delta i_c}{\Delta i_b} = \frac{4.2 \times 10^{-3}}{94 \times 10^{-6}} = 45$$

The change in output voltage Δv_o is given by the change in collector–emitter voltage, i.e.

$$\Delta v_o = v_{ce} = 8.5 - 4.3 = 4.2\ \text{V (peak to peak)}$$

The change in input voltage Δv_i is given by the change in base–emitter voltage, i.e.

$$\Delta v_i = \Delta i_i \times R_i = 39 \times 10^{-6} \times 2.7 \times 10^3 = 0.25\ \text{V (peak to peak)}$$

$$\therefore \qquad G_v = \frac{\Delta v_o}{\Delta v_i} = \frac{4.2}{0.25} = \mathbf{17}$$

$$G_p = G_v G_i = 17 \times 20 = \mathbf{340}$$

Alternatively,

$$\text{RMS output voltage} = \frac{4.2}{2\sqrt{2}} = 1.5\ \text{V}$$

$$\therefore \qquad \text{RMS output current} = \frac{1.5}{2.2 \times 10^3} = 0.68\ \text{mA}$$

and
$$P_o = (0.68 \times 10^{-3})^2 \times 2.2 \times 10^3$$
$$\equiv 1.0\ \text{mW}$$

$$\text{RMS input voltage} = \frac{0.25}{2\sqrt{2}} \equiv 88\ \text{mV}$$

$$\therefore \qquad \text{RMS input current} = \frac{88 \times 10^{-3}}{2.7 \times 10^3} \equiv 33\ \mu\text{A}$$

and
$$P_i = (33 \times 10^{-6})^2 \times 2.7 \times 10^3$$
$$\equiv 2.9\ \mu\text{W}$$

$$\therefore \qquad G_p = \frac{1.0 \times 10^{-3}}{2.9 \times 10^{-6}} = \mathbf{340}, \text{ as before}$$

22.10 Circuit component selection

The simple transistor circuit shown in Fig. 22.17 was translated into an equivalent circuit for the purpose of Example 22.1. However, what would we have done if simply asked to make an amplifier from a given transistor? For instance, given the transistor, the characteristics of which were shown in Fig. 22.21, how were the resistances of the other circuit components selected?

Normally, given the transistor characteristics, it is the practice first to select a quiescent point. This requires selecting a curve somewhere in the middle of the family of characteristics, which in Fig. 22.21 immediately suggests that we should take the curve for $i_b = 60 \ \mu A$.

In order to obtain such a value, we require to forward-bias the base–emitter junction, i.e. being an n–p–n transistor, the base should be positive with respect to the emitter. This is achieved by R_B connected between the positive supply V_S and the base. If we neglect the base–emitter voltage drop, then

$$R_B = \frac{V_s}{I_b} = \frac{12}{60} \times 10^{-6} = 200 \ k\Omega$$

To take the base–emitter voltage as negligible might be rather optimistic since generally it is in the order of 1 V. Even so, the voltage across the bias resistor would be 11 V in which case R_B would require a value of about 183 kΩ.

Having determined R_B, it is also possible to determine an approximate value for R_C. The quiescent point normally takes a value of about half the supply voltage V_S, which in this case suggests a quiescent collector–emitter voltage of about 6 V. From the characteristics, the corresponding collector current is 2.9 mA. If v_{CE} is 6 V then the voltage across R_C is 6 V and the current in R_C is 2.9 mA, hence $R_C = 6/2.9 \times 10^{-3} = 2 \ k\Omega$.

In Example 22.1, it happens that the resistance of R_C was 1.8 kΩ, but a value of 2.2 kΩ, which is reasonably similar in magnitude, would be as acceptable.

This description of the determination of R_B and R_C seems rather off-hand in the manner with which the values have been accepted. Having sought high correlation of input and output waveforms and a linear relationship, it subsequently seems strange that 1.8 kΩ can be taken as much the same as 2 kΩ. However, the circuit components of many electronic circuits are manufactured to a tolerance of ±10 per cent; thus the approximations made above would lie within such limits. The variations in practice result in considerable ranges in performance between supposedly identical amplifiers, although most tend to have the same performance.

When the source of the input alternating signal is connected to the amplifier, consideration has to be given to the problem that the source normally has a low internal resistance. If it were not for the coupling capacitor C_1, some of the current through the bias resistor would be diverted away from the base–emitter junction and thus the bias conditions would be affected. However, the coupling capacitor does not permit any of the direct bias current to pass through the source. Provided the capacitance of C_1 is so chosen that the reactance $1/2\pi f C_1$ is small compared with the input resistance of the transistor, then its effect on the alternating signal is negligible. In this case we would want the reactance to be about 10 per cent of the input resistance in magnitude.

Finally, it should not be thought that the simple bias arrangement considered is the only one used. Among the many other arrangements found in practice, the bias can be effected by a potential divider across the supply or a resistor connected in parallel between the emitter and the 0 V line.

22.11   Equivalent circuits of a transistor

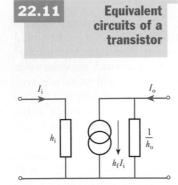

Fig. 22.23 *h*-parameters for a transistor equivalent circuit

So long as a transistor is operated over the linear portions of its characteristics, the actual values of the bias voltages are of no consequence, hence these voltages can be omitted from an equivalent circuit used to calculate the current, voltage and power gains. Instead, the values of the voltages and currents shown on an equivalent circuit diagram refer only to the r.m.s. values of the alternating components of these quantities.

Since the transistor itself is an amplifier, it can be represented by an equivalent circuit of the form shown in Fig. 19.8. In order to indicate that the circuit components and gain ratios apply specifically to the transistor, the values are specified as *h*-parameters as shown in Fig. 22.23.

Here h_i is equivalent to the general input resistance R_i, h_o is equivalent to the general output resistance R_o except that h_o is expressed as a conductance in siemens rather than as a resistance in ohms, and h_f is the current gain of the transistor. Because there is a variety of connections for the transistor, we add a letter to the parameter subscript to indicate the mode of operation, thus h_{ie} is the input resistance of a transistor connected in the common-emitter mode and h_{ib} is the input resistance in the common-base mode.

Consider now the transistor in the amplifier arrangement shown in Fig. 22.24(a). The equivalent circuit for the amplifier is shown in Fig. 22.24(b) and here again the bias resistor is seen to shunt the input to the transistor while the output consists of the load resistor shunted by the collector resistor.

In converting the actual circuit arrangement to the equivalent circuit, it will also be noted that their capacitors have been omitted, it being assumed that their reactances are negligible. In both diagrams, the source has been taken as a voltage source of E_S in series with its source resistance R_S. This contrasts with the equivalent circuit of the transistor which has a current generator, but it has to be remembered that the action of the transistor depends on the input current and therefore it is essentially a current-operated device.

Fig. 22.24 Common-emitter amplifier stage and its equivalent circuit

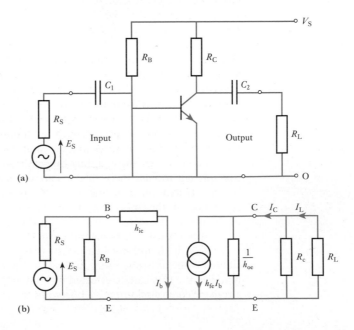

From Fig. 22.24(b), for the transistor

$$G_i = \frac{I_c}{I_b}$$

and

$$I_c = \frac{1/h_{oe}}{1/h_{oe} + R_p} \times h_{fe} I_b$$

where

$$R_p = \frac{R_c R_L}{R_c + R_L}$$

$$\therefore \qquad G_i = \frac{h_{fe}}{1 + h_{oe} R_p} \qquad\qquad\qquad [22.12]$$

This is the current gain of the transistor but not of the amplifier stage. The stage gain is given by

$$G_i' = \frac{I_L}{I_b}$$

and

$$I_L = \frac{R_c}{R_c + R_L} I_c$$

$$\therefore \qquad G_i = \frac{h_{fe}}{1 + h_{oe} R_p} \cdot \frac{R_c}{R_c + R_L} \qquad\qquad [22.13]$$

The voltage gain for the transistor is given by

$$G_v = \frac{V_{ce}}{V_{be}} = \frac{-I_c R_p}{I_b h_{ie}}$$

The minus sign is a consequence of the direction of the collector current.
From relation [22.12]

$$G_v = \frac{-h_{fe} R_p}{h_{ie}(1 + h_{oe} R_p)} \qquad\qquad\qquad [22.14]$$

Since the voltage that appears across the transistor's collector and emitter terminals is the same as that which appears across the load, the voltage gain expression applies both to the transistor and to the amplifier stage. Again the minus sign results from the current directions and also this implies that there is a 180° phase shift between input and output, as shown in Fig. 22.25.

Having examined the approach to analysing the performance of the transistor and also of the amplifier stage, we should now consider a numerical example in which we will see that it is easier to calculate the performance from first principles rather than recall the analyses which resulted in the last three relations.

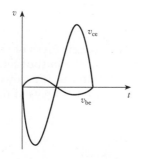

Fig. 22.25 Common-emitter transistor input and output voltage waveforms

Example 22.2

A transistor amplifier stage comprises a transistor of parameters $h_{ie} = 800\ \Omega$, $h_{fe} = 50$ and $h_{oe} = 20\ \mu S$, and bias components and coupling capacitors of negligible effect. The input signal consists of an e.m.f. of 60 mV from a source of internal resistance 2.2 kΩ, and the total load on the stage output is 4 kΩ. Determine the current, voltage and power gains of the amplifier stage.

Fig. 22.26 Amplifier equivalent circuit for Example 22.2

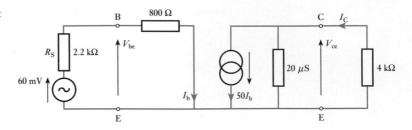

The equivalent circuit is shown in Fig. 22.26.

$$I_b = \frac{60 \times 10^{-3}}{2200 + 800} = 20 \times 10^{-6} \text{ A}$$

$$\frac{1}{h_{oe}} = \frac{1}{20 \times 10^{-6}} \equiv 50 \text{ k}\Omega$$

$$I_c = \frac{50 \times 10^3}{(50 + 4) \times 10^3} \times 50 \times 20 \times 10^{-6} = 926 \times 10^{-6} \text{ A}$$

$$G_i = \frac{926 \times 10^{-6}}{20 \times 10^{-6}} = \mathbf{46.3}$$

$$G_v = \frac{-926 \times 10^{-6} \times 4 \times 10^3}{20 \times 10^{-6} \times 800} = \mathbf{231.5}$$

$$G_p = 46.3 \times 231.5 = \mathbf{10\,720}$$

Example 22.3   A transistor amplifier is shown in Fig. 22.27. The parameters of the transistor are $h_{ie} = 2$ kΩ, $h_{oe} = 25$ μS and $h_{fe} = 55$, and the output load resistor dissipates a signal power of 10 mW. Determine the power gain of the stage and the input signal e.m.f. E. The reactances of the capacitors may be neglected.

Fig. 22.27 Amplifier for Example 22.3

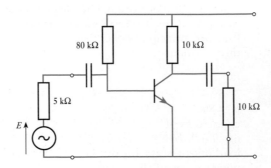

Fig. 22.28 Equivalent circuit for Example 22.3

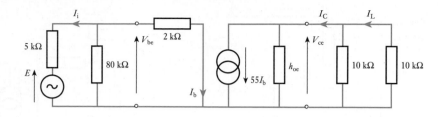

The equivalent circuit is shown in Fig. 22.28.

$$\frac{1}{h_{oe}} = \frac{1}{25 \times 10^{-6}} \equiv 40 \text{ k}\Omega$$

$$P = 10 \times 10^{-3} \text{ W} = \frac{V_{ce}^{2}}{10 \times 10^{3}}$$

$$\therefore \qquad V_{ce} = 10 \text{ V}$$

$$\therefore \qquad 55I_{b} = \frac{10}{40 \times 10^{3}} + \frac{10}{10 \times 10^{3}} + \frac{10}{10 \times 10^{3}}$$

$$\therefore \qquad I_{b} = 40.9 \times 10^{-6} \text{ A}$$

$$\therefore \qquad I_{i} = 40.9 \times 10^{-6} + 40.9 \times 10^{-6} \cdot \frac{2 \times 10^{3}}{80 \times 10^{3}}$$

$$= 41.9 \times 10^{-6} \text{ A}$$

$$E = (41.9 \times 10^{-6} \times 5 \times 10^{3}) + (40.9 \times 10^{-6} \times 2 \times 10^{3})$$

$$= \mathbf{0.29 \text{ V}}$$

$$P_{i} = V_{be}I_{i} = (40.9 \times 10^{-6} \times 2 \times 10^{3}) \times (41.9 \times 10^{-6})$$

$$= 3.43 \times 10^{-6} \text{ W}$$

$$P_{o} = 10 \times 10^{-3} \text{ W}$$

$$G_{p} = \frac{10 \times 10^{-3}}{3.43 \times 10^{-6}} = \mathbf{2920}$$

Before turning our attention to the common-base transistor amplifier, it is worth noting that

$$h_{fe} = \beta \qquad\qquad [22.15]$$

where β is the current amplification factor for a common-emitter circuit as defined in relation [22.2].

The circuit of a common-base transistor amplifier and the corresponding equivalent circuit diagram are shown in Fig. 22.29. The expressions for the gains are identical in form to those derived for the common-emitter

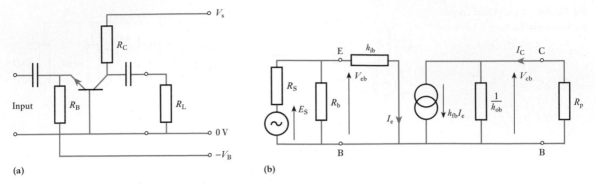

Fig. 22.29 Common-base amplifier stage and its equivalent circuit

arrangement except that the appropriate parameters have to be applied, thus

$$G_i = \frac{I_c}{I_e}$$

$$G_i = \frac{h_{fb}}{1 + h_{ob}R_p} \qquad [22.16]$$

$$G_v = \frac{V_{cb}}{V_{eb}}$$

$$G_v = \frac{-h_{fb}R_p}{h_{ib}(1 + h_{ob}R_p)} \qquad [22.17]$$

$$G_p = G_v G_i$$

Here h_{fb} ($= \alpha$) normally has a value just under unity and is negative, since an increase in the emitter current produces an increase in the collector current which is in the opposite direction, i.e. there is a 180° phase shift between I_e and I_c. There is no phase shift between V_{eb} and V_{cb}.

22.12 Hybrid parameters

For any bipolar junction transistor, there are six possible variables, these being i_b, i_c, i_e, v_{be}, v_{ce} and v_{cb}. However,

$$i_b + i_c = i_e$$

and $$v_{cb} + v_{be} = v_{ce}$$

From these relations, the operation of a transistor can be predicted provided we have the characteristics specifying the relationships between two of the voltages and two of the currents.

It is most convenient to consider input and output quantities. At this point let us restrict our consideration to a common-emitter transistor, in which case the input and output quantities are v_{be}, i_b, v_{ce} and i_c. In general mathematical terms, these relations can be expressed in the forms

$$v_{be} = f_1(i_b, v_{ce}) \qquad [22.18]$$

$$i_c = f_2(i_b, v_{ce}) \qquad [22.19]$$

Fig. 22.30 *h*-parameter equivalent circuit for a common-emitter transistor

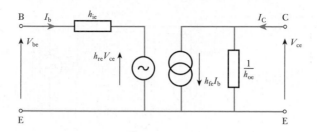

Provided that all the variable quantities are sinusoidal in nature, then these expressions can be expanded as

$$V_{be} = h_{ie}I_{b} + h_{re}V_{ce}$$ [22.20]

$$i_{c} = h_{fe}I_{b} + h_{oe}V_{ce}$$ [22.21]

These parameters are the small-signal hybrid parameters which are generally termed the *h*-parameters. Note that the signals are assumed small in order to ensure that the operation remains within the linear sections of the transistor characteristics.

For the common-emitter transistor:

h_{ie} is the short-circuit input resistance (or impedance)
h_{re} is the open-circuit reverse voltage ratio
h_{fe} is the short-circuit forward current ratio
h_{oe} is the open-circuit output conductance (or admittance)

We have been introduced to all of these parameters in Fig. 22.23 with the exception of the open-circuit reverse voltage ratio h_{re}. When introduced into the full *h*-parameter equivalent circuit, the arrangement becomes that shown in Fig. 22.30.

The voltage generator $h_{re}V_{ce}$ represents a feedback effect which is due to the output signal voltage controlling, to a very limited extent, the input signal current. Owing to the normal voltage gain of a transistor being quite large, it might be expected that as a consequence the feedback effect could be appreciable, but h_{re} is always very small (less than 1×10^{-3}) with the result that the feedback voltage generator can be omitted from the equivalent circuit without introducing any important error. Again we may recall that transistor circuit components are liable to considerable variation in their actual values compared with their anticipated values, and, within the expected range of tolerances, the omission of the feedback generator will have a negligible effect on any circuit analysis. It follows that the analyses undertaken in section 22.11 correspond satisfactorily to the full *h*-parameter equivalent circuit.

Similar equivalent circuits can be used for the common-base and common-collector configurations for a junction transistor.

22.13 **Limitations to the bipolar junction transistor**

For most bipolar transistors, the input resistance, as given by h_{ie}, tends to be about 1 kΩ, which for electronic circuits tends to be rather a small value. It follows that there is always a significant input current which has two effects: (1) the amplifier gain is somewhat limited; (2) the lower levels increase the size and rating of circuit components more than might be desirable.

Fig. 22.31 Effect of input resistance on gain

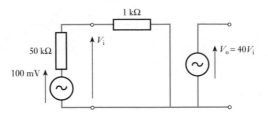

In order to understand the limitation of gain, consider an amplifier with a gain of 40 and an input resistance of 1 kΩ. A source, of e.m.f. 100 mV and internal resistance 50 kΩ, is connected to the input. The circuit is shown in Fig. 22.31. The input voltage is

$$V_i = \frac{1 \times 10^3 \times 100 \times 10^{-3}}{(50 + 1) \times 10^3} \equiv 2 \text{ mV}$$

and the output voltage

$$V_o = 40 \times 2 = 80 \text{ mV}$$

The result is that, in spite of the amplification, the output voltage is less than the input e.m.f. – this is due to the low input resistance when compared with the source internal resistance. Now consider the situation had the input resistance been greater, say 1 MΩ, in which case

$$V_i = \frac{1 \times 10^6 \times 100 \times 10^{-3}}{(50 + 1000) \times 10^3} = 95.2 \text{ mV}$$

and $V_o = 40 \times 95.2 \times 10^{-3} = 3.8 \text{ V}$

It can therefore be seen that the increase of input resistance has increased the gain by more than 47 times. Also, because the input resistance is much higher, the input current is substantially reduced and therefore the size and rating of the input circuit components can be reduced.

In order to obtain the increase in input resistance, we have to look for another type of transistor – the field effect transistor, generally known as the FET.

22.14	Stabilizing voltage supplies

Most electronic equipment requires a constant supply voltage. If the supply voltage varies, the output will not directly respond to the input. It is therefore necessary to provide some means of holding the supply voltage at a reasonably steady value; it is not necessary to ensure that the supply voltage is absolutely constant, only reasonably constant.

The network which provides this steady value is called a stabilization circuit. It maintains a constant output voltage when:

1. the supply voltage varies, and/or
2. the load current changes.

Assuming the supply voltage is derived from a rectified a.c. source, the stabilization circuit is usually inserted between the rectifier with the associated smoothing circuits and the load as shown in Fig. 22.32.

Fig. 22.32 Stability factor

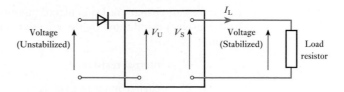

In order to indicate how well a stabilization circuit performs, we use a factor termed the *stability factor*. It assumes that the load requires a constant current, in which case

S = stability factor

$$S = \frac{\text{change in stabilized voltage across load}}{\text{change in unstabilized voltage}}$$ [22.22]

For satisfactory stabilization, S should be small. Effectively the smaller the better.

Example 22.4

Without stabilization, a supply voltage can change from 110 to 80 V even though the load current remains constant. With stabilization effected by a circuit which has a stability factor of 0.06, determine the change in the stabilized voltage.

$$S = \frac{\text{change in stabilized voltage across load}}{\text{change in unstabilized voltage}}$$

$\therefore \qquad 0.06 = \dfrac{\Delta V_s}{110 - 80}$

$\therefore \qquad \Delta V_s = 0.06 \times 30 = 1.8 \text{ V}$

This represents a decrease from the maximum value.

One of the most simple stabilization circuits consists of a resistor in series with the load as well as a Zener diode in parallel with the load. This is shown in Fig. 22.33. The particular characteristics of a Zener diode are that the voltage across it remains nearly constant providing that:

1. the load current remains between set limits;
2. the unstabilized voltage remains between set limits.

Such a simple stabilization circuit cannot handle wide variations of either variable, hence the set limits may not be acceptably far apart.

The stability factor indicates effectiveness provided the load current remains constant. The effectiveness of a stabilization circuit when

Fig. 22.33 A simple stabilization circuit

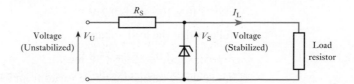

responding to changes of load current (assuming the unstabilized voltage is held constant) is indicated by the effective internal resistance of the supply circuit.

$$R_o = \text{internal resistance}$$

$$= \frac{\text{change in output voltage}}{\text{change in output current}}$$

$$R_o = \frac{\Delta V_s}{\Delta I_L}$$

[22.23]

The change in output current should be taken over the range between the set limits. For instance, let us consider a stabilization circuit in which the load current ranges from 0 to 40 mA. At the same time the input supply voltage remains constant, but the voltage applied to the load falls by 0.25 V as the load increases. If follows that

$$R_o = \frac{0.25}{0.04}$$

$$= 6.25 \ \Omega$$

As with the stabilization factor, the internal resistance should be as small as possible.

Example 22.5

A stabilization circuit has a stability factor of 0.04 and an internal resistance of 5 Ω. The unstabilized voltage can vary between 75 and 100 V, and the load current can vary from 40 to 80 mA. Determine the maximum and minimum values for the stabilized load voltage.

The extreme conditions are low unstabilized voltage with high load current, and high unstabilized voltage with low load current.

In the first case

$$\text{Load voltage} = (\text{lower limit of unstabilized voltage}) - I_1 R_o$$

$$= 75 - (80 \times 10^{-3} \times 5)$$

$$= \textbf{74.6 V}$$

In the second case the unstabilized voltage rises to 100 V and the stabilization circuit allows only a small increase in the load voltage. Change in stabilized voltage is given by

$$S \times \text{change in unstabilized voltage}$$

$$= 0.04 \times (100 - 75) = 1.0 \text{ V}$$

There is a drop in the load voltage due to the internal resistance given by

$$40 \times 10^{-3} \times 5 = 0.2 \text{ V}$$

It follows that the load voltage is

$$75 + 1.0 - 0.2 = \textbf{75.8 V}$$

The stabilized voltage therefore varies from 74.6 to 75.8 V.

Fig. 22.34 A series voltage stabilizer

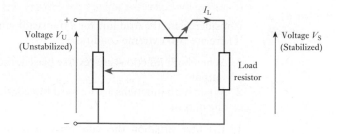

Fig. 22.35 Stabilization circuit incorporating a comparator network

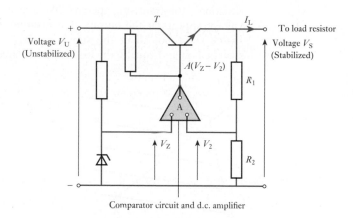

Comparator circuit and d.c. amplifier

A stabilizer which can handle better ranges of variation is effected by placing a transistor in series with the load resistor. The arrangement is shown in Fig. 22.34. The current through the transistor is varied by altering the voltage applied to the base. This in turn is controlled by varying the potentiometer.

The potentiometer is cumbersome and, in more practical arrangements, a comparator network is used in which the output voltage is proportional to the difference between two input voltages, one derived from the unstabilized source voltage and the other from the output voltage to the load. The arrangement is shown in Fig. 22.35.

A reference voltage V is obtained from the unstabilized supply voltage. It is simply derived from a Zener diode as shown. The output voltage is applied to a potentiometer consisting of R_1 and R_2. The voltage across R_2 is V_2 and should be more or less equal to V_Z. The difference between V_2 and V_Z is detected by the comparator network and amplified by a d.c. amplifier. The amplifier output is applied to the base of the transistor.

If the voltage comparison shows that the voltages V_2 and V_Z are equal there is no output from the amplifier, hence there is no control to the transistor and the collector current remains unchanged. Should there be a difference between the voltages V_2 and V_Z, this is amplified and changes the bias condition of the transistor. This either causes the emitter current to increase or to decrease resulting in a change in the output voltage. This change causes V_2 to change until it equals V_Z. At this point the voltages are equal and no further change occurs.

This arrangement can produce very low stabilization factors as well as low internal resistances.

22.15 Transistor as a switch

Consider again the load line for a transistor amplifier as shown in Fig. 22.36. There are two extreme conditions:

1. When both junctions are reverse biased, there is a cut-off condition at the output.
2. When both junctions are forward biased, there is a saturation condition at the output.

In the first situation this conforms to point A on the load line, and the second situation conforms to point B on the load line. In these extreme situations, the transistor operates like a switch in which either the supply voltage appears at the output or it does not. However, it is not a perfect switch since not all of the supply voltage is available to the load when switched on and equally the load is not isolated in the OFF condition.

Fig. 22.36 Transistor as a switch

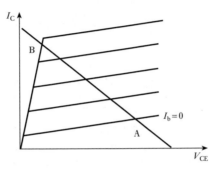

For all its imperfections, the transistor switch is useful because it operates effectively in the transmission of digital signals and also has a quick switching time, thereby responding to very high bit rates.

Summary of important formulae

For a common-base transistor

$$\alpha = \frac{\Delta I_C}{\Delta I_E}\bigg|_{V_{CB}\ \text{constant}}$$

[22.1]

For a common-emitter transistor

$$\beta = \frac{\Delta I_C}{\Delta I_B}\bigg|_{V_{CE}\ \text{constant}}$$

[22.2]

$$\alpha = \frac{\beta}{1+\beta}$$

[22.3]

$$\beta = \frac{\alpha}{1-\alpha}$$

[22.4]

$$i_c = -\frac{1}{R_c}\cdot v_{ce} + \frac{V_s}{R_c}$$

[22.7]

Summary of important formulae continued

For an equivalent circuit with h-parameters, and for common emitter

$$G_i = \frac{h_{fe}}{1 + h_{oe}R_p}$$ [22.12]

$$G_v = \frac{-h_{fe}R_p}{h_{ie}(1 + h_{oe}R_p)}$$ [22.14]

$$h_{fe} = \beta$$ [22.15]

For an equivalent circuit with h-parameters, and for common base

$$G_i = \frac{h_{fb}}{1 + h_{ob}R_p}$$ [22.16]

$$G_v = \frac{-h_{fb}R_p}{h_{ib}(1 + h_{ob}R_p)}$$ [22.17]

Terms and concepts

The **bipolar junction transistor** has three layers – the **emitter**, the **base** and the **collector**. It is connected to two circuits with any one layer common.

The input circuit obtains its power from the signal superimposed on a current supplied from a power source. The output circuit is entirely provided with power from the power source, hence the ability of the transistor to amplify.

The amplified signal can be removed by means of a capacitor which blocks the d.c. power source.

The transistor is essentially a current-operated device and the equivalent circuits are often based on **constant-current generators**.

Consistent operation depends on the stability of the operating conditions. It is common practice to make **supply stabilizers** incorporating transistors.

Transistors are commonly used as switches.

Exercises 22

1. Explain the function of an equivalent circuit.
2. Explain the relationship between h_{ie} and a suitable device characteristic.
3. Define the term h_{ie}.
4. Define the term h_{fe}.
5. Explain the relationship between h_{fe} and a suitable device characteristic.
6. Define the term h_{oe}.
7. Explain the relationship between h_{oe} and a suitable device characteristic.

8. Draw the equivalent circuit for a common-emitter bipolar transistor amplifier and derive suitable formulae for the amplifier current gain, voltage gain and power gain. Neglect bias, decoupling and coupling components.
9. Indicate suitable values of the components employed in the circuit of the previous question if $V_{cc} = 15$ V, $V_{be} = 0.8$ V, $I_b = 200 \mu A$, $I_c = 5$ mA, $V_{ce} = 5$ V.
10. In the simplified amplifier network shown in Fig. A the hybrid parameters of the transistor are $h_{ie} = 650 \Omega$,

Exercises 22 continued

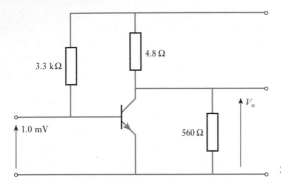

Fig. A

$h_{fe} = 56$ and $h_{oe} = 100\ \mu S$. The input sinusoidal signal is 1.0 mV r.m.s.

(a) Draw the small-signal equivalent network.
(b) Hence calculate: (i) the current in the 560 Ω output resistor; (ii) the voltage gain (in dB); (iii) the current gain (in dB).

11. An amplifier has the following parameters: input resistance, 470 kΩ; output resistance, 8.2 kΩ; open-circuit voltage gain 50. A signal source having an e.m.f. of 75 μV and negligible internal resistance is connected to the input of the amplifier. The output is connected to a 10 kΩ resistive load. Draw the equivalent circuit and hence determine: (a) the output voltage developed across the load; (b) the current gain; (c) the voltage gain.

12. The equivalent circuit of an amplifier is shown in Fig. B. When the load resistance is $R_L = 175\ \Omega$, the voltage gain of the amplifier $A_v = 4375$. Calculate the amplifier output resistance R_o. When the value of R_L is changed to 275 Ω, the current gain of the amplifier is $A_i = 16.7 \times 10^4$. Calculate the value of the amplifier input resistance R_i.

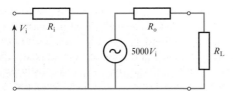

Fig. B

13. The collector characteristics of a p–n–p transistor can be considered as three straight lines through the following points:

(a) $I_b = 300\ \mu A$ ($I_c = 8$ mA, $V_{ce} = 2$ V and $I_c = 9.5$ mA, $V_{ce} = 14$ V)

(b) $I_b = 150\ \mu A$ ($I_c = 4.5$ mA, $V_{ce} = 2$ V and $I_c = 5.3$ mA, $V_{ce} = 14$ V)
(c) $I_b = 0\ \mu A$ ($I_c = 1.5$ mA, $V_{ce} = 2$ V and $I_c = 2$ mA, $V_{ce} = 14$ V)

The supply voltage is 16 V. Draw a load line for a load resistance of 1.5 kΩ and hence find: (i) the collector current and voltage if the base bias current is 150 μA; (ii) the r.m.s. output voltage if the input to the base causes a sinusoidal variation of base current between 0 and 300 μA; (iii) the current gain.

14. An n–p–n transistor, the characteristics of which can be considered linear between the limits shown in the table, is used in an amplifier circuit. A 2.0 kΩ resistor is connected between its collector and the positive terminal of the 9 V d.c. supply and its emitter is connected directly to the negative terminal. Given that the quiescent base current is 40 μA determine the quiescent collector-to-emitter voltage and the quiescent collector current.

If, when a signal is applied to the circuit, the base current varies sinusoidally with time with a peak alternating component of 20 μA, determine the alternating component of the collector current and hence the current gain of the stage. The load across the output terminals of the circuit can be considered very high in comparison with 2.0 kΩ.

$I_b\,(\mu A)$	60		40		20	
$V_{ce}\,(V)$	1	10	1	10	1	10
$I_c\,(mA)$	2.7	3.2	1.8	2.1	0.9	1.1

15. The transistor used in the circuit shown in Fig. C has characteristics which can be considered linear between

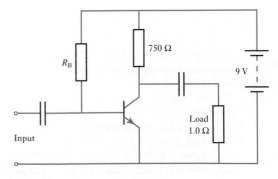

Fig. C

Exercises 22 continued

I_b (μA)	120		100		80		60		40	
V_{cc} (V)	1	12	1	12	1	12	1	12	1	12
I_c (mA)	10.8	13.4	9.0	11.2	7.2	9.0	5.3	6.7	3.7	4.5

the limits shown in the table. Determine the value of R_B to give a quiescent base current of 80 μA.

If the signal base current varies sinusoidally with time and has a peak value of 40 μA determine the r.m.s. value of the signal voltage across the load and hence the signal power in the load. The reactances of the coupling capacitors can be considered zero.

16. The output characteristics, which can be considered linear, for the n–p–n silicon transistor used in the amplifier circuit shown in Fig. D are specified in the table below. The source of signal can be represented by a constant-current generator of 24 μA peak and internal resistance 3.0 kΩ. The stage feeds an identical one. Determine the current, voltage and power gains of the stage. The input resistances of the transistors can be considered constant at 6.0 kΩ and the reactances of the coupling capacitors are negligible.

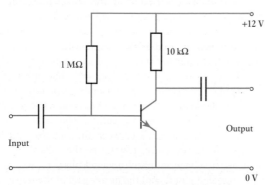

Fig. D

I_b (μA)	0		4		8		12		16		20	
V_{cc} (V)	1	10	1	10	1	10	1	10	1	10	1	10
I_c (mA)	0.0	0.0	0.14	0.18	0.33	0.39	0.49	0.61	0.68	0.83	0.86	1.04

17. A common-emitter, bipolar transistor has the following h parameters: $h_{ie} = 1.5$ kΩ, $h_{fe} = -60$ and $h_{oe} = 12.5$ μS. If the load resistance can vary between 5 kΩ and 10 kΩ, calculate the maximum and minimum values of the amplifier's (a) current gain; (b) voltage gain; (c) power gain.

18. The hybrid parameters for a transistor used in the common-emitter configuration are $h_{ie} = 1.5$ kΩ, $h_{fe} = 70$ and $h_{oe} = 100$ μS. The transistor has a load resistor of 1 kΩ in the collector and is supplied from a signal source of resistance 800 Ω. Calculate: (a) the current gain; (b) the voltage gain; (c) the power gain.

19. The small-signal hybrid parameters for a transistor used in an amplifier circuit are $h_{ie} = 2$ kΩ, $h_{fe} = 60$, $h_{oe} = 20$ μS and $h_{re} = 0$. The total collector-to-emitter load is 10 kΩ and the transistor is supplied from a signal source of e.m.f. 100 mV r.m.s. and internal resistance 3 kΩ. Determine the current, voltage and power gains for the stage and the signal power developed in the load.

20. The transistor in the circuit of Fig. E has the following hybrid parameters: $h_{ie} = 1$ kΩ, $h_{fe} = 50$, $h_{oe} = 100$ μS and $h_{re} = 0$. Draw the equivalent circuit, neglecting the effect of bias resistors and coupling capacitors, and obtain the magnitude of the current gain I_L/I_B and the voltage gain V_o/E_s. What is the power gain from source to load?

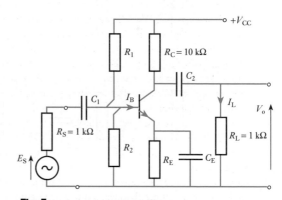

Fig. E

21. A transistor used in the common-emitter configuration has the following small signal parameters: $h_{ie} = 1.0$ kΩ, $h_{fe} = 49$ and $h_{oe} = 80$ μS. The source of signal has an e.m.f. of 10 mV and an internal resistance of 600 Ω. The load resistance is 4.7 kΩ. Estimate the voltage developed across the load and hence the power gain of the transistor, also the power gain in dB.

22. The n–p–n transistor used in the circuit shown in Fig. F has the following small signal parameters: $h_{ie} = 1.4$ kΩ, $h_{fe} = 50$ and $h_{oe} = 25$ μS. The input signal is supplied from a signal source of 30 mV and internal resistance 3 kΩ. Determine the output voltage at mid-band frequencies.

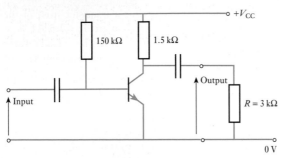

Fig. F

23. The common-emitter stage shown in Fig. G feeds an identical stage, and is fed from a source of constant e.m.f. of 100 mV. $h_{ie} = 2.0$ kΩ, $h_{fe} = 45$, $h_{oe} = 30$ μS. Calculate: (a) the input resistance of the stage; (b) the output resistance of the stage; (c) the input current to the next stage; (d) the input voltage to the next stage; (e) the power gain; (f) the power gain in dB.

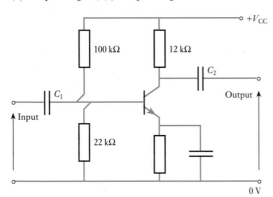

Fig. G

24. The transistor of the amplifier shown in Fig. H has an input resistance of 750 Ω, an output resistance

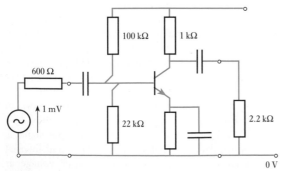

Fig. H

of 100 Ω and a short-circuit current gain of 12 000. Draw the mid-band equivalent circuit of the amplifier. Determine the current, voltage and power gains of the amplifier.

25. In the simplified transistor amplifier network shown in Fig. I, the hybrid parameters are $h_{ie} = 700$ Ω, $h_{fe} = 48$ and $h_{oe} = 80$ μS. The input sinusoidal signal current is 0.1 mA r.m.s. Draw the small-signal equivalent network and hence calculate: (a) the alternating component of the collector current; (b) the current in the 2 kΩ output resistor; (c) the output voltage; (d) the overall voltage amplification.

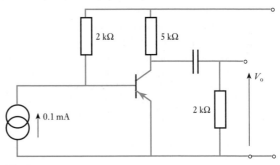

Fig. I

26. A transistor used in the common-base configuration has the following small-signal parameters: $h_{ib} = 75$ Ω, $h_{fb} = -0.95$, $h_{ob} = 0$ S and $h_{rb} = 0$. The effect of the applied sinusoidal signal can be represented by a source of e.m.f. of 30 mV and internal resistance 75 Ω connected between the emitter and base, and the total load connected between the collector and base is 5.0 kΩ resistance. Determine the collector-to-base signal voltage in magnitude and phase relative to the source of e.m.f., and the power gain of the transistor.

27. An amplifier has a voltage gain of 15 at 50 Hz. Its response at 1.0 kHz gives a voltage gain of 40. What is the relative response at 50 Hz in dB to its response at 1.0 kHz?

28. An amplifier has an open-circuit voltage gain of 600 and an output resistance of 15 kΩ and input resistance of 5.0 kΩ. It is supplied from a signal source of e.m.f. 10 mV and internal resistance 2.5 kΩ and it feeds a load of 7.5 kΩ. Determine the magnitude of the output signal voltage and the power gain in dB of the amplifier.

29. An amplifier has a short-circuit current gain of 100, an input resistance of 2.5 kΩ and an output resistance of 40 kΩ. It is supplied from a current generator signal

Exercises 22 continued

source of 12 μA in parallel with a 50 kΩ resistor. The amplifier load is 10 kΩ. Determine the current voltage and power gains in dB of the amplifier.

30. An amplifier operates with a load resistance of 2.0 kΩ. The input signal source is a generator of e.m.f. 50 mV and internal resistance 0.5 kΩ. The parameters of the amplifier are input resistance 800 Ω, output

conductance 80 μS, and short-circuit current gain 47. Draw the small-signal equivalent circuit and determine: (a) the voltage gain; (b) the current gain; (c) the power gain.

31. In the amplifier equivalent circuit shown in Fig. J determine: (a) the short-circuit current gain; (b) the current gain I_L/I_I; (c) the power gain in dB.

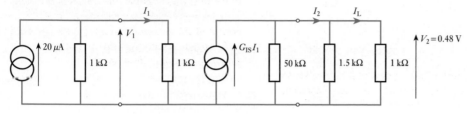

Fig. J

Chapter twenty-three

FET Amplifiers

Objectives

When you have studied this chapter, you should
- have an understanding of how field effect transistors operate
- be aware of the differences between JUGFETs and IGFETs
- be able to describe the advantages of FETs over bipolar transistors

Contents

The bipolar transistor established the transistor as the most significant amplifier component. However, the need to draw current was seen as a drawback and hence there was a desire to produce a better transistor. When it appeared, it was the field effect transistor (FET).

In this chapter, some forms of FET will be introduced together with their manners of operation and characteristics. The application of a FET as a switch will be considered.

By the end of the chapter, you will be familiar with the FET and its actions. You will also be able to undertake simple exercises on circuits containing FETs.

23.1 Field effect transistor (FET)

Unlike the bipolar junction transistors, which are all basically similar in spite of a variety of constructional forms, the FET is more a collective term for a family of transistor devices, of which there are two principal groups: (1) the junction-type FET (JUGFET); and (2) the insulated-type FET (IGFET).

The JUGFET can be in two forms, p-channel and n-channel, depending on the type of semiconductor forming the basis of the transistor. The IGFET has two distinct modes of operation, one known as the depletion mode and the other as the enhancement mode, and again each mode subdivides into p-channel and n-channel.

Like the bipolar junction transistor, all FETs are three-electrode devices; the electrodes are the source, the gate and the drain, which can be taken as corresponding to the emitter, base and collector respectively.

23.2 JUGFET

The more common variety of JUGFET is the n-channel type. This consists of a piece of n-type silicon effectively within a tube of p-type silicon, the interface between the two materials being the same intimate junction as in the bipolar transistor junctions. Non-rectifying contacts are connected to each end of the n-type piece and to the p-type tube, as shown in Fig. 23.1. The two end contacts are termed the source S and the drain D, while that to the tube or shroud is the gate G.

Fig. 23.1 The basic JUGFET

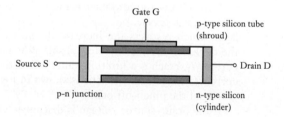

The principle of the action of a JUGFET can be explained by first considering the operation of the source and drain while the gate is left disconnected. Since we are considering an n-channel JUGFET, consider a d.c. supply connected so that the drain is positive with respect to the source, as shown in Fig. 23.2. The application of the drain–source voltage causes a conventional current to flow from the drain to the source; the current consists of electrons, which are the majority carriers in the n-channel, moving from the source to the drain. Being semiconductor material, the voltage/current relationship is almost linear and the n-channel more or less behaves as a resistor.

Fig. 23.2 Application of a drain–source voltage to a JUGFET

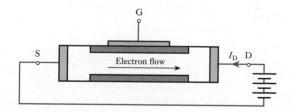

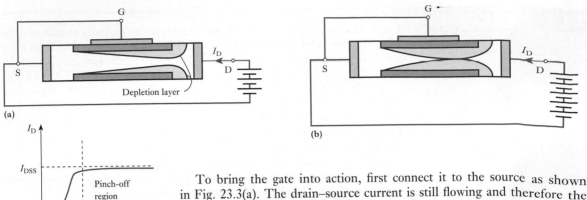

(a)

(b)

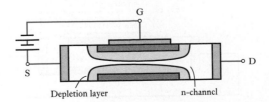

(c)

Fig. 23.3 Depletion layer in a JUGFET

To bring the gate into action, first connect it to the source as shown in Fig. 23.3(a). The drain–source current is still flowing and therefore the n-channel voltage becomes greater as electrons flow from the source towards the drain. It follows that electrons nearer the drain also experience a higher voltage with respect to the gate, hence the voltage across the p–n junction is greater at the right-hand end than at the left. Also the p–n junction is reverse biased and this bias is greater nearer the drain than nearer the source. This causes a depletion layer which becomes greater as the bias increases, i.e. the depletion layer is greater nearer to the drain.

The depletion layer, which acts as an insulator, is shown in Fig. 23.3(a). It reduces the effective cross-section of the n-channel and therefore restricts the flow of electrons. For this reason, the FET is said to be operating in the depletion mode.

If the drain–source voltage is increased, as shown in Fig. 23.3(b), the current might increase, but the increase in voltage also increases the depletion layer which restricts the increase. Eventually a point is reached at which the depletion layer completely absorbs the n-channel, and the drain–source current I_D reaches a limiting value called the saturation current I_{DSS}. The current/voltage characteristic is shown in Fig. 23.3(c). The saturation point is termed the pinch-off point.

If the drain–source voltage is disconnected and a bias voltage applied to the gate, as shown in Fig. 23.4, the depletion layer is evenly set up in the n-channel. This effectively reduces the cross-section and means that when a drain–source voltage is reapplied, the pinch-off point will be experienced at a lower voltage, and the currents, including the saturation current, will all be less.

The combined effect of applying both drain–source voltage and bias voltage is shown in Fig. 23.5. Here the drain current depends on both voltages, and the more negative the bias gate voltage the smaller is the saturation current. The gate voltage could stop the flow of drain current if increased sufficiently but, more importantly, the gate voltage has a more significant effect on the drain current than the drain voltage.

Fig. 23.4 Bias voltage applied to gate of a JUGFET

Fig. 23.5 Combined depletion effects in a JUGFET

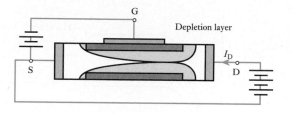

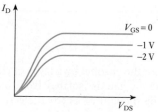

Fig. 23.6 Output characteristics of an n–channel JUGFET

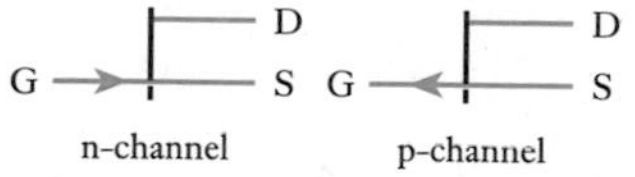

Fig. 23.7 JUGFET symbols

The resulting output characteristics of an n–channel JUGFET are shown in Fig. 23.6. Operation occurs beyond pinch–off so that the control is by the gate voltage. The characteristics compare in form with those shown in Fig. 22.12 for the junction transistor and the amplifier action can be obtained in the manner illustrated in Fig. 22.17.

The symbol used in circuit diagrams for an n–channel JUGFET is shown along with that of the p–channel variety in Fig. 23.7. The operation of the p–channel JUGFET is the converse of the n–channel action, with all polarities and current flow directions reversed.

The gate–source junctions present very high values of resistance (megohms) since they consist of reverse-biased junctions. It follows that the gate currents are very small but, due to the increase in minority charge carriers with temperature, these currents do vary with temperature.

23.3 IGFET

As its name indicates, the IGFET has its gate insulated from the channel. A simplified form of construction is shown in Fig. 23.8, in which the main bulk of the material is low-conductivity silicon; this is termed the *substrate*. For an n–channel IGFET, the substrate is p-type silicon into which is introduced a thin n-channel terminated in the drain and source electrodes. The gate is

Fig. 23.8 Simple IGFET

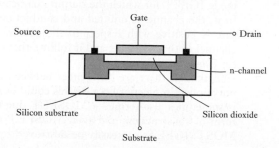

Fig. 23.9 Output characteristics of an n–channel IGFET

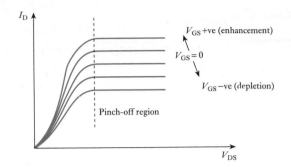

separated from the channel by a layer of silicon dioxide which is an insulator ensuring that the gate is isolated from both source and drain. The silicon dioxide can be thought of as the dielectric in a capacitor consisting of two plates, one being the gate and the other the channel.

In the IGFET, the flow of electrons from drain to source is again the same as in the JUGFET and the pinch-off effect can also be achieved by connecting the gate to the source. However, variation of the gate–source voltage produces a distinctly different effect.

If the gate is made positive with respect to the source and hence to the channel, the source–drain current is increased. Because of this increase, the FET is said to be enhanced. Conversely, if the gate is made negative with respect to the source and hence to the channel, the source–drain current is decreased and the FET is said to be depleted, or in the depletion mode.

The output characteristics of an n–channel IGFET are shown in Fig. 23.9 and the symbols for IGFETs are shown in Fig. 23.10. Again the p–channel IGFET is the converse of the n–channel. Note that a connection to the substrate is available and may be used as the other terminal for the bias arrangement.

The cause of the variation in drain current when bias voltages are applied to the gate lies in the capacitive effect between gate and channel. If the gate is positively charged then the channel is negatively charged and the channel therefore experiences an increase in charge carriers, i.e. it is enhanced. When the bias is reversed, there is a depletion in the number of charge carriers available.

The cause of the pinch-off effect is less obvious, but stems from the p–n junction between channel and substrate, there being no p–n junction between channel and gate as in the JUGFET. Often the substrate is connected to the source, which is also convenient for the gate–source bias.

There is an important derivative of the IGFET which is the enhancement-mode IGFET for which the output characteristics are shown in Fig. 23.11. In it, the channel is omitted and conduction can only take place when the gate is positive with respect to the substrate. This induces a conduction channel in the substrate, and it follows that operation can only take place in the *enhancement mode*.

Although there are differences between the IGFET symbols, these are not applied in practice because this could be confusing. Finally, the IGFET is sometimes also termed a MOSFET due to the metal, oxide and silicon form of construction, and even MOSFET might be further contracted as MOST. IGFETs can easily be destroyed by the build-up of charge across

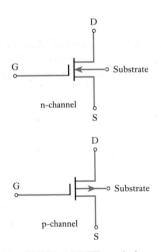

Fig. 23.10 IGFET symbols

Fig. 23.11 Enhancement mode n-channel IGFET

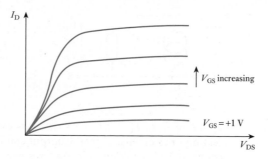

the dielectric separating the gate and substrate. To avoid this, often these are linked at manufacture and the links have to be removed when the IGFETs are installed.

23.4 Static characteristics of a FET

There are three principal characteristics: the drain resistance, the mutual conductance and the amplification factor.

The ratio of the change, ΔV_{DS}, of the drain–source voltage to the change, ΔI_D, of the drain current, for a given gate–source voltage, is termed the drain resistance r_d, i.e.

$$r_d = \frac{\Delta V_{DS}}{\Delta I_D} \quad \text{for a given value } V_{GS} \tag{23.1}$$

The value of r_d is large in most FETs, say in the order of 10–500 kΩ.

The ratio of the change, ΔI_S, of the drain current to the change, ΔV_{GS}, of the gate–source voltage, for a given drain–source voltage, is termed the mutual conductance g_m, i.e.

$$g_m = \frac{\Delta I_D}{\Delta V_{GS}} \quad \text{for a given value } V_{DS} \tag{23.2}$$

where g_m is usually measured in milliamperes per volt.

The product of drain resistance and mutual conductance gives the amplification factor μ, i.e.

$$\mu = r_d g_m \tag{23.3}$$

23.5 Equivalent circuit of a FET

Like the junction transistor, the FET can be represented by a circuit which is equivalent to its response to small signals causing it to operate over the linear part of its characteristics. The FET equivalent circuit, shown in Fig. 23.12, is similar to that for the junction transistor except that the input resistance is so high that it can be considered infinite, i.e. equivalent to an open circuit.

The parameters are termed Y-parameters, Y_{fs} is the forward transconductance (the s indicates that it is for the common-source mode) and is

Fig. 23.12 *Y*-parameter
equivalent circuit for a FET

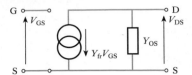

the ratio of I_D to V_{GS}. It follows that it is expressed in amperes per volt, or
sometimes in siemens; Y_{os} is the output conductance (in the common-source
mode) and is also measured in amperes per volt.

Consider the transistor in the amplifier arrangement illustrated in
Fig. 23.13(a). The equivalent circuit is shown in Fig. 23.13(b) and again the
capacitors are taken to have negligible effect.

Fig. 23.13 FET amplifier

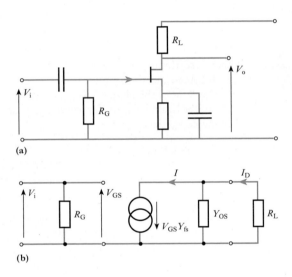

(a)

(b)

From Fig. 23.13(b),

$$I = I_D + V_o Y_{os}$$

$$= I_D + I_D R_L Y_{os}$$

$$I_D = \frac{I}{1 + R_L Y_{os}}$$

$$= \frac{V_{GS} Y_{fs}}{1 + R_L Y_{os}}$$

$$V_o = I_D R_L$$

$$= \frac{V_{GS} Y_{fs} R_L}{1 + R_L Y_{os}}$$

$$A_v = \frac{Y_{fs} R_L}{1 + R_L Y_{os}}$$ [23.4]

| **Example 23.1** | A transistor amplifier stage comprises a FET, of parameters $Y_{fs} = 2.2$ mA/V and $Y_{os} = 20\ \mu S$, and bias components and coupling capacitors of negligible effect. The total load on the output is 2 kΩ. Determine the voltage gain. |

$$A_v = \frac{Y_{fs} R_L}{1 + R_L Y_{os}}$$

$$= \frac{2.2 \times 10^{-3} \times 2 \times 10^3}{1 + (2 \times 10^3 \times 20 \times 10^{-6})} = 4.23$$

23.6 The FET as a switch

In the previous chapter, we have observed that the bipolar transistor can operate as a switch. In particular, it has current flowing from collector to emitter (assuming an n–p–n type) so long as the base–emitter junction is forward biased. Reverse the bias and effectively no current flows. This gives rise to the ON/OFF switching action.

The operation is imperfect because there is a significant input current when the base–emitter junction is forward biased. This gives rise to a collector–emitter voltage known as the offset voltage.

By comparison, the FET has distinct advantages. In particular, the insulated gate ensures that no current is taken by the gate electrode whether the device is effectively ON or OFF. Also, in the OFF condition, effectively no current flows between the drain and the source. This means that the transistor is acting as an open circuit, which is what we would expect of a switch. Finally, when operating in the ON condition, the resistance introduced by the FET is negligible.

There are differences between the operation of JUGFET and IGFET switches. In a JUGFET, the gate voltage must be sufficiently negative to ensure that no drain current flows in the OFF condition. Any suitable voltage between this value and zero results in the device being switched ON. The transfer characteristics of a JUGFET are compared with those of a bipolar transistor in Fig. 23.14.

With an IGFET operating in the depletion mode, the gate voltage must be sufficiently negative to ensure that the switch is OFF. The gate voltage is made positive to switch the device ON. Operation in the enhancement mode avoids the need for a negative signal to ensure the OFF operation since this is its normal condition. Again a positive gate voltage causes the ON

Fig. 23.14 Transfer characteristics

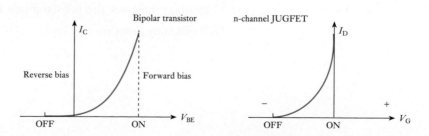

Fig. 23.15 Transfer
characteristics

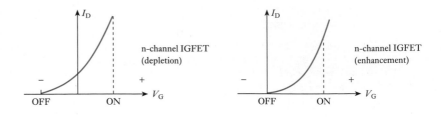

operation to occur. The transfer characteristics for both modes of operation
are shown in Fig. 23.15.

<table>
<tr><td>

**Summary of important
formulae**

</td><td>

For a FET, the drain resistance

$$r_d = \frac{\Delta V_{DS}}{\Delta I_D}\bigg|_{V_{GS} \text{ constant}}$$ [23.1]

The mutual conductance,

$$g_m = \frac{\Delta I_D}{\Delta V_{GS}}\bigg|_{V_{DS} \text{ constant}}$$ [23.2]

The amplification factor,

$$\mu = r_d g_m$$ [23.3]

$$A_D = \frac{Y_{fs} R_L}{1 + R_L Y_{os}}$$ [23.4]

</td></tr>
</table>

Terms and concepts

Field effect transistors (FETs) come in two forms – **JUGFET**s and
IGFETs.

The JUGFET has three connections – the **source**, the **gate** and the
drain. The IGFET also has a **substrate** connection.

An advantage of the FET when compared with the bipolar transistor is
that its input resistance is so high, it is effectively an open circuit.

This high input resistance leads to much better operation as a switch,
there being effectively no current through it when in the OFF condition. With the increasing dependence on digital rather than analogue
signal transmission, this is a significant advantage.

FET switching times are very high.

Exercises 23

1. Without referring to the text, give the symbols for: (a) a p-channel junction FET; (b) an n-channel enhancement and depletion mode insulated gate FET; (c) a p-channel enhancement insulated gate FET; (d) a p–n–p bipolar transistor.

2. In each of the cases in Q. 1, give the polarities of the voltages normally applied to each electrode and state whether the flow of conventional current through each device is primarily by electrons or holes.

3. In a FET the following measurements are noted: with V_{GS} constant at -4 V the drain current is 5 mA when V_{DS} is $+4$ V, and 5.05 mA when V_{DS} is $+15$ V. Find r_d. If the current is restored to 5 mA when the gate voltage is changed to -4.02 V, find g_m.

4. Using the characteristics for an n-channel FET shown in Fig. A find, at $V_{DS} = 12$ V and $V_{GS} = -1.5$ V: (a) r_d; (b) g_m; (c) μ. For a value of $V_{DS} = 14$ V derive the transfer characteristic (I_D/V_{GS}).

   Using the characteristics shown in Fig. A, derive the transfer characteristic for $V_{DS} = 10$ V. Compare the result with that of the previous question. What conclusions do you reach?

5. A bipolar transistor used as a common-emitter amplifier has an input resistance of 2 kΩ and a current gain under operating conditions of 38. If the output current is supplied to a 5 kΩ load resistor, find the voltage gain. If the bipolar transistor is replaced by a FET having g_m 2.6 mA/V and the load resistor is changed to a value of 22 kΩ, find the voltage gain.

6. Find the output voltages for the two devices employed in Q. 5 if the input is derived from a low-frequency generator of e.m.f. 50 mV and output resistance of 6 kΩ.

7. Find the approximate input impedance of a FET at: (a) 10 Hz; (b) 1 MHz; if the input resistance is 5 MΩ shunted by a capacitor of 200 pF.

8. An insulated gate FET operating in the enhancement mode gives a saturation drain current of 4 mA when the gate–source voltage is -2 V and the drain–source voltage is $+25$ V. Is this a p- or n-channel device? What is the value of g_m?

9. The FET in Q. 8 uses as a load a coil of inductance 110 μH and resistance 15 Ω tuned to a frequency of 250 kHz by a capacitor connected in parallel with the coil. What is the value of capacitance needed and what is the stage gain at the resonant frequency if $g_m = 5.5$ mA/V?

10. The tuned load in Q. 9 is replaced by a transformer having 1000 primary turns and 2000 secondary turns and the secondary is connected to a resistor of 5 kΩ resistance. The input to the gate is a sine wave of peak to peak value 80 mV. Find the r.m.s. output voltage across the 5 kΩ resistor.

11. A FET is used as a switch. In the ON state the current flowing is 20 mA with a V_{DS} at 4 V. What is the effective ON resistance? If the FET is used as a series switch with a direct voltage of 40 V and a load resistor of 2.0 kΩ find the voltage across the load in the ON position.

12. What is the average voltage produced across the load in Q. 11 if the FET is switched on by a square wave of mark/space ratio 2:1 (the FET passes no current in the OFF (space) condition)?

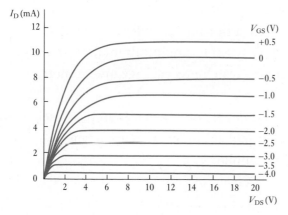

Fig. A

Chapter
twenty-four

Further Semiconductor Amplifiers

Objectives

When you have studied this chapter, you should
- appreciate how amplifiers may be cascaded to achieve required gain
- have an understanding of the operation of operational amplifiers
- be able to determine the component values for inverting, non-inverting and summing operational amplifier circuits
- be aware of common-mode rejection ratio and its significance

Contents

The simple single-stage amplifier, based on either a bipolar transistor or a FET, has already been introduced. Such an amplifier is limited in its effect and for a greater range of performance it is necessary to use other amplifiers.

First, cascaded amplifiers will be considered where a signal is amplified by the first stage and then further amplified by later stages. This permits amplification of a signal by almost as much as required. Then, integrated circuits which pre-package one or more amplifiers will be described.

There are other forms of amplifier known as operational amplifiers (op-amps) which will be considered. These can be used to produce amplifier circuits that are termed inverting, non-inverting, summing and differential amplifiers.

By the end of this chapter, you will be able to cascade amplifiers and will be aware of the functions performed by op-amps. You will also be able to undertake exercises to analyse the performance of such devices.

24.1 **Cascaded amplifiers**

Often a single-transistor amplifier stage is unable to provide the gain necessary for the function to be performed by the system in which the amplifier operates. Alternatively, the stage might be able to provide the necessary gain, but only with the introduction of distortion. In order to overcome the difficulties associated with either of these problems, amplifier stages can be connected in cascade as shown in Fig. 19.10. Cascaded amplifiers are connected in such a manner that the output of one amplifier stage is the input to a second stage, thus providing further gain within the system.

The advantage of cascading is the increase of gain, but this is offset by the increased distortion due to the non-linearity of each amplifier stage.

A simple two-stage n–p–n transistor amplifier could take the form shown in Fig. 24.1.

Fig. 24.1 Two-stage n–p–n transistor amplifier

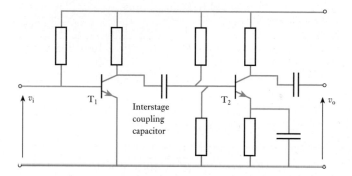

The input signal v_i is amplified in the usual manner by the transistor T_1, and the amplified signal is transferred to the second stage by the interstage coupling capacitor. It should be recalled that the coupling capacitor 'conducts' the a.c. signal but 'blocks' the d.c. component of the supply. The a.c. amplified signal is then further amplified by transistor T_2 and the enhanced output signal emerges as v_o by means of a further coupling capacitor.

The analysis of a two-stage amplifier is similar to that already described in section 22.11 and is best illustrated by means of an example.

Example 24.1

A two-stage amplifier is shown in Fig. 24.2. Draw the mid-band equivalent circuit and hence determine the overall current and power gains of the amplifier. Both transistors have input resistances of 1000 Ω, short-circuit current gains of 60 and output conductances of 20 μS.

Fig. 24.2

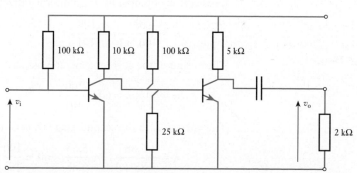

Fig. 24.3

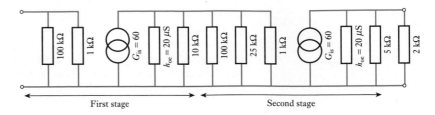

First stage   Second stage

The mid-band equivalent circuit is shown in Fig. 24.3.

$$\frac{1}{h_{oe}} = \frac{1}{2 \times 10^{-6}} \equiv 50 \text{ k}\Omega$$

For the output of the first stage, combine the 50 kΩ, 10 kΩ, 100 kΩ and 25 kΩ resistances thus

$$\frac{1}{R'} = \frac{1}{50} + \frac{1}{10} + \frac{1}{100} + \frac{1}{25} = \frac{17}{100}$$

$$R' = \frac{100}{17} = 5.88 \text{ k}\Omega$$

Also, for the output of the second stage, combine the 50 kΩ and the 5 kΩ resistances thus:

$$\frac{1}{R''} = \frac{1}{50} + \frac{1}{5} = \frac{11}{50}$$

$$R'' = \frac{50}{11} = 4.55 \text{ k}\Omega$$

The network shown in Fig. 24.3 can now be reduced to that shown in Fig. 24.4.

Fig. 24.4

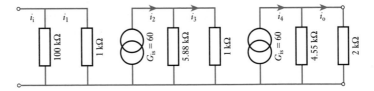

For the input current i_i to the amplifier, the current i_1 to the first-stage transistor is obtained by applying the current-sharing rule

$$i_1 = \frac{100}{100 + 1} i_i = \frac{100}{101} i_i$$

The output of the current generator associated with the first-stage transistor is

$$i_2 = 60 \times i_1 = 60 \times \frac{100}{101} i_i$$

The current i_3 to the second-stage transistor is given by

$$i_3 = \frac{5.88}{5.88 + 1} i_2 = \frac{5.88}{6.88} \times 60 \times \frac{100}{101} i_i$$

The output current of the current generator associated with the second-stage transistor is

$$i_4 = 60 \times i_3 = 60 \times \frac{5.88}{6.88} \times 60 \times \frac{100}{101} i_i$$

The amplifier output current is given by

$$i_o = \frac{4.55}{4.55 + 2} i_4 = \frac{4.55}{6.55} \times 60 \times \frac{5.88}{6.88} \times 60 \times \frac{100}{101} i_i$$

$$\therefore \qquad A_i = \frac{i_o}{i_i} = \frac{4.55}{6.55} \times 60 \times \frac{5.88}{6.88} \times 60 \times \frac{100}{101} = \mathbf{2120}$$

$$A_p = A_i^2 \times \frac{R_o}{R_i} \quad \text{where } R_i = \frac{100 \times 1}{100 + 1} \text{ k}\Omega \equiv 990 \text{ }\Omega$$

$$= 2120^2 \times \frac{2000}{990} = \mathbf{9\ 080\ 000}$$

| Example 24.2 |

Determine the overall voltage gain, current gain and power gain of the two-stage amplifier shown in Fig. 24.5.

Fig. 24.5

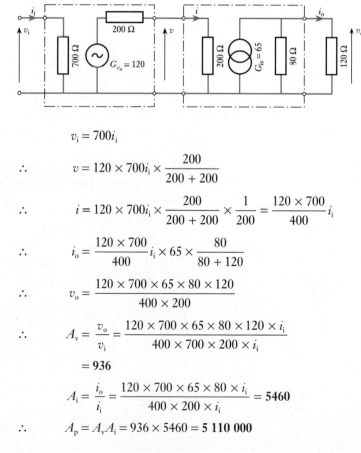

$$v_i = 700 i_i$$

$$\therefore \qquad v = 120 \times 700 i_i \times \frac{200}{200 + 200}$$

$$\therefore \qquad i = 120 \times 700 i_i \times \frac{200}{200 + 200} \times \frac{1}{200} = \frac{120 \times 700}{400} i_i$$

$$\therefore \qquad i_o = \frac{120 \times 700}{400} i_i \times 65 \times \frac{80}{80 + 120}$$

$$\therefore \qquad v_o = \frac{120 \times 700 \times 65 \times 80 \times 120}{400 \times 200}$$

$$\therefore \qquad A_v = \frac{v_o}{v_i} = \frac{120 \times 700 \times 65 \times 80 \times 120 \times i_i}{400 \times 700 \times 200 \times i_i}$$

$$= \mathbf{936}$$

$$A_i = \frac{i_o}{i_i} = \frac{120 \times 700 \times 65 \times 80 \times i_i}{400 \times 200 \times i_i} = \mathbf{5460}$$

$$\therefore \qquad A_p = A_v A_i = 936 \times 5460 = \mathbf{5\ 110\ 000}$$

FET amplifiers can also be connected in cascade. The manner of coupling can be achieved either using the normal resistance–capacitance coupling method or by means of a transformer. The former method is illustrated in Example 24.3.

Example 24.3

A two-stage amplifier is shown in Fig. 24.6. Draw the mid-band equivalent circuit and hence determine the overall voltage, current and power gains of the amplifier. Both transistors have a mutual conductance $g_m = 3$ mA/V and a drain resistance $r_d = 200$ kΩ.

Fig. 24.6 Resistance–capacitance coupling of two FET amplifiers

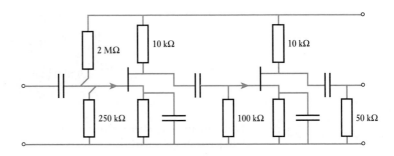

Fig. 24.7

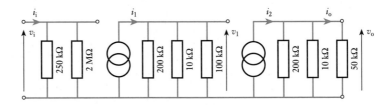

The equivalent circuit for the amplifier is shown in Fig. 24.7.

The parallel resistances have to be combined in order to simplify the circuit analysis; thus

$$R' = \frac{250 \times 2000}{250 + 2000} = 222 \text{ k}\Omega$$

$$\frac{1}{R''} = \frac{1}{200} + \frac{1}{10} + \frac{1}{100} = \frac{23}{100}$$

$$\therefore \qquad R'' = \frac{200}{23} = 8.7 \text{ k}\Omega$$

$$\therefore \qquad \frac{1}{R''} = \frac{1}{200} + \frac{1}{10} + \frac{1}{50} = \frac{25}{200}$$

$$\therefore \qquad R'' = \frac{200}{25} = 8.0 \text{ k}\Omega$$

Thus the mid-band equivalent circuit simplifies to that shown in Fig. 24.8.

Fig. 24.8

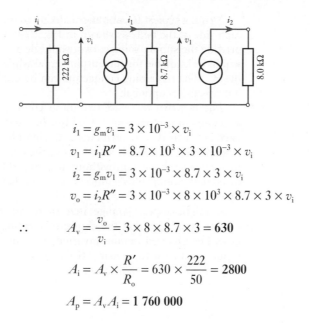

$$i_1 = g_m v_i = 3 \times 10^{-3} \times v_i$$

$$v_1 = i_1 R'' = 8.7 \times 10^3 \times 3 \times 10^{-3} \times v_i$$

$$i_2 = g_m v_1 = 3 \times 10^{-3} \times 8.7 \times 3 \times v_i$$

$$v_o = i_2 R'' = 3 \times 10^{-3} \times 8 \times 10^3 \times 8.7 \times 3 \times v_i$$

$$\therefore \qquad A_v = \frac{v_o}{v_i} = 3 \times 8 \times 8.7 \times 3 = \mathbf{630}$$

$$A_i = A_v \times \frac{R'}{R_o} = 630 \times \frac{222}{50} = \mathbf{2800}$$

$$A_p = A_v A_i = \mathbf{1\ 760\ 000}$$

24.2	Integrated circuits

Thus far, the treatment of semiconductor devices has led to the construction of networks made from a considerable number of components. The consideration of the cascaded amplifier stages indicates that such numbers can rise to 100 or more, even for quite simple arrangements. Apart from the cost of connecting all these pieces together, the space taken up by them is also an important factor when designing electronic equipment.

However, the manufacture of transistors is achieved by a process that makes many transistors on a single slice, which is then cut up to form the individual transistors. Although not simple to effect, it was a logical development to keep the transistors together on the slice, thereby forming the complete amplifier arrangement in one step.

From this development was evolved the integrated circuit, often referred to as an IC. Integrated circuits can be produced to form single amplifier stages or multistage amplifiers, both with the advantage of automatic compensation for temperature drift. The outcome is that one component can replace up to 100 or more with a dramatic reduction in the volume taken up, i.e. the outcome is the miniaturized microcircuit which is very small, cheap and reliable. It cannot be overemphasized that the cost of an IC is significantly less than the circuit components which it can replace.

Typically, an IC might comprise 20–30 transistors with possibly 50–100 other passive components such as resistors, all in the volume formerly taken up by a single bipolar transistor. The form of construction is shown in Fig. 24.9. This is a dual-in-line package; many ICs now come in surface mount packages.

Along each side of the body are mounted a number of connectors. These are considerably larger than would be required for the small currents passing through them, but they have to be of sufficient mechanical strength that they can support the body and can withstand being pushed into the holder, by means of which it is plugged into the network for which it is an amplifier.

IC module

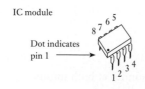

Dot indicates
pin 1

IC connector or holder

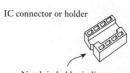

Notch in holder indicates
end at which pin 1 is inserted

Fig. 24.9 An integrated circuit

Again, a typical IC amplifier is likely to have eight connectors, four along each side of the body as shown in Fig. 24.9. Of these, two would be used to provide the supply voltage, two would be available for the input signals, one would be available for the output signal while a sixth would be used to earth the IC. The remaining connectors can be used for frequency compensation, but this use is optional.

The fact that there are two input connectors or terminals and there are also two connectors which need not be used highlights a significant difference between IC arrangements and the individual component networks we have previously considered. Because the complexity of the IC can be varied to a considerable extent without affecting the cost, many ICs are manufactured with optionable facilities; thus a number of functions can be performed by the one IC.

It might appear strange that there are two input terminals, but the versatility is such that the choice of terminal permits the output signal to be either inverted or non-inverted, i.e. out of phase or in phase with the input signal. Further, many ICs can accept two signals at these terminals and operate on the difference between them. This will be discussed in section 24.5.

The component amplifiers which make up a typical integrated-circuit amplifier are directly coupled, rather than having a system within the semiconductor slice of introducing active capacitor components to be used for coupling. It follows that ICs can be used for amplifying direct signals as well as alternating signals.

Although integrated circuits have been discussed in the context of amplifiers, it should be noted that an integrated circuit is a form of construction rather than a device performing a specific function or range of functions. In Chapter 27, we shall look at logic systems, which also are generally manufactured in the form of integrated circuits. However, this chapter is limited to amplifier operation for which the basic integrated circuit usually takes the form of an operational amplifier, often abbreviated to op-amp.

24.3	Operational amplifiers

The common IC operational amplifier is one which has a very high gain and finds widespread use in many areas of electronics. Its applications are not limited to linear amplification systems, but include digital logic systems as well.

There are certain properties common to all operational amplifiers as follows:

1. An inverting input.
2. A non-inverting input.
3. A high input impedance (usually assumed infinite) at both inputs.
4. A low output impedance.
5. A large voltage gain when operating without feedback (typically 10^5).
6. The voltage gain remains constant over a wide frequency range.
7. Relatively free of drift due to ambient temperature change, hence the direct voltage output is zero when there is no input signal.
8. Good stability, being free of parasitic oscillation.

The basic form of operational amplifier is shown in Fig. 24.10. It will be noted that the input terminals are marked + and −. These are not polarity

Fig. 24.10 The basic operational amplifier

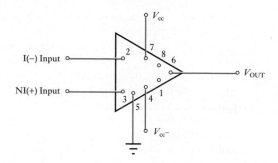

signs; rather the – indicates the inverting input terminal while the + indicates the non-inverting input terminal.

This basic unit performs in a variety of ways according to the manner of the surrounding circuitry. A number of general applications are now considered.

24.4 The inverting operational amplifier

The circuit of an inverting operational amplifier is shown in Fig. 24.11.

Fig. 24.11 Inverting operational amplifier

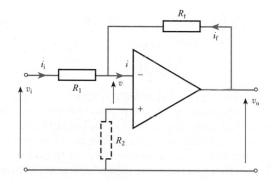

The open-loop gain of the op-amp is A, thus the output voltage $v_o = Av$; R_2 may be assumed negligible, hence

$$v_i - v = i_i R_1$$

If the input impedance to the amplifier is very high then $i \simeq 0$, hence

$$i_i = -i_f$$

but

$$i_f = \frac{v_o - v}{R_f}$$

and

$$i_i = \frac{v_i - v}{R_1}$$

$\therefore$

$$\frac{v_i - v}{R_1} = -\frac{v_o - v}{R_f} = \frac{v - v_o}{R_f}$$

If the output signal is exactly out of phase with the input voltage, the operational amplifier being in its inverting mode, then

$$v_o = -Av$$

and
$$v = -\frac{v_o}{A}$$

hence
$$\frac{v_i + \dfrac{v_o}{A}}{R_1} = \frac{-\dfrac{v_o}{A} - v_o}{R_f}$$

and
$$v_i + \frac{v_o}{A} = -\frac{v_o}{A} \cdot \frac{R_1}{R_f} - v_o \cdot \frac{R_1}{R_f}$$

Generally, R_1 and R_f are of approximately the same range of resistance, e.g. $R_1 = 100 \text{ k}\Omega$ and $R_f = 1 \text{ M}\Omega$, and A is very large, e.g. $A = 10^5$, hence

$$\frac{v_o}{A} \quad \text{and} \quad \frac{v_o}{A} \cdot \frac{R_1}{R_f}$$

can be neglected,

$$\therefore \qquad v_i \simeq -v_o \frac{R_1}{R_f} \qquad\qquad [24.1]$$

It follows that the overall gain is given approximately by

$$A_v = -\frac{R_f}{R_1} \qquad\qquad [24.2]$$

From this relationship it is seen that the gain of the amplifier depends on the resistances of R_1 and R_f, and that the inherent gain of the op-amp, provided it is large, does not affect the overall gain.

Usually, in practice, the non-inverting input is earthed through R_2, thus minimizing the worst effects of the offset voltage and thermal drift. The offset voltage is the voltage difference between the op-amp input terminals required to bring the output voltage to zero. Finally, the output often includes a resistance of about 50–200 Ω in order to give protection in the event of the load being short-circuited.

24.5	The summing amplifier

This is a development of the inverting operational amplifier. Consider the arrangement shown in Fig. 24.12. For the three input signals v_A, v_B and v_C, the currents in the resistors R_A, R_B and R_C are as follows:

$$i_A = \frac{v_A - v}{R_A} \qquad i_B = \frac{v_B - v}{R_B} \quad \text{and} \quad i_C = \frac{v_C - v}{R_C}$$

hence
$$i_i = \frac{v_A - v}{R_A} + \frac{v_B - v}{R_B} + \frac{v_C - v}{R_C} \qquad\qquad [24.3]$$

Again $i_f = \dfrac{v_o - v}{R_f}$

and since $i \simeq 0$

Fig. 24.12 Summing amplifier

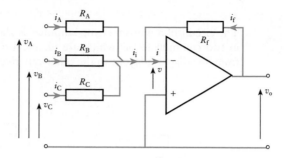

$$\frac{v - v_o}{R_f} = \frac{v_A - v}{R_A} + \frac{v_B - v}{R_B} + \frac{v_C - v}{R_C}$$ [24.4]

In a summing amplifier, usually v is very small compared with the other voltages, hence

$$-\frac{v_o}{R_f} = \frac{v_A}{R_A} + \frac{v_B}{R_B} + \frac{v_C}{R_C}$$

If $\qquad R_f = R_A = R_B = R_C$

then $\qquad -v_o = v_A + v_B + v_C$

or $\qquad v_o = -(v_A + v_B + v_C)$

It can now be seen that, apart from the phase reversal, the output voltage is the sum of the input voltages. From this comes the title – summing amplifier or summer.

It is this form of operation which leads to the general term – operational amplifier. The operation referred to is a mathematical operation and the basic op-amp can be made not only to add but to subtract, integrate, etc.

The summation can be illustrated by the following simple instance. If $v_A = 2$ V, $v_B = -4$ V and $v_C = 6$ V then

$$v_o = -(2 - 4 + 6) = -4 \text{ V}$$

Since instantaneous values have been chosen, it may be inferred that the operation works for alternating voltages as well as for steady voltages.

Example 24.4

Two voltages, +0.6 V and −1.4 V, are applied to the two input resistors of a summation amplifier. The respective input resistors are 400 kΩ and 100 kΩ, and the feedback resistor is 200 kΩ. Determine the output voltage.

$$\frac{v_o}{R_f} = -\left[\frac{v_A}{R_A} + \frac{v_B}{R_B} \right]$$ from [24.4]

$$\therefore \qquad v_o = -200 \left[\frac{0.6}{400} + \frac{-1.4}{100} \right]$$

$$= 2.5 \text{ V}$$

24.6 **The non-inverting amplifier**

The circuit of a non-inverting amplifier is shown in Fig. 24.13. It is shown in two common forms which are identical electrically, but the conversion from one diagram layout to the other can give many readers difficulty.

Fig. 24.13 Non-inverting amplifier

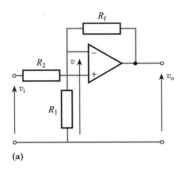

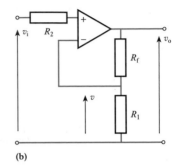

(a)   (b)

Owing to the very high input resistance, the input current is negligible, hence the voltage drop across R_2 is negligible and

$$v_i = v$$

Especially using form (b) of Fig. 24.13, it can be seen that

$$v = \frac{R_1}{R_1 + R_f} v_o$$

$$\therefore \quad v_i = \frac{R_1}{R_1 + R_f} v_o$$

$$\therefore \quad A_v = \frac{v_o}{v_i} = 1 + \frac{R_f}{R_1} \quad\quad [24.5]$$

Again we see that the gain of the amplifier is independent of the gain of the op-amp.

24.7 **Differential amplifiers**

The differential amplifier is the general case of the op-amp, which has been taken only in specific situations so far. The function of the differential amplifier is to amplify the difference between two signals. Being a linear amplifier, the output is proportional to the difference in signal between the two input terminals.

If we apply the same sine wave signal to both inputs, there will be no difference and hence no output signal, as shown in Fig. 24.14. If one of the signals were inverted, the difference between the signals would be twice one of the signals – and hence there would be a considerable output signal.

This at first sight appears to be a complicated method of achieving amplification. Differential amplification has two advantages:

Fig. 24.14 Differential amplification

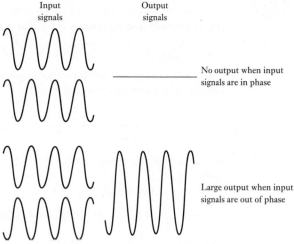

Input signals

Output signals

No output when input signals are in phase

Large output when input signals are out of phase

Fig. 24.15 Effect of interference

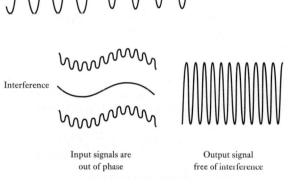

Interference

Input signals are out of phase

Output signal free of interference

1. The use of balanced input signals reduces the effect of interference as illustrated in Fig. 24.15. Here a balanced signal is transmitted by two signals which are identical other than being out of phase. Interference at a lower frequency has distorted each signal, but since the distortion effects are in phase, the amplifier output of the interference signals is zero. It follows that the amplified signal is devoid of interference.

2. The differential amplifier can be used with positive or negative feedback. In an idealized differential amplifier, as shown in Fig. 24.16,

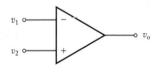

Fig. 24.16 Simple differential amplifier

$$v_o = A_v(v_1 - v_2) \qquad [24.6]$$

In a practical differential amplifier, the relationship is more complicated because the output depends not only on the difference $v_d = v_1 - v_2$ but also upon the average signal – the common-mode signal v_{com} – where

$$v_{com} = \frac{v_1 + v_2}{2} \qquad [24.7]$$

Nevertheless, the gain is still given by

$$A_v = \frac{v_o}{v_1 - v_2} \qquad [24.8]$$

Example 24.5

A differential amplifier has an open-circuit voltage gain of 100. The input signals are 3.25 and 3.15 V. Determine the output voltage.

$$v_o = A_v(v_1 - v_2) = 100(3.25 - 3.15) = 10 \text{ V}$$

The 10 V output is therefore the amplified difference between the input signals. The mean input signal is

$$v_{com} = \frac{v_1 + v_2}{2} = \frac{3.25 + 3.15}{2} = 3.20 \text{ V}$$

It follows that the net input signal $v_1 = 3.25 - 3.20 = 0.05$ V and $v_2 = 3.15 - 3.20 = -0.05$ V.

Using the net signals

$$v_o = A_v(v_1 - v_2) = 100[0.05 - (-0.05)] = \mathbf{10 \text{ V}}$$

thus showing that the mean signal is not amplified. The mean signal is termed the *common-mode signal*.

24.8   **Common-mode rejection ratio**

This is a figure of merit for a differential amplifier, the name usually being abbreviated as CMRR. It is defined as

$$\text{CMRR} = \frac{\text{differential gain}}{\text{common-mode gain}} = \frac{A_v}{A_{com}}$$

The CMRR should be large so that output errors are minimized. For instance, using the figures in Example 24.5, consider the output if the two input signals had both been 3.20 V. In this case

$$v_o = A_v(v_1 - v_2) = 100(3.20 - 3.20) = 0$$

Such a figure is idealistic because, in practice, the circuit components have manufacturing tolerances and it would almost certainly result in there being a very small output. As a result we define the common-mode gain as

$$A_{com} = \frac{v_o}{v_{com}} \tag{24.9}$$

Example 24.6

The differential amplifier used in Example 24.5 has a common input signal of 3.20 V to both terminals. This results in an output signal of 26 mV. Determine the common-mode gain and the CMRR.

$$A_{com} = \frac{26 \times 10^{-3}}{3.20} = \mathbf{0.0081}$$

$$\text{CMRR} = \frac{A_v}{A_{com}} = \frac{100}{0.0081} = \mathbf{12\ 300}$$

$$\equiv 20 \log 12\ 300 = \mathbf{81.8 \text{ dB}}$$

Summary of important formulae

For an inverting operational amplifier,

$$A_v = -\frac{R_f}{R_1}$$ [24.2]

For a non-inverting operational amplifier,

$$A_v = 1 + \frac{R_f}{R_1}$$ [24.5]

For a differential amplifier,

$$A_v = \frac{v_o}{v_1 - v_2}$$ [24.8]

Terms and concepts

It is quite usual that a single amplifier cannot provide the gain which we desire. We therefore use two or more **cascaded amplifiers**.

It does not matter whether the amplifiers incorporate junction transistors or FETs – both can be cascaded.

Amplifiers are often manufactured as **integrated circuits** in which all the components are encapsulated.

A common integrated circuit takes the form of an **operational amplifier** (op-amp) which provides degrees of choice in its applications, e.g. the input can be either **inverting** or **non-inverting**.

Operational amplifiers can be used as **summing amplifiers** in which the output voltage is the sum of the input voltages. Similarly operational amplifiers can be used as **differential amplifiers** in which the output signal is proportional to the difference of two input signals.

Exercises 24

1. Determine the overall voltage and current gains of the two-stage amplifier shown in Fig. A and hence derive the corresponding diagram of an equivalent single-stage amplifier.

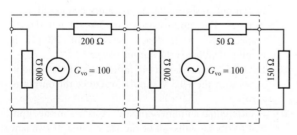

Fig. A

2. Sketch and explain the gain/frequency characteristic of a resistance–capacitance-coupled amplifier.

Determine the overall current and power gains of the two-stage amplifier shown in Fig. B.

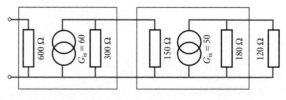

Fig. B

3. Figure C shows the equivalent circuit of two identical amplifiers connected in cascade. The open-circuit voltage gain of each amplifier is 100. Prove that if the output load equals R_i, then the overall power gain of the cascaded system is given by

Exercises 24 continued

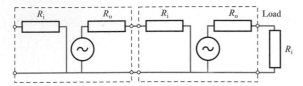

Fig. C

$$10^8 \times \left[\frac{R_i}{(R_o + R_i)} \right]^4$$

Given $R_i = 1000\ \Omega$ and $R_o = 2000\ \Omega$, calculate:
(a) the voltage, current and power gains of each stage;
(b) the overall power gain.

4. Determine the overall voltage and current gains of the two-stage amplifier shown in Fig. D and hence determine the corresponding diagram of an equivalent single-stage amplifier stating all essential values.

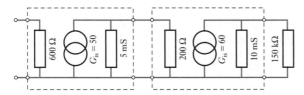

Fig. D

5. The input resistance of each of the identical amplifiers shown in Fig. E is $600\ \Omega$. Calculate V_1, V_2 and the input power.

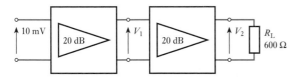

Fig. E

6. The input resistance of each of the amplifiers shown in Fig. F is $600\ \Omega$. Calculate V_i, V_o and the input power P_i.

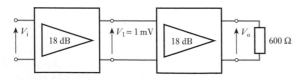

Fig. F

7. Two identical voltage amplifiers, whose input and output impedances are matched, are connected in cascade. If the input signal source is 20 mV and the output voltage of the arrangement is 20 V, determine: (a) the initial gain in dB; (b) the gain if the final stage impedance is reduced by 25 per cent but the power remains constant.

8. (a) Draw the circuit diagram of a non-inverting amplifier which uses an operational amplifier.
(b) Derive the voltage gain. State any assumptions.
(c) Comment on the input resistance.

9. (a) State the function of the circuit shown in Fig. G.
(b) Determine the output voltage for an input of 6 V r.m.s.

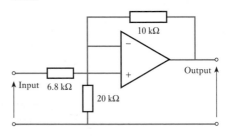

Fig. G

10. (a) Draw the circuit diagram of an op-amp connected as a voltage follower. (b) Derive the voltage gain. State any assumptions. (c) State the characteristics of such an amplifier.

11. (a) Draw the functional diagram of an operational amplifier connected in the inverting mode. (b) Derive the expression for voltage gain in terms of the circuit resistors. (c) Calculate the value of the feedback resistor given that the voltage gain is −1000 and the input resistor is 10 kΩ. (d) State four necessary properties of an operational amplifier.

12. Two voltages, +0.5 V and −1.5 V, are applied to the two input resistors of an op-amp connected as a summer. The input resistors are 500 kΩ and 100 kΩ respectively. A feedback resistor of 200 kΩ is employed. What is the output voltage?

13. In Q. 12, what resistance of input resistor is needed in series with the +0.5 V supply to reduce the output to zero?

14. A differential amplifier employs resistors in the input and feedback paths, all of the same value. If two sine waves $1.5 \sin \omega t$ and $2 \sin \omega t$ are applied simultaneously to the two inputs, what is the equation of the output voltage? If the two waves have equations $1.5 \sin \omega t$ and $1.5 \cos \omega t$ what is the new output?

Exercises 24 continued

15. Calculate the voltage gain of the circuits shown in Figs H(a) and (b).

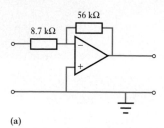

(a)

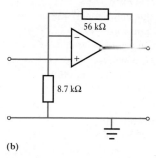

(b)

Fig. H

16. Determine the output voltage of the circuit shown in Fig. I.

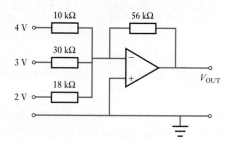

Fig. I

17. Two voltages, +0.1 V and +0.8 V, are applied simultaneously to a differential amplifier of gain 10 and CMRR 60 dB. What is the output voltage?

Chapter twenty-five

Interfacing Digital and Analogue Systems

Objectives

When you have studied this chapter, you should

- be able to explain why there is a need to convert signals from analogue to digital and from digital to analogue
- understand how a digital-to-analogue converter circuit operates
- be able to explain the operation of an analogue-to-digital converter

Contents

We will see the complementary roles of analogue and digital systems in practice in the following chapters. It follows that if information is transmitted by a combination of the two systems, there will be a point at which they have to interface.

This chapter introduces the principles and practice of digital–to–analogue and analogue–to–digital converters. In each case the conversion process is considered, introducing the terms that define the process. Converter circuits based on op–amps are developed, leading to consideration of practical parameters such as voltage stabilization.

By the end of the chapter, you will be aware of the conversion processes and be able to undertake simple exercises based on these processes.

25.1 The need for conversion

Systems can be either digital or analogue. Subsequently we have observed that digital systems have the ability to correct errors which occur and therefore we can expect the output information to be exactly that which was put into the system.

We have also observed that analogue systems are always imperfect in that the output signal is never an exact replica of that which entered the system. Further, we noted that there are many applications which can only operate on an analogue basis, especially in those applications which involve the human senses such as hearing and seeing.

The consequence of these observations is that it is logical to convert an analogue signal as quickly as possible into a digital system, and subsequently to convert back to analogue at the end of the signal transmission.

This makes an important assumption – that such conversion can be achieved with a sufficient degree of freedom from error. It is the object of this chapter to introduce a number of basic conversion processes.

One final general observation. Although much emphasis has been given to the relationship between analogue signals and the human senses, there are many other applications in which analogue signals also apply. These affect control systems and measurements. We will be looking at control systems in Chapter 29 and measurements in Chapter 47. Both involve information which is analogue in nature, but the processing of the signal can be achieved using digital systems.

25.2 Digital-to-analogue conversion

Digital-to-analogue conversion is usually abbreviated to D/A conversion; D/A converters are sometimes abbreviated to DACs. The purpose of such conversion is to take a digital signal and convert it to an analogue signal. Typically a digital signal will be in binary code while the analogue signal will be either a voltage or a current which varies in value over a predetermined range.

We will start by considering the required operation of the system – in this case the system is the converter itself as shown in Fig. 25.1. For simplicity, assume a 4-bit D/A converter with inputs D, C, B and A. These inputs are likely to be taken from the output register of a digital system. Details of the binary number system are given in Chapter 26.

Being a 4-bit system, the input binary numbers can have 16 different values as indicated in Fig. 25.2. Let us now further assume that the output signal – a voltage – has a value numerically equal to that of the equivalent binary number. Thus the binary number 1011 (representing 11) would produce 11 mV.

Fig. 25.1 A D/A converter system: MSB = Most Significant Bit; LSB = Least Significant Bit

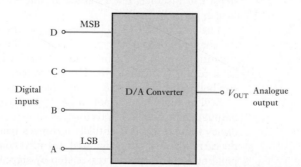

Fig. 25.2 Output waveform from a D/A converter driven by a digital counter

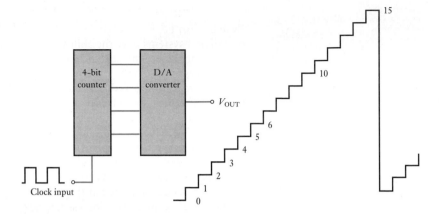

In the same system, 0110 input would produce an output of 6 mV and so on. There is no need for the relationship to be direct. For instance, we could introduce some proportionality factor such as 7. If this were the case, then 1011 would give an output of 77 mV and 0110 would give 42 mV.

Example 25.1

A 5-bit D/A converter has a voltage output. For a binary input of 10100, an output voltage of 12 mV is produced. Determine the output voltage when the binary input is 11100.

The binary input 10100 is equivalent to 20_{10}. The output voltage is 12 mV, hence

$$\text{Proportionality factor} = 12/20$$

$$= 0.6$$

It follows that the output voltage is 0.6 times the binary value. The binary input is equivalent to 28_{10}, hence

$$\text{Output voltage} = 0.6 \times 28$$

$$= \mathbf{16.8 \; mV}$$

The D/A converter in Example 25.1 can provide an output range of voltages between 0 and 18.6 mV, but there is not an infinite range of steps in between these values. In fact, we can only have an output value in 0.6 mV steps. For instance, if we wanted an output of 16.6 mV, the nearest output we can get is 16.8 mV.

The resolution of a D/A converter is the smallest step that can occur in the analogue output due to a change in the digital input. In this case, the resolution would be 0.6 mV and it is derived from a change at the A input, i.e. it results from the LSB. The equivalent of the LSB is known as the step size being the amount the output changes as the input rises by one step.

Still assuming the input to be derived from a 4-bit binary counter, the output can be seen as a waveform as shown in Fig. 25.2. As the counter continues to increase, it eventually advances from 0000 to 1111; assuming that the carry forward can be ignored, the next count reverts to 0000. The corresponding D/A output increases step by step, but eventually reverts to 0.

Generally resolution is expressed as a percentage of the maximum output. For the operation illustrated in Fig. 25.2, a step size is 1 and the maximum output is 15, hence

$$\text{Percentage resolution} = \frac{\text{step size}}{\text{maximum output}} \times 100 \text{ per cent}$$

$$= \frac{1}{15} \times 100 \text{ per cent}$$

$$= 6.67 \text{ per cent}$$

Example 25.2 An 8-bit D/A converter has a step size of 5 mV. Determine the maximum output voltage and hence the percentage resolution.

For an 8-bit system, there are

$$2^8 - 1 = 127 \text{ steps}$$

hence   Maximum voltage $= 127 \times 5$

$$= \textbf{635 mV}$$

$$\text{Percentage resolution} = \frac{5}{635} \times 100 \text{ per cent}$$

$$= \textbf{0.79 per cent}$$

The maximum output value is often known as the full-scale output. The greater the number of input bits, the greater is the number of steps required to reach full-scale output. It follows also that the greater the number of bits, the smaller is the percentage resolution.

The D/A converters have been so far assumed to use binary input codes. However, if we are seeking outputs which can be described by decimal means, then it would seem reasonable that the inputs should also be expressed by decimal codes. This can be achieved by using a binary-coded decimals input code (BCD input code for short). This requires a 4-bit code group for each decimal digit. Thus 8 bits can provide two decimal digits, 12 bits can provide three and so on. To cover the outputs 0–9, the 4-bit code group has to vary from 0000 to 1001.

If we consider a 12-bit D/A converter as shown in Fig. 25.3, the BCD inputs can represent any decimal number from 000 to 999. Within each of the three code groups the steps are the same, but the weighting of group 1 is 10 times that of group 0, and the weighting of group 2 is 10 times that of group 1. It follows that the weighting of group 2 is 100 times that of group 0.

If we apply possible weights to code group 0 of, say, 1 mV, then the LSB $A_0 = 1$ mV, $B_0 = 2$ mV, $C_0 = 4$ mV and $D_0 = 8$ mV, where D is the MSB for code group 0. For code group 1, $A_1 = 10$ mV and so on; for code group 2, $A_2 = 100$ mV and so on.

Code group 0 provides the BCD for the least significant digit (LSD for short), and code group 2 provides the BCD for the most significant digit (MSD for short). Care should be taken not to confuse MSB with MSD.

Fig. 25.3 D/A converter with BCD input code

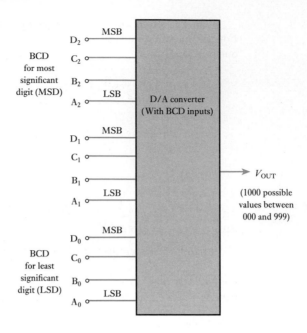

For the D/A converter shown in Fig. 25.3, the LSB has a weighting of 2 mV. The inputs are set as follows:

$$D_2 C_2 B_2 A_2 = 1000$$

$$D_1 C_1 B_1 A_1 = 0101$$

$$D_0 C_0 B_0 A_0 = 1001$$

Determine the output voltage.

$$D_2 C_2 B_2 A_2 = 8_{10}$$

$$D_1 C_1 B_1 A_1 = 5_{10}$$

$$D_0 C_0 B_0 A_0 = 9_{10}$$

thus the input code is 859. With the LSB having a weighting of 2 mV per step, it follows that the output is

$$859 \times 2 = 1718\,\text{mV} \quad\text{or}\quad \textbf{1.718 V}$$

25.3   D/A converter hardware

In practice, D/A converters are manufactured as encapsulated modules, and therefore we cannot directly observe the individual circuit components. However, to obtain an understanding of what might be contained within the module, a common basic network incorporates a summing op-amp as shown in Fig. 25.4.

The input signals are nominally 5 V for a binary input of 1 and 0 V for a binary input of 0. In practice, the input variation can be from about 2 V up to 5.5 V for a binary input of 1, while the low value need not be absolutely 0 V – variation up to 0.8 V is generally acceptable. In section 24.5, we found

Fig. 25.4 D/A converter incorporating a summing op-amp

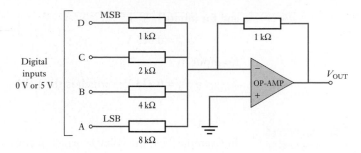

that a summing amplifier multiplies each input voltage by the ratio of the feedback resistance to the appropriate input resistance.

The resistance values are chosen to ensure that the op-amp output is

$$V_{OUT} = V_D + 0.5V_C + 0.25V_B + 0.125V_A \qquad [25.1]$$

For the arrangement shown, the summing op-amp will invert the polarity and therefore the output voltage should be negative. However, we are mostly concerned with the magnitude and therefore we can ignore the polarity – in some arrangements, polarity inversion will not occur.

The output is an analogue voltage representing the weighted sum of the digital inputs. For an input to 0000, the output would be 0 V; for an input of 0001, $V_D = V_C = V_B = 0$ V and

$$V_A = 0.125 \times 5 = 0.625 \text{ V}$$

It follows that the resolution (equal to the weighting of the LSB) is 0.625 V. As the binary input advances one step, so the output voltage increases by 0.625 V.

Example 25.4

For the D/A converter shown in Fig. 25.5, determine the weights for each input bit.

For input D

$$V_D = \frac{1 \times 10^3}{2 \times 10^3} \times 5 = \textbf{2.5 V}$$

hence for input C

$$V_C = \textbf{1.25 V}$$

Fig. 25.5 D/A converter for Example 25.4

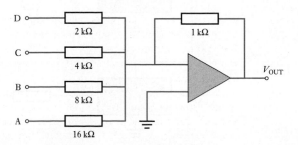

and for input B

$$V_B = \mathbf{0.625\ V}$$

for input A

$$V_A = \mathbf{0.313\ V} \quad \text{which is the LSB}$$

25.4   **D/A converters in practice**

In Example 25.4, the voltages which feature in the answers are idealistic. In practice, the accuracy of the answers would depend on the actual resistances of the resistors and the precision with which the input voltages were maintained at 5 V. Variable resistors could be adjusted to ensure that the resistances are accurate – say to about 0.01 per cent accuracy. However, it has been noted that the input voltage can vary from about 2.0 to 5.5 V. To compensate for variation, precision-level amplifiers are introduced as shown in Fig. 25.6.

Fig. 25.6 D/A converter incorporating precision-level amplifiers

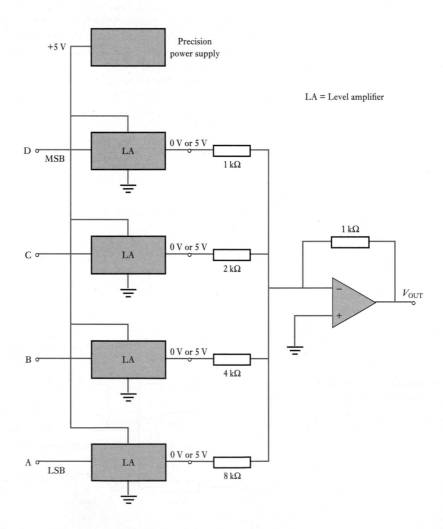

Such precision-level amplifiers ensure that their output levels are exactly 5 and 0 V depending on their inputs. Thus any high input ensures 5 V from the amplifier to the input resistor of the op-amp, and any low input ensures 0 V. In this way, we obtain stable, precise voltage supplies appropriate to the weighted summation op-amp.

By now it will be appreciated that the accuracy of conversion is the most significant factor in determining the suitability of a D/A converter. We specify the accuracy by stating the relative accuracy which is given as a percentage of the full-scale output.

The relative accuracy is the maximum deviation of the D/A converter's output from the ideal value. We can better understand this by considering Example 25.5.

Example 25.5

The D/A converter shown in Fig. 25.7 has a relative accuracy of ±0.02 per cent. Determine the maximum deviation which can occur in the output of the converter.

Fig. 25.7 D/A converter for Example 25.5

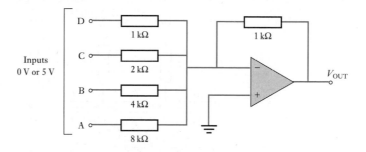

For full-scale deflation (FSD)

$$V_{OUT} = (5 + 0.5 \times 5 + 0.25 \times 5 + 0.125 \times 5) = 9.375 \text{ V}$$

Since relative accuracy is ±0.02 per cent this is equivalent to

$$\frac{0.02 \times 9.375}{100} = \pm 1.875 \text{ mV}$$

Therefore maximum deviation is

±1.875 mV

Another method of specifying accuracy is by giving the differential linearity which is the maximum deviation in step size from the ideal step size. For instance, in Example 25.5, the ideal step size was 625 mV. If the differential linearity is +0.02 per cent then the step size could vary by as much as 1.875 mV.

The previous examples have used such figures as +0.01 and +0.02 per cent which are fairly typical. Better accuracies can be achieved, but this is not usually necessary.

When choosing a D/A converter, we should seek comparable values of resolution and accuracy. Thus if we have a converter with a resolution of 0.05 per cent then we should also expect relative accuracy of about 0.05 per cent.

In conclusion, the performance characteristics of a digital-to-analogue converter may be summarized as follows:

- *Resolution.* The resolution is the reciprocal of the number of discrete steps in the output and can also be given as the number of bits that are converted.
- *Accuracy.* Accuracy is a comparison of the actual output to the expected output and is expressed as a percentage of the maximum output voltage. It should be about 0.2 per cent.
- *Linearity.* A linear error is the deviation from the ideal straight line output.
- *Settling Time.* The settling time is the time taken to settle to the nearest LSB of the final value.

25.5 *R/2R* ladder D/A converter

The binary weighted D/A converter previously considered has one disadvantage in that it requires a variety of component resistors. The *R/2R* D/A converter has the advantage that it uses only two values of resistors. A typical 4-bit arrangement is shown in Fig. 25.8.

Fig. 25.8 *R/2R* ladder D/A converter

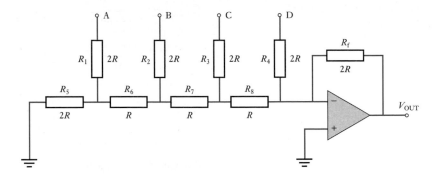

If we assume the D input to be HIGH, i.e. 1, and the others to be LOW, i.e. 0, then this relates to the binary condition 1000. With A effectively at 0 V, i.e. earth or ground, then R_1 is in parallel with R_5 and the equivalent resistance is $R_e = R$, but this is in series with R_6 making a resistance looking left of $2R$. This is in parallel with R_2, and this round of progression continues to reduce the network to that shown in Fig. 25.9.

Fig. 25.9

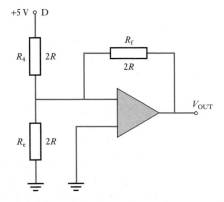

Fig. 25.10

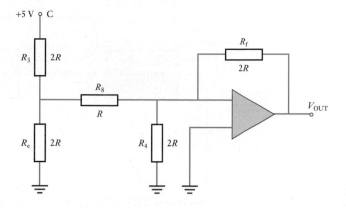

Fig. 25.11

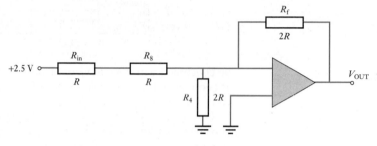

The inverting input is effectively at 0 V (see section 24.5, the summing amplifier) so no current passes through R_e. The current through D is therefore V_D/R_4 which is $5/2R$. The current from D passes through R_4 and R_f and the output voltage is equal to that at D, i.e.

$$I_D R_f = (5/2R) \cdot 2R = +5 \text{ V}$$

If the digital input were changed to 0100, then it is C that becomes HIGH and the network would reduce to that shown in Fig. 25.10.

By applying Thévenin's theorem to the system at the junction between R_3 and R_e, the network reduces to that shown in Fig. 25.11. Note that the resistor R changes the situation which we met in Fig. 25.9 in that R_e is no longer directly connected to the inverting input. However, the inverting input remains effectively at 0 V so no current passes through R_4, hence the current from C passes through R_{in}, R_8 and R_f, and the output voltage is +2.5 V.

Each successive lower weighted input produces an output which is half the previous value. It follows that the output voltage is proportional to the binary weight of the input bits.

25.6   Analogue-to-digital conversion

The analogue-to-digital converter (A/D converter or ADC) takes an analogue signal and produces a digital output code which represents the analogue signal. One of the most significant features of the A/D converter is the time delay between the entry of the input signal and the production of the output code. This arises from the complexity of the process.

Over the time since the first A/D converter was introduced, many methods have appeared. As with the D/A converter, we are unlikely ever to

Fig. 25.12   Block diagram for
an A/D converter

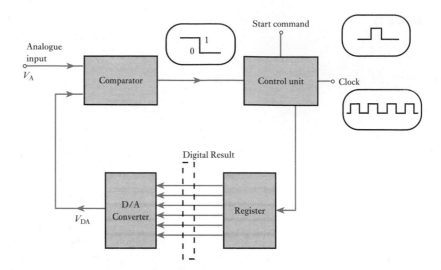

be concerned with the circuitry of such converters since they are produced as complete units. However, in order to understand the function of the A/D converter, we need to consider the operation of appropriate circuits.

One common form of analogue-to-digital converter is the successive approximation one. Its basic operation depends on the difference between two analogue signals being compared with the digital output of the converter. This causes a time delay and the operation involves a D/A converter as part of the system. A block diagram of a successive approximation A/D converter is shown in Fig. 25.12.

The control unit has three inputs which are the output from a voltage comparator, a clock and a start command. The control unit contains a logic system provided it is triggered by a start command. The voltage comparator compares two analogue signals; its output to the control unit depends on which analogue signal is the greater.

Knowing these points, we can better explain the operation by going through the sequence of events when the A/D converter is switched on, as follows:

1.   The start command goes high and this causes the A/D conversion to commence operating.

2.   The clock determines the rate at which data are sent by the control unit to the register. Registers will be detailed in Chapter 27.

3.   The register holds a binary number which is passed on to the D/A converter and we are now familiar with its operation. Its output is an analogue signal which is applied to the comparator.

4.   The input signal to the A/D converter is also fed into the comparator, thus the comparator is comparing the input signal with that provided by the D/A converter. Let the analogue signal be V_A and let the output signal from the D/A converter be V_{DA}.

5.   If $V_{DA} < V_A$, the comparator output is HIGH.
     If $V_{DA} = V_A$, the comparator output is LOW.
     If $V_{DA} > V_A$, the comparator output is LOW.

If the comparator output is LOW, the comparator stops the process of modifying the binary number stored in the register. This is likely to occur when V_{DA} is approximately equal to V_A. At this stage the digital number in the register is also the digital equivalent of the input signal V_A.

6. If the comparator output is HIGH, the comparator continues the process of modifying the binary number stored in the register until the point is reached when the comparator output is LOW.

25.7 Simple comparator

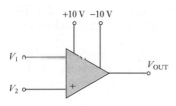

Fig. 25.13 Op-amp comparator

The comparator function can be fairly easily realized by means of an op-amp used in its high-gain differential mode. The gain is important because ideally we are seeking zero difference between the signals being compared. Such a zero difference cannot be amplified so we need a small difference to produce an output. We therefore wish to obtain the required output from the smallest possible input difference and this demands high gain.

A suitable arrangement is shown in Fig. 25.13. Here the device compares V_1 and V_2. The difference is amplified by the gain G of the amplifier. Because of the two supply voltages, we normally obtain output voltage of +10 V or −10 V. If $V_1 > V_2$ by at least a difference, known as the threshold voltage, the comparator output saturates at +10 V. Equally if $V_1 < V_2$ by at least the value of the threshold voltage, the comparator output saturates at −10 V.

If the difference between V_1 and V_2 is less than the value of the threshold voltage, the output voltage of the comparator is G times the difference. It follows that the output/input characteristic, between the stated limits, is a straight line as shown in Fig. 25.14. For the purposes of this characteristic, a gain of 10 000 has been assumed, this being a typical value in practice.

However, in practice an output voltage less than 10 V is required and also we cannot always rely on steady 10 V supplies. However, if we apply the output to a Zener diode in series with a resistor, we can then be sure of the required voltage appearing across the Zener diode.

Fig. 25.14 Output/input characteristic

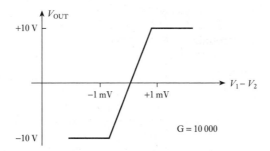

Example 25.6 A comparator has the output/input characteristic shown in Fig. 25.14. Determine:

(a) the threshold voltage;
(b) the output voltage for $V_1 = 6.546$ V and $V_2 = 6.544$ V;
(c) the output voltage for $V_1 = 3.1255$ V and $V_2 = 3.1250$ V.

(a) From Fig. 25.9, V_{OUT} saturates at +10 V when $V_1 - V_2 \geq 1$ mV.

$\therefore$ Saturation voltage = **1 mV**

(b) $V_1 - V_2 = 6.546 - 6.544 \equiv 2$ mV: this is greater than the threshold voltage.

$$\therefore \qquad V_{OUT} = +10 \text{ V}$$

(c) $V_1 - V_2 = 3.1255 - 3.1250 \equiv 0.5$ mV: this is less than the threshold voltage.

$$\therefore \qquad V_{OUT} = G(V_1 - V_2) = +5 \text{ V}$$

25.8   A/D converters

Having established the general principles by which A/D converters operate, let us consider further that which was introduced in Fig. 25.12. It will be recalled that the waveform at the output from the D/A converter is actually that of a step-by-step ramp after the form which was shown in Fig. 25.3. The characteristic looks like a ramp, and for that reason the converter is known as a digital-ramp A/D converter.

The block circuit diagram for a digital-ramp converter is shown in Fig. 25.15. Before explaining its operation, we should note that the previous system assumed that the counter was set to zero. This need not have been the case, hence the arrangement has been developed to ensure that the start pulse resets the counter to zero. Also during the start-up period, there is the risk that a pulse from the clock will affect the counter at a time when it is being reset to zero; the start pulse causes the AND gate to lock out the clock.

Let us assume that V_A has some positive value, hence the operation can be described as follows:

1.   The start pulse is applied. When it becomes high (positive), the counter is reset and clock pulses are inhibited by the AND gate.

2. $V_{DA} = 0$ when the counter is set at zero. It follows that the comparator output goes high.

Fig. 25.15   Digital-ramp A/D converter

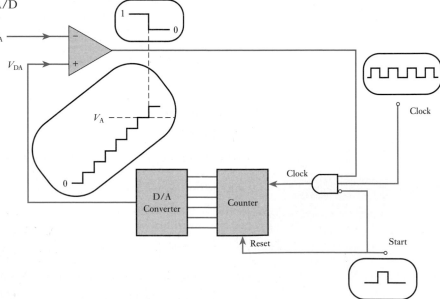

3. Only after the start pulse goes low is the AND gate enabled thus allowing pulses to enter the counter.

4. Each pulse causes the counter to advance. Consequently the output of the D/A converter increases in steps determined by the resolution.

5. This process continues until V_{DA} reaches the step when $V_{DA} > V_A$ by a voltage difference equal to or greater than the threshold voltage. At this step, the comparator output becomes low and this inhibits the pulses going to the counter. The counter has stopped at a count which provides the digital representation of V_A. At this point the conversion process has been completed.

Although this has described the operational process, it has avoided a quantitative consideration of the action. This is better explained by an example as follows.

Example 25.7	An A/D converter has a clock frequency of 1 MHz. The full-scale output of the D/A converter is 8.190 V with a 12-bit input. The comparator threshold voltage is 1 mV.

If $V_A = 4.528$ V, determine:

(a) the digital number obtained from the counter;
(b) the conversion time;
(c) the converter resolution.

(a) The D/A converter has a 12-bit input, thus the total number possible is $2^{12} - 1 = 4095$. Since the full-scale output is 8.192 V, the step size is

$$\frac{8190}{4095} = 2 \text{ mV}$$

This means that V_{DA} increases in steps of 2 mV. Given that $V_A = 4.528$ V and that the threshold voltage is 1 mV, it follows that V_{DA} must be 4.529 V or more before the comparator will switch to low. This will require the voltage to rise further to 4.530 V, hence the number of steps required is

$$\frac{4530}{2} = 2265$$

At the end of the conversion, the counter will hold the binary data equivalent to 2265, this being

100011011001

(b) To complete the conversion 4530 steps are required. The clock frequency is 1 MHz, hence the pulse rate is 1 pulse per microsecond. Therefore the time taken for 4530 pulses is

$$4530 \ \mu s = \textbf{4.53 ms}$$

(c) The resolution is given by the step size of the D/A converter which is 2 mV. This can also be expressed as a percentage which is

$$\frac{2}{8192} \times 100 = \textbf{0.024 per cent}$$

From this example we see that the best approximation we can obtain is the nearest 2 mV which is that of the resolution. This potentially can also be considered as the largest potential error: this inherent error is termed the *quantization error*.

The error can be reduced by further increasing the number of bits in the counter. In Example 25.7, however, with a resolution of 0.025 per cent, it might be thought that the error is less than might be accepted. Against that, the time taken was quite long and therefore we could 'trade in' some of the resolution in order to obtain a faster conversion time.

In the A/D converter, accuracy is not related to the resolution but it is dependent on the components of the circuit. If we are told that a converter has an accuracy of, say, 0.03 per cent, this will be in addition to the quantization error.

Clearly the digital-ramp converter inherently suffers from long conversion times and therefore is unsuitable for high-speed applications. We can improve the performance by omitting the reset procedure on the chance that the level from the previous conversion is nearer to the new analogue signal and therefore a shorter counting sequence will ensue. This requires an up/down counter since it has to count in either direction. Such a converter is called a continuous digital-ramp converter.

Although this arrangement cuts the conversion time, it remains slow and a better arrangement is to be found in the successive-approximation A/D converter. This replaces the counter with a register. The detail of this converter is beyond the coverage of this chapter and indicates the complexity contained within further studies in electronics. No matter how fast the converter, engineers will seek better speeds.

| 25.9 | Converters in action |

Having considered the operation of A/D and D/A converters, we might still be left wondering about their significance in the wider scene of electronic engineering. We shall therefore finish this chapter by considering a simple application for each.

Sitting in a room, you might wish to know its temperature and, in recent times, it is quite likely that your thermometer has a digital display to show the temperature. A simplified block diagram for such a thermometer is shown in Fig. 25.16. The sensor either provides an e.m.f. proportional to the temperature or varies its resistance in proportion to the temperature. Either way an analogue signal is produced and it varies with the temperature. Since the signal is likely to be small, it needs to be amplified by a linear amplifier and then it is applied to an A/D converter.

The output from the converter is applied to a logic controller which samples the signal at regular intervals and converts the binary signal into

Fig. 25.16 Simple electronic thermometer

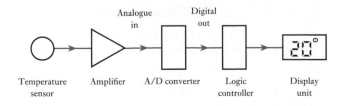

Fig. 25.17 Simple CD player

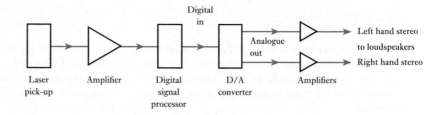

Laser pick-up — Amplifier — Digital signal processor — D/A converter — Amplifiers

Digital in

Analogue out

Left hand stereo
to loudspeakers
Right hand stereo

a form that causes the display to indicate the temperature. The formatting of this display can ensure that it can appear either in degrees centigrade or in degrees Fahrenheit. The logic of such a controller will be explored in Chapter 28.

Still sitting in our room, we might wish to listen to some music and it is quite likely that this will come from a CD player. In order to produce the sound, which is an analogue signal, the CD player incorporates a D/A converter. The basic block diagram for a CD player is shown in Fig. 25.17. The compact disc stores the music by means of a digital recording. This digital information is picked up by means of a laser pick-up device and, like the thermometer, this has to be amplified. Since we are dealing with HIGH and LOW pulses, the quality of amplification is not too important. The digital information is processed and sent on to the D/A converter in a form suitable to that converter. Finally it is amplified and sent to the loudspeakers whereby we can listen to the music.

These examples have demonstrated the roles that can be played by the amplifiers to which we have been introduced as well as the roles of the converters. However, the examples have also shown that there is a need to organize and control the effects of the digital signals. And it is this organization and control which we shall consider in the following chapters.

Terms and concepts

It is convenient to be able to change from digital to analogue systems and vice versa. Digital-to-analogue transfer is achieved by a **D/A converter** and analogue-to-digital transfer by an **A/D converter**.

The **least significant bit** (LSB) determines the step size of a D/A converter. The **resolution** is the ratio of the step size to the maximum output.

A D/A converter is constructed around a **summing op-amp**. In practice D/A converters are limited by the accuracy of the input resistances and the precision with which the input supply voltage is held.

An A/D converter is constructed around a D/A converter comparing the output of the latter with the input to the former. The comparison is undertaken by an **op-amp comparator**.

Exercises 25

1. A 4-bit D/A converter is shown in Fig. A. The binary input 1111 results in an output voltage of −10 V, and the input supplies operate at 24 V. Determine the resistances of the input resistors to produce such an operation.

   Determine the analogue output voltages for input signals of 0001, 0010, 0100 and 1000.
5. For the converter shown in Fig. C, the waveforms represent a sequence of 4-bit binary numbers applied to the inputs. Determine the output voltages for each of the inputs.

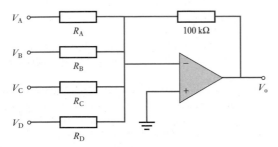

Fig. A

2. A D/A ladder is shown in Fig. B. The value of R is 1 kΩ and the input is 0010. Determine the output voltage.

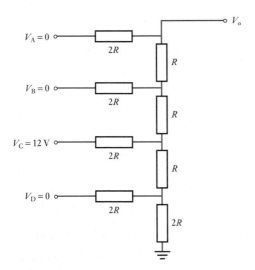

Fig. B

3. Design a D/A converter which will convert a 6-bit binary number to a proportional output voltage.
4. For the ladder shown in Fig. B, the output resistance is 2 kΩ. The input voltages operate at 12 V.

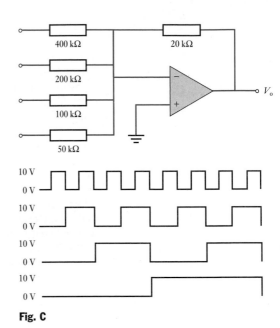

Fig. C

6. Determine the resolution of an 8-bit and a 16-bit D/A converter.
7. An analogue waveform is shown in Fig. D. An A/D converter operates from a 10 V supply. Determine the binary code outputs for each of the indicated intervals.

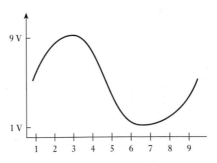

Fig. D

Exercises 25 continued

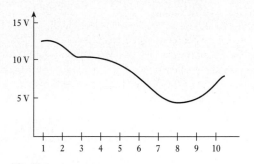

Fig. E

8. An analogue waveform is shown in Fig. E. An A/D converter operates from a 16 V supply. Determine the binary code outputs for each of the indicated intervals.

9. The following series of digital numbers are determined at equal intervals in time. Draw the corresponding analogue waveform.
   0000 0001 0010 0011 0101 0110 0111 1000 1001
   1010 1011 1100 1100 1011 1010 1001 1001 1010
   1011 1100 1101 1110 1111 1111 1110 1100 1010
   1000 0110 0011

10. In a 4-bit D/A converter, the lowest weighted resistor has a value of 15 kΩ. Determine the resistances of the other resistors.

11. In a 4-bit D/A converter, the highest weighted resistor has a value of 200 kΩ. Determine the resistance of the other resistors.

12. A converter has a closed loop voltage gain of 400 with a given inverting amplifier. Determine the feedback resistance if the input resistance is 1 kΩ.

Chapter twenty-six

Digital Numbers

Objectives

When you have studied this chapter, you should

- understand the binary number system

- be able to convert between binary and decimal number systems

- be aware of how negative numbers may be represented in binary format

- have an understanding of the basic arithmetic operations on binary numbers

- appreciate the use of the octal and hexadecimal number systems

Contents

In previous chapters, a wide variety of electronic operations, which are generally associated with analogue signals, have been considered. In this chapter, the nature of digital signals will be introduced.

First, binary numbers and their relation to decimal numbers will be presented. The basic arithmetic operations using binary numbers are explained. The octal and hexadecimal number systems are described. These are easier to use in some situations than binary numbers, which are somewhat cumbersome.

By the end of the chapter, you will be able to manipulate binary numbers and perform exercises using binary, octal and hexadecimal numbers.

26.1 Introduction

In Chapter 25, we were introduced to analogue-to-digital converters. In digital systems, we found that the signals came in two varieties which were designated 1 and 0. This could be quite satisfactory in a very limited range of applications. For instance, if we wish to control a lamp bulb then 1 could conform to the ON condition and 0 to the OFF condition. But what if we wish to use digital signals to define the brilliance of the light emitted? Digital systems can achieve this by coding the information by means of the binary system.

This chapter will therefore address the methods of encoding information by means of binary numbers.

26.2 Binary numbers

We are used to expressing numbers by means of a decimal system. We use 10 digits from 0 to 9, hence we describe the system as being to a base of 10. This means that each digit in a decimal number represents a multiple of a power of 10.

Let us consider the number 458. In this case, we have four hundreds, five tens and eight units. This may be expressed as

$$4 \times 10^2 + 5 \times 10^1 + 8 \times 10^0$$

The position of each digit can be assigned a *weight*. The weights for whole numbers are positive increasing from right to left, the lowest being $10^0 = 1$. For fractional numbers, the weights are negative decreasing from left to right starting with $10^{-1} = 0.1$. Thus the number 256.18 can be expressed as

$$2 \times 10^2 + 5 \times 10^1 + 6 \times 10^0 + 1 \times 10^{-1} + 8 \times 10^{-2}$$

In the binary system, the base is 2 and there are only two digits, 0 and 1. These are known as *bits* and it follows that a binary number can only be expressed in 0's and 1's. For instance, if we have a binary number of 101 then this could be expressed as

$$1 \times 2^2 + 0 \times 2^1 + 1 \times 2^0$$

In decimal terms, this represents $4 + 0 + 1 = 5$. Table 26.1 lists the variations using a four-digit binary system and from it we can see that $1011_2 = 11_{10}$. It can also be seen that the number of bits has to be increased to represent larger numbers, thus 5 bits would double the range, 6 bits would again double that range, and so on.

As with the decimal system, a binary number is a weighted number. The right-hand bit is the least significant bit (LSB) and has a weight $2^0 = 1$. The weights increase from right to left by a power of 2 for each bit. The left-hand bit is the most significant bit (MSB). Its weight depends on the size of the binary number.

Fractional binary numbers can be represented by placing bits to the right of the binary point, thus

Binary	2^6	2^5	2^4	2^3	2^2	2^1	2^0	2^{-1}	2^{-2}	2^{-3}	2^{-4}
Decimal	64	32	16	8	4	2	1	0.5	0.25	0.125	0.0625

Table 26.1 Four-digit binary numbers

Decimal	$2^3 = 8$	$2^2 = 4$	$2^1 = 2$	$2^0 = 1$
0	0	0	0	0
1	0	0	0	1
2	0	0	1	0
3	0	0	1	1
4	0	1	0	0
5	0	1	0	1
6	0	1	1	0
7	0	1	1	1
8	1	0	0	0
9	1	0	0	1
10	1	0	1	0
11	1	0	1	1
12	1	1	0	0
13	1	1	0	1
14	1	1	1	0
15	1	1	1	1

26.3 Decimal to binary conversion

There are two similar approaches to such a conversion, the first being based on observation. We can see this in Example 26.1.

Example 26.1

Convert 24_{10} to binary.

The highest power of 2 not greater than 24 is $2^4 = 16$.
Take 16 from 24 to leave 8.
The highest power of 2 not greater than 8 is $2^3 = 8$.
Take 8 from 8 to leave 0.
The binary for 24_{10} is

$$2^4 + 2^3 \text{ equivalent to } 11000$$

thus $24_{10} = \mathbf{11000_2}$

This approach can easily lead to mistakes, principally because we can overlook the powers which have zero digits. An alternative which is more reliable is repeatedly to divide the decimal number by 2, the remainder indicating the appropriate binary digit. This is known as the *repeated division-by-2 method*.

Example 26.2 Convert 24_{10} to binary.

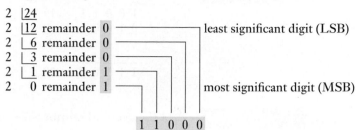

thus $24_{10} = \mathbf{11000_2}$

Example 26.3 Convert 33_{10} to binary.

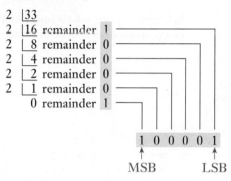

thus $33_{10} = \mathbf{100001_2}$

26.4 **Binary addition**

Binary addition is similar to decimal addition except that we only require to consider two digits, i.e. 1 and 0. There are four simple rule of binary addition as follows:

$$0 + 0 = 0$$
$$0 + 1 = 1$$
$$1 + 0 = 1$$
$$1 + 1 = 0 \quad \text{and carry 1 to the next higher order of significance}$$

This is similar to a decimal addition such as $5 + 6$ giving 11, i.e. the sum is in excess of 9, hence a 1 is carried forward to the next higher order of significance. There is one other rule of addition:

$$1 + 1 + 1 = 1 \quad \text{and carry 1 to the next higher order of significance}$$

Example 26.4 Add the binary numbers 1010 and 0110.

$$
\begin{array}{r}
1\ 0\ 1\ 0 \\
\underline{0\ 1\ 1\ 0} \\
\checkmark \\
\underline{1} \\
1\ 0\ 0\ 0\ 0_2
\end{array}
$$

Example 26.5   Add the decimal numbers 19 and 9 by binary means.

$$
\begin{array}{c}
1\ 0\ 0\ 1\ 1 \quad\quad 19 \\
\underline{0\ 1\ 0\ 0\ 1} \quad\quad \underline{\ 9} \\
\checkmark \\
\underline{1} \\
1\ 1\ 1\ 0\ 0_2 \equiv 28_{10}
\end{array}
$$

Example 26.6   Add the decimal numbers 79 and 31 by binary means.

$$
\begin{array}{c}
1\ 0\ 0\ 1\ 1\ 1\ 1 \quad\quad 79 \\
\underline{0\ 0\ 1\ 1\ 1\ 1\ 1} \quad\quad \underline{31} \\
\checkmark\ \checkmark\ \checkmark\ \checkmark \\
\underline{1\ 1\ 1\ 1} \\
1\ 1\ 0\ 1\ 1\ 1\ 0_2 \equiv 110_{10}
\end{array}
$$

In Chapter 27, we will be introduced to system components which carry out the functions necessary for such additions. Such systems are termed *logic systems*.

26.5 Binary subtraction

Once again, we require some simple rules which are as follows:

$$0 - 0 = 0$$
$$1 - 0 = 1$$
$$1 - 1 = 0$$
$$0 - 1 = 1 \quad \text{and deduct 1 from the next}$$
$$\text{higher order of significance}$$
$$1 - 1 - 1 = 0 \quad \text{and deduct 1 from the next}$$
$$\text{higher order of significance}$$

The deduction again commences with the LSB as illustrated in Example 26.7.

Example 26.7   Subtract 01011_2 from 11001_2.

$$
\begin{array}{c}
1\ 1\ 0\ 0\ 1 \\
\underline{0\ 1\ 0\ 1\ 1} \\
\checkmark \\
\underline{1} \\
1\ 0\ 1\ 1\ 0_2
\end{array}
$$

26.6 Binary multiplication

Digital devices can produce multiplication by the shift–and–add method, which is similar to the decimal multiplication technique with which we should be familiar. There are some basic rules for binary multiplication as follows:

$$1 \times 1 = 1$$
$$1 \times 0 = 0$$
$$0 \times 0 = 0$$

When undertaking multiplication, we take each digit in turn and shift the resulting product one space to the left, i.e. raising it to the next significant power. This is demonstrated in Example 26.8.

Example 26.8 Multiply the binary numbers 1101 and 0101.

$$
\begin{array}{r}
1\ 1\ 0\ 1 \\
\underline{0\ 1\ 0\ 1} \\
1\ 1\ 0\ 1 \\
0\ 0\ 0\ 0 \\
1\ 1\ 0\ 1 \\
\underline{0\ 0\ 0\ 0} \\
\checkmark\ \checkmark\ \checkmark\ \checkmark \\
\underline{1\ 1\ 1\ 1} \\
1\ 0\ 0\ 0\ 0\ 0\ 1
\end{array}
$$

hence $1101 \times 0101 = \mathbf{1000001}$

The steps which involved multiplication by 0 could have been omitted as shown in Example 26.9.

Example 26.9 Multiply the decimal numbers 27 and 10.

$$
\begin{array}{rr}
1\ 1\ 0\ 1\ 1 & 27 \\
\underline{1\ 0\ 1\ 0} & 10 \\
1\ 1\ 0\ 1\ 1 \\
\underline{1\ 1\ 0\ 1\ 1} \\
\checkmark\ \checkmark\ \checkmark \\
\underline{1\ 1\ 1} \\
1\ 0\ 0\ 0\ 0\ 1\ 1\ 1\ 0 \equiv \mathbf{270}
\end{array}
$$

Digital devices such as computers can only add two numbers at a time. When there are more than two intermediate stages as in Example 26.9, then the first two are added, then the third and so on. This is particularly helpful since it saves undue difficulties with the carry-forward digits. This is demonstrated in Example 26.10.

Example 26.10 Multiply 45_{10} by 25_{10} using binary means.

$$
\begin{array}{rr}
1\ 0\ 1\ 1\ 0\ 1 & 45 \\
\underline{1\ 1\ 0\ 0\ 1} & 25 \\
1\ 0\ 1\ 1\ 0\ 1 \\
\underline{1\ 0\ 1\ 1\ 0\ 1} \\
1\ 1\ 0\ 0\ 1\ 0\ 1\ 0\ 1 \\
\underline{1\ 0\ 1\ 1\ 0\ 1} \\
1\ 0\ 0\ 0\ 1\ 1\ 0\ 0\ 1\ 0\ 1 \equiv \mathbf{1125}
\end{array}
$$

26.7 Binary division

Digital devices can produce division by the shift-and-subtract method, which is similar to the binary multiplication technique with which we should now be familiar. There are some basic rules for binary division. To explain them, let us assume a four-digit divisor which we are dividing into the dividend. The basic rules are as follows:

1.   Compare the divisor with the four most significant digits of the dividend, i.e. the four digits furthest to the left. If the divisor is smaller than the dividend, place a 1 in the most significant quotient position; next subtract the divisor from the four most significant digits of the dividend – this produces a partial remainder. This is very similar to the process of long division which we probably met early in our school-days. If the divisor is larger than the four most significant digits of the dividend, place a 0 in the most significant quotient position; do not subtract the divisor.

2.   Draw down the next digit from the dividend and repeat procedure 1.

3.   Continue repeating 2 until the LSB is reached.

Example 26.11   Divide 63_{10} by 9_{10} by means of binary numbers.

$$63_{10} = 111111_2$$

$$9_{10} = 1001_2$$

```
                             1 1 1   quotient
divisor  1 0 0 1 | 1 1 1 1 1 1       dividend
                   1 0 0 1 ↓
                   0 1 1 0 1
                     1 0 0 1
                     0 1 0 0 1
                       1 0 0 1
                       0 0 0 0   remainder
```

hence $111111 \div 1001 = 111$

and $63_{10} \div 9_{10} = 7_{10}$

In this case there was no remainder since 9 divides exactly into 63. Let us consider a situation in which there will be a remainder.

Example 26.12   Divide 61_{10} by 9_{10} by means of binary numbers.

$$61_{10} = 111101_2$$

$$9_{10} = 1001_2$$

```
                           1 1 0
         1 0 0 1 | 1 1 1 1 0 1
                   1 0 0 1 ↓
                   0 1 1 0 0
                   1 0 0 1 ↓
                   0 0 1 1 1
```

hence $111101 \div 1001 = 110$ remainder 111

and $61_{10} \div 9_{10} = 6_{10}$ **remainder** 7_{10}

The above process introduces a method of undertaking the binary division function. In digital systems, it is not possible to compare two numbers, although a digital system can subtract one number from another so that it can inspect the result.

A technique using this form of inspection is called restoring division, but it requires that we can cope with the outcome of a subtraction which results in a negative binary number.

26.8 Negative binary numbers

In order to operate with negative binary numbers, it is necessary to become familiar with the 1's and 2's complements of binary numbers. Usually computers handle negative binary numbers using the 2's complements but it is better to be aware of both systems.

The 1's complement of a binary number is found by changing all the 1's to 0's and all the 0's to 1's as shown below:

$$1 \ 0 \ 1 \ 1 \ 0 \ 1 \ 1 \ 0 \ 1 \qquad \text{Binary number}$$
$$\downarrow \ \downarrow \ \downarrow \ \downarrow \ \downarrow \ \downarrow \ \downarrow \ \downarrow$$
$$0 \ 1 \ 0 \ 0 \ 1 \ 0 \ 0 \ 1 \ 0 \qquad \text{1's complement}$$

The 2's complement can be found by adding 1 to the LSB of the 1's complement.

$$\text{2's complement} = (\text{1's complement}) + 1 \qquad [26.1]$$

This method is illustrated in Example 26.13.

Example 26.13   Determine the 2's complement of 1101100_2 by applying the 1's complement.

$$1111111 - 1101100 + 1 = 0010011 + 1 = \mathbf{0010100}$$

An alternative method of finding the 2's complement is as follows:

1. Start with the LSB and, moving left, write down the bits as they appear up to and including the first 1.
2. Continuing left, write the 1's complements of the remaining bits.

This is demonstrated in Example 26.14.

Example 26.14   Determine the 2's complement of 1101100 directly.

Split the number to the left of the lowest 1.

$$\begin{array}{c|c} 1 \ 1 \ 0 \ 1 & 1 \ 0 \ 0 \\ \text{change} & \text{no change} \\ \text{(invert)} & \\ 0 \ 0 \ 1 \ 0 & 1 \ 0 \ 0 \end{array}$$

hence   2's complement of $1101100 = \mathbf{0010100}$

Digital systems may need to handle both positive and negative numbers. To distinguish between positive and negative, each number can be given a *sign bit*. The left-hand bit of a signed binary number is the sign bit and it indicates the polarity of the number, i.e.

a 0 indicates that the number is positive
a 1 indicates that the number is negative

It follows that in a signed binary number, the left-hand bit is the sign bit and the remaining bits are the magnitude bits. Thus in an 8-bit signed binary number system, the decimal number +30 is expressed as

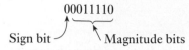

$$00011110$$

Sign bit ⟋   ⟍ Magnitude bits

If the decimal number had been −30 then the signed binary number would have been

$$10011110$$

A binary number need only be signed when there is the possibility of positive and negative numbers appearing in the same system.

Example 26.15   Determine the decimal value of the signed binary number 10011010.

The weights are as follows:

2^6	2^5	2^4	2^3	2^2	2^1	2^0
0	0	1	1	0	1	0

Summing the weights,

$$16 + 8 + 2 = 26$$

The sign bit is 1 therefore the decimal number is **−26**.

26.9   Signed binary addition

In section 26.4, we were introduced to the addition of binary numbers. Let us now consider the addition of signed binary numbers. In Example 26.16, two positive signed binary numbers are added and it can be seen that effectively the outcome is as before. Since these numbers are signed, let us continue to assume the 8-bit format.

Example 26.16   Add 9 and 3 by means of signed binary numbers.

```
  0 0 0 0 1 0 0 1          +9
  0 0 0 0 0 0 1 1          +3
  0 0 0 0 1 1 0 0         +12
```

The first bit is 0 and hence the number is positive as expected.

In the next example 26.17, we shall let the number with the larger magnitude be negative and again carry out the addition process.

Example 26.17   Add −9 and +3 by means of signed binary numbers.

To represent −9, we take the 2's complement thus

```
  1 1 1 1 0 1 1 1          −9      (2's complement)
  0 0 0 0 0 0 1 1          +3
  1 1 1 1 1 0 1 0          −6
```

The sum is negative since the first bit is 1. It is in 2's complement form and hence its magnitude is 0000110 which is 6 and hence the sum is **−6**.

Example 26.18 Add −9 and −3 by means of signed binary numbers.

$$
\begin{array}{llr}
& 1\ 1\ 1\ 1\ 0\ 1\ 1\ 1 & -9 \\
& 1\ 1\ 1\ 1\ 1\ 1\ 0\ 1 & -3 \\
\text{Carry} \to 1 & 1\ 1\ 1\ 1\ 0\ 1\ 0\ 0 & -12
\end{array}
$$

In this solution, we discard the carry bit. The number is negative and its magnitude in true binary form is 0001100 giving a decimal number of **−12**.

Computers carry out most processes by addition. This process is taken one step at a time, thus if a computer were to add three numbers, it would add the first two to give a subtotal and then add the third. Computers can work extremely quickly and therefore this apparently tedious approach need not be as big a problem as it would appear.

Example 26.19 Add the signed numbers 00001000, 00011111, 00001111 and 00101010.

For convenience the decimal equivalent numbers appear in the right-hand column below.

$$
\begin{array}{lll}
0\ 0\ 0\ 0\ 1\ 0\ 0\ 0 & & 8 \\
0\ 0\ 0\ 1\ 1\ 1\ 1\ 1 & & +31 \\
0\ 0\ 1\ 0\ 0\ 1\ 1\ 1 & 1^{\text{st}}\ \text{sum} & 39 \\
0\ 0\ 0\ 0\ 1\ 1\ 1\ 1 & & +15 \\
0\ 0\ 1\ 1\ 0\ 1\ 1\ 0 & 2^{\text{nd}}\ \text{sum} & 54 \\
0\ 0\ 1\ 0\ 1\ 0\ 1\ 0 & & +42 \\
0\ 1\ 1\ 0\ 0\ 0\ 0\ 0 & 3^{\text{rd}}\ \text{sum} & 96
\end{array}
$$

26.10 Signed binary subtraction

Subtraction is easily obtained by using the 2's complement of the number (thereby making it a negative number) to be subtracted, and adding it to the other number taking care to discard any final carry bit. This is demonstrated in Example 26.20.

Example 26.20 Subtract +9 from +12.

$$
\begin{array}{llr}
& 0\ 0\ 0\ 0\ 1\ 1\ 0\ 0 & +12 \\
& 1\ 1\ 1\ 1\ 0\ 1\ 1\ 1 & -9 \quad \text{(2's complement)} \\
1 & 0\ 0\ 0\ 0\ 0\ 0\ 1\ 1 & +3
\end{array}
$$

Discard the carry and we see that the number is positive with a magnitude of 3, i.e. the outcome is **+3**.

Example 26.21 Subtract +19 from −24.

$$
\begin{array}{llr}
& 1\ 1\ 1\ 0\ 1\ 0\ 0\ 0 & -24 \\
& 1\ 1\ 1\ 0\ 1\ 1\ 0\ 1 & -19 \\
1 & 1\ 1\ 0\ 1\ 0\ 1\ 0\ 1 & -43
\end{array}
$$

Discard the carry and we see that the number is negative with a magnitude of 43, i.e. the outcome is **−43**.

26.11   **Signed binary multiplication**

In its most basic form of operation, a computer could achieve multiplication by adding a number to itself by the number of times featured in the other number. For instance, if we were to multiply 6 by 4 then the computer might add 6 to 6 then add a third 6 and finally add a fourth 6. Such a system could become very lengthy depending on the magnitudes of the numbers.

A better system uses partial products. We can use partial products when multiplying decimal numbers, e.g.

$$
\begin{array}{rl}
125 & \\
\times\,31 & \\
\hline
125 & 1^{st}\text{ partial product is }1\times125 \\
375 & 2^{nd}\text{ partial product is }3\times125 \\
\hline
3875 & \text{Final product}
\end{array}
$$

A similar approach can be used for signed binary multiplication using the following procedures:

1. Determine whether the product will be positive or negative – if the MSB's are the same, the product is positive; if different, the product is negative.
2. Change any negative number to its true (uncomplemented) form.
3. Starting with the LSB, generate the partial products using only the magnitude bits. This process is demonstrated in Example 26.22.
4. Add each successive partial product as it occurs.
5. If the sign bit, determined in step 1, is negative, take the 2's complement of the final product. If positive, leave the product in its true (uncomplemented) form. Attach the appropriate sign bit to the final product.

This all appears somewhat complicated but it is really quite simple as will now be demonstrated.

Example 26.22   Multiply the signed binary numbers 01010011 and 11001101.

Step 1: the MSB's are different therefore the product will be negative.

Step 2: change the negative number to its true (uncomplemented) form.

11001101	becomes	00110011

Step 3: multiplication of the magnitude bits

```
                1 0 1 0 0 1 1
              × 0 1 1 0 0 1 1
Step 4:         1 0 1 0 0 1 1      1st partial product
              1 0 1 0 0 1 1        2nd partial product
              1 1 1 1 1 0 0 1
            0 0 0 0 0 0 0          3rd partial product
            0 1 1 1 1 1 0 0 1
          0 0 0 0 0 0 0            4th partial product
          0 0 1 1 1 1 1 0 0 1
        1 0 1 0 0 1 1              5th partial product
        1 1 0 0 0 1 0 1 0 0 1
      1 0 1 0 0 1 1                6th partial product
      1 0 0 0 0 1 0 0 0 1 0 0 1
    0 0 0 0 0 0 0                  7th partial product
    1 0 0 0 0 1 0 0 0 1 0 0 1      Final product
```

Step 5: since the product MSB is 1, take the 2's complement of the final product and attach the sign bit.

$$1000010001001 \quad \text{becomes} \quad 0111101110111$$

and with the sign bit, the product of the given numbers is **10111101110111**.

If you care to check, you will find that the equivalent decimal numbers were 83 and −51 giving a product of −4233.

26.12 Signed binary division

As with multiplication, a computer could achieve division by subtracting the divisor from the dividend until there was no remainder. For instance, dividing 8 by 2 could be achieved by taking the divisor 2 from the dividend 8 to leave 6, with a further 2 subtracted to leave 4 and so on; four subtractions would leave no remainder and therefore we find that 8 divided by 2 goes four times and the quotient is 4. This is a very clumsy method whenever the numbers become large but it serves to introduce the partial remainder system.

When dividing signed binary numbers, both must be in true (uncomplemented) form.

1. Determine whether the quotient will be positive or negative – if the MSB's are the same, the quotient is positive; if different, the quotient is negative.
2. Initialize the quotient to zero.
3. Subtract the divisor from the dividend using the 2's complement to get the first partial remainder and add 1 to the quotient. If the partial remainder is positive, continue to step 4. If the partial remainder is zero or negative, the division is complete.
4. Subtract the divisor from the partial remainder and add 1 to the quotient. Repeat this procedure until the outcome is zero or negative when the division is complete.

Again this all seems somewhat complicated but it is more readily understood when we consider Example 26.23.

Example 26.23

Divide 01111101 by 00011001.

Step 1: the MSB's are the same and therefore the quotient will be positive.

Step 2: set the quotient at 00000000.

Step 3: subtract, using 2's complement addition

$$
\begin{array}{ll}
0\,1\,1\,1\,1\,1\,0\,1 & \\
+\,\underline{1\,1\,1\,0\,0\,1\,1\,1} & \text{(2's complement)} \\
1\,0\,1\,1\,0\,0\,1\,0\,0 & \text{Positive 1}^{\text{st}}\text{ partial remainder}
\end{array}
$$

The final carry is discarded and 1 is added to the quotient, i.e.

$$00000000 + 00000001 = 00000001$$

Step 4:
$$
\begin{array}{ll}
0\,1\,1\,0\,0\,1\,0\,0 & \text{1}^{\text{st}}\text{ partial remainder} \\
\underline{1\,1\,1\,0\,0\,1\,1\,1} & \\
0\,1\,0\,0\,1\,0\,1\,1 & \text{Positive 2}^{\text{nd}}\text{ partial remainder}
\end{array}
$$

Again the final carry has been discarded and 1 is added to the quotient, i.e.

$$00000001 + 00000001 = 00000010$$

0 1 0 0 1 0 1 1	2nd partial remainder
1 1 1 0 0 1 1 1	
0 0 1 1 0 0 1 0	Positive 3rd partial remainder

$$00000010 + 00000001 = 00000011$$

0 0 1 1 0 0 1 0	3rd partial remainder
1 1 1 0 0 1 1 1	
0 0 0 1 1 0 0 1	Positive 4th partial remainder

$$00000011 + 00000001 = 00000100$$

0 0 0 1 1 0 0 1	4th partial remainder
1 1 1 0 0 1 1 1	
0 0 0 0 0 0 0 0	Final remainder is zero or less

$$00000100 + 00000001 = 00000101$$

The MSB is set for a positive quotient and therefore the outcome of the division is **00000101**.

If you care to convert the numbers to decimal, you will find that we have divided 125 by 25 to obtain 5.

26.13	**The octal system**

From Example 26.3, it is clear that as decimal numbers become larger, the binary number takes up more and more digits. This can be reduced by using a variation on the binary system called the octal system which has eight digits, i.e. 0 to 7. It follows that each octal digit represents a power of 8.

Example 26.24 Convert the octal number 236_8 to decimal.

$$236_8 = 2 \times 8^2 + 3 \times 8^1 + 6 \times 8^0$$

$$= 2 \times 64 + 3 \times 8 + 6 \times 1$$

$$= 128 + 24 + 6$$

$$= \mathbf{158}$$

Each octal digit would require to be replaced by three binary digits. This can be observed in Example 26.25.

Example 26.25 Convert 125_8 to binary.

The binary for the first digit is   001
    for the second digit is   010
    for the third digit is   101

hence $125_8 = \mathbf{001010101}_2$ in binary

$$= 2^6 + 2^4 + 2^2 + 2^0$$

$$= 64 + 16 + 4 + 1 = 85_{10}$$

This octal system can be applied to decimal conversion after the manner introduced in Example 26.2.

Example 26.26

Convert decimal 6735 to binary.

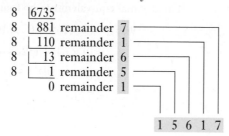

hence $\quad 6735_{10} = 15617_8$

$$= 001\ 101\ 110\ 001\ 111_2$$

We can check that this is correct by converting the solution back to decimal as follows:

$$1101110001111 = 12^{12} + 2^{11} + 2^9 + 2^8 + 2^7 + 2^3 + 2^2 + 2^1 + 2^0$$

$$= 4096 + 2048 + 512 + 256 + 128 + 8 + 4 + 2 + 1$$

$$= 6735$$

26.14 Hexadecimal numbers

Binary numbers are often represented for convenience in hexadecimal code, i.e. base 16 (= 2^4). In the hexadecimal system 16 digits are required: the 10 decimal digits and the first 6 letters of the alphabet. Table 26.2 shows the

Table 26.2

Binary	Decimal	Hexadecimal
0000	0	0
0001	1	1
0010	2	2
0011	3	3
0100	4	4
0101	5	5
0110	6	6
0111	7	7
1000	8	8
1001	9	9
1010	10	A
1011	11	B
1100	12	C
1101	13	D
1110	14	E
1111	15	F

comparison of binary, decimal and hexadecimal numbers. Using hexadecimal base, groups of 4 bits can be represented by a single digit:

$$(1001\ 1100\ 0101)_2 = 9C5_{16}$$
$$(1110\ 0000\ 1010)_2 = E0A_{16}$$

The decimal equivalent can be found by

$$9C5_{16} = (9 \times 16^2) + (C \times 16^1) + (5 \times 16^0)$$
$$= (9 \times 64) + (12 \times 16) + (5 \times 1)$$
$$= (576 + 192 + 5)_{10}$$
$$= 773_{10}$$

Terms and concepts

Digital transmission of data requires that numerical data are available in **binary** form.

We can convert decimal numbers to binary. This quickly gives rise to large numbers of binary digits but these can be reduced by the **octal system**.

We can perform addition, subtraction, multiplication and division using binary digits.

For best use of digital systems, it can be appropriate to apply the **hexadecimal system** of numbers.

Exercises 26

1. Convert 15_{10} to binary.
2. Convert 27_{10} to binary.
3. Convert 86_{10} to binary.
4. Determine the MSB and the LSB of 45_{10} when expressed as a binary number.
5. Convert 11001_2 to decimal.
6. Convert 101011_2 to decimal.
7. Convert 157_8 to decimal.
8. Convert 172_8 to binary.
9. Convert decimal 5632 to binary.
10. Convert decimal 5633 to binary.
11. Add the binary numbers 1001 and 0110.
12. Add the decimal numbers 26 and 24 by binary means.
13. Add the decimal numbers 105 and 62 by binary means.
14. Subtract the binary number 01111 from 11001.
15. Subtract the decimal number 15 from 25 by binary means.
16. Multiply the binary numbers 101 and 010.
17. Multiply the binary numbers 10101 and 01001.
18. Multiply the decimal numbers 6 and 7 by binary means.
19. Multiply the decimal numbers 50 and 20 by binary means.
20. Divide the binary number 0111 into 1110.
21. Divide the decimal number 8 into 48 by use of binary numbers.
22. Divide the decimal number 8 into 62 by use of binary numbers.
23. Determine the 2's complement of 1010110_2.
24. By the use of binary numbers, determine: (a) $25 - 19$; (b) $35 - 19$; (c) $30 - 30$; (d) $18 - 24$; (e) $18 - 35$.

Chapter twenty-seven

Digital Systems

Objectives

When you have studied this chapter, you should
- be aware of the basic Boolean operators
- understand the basic rules of Boolean algebra
- be able to determine logic circuits to implement Boolean functions
- be able to use Karnaugh maps to simplify combinational logic
- understand how circuits with feedback can latch data in registers

Contents

Digital numbers and their arithmetic operations have already been introduced. In this chapter the logic devices that are used in circuits to realize the digital concepts are presented. In particular the use of OR, AND, NOT, NOR and NAND functions to create systems will be explained. There are rules that are used to establish the effects of certain combinations and these rules are termed rules of combinational logic. Karnaugh maps, which may be used for the simplification of combinational logic expressions, are introduced. Bistable circuits and registers, which are building blocks for many digital systems, are introduced.

By the end of the chapter, you will be able to develop and analyse logic networks and undertake exercises including logic network reduction.

27.1 Introduction to logic

It will be noted in Chapter 31 that an analogue amplifier has the inherent problem that the output is not a uniform amplification of the input, due to the non-linear nature of the circuit components and due to the variation of response with frequency. Such variation can be accepted in, say, sound amplification but, when it comes to dealing with signals carrying information, any variations would change the information. However, if the information can be reduced to a simple form of being, say, a circuit switched ON and OFF, and, moreover, if the ON condition could be represented by 10 V and the OFF condition by 0 V, it would not matter if after transmission the ON condition were reduced to 9 V, as this could be applied to a new 10 V source, thereby ensuring the output was exactly the same as the input, i.e. we can ensure that the signal is returned to its original condition. It is by such means that we can see television transmitted from the other side of the world with no apparent loss of quality.

The condition therefore for digital operation is that all information must be capable of being reduced to one or other of two states. In logic systems, these states are ON and OFF, which correspond in human terms to the responses yes and no.

27.2 Basic logic statements or functions

A logic circuit is one that behaves like a switch, i.e. a two-position device with ON and OFF states. This is termed a binary device, in which the ON state is represented by 1 and the OFF state by 0.

We require to devise a logic statement or communication which can be expressed in only one of two forms. For instance, consider the options on boarding a bus with the possibility of it taking you to two towns, A and B.

27.3 The OR function

'The bus will take me to either A or B.' The success of the bus taking you to one or other can be represented by F; thus F occurs when the bus goes to A or B or both. It might travel through A to get to B or vice versa. Thus it can be stated as

$$F = A \text{ OR } B \qquad\qquad [27.1]$$

which in logic symbols is expressed as

$$F = A + B \qquad\qquad [27.2]$$

The positive sign is not the additive function, but means OR in logic. In an electrical system, this statement is equivalent to two switches in parallel, as shown in Fig. 27.1. The lamp F lights when either switch or both switches are closed.

Fig. 27.1 Switch arrangement equivalent to OR logic

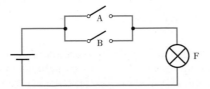

Table 27.1 Combinations of switching options

(a) Switching circuit			(b) Truth table		
A	B	Lamp	A	B	F
Open	Open	Off	0	0	0
Open	Closed	On	0	1	1
Closed	Open	On	1	0	1
Closed	Closed	On	1	1	1

This operation can be better understood by considering the four combinations of switching options as given in Table 27.1(a). The switching circuit of Table 27.1(a) can readily be reduced to the truth table in Table 27.1(b), using a 1 for the closed or ON condition and 0 for the open or OFF condition. A truth table permits a simple summary of the options for any logic function.

27.4 The AND function

'The bus will take me to A and B.' This success F occurs only when the bus goes to both A and B. This can be stated as

$$F = A \text{ AND } B \qquad [27.3]$$

which in logic symbols is expressed as

$$F = A \cdot B \qquad [27.4]$$

The period sign is not the multiplicative function but means AND in logic. In an electric circuit, this statement is equivalent to two switches in series, as shown in Fig. 27.2. The lamp F only lights when both are closed.

The truth table for the AND function is given in Table 27.2.

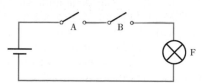

Fig. 27.2 Switch arrangement equivalent to AND logic

Table 27.2

A	B	F
0	0	0
0	1	0
1	0	0
1	1	1

27.5 The EXCLUSIVE-OR function

'The bus will take me either to A or to B but not to both.' In this case the success F can only be achieved either by A or by B but not by both, and the logic statement is

$$F = A \oplus B \qquad [27.5]$$

This situation is typically represented in an electric circuit by the two-way switching associated with a stair light, as shown in Fig. 27.3. The truth table for the EXCLUSIVE-OR function is given in Table 27.3.

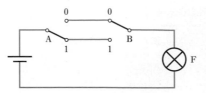

Fig. 27.3 Switch arrangement equivalent to the EXCLUSIVE-OR logic

Table 27.3

A	B	F
0	0	0
0	1	1
1	0	1
1	1	0

27.6 The NOT function

This is a simple function in which the input is inverted; thus if the input represents the ON (1) condition then the output F is the OFF (0) condition, and vice versa. It is stated in logic symbols as

$$F = \overline{A} \qquad\qquad [27.6]$$

and the truth table is given in Table 27.4.

Table 27.4

A	F
0	1
1	0

27.7 Logic gates

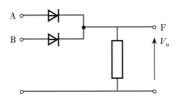

Fig. 27.4 Simple OR gate circuit

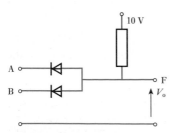

Fig. 27.5 Simple AND gate circuit

Circuits which perform logic functions are called gates. Generally, these are produced as circuit modules in the form associated with integrated circuits, or chips. In order to gain some understanding of the circuitry, consider the following simple OR and AND gates.

A simple diode OR gate is shown in Fig. 27.4. If no voltage is applied to either input then the output voltage V_o is also zero. If, however, a voltage of, say, 10 V is applied to either or both inputs then the respective diodes are forward biased and the output voltage is 10 V.

A simple diode AND gate is shown in Fig. 27.5. If zero voltage is applied to either A or B or to both, then the respective diodes are forward biased and current flows from the source. The result is that the output voltage V_o is 0 V. However, if 10 V is applied to both A and B then the p.d.s across both diodes are zero and the output voltage rises to 10 V from the source. In practice most gates incorporate switching transistors, but these are practical refinements of the principles illustrated by the diode gates.

It is unusual to show the circuitry of a gate; rather a symbol is given to represent the entire gate. The symbols are summarized in Fig. 27.6. Note that there are a number of systems of logic symbols in general use and only the two most common are shown. Gates can have more than two inputs; thus the truth tables in Fig. 27.6 have been extended for three inputs.

Fig. 27.6 Logic gate symbols and their respective truth tables

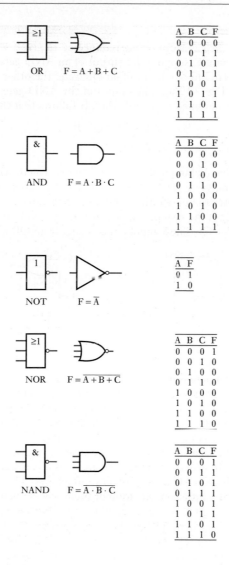

OR $F = A + B + C$

A	B	C	F
0	0	0	0
0	0	1	1
0	1	0	1
0	1	1	1
1	0	0	1
1	0	1	1
1	1	0	1
1	1	1	1

AND $F = A \cdot B \cdot C$

A	B	C	F
0	0	0	0
0	0	1	0
0	1	0	0
0	1	1	0
1	0	0	0
1	0	1	0
1	1	0	0
1	1	1	1

NOT $F = \overline{A}$

A	F
0	1
1	0

NOR $F = \overline{A + B + C}$

A	B	C	F
0	0	0	1
0	0	1	0
0	1	0	0
0	1	1	0
1	0	0	0
1	0	1	0
1	1	0	0
1	1	1	0

NAND $F = \overline{A \cdot B \cdot C}$

A	B	C	F
0	0	0	1
0	0	1	1
0	1	0	1
0	1	1	1
1	0	0	1
1	0	1	1
1	1	0	1
1	1	1	0

27.8 The NOR function

This function is the combination of an OR function and a NOT function. It is realized by connecting a NOT gate to the output of an OR gate. This inverts the output, providing the truth table shown in Fig. 27.6.

27.9 The NAND function

This function is realized by connecting a NOT gate to the output of an AND gate, again inverting the output as indicated by the truth tables shown in Fig. 27.6. Both NOR and NAND gates come as single gates, the combinations of functions having been reduced in each case to a single circuit.

27.10   **Logic networks**

Logic gates can be interconnected to give a wide variety of functions. As an example, suppose the output of an AND gate, with inputs A and C, is connected to one input of an OR gate, the other input to which is B, as shown in Fig. 27.7(a). The output of the AND gate is A · C and this expression is now applied to the OR gate. It follows that the output of the OR gate is

$$F = (A \text{ AND } C) \text{ OR } B$$

i.e. $F = A \cdot C + B$

This technique could have been reversed, whereby we could select a function, say $F = (A + B) \cdot (B + C)$, and we could then produce a combination of gates to operate this function. In this instance, we must start from the output of the system, which must be an AND gate having the inputs $(A + B)$ and $(B + C)$. These inputs must come from OR gates, as shown in Fig. 27.7(b).

Fig. 27.7   Simple logic networks

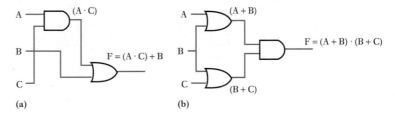

(a)

(b)

There is no end to the variety of such combinations and, as these expand, we quickly realize that we can have difficulties dealing with the calculation of the larger systems; and then we also have to ask whether we are making the best use of the gates. For instance, it will be shown that the combination shown in Fig. 27.7(b) could be achieved using only two gates instead of three, which clearly means that the arrangement shown is costing more than it need.

In order to handle such analyses it is necessary to set up a system of rules, which are known as Boolean identities. These are summarized in Fig. 27.8, and a little consideration of them will show that they are really self-evident.

Fig. 27.8   Boolean identities

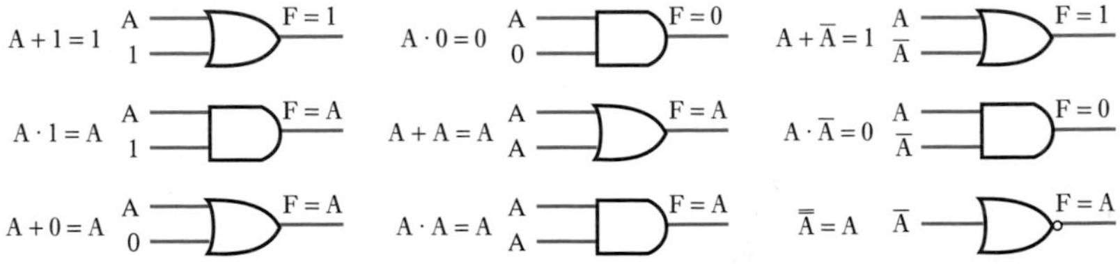

$A + 1 = 1$ $A \cdot 0 = 0$ $A + \bar{A} = 1$

$A \cdot 1 = A$ $A + A = A$ $A \cdot \bar{A} = 0$

$A + 0 = A$ $A \cdot A = A$ $\bar{\bar{A}} = A$

These identities are the logic equivalents of the basic rules of addition and multiplication of numbers. Preferably they should be memorized, although repeated use of them will produce the necessary familiarity. The relationships can easily be derived by considering the logic function included in each identity and then considering the variables to be inputs to the appropriate logic gate.

As an instance, consider the first identity which involves an OR function. The inputs are A and 1, and we know that the output of an OR gate is 1 provided that one of the inputs is 1. In this case, since one of the inputs is 1, it follows that the output must always be 1.

A similar consideration of the other eight identities will show them to be proved.

27.11	**Combinational logic**

In order to deal with more complex logic systems, we need to become familiar with a number of Boolean theorems.

(a) **The commutative rules**

$$A + B = B + A \qquad\qquad [27.7]$$

$$A \cdot B = B \cdot A \qquad\qquad [27.8]$$

i.e. the order of presentation of the terms is of no consequence.

(b) **The associative rules**

$$A + (B + C) = (A + B) + C \qquad\qquad [27.9]$$

$$A \cdot (B \cdot C) = (A \cdot B) \cdot C \qquad\qquad [27.10]$$

i.e. the order of association of the terms is of no consequence.

(c) **The distributive rules**

$$A + B \cdot C = (A + B) \cdot (A + C) \qquad\qquad [27.11]$$

$$A \cdot (B + C) = A \cdot B + A \cdot C \qquad\qquad [27.12]$$

These rules provide the logic equivalents of factorization and expansion in algebra, although we do not use the terms factorization and expansion in Boolean logic. We can satisfy ourselves that relation [27.11] is correct by means of the truth table in Table 27.5. Similarly, we can satisfy ourselves that equation [27.12] is also correct.

There are two interesting cases of the distributive rules in which the inputs are limited to A and B; thus

$$A + A \cdot B = A \qquad\qquad [27.13]$$

$$A \cdot (A + B) = A \qquad\qquad [27.14]$$

Table 27.5

A	B	C	B · C	A + B · C	A + B	A + C	(A + B) · (A + C)
0	0	0	0	0	0	0	0
0	0	1	0	0	0	1	0
0	1	0	0	0	1	0	0
0	1	1	1	1	1	1	1
1	0	0	0	1	1	1	1
1	0	1	0	1	1	1	1
1	1	0	0	1	1	1	1
1	1	1	1	1	1	1	1

These are particularly useful in reducing the complexity of a combinational logic system.

(d)   de Morgan's laws

$$\overline{A \cdot B \cdot C} = \overline{A} + \overline{B} + \overline{C} \qquad\qquad [27.15]$$

$$\overline{A} \cdot \overline{B} \cdot \overline{C} = \overline{A + B + C} \qquad\qquad [27.16]$$

If we consider the truth table for a NAND gate, we can observe the proof of equation [27.15]. A NAND gate with inputs A, B and C can be expressed as

$$F = \text{NOT (A AND B AND C)}$$
$$= \overline{A \cdot B \cdot C}$$

The truth table for the NAND gate is as shown in Table 27.6. From Table 27.6, it can be observed that

$$F = \overline{A} + \overline{B} + \overline{C}$$

hence $\overline{A \cdot B \cdot C} = \overline{A} + \overline{B} + \overline{C}$

A similar proof of equation [27.16] can be observed from the NOR gate.

With the use of the various rules listed above, it is possible to reduce the numbers of gates.

If we consider the output function of the network shown in Fig. 27.7(b) then from the distributive rule [27.11]

$$F = (A + B) \cdot (B + C)$$
$$= (B + A) \cdot (B + C)$$
$$= B + A \cdot C$$
$$= A \cdot C + B$$

However, the output of the network shown in Fig. 27.7(a) is also $F = A \cdot C + B$ and therefore the two networks perform the same function. However, the network in Fig. 27.7(a) uses only two gates and therefore makes better use of the components.

Consider a practical instance, as illustrated by the following example.

Table 27.6 Truth table for three-input NAND gate

A	B	C	$\overline{A}$	$\overline{B}$	$\overline{C}$	F
0	0	0	1	1	1	1
0	0	1	1	1	0	1
0	1	0	1	0	1	1
0	1	1	1	0	0	1
1	0	0	0	1	1	1
1	0	1	0	1	0	1
1	1	0	0	0	1	1
1	1	1	0	0	0	0

Example 27.1

An electrical control system uses three positional sensing devices, each of which produce 1 output when the position is confirmed. These devices are to be used in conjunction with a logic network of AND and OR gates and the output of the network is to be 1 when two or more of the sensing devices are producing signals of 1. Draw a network diagram of a suitable gate arrangement.

If we consider the possible combinations which satisfy the necessary conditions, it will be observed that there are four, i.e. any two devices or all three devices providing the appropriate signals; thus

$$F = A \cdot B \cdot \overline{C} + A \cdot \overline{B} \cdot C + \overline{A} \cdot B \cdot C + A \cdot B \cdot C$$

The term $A \cdot B \cdot C$ can be repeated as often as desired, hence

$$F = A \cdot B \cdot \overline{C} + A \cdot B \cdot C + A \cdot \overline{B} \cdot C + A \cdot B \cdot C$$
$$+ \overline{A} \cdot B \cdot C + A \cdot B \cdot C$$

Using the second distributive rule

$$F = A \cdot B \cdot (\overline{C} + C) + A \cdot C \cdot (\overline{B} + B) + B \cdot C \cdot (\overline{A} + A)$$

but applying the identity illustrated in Fig. 27.8

$$A + \overline{A} = 1, \quad B + \overline{B} = 1 \quad \text{and} \quad C + \overline{C} = 1$$

hence $\quad F = A \cdot B + B \cdot C + C \cdot A$

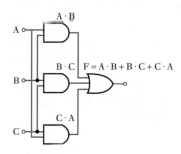

Fig. 27.9

The network which would effect this function is shown in Fig. 27.9.

Example 27.2

Draw the circuit of gates that could effect the function

$$F = \overline{A \cdot B + A \cdot C}$$

Simplify this function and hence redraw the circuit that could effect it.

The gate circuit based on the original function is shown in Fig. 27.10.

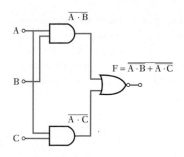

Fig. 27.10

Using de Morgan's theorems, $F = \overline{\overline{A \cdot B} + \overline{A \cdot C}}$

$$= \overline{\overline{A \cdot B}} \cdot \overline{\overline{A \cdot C}}$$

(associative rule) $\quad = A \cdot B \cdot A \cdot C$

$$= A \cdot B \cdot C$$

The simple circuit is shown in Fig. 27.11.

$$A \quad B \quad C \quad F = A \cdot B \cdot C$$

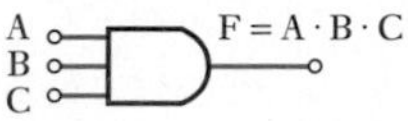

Fig. 27.11

In Example 27.2 it is seen that the original function could be effected by a simple AND gate, which would result in a substantial saving of gate components.

Not all applications of combinational logic rules result in a saving, as shown by Example 27.3.

| **Example 27.3** | Draw the circuit of gates that would effect the function |

$$F = \overline{A + B \cdot C}$$

Simplify this function and hence redraw the circuit that could effect it.

The gate circuit based on the original function is shown in Fig. 27.12. Using de Morgan's theorem

$$F = \overline{A + B \cdot C}$$
$$= \overline{A} \cdot \overline{(B \cdot C)}$$
$$= \overline{A} \cdot (\overline{B} + \overline{C})$$

This can be realized by the network shown in Fig. 27.13, which illustrates that, rather than there being a saving, we have involved the same number of gates with a greater number of inverters.

Fig. 27.12   **Fig. 27.13**

| **27.12** | **Gate standardization** |

In the previous sections it has been assumed that we are free to use any combination of logic gates. Sometimes it is easier to standardize by using a single form of gate. For example, any function can be effected using only NOR gates, and Figs 27.14, 27.15 and 27.16 illustrate how NOT, AND and OR gates can be realized using only NOR gates.

At first sight it would appear that restricting our choice of gates to one type has the advantage of simplicity – of avoiding using a range of gates – but this advantage is apparently offset by the major disadvantage that more gates are being used. However, this need not be as drastic as it might appear. Consider the instance of the NOT-EQUIVALENT gate illustrated by Example 27.4.

Fig. 27.14   NOT gate

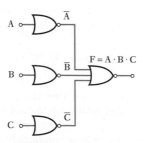

Fig. 27.15   AND gate

Fig. 27.16   OR gate

| Example 27.4 | Draw a logic circuit, incorporating any gates of your choice, which will produce an output 1 when its two inputs are different. Also draw a logic circuit, incorporating only NOR gates, which will perform the same function. |

For such a requirement, the function takes the form

$$F = \overline{A} \cdot B + A \cdot \overline{B}$$

This is the NOT-EQUIVALENT function and the logic circuit is shown in Fig. 27.17.

Fig. 27.17

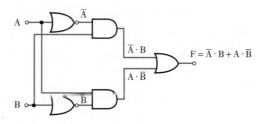

This can be converted directly into NOR logic gate circuitry, as shown in Fig. 27.18. Examination of the circuitry shows that two pairs of NOR gates are redundant since the output of each pair is the same as its input. These gates have been crossed out and the simplified circuit is shown in Fig. 27.19.

Fig. 27.18

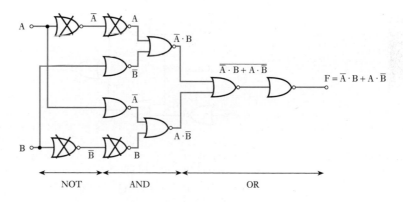

Fig. 27.19

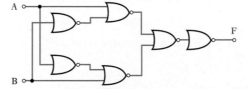

Similar systems to the one developed in Example 27.4 can be created using only NAND gates. Methods specified for NOR or NAND gates will not guarantee the simplest form of circuit, but the methods used in the reduction of circuits to their absolute minimal forms are beyond the scope of an introductory text.

Example 27.5   Draw circuits which will generate the function

$$F = B \cdot (\overline{A} + \overline{C}) + \overline{A} \cdot \overline{B}$$

using

(a) NOR gates;
(b) NAND gates.

$$F = B \cdot (\overline{A} + \overline{C}) + \overline{A} \cdot \overline{B}$$
$$= B \cdot \overline{A} + B \cdot \overline{C} + \overline{A} \cdot \overline{B} \quad \text{(second distributive rule)}$$
$$= \overline{A} \cdot (B + \overline{B}) + B \cdot \overline{C} \quad \text{(second distributive rule)}$$
$$= \overline{A} + B \cdot \overline{C} \quad\quad\quad \text{(first rule of complementation)}$$

(a) For NOR gates, complement of function is

$$\overline{F} = \overline{\overline{A} + B \cdot \overline{C}}$$
$$= A \cdot \overline{(B \cdot \overline{C})} \quad \text{(de Morgan's theorem)}$$
$$= A \cdot (\overline{B} + C) \quad \text{(de Morgan's theorem)}$$
$$= A \cdot \overline{B} + A \cdot C \quad \text{(second distributive rule)}$$

$A \cdot \overline{B}$ and $A \cdot C$ are generated separately, giving the circuit shown in Fig. 27.20.

Fig. 27.20

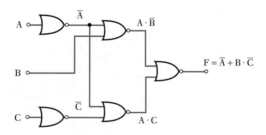

(b) For NAND gates

$$F = \overline{A} + B \cdot \overline{C}$$

Inputs to the final NAND gates are

$$\overline{\overline{A}} = A \quad \text{and} \quad \overline{B \cdot \overline{C}} = \overline{B} + C$$

$\overline{B} + C$ has to be generated separately, giving the circuit shown in Fig. 27.21.

Fig. 27.21

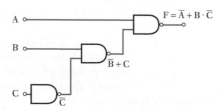

27.13 Karnaugh maps for simplifying combinational logic

Here we look at Karnaugh maps and how these can be used to simplify logic expressions. Minimizing logic expressions using Boolean algebra theory can be quite a tricky process, as it is quite often difficult to tell when the simplest version has been reached. Karnaugh maps are a powerful, widely used, graphical method of logic simplification that will always give the simplest sum-of-products form. Karnaugh maps allow expressions with up to four variables to be handled in a straightforward manner, and can be used with expressions of up to six variables. A Karnaugh map is a way of rearranging a truth table so that terms which can be grouped together are therefore simplified and are more easily identified. Each cell in the Karnaugh map corresponds to one line in the truth table. The map is, in essence, nothing more than the truth table redrawn.

The main stages to creating a logic expression using a Karnaugh map are:

1. Draw the empty Karnaugh map.
2. Fill in the 1's and 0's. (The information could come from different sources, i.e. from a logic expression or a truth table.)
3. Draw the loops on the Karnaugh map.
4. Create the expression for each loop to give the entire logic expression.

We will consider each of these stages here and then look at some complete examples.

(a) Creating an empty Karnaugh map

The number of inputs defines how the empty Karnaugh map should be drawn. For two variables, A and B, there are four possible states. A two-variable Karnaugh map therefore has four cells as shown in Fig. 27.22. These four cells are the equivalent of a truth table (Table 27.7), where the input variables

Fig. 27.22 Two-variable Karnaugh map

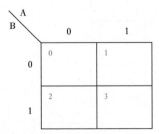

Table 27.7 Truth table for two variables. Binary code for inputs are shown in the right column

B	A	Output	Binary code
0	0		$00_2 = 0_{10}$
0	1		$01_2 = 1_{10}$
1	0		$10_2 = 2_{10}$
1	1		$11_2 = 3_{10}$

of the function are given on the outside of the table (A and B for this case), and the result of the function is given in the appropriate cell of the Karnaugh map. The small numbers in the top left corner of each cell show the binary code for that cell. Sometimes you will be given inputs in terms of the binary code. The input on the left (in this case B) is used as the most significant bit when calculating the binary code. Where binary numbers have A as the least significant bit and B, C and D rising powers of 2 it is easier to enter variables into a Karnaugh map if they are set up as shown here with D, C, B and A going from left to right. However, if you prefer to use A, B, C and D from left to right you will still get the correct answers; both approaches are correct and it is a matter of personal preference. It is useful to have the binary code numbers there as a convenient way of referring to each cell. Each of the cells corresponds to one state from the function's truth table. Each of the cells corresponds to one minterm – a minterm is the ANDing of one combination of all of the inputs of a function.

The appropriate output of the function to be simplified can then be entered in each cell. Karnaugh maps for three (Fig. 27.23) and four (Fig. 27.24) variables have 8 and 16 cells respectively. For five or six variables, three-dimensional maps or map-entered variables have to be used.

Fig. 27.23 Three-variable Karnaugh map

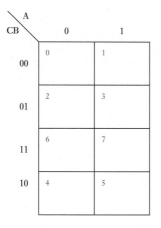

Fig. 27.24 Four-variable
Karnaugh map

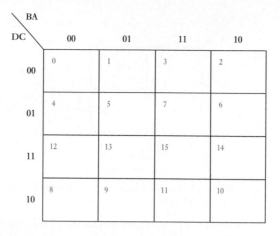

(b)   Filling in the Karnaugh map

Each cell of the Karnaugh map is filled with a 1 or a 0 signifying the output
for that cell. The information supplied enabling you to fill in the 1's and 0's
for the output in the Karnaugh map could be supplied in various formats.
For example:

*   by specifying the binary numbers that correspond to an output of 1;
*   from a logic expression;
*   from a truth table;
*   squares containing logic 0 terms may be left blank.

(c)   Filling in a Karnaugh map given the binary code numbers

The output information may be supplied by giving the binary cell numbers
for an output of 1 and stating what the input variables are.

Example 27.6   Fill out the Karnaugh map given $G = 1, 2, 3$ with input variables BA.

First look at what the question tells us. G is the name for the output in this
case. We are told that there are two inputs, B and A. This specifies the form
of the Karnaugh map, shown in Fig. 27.22. The binary cell numbers have
been shown. Make sure you understand how to get these numbers; they
come from the input variables A and B in this case.

The numbers in the statement $G = 1, 2, 3$ tell us where in the Karnaugh
map logical 1's occur. If we fill in 1's in the specified cells (corresponding to
the numbers 01_2, 10_2, 11_2) and 0's in the other cells we get the Karnaugh map
shown in Fig. 27.25. You may have noticed that this is the Karnaugh map for
the OR function.

Fig. 27.25 Completed
Karnaugh map

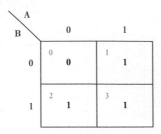

Example 27.7 Fill out the Karnaugh map given F = 0, 5, 6, 8, 12, 13 with input variables DCBA.

What is this problem specification? F is the name for the output in this case. We are told that there are four inputs, D, C, B and A. This specifies the form of the Karnaugh map, shown in Fig. 27.24. The binary cell numbers have been shown. Again, make sure you understand how to get these numbers.

The numbers in the statement F = 0, 5, 6, 8, 12, 13 tell us where in the Karnaugh map logical 1's occur (by converting the numbers to their binary form). If we fill in 1's in the specified cells and 0's in the other cells we get the Karnaugh map shown in Fig. 27.26.

Fig. 27.26 Completed
Karnaugh map for F = 0, 5, 6, 8,
12, 13 with inputs DCBA

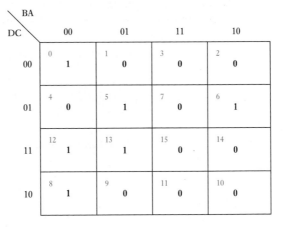

(d) Filling in a Karnaugh map given a logic expression

Sometimes the information required to fill out the Karnaugh map will be given in terms of a logic expression, where the idea of using the Karnaugh map would be to find a simpler logic expression. When you are given a logic expression (for instance, $F = \overline{A} \cdot \overline{B} + \overline{A} \cdot B + A \cdot \overline{B}$), the expression tells us what inputs give an output of 1.

When a line is present over an input this means the input is 0; with no line the input is 1.

Example 27.8

Given the following logic expression create and fill out a Karnaugh map:

$$F = A \cdot B + \overline{A} \cdot B + A \cdot \overline{B}$$

The expression given above tells us we are looking at using a two-input Karnaugh map with inputs A and B (Fig. 27.22), and the individual product terms in the expression tell us which cells have a 1 in them.

For instance, in the expression for F given here, this can be translated as having outputs of 0 for all combinations of inputs for A and B, except when $A = 0$ and $B = 1$ ($\overline{A} \cdot B$), when $A = 1$ and $B = 0$ ($A \cdot \overline{B}$), and when $A = 1$ and $B = 1$ ($A \cdot B$); this is the same as the function shown in the Karnaugh map in Fig. 27.25. You may have noticed that this is the Karnaugh map for the OR function, i.e. $A + B$.

Example 27.9

Given the following logic expression create and fill out a Karnaugh map:

$$G = A \cdot \overline{B} \cdot \overline{C} \cdot \overline{D} + A \cdot B \cdot \overline{C} \cdot \overline{D} + \overline{A} \cdot \overline{B} \cdot C \cdot \overline{D} + A \cdot \overline{B} \cdot C \cdot \overline{D}$$

$$+ A \cdot B \cdot C \cdot \overline{D} + \overline{A} \cdot B \cdot C \cdot \overline{D} + \overline{A} \cdot B \cdot C \cdot D$$

$$+ \overline{A} \cdot \overline{B} \cdot \overline{C} \cdot D + \overline{A} \cdot B \cdot \overline{C} \cdot D$$

The expression given above tells us we are looking at using a four-input Karnaugh map with inputs A, B, C and D (Fig. 27.24), and the individual product terms in the expression tell us which cells have a 1 in them.

For instance, in the expression for G given here, this can be translated as having outputs of 0 for all combinations of inputs for A, B, C and D, except for the nine inputs given. $A \cdot \overline{B} \cdot \overline{C} \cdot \overline{D}$ tells us there is a 1 when $A = 1$, $B = 0$, $C = 0$ and $D = 0$. $A \cdot B \cdot \overline{C} \cdot \overline{D}$ tells us there is a 1 when $A = 1$, $B = 1$, $C = 0$ and $D = 0$. $\overline{A} \cdot \overline{B} \cdot C \cdot \overline{D}$ tells us there is a 1 when $A = 0$, $B = 0$, $C = 1$ and $D = 0$ and so on. The resulting Karnaugh map can be seen in Fig. 27.27.

Fig. 27.27 Karnaugh map for four specified inputs

DC \ BA	00	01	11	10
00	0 **0**	1 **1**	3 **1**	2 **0**
01	4 **1**	5 **1**	7 **1**	6 **1**
11	12 **0**	13 **0**	15 **0**	14 **1**
10	8 **1**	9 **0**	11 **0**	10 **1**

Plotting a Karnaugh map from a truth table for a function involves plotting a '1' in the map if a particular minterm is included in the truth table, i.e. a '1' is plotted for each map square where X = 1.

Example 27.10   Given the truth table for the function G in Table 27.8, fill in the cells of the Karnaugh map.

Table 27.8   Truth table for function G

A	B	C	D	G
0	0	0	0	0
0	0	0	1	1
0	0	1	0	1
0	0	1	1	0
0	1	0	0	0
0	1	0	1	1
0	1	1	0	1
0	1	1	1	1
1	0	0	0	1
1	0	0	1	0
1	0	1	0	1
1	0	1	1	0
1	1	0	0	1
1	1	0	1	0
1	1	1	0	1
1	1	1	1	0

Considering the inputs in the order DCBA, the truth table is shown again in Table 27.9 with binary code so that the cells can be matched with the resulting Karnaugh map shown in Fig. 27.27.

Table 27.9   Truth table for function G

A	B	C	D	G	Binary code DCBA
0	0	0	0	0	0
0	0	0	1	1	8
0	0	1	0	1	4
0	0	1	1	0	12
0	1	0	0	0	2
0	1	0	1	1	10
0	1	1	0	1	6
0	1	1	1	1	14
1	0	0	0	1	1
1	0	0	1	0	9
1	0	1	0	1	5
1	0	1	1	0	13
1	1	0	0	1	3
1	1	0	1	0	11
1	1	1	0	1	7
1	1	1	1	0	15

(e) Creating loops in Karnaugh maps

Each cell of a Karnaugh map must have the appropriate output, for the function being represented, entered into it. When all the values are entered all of the 1's present must be grouped together according to the following rules:

- Groups must contain 2^n adjacent cells, each containing a 1.
- The groups should be made as large as possible, while remaining integer powers of 2.
- The groups have to be square or rectangular.
- The larger the groups chosen, the simpler the function, as with each increase of power of 2 in the size of the group, the group becomes dependent on one less variable. So the minimal form is obtained with the largest possible groups that are integer power of 2.
- All cells containing a 1 must be included in at least one group.
- The groups may overlap, so one cell may be included in several groups, but any group that has all its elements included in other groups is not required.
- There may be several correct minimal forms for a given logic function, dependent upon the particular groupings that are chosen.

The edges of a Karnaugh map are adjacent to each other because of the binary code used for labelling the cells. This means that groups can join from the left side to the right side and from the top to the bottom.

Using these rules the minimal form of a logic expression can be formed. The inverse of a function can be found using the same method, but grouping the 0's instead of the 1's.

Example 27.11   Consider the Karnaugh map shown in Fig. 27.27. Illustrate a possible set of loops.

A typical set of loops can be seen in Fig. 27.28.

Fig. 27.28 A Karnaugh map for four inputs

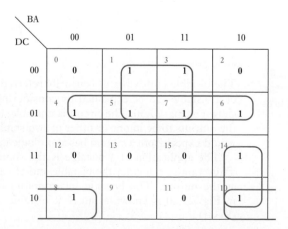

(f)   Going from the loops to a logic expression

Once the loops have been assigned a logic expression can be created. This is done one loop at a time. For each loop write down the inputs that describe that specific loop.

Example 27.12   Consider the loops shown in Fig. 27.28. We will look at generating the expression.

Consider the four loops one at a time, as shown in Fig. 27.29, and calculate the logic expression for each loop.

Fig. 27.29   Karnaugh map with identified loops

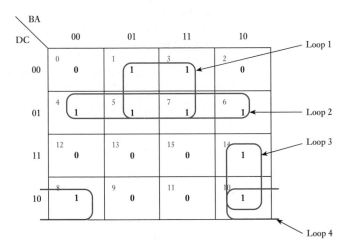

Loop 1 is $A \cdot \overline{D}$, loop 2 is $C \cdot \overline{D}$, loop 3 is $\overline{A} \cdot B \cdot D$ and loop 4 is $A \cdot \overline{C} \cdot D$. Finally we create the simplified logic expression by ORing the individual expressions giving the expression

$$A \cdot \overline{D} + C \cdot \overline{D} + \overline{A} \cdot B \cdot D + \overline{A} \cdot \overline{C} \cdot D$$

27.14   Bistable multivibrator circuits

The bistable multivibrator is usually referred to as a bistable. The operation of a bistable can be explained by considering a typical circuit such as that shown in Fig. 27.30. The circuit is unstable, but at any instant we can expect the outputs to be in one or other of two stable conditions, which is what we would expect from a system based on logic gates.

The application of a bistable is the short-term storage of information. Thus a pulse can enter the bistable and effectively appear a short time later at the output. The bistable incorporating two inverting gates as shown in Fig. 27.30 is known as an S–R bistable. The inputs are S (Set) and R (Reset).

Fig. 27.30

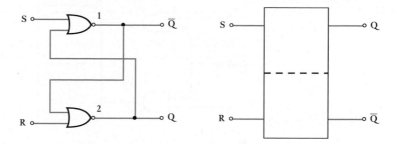

The inverting gates are based on NOR gates. The output of a NOR gate is 1 only if both inputs are 0; therefore if S or R is 1, the output of the respective gate must be 0. For instance if $S = 1$ and $R = 0$ then $\overline{Q} = 0$. However, this output is the second input to the second gate with the result that its output is $Q = 1$. In turn the output Q is the second input to the first gate; both its inputs are now 1 which ensures that Q remains in the 0 state.

Summary: when $S = 1$ and $R = 0$, the circuit is stable with $Q = 1$ and $\overline{Q} = 0$.

Let us now reset S to 0 so that both $S = R = 0$. Gate 1 still has an input set at 1, hence its output remains at $\overline{Q} = 0$. This makes the second input to the second gate 0 so that both its inputs are set at zero and hence its output Q remains at 1.

Summary: when $S = R = 0$, the outputs remain as before.

The next development is to reset R to 1 leaving $S = 0$. The inputs to the second gate are no longer both 0, hence $Q = 0$ and this becomes the second input to the first gate. However, $S = 0$, so both inputs are 0, hence $\overline{Q} = 1$. This output goes to the second input to the second gate so that both inputs are 1 and the output $\overline{Q}$ remains at 0.

Summary: when $S = 0$ and $R = 1$, the circuit is stable with $Q = 0$ and $\overline{Q} = 1$.

Let us return to $S = R = 0$ by resetting R to 0. Gate 2 still has an input set at 1, hence its output remains at $Q = 0$. This makes the second input to the first gate so that both its inputs are set at zero and hence its output $\overline{Q}$ remains at 1.

Summary: when $S = R = 0$, the outputs remain as before.

When $Q = 1$ and $\overline{Q} = 0$, the bistable is said to be in the SET condition. Conversely, when $Q = 0$ and $\overline{Q} = 1$, the bistable is said to be in the RESET condition. It follows that:

1. We set the bistable by setting $S = 1$ and $R = 0$.
2. We reset the bistable by setting $S = 0$ and $R = 1$.

27.15   Registers

We can consider a register to be a group of bistables. Each bistable can respond to an input of a bit. At the end of the bit, it does not change its output which remains as before. Effectively the bistable stores the bit. The group of bistables therefore provides data storage for a short period of time. At the rate of one bistable for every bit to be stored, it is too costly to use bistables extensively.

An S–R bistable is said to be an asynchronous device. This is a way of defining the system in which Q and $\overline{Q}$ immediately take up logic states which are set by inputs S and R. However, we need to ensure that bistable operation is synchronized to an external clock which provides a series of regular pulses which regulate when changes of output may occur.

Fig. 27.31

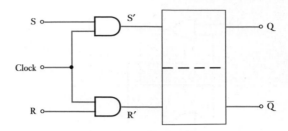

Let us consider a clocked bistable shown in Fig. 27.31. Before the clock provides a pulse its outputs are set to 0, hence the AND gates' outputs remain at 0. When the clock provides a pulse, then either S′ or R′ is set to 1, the respective AND gate has an output of 1 and the bistable reacts as before. Subsequent changes of R or S have no effect until the next clock pulse appears.

This is a very simple introduction to registers, but it is sufficient for the purposes of analogue-to-digital converters in Chapter 25 and microprocessor systems in Chapter 28. Electronics engineers require to consider a variety of bistables and registers in order to be familiar with those in common use.

27.16 Timing diagrams

So far, most of this chapter has assumed steady input conditions. In the sections dealing with bistable circuits and registers we have inferred that the inputs might vary. In practice, most digital systems are responding to a digitally varied set of inputs.

If an input signal is 1 then it is said to be HIGH and if it is 0 then the input is LOW. Let us consider a simple AND gate, as shown in Fig. 27.32, with varying inputs to A, B and C. In this example we wish to determine the output waveform.

Fig. 27.32 AND gate timing diagram

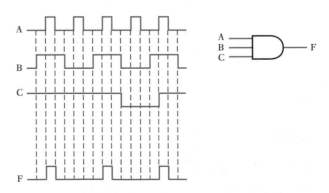

The output waveform can only be HIGH when all three input signals are HIGH, hence we can draw the output waveform as shown.

A similar output waveform can be derived for an OR gate as shown in Fig. 27.33.

Fig. 27.33 OR gate timing diagram

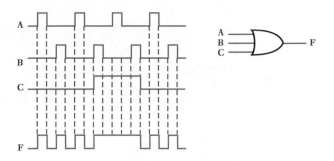

27.17 Integrated circuit logic gates

There are two commonly used digital integrated circuit formats used for basic logic gates. These are CMOS and TTL.

CMOS is the abbreviated form of Complementary Metal Oxide Semiconductor and it is the favoured form for many manufacturers. Its principal advantage is that of lower power dissipation compared with TTL. One reason for lower power dissipation is that the operating voltage for many gates was 5 V and this has been reduced to 3.3 V. The disadvantage with CMOS is that it is sensitive to electrostatic discharge. It also suffered from slower switching speeds. But this has largely been overcome and its speeds compare with TTL.

TTL stands for Transistor–Transistor Logic and is constructed using bipolar transistors. Although not as popular as it once was, it is useful when developing prototypes because of not being sensitive to electrostatic discharge. Many CMOS IC devices are pin compatible with the same devices in TTL so that development work can proceed using TTL and then be finished by transferring and manufacturing using CMOS ICs.

A typical logic gate IC contains four gates mounted in a 14-pin dual-in-line package (DIP) which is intended for plug-in applications. This format is shown in Fig. 27.34. An alternative format is the small outline package (SOIC) which is intended for surface mounting.

Fig. 27.34 14-pin DIP

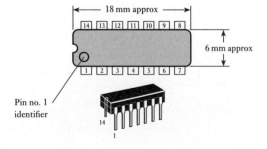

The gate arrangements come in many forms. Figure 27.35 illustrates only four of the many common IC gate configurations.

Fig. 27.35 Pin configuration diagrams for common IC gates

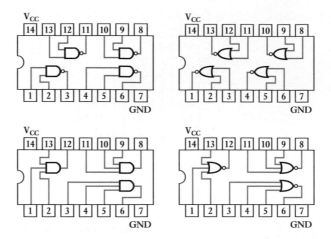

Summary of important formulae

The OR function,

$$F = A + B \qquad\qquad [27.2]$$

The AND function,

$$F = A \cdot B \qquad\qquad [27.4]$$

The NOT function,

$$F = \overline{A} \qquad\qquad [27.6]$$

The commutative rules,

$$A + B = B + A \qquad\qquad [27.7]$$

$$A \cdot B = B \cdot A \qquad\qquad [27.8]$$

The associative rules,

$$A + (B + C) = (A + B) + C \qquad\qquad [27.9]$$

$$A \cdot (B \cdot C) = (A \cdot B) \cdot C \qquad\qquad [27.10]$$

The distributive rules,

$$A + B \cdot C = (A + B) \cdot (A + C) \qquad\qquad [27.11]$$

$$A \cdot (B + C) = A \cdot B + A \cdot C \qquad\qquad [27.12]$$

de Morgan's laws,

$$\overline{A \cdot B \cdot C} = \overline{A} + \overline{B} + \overline{C} \qquad\qquad [27.15]$$

$$\overline{A} \cdot \overline{B} \cdot \overline{C} = \overline{A + B + C} \qquad\qquad [27.16]$$

Terms and concepts

Logic gates operate in two conditions – ON and OFF, i.e. they are digital.

Logic gates can be made from diode circuits but most incorporate transistors, usually FETs since they can effectively operate as though they were open circuits when in the OFF condition.

The basic logic gates are AND, OR, NAND and NOR. They can be connected in any group required but it is common practice to create circuits which only incorporate one form of gate, e.g. the NOR gate.

The number of gates required can often be reduced by the application of **combinational logic**.

Karnaugh maps are a convenient way of simplifying combinational logic expressions.

A common gate arrangement is the **S–R bistable** which locates its output in either of two conditions until it is intentionally reset.

A group of bistables can be made to form a **register** which is driven by a clock, i.e. its condition can only be set or reset by the application of pulses which are applied at regular intervals.

Exercises 27

1. Simplify the following Boolean expressions:

(a) $F = A + B \mid 1$
(b) $F = (A + B) \cdot 1$
(c) $F = A + B \cdot 1$

2. Simplify the following Boolean expressions:

(a) $F = A + B + 0$
(b) $F = (A + B) \cdot 0$
(c) $F = A + B \cdot 0$

3. Simplify the following Boolean expressions:

(a) $F = (A + B) \cdot (A + B)$
(b) $F = A \cdot B + B \cdot A$
(c) $F = (A + B \cdot \overline{C}) \cdot (A + B \cdot \overline{C})$

4. Simplify the following Boolean expressions:

(a) $F = (A + B) \cdot \overline{(A + B)}$
(b) $F = A \cdot B + \overline{A \cdot B}$
(c) $F = (A + \overline{B} \cdot C) \cdot \overline{(A + \overline{B} \cdot C)}$

5. Draw a network to generate the function
$F = \overline{A \cdot B} + C$.

6. Draw a network to generate the function
$F = \overline{\overline{A \cdot B} + \overline{A} \cdot C}$. Using de Morgan's laws, simplify this expression and hence draw the diagram of a simpler network which would produce the same result.

7. Draw a network to generate the function
$F = \overline{A} \cdot \overline{C} + A \cdot \overline{B} \cdot C$. Simplify this expression and

hence draw the diagram of a simpler network which produces the same result.

8. Simplify the following Boolean expression:
$F = A \cdot B \cdot \overline{C} + A \cdot \overline{B} \cdot C + A \cdot \overline{B} \cdot \overline{C} + A \cdot B \cdot C$

9. Simplify the following logic functions and hence draw diagrams of circuits which will generate the functions using: (a) AND, OR and NOT gates; (b) NAND gates; (c) NOR gates.

(i) $F = A \cdot \overline{B} \cdot \overline{C} + A \cdot \overline{B} \cdot C + \overline{A} \cdot \overline{B} \cdot \overline{C} + \overline{A} \cdot \overline{B} \cdot C$
(ii) $F = \overline{A} \cdot B \cdot \overline{C} + \overline{A} \cdot B \cdot C + A \cdot \overline{B} \cdot C + \overline{A} \cdot \overline{B} \cdot C$
(iii) $F = A \cdot \overline{B} \cdot \overline{C} + \overline{A} \cdot \overline{B} \cdot \overline{C} + \overline{A} \cdot B \cdot C$
(iv) $F = A \cdot B \cdot C + \overline{A} \cdot \overline{B} \cdot \overline{C}$
(v) $F = B \cdot \overline{C} \cdot \overline{D} + A \cdot \overline{B} \cdot D + B \cdot C \cdot \overline{D} + \overline{A} \cdot \overline{B} \cdot D$

10. Draw a circuit containing AND, OR and NOT gates to generate the function specified in the truth table below.

A	B	C	Z
0	0	0	0
0	0	1	1
0	1	0	1
0	1	1	0
1	0	0	1
1	0	1	0
1	1	0	0
1	1	1	0

Exercises 27 continued

11. A circuit is required which will produce a logical 1 when its two inputs are identical. Indicate how such a circuit can be constructed using: (a) AND, OR and NOT gates; (b) NAND gates; (c) NOR gates.

12. (a) The gate network shown in Fig. A has three inputs A, B and C. Find an expression for the output Z and simplify this expression.

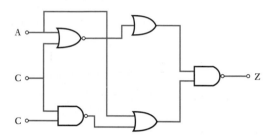

Fig. A

(b) From the simplified expression for the output Z, determine an equivalent network which does *not* contain NAND gates.

13. Three inputs A, B and C are applied to the inputs of AND, OR, NAND and NOR gates. Give the algebraic expression for the output in each case.

Assuming that AND, OR, NAND, NOR and NOT gates are available, sketch the combinations that will realize the following:

(a) $A \cdot B \cdot \overline{C} + \overline{A \cdot (\overline{B + C})}$
(b) $B \cdot (A + \overline{B} + \overline{C}) + \overline{A} + B \cdot C$

14. For the circuit shown in Fig. B, determine the relationship between the output Z and the inputs A, B and C. Construct a truth table for the function.

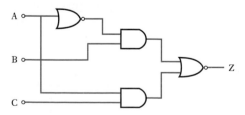

Fig. B

15. Simplify the following logic expressions:

(a) $A \cdot (B + C) + A \cdot B \cdot (\overline{A} + \overline{B} + C)$
(b) $\overline{A} \cdot B \cdot C \cdot (B \cdot C + A \cdot B)$
(c) $(\overline{A} \cdot B + \overline{B} \cdot A) \cdot (A + B)$

Using AND, NAND, OR, NOR or NOT gates as required, develop circuits to generate each of the functions (a), (b) and (c) above.

16. The gate network shown in Fig. C has three inputs A, B and C. Find an expression for the output Z and simplify this expression.

From the simplified expression, draw a simpler network that would produce the same output. Draw the truth table for the simplified network.

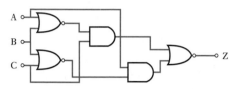

Fig. C

17. Explain with the aid of truth tables the functions of two-input AND, OR, NAND and NOR gates. Give the circuit symbol for each.

The gate network shown in Fig. D has three inputs, A, B and C. Find an expression for the output Z, and obtain a truth table to show all possible states of the network. From the truth table suggest a simpler network with the same output.

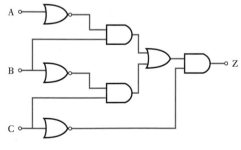

Fig. D

18. Use a Karnaugh map to simplify each of these functions:

(a) $X = \overline{A} \cdot \overline{B} \cdot \overline{C} + A \cdot \overline{B} \cdot C + \overline{A} \cdot B \cdot C + A \cdot B \cdot \overline{C}$
(b) $X = A \cdot C[\overline{B} + A \cdot (B + \overline{C})]$
(c) $X = D \cdot E \cdot \overline{F} + \overline{D} \cdot E \cdot \overline{F} + \overline{D} \cdot \overline{E} \cdot \overline{F}$

19. Using Karnaugh maps produce simplified logic functions for the functions F and G defined in the truth table (Table A).

Table A

A	B	C	F	G
0	0	0	1	0
0	0	1	0	1
0	1	0	0	1
0	1	1	0	1
1	0	0	0	0
1	0	1	1	1
1	1	0	0	1
1	1	1	1	0

20. Obtain simplified logic expressions for the Karnaugh maps shown in Fig. E.

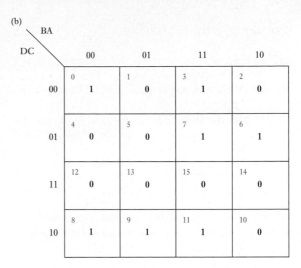

(b)

(a)

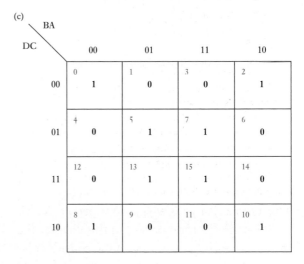

Fig. E

Fig. E (Continued)

Chapter twenty-eight

Microprocessors and Programs

The application with which most people associate digital systems is the computer, at the heart of which is the microprocessor. In this chapter the microprocessor and its principles of operation will be presented. This will lead to the consideration of ROMs and RAMs, ports, decoders and buses.

To gain an improved understanding of microprocessor operation the process of programming is introduced. There are many programming systems and here a restricted and simplified set of instructions, which can be used to design simple programs, is used. This allows an understanding of the concepts to be developed without requiring to learn all of the details of a particular language.

Programmable logic controllers (PLCs) are rapidly becoming more common. They are frequently termed microcontrollers and are widely used in all types of electronic goods. They are a type of microprocessor and, as such, can perform different tasks depending upon the program that they have stored. The tasks these devices are used for, as their name suggests, are simple control tasks where there is a need to be able to alter the task specifications. These tasks may involve interlocking, counting, sequencing, timing and monitoring. The PLCs can be used to directly control certain types of output.

By the end of the chapter, you will be aware of the basics of microprocessors and PLCs and of their operation and will be able to prepare short programs.

28.1 Microprocessors

The microprocessor is an advanced electronic device which has arisen out of logic integrated circuits. The rate of development has arguably been greater than with any other electrical device. Its introduction commenced with the microelectronic developments for integrated circuits, but it has reached the point of creating circuitry with a density of several million transistors per square centimetre of the semiconductor slice.

Logic integrated circuits are capable of providing a range of functions within a fairly well-defined sphere of operation for each IC; by comparison, the microprocessor seems almost to be without limit in its operation, although this is really a slight overstatement. Its greatest advantage is the vast range of functions to which a single microprocessor can be applied. In particular, the microprocessor can be applied to three principal ranges of operation:

1. Control (anything from a washing machine program to an oil refinery).
2. Calculation (anything from a pocket calculator to a simple computer).
3. Administration (anything from a list of names and addresses to commercial control).

The term *microprocessor* has become somewhat ambiguous in its development. The device is specifically a small slice of silicon on which large-scale integrated circuits have been created, the slice being mounted in a 20-pin, 40-pin or even larger packaging unit. However, many people would associate the term with the microprocessor computer, which is sold in large numbers through many stores. The name of this device has been generally curtailed to 'microprocessor', which is inappropriate to this text. In this chapter, the microprocessor is the package and not the application. In particular, it is important not to associate a keyboard with the term 'microprocessor'.

The microprocessor requires to interact with other electrical devices and cannot operate on its own. It is therefore only part of a system, in much the same manner that a transistor is only part of an amplifier.

28.2 Microprocessor operation

The microprocessor consists of thousands of electronic switching circuits. As with any logic operation, each circuit can be either ON or OFF, i.e. HIGH or LOW. It follows that the operation that is effected by a microprocessor must be a binary one. Instructions fed into a microprocessor are in binary digits known as *bits*. The form of the instruction must be a series of ones and zeros which relate to circuits being closed or open, i.e. ON or OFF, or HIGH or LOW. Typically, a circuit which is ON would supply a signal of about 5 V and a circuit which is OFF would supply a signal of 0 V.

A typical instruction could be 01010110. This is an 8-bit instruction and is the standard form used for the first generation of microprocessors (later generations have increased to 32 and more). The 8-bit instruction is termed a *byte*. Note that for the instructions given above, the first digit appears superfluous, but all 8 bits must be given to complete an instruction, even to the point of 00000001 or 00000000.

The byte is fed into a microprocessor either by a sequence of pulses, which is known as asynchronous action, or all pulses at one time, which is known as synchronous action. The former is a serial operation while the latter is a parallel operation. The rate at which bits are fed into a microprocessor is determined by a clock which typically, for low-speed microcontrollers, operates at 16 MHz. Thus in serial operation 16 000 000 bits can be handled every second, which is equivalent to 2 000 000 instructions or bytes per second.

In order to make sense of a sequence of input information, the micro-processor has to be able to obey a number of instructions. It also requires a memory in which input information can be stored. The memory requires an addressing system so that it is known where the information is stored and, equally, from whence it can be recalled.

The input information comes in two distinct forms: instructions and data. Both are expressed in binary form but the microprocessor must be capable of distinguishing between them.

The instructions relate to a particular section of the memory arrange-ment. This section is termed the *read-only memory*, or ROM. This has a set of instructions manufactured into it and they cannot be changed. Also, when the microprocessor is switched off, the instructions remain permanently in the ROM and are called non-volatile. Thus if a microprocessor is given the instructions to place a byte of data in store, the instruction code introduced causes a number of instructions retained in the ROM to be recalled and the resulting effect is that the data are placed in the next available store.

Implicit in this observation is the alternative form of memory, i.e. one in which information can be temporarily stored. This form of memory is termed a *random-access memory* or RAM. A RAM comprises a large number of stores, each of which has an address; therefore, when we insert an information byte, we require an associated address indicating the particular location in which it is to be stored. Similarly, we need to know the address should we wish to recall the information byte. Unlike a ROM, the information stored in a RAM is lost when it is de-energized. The RAM is therefore said to be volatile.

You may well have come across the effects of a ROM or a RAM if you have operated a personal computer (p.c.). When you switch on a p.c., it performs a program which is recalled from the ROM and eventually it gives you access to that program. Consequently you enter data, but if while you are working on it, you switch off the p.c. by mistake, the result is that all your entries are lost. This is because the data were held in the RAM prior to being saved.

The microprocessor therefore can be seen to operate on the interaction of a number of interacting processes. At the heart of these interactions is the *accumulator*. This can be considered as the section where the main activity takes place. Thus, when it is proceeding through a series of operations, the changes in the information process take place in the accumulator. Typically, a sequence of events could require two or three changes, at which stage the processor would have gone as far as it could. The result can then be stored in the RAM, clearing the accumulator ready for the next series of operations. The control of the sequence may come either from the ROM or from another section of the RAM in conjunction with the ROM.

By this stage we have introduced a number of terms, and probably confusion is setting in. However, a diagram of the microprocessor, shown in Fig. 28.1, helps to relate the terms.

The central processing unit is the microprocessor chip containing the accumulator. It is connected to the ROM and to the RAM by three sets of circuits, or *buses*. The address bus relays the direction of the data to be stored or recalled from memory in order that the correct storage system is used. The data, however, are transferred through the data bus. We can think of this as being like a railway where the address bus sends the information that

Fig. 28.1 A simple block diagram of a microprocessor system

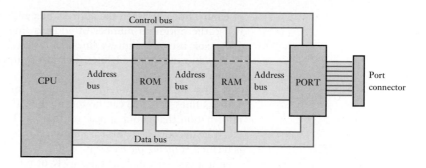

controls the track points, thus ensuring that the train of data arrives at the appropriate destination. A subsequent changing of the points permits the following train of data to be directed to a different destination.

The diagram also shows the control bus, which carries the instructions for the organization of the sequence of operations including the commencement and termination of the sequences.

Finally, the *port* is the circuitry which connects the microprocessor system to the world around it.

In the limited content of a general book of electrical and electronic engineering, it is not possible to explain fully the operation of each of the parts of a microprocessor system. Should a reasonably detailed understanding be required, it is necessary to refer to a text relating specifically to microprocessor systems, but this introduction would not be complete without slightly expanding the simplified system shown in Fig. 28.1.

Fig. 28.2 Expanded block diagram of a microprocessor system

In particular, Fig. 28.2 has added in the clock, which provides the timing pulses. Also decoders have been added, and require introduction. Each set

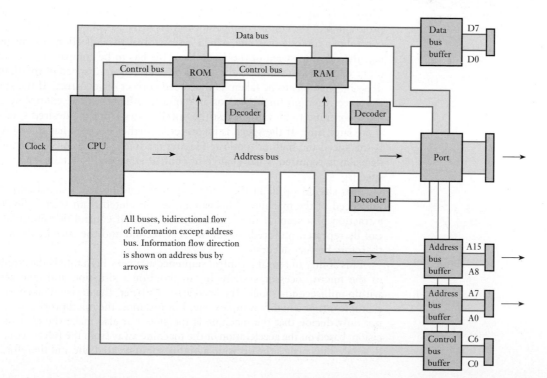

All buses, bidirectional flow of information except address bus. Information flow direction is shown on address bus by arrows

of connections between the address bus and the various chips connected to it use the optimum number of address lines. A conventional 8-bit microprocessor uses 16 address lines, numbered A0–A15. However, a typical ROM chip uses only 11 lines, i.e. A0–A10, thus lines A11–A15 cannot be connected.

Using the binary system, the range of information that can be carried by address lines A0–A10 can be repeated 32 times appropriate to the range of information that could appear on lines A11–A15, i.e. 00000, 00001, 00010, etc. The decoder uses these upper lines to control the memory chips and this is termed *mapping the memory*.

For a system which handles 64 K of memory (equivalent to 65 526 bytes since K = 1024), typical memory maps are shown in Fig. 28.3.

Fig. 28.3 Typical memory mappings

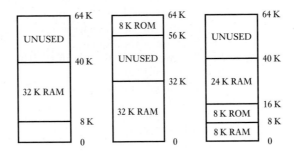

28.3 Microprocessor control

Most uses of a microprocessor involve data handling and, if suitably programmed, the control of machines. A typical example would be the speed control of a rotating machine, which first involves the collection of some data, e.g. the existing speed of the machine. This has to be compared with some other data already stored in the RAM, i.e. the required speed of rotation. A decision then has to be taken: is the speed correct or incorrect? If the speed is correct then no further action is required other than the control system must continue to be vigilant lest the speed varies from the desired speed at some later time. If the speed is incorrect, a further decision has to be taken: is the speed too high or too low? Having made these decisions, yet another decision is required: what change of control is needed to attain the desired speed?

This should result in the control system sending out a signal adjusting the speed of the machine. This can all be completed from start to finish in a fraction of a second; thus the complete cycle of control decision-making can be repeated hundreds of times every second and the speed control can therefore be very accurate.

This ability to repeat a control sequence at a very high rate is the strength of the microprocessor system; it can monitor a situation and take action within milliseconds should the need arise. Further, the decision taken can be a simple one or a highly complex one. For instance, the microprocessor might not only decide that the machine is too slow but also make the control decision, based on the acceleration of the machine away from the reference speed desired. This makes the decision a highly sophisticated one and therefore we

Terminal point of
flowchart, e.g. start,
stop, halt, delay or
interrupt

Process of executing
a defined operation
or group of operations
resulting in a change
in value, form or
location of information

Decision operation
that determines
which of a number of
alternative paths is
followed

Input/Output function
where the data
medium is not
specified

Connector represents
an exit to, or entry
from, another part of
the flowchart

Fig. 28.4 Flowchart symbols

have to plan very carefully the program of events which we wish the micro-processor system to execute.

We also have to remember that the microprocessor can basically only make decisions of the yes/no variety. In order to obtain a decision by this means, it might be necessary to make tens or even hundreds of yes/no decisions to result in one apparently complex decision, but such is the rapidity of microprocessor operation that this is still achieved very quickly. A seemingly simple stage in a program might require many instructions before it is completed, and therefore we have to appreciate that the approach to programming a microprocessor involves two steps:

1. The first step is to determine the interrelated stages, each of which performs a decision-making routine.
2. The second step is to write the program for each of the above stages.

Once the various stage programs have been written, they have to be combined to make the entire operational program for the microprocessor system.

A typical method of determining the required stages of a program is the drawing of a flowchart. The principal symbols used in flowcharts are shown in Fig. 28.4.

For the example of the control of the speed of a rotating machine, the flowchart is shown in Fig. 28.5.

The control process need not limit itself to one action. In fact, the problem about the suggested arrangement for the machine speed control is that it is repeating too often, i.e. there is no possibility that within about a millisecond the machine will have corrected any deviation in speed. Rather it might be better to give it a chance to take some action before reassessing the need for change of speed control. This delay could be achieved by switching off the control system for a number of milliseconds before repeating the sequence.

This off time could be used for other applications, however; for instance, let us consider a number of machines which are independent of one another. The microprocessor system could consider the speed of number 1, then go on to consider number 2 and so on until, having reviewed them all and having taken action on all, it could come back to the first and start all over again. Typically, control of 40 or more machines could be kept under surveillance by one system, while at the same time other information could be obtained to promote the more effective use of the machines.

Microprocessor systems, however, are rarely used to their full capacity and it is interesting to note that again we have to consider the change of approach to engineering design which they have brought, i.e. the micro-processor is a device capable of an almost infinite number of functions and we have to decide how many of these we use in any given application. It makes little odds whether we use many of its possibilities or few, because the initial cost of the chip remains the same and that cost is relatively small.

We have looked at the microprocessor system in its widest sense. However, there is a significant application in which the microprocessor appears in programmable logic controllers (PLCs). This term is sometimes abbreviated to programmable controllers but we shall continue with the PLC version in this text. In PLCs, the inputs and outputs are generally single-digit quantities. By considering applications more closely related to the PLC, we can further our study of the microprocessor system in this chapter and continue to look further at PLCs later in the chapter.

Fig. 28.5 Flowchart for
machine speed control

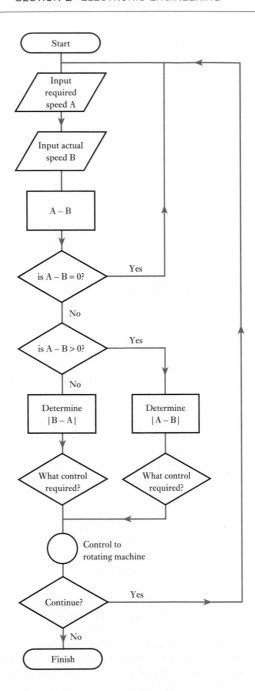

Example 28.1   Draw the flowchart of a program controlling the supply valve to a
water tank. The valve should be energized when the water level falls to
1.0 m and should be de-energized when the water level rises to 1.5 m.

   The flowchart is shown in Fig. 28.6. From it, you will see that there are
two opportunities to discontinue the sequence. Unless it is effectively

Fig. 28.6 Flowchart for
Example 28.1

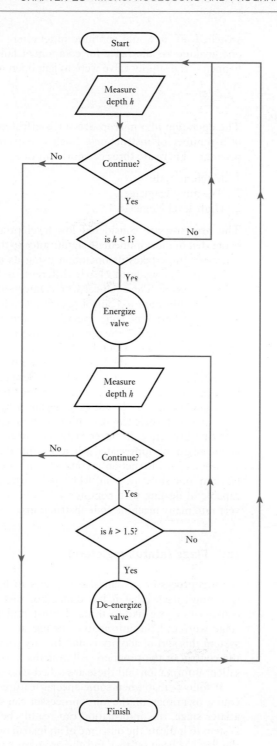

switched off, it would continue indefinitely. The arrangement is a simple one, making no allowance for component failure; for example, what would happen if the supply valve were to jam open or shut?

28.4   Programs

The operation of a microprocessor is effected by the sequential application of a number of instructions. Such a sequence of instructions is called a program. There are three categories of programming:

1. Machine code.
2. Assembly language.
3. High-level language.

The first two are considered low-level programming, the machine code being that of most immediate significance to the microprocessor. The microprocessor undertakes its operation using its own machine code and it is at this point that we must clearly differentiate between the two views of the microprocessor. The engineer who is interested in the circuit operation, and even its application to control systems which are preprogrammed, requires to know the machine code operation of the microprocessor. The computer operator who controls the microprocessor by means of a keyboard will use a high-level language, which has to be translated into the machine code using instructions stored in the ROM. Equally, after the microprocessor is ready to relate the result of its work, this has to be translated back into the high-level language recognized by the computer operator. The translation of assembly language into machine code is undertaken by an *assembler*, while a *compiler* translates high-level language into machine code.

The difference in level between assembly language and high-level language can be appreciated when it is noted that in assembly language each symbolic code instruction is connected to one machine code instruction and takes up one store location, whereas in high-level languages the compiler is capable of dealing with complex symbolic instructions, each of which convert into many machine code instructions.

(a)   Flags (status registers)

A microprocessor system also requires to have its own form of note-pad to remind itself what it has done. For instance, often in the course of a conversation, we like to think 'I must remember that!' and at some later stage we recall the information to use at that time. To a microprocessor system this sort of activity is not directly part of the program, but is part of the control of the program. All systems have some means of remembering salient information and these are called *flags* or *status registers*.

It follows that when some special feature has been achieved during the course of a program, the microprocessor can set its flag either to 0 or to 1. At a later stage, this information can readily be made available to the control system to indicate the outcome of an instruction. For instance, let us say that the flag has been set to 1 and the program has reached the BNZ instruction (see below). This instruction considers the status of the flag: because the flag is at 1 the program continues, but had the flag been at 0 it would have turned back to an earlier stage in the program. The information held by the flag is control information and therefore it does not enter the accumulator which holds the program data. Also because it does not have to be stored in the memory, the flag status can be obtained much more quickly. Most systems incorporate a number of flags, but for this introduction to programs we shall assume there is only one.

(b) Simplified instruction set

For the purpose of this introduction to programming, it is necessary to select a number of simplified instructions. While the code letters used are similar to those used in a variety of applications, it is worth noting that each micro–processor system has its own instruction set with which one has to become familiar.

ADD Add memory contents to the accumulator
This instruction adds the contents of a stated memory location to the contents of the accumulator, and stores the result in the accumulator. If several successive additions are carried out, no account is taken of any carries that are generated.

AND Logical AND memory with accumulator
This instruction performs a logical AND between the contents of a stated memory location and the contents of the accumulator. The AND is between the equivalent bits of the two bytes; the result is stored in the accumulator. If there are no bit pairs which are both logic 1, the result will be zero and the zero flag will be set to logic 1.

BRK Break
This instruction stops the program. No operand is required.

BNZ Branch if not equal to zero
When this instruction is encountered, the zero flag is tested. Thus the BNZ instruction should follow an instruction which is designed to affect the zero flag. If the flag is set to logic 0, the branch is taken; otherwise the program continues to the next instruction in sequence.

BZE Branch if equal to zero
When this instruction is encountered, the zero flag is tested. Thus the BZE instruction should follow an instruction which is designed to affect the zero flag. If the flag is set to logic 1, then the branch is taken; otherwise the program continues to the next instruction in sequence.

CMP Compare memory with accumulator
This instruction compares the contents of a stated memory location with the contents of the accumulator. This is done by subtracting the memory contents from the accumulator contents. The answer to the subtraction is not stored so the data in the accumulator are not destroyed. The zero flag will be set to logic 1 if the answer is zero; it will be set to logic 0 otherwise. The CMP instruction will normally be followed by a conditional branch.

DEC Decrement memory
This instruction decrements the contents of a stated memory location by 1. If the contents of the memory become zero as a result of this operation, the zero flag will be set to logic 1.

EOR EXCLUSIVE-OR memory with accumulator

This instruction performs the EXCLUSIVE-OR between the contents of a stated memory location and the contents of the accumulator. The EXCLUSIVE-OR is between the equivalent bits of the two bytes; the result is stored in the accumulator.

JMP Jump

This instruction causes the program to jump to a stated memory location (and hence instruction) rather than proceed to the next instruction in sequence. The jump is unconditional and is always taken.

LD Load accumulator

This instruction has two modes:

1. To load the accumulator with fixed data, the data value being held in memory in the location immediately following the instruction.
2. To load the accumulator with data contained in a stated memory location. The instruction is followed by the address in memory from which data are to be loaded.

OR Logical OR memory with accumulator

This instruction performs a logical OR between the contents of a stated memory location and the contents of the accumulator. The OR is between the equivalent bits of the two bytes; the result is stored in the accumulator.

SHL Shift memory contents left

This instruction shifts the contents of the stated memory location to the left by 1 bit. The most significant bit is lost; the least significant bit is replaced by a 0.

SHR Shift memory contents right

This instruction shifts the contents of the stated memory location to the right by 1 bit. The least significant bit is lost; the most significant bit is replaced by a 0.

ST Store accumulator in memory

This instruction stores the contents of the accumulator in a stated memory location. The memory location is held in the program in the two bytes following the instruction.

SUB Subtract memory contents from the accumulator

This instruction subtracts the contents of a stated memory location from the contents of the accumulator and stores the result in the accumulator. If several successive subtractions are carried out no account is taken of borrows. If the subtraction results in a zero answer, this is indicated by the zero flag being set to logic 1.

To this list, we can add four other instructions. These are intended for PLC programs in which the input is either ON or OFF.

ADDN   Logical AND accumulator with NOT content at stated location.

This instruction answers the question – is the signal in the accumulator AND NOT in the stated location?

LDN   Load accumulator with NOT content at stated location.
ORN   Logical OR accumulator with NOT content at stated location.

This instruction answers the question – is the signal in the accumulator OR NOT in the stated location?

STN   Send NOT accumulator to stated location.

Each of the above four instructions inverts the signal taken from or sent to the stated location.

Finally, for the sake of easy identification, we shall use the prefix I for all input locations and the prefix Q for all output locations. The numbers which follow have been chosen at random and it is only for consistency that the following examples indicate the input as I001 and the output as Q010.

28.5 Simple programs

The most fundamental programs are almost self-evident, but they represent an opportunity to become familiar with the terms.

Example 28.2

In a microprocessor system, a single-byte number is located at 0060 and also a single-byte number is available at the input I001. The program of this system is given below.

Explain each instruction of the following program and describe the purpose of the program.

```
LD    0060
ST    0040
LD    I001
SUB   0040
ST    0041
BRK
```

The instructions are as follows:

LD   0060   Load the accumulator with the number stored at location 0060 of the RAM.

ST   0040   Store this number in location 0040 of the RAM.

LD   I001   Load the accumulator with the number available at input I001, thus replacing the previous number held in the accumulator.

SUB 0040   Subtract the number held in location 0040 from the number held in the accumulator.

ST   0041   Store this new number (i.e. the difference resulting from the subtraction) in location 0041.

BRK   End the program.

The program determines the difference between the input number available at I001 and the number stored in 0060, and subsequently stores the difference in 0041.

Note that there is no apparent need to shift the number held at 0060 to 0040 in order to undertake the subtraction, but this step illustrates two points. First, the loading of the accumulator replaces any previous information held there, and second, this program might be a stage in a greater program in which the information held at 0060 could later be modified for other purposes, leaving the original information still intact at 0040.

Example 28.3

The input byte to a microprocessor system is addressed at I001 and represents a hexadecimal number less than 30. The program is given below, the output being addressed Q010.

Explain each instruction of the program and hence explain the function of the program.

```
LD    I001
ST    0041
SHL   0041
LD    0041
ST    Q010
BRK
```

The program is modified to that shown below by the addition of three instructions. Explain these additional instructions and hence explain the function of the new program.

```
LD    I001
ST    0040
ST    0041
SHL   0041
SHL   0041
LD    0041
ADD   0040
ST    Q010
BRK
```

Before giving the answer to this question, we require to consider the limitation given, i.e. a hexadecimal number less than 30 and its significance relative to the instruction SHL. The number 30 would be given in digital form as 0011 0000 and the SHL instruction moves each digit one place to the left. The application of the SHL instruction would thus give 0110 0000, which represents the hexadecimal number 60, i.e. the SHL instruction is equivalent to multiplying by 2. Similarly, the SHR instruction is equivalent to dividing by 2. In this example, we apply the SHL instruction twice, which results in 1100 0000. A further application would shift the left-hand digit outside the 8-bit register and we would require to involve an overflow byte, which is in advance of this simple introduction to programming so we must commence with a number which ensures that, initially, the first two bits are 0.

LD   I001   Load the accumulator with the information at input I001.
ST   0041   Store the information in the accumulator in 0041.
SHL 0041   Using the information stored in 0041, shift each bit one place to the left and store this new information in 0041. Note that the accumulator operated on the information initially in 0041 and not on that in the accumulator, even though it happened to be the same. Also, after the operation, the information stored in 0041 is changed and the original formation taken from I001 is lost, unless it happens still to be available at I001.
LD   0041   Load the accumulator with the information stored at 0041.
ST   Q010   Store the information held in the accumulator at output point Q010.
BRK   End program.

Had it been desirable to retain the original input information, this should have been stored somewhere other than 0041 as illustrated in the modified program. However, it will be noted that the previous program had the function of taking in a number from input port I001, multiplying it by 2 and making it available to output port Q010.

The modified program performs as follows:

LD	I001	Load the accumulator with the information at input I001.
ST	0040	Store the information in the accumulator in 0040.
ST	0041	Store the information in the accumulator in 0041.
SHL	0041	Shift the memory content of 0041 one digit to the left.
SHL	0041	Shift the memory content of 0041 one digit to the left (this results in the shift having been undertaken twice).
LD	0041	Load the accumulator with the information stored in 0041.
ADD	0040	Add the information stored in 0040 to the information held in the accumulator.
ST	Q010	Store the information held in the accumulator at output port Q010.
BRK		End program.

In this example, we saw the need to store information twice because, in the second instance, the stored information was modified by the SHL instruction and was therefore effectively lost.

The program had the formation of taking in a number from input port I001, storing it, multiplying it by 4, adding the original number (which is equivalent in total to multiplying by 5) and making the resultant available to output port Q010.

These examples have illustrated the most simple instructions which are generally associated with calculations rather than with control systems. Control systems depend on decision-making, and for that we require to consider control programming.

28.6 Control programs

Control programs depend on decisions which usually can be thought of as yes/no decisions. However, there are terms unknown to a logic system that require the rephrasing of a question as a comparison, e.g. is A = B? If the answer is yes then the program might continue, otherwise it loops back to an earlier step in the program. Should the decision be to loop back, then the program will repeat until eventually the desired answer is obtained and then the program can proceed.

Usually the comparison question is phrased in one of two ways. Either the accumulator should hold a number which must be equal to zero for the program to proceed, in which case the instruction is BNZ, or the accumulator should hold a number other than zero for the program to proceed, in which case the instruction is BZE. Both are used in Example 28.4.

Example 28.4

A hoist is fitted with a protective device which operates at a height of 2.0 m. Above 2.0 m the protective device inputs logic 00 to the controlling microprocessor system.

As the hoist descends towards the 2.0 m limit, the input changes to logic 01. Should the hoist descend past the 2.0 m limit, the input returns to logic 00. Write a program that continually checks the state of the input signal at 001 and which outputs the byte 20 (which stops the hoist) at 00 after the protection device input has changed from 00 to 01 and back to 00.

The program could take the following form:

LD	#01	Load the accumulator with the data 01 to ensure the data are available for comparison.
ST	0040	Store the data in 0040.
LD	#20	Load the accumulator with the data 20 to ensure the data are available for the protective device.
ST	0041	Store the data in 0041.
→ LD	I001	Load the accumulator with the information at input port I001.
AND	0040	Logical AND this memory with the accumulator information to set the zero flag required for the BZE instruction.
└ BNZ		Check the zero flag and branch back if equal to zero. (If the hoist is descending, the input is 00 and the reference is 01, hence the AND operation gives a zero-flag setting of logic 1 and the branch is taken, thus reflecting the sequence awaiting the further descent of the hoist. If the hoist reaches the 2.0 m limit the I001 changes to 01, hence the AND operation has a result of 1 and the flag is set to logic 0. If the flag is set to 0, the program moves to the next instruction.)
→ LD	I001	Load the accumulator with the information at input port I001.
AND	0040	Logical AND this memory with the accumulator information to set the zero flag required for the BNZ instruction.
└ BZE		Check the zero flag and branch back if not equal to zero. (If the hoist is around the 2.0 m limit, then I001 gives 01, hence the AND operation has a result of 1 and the flag is set to logic 0. If the flag is set to 0 then the branch is taken, thus repeating the sequence awaiting the action by the hoist. Should the hoist continue then I001 changes to 00, hence the AND operation has a result of 0 and the flag is set to logic 1, in which case the program moves to the next instruction.)
LD	0041	Load the accumulator with the data in 0041.
ST	Q010	Make the accumulator data available to output port Q010.
BRK		End program.

In this program it is clear that the BZE and BNZ instructions depend on a previous logic decision, in this instance taken by the AND operation which has set the zero flag. Each instruction we program can carry a reference number, hence the BZE and BNZ instructions can include details of the instructions to which they have to branch; however, at this introductory stage an arrow is sufficient to indicate the intention.

Example 28.5

The level of liquid in a tank is measured by a float transducer which provides an input signal to a microprocessor system, the signal being proportional to the level of the liquid. It is required that the microprocessor should output a signal, stored at address 0080, to switch on an indicator lamp when the input falls to 05, and it is required that it should output a signal, stored at address 0081, to switch off the lamp when the input rises to 25. The input is addressed at I001 and the output at Q010.

Write a program setting up the operating conditions and continuously monitoring the input signal. Assume that the initial level of liquid is between the limits and falling.

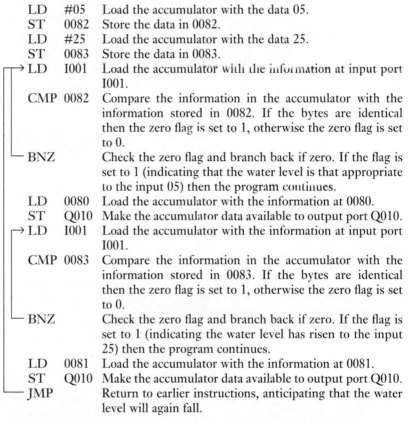

LD	#05	Load the accumulator with the data 05.
ST	0082	Store the data in 0082.
LD	#25	Load the accumulator with the data 25.
ST	0083	Store the data in 0083.
LD	I001	Load the accumulator with the information at input port I001.
CMP	0082	Compare the information in the accumulator with the information stored in 0082. If the bytes are identical then the zero flag is set to 1, otherwise the zero flag is set to 0.
BNZ		Check the zero flag and branch back if zero. If the flag is set to 1 (indicating that the water level is that appropriate to the input 05) then the program continues.
LD	0080	Load the accumulator with the information at 0080.
ST	Q010	Make the accumulator data available to output port Q010.
LD	I001	Load the accumulator with the information at input port I001.
CMP	0083	Compare the information in the accumulator with the information stored in 0083. If the bytes are identical then the zero flag is set to 1, otherwise the zero flag is set to 0.
BNZ		Check the zero flag and branch back if zero. If the flag is set to 1 (indicating the water level has risen to the input 25) then the program continues.
LD	0081	Load the accumulator with the information at 0081.
ST	Q010	Make the accumulator data available to output port Q010.
JMP		Return to earlier instructions, anticipating that the water level will again fall.

No indication of the manner in which the program might terminate has been given as it has been assumed that the monitoring of the water level will be an on-going process. However, the reader can write a further program to terminate the process as considered appropriate.

28.7 Programming in hexadecimal representation

During the execution of each instruction, the address specified by the program counter is passed along the address bus to the memory. Since each address refers to a single register, the number of bits in the program counter limits the maximum number of memory locations. For a 16-bit program counter 2^{16} or 65 536 locations can be addressed.

The program counter, however, stores 16 binary digits, usually represented in hexadecimal for convenience:

	0100 1011 1001 0011	in binary
or	4 B 9 3	in hexadecimal

When using the instruction set listed in section 28.4, each instruction requires a 4-digit hex address. For example:

> LD $0240   ($ indicates a hexadecimal number)

This causes the contents of the register at $0240 to be transferred to the accumulator.

The accumulator is an 8-bit register, which implies that each memory register is also 8-bit. Hence, a memory address can store:

> either: an 8-bit instruction code
>   or: the lower half of a 16-bit address
>   or: the upper half of a 16-bit address
>   or: an 8-bit dataword

In a microprocessor, it is usual for the lower order of the address to precede the upper order.

The use of hexadecimal numbers is best appreciated by considering a series of instructions and their explanations as follows:

1. LD   #$26
   AND $0247

If the contents of $0247 are $0F the bit-by-bit AND result will be $06:

$$0010 \quad 0110 = \$26$$
$$\underline{0000 \quad 1111} = \$0F$$
$$0000 \quad 0110 = \$06$$

Since the result is non-zero the zero flag will be cleared.

However, if $0247 stores $91 the result is $00:

$$0010 \quad 0110 = \$26$$
$$\underline{1001 \quad 0001} = \$91$$
$$0000 \quad 0000 = \$00$$

The result is zero and so the zero flag will be set.

2. DEC $025D
   BZE $0213

When the DEC instruction is executed the contents of $025D are reduced by 1. If the result is zero the zero flag is set to 1. Otherwise it will be zero.

When the BZE instruction is executed the zero flag is tested. If the flag is set to 0 (i.e. a non-zero result) the branch is not taken and the normal sequence is continued.

However, if the result was zero, the flag is set to 1 and the branch is taken. The four hex characters ($0213) indicate the address of the first instruction in the new sequence.

3. LD   #$04
   CMP $0283
   BNZ $022E

The LD instruction causes the accumulator to be loaded with $04. The CMP instruction compares the dataword in $0283 with the contents of the accumulator and the zero flag is set to 1 if the dataword is $04, otherwise it is cleared (set to 0).

When the BNZ instruction is executed the branch is taken only if the zero flag is set to 0 (i.e. a non-zero result).

Example 28.6 Write a program to add two numbers, stored at $0300 and $0301, and store the result at $0302.

Location	Instruction code	Memory address	Comment
0200	LD	0300	Get first number
0203	ADD	0301	Add second
0206	ST	0302	Store result
0209	BRK		Stop

Except for BRK, each instruction occupies three 8-bit memory locations:

```
0200  LD      Load accumulator
   1     00 ⎫
   2     03 ⎭  Address of data (low byte first)
   3  ADD     Add to accumulator
   4     01 ⎫
   5     03 ⎭  Address of data (low byte first)
   6  ST      Store accumulator
   7     02 ⎫
   8     03 ⎭  Memory address
   9  BRK     Halt
```

If $0300 contains $02 and $0301 contains $06 the result will be $08 stored in $0302. (Note that $0300 and $0301 will not be changed by this program.)

The total should not exceed $FF, since this generates a carry: thus if

$$\$0300: \$D4 = 1101 \quad 0100$$
and
$$\$0301: \$72 = \underline{0111 \quad 0010}$$
the total is
$$1 \quad \underline{0100 \quad 0110}$$
$$\text{carry}$$

The carry indicates that a further byte is required.

Example 28.7 Write a program to set the four most significant bits of the data in $0303 to zeros, leaving the remaining bits unchanged.

Location	Code	Address	Comment
0200	LD#	0F	Get $0F to acc. ($0F: 0000 1111)
0202	AND	0303	AND with ($0303)
0205	ST	0303	Store new data
0208	BRK		

Notice that LD# uses only one address byte.

> 0200   LD# Load accumulator, immediate dataword
> 0201 $0F
> 0202   AND
> etc.

If $0303 initially stores $6D:

$$\begin{array}{lll} & \$6D: 0110 & 1101 \\ \text{AND} & \$0F: \underline{0000 \quad 1111} \\ & 0000 & 1101 \Rightarrow \$0D \end{array}$$

i.e. $1.0 = 0$ and $1.1 = 1$. Hence, the lower bits are unchanged, whereas the upper bits are made 0.

This is a common operation called masking.

Example 28.8 Write a program to multiply the contents of $0309 by 5 and store the result at $030A. (Assume data are less than $33.)

Location	Code	Address	Comment
0200	LD	0309	Get number
0203	SHL	0309	Multiply number by 2
0205	SHL	0309	Multiply number by 2
0208	ADD	0309	Add to number
020B	ST	0309	Store result
020E	BRK		

The SHL instruction moves each bit one place towards the most significant bit (MSB) and therefore doubles its value:

$$\begin{array}{ll} \$01 & 0000\ 0001 \\ \text{SHL} \Rightarrow & 0000\ 0010 = \$02 \end{array}$$

If data = $26 (= 38 in decimal)

$$\begin{array}{lll} \text{SHL} & 0100 & 1100 \ (= 76_{10}) \\ \text{SHL} & 1001 & 1000 \ (= 152_{10}) \\ \$26 & \underline{0010 \quad 0110} \ (= 38_{10}) \\ & 1011 & 1110 \ (= 190_{10}) \end{array}$$

Note: $0010\ 0110 = (0 \times 2^7) + (0 \times 2^6) + (1 \times 2^5) + (0 \times 2^4) + (0 \times 2^3)$

$$+ (1 \times 2^2) + (1 \times 2^1) + (0 \times 2^0)$$

$$= 0 + 0 + 32 + 0 + 0 + 4 + 2 + 0 = 38_{10}$$

28.8 The programmable logic controller

The programmable logic controller (PLC) or microcontroller is a device built around a microprocessor and its functions are determined by a program which it stores. Such programs are relatively small compared with those envisaged for a computer. The programs can be fairly readily changed, but generally, once they have been established, the program is set to operate and left to perform on its own. For example, having set up a PLC to control a lift, it would be switched on 24 hours a day ready to respond to any call.

PLCs are becoming more and more common because they are comparatively cheap. Most are manufactured with a program already established so that the potential flexibility might not be required yet they are favoured because of cost. For instance, most lifts will operate to the same basic style of program but if a particular application calls for, say, additional indicator lights then the program can be readily adapted to suit the specification. PLCs, sometimes known as 'microcontrollers', are so cheap that they are found in a wide range of domestic applications as well as non-specialized industrial applications.

The principal tasks undertaken are the performance of a control program which involves interlocking, counting, sequencing, timing and monitoring. More recent applications involve communication whereby a PLC can interact with another control system or report on its monitoring activity. For example, a ship could automatically report its progress from the South Atlantic to its European headquarters where decisions could be made about its further voyage. A decision might be made to slow down and save on fuel costs, or to speed up to catch a berth in Rotterdam.

The PLC directly controls the output provided it is not too heavily rated. Lamp bulbs, thyristor control gates, contactor coils and solenoid valves are all typical output loads. These are supplied at 115 or 230 V a.c. or 24 or 48 V d.c. The fact that a PLC can supply such power loads directly makes it distinct from a computer. So that raises the question – when is a control system suitable for PLC operation?

28.9 Control system characteristics

There are three basic characteristics:

1. There must be definite actions to be taken, e.g. turning a signal lamp ON or OFF.
2. There must be definite RULES governing these actions, e.g. the lamp being ON indicates a given state and being OFF excludes that state.
3. The RULES must take into account the relevant conditions, e.g. an outer door to a passenger lift ought not to be able to open unless the lift is present.

If a system meets these characteristics then it can be controlled by a PLC. However, if, say, the second characteristic found that the action could be performed in a random fashion then a PLC would not be suitable.

28.10 Flexibility of PLCs

Let us look at a simple conventional control system. A lamp is to light only when all three control switches are closed. Using conventional mechanical equipment, the system would be that shown in Fig. 28.7

In this instance, the wiring goes from switch to switch and then to the lamp. The arrangement is therefore entirely determined by the wiring of the system. Now let us look at the equivalent PLC system shown in Fig. 28.8.

At first, the wiring seems more complicated because every unit is connected directly to the PLC. However, the system is no longer determined by

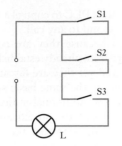

Fig. 28.7 Simple control system

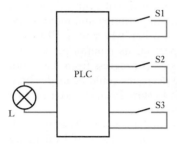

Fig. 28.8 Simple PLC control system

the wiring – instead it is determined by the program which, in this instance, effectively says: if S1 AND S2 AND S3 are ON (or HIGH) then the lamp lights.

Let us decide to change the conditions for the lamp to light. Let us say that S1 and S2 should be closed or ON and that S3 should be open or OFF. In Fig. 28.7, a relay would have to be inserted so that the S3 function is inverted. (Alternatively the S3 action should be changed or inverted.) With the PLC in Fig. 28.8, no change is required to the wiring but the program needs to effectively say: if S1 AND S2 AND NOT S3 are ON then the lamp goes ON. In fact, we could have any combination of S1, S2 and S3 being ON or OFF simply by changing the program. What a difference to rewiring the traditional control system!

You might wonder if systems are changed all that often. Consider a new control system. You think that you have designed its program to perform all the given functions, that you have built in all the safeguards and that it is foolproof – and a week after it starts, someone finds a way to break the system. There has been a combination of events which you had not envisaged. At this stage, all you have to do is modify the program and you have the missing safeguard in place. We would all like to think that we would produce a perfect system first time but the practical engineer knows that there is always the unforeseen waiting to trip him or her up. How much better to have a fallback position by being able to change the program.

28.11   Inside a PLC

Unlike the integrated circuit, the PLC comes in a variety of forms which are either of the block-type construction or the rack-type construction. The block is used for small controllers and comes as a complete package, i.e. all the inputs, outputs and power supplies are accessed through its surface terminals. Also it has a shallow mounting depth which can be advantageous. The rack type of construction is used for larger controllers. It consists of the base rack, which is plugged into a number of functional units or modules such as the power supply, the microprocessor, the input(s) and the output(s). It is the rack which provides the interconnection between the modules. This form of construction allows a degree of flexibility in the choice of module and also the number of modules, thereby permitting a greater number of, say, inputs. Although somewhat sizeable, the modules have the advantage of rapid replacement should a fault occur.

We shall look at devices which can be connected to the inputs and outputs throughout the remainder of this chapter but, at this stage, it is useful to note the symbols shown in Fig. 28.9 as they shall be appearing in the sections which follow.

Each controller has separate sets of terminals for the power functions, the input functions and the output functions. The power terminals are normally 115 V (or 230 V) a.c. and the inputs operate from 24 V (or 48 V) d.c. The output switches are electronic switches integral to the PLC.

At this point theory and practice become less consistent because, although there are IEC recommendations, there is no standard form of physical construction or of addressing the input and output terminals. Consequently the user has to become familiar with the arrangements specific to the manufacturer of the equipment used.

Fig. 28.9 Symbols of input and
output devices

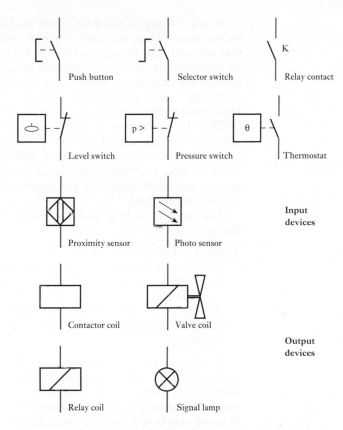

Push button   Selector switch   Relay contact

Level switch   Pressure switch   Thermostat

Proximity sensor   Photo sensor

Input
devices

Contactor coil   Valve coil

Output
devices

Relay coil   Signal lamp

IEC 1131–3 recommends that input terminals be given the prefix I and
the output terminals the prefix Q. These were the ones we used earlier in the
chapter. However some manufacturers use X for inputs and Y for outputs,
whilst others use O instead of Q for their outputs. The last can be quite confusing when dealing with an output O0.0.

A small PLC might have up to 128 input/output (I/O) terminals, a
medium PLC up to 1024 I/O terminals and a large one up to 4096 I/O terminals. Not all input and output demands are equal and care has to be taken
that the sum of the loads, should they all be applied simultaneously, does not
overload the rating of the PLC.

28.12	The PLC program

We are familiar with programming and the manner in which programs
are written. Using the instruction list used earlier in the chapter, let us
consider a program appropriate to the three switches and the lamp shown in
Fig. 28.8.

LD   I400

AND   I401

AND   I402

OUT   Q430

In this instance, if the three inputs are ON then the output will switch on the lamp. When operating, the PLC program continually goes to the first step and runs through to the last. Each step is taken in sequence and therefore the processor is dealing with only one step at a time. However, it goes through the entire sequence so quickly that the total time is almost negligible – even a reasonably long program is repeated within 10 ms. Having completed the program, it is rerun time after time.

The PLC therefore scans the program updating the outputs at frequent intervals – in this case, at the end of each cycle. It is therefore necessary to have a choice of operating modes which are STOP and RUN. STOP indicates that, provided power input is available, the PLC is ready to operate, and RUN makes the PLC operate by repeatedly executing all the program instructions.

The program is held in the memory which can be in one of three forms:

1. the RAM which as we have seen previously can be programmed quite readily but it clears should the power supply be interrupted;
2. the ROM which is secure against loss of power but which cannot be reprogrammed;
3. the electrically erasable programmable read-only memory EPROM which can be reprogrammed quite readily but which is secure against loss of power.

It follows that in most instances, PLCs incorporate EPROMs but microcontrollers may incorporate PROMs. RAMs are often used at the development stage of new programs. Usually the RAMs are specially adapted for this application and incorporate batteries which limit the dangers of power supply interruption. Once the program is established, it is copied to an EPROM which then replaces the RAM.

The memory size associated with PLCs is much smaller than that associated with computers. The memory of a small PLC might be 2 or 4 kbytes. A large PLC might have 96 kbytes and a medium PLC would be somewhere in between, say 32 kbytes.

In order to minimize the size of program, many PLCs have a number of built-in functions. For instance, it might be that a particular series of instructions has to be repeated many times in the program cycle. To avoid this, we can scan the series of instructions and load the outcome into an internal relay. Basically this is an output to which there is no terminal access. However, the result stored by the internal relay can be used in subsequent steps of the program. For instance, we might test six inputs and store the result in the internal relay. Subsequently we need only test the relay instead of the six inputs. This greatly increases the programming power of the controller.

Other internal functions can include flip-flops, counters, timers and sequencers.

To consider a program further, it is best that we review the practical possibilities available to us. For example, if we wish to operate a lift, can we measure its position? or can we know if someone has entered the lift? or can we indicate the position of the lift? In practice, we can produce information satisfying the requirements of these questions and we can supply it to the inputs of a PLC. However, the inputs have to be digital in concept so that the information comes in the form of an input being ON or OFF. There we need to look further at the input devices which make the information available to a PLC.

28.13 Input devices

There is a very large range of input devices, the most simple of which is the push button. Even such a simple device comes in two forms. In the first, the contacts make when the button is pushed, i.e. the contacts are normally open until operated. In the second, the contacts break when the button is pushed, i.e. the contacts are normally closed until operated. It is normal practice to apply the make operation to give START signals and the break operation to give STOP signals. Thus a fault in a STOP circuit has the same effect as the break action and will cause the system to STOP for safety.

When a push-button switch is pressed, it may only be for a short time – almost certainly less than a second. However, the program should scan the switch state every 10 ms so that even a short closure will be picked up by the PLC. There are other manual switches such as the selector switch, which is rotated to two or more positions, and the thumb-wheel switch, which is rotated to show a numerical selection.

The simplest non-manual switch is the limit switch in which a moving object (our lift for example) comes into contact with a lever or cam which makes or breaks the switch contacts. A level switch operates in much the same manner with the lever attached to a float mechanism. The limitations of such devices are that moving parts are prone to wear and thus may need to be replaced or the electrical switch contacts may mechanically wear or even burn.

In more recent times, a family of sensors has been introduced. An inductive proximity sensor can detect the approach or passing of an object by responding to the movement of a ferromagnetic plate attached to the object. The approach of the plate changes the magnetic field linkage of the sensor and can thus trigger a transistor switch. By this means, there is no need for any moving parts in either the sensor or the switch, hence the system is more reliable in its operation. A capacitive proximity sensor works in a similar fashion but it can respond to non-metallic objects. We meet this type of device in keypads in which we touch a number without having to push it.

Photosensors are used to indicate the passage of an object. In one common form, a beam of light is directed to a reflector and returned to a receiver. If the beam is interrupted then we know that an object has passed, e.g. a person entering a lift. An alternative arrangement mounts the receiver opposite the light transmitter. Photosensors are more demanding in power than other sensors and we have to be careful not to make too great a total power demand on a PLC.

Temperature can be determined by thermostats many of which are still basically mechanical, i.e. bimetallic strips open and close the switch contacts. Transistorized versions are becoming available but their operating times are slow. Pressure sensors operate with bellows which control make or break contacts.

All the input devices have the common factor that the inputs are digital, i.e. they are either ON or OFF. It is possible to provide numerical information to a PLC but again this needs to be in a digital form. Generally we design our systems only to answer questions of the yes/no variety, i.e. is the switch open (yes) or closed (no)?

28.14 Outputs

Most controllers are intended to control at least one significant power output. For example, if we consider our lift, then we shall be controlling a motor to raise and lower the cabin. In such applications, the current to the motor is normally controlled by a contactor or by some power electronic equivalent. In a contactor, an electromagnetic coil energized by a small current causes the contacts to close thus supplying current from a separate power source to the motor. Contactors are used for currents in excess of 10 A. A lesser variety of contactor is the relay which operates on the same principle but is rated up to about 10 A. With either device, the currents controlled are in excess of those which can be fed directly from a PLC. Contactors and relays are further explained in Chapter 36.

Contactors and relays are mechanisms which contain moving parts and therefore are prone to wear. Thyristors, IGBTs and GTOs are all electronic devices which perform similar functions in that a small current (or voltage) from one source can control a large current from another source. Having no moving parts, they require less maintenance and are progressively replacing the contactor and the relay. Such electronic switches are further explained in Chapter 45.

The most common outputs from PLCs are indicator lights such as incandescent lamps, neon lamps and l.c.d.s. The last mentioned is increasingly popular because of its low power rating. However, the demand for indicator lamps is declining, to be replaced by message displays or operator interfaces. These provide the operator with a variety of messages which can be quite explicit rather than the need to interpret the signals from the indicator lamps. For example, the message might be 'Fuel level low' as opposed to a red lamp illuminated on a panel.

As PLCs have developed, the need to keep an operator informed has become more significant. There is a limit to how readily an operator can interpret a display of indicator lamps especially if there are too many to be observed. A solution, of which mention has already been made, is the use of message displays although it should be noted that the range of messages is fairly limited for any one PLC. In such instances, a display unit holds the messages in its memory and each has a recall number. The PLC activates the display unit by sending an appropriate array of outputs which are interpreted by the display unit interpreting the relevant recall number. By using five PLC outputs, it is possible to control up to 32 messages.

A simple operation interface should be capable of being used with any PLC. It operates with only the outputs and does not require a special communications port: however, there is usually a requirement for some input from the operator, thus there is a need to use some of the input terminals as well as the outputs. Finally such an operation can easily be applied to an existing system with only the need to make minor changes to the program.

28.15 A practical application

We have looked at the basic operation of a PLC. In order to obtain a more practical understanding of the operation, let us finish by considering a possible application. Throughout the chapter, we have often instanced the application of a PLC to a lift and it would seem appropriate to investigate such an application further.

Let us consider a lift intended to raise people in wheel chairs from one level to another thereby avoiding a short flight of steps. The height of the rise is about 2 m and entry to the lift is attained in response to a push button. The gates operate in pairs (one on the lift and the other on the fixed entry) and they must only open provided the lift is in the appropriate position. A motor at each level can move the relevant pairs of gates. The lift is raised and lowered by a three-phase motor and is also fitted with a brake.

Having outlined the project, we can draw up a list of system requirements as follows:

A. The lift control system will normally remain in the RUN mode at all times and will respond to call push-button switches only sited at each level; it will also respond to raise/lower push-button switches within the lift.

B. The lift may only move when all four gates are closed and a call or raise/lower command is given.

C. Once the lift starts to move, it must complete its travel

D. Signal lamps, sited at each level and indicating 'LIFT IN USE', will operate should either of the pairs of gates be open or should the lift be moving following the raise/lower command.

E. The gates will close after the lift has been used.

F. An alarm bell will sound should the lift overrun or should an emergency push button sited in the lift be operated.

We can now draw up a schedule for the equipment to be controlled by the PLC as follows:

1. The lift motor has an UP contactor MC1 and a DOWN contactor MC2 for normal operation. For protection, there is a thermal overload TR1.

2. The lift brake operates when the lift motor is OFF and therefore does not involve the PLC.

3. The lower gate motor has an OPEN contactor MC3 and a CLOSE contactor MC4. For protection, there is a thermal overload TR2.

4. The upper gate motor has an OPEN contactor MC5 and a CLOSE contactor MC6. For protection, there is a thermal overload TR3.

5. Signal lamps L1 and L2 indicate 'LIFT IN USE'.

6. The lift is called to the lower level by push-button switch S1 and to the upper level by push-button switch S2.

7. The gates are opened in response to push-button switch S3 at the lower level and to push-button switch S4 at the upper level.

8. On entry, the gates are closed by push-button switch S5 and are opened at either level by push-button S8.

9. Once the gates are closed, the motion of the lift is initiated by push-button switch S6 to go UP and by push-button switch S7 to go DOWN.

10. Limit switches S9 and S10 detect the gates being CLOSED at the lower level.

11. Limit switches S11 and S12 detect the gates being CLOSED at the upper level.

12. Limit switches S13 and S14 detect the outer gates being OPEN at the lower and upper levels respectively.

13. Limit switches S15 and S16 detect the lift being at the lower and upper levels respectively.

14. Safety limit switches S17 and S18 detect overrun by the lift at the lower and upper levels respectively thereby ensuring the control to the lift motor is interrupted.
15. Alarm push-button switch S19 is in the lift for emergency calls.
16. Alarm bell A1 is to ring should S17, S18 or S19 be operated.

This schedule is not exhaustive but it is a typical initial attempt. In particular, no attempt has been made to make the gate operation automatic in view of the specialized application. Once we have completed an operational program, we shall consider some of its practical limitations and hence gain an appreciation of the critical approach required to program development.

In order to produce a program, it is necessary to assign the inputs and outputs as follows:

Device	Component	Input/output
Lower call push button	S1	I001
Upper call push button	S2	I002
Lower outer gate OPEN push button	S3	I003
Upper outer gate OPEN push button	S4	I004
CLOSE gate push button in lift	S5	I005
Raise (UP) push button in lift	S6	I006
Lower (DOWN) push button in lift	S7	I007
OPEN gate push button in lift	S8	I008
Lower outer gate CLOSED limit switch	S9	I009
Lower inner gate CLOSED limit switch	S10	I010
Upper outer gate CLOSED limit switch	S11	I011
Upper inner gate CLOSED limit switch	S12	I012
Lower outer gate OPEN limit switch	S13	I013
Upper outer gate OPEN limit switch	S14	I014
Lift at lower level position switch	S15	I015
Lift at upper level position switch	S16	I016
Lower safety limit switch (overrun)	S17	I017
Upper safety limit switch (overrun)	S18	I018
Emergency push button	S19	I019
Lift motor thermal overload	TR1	I020
Lower gate motor thermal overload	TR2	I021
Upper gate motor thermal overload	TR3	I022
Motor control to raise lift	MC1	Q001
Motor control to lower lift	MC2	Q002
Motor control to open lower gates	MC3	Q003
Motor control to close lower gates	MC4	Q004
Motor control to open upper gates	MC5	Q005
Motor control to close upper gates	MC6	Q006
'LIFT IN USE' lamps	L1 & L2	Q007
Alarm bell	A1	Q008

Having assigned the inputs and outputs, we can now write the program. The program would normally consist only of the instructions but in order to make it easier to follow, comment is made in italic print so that you can understand the significance of each step. Also reference has already been made to the use of an internal relay whereby the program can avoid repeating a number of steps. This program commences by setting up such a relay in order to avoid repeatedly checking that the gates are closed and that the lift has not overrun.

Relay	LD	I009	*Check lower outer gate closed*
	AND	I010	*Check lower inner gate closed*
	AND	I011	*Check upper outer gate closed*
	AND	I012	*Check upper inner gate closed*
	ANDN	I017	*Check lift has not overrun downwards*
	ANDN	I018	*Check lift has not overrun upwards*
	ANDN	Q008	*Check alarm has not been operated*
	OUT	C100	*Output from internal relay*
Operate lift down	LD	I001	*Lower call push button operated*
	OR	I007	*DOWN push button in lift operated*
	OR	Q002	*Continue motor operation downwards*
	AND	I015	*Check that lift is not in lower position*
	AND	C100	*Check safety of operation*
	ANDN	Q001	*Check lift is not moving upwards*
	OUT	Q002	*Motor control to lower lift*
Operate lift up	LD	I002	*Upper call push button operated*
	OR	I006	*UP push button in lift operated*
	OR	Q001	*Continue motor operation upwards*
	AND	I016	*Check that lift is not in upper position*
	AND	C100	*Check safety of operation*
	ANDN	Q001	*Check lift is not moving upwards*
	OUT	Q001	*Motor control to raise lift*
LIFT IN USE lamps	LDN	I009	*Check lower gates open*
	ORN	I010	*Check upper gates open*
	OR	Q001	*Check lift in motion upwards*
	OR	Q002	*Check lift in motion downwards*
	OUT	Q007	*Illuminate LIFT IN USE lamps*
Alarm bell	LDN	I017	*Check lower safety limit switch*
	ORN	I018	*Check upper safety limit switch*
	ORN	I019	*Check emergency push button*
	OR	I020	*Check lift motor thermal overload*
	OR	I021	*Check lower gate motor thermal overload*
	OR	I022	*Check upper gate motor thermal overload*
	OUT	Q008	*Sound alarm bell*
Open lower gates	LD	I003	*Lower outer gate OPEN push button*
	OR	I008	*OPEN gate push button in lift*
	OR	Q003	*Continue opening motor operation*
	AND	I015	*Check lift at lower level*
	ANDN	I013	*Check lower gate is not open*
	ANDN	Q004	*Check gate motor is not closing*
	OUT	Q003	*Motor control to open lower gate*

Close lower gates	LD	I005	*CLOSE push button in lift*
	OR	I002	*Upper call push button*
	OR	Q004	*Continue closing motor operation*
	AND	I015	*Check lift at lower level*
	ANDN	I009	*Check lower gate is not closed*
	ANDN	Q003	*Check gate motor is not opening*
	OUT	Q004	*Motor control to close lower gate*
Open upper gates	LD	I004	*Upper outer gate OPEN push button*
	OR	I008	*OPEN gate push button in lift*
	OR	Q005	*Continue opening motor operation*
	AND	I016	*Check lift at upper level*
	ANDN	I014	*Check upper gate is not open*
	ANDN	Q006	*Check gate motor is not closing*
	OUT	Q005	*Motor control to open lower gate*
Close upper gates	LD	I005	*CLOSE push button in lift*
	OR	I001	*Lower call push button*
	OR	Q006	*Continue closing motor operation*
	AND	I016	*Check lift at lower level*
	ANDN	I013	*Check lower gate is not closed*
	ANDN	Q005	*Check gate motor is not opening*
	OUT	Q006	*Motor control to close lower gate*

Having followed the program, we should now have a better understanding of the manner in which a PLC can be applied to the control of a system. The above program would work but, because it has been kept relatively simple, it might not work as well as we would wish. This is always the case when developing a program – so here are some questions for you to consider:

1. What if the person in the wheel chair cannot reach the lift push buttons?
2. What if the lift were called just as someone was passing through the gates – might not the gates close on them?
3. What if someone calls the lift when it is already in use? – there is no memory to record the call signal.
4. Is there a need to make the gates close automatically bearing in mind the variety of wheel chairs that might use the lift and their various speeds of entry?
5. What would happen if the lift were inadvertently to stop half way?

The first question can be readily answered by positioning duplicate push buttons with the lift bearing in mind that the person operating the lift might not be the person sitting in the wheel chair but rather could be someone pushing the chair. The duplicate switches could be in parallel and therefore there is no need to increase the number of inputs to the PLC.

For the second question, we might wish to introduce a timer, and a simple addition to the program might be as follows:

Upper gate to remain open	LD	I014	*Check outer gate OPEN*
	TMR	V001	*Timer operation*
	OUT	C101	*Output from second internal relay*

The output from this second relay can now be introduced to the upper gate closure program thus preventing gate closure for a period commencing with the complete opening of the outer gate.

With regard to the remaining questions, think of them as an exercise on which to try out both your programming skills and also your design skills. Remember that the lift only has a short rise and may not require the sophistication associated with multi-level lifts which disappear from sight.

Terms and concepts	A **microprocessor** is a device containing many logic circuits. It can operate at very high speeds.

In order to control the processing unit, we require memories to store instructions and data. The instructions are held in the **read-only memory** (ROM) and the temporary data are in the **random-access memory** (RAM). The interchange of information between the **accumulator** (CPU), the RAM and the ROM is achieved by **buses**.

Information into and out from the system is transferred through **ports**.

Sequences of instructions are called **programs** which may be in machine code, assembly language or high-level language. The use of such languages is a study of its own.

A **programmable logic controller (PLC)** is a device built around a microprocessor and its functions are determined by a program stored within the PLC.

The tasks of a PLC involve interlocking, counting, sequencing, timing and monitoring.

The characteristics of a PLC control system are that there are definite actions to be taken, definite rules governing these actions and the rules must take into account the relevant conditions.

PLCs come either in block-type form or in rack-type depending on the size of the controller. A small controller might have around 100 inputs and outputs whereas a large one would have over 1000.

There is a large variety of input devices all of which must be digital. The outputs can be directly powered from the PLC but the power ratings are not high. In those instances in which a high power rating is involved, the PLC output can supply an electronic switch or a contactor.

Exercises 28

1. (a) Distinguish between a *register* and a *memory* in a microcomputer.

    (b) Describe the need for analogue-to-digital interfaces and digital-to-analogue interfaces within a microcomputer system.

2. (a) (i) Draw a block diagram of a bus-organized microprocessor system illustrating the microprocessor, the memory, the input and output. (ii) Explain the function of the various buses in the above system.

    (b) Distinguish between the serial and parallel transfer of digital signals, and give an example of a device employing each.

3. (a) In a microprocessor system distinguish between the following: (i) operand and instruction set; (ii) conditional and unconditional branch instructions; (iii) the two address modes for the LDA instruction LD# xx and LD yyyy.

    (b) It is required to interface a microcomputer (via the output port) to a power circuit having an a.c. rating of 230 V, 10 A. List two examples of a suitable interface between the microcomputer and the power circuit.

(c) A flowchart can be a valuable aid in programming development. List four of the symbols used.

4. Write a program to add two numbers, stored at 0260 and 0261, and store the result at 0262.

5. Using the instruction set provided in section 28.4, write a program which takes the number 27 from location 0030 and converts this number to 72 (i.e. digit interchange) which it finally stores in location 0031. The program should be accompanied by comments explaining each instruction.

6. In a microprocessor system, a single-byte number is located at 0060 and also a single-byte number is available at the input I001. The program of this system is given below. Explain each instruction and describe the purpose of the program.

```
LD    0060
ST    0040
LD    I001
SUB   0040
ST    0041
BRK
```

7. (a) Describe the role of the microprocessor system in a typical application indicating the origin of the data, the processing function and the destination of the output signal.

(b) A program using the instruction set in section 28.4 is given below. (i) Describe each instruction and the operation of the program. (ii) State the value stored in location 0042 after running the program.

```
    LD    #08
    ST    0040
    LD    #04
    ST    0041
    LD    #00
┌→  ADD   0041
│   DEC   0040
└─  BNZ
    ST    0042
    BRK
```

8. (a) Using the instruction set in section 28.4, provide the instructions for the following: (i) to load the accumulator immediately with the operand 55; (ii) to subtract the contents of memory location 0020 from the accumulator without affecting the data in the accumulator.

(b) A part of a program using the instruction set listed in section 28.4 is given below. Describe each instruction and the operation of this sequence of instructions.

```
    LD    #04
    ST    0040
┌→  DEC   0040
└─  BNZ
    LD    0040
    ST    I000
```

9. A hoist is fitted with a protective device which operates at a height of 3.0 m. Above 3.0 m, the protective device inputs logic 0 to the controlling microprocessor system. As the hoist descends towards the 3.0 m limit, the input changes to logic 1. Should the hoist descend past the 3.0 m limit, the input returns to logic 0. Write a program that continually checks the state of the input signal at I001 and which outputs the byte 2F (which stops the hoist) at Q010 after the protection device input has changed from 0 to 1 back to 0.

10. A microcomputer has eight switches connected to the input port I001 and eight LEDs connected to the output port Q010. A logic 1 is applied to the input from a closed switch and an LED is lit by the application of a logic 1.

(a) Write a program which performs the following: (i) reads the state of the switches and tests this state against a pattern of FF; (ii) if the state of the switches matches this pattern, outputs a pattern AA which lights alternate LEDs.

(b) Comment on each instruction of the program.

11. Write a program to set the four most significant bits of the data in $0273 to zeros, leaving the remaining bits unchanged.

12. Write a program to set the four most significant bits of the data in $0284 to ones, leaving the remaining bits unaffected.

13. Write a program to invert each bit of the data stored in $0266.

14. Write a program to multiply the contents of $0266 by 5 and store the result in $0268 without changing the data in $0266. Assume that the data are less than 33.

15. Write a program to multiply the contents of $028A by 7 and store the result in $028B without changing the data in $028A.

16. In an 8-bit dataword, the bits are numbered 0, 1, 2, 3, 4, 5, 6, 7, where the highest bit is 7. Write a program to set (make 1) bits 4, 3, 2 and 1 of $0293 and store the result in $0202.

17. Write a program to clear (make 0) the four least significant bits of $0202 and invert the remaining bits.

18. Write a program to move each bit of $0292 one place to the right and set the two most significant bits.

19. Write a program to form a dataword from the four most significant bits of $024A and the four least significant bits of $024B and store the result in $024C, leaving the data in $024A and in $024B unaffected.

20. Write a program to move the four most significant bits of $0291 to the four least significant bits of $0202, leaving $0291 unaffected and set bits 7 and 5 to logic 1.

21. Write a program which tests the most significant bit of the data in $0260 and stores data at $0900 so that: if the MSB=1 then the data stored at $0900 are those already stored in $0700; if the MSB=0 then the data stored at $0900 are those already stored in $0701. Write the program first using the BZE instruction and then using the BNZ instruction.

22. Write a program which sets the MSB of $0800 if both bits 7 and 6 of the data in $0250 are zeros and clears the MSB of $0800 if either bit is a one.

23. Write a program to store the data in $0250 at $0900 if the data in $0901 are equal to $0018 and store the data at $0251 at $0900 if the data in $0901 are equal to $0019. Write the program as a continuous loop.

24. What are the characteristics that a system must meet if it is to be controlled by a PLC?

25. What are the advantages that using a PLC for a control task will give over conventional mechanical equipment?

26. How is the program for a PLC held?

27. What built-in functions do many PLCs have to help minimize program size?

28. Describe the types of input and output device that can be used by PLCs.

29. In section 28.15 a program for a PLC was developed. How would you modify this program to ring the alarm if there is a weight overload that is indicated by TR4?

Control Systems

Objectives

When you have studied this chapter, you should
- be aware of the differences between open-loop and closed-loop systems
- understand how feedback signals are compared to input signals to give the error signal
- have an appreciation of transients and the concept of damping
- be able to explain the settling time for a system

Contents

Although microprocessors are used in computers they are also widely used in control systems, particularly in the form of PLCs as described in the previous chapter. These are systems that automatically cause a series of events to occur.

In this chapter examples of control systems will be explored, splitting them into two categories. Regulators are associated with varying operational speeds while servomechanisms are associated with varying position. In either case there is a need for feedback which informs the control system of how well it is succeeding. The transient period of these systems will also be considered.

To shorten the transient period, damping is introduced. The relative value of damping gives rise to a variety of transient effects that are explained. The choice of damping effects is made dependent upon the application requirements.

By the end of the chapter, you will be familiar with the concepts associated with simple control systems and the principles of feedback and damping. You will also be able to undertake simple exercises applying these principles.

29.1 Introduction

A control system is one which undertakes some function, checks its success and takes further action until the objective is attained. And if in the meantime the objective varies then the system will respond to the change. This sounds very complicated so let us consider some simple examples.

For instance, Example 28.5 involved a basic control system. In it, we were concerned with the level of liquid in a tank. If the level rose to a given level, a lamp was switched on, and conversely if it fell to a lower given level, the lamp was switched off. We might just as well have arranged to switch on a pump when the level fell (thereby filling the tank) and switch off the pump when the liquid level had risen.

Such control systems are basically digital in nature, i.e. they require an action to be either ON or OFF. They make no allowance for the amount of water needed to fill the tank and it might take a minute or an hour to refill it. So perhaps control systems not only should react to need but they should also respond to the speed of response required. If that is what we want, we require a control system which either is analogue in principle or is digital as a consequence of a D/A converter.

A common example of an analogue control system is that of a person driving a car. The car is guided by the steering wheel and the driver assesses whether an adjustment of the steering wheel is appropriate. If the driver decides that the car is moving too far to the left, he or she will turn the steering wheel clockwise. Eventually this will bring the car back on line, but then it would be continuing to swing towards the right unless a second adjustment is made to stop the swing and maintain the car on its new course.

A further complication can be anticipated by the driver when it is seen that the road ahead is entering a bend. The driver can estimate the necessary adjustment to the steering so that the steering wheel moves to keep the car on course rather than waiting to find that it is going off course.

If we wish to replace the driver with a totally automatic control system, we need to determine that there is a problem (the car is going off the road), the extent of the problem (how far is the car off its lane) and the extent of the response required (how quickly must the car return to its lane).

A typical development for this situation is to be found in larger passenger-carrying aircraft which land on automatic pilot. In particular, such a system has to continually make allowance for changes in the wind as the aircraft approaches the runway. A side wind could drive the aircraft to the side of the runway while changes in head wind could bring the aircraft down too early or too late.

Control systems depend on the quantities which we can measure. The five basic quantities are as follows:

1. Displacement.
2. Force.
3. Pressure.
4. Temperature.
5. Velocity.

There are other more complex quantities such as light intensity, chemical composition, rate of flow and conductivity.

When we look at further instances of control systems, we will see these quantities being used to effect. However, we must first consider the loops which are fundamental to control systems.

29.2 Open-loop and closed-loop systems

Let us assume we have a mechanical arm which lifts boxes from a conveyor belt to fill a container. The arm has been instructed to repeat its action 10 times, after which the container should be full. This works well so long as the supply of boxes on the conveyor belt is uninterrupted. However, if there is an interruption, the arm keeps on repeating its action even though there are no boxes for it to move. The result is that the cycle of 10 movements has been completed, but the container has not received 10 boxes – perhaps it has received none.

A control system of this type is called an open-loop system because it has no feedback to tell the control that some of the movements have failed in their intended action to move boxes.

We could have improved the system by one or other of two simple modifications. One would have been that the device lifting a box could register that it has failed to find a box. This could be discovered because the gripping mechanism closed too far showing the absence of a box (this is an example of displacement measurement). Thus only a movement with a box in the gripping mechanism would count towards the 10 effective movements.

Alternatively the container would gain weight every time a box was dropped into it. Only when the weight was achieved would the mechanical arm stop repeating its action on the understanding that 10 effective movements had been completed. The measurement of the weight is an example of force measurement.

A control system with such feedback arrangements is known as an automatic closed-loop system. If we think back to the driver in a car, that would represent a closed-loop system which is not automatic. In general, however, when we talk of closed-loop systems, we are inferring that they are automatic, i.e. there is no human intervention. We can represent the actions of both the open-loop and closed-loop systems described above by means of block diagrams as shown in Fig. 29.1.

A closed-loop control system therefore depends on there being two significant functions:

Fig. 29.1 Open-loop and closed-loop systems

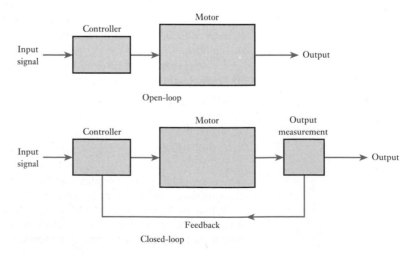

1. Feedback from the output to the input.
2. A difference (called the error) between the feedback and the desired outcome specified by the input control signal.

For instance, if we were weighing the container, the feedback would be equivalent to the weight, and the difference between the actual weight and the desired weight provides the error. A device which converts a quantity measurement such as force into an electrical signal is called a transducer. Although the preferred term for error is deviation, we have to accept that most engineers talk of error.

29.3 Automation

There are many automatic control systems in general everyday use, but most of the sophisticated control systems are to be found in industrial applications. These applications are often referred to as automation.

There are many advantages to be found in automation. If we think of the mechanical arm loading the container, only the automatic system ensured that the container was completely filled. In industry, it could be an offence to sell a container claiming to hold 10 units yet loaded with fewer units. Even if it were not an offence, customers would become dissatisfied and the effect could be just as bad for the manufacturer.

So we can think of advantages as being:

1. Consistent production.
2. Release of production operators for more useful work.
3. Improved conditions for the operators.

If the work is more complex, other advantages would include:

4. Improved accuracy of manufacture.
5. Economic use of expensive plant.

29.4 Components of a control system

Let us consider the basic control system shown in Fig. 29.2. The system has four significant features:

1. **INPUT SIGNAL** – this sets the condition which it is intended to achieve. It is sometimes called the reference signal or even the set signal.
2. **OUTPUT SIGNAL** – this represents a measurement of the outcome achieved and hence it provides the feedback signal. It is sometimes called the reset signal.

Fig. 29.2 Basic control system

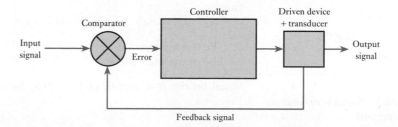

3. COMPARATOR – this is the part of the system in which the input and output signals are compared. The difference is the error which is fed to the controller (see below).

4. CONTROLLER – this is the device which causes the required activity to happen. This is rather like the amplifier in that the controller controls the supply of energy from a separate power supply.

The basic system assumes a degree of stability which might occur in our mechanical arm placing boxes in a container. It would not matter whether or not the power supply to the arm was interrupted. And it would not matter if the supply of boxes on the conveyor was interrupted. Such problems merely delay the filling of the container. Only permanent interruption of power supply or box supply would prevent the container being filled sooner or later.

However, let us consider the automatic pilot on the aircraft which was landing on a runway. If the wind were to move the aircraft off line, it would not be sufficient just to let the automatic pilot get the aircraft back on line sooner or later. Now we have a limited time in which to take action – and for that action to be completed. The nearer to time of touchdown, the quicker the required action.

It follows that many control systems need to operate not only on the error, but also on the rate of reaction which is necessary. However, this introduces a further complication. If the feedback circuit detects a change of error, it may be that it will not properly reflect this because it does not respond directly. We can describe this indirect response in terms of the transfer function.

29.5 Transfer function

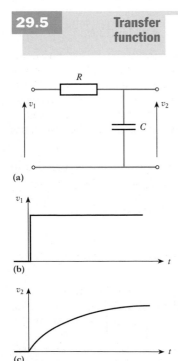

Fig. 29.3 Signal transfer in an *R–C* network

Let us look at a feedback circuit which comprises a resistor *R* and a capacitor *C* as shown in Fig. 29.3(a). Also let us assume the introduction of a sudden error as shown in Fig. 29.3(b) – this type of signal is described as a step change. From our understanding of *R–C* transients, we can anticipate that the output signal will take the form shown in Fig. 29.3(c).

We can predict such a reaction, therefore we should be able to specify it by some function. In fact we can specify the reaction by means of a transfer function. The transfer function is the ratio of the output signal to the input signal making due allowance for the time element.

We have considered the application of a step change, i.e. one which rises from zero to a finite value instantaneously. However, we could have predicted the outcome in response to, say, a sinusoidal input. The output would have been sinusoidal, of a smaller peak value, and would lag the input by up to quarter of a period.

The consideration of sinusoidal or even more complex signals requires quite advanced mathematical techniques – usually we apply Laplace transforms. At this introductory stage, we will limit ourselves to considering step changes.

For a step change, the time variation can be considered in two parts:

1. The transient period.
2. The steady-state period.

Based on the *R–C* circuit of Fig. 29.3, we can illustrate these periods in Fig. 29.4.

Fig. 29.4 Response periods

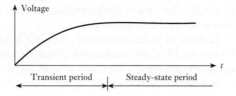

Fig. 29.5 A basic regulator system

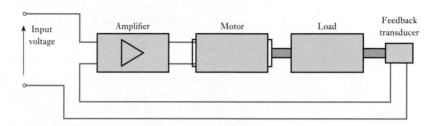

It is not only the *R–C* circuit which will provide the responses shown in Fig. 29.4. An alternative would be the sudden application of an electromagnetic field to a rotating coil. Switching on the current to create the field as shown in Fig. 29.5 does not immediately create the field, instead it builds up exponentially. It follows that the e.m.f. induced in the rotating coil also builds up, thus the transfer function is of the same form as that for the *R–C* circuit.

Once again it is a simple step forward to replacing the circuit details by a block diagram. For most purposes, the significant factor is the transfer function which relates the output to the input: the manner in which the transfer is effected is of little importance.

Before leaving transfer functions, it is necessary to note certain assumptions. In particular, just as with amplifiers, we assume that the system will not saturate and that the relation is derived from linear components. This is not always the case, especially when we introduce devices with ferromagnetic cores. This involves hysteresis which has to be minimized to obtain a performance which is almost linear.

29.6 Regulators and servomechanisms

Automatic control systems usually require some outcome which is not electrical. Most outcomes are mechanical in nature and these can be divided into two groups:

1. Regulators.
2. Servomechanisms (also known as remote position controllers or r.p.c.s for short).

The difference between the two groups is the form in which the error appears. For steady-state conditions, regulators require an error signal, whereas servomechanisms require zero error. A system incorporating a servomechanism is known as a servosystem.

In either case, the mechanism will have an inertia to be overcome when motion is required. This infers that the mass of the mechanism has to be accelerated to produce movement and this will probably be followed by deceleration so that the mechanism stops at the required position. This is especially the case in servomechanisms, but regulators also have to respond to the demand for movement.

If there is to be movement, then we have to introduce a motor into the system to drive the mechanism which we refer to as the load. Having added these components, a regulator system takes the form shown in Fig. 29.5.

We can use this regulator system to explore the effect of the error. Let us assume that the regulator is a speed controller ensuring that the load is driven at a constant speed set by an input signal of 80 V. To do this, let us assume the operation to be in the steady-state condition, i.e. the load being driven at constant speed, and that we know the transfer functions for the system blocks.

For the feedback transducer, the transfer function is 40 mV per r/min.

For the amplifier, the transfer function is 20 V per volt.

For the motor/load, the transfer function is 100 r/min per volt.

The error signal fed into the amplifier is the difference between the input signal and the output signal supplied by the feedback transducer. If the steady-state speed is N_r, then the feedback signal is

$$40 \times 10^{-3} \times N_r$$

Given that the input voltage is 80 V, the error signal is

$$80 - (0.04 N_r)$$

The amplifier transfer function is 20 V per volt, hence the amplifier output voltage is

$$20(80 - (0.04 N_r))$$
$$= 1600 - 0.8 N_r$$

The motor speed is given by the transfer function being applied to the amplifier output, hence

$$N_r = 100(1600 - 0.8 N_r)$$
$$81 N_r = 160\ 000$$
$$N_r = 1975.3 \text{ r/min}$$

However, if we recall that the input signal was 80 V, we might determine that the desired load speed to give a feedback of 80 V is 2000 r/min. The error measured in speed is therefore 24.7 r/min and the error voltage is 0.988 V.

In its simple form, the regulator therefore required an error of 24.7 r/min to ensure steady speed. If we wished actually to attain a steady speed of 2000 r/min, we would require to offset the input signal by the error voltage, thus an input signal of 80.988 V would cause the load to be driven at a steady speed of 2000 r/min.

The difference between the feedback signal and the input signal is the error not only under steady-state conditions but also during the transient period. Once the system has settled down, the difference between the desired speed and the actual speed developed under steady-state conditions is called the accuracy. Usually it is expressed as a percentage of the desired speed which in the situation just investigated is 1.235 per cent.

In the regulator, if we had sought a different speed, it would have been necessary to adjust the input signal voltage. A simple method of varying the input signal is to derive it from a potentiometer.

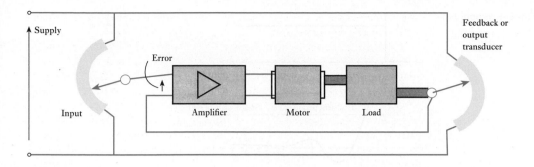

Fig. 29.6 Simple
servomechanism or r.p.c.

This arrangement was suitable for controlling speed – had we desired to control the position of the load, we would need to modify the arrangement slightly. Since any movement will result in positioning the load, at which point no further motion is required, we must change the transducer as shown in Fig. 29.6. Here a further potentiometer can be used to indicate the position, so by comparing the input potentiometer with the output potentiometer, we can determine if the load has moved to the desired position.

If the two potentiometers are aligned, there will be zero error and hence there will be no drive to the motor from the amplifier. If there is an error signal, it will be amplified to drive the motor and produce the desired movement.

Lest there be any confusion, once the load is aligned to the position determined by the input signal, no further motion will occur – unless the input signal is varied. Should it be varied then further action is required to reposition the load.

So in the r.p.c. or servomechanism, the steady-state condition occurs when the error is zero, which is distinctly different from the regulator which required an error.

29.7 In transient periods

We have considered both regulators and servosystems under steady-state conditions. However, there are two transient conditions of which we need to be aware. First there is the situation which arises when the system is switched on.

If we consider the constant-speed drive shown in Fig. 29.5, at the time of switch-on, the motor will be at rest. Therefore it has to accelerate towards the operational speed required. In practice we find that it accelerates and is still gaining speed when it reaches the operational speed, the result being that it accelerates beyond the desired speed and subsequently has to slow down. This response is shown in Fig. 29.7.

It is quite likely that the deceleration will give rise to a further overshoot and the speed may fluctuate several times before a steady state is reached.

In the r.p.c., a similar situation can arise. Here the load has an inertia which carries it beyond the desired position. As soon as it has passed the desired position, the error reverses and the motor is thrown into reverse. This helps stop the load, but starts it off in the opposite direction. Again an overshoot can take place, but each time it is smaller until the load stops in the desired position, there being no error signal to drive the motor further.

Fig. 29.7 Transient response of a simple regulator

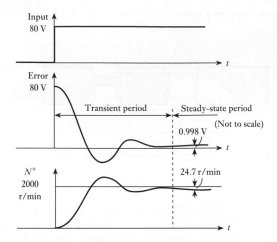

The other transient condition occurs if there is a change in condition after the steady-state condition has been achieved. Typically, the constant-speed regulator might be disturbed if the torque of the load were to change. Say it were increased, the motor would slow down, creating an error signal. This would increase the speed back to the original value – but this would give rise to a transient variation in speed for a short period of time.

Regulators quite often find themselves chasing a continuously varying objective. For instance a generator might find its load always varying as consumers switch off and on loads such as heating, lighting and motors – consider the generator chasing the varying demand of a motor as it accelerates its load.

Servosystems can experience similar problems if the input signal is continually varying, thus causing the load to be continuously chasing the desired moving target. It can also be difficult if the load experiences a force trying to displace it – this causes it to move from the desired position and hence an error is detected. This causes the motor to produce a force seeking to return the load to the desired position. However, a point of stability is achieved when the displacement force is equal but opposite to the driving torque. This is not at the desired position but one at which the error causes the motor to offset the displacement force.

29.8	Damping

Let us again consider the transient response shown in Fig. 29.8. The output position moves towards the required objective until it is reached and then the objective is passed, i.e. there is overshoot. The output then approaches the objective from the other side and again there is overshoot. This happens two, three or more times. Such a response is said to be undamped.

We can reduce the overshoot by introducing a second form of feedback. This depends on the velocity of approach of the load position to the objective. For instance, at the point where the error is zero, there will be a feedback signal due to the overshoot of the position characteristic. The feedback therefore has to oppose the error positional signal so that at the given instant it will cause the load to be braking. If we introduced sufficient

Fig. 29.8 Undamped system response to step change

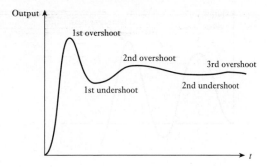

Fig. 29.9 Response with critical damping

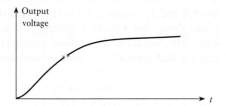

so-called velocity feedback, the characteristic would take the form shown in Fig. 29.9.

In this response, sufficient velocity feedback has been introduced to ensure that no overshoot takes place. It will be noted that the velocity feedback ceases when the steady-state condition has been reached, the velocity being zero. Sometimes the term *velocity* can appear ambiguous. For instance in the constant-speed drive previously considered, the velocity would be the rate of change of speed relative to the target speed, e.g. the rate of the rotational speed approaching 1975.3 r/min.

The introduction of velocity feedback is known as *damping*. When sufficient velocity feedback ensures that there is no overshoot, this is known as *critical damping*. If there were a slight reduction in velocity feedback, overshoot would occur and the system would be *underdamped*. If there were an increase in velocity feedback, no overshoot would take place, but the transient time would be even longer: such a system is said to be *overdamped*.

System responses which are underdamped, overdamped and critically damped are shown in Fig. 29.10.

Care has to be taken when introducing velocity damping – if too much is introduced, a system can become unstable, i.e. instead of approaching a steady-state condition, the overshooting exceeds the initial error and

Fig. 29.10 Damped system response

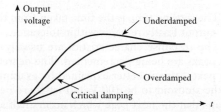

Fig. 29.11 Response of an unstable system

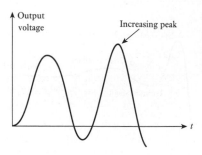

progressively the errors become larger and larger as the load swings backwards and forwards. This is illustrated in Fig. 29.11.

When designing a control system, it is essential that not only should it be stable but also it should respond as quickly as possible. From Fig. 29.10, we can see that critically damped or overdamped systems are slow to move the load to the desired objective. The underdamped system arrives at the objective more quickly, but then overshoot occurs.

Now if we were driving a car up to the edge of a cliff, underdamping would be an unsatisfactory condition since it infers that the car would have to pass over the edge. Perhaps less spectacular would be the control of a cutting tool, yet here again it would travel too far and damage the workpiece.

However, most systems can afford a degree of overshoot, and therefore most control systems are designed with underdamping. This requires that we have two design parameters – the tolerance limits and the settling time.

The relationship between these parameters is illustrated by Fig. 29.12. The tolerance limits define the extent to which we can accept variation between the load position and the objective position. This is usually expressed as a percentage of a step function giving rise to the change. Values of 5 or 2 per cent are common.

Fig. 29.12 Settling time and tolerance limits

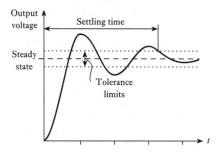

The settling time is the time taken from the start of the step change until the output finally remains within tolerance.

The response characteristics are usually sinusoidal in form except that the peaks are being decremented. The decrement is the ratio of one peak to the next – this is better explained by example. Let us assume the first peak of the sinusoid to be 100 V and let there be a decrement factor of 4. This means that the next peak which occurs half a cycle later will be 100/4 volts,

Fig. 29.13 A decremented
response (decrement equal to 4)

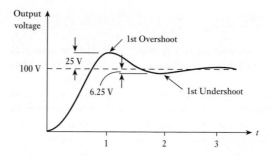

i.e. 25 V. Applying the same approach, it follows that the next peak is
25/4 = 6.25 V and so on. This is illustrated in Fig. 29.13.

A common decrement is 10 so that the first overshoot is limited to 10 per
cent of the step change. If the tolerance is also 10 per cent, this provides a
very quick response by the system yet remaining within a reasonable proximity of the objective. Further, the return motion will be limited to a 1 per
cent overshoot. Similarly a decrement of 5 will take just over two half-cycles
before the settling time is achieved for a tolerance limit of 1 per cent.

The determination of damping is a most complex one, and as previously
noted requires the application of Laplace transforms. Further, the complexity varies with the demands being made on the control system. However,
for engineers, control systems provide the greatest and most interesting
challenges and are worthy of much greater attention than can be contained
in a general introductory text such as this.

Terms and concepts

Control systems can be either **open loop** or **closed loop**. An open-loop
system takes no recognition of the output in the belief that the input
will be achieved. The closed-loop system feeds back information of the
output to ensure that the input intention is achieved.

The difference between the feedback signal and the input signal is
referred to as the **error**.

The device which produces the feedback signal is called a **transducer**.

The error is produced by the comparator and is supplied to the controller.
In turn, the controller causes a motor to cause the desired output
movement.

Control systems can be divided into **regulator** and **remote position
controllers** (r.p.c.s). Typically regulators control speed whilst r.p.c.s
control position.

A change of input signal gives rise to a period of transient change prior
to a new steady-state condition being achieved. Generally the steady-
state condition is achieved with an error which is not zero but is
insufficient to cause further change.

To reduce the transient period, **damping** is introduced. This may be
critical damping, overdamping or underdamping. Excessive damping
can make the system unstable.

The **settling time** is the time taken for the transient to reduce within
given tolerance limits.

Exercises 29

1. Explain the concept of a closed-loop control system. Describe the manner in which it differs from an open-loop system.

2. Describe the essential components of a closed-loop control system making reference to a system with which you are familiar. In such a control system, what are the advantages and disadvantages associated with the use of feedback?

3. For the speed control system shown in Fig. 29.5, the tachogenerator (feedback transducer) has a transfer function of 50 mV per r/min and the transfer function for the motor/load is 100 r/min per volt. The reference signal applied to the system is 100 V. Given that the allowable speed error is 40 r/min determine the amplifier gain (transfer function).

4. For the system in Q. 3, the reference voltage is reduced to 80 V. Determine the corresponding reference speed.

5. For the system in Q. 3, the load speed is adjusted to 1900 r/min. Determine the corresponding reference voltage.

6. Describe the settling time in a servosystem.

7. A servosystem has a periodic oscillation time of 0.4 s and its first overshoot is 25 per cent. Determine the settling time if the allowable tolerance limits are ±5 per cent.

8. For the servosystem in Q. 7, the settling time is 0.81 s for a tolerance limit of ±1 per cent. Determine the first overshoot.

9. Explain the terms underdamped, overdamped and critically damped.

10. Explain the manner in which a transducer can produce a signal proportional to angular displacement.

Chapter thirty

Signals

Signals are central to the analysis and design of most electrical and electronic systems. Signals form both the inputs to and outputs from these systems. To be able to process and analyse the signals properly it is necessary to gain an understanding of the different signal types and their properties. The way the signal is altered between the input and output of a system is a function of the system transfer function. Given the impulse response of a system and the input signal it is possible to determine what the output signal will be using the process of convolution.

This chapter explains how signals are classified. The way a signal can be represented as a series of impulses is introduced, leading to the presentation of the process of convolution, and convolution techniques for both discrete-time and continuous–time signals are presented.

By the end of the chapter you will be able to classify signals into different types. You will be capable of performing simple convolution exercises for both continuous and discrete–time signals and understand how the impulse response may be obtained from the step response.

<table>
<tr><td>**30.1**</td><td>**Classification of signals**</td></tr>
</table>

Signals arise from an infinite number of sources with a wide variety of characteristics. The inputs and outputs to any system are termed signals. These signals can be thought of as a representation of a particular variable of a system. This representation when dealing with electrical systems will be a current or a voltage. However, in other systems, signals such as temperature and pressure may be considered, which may be converted by appropriate transducers to electrical signals. These signals vary (or in some cases remain constant) as a function of time. The signals of interest here are functions of time and, depending on their characteristics, are classified in a number of different ways.

(a) Continuous-time and discrete-time signals

A continuous-time signal is one that is defined for each and every instant of time within a given range. This does not imply that the signal itself is a mathematically continuous function of time, but that it has values at all instances of time. The step function shown in Fig. 30.1 is not a mathematical continuous function because of the discontinuity at $t = t_0$. It is, however, a continuous-time function.

Fig. 30.1   A step function

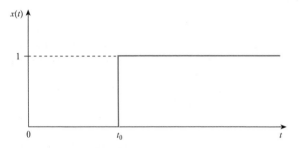

Another example of a continuous-time signal is shown in Fig. 30.2. This figure represents the speech signal produced by a microphone that senses variations in acoustic pressure from sound and converts them into an electrical voltage.

Fig. 30.2   A section of speech signal

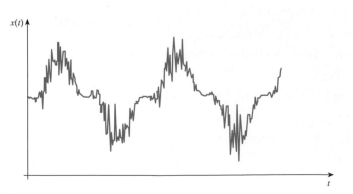

A discrete-time signal has values only at discrete instances of time. That is, the domain of t is a set of discrete numbers. Consequently such a signal will usually be denoted as $x(n)$, instead of $x(t)$, where n is an integer. Quite often, a discrete-time signal is obtained by sampling a continuous-time signal. When this is the case n may be replaced by kT, where k is an integer and T is the sampling interval. If T is kept constant, the values of x are obtained at equally spaced values of time. Such is the case in digital telephony, where the speech signal is sampled to obtain a discrete-time signal. Discrete-time signals can also occur without a continuous-time signal being sampled. This happens in many areas of life including business, science and engineering. An example is the daily maximum temperature in a city. A discrete-time signal $x(n)$ is often referred to as a sequence.

Discrete-time and quantized signals are different. Quantized signals arise when the values of a signal are limited to a countable set, e.g. when these values must be in the form of eight binary digits; this process is considered in Chapter 25. All discrete-time signals are not necessarily quantized, but if they are to be processed by a digital computer they must be quantized.

(b) Periodic and aperiodic signals

A continuous-time signal $x(t)$ is periodic if and only if

$$x(t + T) = x(t) \qquad \text{for all } t \in (-\infty, \infty) \qquad [30.1]$$

The smallest positive value of T for which the above equation holds true is termed the period of the signal. This equation will also hold when replacing T by kT, where k is an integer.

Similarly, a discrete-time signal $x(n)$ is periodic if and only if

$$x(n + N) = x(n) \qquad \text{for all } n \in (-\infty, \infty) \qquad [30.2]$$

where the smallest positive value of integer N for which the above holds is termed the period of the signal.

Any signal that is not periodic is aperiodic.

The most common example of a class of continuous-time periodic functions is

$$x(t) = A \cos(2\pi f t + \phi) \qquad [30.3]$$

where A, f and ϕ are constants (amplitude, frequency in hertz and phase in radians).

Angular frequency, which is expressed in radians per second, is defined as

$$w = 2\pi f \qquad [30.4]$$

The period T, in seconds, is

$$T = \frac{1}{f} = \frac{2\pi}{w} \qquad [30.5]$$

Two sinusoids when summed together will give a periodic function if and only if the ratio of their respective periods is a rational number. For example, consider the following sinusoids:

$$x_1(t) = 10 \sin 3\pi t$$

$$x_2(t) = 20 \cos 5\pi t$$

$$x_3(t) = 10 \sin 16t$$

The sum of x_1 and x_2 will yield a periodic function, since the period of x_1 is $\frac{2}{3}$ s and that of x_2 is $\frac{2}{5}$ s. The ratio of the two periods is a rational fraction. As a result, the period of the sum will be 2 s, which comprises 3 complete cycles of x_1 and 5 complete cycles of x_2. On the other hand the sum of x_1 and x_3 will not be periodic since the ratio of their respective periods, $\frac{2}{3}$ and $\frac{\pi}{8}$, is an irrational number.

Figure 30.3 shows two examples of periodic continuous-time signals.

Fig. 30.3   Two examples of periodic continuous-time signals: (a) a square wave; (b) a triangular wave

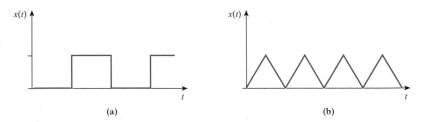

(a)   (b)

A discrete-time periodic signal is given by the sinusoidal sequence period N

$$x(n) = A \sin\left(\frac{2\pi n}{N}\right) \qquad [30.6]$$

where n and N are integers. Figure 30.4 shows a sinusoidal sequence for $N = 8$.

Euler's relationship can be used to express sinusoids in terms of complex exponentials:

$$\exp(j\theta) = \cos\theta + j \sin\theta$$

$$x(n) = \text{Real part} \left\{ A \exp\left(\frac{j2\pi n}{N}\right) \right\} \qquad [30.7]$$

Fig. 30.4   A discrete-time sinusoidal sequence

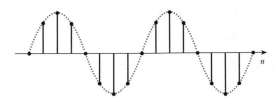

The angular frequency is defined as

$$\Omega = \frac{2\pi}{N} \qquad [30.8]$$

Because N is dimensionless, the units for angular frequency Ω are radians, whereas for the continuous-time case, the units of angular frequency w are radians per second.

A discrete-time step sequence $x(n)$, which is zero for $n < 0$ and 1 for $n \geq 0$, is shown in Fig. 30.5.

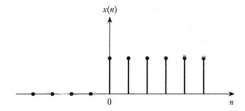

Fig. 30.5 A discrete-time step sequence

(c) Deterministic and random signals

If a signal can be represented by a set of equations it is said to be deterministic. That is, a deterministic signal is fully characterized by its defining equation. This means that all future values of the signal can be predicted from this equation. If the future value of a signal cannot be predicted with certainty then it is referred to as random. A common example of a random signal is the noise voltage generated in an amplifier. Such signals can be described only in terms of probabilistic functions such as their mean, variance and distribution functions. There are different classes of random signal, such as random ergodic signals. The term ergodic, amongst other things, means that the time and the ensemble averages will be the same for that signal.

(d) Singularity functions

Some aperiodic signals have unique properties and are known as singularity functions because they are discontinuous or have discontinuous derivatives. Figure 30.6 shows the simplest of these, the unit step function. This is defined as

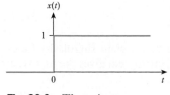

Fig. 30.6 The unit step function

$$x(t) = \gamma(t) = \begin{cases} 0 & t < 0 \\ 1 & t > 0 \end{cases} \qquad [30.9]$$

At $t = 0$ the value of the function jumps from 0 to 1, therefore it is discontinuous at that instant. This function is often represented by $u_{-1}(t)$ or $u(t)$. Here, however, it will be denoted by the symbol $\gamma(t)$. A delayed step

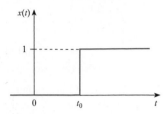

Fig. 30.7 The delayed unit step function

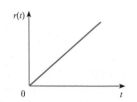

Fig. 30.8 The unit ramp function

function, i.e. a function that changes from 0 to 1 at $t = t_0$, is denoted by $\gamma(t - t_0)$, as shown in Fig. 30.7.

The process of integration can produce other singularity functions:

$$u_{i-1}(t) = r(t) = \int_{-\infty}^{t} u_i(\tau)\, d\tau \qquad [30.10]$$

This shows that $u_{i-1}(t)$ is obtained by integrating $u_i(t)$. Integration of the unit step function gives the unit ramp function, $r(t)$. This can be seen from the above equation to be $u_{-2}(t)$. The slope of this is unity. A ramp function with slope m will be denoted by $mr(t)$ or $mu_{-2}(t)$. Figure 30.8 shows a unit ramp function.

Linear combinations of various singularity functions may be used to generate a large number of functions. Figure 30.9 illustrates two signals formed from linear combinations of step and ramp functions.

The rectangular pulse is expressed as

$$x(t) = A\gamma(t - t_0) - A\gamma(t - t_0 - T) \qquad [30.11]$$

The triangular pulse is expressed as

$$x(t) = \frac{2A}{T}\left[r(t - t_0) - 2r\left(t - t_0 - \frac{T}{2}\right) + r(t - t_0 - T) \right] \qquad [30.12]$$

Fig. 30.9 Two signals formed from linear combinations of step and ramp functions

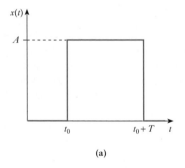

(a)

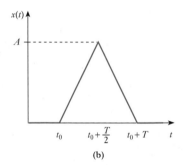

(b)

(e) The impulse function

The impulse (or delta) function, $\delta(t)$, is a very important singularity function. It is defined as the function which, when integrated, gives the unit step:

$$\gamma(t) = \int_{-\infty}^{t} \delta(\tau)\, d\tau \qquad [30.13]$$

Alternatively the impulse can be considered as the time derivative of the unit step. The impulse function satisfies the following:

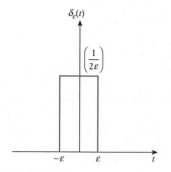

Fig. 30.10 The rectangular pulse

(i) $\delta(t) = 0$ for $t \neq 0$ [30.14]

(ii) $\displaystyle\int_{-\infty}^{\infty} \delta(t)\,dt = 1$

That is, the area of the impulse function is unity and it is obtained over an infinitesimal interval around $t = 0$. This implies the height of the function approaches infinity at $t = 0$, with width approaching zero. Figure 30.10 illustrates a rectangular pulse which, in the limit of $\varepsilon \to 0$, gives an impulse.

The impulse function may appear to be an artificial function at first, but several physical phenomena can be modelled approximately with the impulse. The main property of these phenomena is that they happen in time intervals that are very short compared with the resolution of the measuring apparatus used, but they produce a measurable change in a physical quantity that is almost instantaneous. An example is connecting an ideal capacitor across the terminals of a battery; then the voltage will change almost instantaneously. This is the result of a very large current flowing into the capacitor for a very small instant of time, i.e. the current flowing into the capacitor is an impulse function.

(f) The discrete-time unit impulse

This is defined in a similar manner to the definition of the unit impulse for the continuous–time case. The essential difference is that because of the discrete nature of the signal, the process of integration must be replaced by summation

$$\gamma(n) = \sum_{k=-\infty}^{n} \delta(k)$$ [30.15]

It follows that

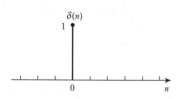

Fig. 30.11 The discrete-time unit impulse

$$\delta(n) = \begin{cases} 0 & n \neq 0 \\ 1 & n = 0 \end{cases}$$ [30.16]

This is shown in Fig. 30.11.

(g) Power and energy signals

Another method for classifying signals is based on whether they contain finite energy or finite average power. Some signals have neither finite energy nor finite average power.

The energy content of a signal $x(t)$ is

$$E = \lim_{T \to \infty} \int_{-T}^{T} x^2(t)\,dt$$ [30.17]

The average power content of a signal is

$$P = \lim_{T \to \infty} \frac{1}{2T} \int_{-T}^{T} x^2(t) \, dt \qquad\qquad [30.18]$$

Note: If $x(t)$ is the current (in amperes) flowing through a $1 \, \Omega$ resistor, then E is the total energy (in joules) dissipated in the resistor and P is the average power (in watts) content in the signal. If the average power of the signal is finite and greater than zero then its energy content will be infinite. Also, if the energy content of a signal is finite, then its average power content must be zero.

- $x(t)$ is a power signal if and only if its average power content P is finite and greater than zero, implying that its energy content is infinite.
- $x(t)$ is an energy signal if and only if its energy content E is finite, implying that its power content is zero.
- If $x(t)$ does not satisfy either of the two conditions above it is neither a power nor an energy signal.

For example,

$$x(t) = A^{\alpha t} \gamma(t)$$

where $\gamma(t)$ is the unit step function and A and α are constants with $\alpha < 0$. For this case the energy content of the signal is given by

$$E = \frac{A^2}{2\alpha}$$

Thus it is an energy signal.

However, if $\alpha = 0$ we get the step function which has E being infinite, but P is finite and it is a power signal.

The average power content of a periodic signal $x(t)$ with period T can be determined by performing the integration over one complete cycle, i.e.

$$P = \lim_{T \to \infty} \int_{-T}^{T} x^2(t) \, dt \qquad\qquad [30.19]$$

30.2 Representation of a signal by a continuum of impulses

The impulse function can be used to represent any arbitrary continuous–time signal as a continuum of impulses. Consider an arbitrary continuous–time function of time $x(t)$, which is defined over the interval $[-T, T]$. One possible approximation to the function can be obtained by representing it by a sequence of rectangular pulses of width Δ, where the height of each pulse is made equal to the value of $f(t)$ at the centre of the pulse. This is shown in Fig. 30.12.

This approximation will improve as the width of the pulses Δ is decreased. Defining the pulse duration Δ at time $t = k\Delta$ as $P_\Delta(t - k\Delta)$, the approximation can be represented as

$$x(t) \approx \sum_{k=-N}^{N} x(k\Delta) P_\Delta(t - k\Delta)$$

Fig. 30.12 Representation of
signal as sequence of rectangular
pulses of width Δ

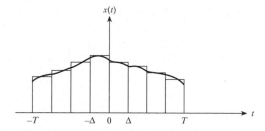

The total number of pulses is

$$2N + 1 = \frac{2T}{\Delta}$$

Rearranging gives

$$x(t) \approx \sum_{k=-N}^{N} x(k\Delta) \frac{P_{\Lambda}(t - k\Lambda)}{\Delta} \Delta$$

As $\Delta \to 0$ the fractional part of the equation approaches the unit impulse $\delta(t - k\Delta)$, an impulse occurring at $t = k\Delta$. Also, the value of N approaches infinity.

The equation can be rewritten as

$$x(t) \approx \sum_{k=-N}^{N} x(k\Delta)\delta(t - k\Delta)\Delta$$

As $\Delta \to 0$ the summation can be replaced by an integral:

$$x(t) = \int_{-T}^{T} x(\tau)\delta(t - \tau)\, d\tau \qquad [30.20]$$

where Δ is replaced by $d\tau$ and $k\Delta$ by τ. This can be understood by considering the graphical interpretation of the integral in the equation as the area under $x(\tau)\delta(t - \tau)$, which would approach the area under the right-hand side of the first summation equation as $\Delta \to 0$. Finally, the complete time function is obtained as $T \to \infty$, so that

$$x(t) = \int_{-\infty}^{\infty} x(\tau)\delta(t - \tau)\, d\tau \qquad [30.21]$$

This equation represents the function $x(t)$ as the summation (integral) of a continuum of impulses, where the strength of each impulse is equal to the value of the function $x(t)$ at that instant. This equation is the result of a property of the impulse function termed the sifting property, which states that the integral of the product of a signal and a unit impulse will be equal to the value of the signal at the time when the impulse occurs. This is because the impulse is zero for all other values of time and the area under the impulse itself is one. The result of the equation is illustrated in Fig. 30.13, where the length of each vertical line represents the strength of the corresponding impulse. Although in the figure the spacing has been shown as non-zero for

Fig. 30.13 Representation of
signal as a continuum of
impulses

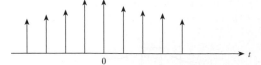

clarity, it should really be for the interval Δ, which is approaching zero. This
ability to represent a signal as a continuum of impulses is important when
trying to determine system responses.

30.3 Impulse response

(a) Continuous-time impulse response

Since any continuous-time signal can be represented by a continuum of
impulses, the Superposition theorem can be used to determine the response
to any input, given the response of the system to a unit impulse. This
approach will also work for discrete-time systems, where the input is exactly
a sequence of discrete-time impulses. Thus, it is important to be able to
calculate the response of a linear system to a unit pulse. The impulse
response of a system is normally denoted $h(t)$.

(b) Discrete-time impulse response

Consider now the response of a discrete-time system to a unit impulse.
Since this will also be a discrete-time function, it will be called the impulse
response sequence and will be denoted as $h(k)$.

30.4 Convolution sum for discrete-time systems

An important problem in system theory is to determine the response of a system to a given input function which cannot be expressed in terms of standard
functions like exponentials and sinusoids. There may even be instances
where it may not be possible to write an analytical expression for the input
function. Here, we study an approach that enables us to obtain the response
of the system to any given arbitrary excitation, if we know its response to
either a unit impulse or a unit step input. This result, called the Convolution
theorem, is a consequence of the application of the principle of superposition
to linear systems. It may be noted that although the theorem is usually
expressed in terms of the impulse response of the system, it can be easily
related to the step response, which can be obtained experimentally without
much difficulty.

Consider a discrete-time system with the impulse response sequence
given by $h(k)$. For convenience, we denote the sequence as

$$\{h_0, h_1, h_2, \ldots, h_k, h_{k+1}, \ldots\}$$

where h_i represents $h(i)$. Our object is to determine the output sequence of
the system for the input sequence

$$\{u_0, u_1, u_2, \ldots, u_k, u_{k+1}, \ldots\}$$

where u_i represents $u(i)$.

First we note that the input sequence can be expressed as a sequence of impulses, of strength (or area) u_i, as below:

$$u(k) = \sum_{i=0}^{\infty} u_i \delta(k - i)$$

Consequently we can use the superposition property to determine the total output at the kth sampling instant as the sum of the responses to each of the individual impulses that have occurred up to that instant. Thus we obtain

$$y_k = u_0 h_k + u_1 h_{k-1} + \ldots + u_k h_0$$

This is called the convolution sum. Note that the first term on the right-hand side is the contribution of the impulse that occurred at $i = 0$ and is the product of u_0 and h_k. Similarly, the second term is the effect of the impulse that occurred at $i = 1$ and is the product of u_1 and h_{k-1}, and so on. The resultant sequence y will be the length of the lengths of u and h added together minus 1. The compact notation for convolution is

$$y_k = \sum_{i=0}^{k} u_k h_{k-i} \qquad [30.22]$$

Note that for a causal system, h_i is zero for $i < 0$. Therefore, we can change the upper limit of the summation to ∞ without affecting the result. Furthermore, it is not necessary that the input sequence be also causal, i.e. u_i may not be zero for $i < 0$. In such cases we must change the lower limit from 0 to $-\infty$. Accordingly, for the general case, we obtain

$$y_k = \sum_{i=-\infty}^{\infty} u_k h_{k-i} \qquad [30.23]$$

These two equations are known as the convolution sum. It is often convenient to denote this as

$$y(k) = h(k) * u(k) \qquad [30.24]$$

which is read as $y(k)$ equals $h(k)$ convolution with $u(k)$. It should be noted that because of the symmetry of these equations we may write

$$h(k) * u(k) = u(k) * h(k) \qquad [30.25]$$

i.e. convolution is commutative. This follows from the fact that the original equation could have been written as

$$y_k = \sum_{i=0}^{k} h_i u_{k-i} \qquad [30.26]$$

Example 30.1

The impulse response sequence of a given discrete-time system is

$$h(k) = 0.5^k = \{1, 0.5, 0.25, 0.125, 0.0625, \ldots\}$$

It is required to find the output sequence if the input sequence is given by

$$u(k) = \{1, 0, -1, -1, 1, 0\}$$

In this case, since we have an input sequence of finite length, we set the initial input at $k = 0$. Hence, we can apply the sum equation to obtain the output sequence for various values of k. The following values are obtained:

$$y(0) = h(0)u(0) = 1$$

$$y(1) = h(0)u(1) + h(1)u(0) = 0.5$$

$$y(2) = h(0)u(2) + h(1)u(1) + h(2)u(0) = -0.75$$

$$y(3) = h(0)u(3) + h(1)u(2) + h(2)u(1) + h(3)u(0) = -1.375$$

and so on

The importance of the convolution sum is obvious immediately, because it allows us to determine the response to any arbitrary input sequence if we know the response of the system to a unit impulse. A better understanding of convolution is therefore very important, especially for the most common situation when both the input and the system are causal.

Two methods will be described. The first method, termed the sliding-strip, involves writing out on separate strips of paper the input sequence $u(k)$ and the impulse response $h(k)$. These sequences should be written out with an even spacing. Then, reverse the h strip and align the entry $h(0)$ with $u(i)$, to determine the output $y(i)$. That is for $y(0)$, $h(0)$ is aligned with $y(0)$, and so on. The output $y(i)$ is then obtained by multiplying the aligned elements of the strip and adding the results. Those elements that do not align do not contribute to the result for that point. The result obtained is identical to the convolution sum equation.

The second algorithm to perform convolution is based on arranging the data in an array. The input sequence and the impulse sequence are arranged in horizontal and vertical arrays as shown in Table 30.1. This is followed by multiplying each of the elements in the various rows and columns to get a table as shown in Table 30.1. The values held in the array will be those that are required to be summed to give the convolution result.

Table 30.1 Data array for convolution

	$u(0)$	$u(1)$	$u(2)$	$u(3)$	$u(4)$
$h(0)$	$h(0)\,u(0)$	$h(0)\,u(1)$	$h(0)\,u(2)$	$h(0)\,u(3)$	$h(0)\,u(4)$
$h(1)$	$h(1)\,u(0)$	$h(1)\,u(1)$	$h(1)\,u(2)$	$h(1)\,u(3)$	$h(1)\,u(4)$
$h(2)$	$h(2)\,u(0)$	$h(2)\,u(1)$	$h(2)\,u(2)$	$h(2)\,u(3)$	$h(2)\,u(4)$
$h(3)$	$h(3)\,u(0)$	$h(3)\,u(1)$	$h(3)\,u(2)$	$h(3)\,u(3)$	$h(3)\,u(4)$
$h(4)$	$h(4)\,u(0)$	$h(4)\,u(1)$	$h(4)\,u(2)$	$h(4)\,u(3)$	$h(4)\,u(4)$

To obtain $y(k)$ for any value of k, we add the terms on a diagonal of the matrix starting from the entry to the right of the corresponding value of $h(k)$ and proceeding toward the right and top. For instance, to find $y(3)$, we start with the entry $h(3)u(0)$ and proceed along the diagonal that contains the elements $h(2)u(1)$, $h(1)u(2)$ and $h(0)u(3)$. It is easily verified that the sum of these elements is, in fact, $y(3)$.

Using the array method we can recalculate the results of Example 30.1. From the values of $u(k)$ and $h(k)$ given, we obtain the matrix shown in Table 30.2.

Table 30.2 Data array for convolution example

	1	0	-1	-1	1	0
1	1	0	-1	-1	1	0
0.5	0.5	0	-0.5	-0.5	0.5	0
0.25	0.25	0	-0.25	-0.25	0.25	0
0.125	0.125	0	-0.125	-0.125	0.125	0
0.0625	0.0625	0	-0.0625	-0.0625	0.0625	0

From this array we can obtain the following values of the output sequence. The values to be summed are found by taking a diagonal line through the terms of the array:

$$y(0) = 1.0$$

$$y(1) = 0.5 + 0.0 = 0.5$$

$$y(2) = 0.25 + 0.00 - 1.00 = -0.75$$

$$y(3) = 0.125 + 0.000 - 0.500 + (-1.000) = -1.375$$

and so on

These values are identical to those calculated previously.

30.5 Convolution integral for continuous-time systems

We now extend the concept of convolution to continuous-time systems. In this instance we do not have a sequence of impulses. However, in view of our earlier discussion, a continuous-time signal $u(t)$ can be approximated by a sum of impulses as

$$u(t) \approx \sum_{k=-\infty}^{\infty} u(k\Delta)\delta(t - k\Delta)\Delta$$

This approximation improves as $\Delta \to 0$ and in the limit we obtain the integral

$$u(t) = \int_{-\infty}^{\infty} u(\tau)\delta(t - \tau)\, d\tau$$

Applying the Superposition theorem, the following approximation is obtained for $y(t)$:

$$y(t) \approx \sum_{k=-\infty}^{\infty} u(k\Delta)h(t - k\Delta)\Delta$$

This equation states that the total response $y(t)$ is the sum of the responses to the various impulses $u(k\Delta)\delta(t - k\Delta)$. In the limit, as $\Delta \to 0$, the convolution sum expressed in the equation becomes the convolution integral as we replace Δ with $d\tau$, $k\Delta$ with τ, and $u(k\Delta)\delta(\tau - k\Delta)$ with $u(\tau)$. Thus, the output is given by

$$y(t) = \int_{-\infty}^{\infty} u(\tau)\delta(t - \tau)\, d\tau \qquad [30.27]$$

or

$$y(t) = u(t) * h(t) = h(t) * u(t) \qquad\qquad [30.28]$$

If both the input $u(t)$ and the impulse response $h(t)$ are causal there are important simplifications. As a result of $u(t)$ being causal the lower limit can be changed from $-\infty$ to 0, since $u(\tau)$ will be zero for negative τ. Similarly, causality of the impulse response can be utilized to change the upper limit from ∞ to t, since $h(t - \tau)$ will be zero for $\tau > t$. This fact is especially useful in giving a graphical interpretation to convolution in a manner similar to that used for the discrete-time case.

We can now look at a graphical interpretation of the convolution integral. The integral in the convolution integral equation is the product of two functions $u(\tau)$ and $h(t - \tau)$. To obtain the second term, first we fold back the impulse response $h(\tau)$ on the τ-axis and then shift it to the right to the value of t at which the output is desired. Then we multiply the functions $u(\tau)$ and the folded and shifted impulse response $h(t - \tau)$ and integrate the product over the period 0 to t. Note that both the convolution sum and the convolution integral require multiplication of the forcing function with the folded impulse response, but summation in the former is replaced with integration in the latter. This is illustrated in Fig. 30.14. The output $y(t)$ is obtained by integrating this final product over the interval 0 to t.

Fig. 30.14 A graphical interpretation of convolution

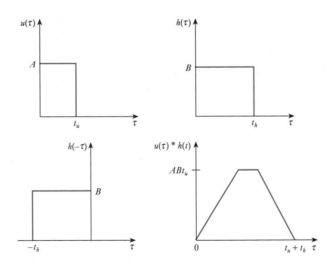

30.6 Deconvolution

The convolution sum is represented by

$$y(k) = u(k) * h(k)$$

The convolution integral by

$$y(t) = u(t) * h(t)$$

In both instances, the objective is to determine the output of the system, given its impulse response and the input. Often there are situations where we need to go the other way, that is, given the input and the output, we need to find the impulse response, or given the output and the impulse response we need to find the input. This is termed deconvolution, which may be regarded as the inverse of convolution. Note that owing to the commutative nature of convolution these two problems are not different as far as the mathematical procedure is concerned.

There are many applications of deconvolution. For instance, one way to obtain a model for a linear system is to measure its output for a given input. From these input–output data, we can use deconvolution to determine the impulse response of the system and, hence, the system model. This is termed system identification. Deconvolution is used when trying to determine the input for a given impulse response and output. This is very important for control systems, where the objective is to force the system to have a specified output by applying the appropriate input that will cause that output.

30.7 Relation between impulse response and unit step response

The practical usefulness of convolution methods may appear to be limited because it is not convenient to determine the response of a system to a unit impulse experimentally. Not only is it not possible to generate a true impulse, but it is also undesirable to apply an impulse to an actual system. For example, applying an impulse voltage function to an electrical circuit may cause some serious problems.

On the other hand, it is a lot easier to apply an input that will be a close approximation to a unit step input. Therefore, it would be desirable to find some way to relate the step response of a linear system to its impulse response. This is done readily by applying the Convolution theorem.

Consider a linear continuous-time system with impulse response given by $h(t)$. Also, let the response of the same system to a unit step input be denoted by $g(t)$. Then using convolution, we get

$$g(t) = \int_{-\infty}^{\infty} \gamma(\tau)h(t-\tau)\, d\tau$$

where $\gamma(t)$ represents the unit step. Owing to the causal nature of the input and the impulse response, we can replace the lower limit by 0 and the upper limit by t. Furthermore, because of the commutative nature of convolution, we can interchange the roles of the input and the impulse response. Consequently we get

$$g(t) = \int_{0}^{t} h(\tau)\gamma(t-\tau)\, d\tau$$

Finally, since between the limits of integration, that is, for $0 \leq \tau \leq t$, the function $\gamma(t-\tau)$ is equal to 1 we can simplify to give

$$g(t) = \int_{0}^{t} h(\tau)\, d\tau$$

Thus the response of a system to a unit step at a particular instant of time t is the integral of its response to a unit impulse evaluated between 0 and t. This integral is often called the running integral. Alternatively, it may be said that the impulse response can be obtained by differentiating the step response with respect to time. Thus we have a practical way of determining the impulse response of a system experimentally from its response to a unit step. This important result is intuitively obvious if we recall that the unit impulse is the derivative of the unit step.

The input–output relationship can also be expressed directly in terms of the step response $g(t)$. Let us first rewrite the convolution integral

$$y(t) = \int_{-\infty}^{\infty} h(\tau)u(t - \tau) \, d\tau$$

and then integrate it by parts, using the relationship in the previous equation. This yields

$$y(t) = g(\tau)u(t - \tau) \Big|_{-\infty}^{\infty} - \int_{-\infty}^{\infty} g(\tau) \frac{du(t - \tau)}{d\tau} \, d\tau$$

If we assume that both the input and the system are causal, then the first term vanishes, since $g(-\infty) = 0$ and $u(t - \tau) = 0$ for $\tau > t$. The following result is then obtained:

$$y(t) = -\int_{-\infty}^{\infty} g(\tau) \frac{du(t - \tau)}{d\tau} \, d\tau \qquad [30.29]$$

or by interchanging variables

$$y(t) = \int_{-\infty}^{\infty} g(t - \tau) \frac{du(\tau)}{d\tau} \, d\tau \qquad [30.30]$$

The equations are useful since they enable us to obtain the output of the system to any input if its response can be measured experimentally. In practice it may be more convenient to use the second of the two equations above, except when the input function has jump discontinuities for some values of t. In such cases, differentiation of the input will give rise to impulses. This is not a problem if one recalls the sifting property of the impulse:

$$\int_{-\infty}^{\infty} f(\tau)\delta(t - \tau) \, d\tau = f(t)$$

Therefore, the corresponding integral can be easily evaluated for this case.

30.8 Step and impulse responses of discrete-time systems

The discrete-time equivalent of the step input is the step sequence $\gamma(k) = 1$ for all $k \geq 0$. Denoting the impulse response sequence of such a system by $h(k)$ and the step response by $g(k)$ and applying the convolution theorem, we obtain

$$g(k) = \gamma(0)h(k) + \gamma(1)h(k - 1) + \ldots + \gamma(k)h(0) = \sum_{i=0}^{k} h(i)$$

since $\gamma(i) = 1$ for all $i \geq 0$.

This equation is the discrete-time equivalent of the equation developed for the continuous-time case. Note that the operation of integration for $0 \leq \tau \leq t$ has been replaced by the operation of summation for $0 \leq i \leq k$, which is called the running sum.

Similarly, just as the impulse response of a continuous-time system can be obtained by differentiating the response to a unit step, for a discrete-time system the impulse response sequence can be determined by taking successive differences of the step response sequence.

Example 30.2 The impulse response sequence of a discrete-time system is given by

$$h(k) = 10(0.6)^k$$

Using the above equation we will determine the step response:

$$g(k) = \sum_{i=0}^{k} 10(0.6)^i$$

$$= 10 \frac{1 - 0.6^{k+1}}{1 - 0.6}$$

$$= 50(1 - 0.6^{k+1})$$

$$= 50 - 30(0.6)^k$$

Summary of important formulae

Periodic signals are defined as

$$x(t + T) = x(t) \qquad \text{for all } t \in (-\infty, \infty) \qquad [30.1]$$

$$x(n + N) = x(n) \qquad \text{for all } n \in (-\infty, \infty) \qquad [30.2]$$

The unit step is defined as

$$x(t) = \gamma(t) = \begin{cases} 0 & t < 0 \\ 1 & t > 0 \end{cases} \qquad [30.9]$$

The impulse function is defined as

(i) $\delta(t) = 0$ for $t \neq 0$ $\qquad [30.14]$

(ii) $\displaystyle\int_{-\infty}^{\infty} \delta(t) \, dt = 1$

The discrete-time unit impulse is given as

$$\delta(n) = \begin{cases} 0 & n \neq 0 \\ 1 & n = 0 \end{cases} \qquad [30.16]$$

The energy content of a signal $x(t)$ is

$$E = \lim_{T \to \infty} \int_{-T}^{T} x^2(t) \, dt \qquad [30.17]$$

**Summary of important
formulae continued**

The average power content of a signal is

$$P = \lim_{T \to \infty} \frac{1}{2T} \int_{-T}^{T} x^2(t)\, dt \qquad [30.18]$$

The notation for convolution is

$$y_k = \sum_{i=-\infty}^{\infty} u_k h_{k-i} \qquad [30.23]$$

$$y(t) = \int_{-\infty}^{\infty} u(\tau)\delta(t - \tau)\, d\tau \qquad [30.27]$$

Terms and concepts

Signals may be classified as: continuous-time, discrete-time, periodic, aperiodic, deterministic, random, energy and power signals.

Continuous-time signals may be represented as a continuum of impulses.

The impulse response of a system characterizes how a system will behave when an impulse is applied to its input. Signals may be represented as a continuum of impulses, which leads to being able to determine the system output for a given input signal and impulse response using the process of convolution.

Exercises 30

1. Explain clearly the difference between continuous-time and discrete-time signals.
2. Consider the following sinusoids:

$x_1(t) = 5 \sin 5\pi t$
$x_2(t) = 7 \sin 3\pi t$
$x_3(t) = 8 \sin 5t$

Which of the following signals will be periodic and if so what is their period?

(a) $x_1(t) + x_2(t)$
(b) $x_2(t) + x_3(t)$

3. What is the relation of the unit step function to the impulse function $\delta(t)$?
4. Express each of the signals shown in Fig. A in terms of singularity functions.
5. The equations for a number of signals are given below. Determine if they are (a) periodic or non-periodic signals and (b) energy or power signals.

(i) $x(t) = 20\, e^{-2t} \cos 100t$
(ii) $x(t) = 20 \cos 2t + 10 \sin 3\pi t$
(iii) $x(t) = 30 \cos 4\pi t [\gamma(t) - \gamma(t-1)]$

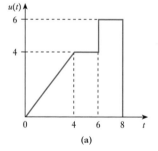 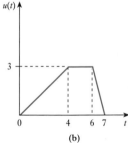

Fig. A

(iv) $x(t) = 4 \cos 4\pi t + 6 \sin 5\pi t$
(v) $x(t) = 10\, e^{3t} \gamma(t)$
(vi) $x(t) = (4 \cos 4\pi t + 6 \sin 5\pi t)\gamma(t)$

6. The impulse response sequence $h(k)$ of a linear discrete-time system is non-zero only for $0 \le k \le 5$ and is given below:

k	0	1	2	3	4	5
$h(k)$	6	4	2	0	-2	-4

Determine its response to the input sequence $u(k) = 1$.

Exercises 30 continued

7. Repeat Q. 6 for the following input sequence:

k	0	1	2	3	4	5	6	7
$h(k)$	1	1	0	−1	−1	0	0	0

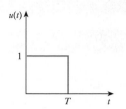

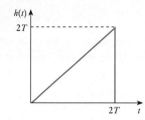

Fig. B

8. The impulse response $h(t)$ of a continuous-time system is shown in Fig. B. Determine the response of this system to the input shown.

9. The impulse response of a system is given by

$$h(t) = 3\,e^{-4t}\cos 3t \cdot \gamma(t)$$

Use the convolution integral to determine its response to the input

$$u(t) = \gamma(t)$$

10. The response of a system to a unit step is given by

$$g(t) = (0.48 - 0.48\,e^{-4t}\cos 3t + 0.36\,e^{-4t}\sin 3t)\gamma(t)$$

Determine its impulse response.

Chapter thirty-one

Data Transmission and Signals

Objectives

When you have studied this chapter, you should

- be familiar with the concept of analogue and digital signals
- understand the term bandwidth
- be able to explain modulation and demodulation
- have an understanding of frequency division multiplexing

Contents

Electronic systems are used to transmit information. The process by which this is achieved can be classified as analogue or digital. Analogue systems use waveforms, which are directly proportional to the information being transmitted. Digital systems operate by converting the information into some form of pulse for transmission.

In this chapter, the relative advantages and disadvantages of such systems will be presented. When considering analogue systems the concepts of bandwidth and modulation will be introduced.

By the end of the chapter you should be familiar with the concepts underlying analogue and digital systems and be aware of the need to develop electronic techniques, which can handle one or other system. The following chapters will further develop these techniques.

31.1 Transmission of information

Electronic systems are frequently associated with the transmission of information. This might be taking the information keyed into a pocket calculator through to the microprocessor and finally to the display. Alternatively, a system might collect information about the pressure of gas in a pipe and transmit it to a control room. Yet again a system might take a television picture from one side of the world to the other.

Information can come in many forms. For instance we could indicate light by sending a voltage when a lamp is switched on and sending no voltage when the lamp is switched off. This is quite crude but effective. But information can be complex, especially when it is visual as in a television picture.

Information transmission need not always be instantaneous. A sound cassette stores information to be retrieved at some time in the future – and as often as we like. In this way we can recall music on demand. Another form of information recall is given by the floppy discs we use in computers to save the data information entered into the system.

In electronic systems it is implicit that information has to be transmitted to a point some distance away from the source. How this is achieved depends on the complexity of the information, the distance over which the information has to be transmitted and the speed at which the information is needed.

When we transmit information, we require it to be translated into some equivalent electrical value. If we wish to transmit sound, it is necessary to convert it into an electrical energy equivalent. In practice, we do not describe this in terms of energy but in terms of one of electrical energy's component parts – voltage or current.

When we vary a voltage to represent, say, sound, then the varying voltage is called a signal. We looked at some different types of signal in Chapter 30. Equally, if we were to vary a current for the same purpose, we would describe the varying current as a signal. In either case, the signal is transmitted to a receiving component which converts it to an appropriate energy form.

As an example, we could speak into a telephone, i.e. the input is sound. The signal is transmitted electrically to a receiver at the far end of the line where it is converted back to sound, hence the person at the far end can listen to the input information.

A different form of data transmission occurs in a security light. Here movement is sensed by a change of heat intensity. This is converted to an electrical signal which switches on a lamp, thus providing a final output of light.

The basic system has three parts. The first part concerns the conversion of the input into an electrical signal, while the final part concerns the conversion of the signal to the output. The middle part concerns the transmission of the signal – and this signal operation involves most of our interest in electronics. Let us therefore make further investigation of electronic signals.

31.2 Analogue signals

Electronic signals fall into two categories, analogue and digital. Both types of signal have been used from the earliest days of electrical communication. Analogue systems first came to prominence with the telephone, whereas digital systems first appeared when telegraph systems introduced Morse code.

In analogue systems the information or data is given as an electrical signal that varies in direct proportion to the information or data. It follows that the variation must be continuous and, between the limits of operation of the system, the variation can have any value from an infinite number of values. Such variation is associated with the production of sound (as in the telephone) in radio receivers or vision in television sets.

An analogue system in its most basic form has an input electrical signal which is either a voltage or a current varying directly in proportion to the input information. The input information is converted into the electrical signal by a transducer. Typical transducers include a microphone (converting sound to electrical e.m.f.), pressure transducers (converting pressure, e.g. air pressure, to electrical e.m.f.), tachometers (converting speed to an e.m.f. using the principle $e = Blu$) and light detectors (converting light intensity to e.m.f.).

An example of an analogue signal is shown in Fig. 31.1. Here we can observe the continuing variation in a complicated pattern, which contrasts with the regulated sinusoidal a.c. signals to which we have previously been introduced.

Fig. 31.1   An analogue signal

The analogue system frequently takes the input signal and enlarges it. In practice, this is achieved by a variety of means, but all have the common feature that the input signal is used to control the flow of energy from a more powerful source. This is the process of amplification to which we were introduced in Chapter 17. Therefore when we give further consideration to amplifiers, we have to recognize that the amplifier is handling a signal which is varying.

In practice, analogue systems are limited by the linearity of the amplifiers. A linear amplifier should amplify all signals equally. For instance, an amplifier with a gain of 10 should take an input signal of 1 mV and provide an output signal of 10 mV. Similarly an input signal of 7 mV should result in an output of 70 mV. However, it is quite possible that the output signal could be 69.5 mV. The result of this slight change of output is that the output signal will not have exactly the same shape as that of the input. This effect is called distortion – the sort of thing which results in a singer sounding either unclear or even as though he or she were drowning.

31.3   Digital signals

Digital signals have one of a limited number of discrete values. In many applications, there are two values which in crude terms could be explained as ON and OFF. These values are usually described as 1 and 0, being the presence and absence of the supply voltage or current.

The first form of digital signal in common use was that associated with the Morse code. This uses short and long pulses to make up a code. Examples of

Fig. 31.2   The Morse code

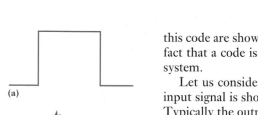

(a)

(b)

Fig. 31.3   A transmitted pulse.
(a) Input transmitted; (b) output
received

this code are shown in Fig. 31.2; this is discussed further in Chapter 32. The fact that a code is required can be an obstacle to the application of a digital system.

Let us consider the transmission of a simple pulse over a long line. The input signal is shown in Fig. 31.3(a) and is seen to be a simple square pulse. Typically the output signal which could result is shown in Fig. 31.3(b); here we find the pulse has become distorted. This is not quite the same distortion which was caused by a non-linear amplifier with analogue signals, but it is another instance of the difficulty of the output not being the same as the input.

At this point, we remember that the input signal can only be a pulse. It follows that the distorted output signal must represent a square pulse. We can therefore arrange for the distorted signal to be replaced by a square pulse, thus generating an output which exactly represents the input signal. For this reason, we find that many data and information transmission systems now use digital techniques, even to finally produce analogue signals. A typical example is the compact disc or even the digital cassette replacing the traditional music cassette.

Following the Morse telegraph systems, most digital systems have progressed to being a series of pulses of identical duration. The period of a pulse is defined as the time between any two similar points of the pulse train.

Pulse period   **Symbol:** T **Unit:** second (s)

The pulse repetition rate or pulse frequency is the number of pulses per second.

A pulse is called a binary digit or bit for short. The bit rate is the number of pulses per second, bearing in mind that a pulse can be a 1 or a 0. Each pulse is separated by a zero period so that subsequent pulses can be identified. A typical series of bits is shown in Fig. 31.4.

A system of the form illustrated is a return–to–zero binary system. The period between pulses is 10 ms, in which case the number of bits per second would be 100 000, hence giving a bit rate of 100 kilobits per second.

Digital systems are commonly associated with calculators, computers and microprocessors. However, most other systems interact with analogue systems for the transmission of sound and vision.

At this introductory stage we will treat the digital and analogue systems separately; in Chapter 25 we have already considered analogue-to-digital converters and digital-to-analogue converters.

Fig. 31.4   A series of bits

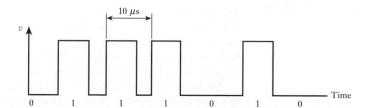

Not all of the period of a pulse is occupied by the pulse. The duty cycle is the ratio of the pulse width to the period. This is best explained by means of an example.

Example 31.1   Determine the pulse repetition frequency and the duty cycle for the pulse train shown in Fig. 31.5.

Fig. 31.5 Pulse train for Example 31.1

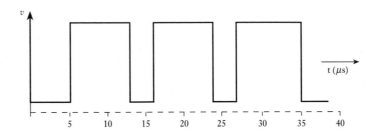

$$T = (16 - 5) = 11 \ \mu s$$

Pulse repetition rate is

$$\frac{1.0 \times 10^{6}}{11} = 88 \text{ kHz}$$

Pulse width is

$$(13 - 5) = 8 \ \mu s$$

Duty cycle is

$$\frac{8}{11} = 73 \text{ per cent}$$

The fact that a digital system can be self-correcting seems so attractive that we might well ask if it is necessary to bother with analogue systems at all. In fact more and more communications systems are now becoming predominantly digital, with the reducing costs of digital microelectronics. The problem is that what we see and what we hear and what we say all occur by analogue. Therefore any system which involves our senses requires to be analogue at least at the input and output stages; the intervening transmission system, however, can be either analogue or digital. Let us look at some of the factors which affect signal transmission.

31.4   **Bandwidth**

The range of frequencies over which a signal can operate is known as the bandwidth. There is usually a limit to the range of frequencies available to the bandwidth of a system, hence it is necessary to ensure that a signal being transmitted is within the limited bandwidth. If components of the signal lie outside the bandwidth, they will be partially if not entirely lost.

Analogue systems can occupy a significant bandwidth. In section 19.4 we saw how to determine normal limits of bandwidth.

We are familiar with recording sound on audio cassettes. Most people can hear frequencies between the range 25 Hz and 16 kHz, although those with better hearing might reach higher values. For many purposes, it is sufficient to record the range up to 16 kHz on a tape running at 5 cm/s. However, if we wished to transfer the recording from one tape to another (as is done in the production of prerecorded tapes), then we could run the tapes at, say, 20 cm/s. The information being transferred from one tape to the other now comes with frequencies four times the original. Hence the system linking one tape to the other would require a bandwidth up to 16×4, i.e. 64 kHz.

More briefly, a television picture requires that 25 images are transmitted every second and each image takes up 625 display lines. Each line requires a complex analogue signal to provide the variation of picture density. For such a rate of information delivery, a very high bandwidth is required, preferably about 5 MHz. We can also reproduce a single picture over a telephone line. Telephone lines are very limited in bandwidth, say, 3.4 kHz, hence the transmission of an image takes seconds or even minutes.

We can summarize these observations as follows. The bandwidth required for a signal depends upon the complexity of the information being transmitted and the speed with which it is to be transmitted. If either or both increase, the bandwidth needs to be extended.

31.5 Modulation

Let us once again consider the transmission of speech by telephone line. We have noted that we require only a bandwidth of 3.4 kHz – but what if we wished to send two signals at the one time? A possible solution would be to shift the frequency range. For instance, we could add 10 kHz to the second signal, thus providing it with the frequency bandwidth of 10 kHz up to 13.4 kHz.

By this method we can transmit several signals along a single pair of wires. It follows that if we had a total bandwidth of 100 kHz, we could have several speech channels each requiring 3.4 kHz bandwidth. Each channel requires separation from the next; this can be achieved by setting the lower limits at 10 kHz intervals. This would permit 10 channels along the same pair of wires. This process is known as frequency division multiplexing and is illustrated in Fig. 31.6.

The process of frequency shifting is known as modulation. The signal which has its frequency spectrum shifted is said to be modulated. There are a number of ways in which this can be achieved and one of the most common is amplitude modulation.

Let us take a relatively high frequency which we will use to shift the signal frequencies. For instance we might use a high frequency of 120 kHz

Fig. 31.6 Frequency division multiplexing

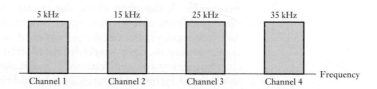

Fig. 31.7 Waveforms associated with amplitude modulation

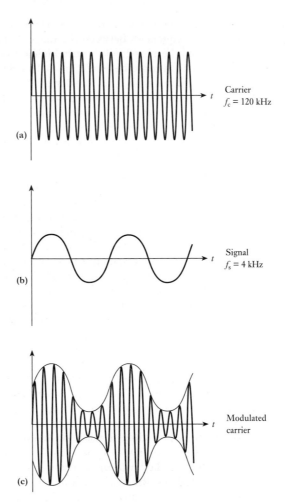

(a) Carrier $f_c = 120$ kHz

(b) Signal $f_s = 4$ kHz

(c) Modulated carrier

as the carrier frequency for a speech bandwidth of say 4 kHz. The carrier and signal are shown in Fig. 31.7(a) and (b). By using the signal to control the carrier, the resulting modulated signal appears in the form shown in Fig. 31.7(c). For the figures given, a 120 kHz carrier with a signal frequency of 4 kHz would result in frequencies of 116–124 kHz. This has doubled the original bandwidth and shows the reason that we left large intervals between the carrier frequencies in frequency division multiplex systems.

When we tune a radio receiver to a given frequency, we are in fact tuning to the carrier frequency. The receiver circuit is designed to have a frequency response which takes in the signal frequencies above and below that carrier.

Let the carrier frequency be f_c and the signal frequency be f_s. The frequencies in the band above the carrier frequency are said to be in the upper sideband. Hence the upper sideband ranges from f_c to $(f_c + f_s)$. The frequencies between $(f_c - f_s)$ and f_c are said to be in the lower sideband. It is possible to have single sideband transmission. This permits more channels within a given bandwidth, but it also requires a significantly more complicated system. The range of sidebands is illustrated in Fig. 31.8.

Fig. 31.8 The sidebands associated with amplitude modulation

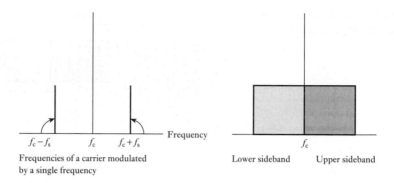

Frequencies of a carrier modulated by a single frequency

Lower sideband   Upper sideband

31.6 Filters

Having transmitted a number of signals along a line, there remains the problem of separating them. The devices which are used for this purpose are called filters.

We can recall that in sections 10.11 and 11.3, we were introduced to series and parallel RLC networks. Each demonstrated a variation of impedance with frequency, with an extreme condition arising at resonance. Basically this type of response is that required from filters.

However, it is important that the filter should accept the entire bandwidth appropriate to a channel. As an example, a speech channel using a carrier frequency of 50 kHz requires a bandwidth of 46–54 kHz. The filter for that channel should react in such a way that a signal at any frequency within the bandwidth should pass freely yet any outside the bandwidth should be rejected. Such a filter is called a bandpass filter.

The ideal response of a bandpass filter is shown in Fig. 31.9(a) while that of an RLC parallel network is shown in Fig. 31.9(b). Clearly the latter is not going to treat all frequencies within the bandwidth equally and distortion will result. More complex designs for filter networks are therefore required, although none can be described as ideal. This indicates another reason for the preference given to transmission by digital techniques rather than by analogue methods.

Given that speech channels require a bandwidth of about 8 kHz and assuming that the carrier frequencies are at 10 kHz intervals, the resulting frequency division multiplex system would be of the form shown in

Fig. 31.9 Bandpass filter characteristics

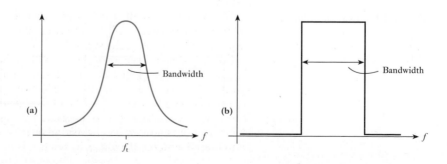

Fig. 31.10 A frequency division multiplex system

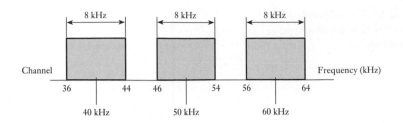

Fig. 31.10. It can be seen that there are gaps of only 2 kHz between the channels. It follows that a filter therefore should ideally totally accept a frequency of, say, 54 kHz yet totally reject a frequency of 56 kHz. This again leads to more complex network design, although it cannot be totally achieved.

We have introduced the function of filters by considering signals associated with speech. We are not limited to the transmission of sound – any form of data transmission would be applicable, but depending on the information greater or lesser bandwidths would apply. Even replacing speech by music would require the bandwidth to be raised to about 40 kHz.

31.7 Demodulation

Once again let us consider the speech transmission system with which we were introduced to filters. After having separated the various channels, it remains to separate the carrier from the original speech signal. The original signal occupied a bandwidth from a few hertz up to 4 kHz, and it is necessary to remove the carrier from the received signal to regain this original signal. This can be achieved by a reasonably simple network incorporating a rectifier.

The process of separating the original signal from the received signal is called demodulation. The modulated carrier is shown in Fig. 31.11(a) and the demodulated signal is shown in Fig. 31.11(b).

Before leaving demodulation, let us look at an example of its action in practice. Figure 31.12 shows the block system diagram for a simple radio receiver. The aerial picks up the transmitted signal and a bandpass filter is used to select the station to which we wish to listen. The output from the filter is the modulated carrier which generally has to be amplified before the demodulation process can be effected. The output from the demodulator leaves the required signal; it then only remains to amplify the signal so that it has sufficient power to drive a loudspeaker.

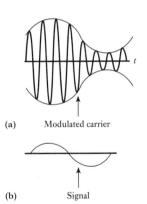

(a)   Modulated carrier

(b)   Signal

Fig. 31.11 Demodulation

Fig. 31.12 A simple radio receiver

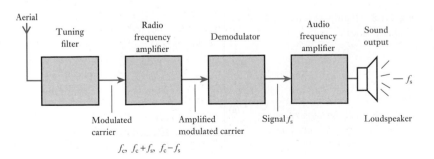

31.8 Amplifying signals

Throughout this chapter there has been an underlying assumption – the linearity of response. Given appropriate design, most systems can achieve this reasonably well. However, we have also noted that failure to do so does not dramatically affect digital systems because they can interpret the output signal to be identical to the input signal. The problems of not being linear are therefore associated with analogue systems – and the principal problem is that of distortion. Let us therefore look to some of the causes.

We have noted that amplifiers work by using a small input signal to control the release of energy from a suitable source. For instance, we could use a signal of 3 mV to control the release of energy from a 3 V source. The input signal is an a.c. one, whereas the source voltage is a d.c. one. It is important that the peak-to-peak variation of the input signal when amplified does not exceed 3 V. Approximately this would occur if the signal gain were greater than 300.

This means that there is a maximum value of input signal that will produce an output signal, the waveform of which remains an acceptable replica of the input signal waveform. Increasing the input signal beyond this level will produce clipping of the input signal waveform as illustrated in Fig. 31.13, although the clipping levels need not be symmetrical about the zero level.

Fig. 31.13 Clipping of output signal waveform

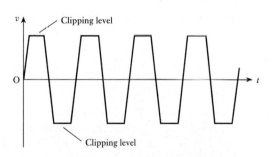

Examination of the gain of an amplifier will show that it does not remain constant with the frequency of the input signal. Some amplifiers exhibit a reduction in gain at both high and low frequencies while others have a reduction at high frequencies only. In both cases there is a considerable reduction in gain at both high and low frequencies while others have a frequency range over which the gain remains essentially constant. It is within this frequency range that the amplifier is designed to operate. The effect of the unequal gain is to produce another form of waveform distortion since the harmonics present in a complex input signal waveform may not be amplified by the same amount. Figure 31.14 shows typical gain/frequency characteristics where it should be noted that the frequency scales are logarithmic. The advantage of an amplifier with a characteristic as illustrated in Fig. 31.14(b) is that it is capable of amplifying signals at the very low frequencies met with in many industrial applications. It is known as a direct-coupled (d.c.) amplifier and integrated circuit amplifiers are of this type. Figure 31.15 shows the responses obtained from both types of amplifier for an input signal which changes instantaneously from one level to another. Lack of low-frequency gain in the first type prevents faithful reproduction

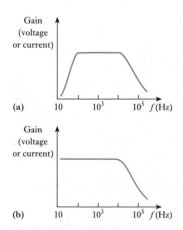

Fig. 31.14 Gain/frequency characteristics. (a) Capacitance-coupled amplifier; (b) direct-coupled amplifier

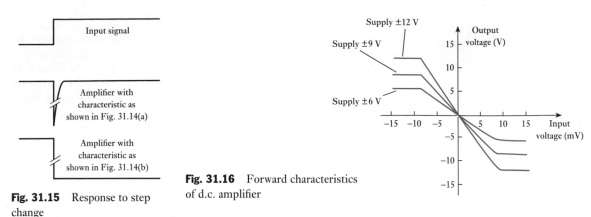

Fig. 31.15 Response to step change

Fig. 31.16 Forward characteristics of d.c. amplifier

of the steady parts of the input signal, the output being merely a 'blip' as the input signal changes from one level to the other. The characteristics showing output voltage against input voltage for several d.c. supply voltages as given in Fig. 31.16 are representative of integrated circuit d.c. amplifiers. Output voltage variation on either side of zero volts is obtained by the use of power supply voltages which are spaced on both sides of zero, e.g. +12 V and −12 V as opposed to +24 V and 0 V.

31.9   Digital or analogue?

The public first became generally aware of transmission systems thanks to the telephone. All the original development grew up around telephone systems and they remain pre-eminent in the minds of people as the basis of communications. That has been shown by the number of times we have used the telephone to explain the concepts of data transmission. Even today, we still give great emphasis to being able to talk to one another on the phone. It follows that the human aspects of speaking and hearing play an extremely important role in our requirements for electronics systems.

Over the last few years we have seen communications systems involving speech, sound and vision become the lesser part of the overall picture. Most communications systems transmit commercial information in digital data form. It is not so much that people make less use of analogue systems but rather that machines will demand more and more data transmission over and above that which we already use. In addition, the rise of text messaging has acted as an increase in communications in a digital format. Also, there are now more affordable (due to reduced cost of electronics) digital consumer wants. These include music, images and digitally encoded movies, which will all continue to increase the amount of digital communication taking place.

Digital systems now predominate but we humans will still need analogue input and outputs to meet our needs.

Terms and concepts

When information is transmitted electrically, the relevant energies are contained in **signals**.

Signals are either **analogue signals** or **digital signals**.

Analogue signals have continuous variation in direct proportion to the information being carried. In practice, it is difficult to maintain direct proportionality and the consequent differences give rise to **distortion**.

Digital signals normally come in two forms – ON and OFF. The length of the signal can be varied, as in Morse code, but it is by far the most common practice to use binary digits of equal duration. As a consequence, any distortion can be removed since we anticipate equal digits which are either ON or OFF.

Transmission systems can normally operate only over a limited range of frequencies. Outside these frequencies, their performance is unacceptably poor. The range of satisfactory operation is called the **bandwidth**.

Modulation permits us to change the frequency of operation so that we can have a number of signals being transmitted in a system at one time. Such an operation is known as **frequency division multiplexing**.

The **carrier frequency** is the change of frequency applied.

The **signal frequency** can be added to or subtracted from the carrier frequency. When added, we obtain the **upper sideband** and when subtracted we obtain the **lower sideband**. When only one sideband is used, we have **single sideband transmission**.

Filters permit the separation of one signal from a number which are all using the same transmission system. **Passband filters** permit the passage of one signal and reject signals at all other frequencies.

Demodulation is the reversal of modulation, and is the process whereby the carrier frequency is removed in order that the signal returns to its original range of frequency operation.

Amplifiers do not always amplify equally either at all frequencies or at all magnitudes. When applying amplifiers, we have to take into consideration these practical limitations.

Exercises 31

1. Explain the difference between an analogue and a digital signal.
2. It is essential that an analogue amplifier should be as linear as possible. Explain why linear amplification is necessary.
3. A digital system has a pulse period of 8 μs and a duty cycle of 75 per cent. Determine: (a) the pulse repetition rate; (b) the pulse width.
4. Describe the process of frequency division multiplexing.
5. Draw the attenuation/frequency characteristic of a bandpass filter and hence describe a possible application.
6. Describe the operation of a frequency division multiplex system.
7. Explain the functions of modulation and demodulation.
8. Draw the gain/frequency characteristic of a capacitance-coupled amplifier and hence explain the consequences of the gain varying with frequency.
9. Describe the effect known as clipping.
10. Describe the difference between a capacitance-coupled amplifier and a direct-coupled amplifier.

Chapter thirty-two

Communications

Objectives

When you have studied this chapter, you should

- be aware of basic communication concepts

- be able to calculate the theoretical maximum channel capacity

- be able to perform calculations to determine information content

- understand why source coding is useful

- be able to generate a Huffman code

Contents

Communications is a rapidly developing field, with ever-greater numbers of pieces of consumer electronics requiring communications technology. These include mobile telephones, WAP devices, internet communication and digital television. To support these developments an understanding of some basic principles is required. These basic principles can be used to determine the requirements from circuits and systems to tackle the different applications.

In this chapter, channel capacity, some basic information theory and simple source coding will be discussed. These topics form the core of understanding to help determine parameters like required signal to noise ratio, code bit rate and bandwidth. These parameters then form the basis for part of the system specifications when analysing or designing a communications system.

By the end of this chapter you will be aware of basic communication concepts. You will be able to calculate the theoretical maximum channel capacity given the bandwidth and the signal to noise ratio. Given the probabilities of occurrence of symbols, you will be able to perform calculations to determine information content and to generate a Huffman code.

32.1 Basic concepts

Data communication is the process of transferring information between two places in a form that can be handled by data communication equipment. Here we are going to look at some of the basics of data communication. Information, in data communication terms, loosely refers to the symbols or signals used to make up a message. In data communications, the meaning of the message is not of particular relevance, it is only the actual data that are important. One definition of information uses the meaning as a definition of information; this will be considered later.

Messages may be sent in a variety of ways using many transmission media from the information source to its destination. Speech, for example, enables people to communicate using sounds, with air as the transmission medium. A telephone line is an electrical communication line, which allows us to hold voice conversations over a long distance or to transmit computer data over long distances. A standard telephone line uses a pair of copper wires to transmit the electrical signals which contain the information (speech or data) being transmitted.

Interface devices are required to transfer data over telephone lines (the handset acts as an interface device for voice communications). Modems are used to permit electronic devices, in general, and computers in particular, to transmit data over communication links, which for many are simply telephone lines.

Different types of data that may be transmitted include text, numeric data and graphical data. Other forms of data require to be sent and these can all be accommodated with appropriate coding.

(a) Bandwidth

The bandwidth is the range of frequencies that are available for use in a communications channel, expressed in hertz (Hz). The bandwidth of a communications system is one of the key parameters as it partially determines the channel capacity, which puts an upper limit on the amount of information a channel may transmit.

(b) Channel capacity

The channel capacity is the upper limit of the amount of data that can be passed through a communications channel in a given time, normally calculated in bits per second (bps). This theoretical maximum capacity is never reached but with improving technology some communication systems are approaching the limit.

(c) Signal to noise ratio

The signal to noise ratio is the ratio of user signal to background noise (S/N), usually expressed in decibels (dB). This, together with the bandwidth of the system, determines the theoretical channel capacity using the Shannon–Hartley law.

(d)   Shannon–Hartley law

The Shannon–Hartley law can be expressed as

$$C = w \log_2(1 + S/N) \qquad [32.1]$$

where C is the capacity of the channel, w is the bandwidth and S/N is the signal to noise ratio. The relationship between channel capacity and signal to noise ratio is shown in Fig. 32.1.

Fig. 32.1   Relationship between channel capacity and signal to noise ratio

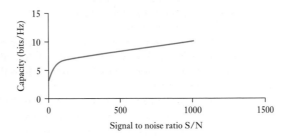

Signal to noise ratio S/N

Example 32.1   For a given communications system with a bandwidth of 5 kHz and a signal to noise ratio of 20 dB, it is required to calculate the channel capacity. This is done using the Shannon–Hartley law. Using equation [32.1]:

$$C = 5000 \log_2(1 + 20) \text{ bps}$$

Such a system's channel capacity will therefore be approximately 22 kbps.

(e)   Binary data

To be able to determine if the data that we wish to transfer can be communicated over the channel we have, we need to have a measure of the information in the data. Data communication devices operate on digital signals. There are a variety of types of digital signals, from simple on–off binary signals through to complex multilevel multiphase signals. The detail of these signalling types is beyond the boundaries of this text (there are many excellent texts on digital signalling techniques). It should be noted that digital and binary do not mean the same thing. A binary signal is digital, but digital is not necessarily binary. In Chapter 26 binary numbers were discussed; here we look at extending the range of binary representations.

A BInary digiT is termed a *bit* and this is also used as the unit for measuring information. A single bit has 2^1 states, that is 2 states. These states can represent on/off, true/false, high/low or a variety of other binary pairs. This is not much use for conveying a lot of information. To represent more than two states it is normal to group several bits together to give the opportunity to code information. If 8 bits are grouped together (this is termed a byte) there are 2^8 (that is 256) possible combinations of 0s and 1s. By allocating symbols to each pattern we can use the bytes to transfer useful data. The process of allocating symbols to unique patterns of bits is called coding. If groups of 4 bits are used, we have a nibble, which has 16 possible combinations (used with hexadecimal). Computers generally work with many bits in parallel; some microprocessors have a word length of 1 byte (8 bits), but most have much longer word lengths such as 32 bits (4 bytes) or 64 bits. Normally the word length will be an integral number of bytes.

(f) Character codes

These are used to encode symbols or characters such as keyboard characters. To ensure compatibility when communicating, several standard codes have been developed. ASCII is one of the most widely used character codes. ASCII stands for American Standard Code for Information Interchange; it is now an internationally accepted standard.

ASCII uses 7 bits to represent the letters of the alphabet (upper and lower case), numbers, punctuation symbols and control characters. The code has been arranged so that conversion between upper and lower case is performed by only changing one bit (e.g. C – 1000011 and c – 1100011). This is for ease of implementation for keyboard encoders and character manipulation.

Although ASCII is a 7-bit code there are variants that are 8-bit codes. These have the extra bit giving an extra 128 codes that are used for graphic and other language characters. Other extensions can be set up for particular computers but the standards are reverted to when communicating between different types of machines. Another binary character code is EBCDIC (Extended Binary Coded Decimal Interchange Code) which uses all 8 bits to define the characters. This started as a proprietary standard with IBM and has now been accepted as an open standard. However, it means computers using ASCII and EBCDIC have to convert codes when communicating with each other. A more recent code is Unicode, which is a 16-bit character set that supports a range of alphabets and is used to ease problems with internationalization.

32.2   Information theory for source coding

Here the basics of information theory will be covered, leading to the idea of source coding (Huffman coding). All messages do not have the same importance, nor are all messages equally likely. These ideas will be used to help in coding. There is a link between information transfer and the likelihood of occurrence (or surprise). In terms of probability, a message that is expected has a high probability of arrival, whilst a message not expected has a low probability of arrival. The probability (P) of an event occurring lies in the range $0 \le P \le 1$, where if $P = 0$ the event is impossible, if $P = 1$ the event is certain and if $P = 0.5$ there is a 50:50 chance of it happening.

Intuitively, if something is certain, then the information from that having occurred is zero: if $P = 1$ then $I = 0$. Conversely, if something is impossible, then the information from that occurring is infinite: if $P = 0$ then $I = \infty$. If the probability lies between these two extremes, a measure of the information is given by

$$I = \log_2 \frac{1}{P} \text{ bits} \tag{32.2}$$

note
$$\log_2 x \approx 3.32 \log_{10} x$$

The information is measured in bits. If there are two possible outcomes A and B, where the probability of A occurring is $P(A)$ and the probability of B occurring is $P(B)$, the sum of these probabilities occurring must be 1. That is, $P(A) + P(B) = 1$. This means that $P(A) = P(B) = 0.5$. Information contained in a message is given by $I = \log_2 \frac{1}{0.5}$ bits = 1 bit. The message is that which is transmitted by a single equiprobable binary digit. When binary digits are not equiprobable then the information is no longer 1 bit.

The average information contained in a message is termed the entropy H:

$$H = P(A) \log_2 \frac{1}{P(A)} + P(B) \log_2 \frac{1}{P(B)}$$

or   [32.3]

$$H = P(A) \log_2 \frac{1}{P(A)} + (1 - P(A)) \log_2 \frac{1}{(1 - P(A))}$$

The average information in a binary source is shown in Fig. 32.2. The average information is greatest when both events are equiprobable. This is the case for 1 bit but how is the information determined for n equiprobable messages? The probability of occurrence of each one is

$$P = 1/n$$

Fig. 32.2 The average information for a binary source

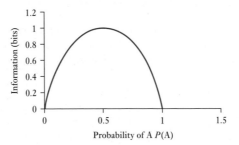

The information is therefore given by

$$I = \log_2 n \text{ bits}$$ [32.4]

If the messages are not equiprobable, as is frequently the case, then the information associated with each message is

$$I_i = \log_2 \frac{1}{P_i} \text{ bits}$$ [32.5]

The expected information gained from any message, the entropy, is

$$H = \sum_{i=1}^{n} P_i I_i = \sum_{i=1}^{n} P_i \log_2 \frac{1}{P_i}$$ [32.6]

The maximum for equiprobable messages is

$$H = \log_2 n \text{ bits} \qquad [32.7]$$

Note that the number of bits needed to send a message from a set of equiprobable messages can be found from

$$N = \log_2 n \text{ bits, where } N \text{ is an integer} \qquad [32.8]$$

For example, if there are eight equiprobable messages, then the information gained by the arrival of any message is $\log_2 8 = 3$ bits. This is also the minimum number of bits needed to code eight messages, with the same length of code for each message. For example the octal coding scheme is given in Table 32.1.

Table 32.1 Octal coding scheme requiring 3 bits

message	code (bits)
1	000
2	001
3	010
4	011
5	100
6	101
7	110
8	111

32.3 Data communications systems

By far the most widely used system is the PSTN (Public Switched Telephone Network) which was designed for analogue voice traffic and, even for digital traffic, offers fairly low data rates (e.g. V34 standard for modems is 28.8 kbps). This is a measure of the number of bits per second that can be transmitted. Even for a situation where the information in the messages that require to be transmitted is below this level, if the coding scheme used does not code the data efficiently, then the channel capacity will not be high enough. In the next section coding schemes to permit information rates to approach data rates will be discussed.

The fully digital version of PSTN is ISDN (Integrated Services Digital Network) which comes in a number of versions based on a basic channel (B-channel) of 64 kbps. The standard offering is 2B+D, which is two B-channels plus a 16 kbps D-type signalling channel, over copper cable, and the two channels can be integrated to offer a maximum capacity of 128 kbps. Another common option is 30B+D, 30 B-channels plus a D-channel, over fibreoptic cable, offering just under 2 Mbps.

LANs (Local Area Networks) and WANs (Wide Area Networks) provide direct digital linking for computer systems. The difference originally was that LANs were only available in relatively small configurations, limited by cable length, whereas WANs were able to provide coverage over a much wider area. This distinction no longer applies as bridges, switches, routers and repeaters have enabled LANs to extend over very wide areas. The distinction now is that LANs use access methods and technology applicable to high-speed networks in localized environments whereas WANs use technology that is generally available through a service provider, such as a PSTN, over a wide area. The actual capacity on a LAN or WAN is governed by the communications protocol that manages communication on the system. Current developments in technology mean that the available channel capacity is rising.

32.4 Coding for efficient transmission

What is the best way to set up the information to be transmitted to give the maximum information transfer across a given channel? This is another way of asking, how does the data rate get maximized? This is of interest in many applications; an example is sending a large document by email. The document may be several hundred kilobytes in size and to send it by email over a telephone line using a modem operating at 28.8 kbps may take a considerable time. If the data are compressed, the file can in certain cases be reduced in size by a factor of over 10, which would reduce the time taken to transmit the file, thus improving the convenience and minimizing cost.

If the source information contains redundant information, data compression can shorten the message by removing redundancy. An example would be sending a text file over a communications link where the source device removes redundant characters before transmission. If the file contains sequences that are repeated, such as several space characters in a row, then instead of transmitting each of them the number of spaces could be transmitted. At the receiver the reverse process would allow the original document to be recovered. This process can be done with any character and can be used on simple binary data. This is termed run length encoding and is widely used in a variety of applications including fax machines. Try sending a fax of plain paper and compare the time taken with that taken for a page of typed notes.

More powerful algorithms are available but they all work by identifying redundancy and removing it. Note that the information content of the message is not altered and these techniques are termed compaction. Data compression may lose some information, which is acceptable in certain situations such as digital TV. The term data compression is used both for those processes that may lose information and for those that do not (i.e. compaction). Another example of coding source information is the Morse code.

(a) A brief history of the Morse code

In the early 19th century, all of the essential components necessary to construct an electrical communication system had been discovered. The most important of these were the battery by Volta, the relationship between electric current and magnetism by Oersted, and the electromagnet by Henry. It now remained for someone to find a practical method to combine these technologies into a working communication system.

Some commercial electrical communications systems existed in Europe as early as the 1830s. A classic example of this is the English 'Needle Telegraph'. The needle telegraph required two or more lines to form a complete circuit. It was relatively slow and the design of the transmitting and receiving instruments was complex. Something simple and efficient was needed.

The Morse system of telegraphy was invented by Samuel Finley Breese Morse (an American painter and founder of the National Academy of Design in New York) in the 1840s in the United States. Morse code is essentially a simple way to represent the letters of the alphabet using patterns of long and short pulses. A unique pattern is assigned to each character of the alphabet, as well as to the ten numerals. An operator using a telegraph key translates these long and short pulses into electrical signals, and a skilled operator, at the distant receiving instrument, translates the electrical signals back into the alphabetic characters. This was demonstrated in 1844 sending the message 'what hath God wrought' via an experimental telegraph from Washington DC to Baltimore.

In the 1920s automated teleprinter technology had become reliable enough to begin to replace the Morse operator. Manual landline telegraphy was slowly phased out until the 1960s when Western Union and the railroads discontinued use of their last Morse circuits. Morse continued to be used in Canada until the mid-1970s, and railroads in Mexico were still using the wire at least until 1990. A small but hardy group of retired telegraphers and telegraph enthusiasts continues to keep landline Morse alive in the US via a mode called 'dial-up' telegraphy.

A dot is the basic timing element. A dash is equivalent to three dots. A space between the dots and dashes in a character is equivalent to one dot and the spaces between characters are three dots long. Words are separated by seven dots in length. The character length in Table 32.2 shows the length, including space between characters in terms of dot length. Note that the most frequently occurring characters have the shortest length and vice versa. This relationship is similar to the relationship between frequency of occurrence of letters and their points value in the game Scrabble. From Table 32.2, you can work out that the average length of a Morse character is 11.2 dots; however, if the letters are weighted by their percentage occurrence, the average character length is 9.04 dots, giving a real saving in overall length. Morse matched the information source (a piece of newspaper text) to the telegraph channel, eliminating some redundancy and efficiently coding the alphabet. This is one of the earliest forms of source coding where the code is matched to the source data. Source coding will be discussed later in the chapter.

(b) Parity checking

Parity checking is used to help detect and/or correct errors that may occur between a transmitter and receiver. The use of parity checking does not improve the efficiency of the source data (it actually reduces the efficiency) but it does permit systems to operate with low error rates. Basic parity checking will detect an error requiring retransmission, to correct the error. Parity checking, as with all other coding techniques for error detection and correction, adds redundancy to the message. It, in effect, averages the noise effects over the message. Parity also forms the basis for many more complex error correcting codes. For a scheme with no redundancy, every message is valid so it is not possible to establish which messages contain

Letter	Morse code	Length	% Occurrence	Character	Code
A	.-	8	8.25	1	.----
B	-...	12	1.78	2	..---
C	-.-.	14	3.14	3	...--
D	-..	10	3.38	4	....-
E	.	4	12.77	5	.....
F	..-.	12	2.38	6	-....
G	--.	12	2.04	7	--...
H	....	10	5.06	8	---..
I	..	6	7.03	9	----.
J	.---	16	0.19	0	-----
K	-.-	12	0.58	Period (Break)	.-.-.-
L	.-..	12	4.30	Comma	--..--
M	--	10	2.29	Question Mark	..--..
N	-.	8	7.02	Double Dash (BT)	-...-
O	---	14	7.13	Fraction Bar	-..-.
P	.--.	14	2.03	End of Message (AR)	.-.-.
Q	--.-	16	0.14	End of Contact (SK)	...-.-
R	.-.	10	6.30		
S	...	8	7.06		
T	-	6	9.17		
U	..-	10	2.83		
V	...-	12	1.20		
W	.--	12	1.80		
X	-..-	14	0.28		
Y	-.--	16	1.76		
Z	--..	14	0.09		

Table 32.2 International Morse code

errors. Natural language is an example of a code with considerable in-built redundancy. The number of errors that can be tolerated varies considerably depending on the situation. An example is 'Thr dog barked'; this is normally recognized as 'The dog barked'. This would not be the case if we were not dealing with language but just lists of characters. It is the redundancy in language which permits this error correction.

In parity error detection schemes, the parity bit (or bits) is used to ensure that the message has either an odd or even number of ones (odd or even parity).

Example 32.2

For example, consider sending the ASCII character 'K' using odd parity. Character 'K' is 1001011 in binary so the parity bit is set to 1 to make the overall number of 1s odd. With the parity bit in the MSB, the transmitted message would be as shown in Table 32.3.

Table 32.3 Transmitted message

Parity	Bit 7	Bit 6	Bit 5	Bit 4	Bit 3	Bit 2	Bit 1
1	1	0	0	1	0	1	1

Table 32.4 Received message

Parity	Bit 7	Bit 6	Bit 5	Bit 4	Bit 3	Bit 2	Bit 1
1	1	1	0	1	0	1	1

Parity can detect any odd number of errors (irrespective of whether it is odd or even parity). If the message is received as shown in Table 32.4, we can see the error is in bit 6 position.

The number of bits is even, so an error must have occurred. This form of coding does not identify which bit is in error, so for correct reception, retransmission is required. Any odd number of errors will be detected but any even number of errors will remain undetected. This is useful for many applications where errors are randomly occurring and normally occur singly. However, in many situations burst errors are prevalent. In this case, other approaches are required. Interleaving of simple codes is one approach to tackling burst errors but more commonly cyclic redundancy check (CRC) codes are used for these types of situation. This is a scheme that is used for longer messages but is still based on parity checking. Error correction and detection codes are an extensive subject in their own right and there are many specialist textbooks on the subject.

32.5 Source coding

(a) Huffman coding

Huffman coding is a statistical data compression technique, which gives a reduction in the average code length, used to represent the symbols of an alphabet. The Huffman code is an example of a code that is optimal in the case where all symbol probabilities are integral powers of 0.5. A Huffman code can be built in the following manner:

1. Rank all symbols in order of probability of occurrence.
2. Successively combine the two symbols of the lowest probability to form a new composite symbol; eventually to build a binary tree where each node is the probability of all nodes beneath it.
3. Trace a path to each leaf, noticing the direction at each node.

For a given probability distribution, there are many possible Huffman codes, but the total compressed length will be the same. This technique is used in most archivers. This is a means of producing a variable length code which has an average length $\bar{L}$ close to the average information contained in a message. $\bar{L}$ will always be greater than or equal to H.

Thus with known probabilities of occurrence, we can produce an optimum solution using Huffman coding. Huffman coding requires a priori probabilities, which can be a problem in some situations. The Lempel–Ziv algorithm learns the source characteristics whilst coding and is used in situations where the a priori probabilities are not available.

Example 32.3

The process for producing a Huffman code is illustrated here. For example, suppose we have the six symbols with the probabilities given in Table 32.5. Note that the total sum of the probabilities is 1.

Table 32.5 Six symbols and their associated probabilities

Symbol	A	B	C	D	E	F
Probability	0.35	0.2	0.15	0.1	0.1	0.1

Firstly, the symbols are listed in descending order of probability (Fig. 32.3). Combine the two least probable symbols and reorder in the next column, again, in terms of descending probability, i.e. the first stage combines E and F. Use arrows to help to keep track of where each column comes from. When two probabilities are the same, place the newly combined one higher as this reduces the difference in length of codewords. Keep doing this until they are all combined.

Fig. 32.3 Huffman coding for Example 32.3

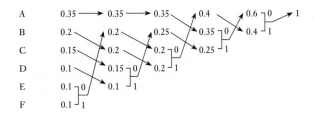

Work back along the tree, labelling each combination with a 0 or a 1, depending on whether it is the upper or lower route. Read out the codeword for each symbol. Read the bits of the codeword from right to left following back the arrows. This ensures that codewords are not the prefix for other codewords. The codewords are generated as shown in Table 32.6.

Table 32.6 Codewords for Huffman coding example

Symbol	A	B	C	D	E	F
Probability	0.35	0.2	0.15	0.1	0.1	0.1
Codeword	00	11	010	011	100	101

The average length of the symbol set is

$$\bar{L} = \sum_{i=1}^{n} P_i L_i$$

$$= 2.45 \text{ bits}$$

The average information/symbol is

$$H = \sum_{i=1}^{n} P_i \log_2 \frac{1}{P_i}$$

$$= 2.4 \text{ bits}$$

The efficiency is defined as

$$\eta = \frac{H}{\bar{L}} \approx 0.98 \tag{32.9}$$

i.e. 98% efficient, which is close to the optimum that could be obtained if the probabilities were integer powers of 0.5. The redundancy r is defined as

$$r = 1 - \eta \qquad r = 0.02 \qquad\qquad [32.10]$$

(b)   Shannon–Fano coding

A technique related to Huffman coding is Shannon–Fano coding, which works as follows:

1. Divide the set of symbols into two equal or almost equal subsets based on the probability of occurrence of characters in each subset. The first subset is assigned a binary zero, the second a binary one.
2. Repeat step (1) until all subsets have a single element.

The algorithm used to create the Huffman codes is bottom-up, and the one for the Shannon–Fano codes is top-down. Huffman encoding always generates optimal codes, Shannon–Fano sometimes uses a few more bits.

Summary of important formulae

The Shannon–Hartley law can be expressed as

$$C = w \, \log_2(1 + S/N) \qquad\qquad [32.1]$$

A measure of the information is given by

$$I = \log_2 \frac{1}{P} \text{ bits} \qquad\qquad [32.2]$$

The expected information gained from any message, the entropy, is

$$H = \sum_{i=1}^{n} P_i I_i$$

$$= \sum_{i=1}^{n} P_i \log_2 \frac{1}{P_i} \qquad\qquad [32.6]$$

The maximum entropy is for equiprobable messages and is

$$H = \log_2 n \text{ bits} \qquad\qquad [32.7]$$

Terms and concepts

The relationship between available bandwidth, signal to noise ratio and channel capacity has been presented and is given by the Shannon–Hartley law.

The unit of information is the bit and given the probability of occurrence of an event a measure of the information is given by equation [32.2].

The expected information gained from any message is termed the entropy. With appropriate source coding the average length of a set of codewords can approach the entropy.

Exercises 32

1. For a channel with an available bandwidth of 3.5 kHz and a signal to noise ratio of 20 dB, what is the theoretical maximum channel capacity?

2. (a) How much information does the receipt of a symbol (from a symbol set of 10 equally probable symbols) convey?
   (b) Determine the information gained when one of 26 equiprobable symbols is received.

3. Consider a source having an $M = 3$ symbol alphabet where: $P(x_1) = 0.5$, $P(x_2) = P(x_3) = 0.25$ and the symbols are statistically independent. Calculate the information conveyed by the receipt of the symbol x_1, x_2, x_3. What is the source entropy for this source alphabet?

4. Consider a source whose (statistically independent) symbols consist of all possible binary sequences of length k. Assume all symbols are equiprobable. How much information is conveyed on the receipt of any symbol?

5. Determine the information conveyed by the message x_1 x_3 x_2 x_1 when it emanates from the following statistically independent sources: (a) $P(x_1) = \frac{1}{2}$, $P(x_2) = \frac{1}{4}$, $P(x_3) = P(x_4) = \frac{1}{8}$, (b) $P(x_1) = P(x_2) = P(x_3) = P(x_4) = \frac{1}{4}$.

6. What is the maximum entropy of an 8-symbol source and under what conditions is this situation achieved?

7. An information source contains 100 different statistically independent equiprobable symbols. Find the maximum code efficiency if for transmission all of the symbols are represented by binary codewords of equal length.

8. (a) Write the 8-bit code for the data 1010110 with an even parity bit.
   (b) Write the 8-bit code for the data 1010110 with an odd parity bit.

9. Can a parity check determine if there are three bits in error?

10. State the types of error that parity checks can and cannot detect. Does using even or odd parity make any difference to this?

11. Consider a source having an $M = 3$ symbol alphabet where: $P(x_1) = \frac{1}{2}$, $P(x_2) = P(x_3) = \frac{1}{4}$ and the symbols are statistically independent.
    (a) Apply Huffman's algorithm to this source to deduce an optimal code.
    (b) What is the efficiency of the code?
    (c) If equal length binary codewords were used what would the efficiency of the code have been?

12. Using the data in Table 32.1 to work out the probabilities for the letter occurrences, calculate the information transferred by the receipt of your full name assuming that the letters are independent of each other. Why is this measure of information about your name not correct?

13. A communications system uses a set of eight symbols. The list of symbols and their probability of occurrence are given in Table A. Construct a code such that individual symbols can be transmitted with the fewest number of bits. Calculate the theoretical minimum average number of bits per symbol and the average number of bits per symbol.

Table A Eight symbols and their probabilities of occurrence

A	B	C	D	E	F	G	H
0.04	0.08	0.05	0.11	0.06	0.5	0.07	0.09

Chapter thirty-three

Fibreoptics

Objectives

When you have studied this chapter, you should

- have a knowledge of the frequency range of fibreoptic systems

- have the ability to describe using the ideas of refraction and reflection how light travels along optical fibres

- be able to calculate power output given power input and attenuation

- be aware of the benefits of the greater bandwidth available from optical systems

Contents

In this chapter, the transmission of information between transmitter and receiver will be considered. The effects of refraction and the angles of acceptance will be looked at. The ideas of attenuation, modulation and bandwidth in optical systems are described together with some of the practical considerations of fibreoptic systems.

By the end of the chapter, you will be aware of the uses of fibreoptic technology in linking electronic systems and of the benefits of the greater bandwidth available from optical systems.

33.1 Introduction

We have been introduced in Fig. 9.22 to the ranges of frequency normally associated with electrical signals. The highest frequency was of the order of 1000 GHz. If the frequency is increased to about 10^{15} Hz, we would find that the signal would literally appear as visible light. Signals in the range 10^{12}–10^{15} Hz appear as infra-red light, while those in the range 10^{15}–10^{17} Hz appear as ultraviolet light.

This raises the question, if we can transmit data using the lower frequency ranges, why should we not communicate at the higher light frequencies? In fact there is no reason why we should not, but there are difficulties in taking the light from a source to some form of receiving unit within which the carrier could be separated from the basic information signal.

There are plenty of materials such as glass through which we can pass light. A glass fibre would make a simple conductor of light. The problem is to retain the light within the glass, but it is possible by simply placing the fibre within a cladding.

33.2 Fibre loss

Let us consider a fibre which has been bent as shown in Fig. 33.1. Light moves in straight lines; therefore, when the beam of light arrives at the bend, it hits the side and bounces off the side at the same angle as that of impact. This is appropriate so long as the fibre is held in air, but if the fibre is held against another material, the light may be partially or totally lost from the fibre.

The need was therefore to introduce a cladding which would prevent the fibre touching such unsuitable external materials. The cladding which surrounds the fibre must have a reaction to light similar to that of air, i.e. an insulator for light. A clad optical fibre is shown in Fig. 33.2 and here we see the light continues through the fibre even though it has been bent.

Fig. 33.1 Unclad optical fibre

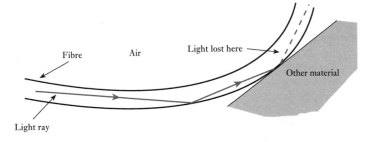

Fig. 33.2 Clad optical fibre

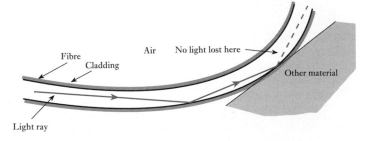

Let us now consider passing light through a pane of glass. We tend to think that all the light has passed through, but in fact there is a small loss. We could demonstrate this by producing thicker and thicker panes of glass because we would find that the image on the other side would become progressively less easy to see.

Even with a reasonably good form of glass, we would find that after passing through 10 m of glass, the signal strength would be down to about 10 per cent of the input, and 10 per cent of 10 per cent means that we would be down to a signal strength of only 1 per cent after 20 m. It follows that we could not transmit a signal along any appreciable length of fibre made from ordinary glass.

Therefore we have to look to removing the impurities in the glass, rather like removing the impurities in semiconductor materials. Once the impurity level is sufficiently low, we find we achieve outcomes which otherwise were impossible. In optical fibres, the 10 m for a 10 per cent signal has been extended to 50 km, and no doubt this will continue to be improved upon.

Although we have introduced light signals in terms of frequency, it is also common to define the operation by its wavelength. It will be recalled that the speed of light is given by

$$c = 3 \times 10^8 \text{ m/s}$$

Wavelength   Symbol: λ Unit: metre (m)

The wavelength, frequency and speed of light are related by

$$c = \lambda f \qquad\qquad [33.1]$$

Usually we find that the wavelength of light is measured in micrometres.

Most optical fibre systems operate with wavelengths in the range 0.8 to 1.6 μm which are the infra-red wavelengths. The type of glass used is silica glass.

33.3   Refraction

In Figs 33.1 and 33.2, we have observed that light, which travels in straight lines, can pass through fibres which are curved. Also in Fig. 33.1, we were shown that losses could occur at the edge of the fibre according to the material which it was touching. To explain these occurrences, we need to briefly consider refraction.

The significant measurement of any transparent material is its refractive index which is the ratio of the speed of light in free space (c_0) to the speed of light in the transparent material (c_{mat}).

Refractive index   Symbol: n Unit: none

$$n = \frac{c_0}{c_{mat}} \qquad\qquad [33.2]$$

The speed of light in a material is always slower than that in free space.

Light rays travel in straight lines through optical materials, but when a light ray passes from one material to another, the refractive index changes and this causes the ray to bend. The amount of bending depends on the

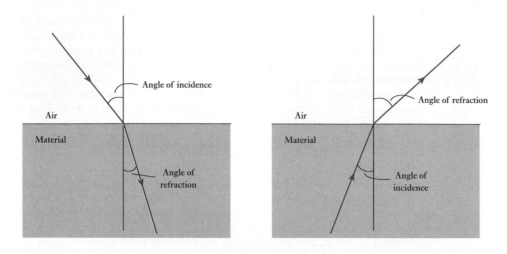

Fig. 33.3 Light refraction

choice of materials and the angle of incidence as shown in Fig. 33.3. The angle of refraction is that relating to the ray of light after bending.

In the instance shown, we see that the ray of light when entering the glass bends towards the normal. However, if leaving the glass, the light bends away from the normal. If we increase the angle of incidence within the glass eventually we reach the point that the angle of refraction is 90°. Increase the angle of incidence and the ray is then reflected as shown in Fig. 33.4, as total internal reflection.

Fig. 33.4 Reflection

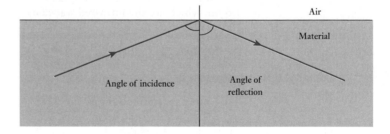

This is what we observed in Fig. 33.1 when the glass fibre was in space. However, when the fibre touched some other material it is possible that reflection would not take place and refraction would operate, hence the loss of the light from within the fibre. We therefore have to select cladding which has an index of refraction less than that of the fibre. It is not necessary for the difference to be large – even 1 per cent is sufficient. A difference of 1 per cent permits an angle of 8° between the ray of light and the fibre surface. This is well within the limit needed when the fibre is more or less straight, but it does limit the extent to which the fibre can be bent round corners, after the fashion indicated in Fig. 33.2. The greatest angle between the ray and the fibre surface without refraction is known as the angle of confinement. This is illustrated in Fig. 33.5.

Fig. 33.5 Angle of confinement

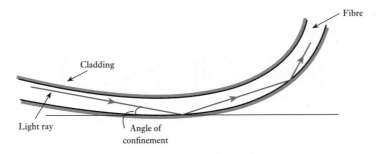

| 33.4 | Light acceptance |

So far, we have considered the effect of a ray of light passing along an optical fibre. This assumes that we can take it from a suitable light source and ensure that it passes along the fibre. Ideally the light source would produce parallel rays of light which move directly into the fibre, but in practice some of the rays will enter the end of the fibre at an angle. The limiting angle is the acceptance angle which is illustrated in Fig. 33.6.

In practice, optical fibres are about 0.25–0.5 mm in diameter. This includes a plastic protective coating and the cladding so that the light-transmitting core is only about 10–80 μm. This means that the light source has to be very small to remain within the angle of acceptance. Such a source is obtainable from a semiconductor diode laser. If we accept larger core sizes, say 100 μm up to 1.0 mm, light-emitting diodes (LEDs) can be used. These are cheaper and last longer than the diode lasers, but they cannot operate so quickly.

Fig. 33.6 Angle of acceptance

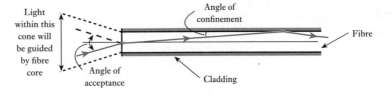

| 33.5 | Attenuation |

When light passes along an optical fibre, losses occur and the light signal is attenuated. There are a variety of reasons for this including absorption of light energy by impurities in the fibre material. Light can also be scattered out of the fibre by impurities. An attenuation/wavelength curve for a typical fibre is shown in Fig. 33.7. The unexpected hump in the characteristic is due to traces of water impurity in the fibre. The characteristic is limited by ever-increasing attenuation at smaller wavelengths and by rapidly increasing attenuation due to absorption by the silica in the glass at longer wavelengths. For the characteristic shown, we are limited in the choice of operating condition by the light sources available. However, it should be noted that the attenuation values are low when compared with conventional electrical conductor and waveguide systems. This means that we can transmit over relatively long distances without amplifiers or repeaters.

Fig. 33.7 Attenuation/ wavelength characteristic

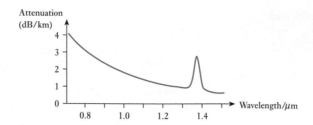

33.6 Bandwidth

When introduced to bandwidth in Chapter 31, we considered the importance of being able to transmit a number of signals at different frequencies within one system. In digital systems, we seek high bandwidth associated with high information capacity.

Let us consider a system capable of handling 100 Mbps. This is equivalent to 100 systems operating at 1 Mbps – and predictably it is cheaper to build the 100 Mbps system. The same would be true for analogue signals except that we would measure the effect in megahertz.

If, for a moment, we were to return to more conventional electrical transmission cables, we find that as frequency or information capacity increases, so the transmission losses increase. In optical fibres the losses are almost independent of frequency or information capacity – and this is the significant advantage of using optical fibre systems instead of conventional cable systems. The comparison of loss/frequency characteristics is shown in Fig. 33.8.

Fig. 33.8 Loss/frequency characteristics

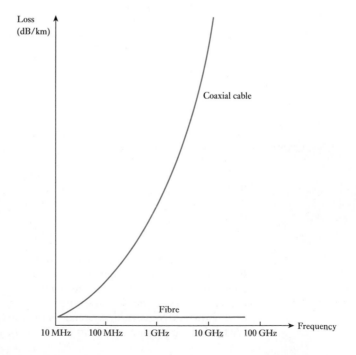

33.7   Modulation

In conventional electrical systems, we produce the carrier signal and then modulate it. Semiconductor lasers and LEDs combine the two functions so that they are modulated by the same current which energizes them. It follows that, by imposing the signal on the energizing current, the light varies in proportion to the signal.

For a digital system involving a diode laser source, the output/input characteristic is shown in Fig. 33.9. Analogue transmission can also take place, in which case the light intensity will rise and fall in proportion to the analogue input signal.

Fig. 33.9 Output/input characteristic of a diode laser

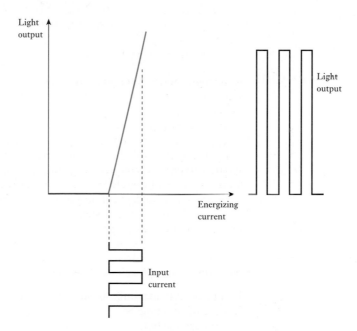

33.8   Optical fibre systems

In principle, an optical fibre system takes the form shown in Fig. 33.10 with a light source providing the conversion of an electrical signal into a light signal which passes through the fibre. At the receiving end, we require a detector which will convert the light signal back into an electrical signal.

A receiver is likely to consist of a semiconductor photodetector. This will not produce a particularly strong electrical signal and therefore an electronic amplifier will be incorporated with the photodetector to make up the receiver. The photodetectors can use a variety of materials depending on the system wavelengths. For instance a silicon diode would be used if the

Fig. 33.10 An optical fibre transmission system

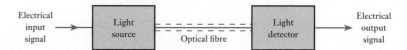

Fig. 33.11   Common forms of optical fibres

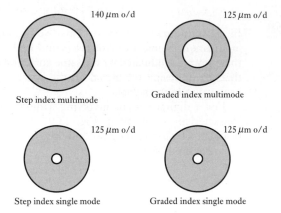

140 μm o/d

125 μm o/d

Step index multimode

Graded index multimode

125 μm o/d

125 μm o/d

Step index single mode

Graded index single mode

wavelength were about $0.8\,\mu$m, whereas a germanium diode would be suitable for $1.2\,\mu$m. For digital systems, the amplifier will also clean up the signal.

In certain respects optical fibre systems are specialized in their applications, and for that reason it is not possible to give them further detailed consideration within this book. However, many communication transmission systems have been converted to optical transmission instead of conductor transmission, and therefore even at this early introductory stage we need to be aware of the concepts of optical fibre systems.

Before leaving such systems, some mention should be made of the variety of optical fibres. Examples of the more common types are shown in Fig. 33.11. The single-mode fibres are now the most favoured because of their high information capacity, although the multimode fibres are often used in short-distance applications. The choice of fibre depends on such factors as

1. The need to maximize bandwidth and capacity.
2. Ease of coupling.
3. Low attenuation.
4. Minimum number of repeaters.
5. Flexibility of fibre and ability to be bent.

The rate of progress in fibre design is such that this year's achievements will be next year's rather ordinary performance. By the early 1990s, fibres were transmitting capacities of 400 Mbps over 50 km at $1.3\,\mu$m wavelength without needing to resort to repeaters. We can look forward to capacities continuing to increase.

Summary of important formulae

Wavelength (in metres)

$$c = \lambda f \qquad\qquad\qquad\qquad [33.1]$$

Refractive index

$$n = \frac{c_0}{c_{\text{mat}}} \qquad\qquad\qquad\qquad [33.2]$$

Terms and concepts

Fibreoptic systems operate at frequencies between 10^{10} and 10^{17} Hz.

The light passes along a fibre which is clad with a material of suitable refractive index. **Refraction** is observed when light passes from one material to another and changes direction as a consequence. **Reflection** occurs when the light requires to deflect by more than $90°$ at such a transfer.

Light can be modulated by the suitable control of a **diode laser** or **LED**.

Optical fibres include **multimode** and **single-mode** types.

Exercises 33

In questions 1 to 6, select the appropriate statement.
1. Light can be guided around corners most efficiently in:
   (a) hollow pipes with gas lenses;
   (b) clad optical fibres;
   (c) bare glass fibres;
   (d) reflective pipes.
2. Optical fibres are made of:
   (a) glass coated with plastic;
   (b) ultrapure glass;
   (c) plastic;
   (d) all of these.
3. Essential components of any fibreoptic communication system are:
   (a) light source, fibre, and receiver;
   (b) light source and cable;
   (c) fibre and receiver;
   (d) fibre only.
4. If light passes from air into glass, it is:
   (a) reflected;
   (b) refracted;
   (c) absorbed;
   (d) scattered.
5. Light is confined within the core of a simple optical fibre by:

(a) total internal reflection at the outer edge of the cladding;
(b) total internal reflection at the core–cladding boundary;
(c) reflection from the fibre's plastic coating;
(d) refraction.

6. The input power to a fibreoptic cable is 1 mW. The cable's loss is 20 dB. What is the output power, assuming there are no other losses?
   (a) 100 μW;
   (b) 50 μW;
   (c) 10 μW;
   (d) 1 μW.
7. The wavelength of a light signal is 1.2 μm. Determine the frequency of the signal.
8. Describe, in terms of improved bandwidth, the benefits to be derived from using fibre systems compared with coaxial cable systems.
9. Describe the operation of a light detector in a fibre system.
10. The refractive index of an optical fibre is 2.25. Determine the speed of light in the fibre material.

Section three Power Engineering

Chapter thirty-four

Multiphase Systems

Objectives

When you have studied this chapter, you should
- have an understanding of the generation of three phase e.m.f.s
- be familiar with the delta or mesh connection of three phases
- be familiar with the star or wye connection of three phases
- be able to calculate voltages and currents in the delta connection
- be able to calculate voltages and currents in the star connection
- understand the relationship between line and phase values
- be able to construct phasor diagrams for delta and star connections
- be able to calculate power in a three-phase system
- be familiar with the measurement of power

Contents

Electricity supply systems have to deliver power to many types of load. The greater the power supplied, for a given voltage, the greater the current. Three-phase systems are well suited to electricity supply applications because of their ability to transmit high powers efficiently and to provide powerful motor drives.

This chapter introduces the principles associated with three-phase systems, including the two methods of connection, delta and star. The relationship between phase and line currents and voltages for both forms of connection will be developed. Finally, the calculation and measurement of power will be introduced.

By the end of the chapter, you will be familiar with the terms and relationships which apply to all basic three-phase configurations.

34.1 Disadvantages of the single-phase system

The earliest application of alternating current was for heating the filaments of electric lamps. For this purpose the single-phase system was perfectly satisfactory. Some years later, a.c. motors were developed, and it was found that for this application the single-phase system was not very satisfactory. For instance, the single-phase induction motor – the type most commonly employed – was not self-starting unless it was fitted with an auxiliary winding. By using two separate windings with currents differing in phase by a quarter of a cycle or three windings with currents differing in phase by a third of a cycle, it was found that the induction motor was self-starting and had better efficiency and power factor than the corresponding single-phase machine.

The system utilizing two windings is referred to as a *two-phase system* and that utilizing three windings is referred to as a *three-phase system*. We shall now consider the three-phase system in detail.

34.2 Generation of three-phase e.m.f.s

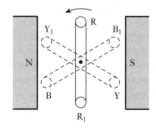

Fig. 34.1 Generation of three-phase e.m.f.s

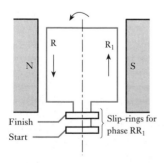

Fig. 34.2 Loop RR_1 at instant of maximum e.m.f.

In Fig. 34.1, RR_1, YY_1 and BB_1 represent three similar loops fixed to one another at angles of 120°, each loop terminating in a pair of slip-rings carried on the shaft as indicated in Fig. 34.2. We shall refer to the slip-rings connected to sides R, Y and B as the 'finishes' of the respective phases and those connected to R_1, Y_1 and B_1 as the 'starts'.

The letters R, Y and B are abbreviations of 'red', 'yellow' and 'blue', namely the colours used to identify the three phases. Also, 'red–yellow–blue' is the sequence that is universally adopted to denote that the e.m.f. in the yellow phase lags that in the red phase by a third of a cycle, and the e.m.f. in the blue phase lags that in the yellow phase by another third of a cycle.

Suppose the three coils are rotated anticlockwise at a uniform speed in the magnetic field due to poles NS. The e.m.f. generated in loop RR_1 is zero for the position shown in Fig. 34.1. When the loop has moved through 90° to the position shown in Fig. 34.2, the generated e.m.f. is at its maximum value, its direction round the loop being from the 'start' slip-ring towards the 'finish' slip-ring. Let us regard this direction as positive; consequently the e.m.f. induced in loop RR_1 can be represented by the full-line curve of Fig. 34.3.

Since the loops are being rotated anticlockwise, it is evident from Fig. 34.1 that the e.m.f. generated in side Y of loop YY_1 has exactly the same amplitude as that generated in side R, but lags by 120° (or one-third of a cycle). Similarly, the e.m.f. generated in side B of loop BB_1 is equal to but lags that in side Y by 120°. Hence the e.m.f.s generated in loops RR_1, YY_1 and BB_1 are represented by the three equally spaced curves of Fig. 34.3, the e.m.f.s being assumed positive when their directions round the loops are from 'start' to 'finish' of their respective loops.

If the instantaneous value of the e.m.f. generated in phase RR_1 is represented by $e_R = E_m \sin \theta$, then instantaneous e.m.f. in YY_1 is

$$e_Y = E_m \sin(\theta - 120°)$$

and instantaneous e.m.f. in BB_1 is

$$e_B = E_m \sin(\theta - 240°)$$

Fig. 34.3 Waveforms of three-phase e.m.f.s

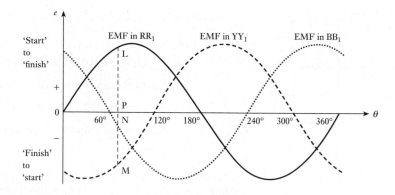

34.3 **Delta connection of three-phase windings**

The three phases of Fig. 34.1 can, for convenience, be represented as in Fig. 34.4 where the phases are shown isolated from one another; L_1, L_2 and L_3 represent loads connected across the respective phases. Since we have assumed the e.m.f.s to be positive when acting from 'start' to 'finish', they

Fig. 34.4 Three-phase windings with six line conductors

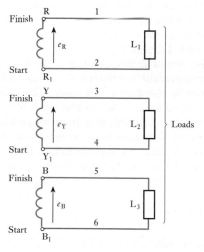

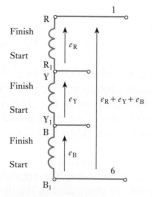

Fig. 34.5 Resultant e.m.f. in a delta-connected winding

can be represented by the arrows e_R, e_Y and e_B in Fig. 34.4. This arrangement necessitates six line conductors and is therefore cumbersome and expensive, so let us consider how it may be simplified. For instance, let us join R_1 and Y together as in Fig. 34.5, thereby enabling conductors 2 and 3 of Fig. 34.4 to be replaced by a single conductor. Similarly, let us join Y_1 and B together so that conductors 4 and 5 may be replaced by another single conductor. If we join 'start' B_1 to 'finish' R, there will be three e.m.f.s chasing each other around the loop and these would produce a circulating current in that loop. However, we can show that the resultant e.m.f. between these two points is zero and that there is therefore no circulating current when these points are connected together.

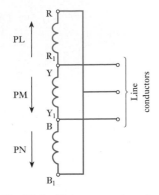

Fig. 34.6 Delta connection of three-phase winding

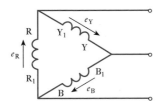

Fig. 34.7 Conventional representation of a delta or mesh-connected winding

Instantaneous value of total e.m.f. acting from B_1 to R is

$$e_R + e_Y + e_B$$
$$= E_m\{\sin\theta + \sin(\theta - 120°) + \sin(\theta - 240°)\}$$
$$= E_m(\sin\theta + \sin\theta \cdot \cos 120° - \cos\theta \cdot \sin 120°$$
$$+ \sin\theta \cdot \cos 240° - \cos\theta \cdot \sin 240°)$$
$$= E_m(\sin\theta - 0.5\sin\theta - 0.866\cos\theta - 0.5\sin\theta + 0.866\cos\theta)$$
$$= 0$$

Since this condition holds for every instant, it follows that R and B_1 can be joined together, as in Fig. 34.6, without any circulating current being set up around the circuit. The three line conductors are joined to the junctions thus formed.

It might be helpful at this stage to consider the actual values and directions of the e.m.f.s at a particular instant. For instance, at instant P in Fig. 34.3 the e.m.f. generated in phase R is positive and is represented by PL acting from R_1 to R in Fig. 34.6. The e.m.f. in phase Y is negative and is represented by PM acting from Y to Y_1, and that in phase B is also negative and is represented by PN acting from B to B_1. But the sum of PM and PN is exactly equal numerically to PL; consequently the algebraic sum of the e.m.f.s round the closed circuit formed by the three windings is zero.

It should be noted that the directions of the arrows in Fig. 34.6 represent the directions of the e.m.f. at a *particular instant*, whereas arrows placed alongside symbol e, as in Fig. 34.5, represent the *positive* directions of the e.m.f.s.

The circuit derived in Fig. 34.6 is usually drawn as in Fig. 34.7 and the arrangement is referred to as *delta* (from the Greek capital letter Δ) connection, also known as a mesh connection.

It will be noticed that in Fig. 34.7, R is connected to Y_1 instead of B_1 as in Fig. 34.6. Actually, it is immaterial which method is used. What is of importance is that the 'start' of one phase should be connected to the 'finish' of another phase, so that the arrows representing the positive directions of the e.m.f.s point in the same direction round the mesh formed by the three windings.

34.4 Star connection of three-phase windings

Let us go back to Fig. 34.4 and join together the three 'starts', R_1, Y_1 and B_1 at N, as in Fig. 34.8, so that the three conductors 2, 4 and 6 of Fig. 34.4 can be replaced by the single conductor NM of Fig. 34.8.

Since the generated e.m.f. has been assumed positive when acting from 'start' to 'finish', the current in each phase must also be regarded as positive when flowing in that direction, as represented by the arrows in Fig. 34.8. If i_R, i_Y and i_B are the instantaneous values of the currents in the three phases, the instantaneous value of the current in the common wire MN is $(i_R + i_Y + i_B)$, having its positive direction from M to N.

This arrangement is referred to as a *four-wire star-connected* system and is more conveniently represented as in Fig. 34.9, and junction N is referred to as the *star* or *neutral point*. Three-phase motors are connected to the line conductors R, Y and B, whereas lamps, heaters, etc. are usually connected between the line and neutral conductors, as indicated by L_1, L_2 and L_3, the

Fig. 34.8 Star connection of three-phase winding

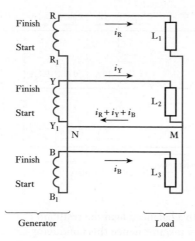

Fig. 34.9 Four-wire star-connected system

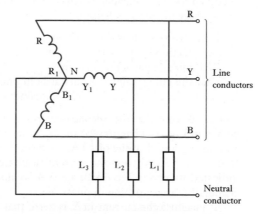

total load being distributed as equally as possible between the three lines. If these three loads are exactly alike, the phase currents have the same peak value, I_m, and differ in phase by 120°. Hence if the instantaneous value of the current in load L_1 is represented by

$$i_1 = I_m \sin \theta$$

instantaneous current in L_2 is

$$i_2 = I_m \sin(\theta - 120°)$$

and instantaneous current in L_3 is

$$i_3 = I_m \sin(\theta - 240°)$$

Hence instantaneous value of the resultant current in neutral conductor MN (Fig. 34.8) is

$$i_1 + i_2 + i_3$$
$$= I_m\{\sin \theta + \sin(\theta - 120°) + \sin(\theta - 240°)\}$$
$$= I_m \times 0 = 0$$

Fig. 34.10 Three-wire star-connected system with balanced load

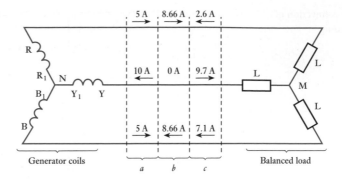

i.e. with a balanced load the resultant current in the neutral conductor is zero at *every* instant; hence this conductor can be dispensed with, thereby giving us the *three-wire star-connected* system shown in Fig. 34.10.

When we are considering the distribution of current in a three-wire, three-phase system it is helpful to bear in mind:

1. That arrows such as those of Fig. 34.8, placed alongside *symbols*, indicate the direction of the current when it is assumed to be *positive* and not the direction at a particular instant.
2. That the current flowing outwards in one or two conductors is equal to that flowing back in the remaining conductor or conductors.

Let us consider the second statement in greater detail. Suppose the curves in Fig. 34.11 represent the three currents differing in phase by 120° and having a peak value of 10 A. At instant *a*, the currents in phases R and B are each 5 A, whereas the current in phase Y is −10 A. These values are indicated above *a* in Fig. 34.10, i.e. 5 A are flowing outwards in phases R and B and 10 A are returning in phase Y.

At instant *b* the current in Y is zero, that in R is 8.66 A and that in B is −8.66 A, i.e. 8.66 A are flowing outwards in phase R and returning in phase B. At instant *c*, the currents in R, Y and B are −2.6, 9.7 and −7.1 A respectively,

Fig. 34.11 Waveforms of current in a balanced three-phase system

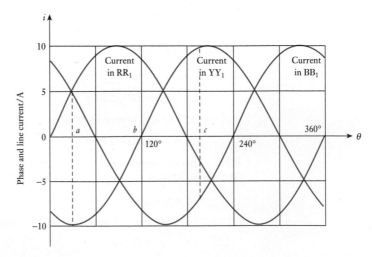

i.e. 9.7 A flow outwards in phase Y and return via phases R (2.6 A) and B (7.1 A).

It will be seen that the distribution of currents between the three lines is continually changing, but at every instant the algebraic sum of the currents is zero.

34.5 Voltages and currents in a star-connected system

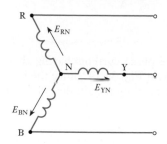

Fig. 34.12 Star-connected generator

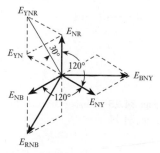

Fig. 34.13 Phasor diagram for Fig. 34.12

Let us again assume the e.m.f. in each phase to be positive when acting from the neutral point outwards, so that the r.m.s. values of the e.m.f.s generated in the three phases can be represented by E_{NR}, E_{NY} and E_{NB} in Figs 34.12 and 34.13.

When the relationships between line and phase quantities are being derived for either the star- or the delta-connected system, it is essential to relate the phasor diagram to a circuit diagram and to indicate on each phase the direction in which the voltage or current is assumed to be positive. A phasor diagram by itself is meaningless.

The value of the e.m.f. acting from Y via N to R is the phasor *difference* of E_{NR} and E_{NY}. Hence E_{YN} is drawn equal and opposite to E_{NY} and added to E_{NR}, giving E_{YNR} as the e.m.f. acting from Y to R via N. Note that the *three* subscript letters YNR are necessary to indicate unambiguously the *positive* direction of this e.m.f.

Having decided on YNR as the positive direction of the line e.m.f. between Y and R, we must adhere to the same sequence for the e.m.f.s between the other lines, i.e. the sequence must be YNR, RNB and BNY. Here E_{RNB} is obtained by subtracting E_{NR} from E_{NB}, and E_{BNY} is obtained by subtracting E_{NB} from E_{NY}, as shown in Fig. 34.13. From the symmetry of this diagram it is evident that the line voltages are equal and are spaced 120° apart. Further, since the sides of all the parallelograms are of equal length, the diagonals bisect one another at right angles. Also, they bisect the angles of their respective parallelograms; and, since the angle between E_{NR} and E_{YN} is 60°,

$$\therefore \qquad E_{YNR} = 2E_{NR} \cos 30° = \sqrt{3}E_{NR}$$

i.e.   Line voltage = 1.73 × star (or phase) voltage

From Fig. 34.12 it is obvious that in a star-connected system the current in a line conductor is the same as that in the phase to which that line conductor is connected.

Hence, in general, if

V_L = p.d. between any two line conductors

= line voltage

and

V_P = p.d. between a line conductor and the neutral point

= star voltage (or voltage to neutral)

and if I_L and I_P are line and phase currents respectively, then for a star-connected system,

$$V_L = 1.73V_P \qquad\qquad [34.1]$$

and

$$I_\mathrm{L} = I_\mathrm{P} \qquad [34.2]$$

In practice, it is the voltage between two line conductors or between a line conductor and the neutral point that is measured. Owing to the internal impedance drop in the windings, this p.d. is different from the corresponding e.m.f. generated in the winding, except when the generator is on open circuit; hence, in general, it is preferable to work with the potential difference, V, rather than with the e.m.f., E.

The voltage given for a three-phase system is always the line voltage unless it is stated otherwise.

34.6 Voltages and currents in a delta-connected system

Let I_1, I_2 and I_3 be the r.m.s. values of the phase currents having their positive directions as indicated by the arrows in Fig. 34.14. Since the load is assumed to be balanced, these currents are equal in magnitude and differ in phase by 120°, as shown in Fig. 34.15.

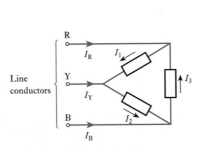

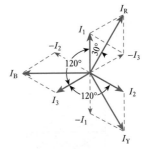

Fig. 34.14 Delta-connected system with balanced load

Fig. 34.15 Phasor diagram for Fig. 34.14

From Fig. 34.14 it will be seen that I_1, when positive, flows away from line conductor R, whereas I_3, when positive, flows towards it. Consequently, I_R is obtained by subtracting I_3 from I_1, as in Fig. 34.15. Similarly, I_Y is the phasor difference of I_2 and I_1, and I_B is the phasor difference of I_3 and I_2. From Fig. 34.15 it is evident that the line currents are equal in magnitude and differ in phase by 120°. Also

$$I_\mathrm{R} = 2I_1 \cos 30° = \sqrt{3}I_1$$

Hence for a delta-connected system with a balanced load

Line current = 1.73 × phase current

i.e. $$I_\mathrm{L} = 1.73I_\mathrm{P} \qquad [34.3]$$

From Fig. 34.14 it can be seen that in a delta-connected system, the line and the phase voltages are the same, i.e.

$$V_\mathrm{L} = V_\mathrm{P} \qquad [34.4]$$

Example 34.1 In a three-phase four-wire system the line voltage is 400 V and non-inductive loads of 10 kW, 8 kW and 5 kW are connected between the three line conductors and the neutral as in Fig. 34.16. Calculate:

(a) the current in each line;
(b) the current in the neutral conductor.

Fig. 34.16 Circuit diagram for Example 34.1

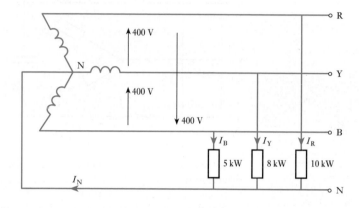

(a)   Voltage to neutral $= \dfrac{\text{line voltage}}{1.73} = \dfrac{400}{1.73} = 230$ V

If I_R, I_Y and I_B are the currents taken by the 10 kW, 8 kW and 5 kW loads respectively,

$$I_R = 10 \times 1000/230 = \mathbf{43.5 \ A}$$

$$I_Y = 8 \times 1000/230 = \mathbf{34.8 \ A}$$

and $I_B = 5 \times 1000/230 = \mathbf{21.7 \ A}$

These currents are represented by the respective phasors in Fig. 34.17.

(b) The current in the neutral is the phasor sum of the three line currents. In general, the most convenient method of adding such quantities is to calculate the resultant horizontal and vertical components thus: horizontal component is

$$I_H = I_Y \cos 30° - I_B \cos 30°$$

$$= 0.866(34.8 - 21.7) = 11.3 \ A$$

and vertical component is

$$I_V = I_R - I_Y \cos 60° - I_B \cos 60°$$

$$= 43.5 - 0.5(34.8 + 21.7) = 13.0 \ A$$

These components are represented in Fig. 34.18.

Current in neutral $= I_N = \sqrt{\{(11.3)^2 + (13.0)^2\}}$

$$= \mathbf{17.2 \ A}$$

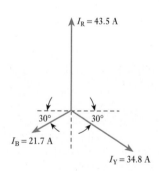

Fig. 34.17 Phasor diagram for Fig. 34.16

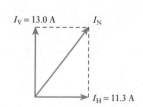

Fig. 34.18 Vertical and horizontal components of I_N

| Example 34.2 | A delta-connected load is arranged as in Fig. 34.19. The supply voltage is 400 V at 50 Hz. Calculate: |

(a) the phase currents;
(b) the line currents.

Fig. 34.19 Circuit diagram for Example 34.2

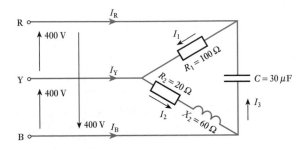

(a) Since the phase sequence is R, Y, B, the voltage having its positive direction from R to Y leads 120° on that having its positive direction from Y to B, i.e. V_{RY} is 120° in front of V_{YB}.

Similarly, V_{YB} is 120° in front of V_{BR}. Hence the phasors representing the line (and phase) voltages are as shown in Fig. 34.20.

If I_1, I_2, I_3 are the phase currents in loads RY, YB and BR respectively:

$$I_1 = 400/100 = \textbf{4.0 A}, \quad \text{in phase with } V_{RY}$$

$$I_2 = \frac{400}{\sqrt{(20^2 + 60^2)}} = \frac{400}{63.3} = \textbf{6.32 A}$$

Fig. 34.20 Phasor diagram for Fig. 34.19

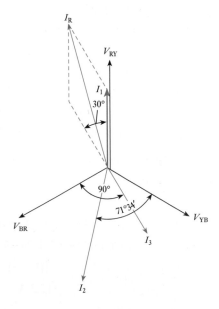

Fig. 34.21 Phasor diagram for deriving I_Y

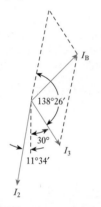

Fig. 34.22 Phasor diagram for deriving I_B

I_2 lags V_{YB} by an angle ϕ_2 such that

$$\phi_2 = \tan^{-1} \frac{60}{20} = 71°34'$$

Also $\quad I_3 = 2 \times 3.14 \times 50 \times 30 \times 10^{-6} \times 400$

$$= 3.77 \text{ A}, \quad \text{leading } V_{BR} \text{ by } 90°$$

(b) If the current I_R in line conductor R is assumed to be positive when flowing towards the load, the phasor representing this current is obtained by subtracting I_3 from I_1, as in Fig. 34.20.

$\therefore \qquad I_R^2 = (4.0)^2 + (3.77)^2 + 2 \times 4.0 \times 3.77 \cos 30° = 56.3$

$\therefore \qquad I_R = 7.5 \text{ A}$

The current in line conductor Y is obtained by subtracting I_1 from I_2, as shown separately in Fig. 34.21. But angle between I_2 and I_1 reversed is

$$\phi_2 - 60° = 71°34' - 60° = 11°34'$$

$\therefore \qquad I_Y^2 = (4.0)^2 + (6.32)^2 + 2 \times 4.0 \times 6.32 \times \cos 11°34'$

$$= 105.5$$

$\therefore \qquad I_Y = 10.3 \text{ A}$

Similarly, the current in line conductor B is obtained by subtracting I_2 from I_3, as shown in Fig. 34.22. Angle between I_3 and I_2 reversed is

$$180° - 30° - 11°34' = 138°26'$$

$\therefore \qquad I_B^2 = (6.32)^2 + (3.77)^2 + 2 \times 6.32 \times 3.77 \times \cos 138°26'$

$$= 18.5$$

$\therefore \qquad I_B = 4.3 \text{ A}$

This problem could be solved graphically, but in that case it would be necessary to draw the phasors to a large scale to ensure reasonable accuracy.

34.7

Power in a three-phase system with a balanced load

If I_P is the r.m.s. value of the current in each phase and V_P the r.m.s. value of the p.d. across each phase,

$$\text{Active power per phase} = I_P V_P \times \text{power factor}$$

and $\qquad$ Total active power $= 3I_P V_P \times$ power factor

$$P = 3I_P V_P \cos \phi \qquad\qquad\qquad [34.5]$$

If I_L and V_L are the r.m.s. values of the line current and voltage respectively, then for a *star-connected system*,

$$V_P = \frac{V_L}{1.73} \quad \text{and} \quad I_P = I_L$$

Substituting for I_P and V_P in equation [34.5], we have

Total active power in watts $= 1.73 I_L V_L \times$ power factor

For a *delta-connected system*

$$V_P = V_L \quad \text{and} \quad I_P = \frac{I_L}{1.73}$$

Again, substituting for I_P and V_P in equation [34.5], we have

Total active power in watts $= 1.73 I_L V_L \times$ power factor

Hence it follows that, for any balanced load,

Active power in watts $= 1.73 \times$ line current $\times$ line voltage

$\times$ power factor

$= 1.73 I_L V_L \times$ power factor

$$P = \sqrt{3} V_L I_L \cos \phi \qquad\qquad [34.6]$$

Example 34.3   A three-phase motor operating off a 400 V system is developing 20 kW at an efficiency of 0.87 p.u. and a power factor of 0.82. Calculate:

(a) the line current;
(b) the phase current if the windings are delta-connected.

(a) Since

$$\text{Efficiency} = \frac{\text{output power in watts}}{\text{input power in watts}}$$

$$\eta = \frac{\text{output power in watts}}{1.73 I_L V_L \times \text{p.f.}}$$

$\therefore\qquad 0.87 = \dfrac{20 \times 1000}{1.73 \times I_L \times 400 \times 0.82}$

and   Line current $= I_L = \textbf{40.0 A}$

(b) For a delta-connected winding

$$\text{Phase current} = \frac{\text{line current}}{1.73} = \frac{40.0}{1.73} = \textbf{23.1 A}$$

34.8 **Measurement of active power in a three-phase, three-wire system**

(a) Star-connected balanced load, with neutral point accessible

If a wattmeter W is connected with its current coil in one line and the voltage circuit between that line and the neutral point, as shown in Fig. 34.23, the reading on the wattmeter gives the power per phase:

$\therefore\qquad$ Total active power $= 3 \times$ wattmeter reading

Fig. 34.23 Measurement of active power in a star-connected balanced load

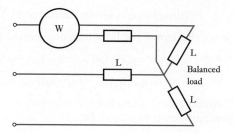

(b) Balanced or unbalanced load, star- or delta-connected. The two-wattmeter method

Suppose the three loads L_1, L_2 and L_3 are connected in star, as in Fig. 34.24. The current coils of the two wattmeters are connected in any two lines, say the 'red' and 'blue' lines, and the voltage circuits are connected between these lines and the third line.

Fig. 34.24 Measurement of power by two wattmeters

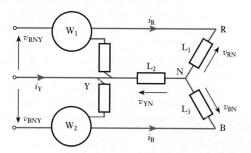

Suppose v_{RN}, v_{YN} and v_{BN} are the instantaneous values of the p.d.s across the loads, these p.d.s being assumed positive when the respective line conductors are positive in relation to the neutral point. Also, suppose i_R, i_Y and i_B are the corresponding instantaneous values of the line (and phase) currents. Therefore instantaneous power in load $L_1 = i_R v_{RN}$, instantaneous power in load $L_2 = i_Y v_{YN}$, and instantaneous power in load $L_3 = i_B v_{BN}$. Therefore

$$\text{Total instantaneous power} = i_R v_{RN} + i_Y v_{YN} + i_B v_{BN}$$

From Fig. 34.24 it is seen that instantaneous current through current coil of $W_1 = i_R$ and instantaneous p.d. across voltage circuit of $W_1 = v_{RN} - v_{YN}$. Therefore

$$\text{Instantaneous power measured by } W_1 = i_R(v_{RN} - v_{YN})$$

Similarly, instantaneous current through current coil of $W_2 = i_B$ and instantaneous p.d. across voltage circuit of $W_2 = v_{BN} - v_{YN}$.

It is important to note that this p.d. is not $v_{YN} - v_{BN}$. This is due to the fact that a wattmeter reads positively when the currents in the current and voltage coils are *both* flowing from the junction of these coils or *both* towards that junction; and since the positive direction of the current in the current coil of W_2 has already been taken as that of the arrowhead alongside i_B in Fig. 34.24 it follows that the current in the voltage circuit of W_2 is positive when flowing from the 'blue' to the 'yellow' line.

$\therefore$ Instantaneous power measured by $W_2 = i_B(v_{BN} - v_{YN})$

Hence the sum of the instantaneous powers of W_1 and W_2 is

$$i_R(v_{RN} - v_{YN}) + i_B(v_{BN} - v_{YN})$$

$$= i_R v_{RN} + i_B v_{BN} - (i_R + i_B)v_{YN}$$

From Kirchhoff's first law (section 3.4), the algebraic sum of the instantaneous currents at N is zero, i.e.

$$i_R + i_Y + i_B = 0$$

$\therefore$ $$i_R + i_B = -i_Y$$

so that sum of instantaneous powers measured by W_1 and W_2 is

$$i_R v_{RN} + i_B v_{BN} + i_Y v_{YN}$$

$$= \text{total instantaneous power}$$

Actually, the power measured by each wattmeter varies from instant to instant, but the inertia of the moving system causes the pointer to read the average value of the power. Hence the sum of the wattmeter readings gives the average value of the total power absorbed by the three phases, i.e. the active power.

Since the above proof does not assume a balanced load or sinusoidal waveforms, it follows that the sum of the two wattmeter readings gives the total power under all conditions. The above proof was derived for a star-connected load, and it is a useful exercise to prove that the same conclusion holds for a delta-connected load.

34.9 Power factor measurement by means of two wattmeters

Suppose L in Fig. 34.25 to represent three similar loads connected in star, and suppose V_{RN}, V_{YN} and V_{BN} to be the r.m.s. values of the phase voltages and I_R, I_Y and I_B to be the r.m.s. values of the currents. Since these voltages and currents are assumed sinusoidal, they can be represented by phasors, as in Fig. 34.26, the currents being assumed to lag the corresponding phase voltages by an angle ϕ. Current through current coil of W_1 is I_R. Potential difference across voltage circuit of W_1 is

Phasor difference of V_{RN} and $V_{YN} = V_{RNY}$

Phase difference between I_R and $V_{RNY} = 30° + \phi$. Therefore reading on W_1 is

$$P_1 = I_R V_{RNY} \cos(30° + \phi)$$

Fig. 34.25 Measurement of active power and power factor by two wattmeters

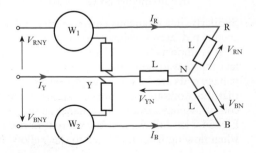

Fig. 34.26 Phasor diagram for Fig. 34.25

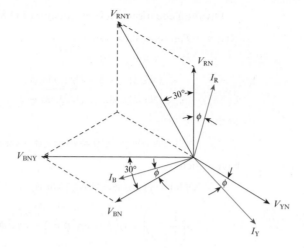

Current through current coil of $W_2 = I_B$. Potential difference across voltage circuit of W_2 is

Phasor difference of V_{BN} and $V_{YN} = V_{BNY}$

Phase difference between I_B and $V_{BNY} = 30° - \phi$. Therefore reading on W_2 is

$$P_2 = I_B V_{BNY} \cos(30° - \phi)$$

Since the load is balanced,

$$I_R = I_Y = I_B = \text{(say) } I_L, \text{ numerically}$$

and

$$V_{RNY} = V_{BNY} = \text{(say) } V_L, \text{ numerically}$$

Hence

$$P_1 = I_L V_L \cos(30° + \phi) \qquad [34.7]$$

and

$$P_2 = I_L V_L \cos(30° - \phi) \qquad [34.8]$$

$$P_1 + P_2 = I_L V_L \{\cos(30° + \phi) + \cos(30° - \phi)\}$$
$$P_1 + P_2 = I_L V_L (\cos 30° \cdot \cos \phi - \sin 30° \cdot \sin \phi$$
$$+ \cos 30° \cdot \cos \phi + \sin 30° \cdot \sin \phi)$$

$$P_1 + P_2 = 1.73 I_L V_L \cos \phi \qquad [34.9]$$

namely the expression deduced in section 34.7 for the total active power in a balanced three-phase system. This is an alternative method of proving that the sum of the two wattmeter readings gives the total active power, but it should be noted that this proof assumed a balanced load and sinusoidal voltages and currents.

Dividing equation [34.7] by equation [34.8], we have

$$\frac{P_1}{P_2} = \frac{\cos(30° + \phi)}{\cos(30° - \phi)} = (\text{say})\, y$$

$$y = \frac{(\sqrt{3}/2)\cos\phi - (1/2)\sin\phi}{(\sqrt{3}/2)\cos\phi + (1/2)\sin\phi}$$

so that

$$\sqrt{3}y\cos\phi + y\sin\phi = \sqrt{3}\cos\phi - \sin\phi$$

from which

$$\sqrt{3}(1 - y)\cos\phi = (1 + y)\sin\phi$$

$$\therefore \quad 3\left(\frac{1-y}{1+y}\right)^2 \cos^2\phi = \sin^2\phi = 1 - \cos^2\phi$$

$$\left\{1 + 3\left(\frac{1-y}{1+y}\right)^2\right\}\cos^2\phi = 1$$

$$\therefore \quad \text{Power factor} = \cos\phi = \frac{1}{\sqrt{\left\{1 + 3\left(\dfrac{1-y}{1+y}\right)^2\right\}}} \qquad [34.10]$$

Since y is the ratio of the wattmeter readings, the corresponding power factor can be calculated from expression [34.9], but this procedure is very laborious. A more convenient method is to draw a graph of the power factor for various ratios of P_1/P_2; and in order that these ratios may lie between $+1$ and -1, as in Fig. 34.27, it is always the practice to take P_1 as the smaller of the two readings. By adopting this practice, it is possible to derive reasonably accurate values of the power factor from the graph.

Fig. 34.27 Relationship between power factor and ratio of wattmeter readings

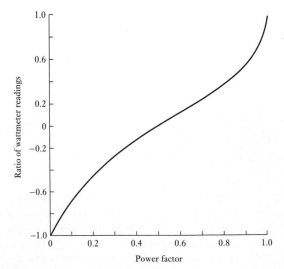

When the power factor of the load is 0.5 lagging, ϕ is 60°; and from equation [34.7], the reading on $W_1 = I_L V_L \cos 90° = 0$. When the power factor is less than 0.5 lagging, ϕ is greater than 60° and $(30° + \phi)$ is therefore greater than 90°. Hence the reading on W_1 is negative. To measure this active power it is necessary to reverse the connections to either the current or the voltage coil, but the reading thus obtained must be taken as negative when the total active power and the ratio of the wattmeter readings are being calculated.

An alternative method of deriving the power factor is as follows:

From equations [34.7], [34.8] and [34.9]

$$P_2 - P_1 = I_L V_L \sin \phi$$

and

$$\tan \phi = \frac{\sin \phi}{\cos \phi} = 1.73\left(\frac{P_2 - P_1}{P_2 + P_1}\right) \qquad [34.11]$$

Hence, ϕ and $\cos \phi$ can be determined with the aid of trigonometrical tables.

Example 34.4

The input power to a three-phase motor was measured by the two-wattmeter method. The readings were 5.2 kW and −1.7 kW, and the line voltage was 400 V. Calculate:

(a) the total active power;
(b) the power factor;
(c) the line current.

(a) Total power = 5.2 − 1.7 = **3.5 kW**.

(b) Ratio of wattmeter readings is

$$\frac{-1.7}{5.2} = -0.327$$

From Fig. 34.27, power factor = 0.28. Or alternatively, from equation [34.11],

$$\tan \phi = 1.73\left\{\frac{5.2 - (-1.7)}{5.2 + (-1.7)}\right\} = 3.41$$

$$\phi = 73°39'$$

and   Power factor = $\cos \phi$ = **0.281**

From the data it is impossible to state whether the power factor is lagging or leading.

(c) From equation [34.6]

$$3500 = 1.73 \times I_L \times 400 \times 0.281$$

$$I_L = \textbf{18.0 A}$$

34.10   Two-phase systems

Two-phase systems were never as common as d.c. systems or three-phase systems. In fact, their applications were mainly to be found where d.c. distribution systems were being uprated without replacing the underground cables, i.e. improvements being done on the cheap. Generally their limitations soon lead to further upgrading to three-phase systems so that almost all distribution systems are now three-phase.

The waveforms of a two-phase system are shown in Fig. 34.28. In it, we see that the two phases are separated by 90°. However, two-phase systems are used in some control systems. By varying one of the phases, it is possible to control the torque of a motor and hence, over a limited range, its speed. The advantages of such an arrangement have been superseded, but instances of two-phase systems are by no means obsolete.

Fig. 34.28 Waveforms for a two-phase system

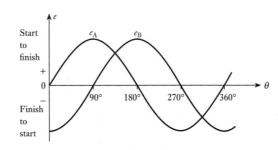

Summary of important formulae

For a star-connected system,

$$V_L = 1.73 V_P \qquad [34.1]$$

$$I_L = I_P \qquad [34.2]$$

For a delta-connected system,

$$I_L = 1.73 I_P \qquad [34.3]$$

$$V_L = V_P \qquad [34.4]$$

For star- or delta-connected system with a balanced load,

$$P = 3 V_P I_P \cos \phi \qquad [34.5]$$

$$= 1.73 V_L I_L \cos \phi \qquad [34.6]$$

If P_1 and P_2 are the indications obtained from the two wattmeters applied to a three-wire, three-phase system,

$$P = P_1 + P_2 \quad \text{under all conditions}$$

For a balanced load operating with sinusoidal waveforms

$$\cos \phi = \frac{1}{\sqrt{\left\{1 + 3\left(\dfrac{P_2 - P_1}{P_2 + P_1}\right)^2\right\}}} \qquad [34.10]$$

and

$$\tan \phi = 1.73 \left(\frac{P_2 - P_1}{P_2 + P_1}\right) \qquad [34.11]$$

Terms and concepts	**Multiphase systems** are best noted for their general ability to transmit high powers efficiently and also to provide powerful motor drives.

Multiphase systems are best noted for their general ability to transmit high powers efficiently and also to provide powerful motor drives.

Most multiphase systems operate with three phases, although others operate with two, six and even twelve phases.

Three-phase systems often identify the phases by the colours red, yellow and blue, although higher power systems use the numbers 1, 2 and 3.

Three phases can be connected either in **star** or in **delta**. Star connection is sometimes called a wye connection while the delta connection is sometimes called a mesh connection.

The voltages across, and the currents in, the components of the load or source are termed the **phase values**. The voltages between the supply conductors and the currents in these supply conductors are termed the **line values**.

In the star connection, the phase and line currents are identical.

In the delta connection, the phase and line voltages are identical.

In both star- and delta-connected systems, the line voltages are mutually displaced by 120°.

The sum of the currents in the supply conductors in a three-wire system is always zero.

The active power can be given by measuring the active power in one phase and multiplying by 3 provided the load is balanced. However, two wattmeters can be used to measure the total active power whether the load is balanced or not.

Two-phase systems have phase voltages displaced by 90°.

Exercises 34

1. Deduce the relationship between the phase and the line voltages of a three-phase star-connected generator.

   If the phase voltage of a three-phase star-connected generator is 200 V, what will be the line voltages: (a) when the phases are correctly connected; (b) when the connections to one of the phases are reversed?

2. Show with the aid of a phasor diagram that for both star- and delta-connected balanced loads, the total active power is given by $\sqrt{3}VI\cos\phi$, where V and I are the line values of voltage and current respectively and ϕ is the angle between phase values of voltage and current.

   A balanced three-phase load consists of three coils, each of resistance 4 Ω and inductance 0.02 H. Determine the total active power when the coils are (a) star-connected, (b) delta-connected to a 400 V, three-phase, 50 Hz supply.

3. Derive, for both star- and delta-connected systems, an expression for the total power input for a balanced three-phase load in terms of line voltage, line current and power factor.

   The star-connected secondary of a transformer supplies a delta-connected motor taking a power of 90 kW at a lagging power factor of 0.9. If the voltage between lines is 600 V, calculate the current in the transformer winding and in the motor winding. Draw circuit and phasor diagrams, properly labelled, showing all voltages and currents in the transformer secondary and the motor.

4. A three-phase delta-connected load, each phase of which has an inductive reactance of 40 Ω and a resistance of 25 Ω, is fed from the secondary of a three-phase star-connected transformer which has a phase voltage of 230 V. Draw the circuit diagram of the system and calculate: (a) the current in each phase of the load; (b) the p.d. across each phase of the load; (c) the current in the transformer secondary windings; (d) the total active power taken from the supply and its power factor.

5. Three similar coils, connected in star, take a total power of 1.5 kW, at a power factor of 0.2, from a three-phase, 400 V, 50 Hz supply. Calculate: (a) the

resistance and inductance of each coil; (b) the line currents if one of the coils is short-circuited.

6. (a) Three 20 μF capacitors are star-connected across a 400 V, 50 Hz, three-phase, three-wire supply. Calculate the current in each line. (b) If one of the capacitors is short-circuited, calculate the line currents. (c) If one of the capacitors is open-circuited, calculate: (i) the line currents; (ii) the p.d. across each of the other two capacitors.

7. (a) Explain the advantages of three-phase supply for distribution purposes.

  (b) Assuming the relationship between the line and phase values of currents and voltages, show that the active power input to a three-phase balanced load is $\sqrt{3}VI \cos \phi$, where V and I are line quantities.

  (c) Three similar inductors, each of resistance 10 Ω and inductance 0.019 H, are delta-connected to a three-phase, 400 V, 50 Hz sinusoidal supply. Calculate: (i) the value of the line current; (ii) the power factor; (iii) the active power input to the circuit.

8. A three-phase, 400 V, star-connected motor has an output of 50 kW, with an efficiency of 90 per cent and a power factor of 0.85. Calculate the line current. Sketch a phasor diagram showing the voltages and currents. If the motor windings were connected in mesh, what would be the correct voltage of a three-phase supply suitable for the motor?

9. Derive the numerical relationship between the line and phase currents for a balanced three-phase delta-connected load.

  Three coils are connected in delta to a three-phase, three-wire, 400 V, 50 Hz supply and take a line current of 5 A 0.8 power factor lagging. Calculate the resistance and inductance of the coils. If the coils are star-connected to the same supply, calculate the line current and the total power. Calculate the line currents if one coil becomes open-circuited when the coils are connected in star.

10. The load connected to a three-phase supply comprises three similar coils connected in star. The line currents are 25 A and the apparent and active power inputs are 20 kVA and 11 kW respectively. Find the line and phase voltages, reactive power input and the resistance and reactance of each coil. If the coils are now connected in delta to the same three-phase supply, calculate the line currents and the active power taken.

11. Non-reactive loads of 10, 6 and 4 kW are connected between the neutral and the red, yellow and blue phases respectively of a three-phase, four-wire system. The line voltage is 400 V. Find the current in each line conductor and in the neutral.

12. Explain the advantage of connecting the low-voltage winding of distribution transformers in star.

  A factory has the following load with power factor of 0.9 lagging in each phase. Red phase 40 A, yellow phase 50 A and blue phase 60 A. If the supply is 400 V, three-phase, four-wire, calculate the current in the neutral and the total active power. Draw a phasor diagram for phase and line quantities. Assume that, relative to the current in the red phase, the current in the yellow phase lags by 120° and that in the blue phase leads by 120°.

13. A three-phase, 400 V system has the following load connected in delta: between the red and yellow lines, a non-reactive resistor of 100 Ω; between the yellow and blue lines, a coil having a reactance of 60 Ω and negligible resistance; between the blue and red lines, a loss-free capacitor having a reactance of 130 Ω. Calculate: (a) the phase currents; (b) the line currents. Assume the phase sequence to be R–Y, Y–B and B–R. Also, draw the complete phasor diagram.

14. The phase currents in a delta-connected three-phase load are as follows: between the red and yellow lines, 30 A at p.f. 0.707 leading; between the yellow and blue lines, 20 A at unity p.f.; between the blue and red lines, 25 A at p.f. 0.866 lagging. Calculate the line currents and draw the complete phasor diagram.

15. (a) If, in a laboratory test, you were required to measure the total power taken by a three-phase balanced load, show how to do this, using two wattmeters. Explain the principle of the method. Draw the phasor diagram for the balanced-load case with a lagging power factor and use this to explain why the two wattmeter readings differ.

  (b) The load taken by a three-phase induction motor is measured by the two-wattmeter method, and the readings are 860 W and 240 W. What is the active power taken by the motor and at what power factor is it working?

16. With the aid of a circuit diagram, show that two wattmeters can be connected to read the total power in a three-phase, three-wire system.

  Two wattmeters connected to read the total power in a three-phase system supplying a balanced load read 10.5 kW and −2.5 kW respectively. Calculate the total active power.

  Drawing suitable phasor diagrams, explain the significance of: (a) equal wattmeter readings; (b) a zero reading on one wattmeter.

17. Two wattmeters are used to measure power in a three-phase, three-wire network. Show by means of connection and complexor (phasor) diagrams that the sum of the wattmeter readings will measure the total active power.

Two such wattmeters read 120 W and 50 W when connected to measure the active power taken by a balanced three-phase load. Find the power factor of the load. If one wattmeter tends to read in the reverse direction, explain what changes may have occurred in the circuit.

18. Each branch of a three-phase star-connected load consists of a coil of resistance 4.2 Ω and reactance 5.6 Ω. The load is supplied at a line voltage of 400 V, 50 Hz. The total active power supplied to the load is measured by the two-wattmeter method. Draw a circuit diagram of the wattmeter connections and calculate their separate readings. Derive any formula used in your calculations.

19. Three non-reactive loads are connected in delta across a three-phase, three-wire, 400 V supply in the following way: (i) 10 kW across R and Y lines; (ii) 6 kW across Y and B lines; (iii) 4 kW across B and R lines. Draw a phasor diagram showing the three line voltages and the load currents and determine: (a) the current in the B line and its phase relationship to the line voltage V_{BR}; (b) the reading of a wattmeter whose current coils are connected in the B line and whose voltage circuit is connected across the B and R lines. The phase rotation is R–Y–B. Where would a second wattmeter be connected for the two-wattmeter method and what would be its reading?

20. Two wattmeters connected to measure the input to a balanced three-phase circuit indicate 2500 W and 500 W respectively. Find the power factor of the circuit: (a) when both readings are positive; (b) when the latter reading is obtained after reversing the connections to the current-coil of one instrument. Draw the phasor and connection diagrams.

21. With the aid of a phasor diagram show that the active power and power factor of a balanced three-phase load can be measured by two wattmeters.

For a certain load, one wattmeter indicated 20 kW and the other 5 kW after the voltage circuit of this wattmeter had been reversed. Calculate the active power and the power factor of the load.

22. The current coil of a wattmeter is connected in the red line of a three-phase system. The voltage circuit can be connected between the red line and either the yellow line or the blue line by means of a two-way switch.

Assuming the load to be balanced, show with the aid of a phasor diagram that the sum of the wattmeter indications obtained with the voltage circuit connected to the yellow and the blue lines respectively gives the total active power.

23. A single wattmeter is used to measure the total active power taken by a 400 V, three-phase induction motor. When the output power of the motor is 15 kW, the efficiency is 88 per cent and the power factor is 0.84 lagging. The current coil of the wattmeter is connected in the yellow line. With the aid of a phasor diagram, calculate the wattmeter indication when the voltage circuit is connected between the yellow line and (a) the red line, (b) the blue line. Show that the sum of the two wattmeter indications gives the total active power taken by the motor. Assume the phase sequence to be R–Y–B.

24. A wattmeter has its current coil connected in the yellow line, and its voltage circuit is connected between the red and blue lines. The line voltage is 400 V and the balanced load takes a line current of 30 A at a power factor of 0.7 lagging. Draw circuit and phasor diagrams and derive an expression for the reading on the wattmeter in terms of the line voltage and current and of the phase difference between the phase voltage and current. Calculate the value of the wattmeter indication.

25. Discuss the importance of power-factor correction in a.c. systems.

A 400 V, 50 Hz, three-phase distribution system supplies a 20 kVA, three-phase induction motor load at a power factor of 0.8 lagging, and a star-connected set of impedances, each having a resistance of 10 Ω and an inductive reactance of 8 Ω. Calculate the capacitance of delta-connected capacitors required to improve the overall power factor to 0.95 lagging.

26. State the advantages to be gained by raising the power factor of industrial loads.

A 400 V, 50 Hz, three-phase motor takes a line current of 15.0 A when operating at a lagging power factor of 0.65. When a capacitor bank is connected across the motor terminals, the line current is reduced to 11.5 A. Calculate the rating (in kVA) and the capacitance per phase of the capacitor bank for: (a) star connection; (b) delta connection. Find also the new overall power factor.

Chapter thirty-five

Transformers

Objectives

When you have studied this chapter, you should

- have an understanding of transformer core material
- be familiar with the principles of transformer action
- understand the e.m.f. equation of the transformer
- have an understanding of no-load losses
- understand the concept of leakage flux
- understand the transformer equivalent circuit
- be familiar with the transformer phasor diagram
- be capable of calculating voltage regulation
- be capable of calculating efficiency
- understand the significance of open-circuit and short-circuit testing

Contents

One of the most important and ubiquitous electrical machines is the transformer. It receives power at one voltage and delivers it at another. This conversion aids the efficient long-distance transmission of electrical power from generating stations. Since power lines incur significant I^2R losses, it is important to minimize these losses by the use of high voltages. The same power can be delivered by high-voltage circuits at a fraction of the current required for low-voltage circuits.

In this chapter, the design of the magnetic circuit, the core of the transformer, will first be considered. Thereafter, the principle of transformer action is introduced and developed by consideration of the transformer's phasor diagram. The significance of the no-load behaviour of the transformer is explained and of the magnetizing current which exists under all operating conditions. The concept of leakage flux is examined.

The transformer equivalent circuit is a powerful analytical tool which facilitates an understanding of the behaviour of the transformer on load and of the losses which occur in the transformer windings. The losses cause the ratio of input to output voltage to vary – they usually cause the output voltage to fall but it can, under certain circumstances, rise. The important concepts of voltage regulation and efficiency are developed.

Finally, a variety of specialist transformers, the current transformer, the auto-transformer and the air-cored transformer, are introduced.

35.1 Introduction

One of the main advantages of a.c. transmission and distribution is the ease with which an alternating voltage can be increased or reduced. For instance the general practice is to generate at voltages about 22 kV, then step up by means of transformers to higher voltages for the transmission lines. At suitable points, other transformers are introduced to step the voltage down to values suitable for motors, lamps, heaters, etc. A medium-sized transformer has a full-load efficiency of about 97–98 per cent, so that the loss at each point of transformation is small (although 2 per cent of 100 MW is not insignificant!). Since there are no moving parts in a transformer, the amount of supervision is practically negligible.

Although transformers are generally associated with power system applications, they also occur in many low-power applications including electronic circuits. However, it is best to first consider the common power-system transformer.

The common form of transformer involves a ferromagnetic core in order to ensure high values of magnetic flux linkage. This is also true of the rotating machines which we shall meet in the following chapters. There are factors about the ferromagnetic core which affect the construction of transformers and rotating machines; these factors are responsible for part of the loss associated with power transfer and require a brief explanation before considering the principle of action of a transformer.

35.2 Core factors

We have observed in Chapter 8 that the flux linking coils can be greatly improved by the introduction of a ferromagnetic core. Although this improved action has been explained by reference to the B/H characteristic, the explanation did not continue to consider the effect of varying the magnetizing force. When the core is energized from an a.c. source, the magnetizing force rises and falls in accordance with the magnetizing current which is basically sinusoidal. This variation does not cause B and H to vary according to the magnetic characteristic, but rather as shown in Fig. 35.1. This loop is called the *hysteresis loop*.

Fig. 35.1 Hysteresis loop

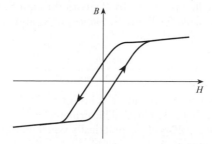

It will be shown in section 44.9 that, when drawn to scale, the larger the loop the greater the energy required to create the magnetic field – and this energy has to be supplied during each cycle of magnetization. This requirement of supplying energy to magnetize the core is known as the *hysteresis loss*. We need not explore it further at this stage, and it is sufficient to think of it as the cost we pay to get better magnetic linkage.

There is an unfortunate loss which is associated with the hysteresis loss. The varying flux in the core induces e.m.f.s and hence currents in the core material. These give rise to I^2R losses. These losses are called *eddy-current losses* and will also be further considered in section 44.10.

The sum of the hysteresis loss and the eddy-current losses is known as the *core loss*. As with the I^2R losses in conductors, they are the imperfections which we have to accept, but it is better to consider them in detail after we understand the principle of machine action.

35.3 Principle of action of a transformer

Figure 35.2 shows the general arrangement of a transformer. A steel core C consists of laminated sheets, about 0.35–0.7 mm thick, insulated from one another. The purpose of laminating the core is to reduce the eddy-current loss. The vertical portions of the core are referred to as *limbs* and the top and bottom portions are the *yokes*. Coils P and S are wound on the limbs. Coil P is connected to the supply and is therefore termed the *primary*; coil S is connected to the load and is termed the *secondary*.

Fig. 35.2 A transformer

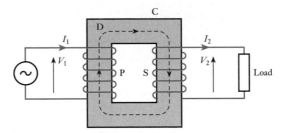

An alternating voltage applied to P circulates an alternating current through P and this current produces an alternating flux in the steel core, the mean path of this flux being represented by the dotted line D. If the whole of the flux produced by P passes through S, the e.m.f. induced in each turn is the same for P and S. Hence, if N_1 and N_2 are the number of turns on P and S respectively,

$$\frac{\text{Total e.m.f. induced in S}}{\text{Total e.m.f. induced in P}} = \frac{N_2 \times \text{e.m.f. per turn}}{N_1 \times \text{e.m.f. per turn}} = \frac{N_2}{N_1}$$

When the secondary is on open circuit, its terminal voltage is the same as the induced e.m.f. The primary current is then very small, so that the applied voltage V_1 is practically equal and opposite to the e.m.f. induced in P. Hence:

$$\frac{V_2}{V_1} \simeq \frac{N_2}{N_1} \tag{35.1}$$

Since the full-load efficiency of a transformer is nearly 100 per cent,

$$I_1 V_1 \times \text{primary power factor} \simeq I_2 V_2 \times \text{secondary power factor}$$

But the primary and secondary power factors at full load are nearly equal,

$$\therefore \qquad \frac{I_1}{I_2} \simeq \frac{V_2}{V_1} \qquad\qquad [35.2]$$

An alternative and more illuminating method of deriving the relationship between the primary and secondary currents is based upon a comparison of the primary and secondary ampere-turns. When the secondary is on open circuit, the primary current is such that the primary ampere-turns are just sufficient to produce the flux necessary to induce an e.m.f. that is practically equal and opposite to the applied voltage. This magnetizing current is usually about 3–5 per cent of the full-load primary current.

When a load is connected across the secondary terminals, the secondary current – by Lenz's law – produces a demagnetizing effect. Consequently the flux and the e.m.f. induced in the primary are reduced slightly. But this small change can increase the difference between the applied voltage and the e.m.f. induced in the primary from, say, 0.05 per cent to, say, 1 per cent, in which case the new primary current would be 20 times the no-load current. The demagnetizing ampere-turns of the secondary are thus nearly neutralized by the increase in the primary ampere-turns; and since the primary ampere-turns on no load are very small compared with the full-load ampere-turns, full-load primary ampere-turns are approximately equal to full-load secondary ampere-turns, i.e.

$$I_1 N_1 \simeq I_2 N_2$$

so that

$$\frac{I_1}{I_2} \simeq \frac{N_2}{N_1} \simeq \frac{V_2}{V_1} \qquad\qquad [35.3]$$

It will be seen that the magnetic flux forms the connecting link between the primary and secondary circuits and that any variation of the secondary current is accompanied by a small variation of the flux and therefore of the e.m.f. induced in the primary, thereby enabling the primary current to vary approximately proportionally to the secondary current.

This balance of primary and secondary ampere-turns is an important relationship wherever transformer action occurs.

35.4 EMF equation of a transformer

Suppose the maximum value of the flux to be Φ_m webers and the frequency to be f hertz. From Fig. 35.3 it is seen that the flux has to change from $+\Phi_m$ to $-\Phi_m$ in half a cycle, namely in $\frac{1}{2f}$ seconds.

$$\therefore \qquad \text{Average rate of change of flux} = 2\Phi_m \div \frac{1}{2f}$$

$$= 4f\,\Phi_m \text{ webers per second}$$

and average e.m.f. induced per turn is

$$4f\,\Phi_m \text{ volts}$$

Fig. 35.3 Waveform of flux variation

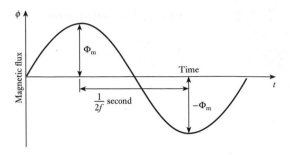

But for a sinusoidal wave the r.m.s. or effective value is 1.11 times the average value,

∴   RMS value of e.m.f. induced per turn $= 1.11 \times 4f\,\Phi_m$

Hence, r.m.s. value of e.m.f. induced in primary is

$$E_1 = 4.44N_1 f\,\Phi_m \quad \text{volts} \tag{35.4}$$

and r.m.s. value of e.m.f. induced in secondary is

$$E_2 = 4.44N_2 f\,\Phi_m \quad \text{volts} \tag{35.5}$$

An alternative method of deriving these formulae is as follows:

If $\phi =$ instantaneous value of flux in webers

$$= \Phi_m \sin 2\pi ft$$

therefore instantaneous value of induced e.m.f. per turn is $d\phi/dt$ volts

$$d\phi/dt = 2\pi f\,\Phi_m \times \cos 2\pi ft \text{ volts}$$

$$d\phi/dt = 2\pi f\,\Phi_m \times \sin(2\pi ft + \pi/2) \tag{35.6}$$

therefore maximum value of induced e.m.f. per turn $= 2\pi f\,\Phi_m$ volts and r.m.s. value of induced e.m.f. per turn is

$$0.707 \times 2\pi f\,\Phi_m = 4.44 f\,\Phi_m \text{ volts}$$

Hence r.m.s. value of primary e.m.f. is

$$E_1 = 4.44N_1 f\,\Phi_m \text{ volts}$$

and r.m.s. value of secondary e.m.f. is

$$E_2 = 4.44N_2 f\,\Phi_m \text{ volts}$$

Example 35.1

A 250 kVA, 11 000 V/400 V, 50 Hz single-phase transformer has 80 turns on the secondary. Calculate:

(a) the approximate values of the primary and secondary currents;
(b) the approximate number of primary turns;
(c) the maximum value of the flux.

(a) Full–load primary current

$$\simeq \frac{250 \times 1000}{11\,000} = \textbf{22.7 A}$$

and full-load secondary current

$$= \frac{250 \times 1000}{400} = \textbf{625 A}$$

(b) No. of primary turns

$$\simeq \frac{80 \times 11\,000}{400} = \textbf{2200}$$

(c) From expression [35.5]

$$400 = 4.44 \times 80 \times 50 \times \Phi_{\mathrm{m}}$$

$$\Phi_{\mathrm{m}} = \textbf{22.5 mWb}$$

<table>
<tr><td>35.5</td><td>Phasor diagram for a transformer on no load</td></tr>
</table>

It is most convenient to commence the phasor diagram with the phasor representing the quantity that is common to the two windings, namely the flux Φ. This phasor can be made any convenient length and may be regarded merely as a reference phasor, relative to which other phasors have to be drawn.

In the preceding section, expression [35.6] shows that the e.m.f. induced by a sinusoidal flux leads the flux by a quarter of a cycle. Consequently the e.m.f. E_1 induced in the primary winding is represented by a phasor drawn $90°$ ahead of Φ, as in Fig. 35.4. Note that expression [35.6] is comparable to expression [10.5]; in the latter, we can replace Li by $N\phi$ and hence obtain [35.6].

The e.m.f. E_2 also leads the flux by $90°$, but the effect which this produces at the terminals of the transformer depends on the manner in which the secondary winding is constructed and, more importantly, the manner in which the ends of the winding are connected to the transformer terminals. In practice, the normal procedure ensures that V_2 is in phase with V_1 and only a very few transformers depart from this arrangement.

However, if V_2 and V_1 were drawn in phase with one another on Fig. 35.4, the diagram would become cluttered and therefore, for convenience, it is usual to show E_1 and E_2 in phase opposition with one another, thus ensuring that V_2 appears in the opposite quadrant of the phasor diagram from V_1. This gives the appearance that the voltages are in antiphase and it should be remembered that the manner of drawing is for convenience only and that the voltages are in fact in phase.

The values of E_2 and E_1 are proportional to the number of turns on the secondary and primary windings, since practically the whole of the flux set up by the primary winding is linked with the secondary winding when the latter is on open circuit. Another matter of convenience in drawing these transformer phasor diagrams is that it has been assumed that N_2 and N_1 are equal so that $E_2 = E_1$, as shown in Fig. 35.4.

Since the difference between the value of the applied voltage V_1 and that of the induced e.m.f. E_1 is only about 0.05 per cent when the transformer is

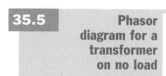

Fig. 35.4 Phasor diagram for transformer on no load

on no load, the phasor representing V_1 can be drawn equal to that representing E_1. Since we have accepted that back e.m.f.s are treated as volt drops, V_1 and E_1 are drawn in phase with one another.

The no-load current, I_0, taken by the primary consists of two components:

1. A reactive or magnetizing component,* I_{0m}, producing the flux and therefore in phase with the latter.
2. An active or power component, I_{0l}, supplying the hysteresis and eddy-current losses in the core and the negligible I^2R loss in the primary winding. Component I_{0l} is in phase with the applied voltage, i.e. $I_{0l}V_1$ = core loss. This component is usually very small compared with I_{0m}, so that the no-load power factor is very low.

From Fig. 35.4 it will be seen that no-load current is

$$I_0 = \sqrt{(I_{0l}^2 + I_{0m}^2)} \qquad [35.7]$$

and power factor on no load is

$$\cos \phi_0 = I_{0l}/I_0 \qquad [35.8]$$

Example 35.2   A single-phase transformer has 480 turns on the primary and 90 turns on the secondary. The mean length of the flux path in the core is 1.8 m and the joints are equivalent to an airgap of 0.1 mm. The value of the magnetic field strength for 1.1 T in the core is 400 A/m, the corresponding core loss is 1.7 W/kg at 50 Hz and the density of the core is 7800 kg/m^3.

If the maximum value of the flux density is to be 1.1 T when a p.d. of 2200 V at 50 Hz is applied to the primary, calculate:

(a) the cross-sectional area of the core;
(b) the secondary voltage on no load;
(c) the primary current and power factor on no load.

(a) From [35.4],

$$2200 = 4.44 \times 480 \times 50 \times \Phi_m$$

$$\Phi_m = 0.0206 \text{ Wb}$$

and cross-sectional area of core is

$$\frac{0.0206}{1.1} = 0.0187 \text{ m}^2$$

This is the net area of the core; the gross area of the core is about 10 per cent greater than this value to allow for the insulation between the laminations.

(b) Secondary voltage on no load is

$$2200 \times \frac{90}{480} = 413 \text{ V}$$

---

* The waveform of this component is discussed in section 35.23.

(c) Total magnetomotive force for the core is

$$400 \times 1.8 = 720 \text{ A}$$

and magnetomotive force for the equivalent airgap is

$$\frac{1.1}{4\pi \times 10^{-7}} \times 0.0001 = 87.5 \text{ A}$$

Therefore total m.m.f. to produce the maximum flux density is

$$720 + 87.5 = 807.5 \text{ A}$$

therefore maximum value of magnetizing current is

$$\frac{807.5}{480} = 1.682 \text{ A}$$

Assuming the current to be sinusoidal, r.m.s. value of magnetizing current is

$$I_{0m} = 0.707 \times 1.682 = 1.19 \text{ A}$$

$$\text{Volume of core} = 1.8 \times 0.0187$$

$$= 0.0337 \text{ m}^3$$

$$\therefore \qquad \text{Mass of core} = 0.0337 \times 7800$$

$$= 263 \text{ kg}$$

and $\qquad$ Core loss $= 263 \times 1.7 = 447 \text{ W}$

Therefore core-loss component of current is

$$I_{0l} = \frac{447}{2200} = 0.203 \text{ A}$$

From equation [35.7],

$$\text{No-load current} = I_0$$

$$= \sqrt{\{(1.19)^2 + (0.203)^2\}}$$

$$= \mathbf{1.21 \text{ A}}$$

and from equation [35.8], power factor on no load is

$$\frac{0.203}{1.21} = \mathbf{0.168 \text{ lagging}}$$

35.6 Phasor diagram for an ideal loaded transformer

With this assumption, it follows that the secondary terminal voltage V_2 is the same as the e.m.f. E_2 induced in the secondary, and the primary applied voltage V_1 is equal to the e.m.f. E_1 induced in the primary winding. Also, if we again assume equal number of turns on the primary and secondary windings, then $E_1 = E_2$.

Let us consider the general case of a load having a lagging power factor $\cos \phi_2$; hence the phasor representing the secondary current I_2 lags V_2 by an angle ϕ_2, as shown in Fig. 35.5. Phasor $I_{2'}$ represents the component of the primary current to neutralize the demagnetizing effect of the secondary current and is drawn equal and opposite to I_2. Here $I_{2'}$ is described as

Fig. 35.5 Phasor diagram for
a loaded transformer having
negligible voltage drop in
windings

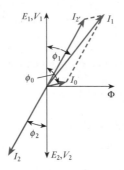

'I_2 referred'; I_0 is the no-load current of the transformer, already discussed
in section 35.5. The phasor sum of $I_{2'}$ and I_0 gives the total current I_1 taken
from the supply, and the power factor on the primary side is $\cos \phi_1$, where ϕ_1
is the phase difference between V_1 and I_1.

In Fig. 35.5 the phasor representing I_0 has, for clarity, been shown far
larger relative to the other current phasors than it is in an actual transformer.

Example 35.3

A single-phase transformer has 1000 turns on the primary and 200
turns on the secondary. The no-load current is 3 A at a power factor
0.2 lagging when the secondary current is 280 A at a power factor of
0.8 lagging. Calculate the primary current and power factor. Assume
the voltage drop in the windings to be negligible.

If $I_{2'}$ represents the component of the primary current to neutralize the
demagnetizing effect of the secondary current, the ampere-turns due to $I_{2'}$
must be equal and opposite to those due to I_2, i.e.

$$I_{2'} \times 1000 = 280 \times 200$$

$\therefore$ $I_{2'} = 56$ A

$\cos \phi_2 = 0.8$ $\therefore \sin \phi_2 = 0.6$

and $\cos \phi_0 = 0.2$ $\therefore \sin \phi_0 = 0.98$

From Fig. 35.5 it will be seen that

$$I_1 \cos \phi_1 = I_{2'} \cos \phi_2 + I_0 \cos \phi_0$$

$$= (56 \times 0.8) + (3 \times 0.2) = 45.4 \text{ A}$$

and $I_1 \sin \phi_1 = I_{2'} \sin \phi_2 + I_0 \sin \phi_0$

$$= (56 \times 0.6) + (3 \times 0.98) = 36.54 \text{ A}$$

Hence, $I_1^2 = (45.4)^2 + (36.45)^2 = 3398$

so that $I_1 = \textbf{58.3 A}$

Also, $\tan \phi_1 = \dfrac{36.54}{45.4} = 0.805$

so that $\phi_1 = 38°50'$

Hence primary power factor is

$$\cos \phi_1 = \cos 38°50' \equiv \textbf{0.78 lagging}$$

35.7 | **Useful and leakage fluxes in a transformer**

When the secondary winding of a transformer is on open circuit, the current taken by the primary winding is responsible for setting up the magnetic flux and providing a very small power component to supply the loss in the core. To simplify matters in the present discussion, let us assume:

1. The core loss and the I^2R loss in the primary winding to be negligible.
2. The permeability of the core to remain constant, so that the magnetizing current is proportional to the flux.
3. The primary and secondary windings to have the same number of turns, i.e. $N_1 = N_2$.

Figure 35.6 shows all the flux set up by the primary winding passing through the secondary winding. There is a very small amount of flux returning through the air space around the primary winding, but since the relative

Fig. 35.6 Transformer on no load

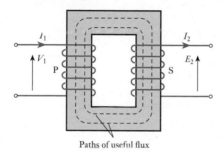

Paths of useful flux

Fig. 35.7 Waveforms of voltages, current and flux of a transformer on no load

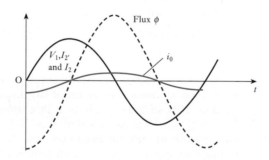

permeability of transformer core is of the order of 1000 or more, the reluctance of the air path is 1000 times that of the parallel path through the limb carrying the secondary winding. Consequently the flux passing through the air space is negligible compared with that through the secondary. It follows that the e.m.f.s induced in the primary and secondary windings are equal and that the primary applied voltage, V_1, is equal to the e.m.f., E_1, induced in the primary, as shown in Figs 35.7 and 35.8.

Next, let us assume a load having a power factor such that the secondary current is in phase with E_2. As already explained in section 35.5 the primary current, I_1, must now have two components:

1. I_{0m} to maintain the useful flux, the maximum value of which remains constant within about 2 per cent between no load and full load.
2. A component, $I_{2'}$, to neutralize the demagnetizing effect of the secondary current, as shown in Figs 35.9(a) and 35.10.

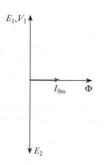

Fig. 35.8 Phasor diagram for Fig. 35.7

Fig. 35.9 Waveforms of
induced e.m.f.s, currents and
fluxes in a transformer on load

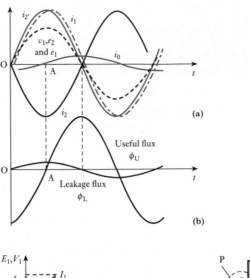

(a)

(b)

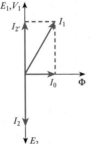

Fig. 35.10 Phasor diagram for
Fig. 35.9

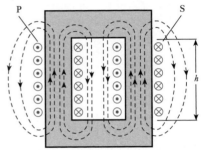

Fig. 35.11 Paths of leakage flux

At instant A in Fig. 35.9(a), the magnetizing current is *zero*, but I_2 and $I_{2'}$ are at their *maximum* values; and if the direction of the current in primary winding P is such as to produce flux upwards in the left-hand limb of Fig. 35.11, the secondary current must be in such a direction as to produce flux upwards in the right-hand limb, and the flux of each limb has to return through air. Since the flux of each limb is linked only with the winding by which it is produced, it is referred to as *leakage* flux and is responsible for inducing an e.m.f. of self-inductance in the winding with which it is linked. The reluctance of the paths of the leakage flux, Φ_L, is almost entirely due to the long air paths and is therefore practically constant. Consequently the value of the leakage flux is proportional to the load current, whereas the value of the useful flux remains almost independent of the load. The reluctance of the paths of the leakage flux is very high, so that the value of this flux is relatively small even on full load when the values of $I_{2'}$ and I_2 are about 20–30 times the magnetizing current I_{0m}.

From the above discussion it follows that the actual flux in a transformer can be regarded as being due to the two components shown in Fig. 35.9(b), namely:

1. The useful flux, Φ_U, linked with both windings and remaining practically constant in value at all loads.
2. The leakage flux, Φ_L, half of which is linked with the primary winding and half with the secondary, and its value is proportional to the load.

The case of the secondary current in phase with the secondary induced e.m.f. has been considered because it is easier to see that the useful and the leakage fluxes can be considered independently of each other for this condition than it is for loads of other power factors.

35.8 Leakage flux responsible for the inductive reactance of a transformer

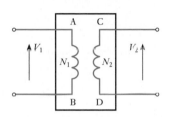

Fig. 35.12 Ideal transformer enclosed in a box

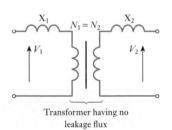

Transformer having no leakage flux

Fig. 35.13 Transformer with leakage reactances

When a transformer is on no load there is no secondary current, and the secondary winding has not the slightest effect upon the primary current. The primary winding is then behaving as an inductor, having a very high inductance and a very low resistance. When the transformer is supplying a load, however, this conception of the inductance of the windings is of little use and it is rather difficult at first to understand why the useful flux is not responsible for any inductive drop in the transformer. So let us first of all assume an ideal transformer, namely a transformer the windings of which have negligible resistance and in which there is no core loss and no magnetic leakage. Also, for convenience, let us assume unity transformation ratio, i.e. $N_1 = N_2$.

If such a transformer were enclosed in a box and the ends of the windings brought to terminals A, B, C and D on the lid, as shown in Fig. 35.12 the p.d. between C and D would be equal to that between A and B; and, as far as the effect upon the output voltage is concerned, the transformer behaves as if A were connected to C and B to D. In other words, the useful flux is not responsible for any voltage drop in a transformer.

In the preceding section it was explained that the leakage flux is proportional to the primary and secondary currents and that its effect is to induce e.m.f.s of self-induction in the windings. Consequently the effect of leakage flux can be considered as equivalent to inductive reactors X_1 and X_2 connected in series with a transformer having no leakage flux, as shown in Fig. 35.13; these reactors being such that the flux-linkages produced by the primary current through X_1 are equal to those due to the leakage flux linked with the primary winding, and the flux-linkages produced by the secondary current through X_2 are equal to those due to the leakage flux linked with the secondary winding of the actual transformer. The straight line drawn between the primary and secondary windings in Fig. 35.13 is the symbol used to indicate that the transformer has a ferromagnetic core.

35.9 Methods of reducing leakage flux

The leakage flux can be practically eliminated by winding the primary and secondary, one over the other, uniformly around a laminated ferromagnetic ring of uniform cross-section. But such an arrangement is not commercially practicable except in very small sizes, owing to the cost of threading a large number of turns through the ring.

The principal methods used in practice are:

1. Making the transformer 'window' long and narrow.
2. Arranging the primary and secondary windings concentrically (see Fig. 35.14).
3. Sandwiching the primary and secondary windings (see Fig. 35.15).
4. Using shell-type construction (see Fig. 35.16).

Fig. 35.14 Concentric
windings.
HVW = high-voltage winding;
LVW = low-voltage winding

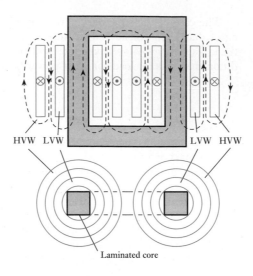

HVW LVW   LVW HVW

Laminated core

Fig. 35.15 Sandwiched
windings

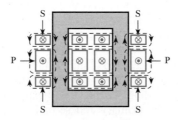

Fig. 35.16 Shell-type
construction

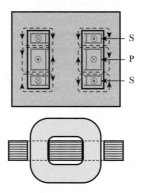

35.10   Equivalent circuit of a transformer

The behaviour of a transformer may be conveniently considered by assuming it to be equivalent to an ideal transformer, i.e. a transformer having no losses and no magnetic leakage and a ferromagnetic core of infinite permeability requiring no magnetizing current, and then allowing for the imperfections of the actual transformer by means of additional circuits or impedances inserted between the supply and the primary winding and between the secondary and the load. Thus, in Fig. 35.17 P and S represent

Fig. 35.17 Equivalent circuit of a transformer

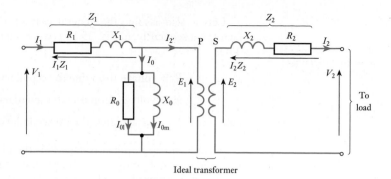

Ideal transformer

the primary and secondary windings of the ideal transformer, R_1 and R_2 are resistances equal to the resistances of the primary and secondary windings of the actual transformer. Similarly, inductive reactances X_1 and X_2 represent the reactances of the windings due to leakage flux in the actual transformer, as already explained in section 35.8.

The inductive reactor X_0 is such that it takes a reactive current equal to the magnetizing current I_{0m} of the actual transformer. The core losses due to hysteresis and eddy currents are allowed for by a resistor R_0 of such value that it takes a current I_{0l} equal to the core-loss component of the primary current, i.e. $I_{0l}^2 R_0$ is equal to the core loss of the actual transformer. The resultant of I_{0m} and I_{0l} is I_0, namely the current which the transformer takes on no load. The phasor diagram for the equivalent circuit on no load is exactly the same as that given in Fig. 35.4.

35.11 Phasor diagram for a transformer on load

For convenience let us assume an equal number of turns on the primary and secondary windings, so that $E_1 = E_2$. As shown in Fig. 35.18 E_1 leads the flux by 90°, and represents the voltage across the primary of the ideal transformer.

Fig. 35.18 Phasor diagram for a transformer on load

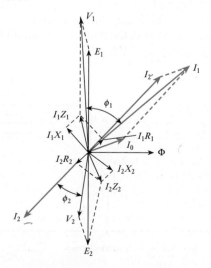

Let us also assume the general case of a load having a lagging power factor; consequently, in Fig. 35.18 I_2 has been drawn lagging E_2 by about 45°. Then

$$I_2 R_2 = \text{voltage drop due to secondary resistance}$$

$$I_2 X_2 = \text{voltage drop due to secondary leakage reactance}$$

and $I_2 Z_2 = \text{voltage drop due to secondary impedance}$

The secondary terminal voltage V_2 is the phasor difference of E_2 and $I_2 Z_2$; in other words, V_2 must be such that the phasor sum of V_2 and $I_2 Z_2$ is E_2, and the derivation of the phasor representing V_2 is evident from Fig. 35.18. The power factor of the load is $\cos \phi_2$, where ϕ_2 is the phase difference between V_2 and I_2, $I_{2'}$ represents the component of the primary current to neutralize the demagnetizing effect of the secondary current and is drawn equal and opposite to I_2. Here I_0 is the no-load current of the transformer (section 35.5). The phasor sum of $I_{2'}$ and I_0 gives the total current I_1 taken from the supply.

$$I_1 R_1 = \text{voltage drop due to primary resistance}$$

$$I_1 X_1 = \text{voltage drop due to primary leakage reactance}$$

$$I_1 Z_1 = \text{voltage drop due to primary impedance}$$

and $\mathbf{V_1} = \mathbf{E_1} + \mathbf{I_1 Z_1} = \text{supply voltage}$

If ϕ_1 is the phase difference between V_1 and I_1, then $\cos \phi_1$ is the power factor on the primary side of the transformer. In Fig. 35.18 the phasors representing the no-load current and the primary and secondary voltage drops are, for clearness, shown far larger relative to the other phasors than they are in an actual transformer.

35.12   **Approximate equivalent circuit of a transformer**

Since the no-load current of a transformer is only about 3–5 per cent of the full-load primary current, we can omit the parallel circuits R_0 and X_0 in Fig. 35.17 without introducing an appreciable error when we are considering the behaviour of the transformer on full load. Thus we have the simpler equivalent circuit of Fig. 35.19.

Fig. 35.19 Approximate equivalent circuit of a transformer

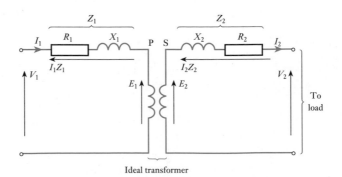

Ideal transformer

35.13 Simplification of the approximate equivalent circuit of a transformer

We can replace the resistance R_2 of the secondary of Fig. 35.19 by inserting additional resistance $R_{2'}$ in the primary circuit such that the power absorbed in $R_{2'}$ when carrying the primary current is equal to that in R_2 due to the secondary current, i.e.

$$I_1^2 R_{2'} = I_2^2 R_2$$

$$R_{2'} = R_2 \left(\frac{I_2}{I_1}\right)^2 \simeq R_2 \left(\frac{V_1}{V_2}\right)^2$$

Hence if R_e is a single resistance in the primary circuit equivalent to the primary and secondary resistances of the actual transformer then

$$R_e = R_1 + R_{2'} = R_1 + R_2 \left(\frac{V_1}{V_2}\right)^2 \qquad [35.9]$$

Similarly, since the inductance of a coil is proportional to the square of the number of turns, the secondary leakage reactance X_2 can be replaced by an equivalent reactance $X_{2'}$ in the primary circuit, such that

$$X_{2'} = X_2 \left(\frac{N_1}{N_2}\right)^2 \simeq X_2 \left(\frac{V_1}{V_2}\right)^2$$

If X_e is the single reactance in the primary circuit equivalent to X_1 and X_2 of the actual transformer

$$X_e = X_1 + X_{2'} = X_1 + X_2 \left(\frac{V_1}{V_2}\right)^2 \qquad [35.10]$$

If Z_e is the equivalent impedance of the primary and secondary windings referred to the primary circuit

$$Z_e = \sqrt{(R_e^2 + X_e^2)} \qquad [35.11]$$

If ϕ_e is the phase difference between I_1 and $I_1 Z_e$, then

$$R_e = Z_e \cos \phi_e \quad \text{and} \quad X_e = Z_e \sin \phi_e$$

The simplified equivalent circuit of the transformer is given in Fig. 35.20, and Fig. 35.21(a) is the corresponding phasor diagram.

Fig. 35.20 Simplified equivalent circuit of a transformer

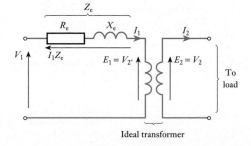

Ideal transformer

Fig. 35.21   Phasor diagram for
Fig. 35.20

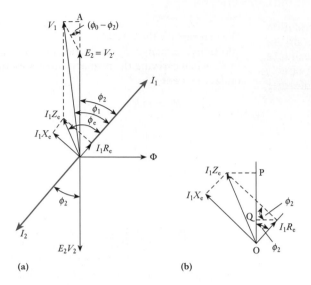

(a)   (b)

35.14	Voltage regulation of a transformer

The voltage regulation of a transformer is defined as the variation of the secondary voltage between no load and full load, expressed as either a per-unit
or a percentage of the *no-load* voltage, the primary voltage being assumed
constant, i.e.

$$\text{Voltage regulation} = \frac{\text{no-load voltage} - \text{full-load voltage}}{\text{no-load voltage}} \qquad [35.12]$$

If V_1 is primary applied voltage

$$\text{Secondary voltage on no load} = V_1 \times \frac{N_2}{N_1}$$

since the voltage drop in the primary winding due to the no-load current is
negligible.

If V_2 is secondary terminal voltage on full load,

$$\text{Voltage regulation} = \frac{V_1 \dfrac{N_2}{N_1} - V_2}{V_1 \dfrac{N_2}{N_1}}$$

$$= \frac{V_1 - V_2 \dfrac{N_1}{N_2}}{V_1} \text{ per unit}$$

$$= \frac{V_1 - V_2 \dfrac{N_1}{N_2}}{V_1} \times 100 \text{ per cent}$$

In the phasor diagram of Fig. 35.21, N_1 and N_2 were assumed equal, so that $V_{2'} = V_2$. In general, $V_{2'} = V_2(N_1/N_2)$,

$$\therefore \quad \text{Per-unit voltage regulation} = \frac{V_1 - V_{2'}}{V_1} \qquad [35.13]$$

In Fig. 35.21(a), let us draw a perpendicular from V_1 to meet the extension of $V_{2'}$ at A. Let the vertical extension of $V_{2'}$ be $V_{2'}A$ and the perpendicular line be V_1A; then

$$V_1^2 = (V_{2'} + V_{2'}A)^2 + (V_1A)^2$$
$$= \{V_{2'} + I_1Z_e \cos(\phi_e - \phi_2)\}^2 + \{I_1Z_e \sin(\phi_e - \phi_2)\}^2$$

In actual practice, $I_1Z_e \sin(\phi_e - \phi_2)$ is very small compared with $V_{2'}$, so that

$$V_1 \simeq V_{2'} + I_1Z_e \cos(\phi_e - \phi_2)$$

Hence

$$\text{Per-unit voltage regulation} = \frac{V_1 - V_{2'}}{V_1}$$

$$\text{Per-unit voltage regulation} = \frac{I_1Z_e \cos(\phi_e - \phi_2)}{V_1} \qquad [35.14]$$

Since

$$Z_e \cos(\phi_e - \phi_2) = Z_e(\cos \phi_e \cdot \cos \phi_2 + \sin \phi_e \cdot \sin \phi_2)$$
$$= R_e \cos \phi_2 + X_e \sin \phi_2$$

therefore

$$\text{Per-unit voltage regulation} = \frac{I_1(R_e \cos \phi_2 + X_e \sin \phi_2)}{V_1} \qquad [35.15]$$

This expression can also be derived by projecting I_1R_e and I_1Z_e on to OA, as shown enlarged in Fig. 35.21(b), from which it follows that

$$V_{2'}A \text{ in Fig. 35.21(a)} = OP \text{ in Fig. 35.21(b)}$$
$$= OQ + QP$$
$$= I_1R_e \cos \phi_2 + I_1X_e \sin \phi_2$$

therefore per-unit voltage regulation is

$$\frac{V_1 - V_{2'}}{V_1} \simeq \frac{V_{2'}A}{V_1} = \frac{OP}{V_1}$$
$$= \frac{I_1(R_e \cos \phi_2 + X_e \sin \phi_2)}{V_1}$$

The above expressions have been derived on the assumption that the power factor is lagging. Should the power factor be leading, the angle in expression [35.14] would be $(\phi_e + \phi_2)$ and the term in brackets in expression [35.15] would be $(R_e \cos \phi_2 - X_e \sin \phi_2)$.

Example 35.4

A 100 kVA transformer has 400 turns on the primary and 80 turns on the secondary. The primary and secondary resistances are 0.3 Ω and 0.01 Ω respectively, and the corresponding leakage reactances are 1.1 Ω and 0.035 Ω respectively. The supply voltage is 2200 V. Calculate:

(a) the equivalent impedance referred to the primary circuit;
(b) the voltage regulation and the secondary terminal voltage for full load having a power factor of (i) 0.8 lagging and (ii) 0.8 leading.

(a) From equation [35.9], equivalent resistance referred to primary is

$$R_e = 0.3 + 0.01\left(\frac{400}{80}\right)^2 = 0.55\ \Omega$$

From equation [35.10], equivalent leakage reactance referred to primary is

$$X_e = 1.1 + 0.035\left(\frac{400}{80}\right)^2 = 1.975\ \Omega$$

From equation [35.11], equivalent impedance referred to primary is

$$Z_e = \sqrt{\{(0.55)^2 + (1.975)^2\}} = 2.05\ \Omega$$

(b) (i) Since $\cos\phi_2 = 0.8$, therefore $\sin\phi_2 = 0.6$.

$$\text{Full-load primary current} \simeq \frac{100 \times 1000}{2200} = 45.45\ \text{A}$$

Substituting in equation [35.15], we have voltage regulation for power factor 0.8 lagging is

$$\frac{45.45(0.55 \times 0.8 + 1.975 \times 0.6)}{2200} = 0.0336\ \text{per unit}$$

$$= 3.36\ \text{per cent}$$

Secondary terminal voltage on no load is

$$2200 \times \frac{80}{400} = 440\ \text{V}$$

Therefore decrease of secondary terminal voltage between no load and full load is

$$440 \times 0.0336 = 14.8\ \text{V}$$

Therefore secondary terminal voltage on full load is

$$440 - 14.8 = 425\ \text{V}$$

(ii) Voltage regulation for power factor 0.8 leading is

$$\frac{45.45(0.55 \times 0.8 - 1.975 \times 0.6)}{2200} = -0.0154\ \text{per unit}$$

$$= -1.54\ \text{per cent}$$

Increase of secondary terminal voltage between no load and full load is

$$440 \times 0.0154 = 6.78 \text{ V}$$

Therefore secondary terminal voltage on full load is

$$440 + 6.78 = \mathbf{447 \text{ V}}$$

Example 35.5

Calculate the per-unit and the percentage resistance and leakage reactance drops of the transformer referred to in Example 35.4.

Per-unit resistance drop of a transformer

$$= \frac{\left(\begin{array}{c}\text{full-load primary} \\ \text{current}\end{array}\right) \times \left(\begin{array}{c}\text{equivalent resistance} \\ \text{referred to primary circuit}\end{array}\right)}{\text{primary voltage}}$$

$$= \frac{\left(\begin{array}{c}\text{full-load secondary} \\ \text{current}\end{array}\right) \times \left(\begin{array}{c}\text{equivalent resistance} \\ \text{referred to secondary circuit}\end{array}\right)}{\text{secondary voltage on no load}}$$

Thus, for Example 35.4, full-load primary current $\simeq 45.45$ A, and equivalent resistance referred to primary circuit $= 0.55 \ \Omega$,

$$\text{Resistance drop} = \frac{45.45 \times 0.55}{2200} = \mathbf{0.0114 \text{ per unit}}$$

$$= \mathbf{1.14 \text{ per cent}}$$

Alternatively, full-load secondary current

$$\simeq 45.45 \times 400/80$$

$$= 227.2 \text{ A}$$

and equivalent resistance referred to secondary circuit is

$$0.01 + 0.3\left(\frac{80}{400}\right)^2 = 0.022 \ \Omega$$

Secondary voltage on no load $= 440$ V

$$\therefore \qquad \text{Resistance drop} = \frac{227.2 \times 0.022}{440} = \mathbf{0.0114 \text{ per unit}}$$

$$= \mathbf{1.14 \text{ per cent}}$$

Similarly, leakage reactance drop of a transformer

$$= \frac{\left(\begin{array}{c}\text{full-load primary} \\ \text{current}\end{array}\right) \times \left(\begin{array}{c}\text{equivalent leakage resistance} \\ \text{referred to primary circuit}\end{array}\right)}{\text{primary voltage}}$$

$$= \frac{45.45 \times 1.975}{2200} = \mathbf{0.0408 \text{ per unit}} = \mathbf{4.08 \text{ per cent}}$$

It is usual to refer to the per-unit or the percentage resistance and leakage reactance drops on full load as merely the per-unit or the percentage resistance and leakage reactance of the transformer; thus, the above transformer has a per-unit resistance and leakage reactance of 0.0114 and 0.0408 respectively or a percentage resistance and leakage reactance of 1.14 and 4.08 respectively.

35.15 Efficiency of a transformer

The losses which occur in a transformer on load can be divided into two groups:

1. I^2R losses in primary and secondary windings, namely $I_1^2R_1 + I_2^2R_2$.
2. Core losses due to hysteresis and eddy currents. The factors determining these losses have already been discussed in sections 35.1 and 35.2.

Since the maximum value of the flux in a normal transformer does not vary by more than about 2 per cent between no load and full load, it is usual to assume the core loss constant at all loads.

Hence, if P_c = total core loss, total losses in transformer are

$$P_c + I_1^2R_1 + I_2^2R_2$$

and   Efficiency $= \dfrac{\text{output power}}{\text{input power}} = \dfrac{\text{output power}}{\text{output power} + \text{losses}}$

$$\text{Efficiency} = \frac{I_2V_2 \times \text{p.f.}}{I_2V_2 \times \text{p.f.} + P_c + I_1^2R_1 + I_2^2R_2} \qquad [35.16]$$

Greater accuracy is possible by expressing the efficiency thus:

$$\text{Efficiency} = \frac{\text{output power}}{\text{input power}} = \frac{\text{input power} - \text{losses}}{\text{input power}}$$

$$\eta = 1 - \frac{\text{losses}}{\text{input power}} \qquad [35.17]$$

Example 35.6

The primary and secondary windings of a 500 kVA transformer have resistances of 0.42 Ω and 0.0019 Ω respectively. The primary and secondary voltages are 11 000 V and 400 V respectively and the core loss is 2.9 kW, assuming the power factor of the load to be 0.8. Calculate the efficiency on

(a) full load;
(b) half load.

(a) Full-load secondary current is

$$\frac{500 \times 1000}{400} = 1250 \text{ A}$$

and   Full-load primary current $\simeq \dfrac{500 \times 1000}{11\ 000} = 45.5 \text{ A}$

Therefore secondary I^2R loss on full load is

$$(1250)^2 \times 0.0019 = 2969 \text{ W}$$

and primary I^2R loss on full load is

$$(45.5)^2 \times 0.42 = 870 \text{ W}$$

$\therefore$ Total I^2R loss on full load $= 3839 \text{ W} = 3.84 \text{ kW}$

and   Total loss on full load $= 3.84 + 2.9 = 6.74 \text{ kW}$

Output power on full load $= 500 \times 0.8 = 400 \text{ kW}$

$\therefore$ Input power on full load $= 400 + 6.74 = 406.74 \text{ kW}$

From equation [35.17], efficiency on full load is

$$\left(1 - \frac{6.74}{406.74}\right) = 0.983 \text{ per unit}$$

$$= \textbf{98.3 per cent}$$

(b) Since the I^2R loss varies as the square of the current,

$\therefore$ Total I^2R loss on half load $= 3.84 \times (0.5)^2 = 0.96 \text{ kW}$

and   Total loss on half load $= 0.96 + 2.9 = 3.86 \text{ kW}$

$\therefore$ Efficiency on half load $= \left(1 - \dfrac{3.86}{203.86}\right) = 0.981 \text{ per unit}$

$$= \textbf{98.1 per cent}$$

35.16 Condition for maximum efficiency of a transformer

If R_{2e} is the equivalent resistance of the primary and secondary windings referred to the *secondary* circuit,

$$R_{2e} = R_1 \left(\frac{N_2}{N_1}\right)^2 + R_2$$

$$= \text{a constant for a given transformer}$$

Hence for any load current I_2

Total I^2R loss $= I_2^2 R_{2e}$

and $\text{Efficiency} = \dfrac{I_2 V_2 \times \text{p.f.}}{I_2 V_2 \times \text{p.f.} + P_c + I_2^2 R_{2e}}$

$$\text{Efficiency} = \frac{V_2 \cos \phi}{V_2 \cos \phi + (P_c / I_2) + I_2 R_{2e}} \qquad [35.18]$$

For a normal transformer, V_2 is approximately constant, hence for a load of given power factor the efficiency is a maximum when the denominator of equation [35.18] is a minimum, i.e. when

$$\frac{d}{dI_2}\left(V_2 \times \text{p.f.} + \frac{P_c}{I_2} + I_2 R_{2e}\right) = 0$$

$$\therefore \qquad -\frac{P_c}{I_2^2} + R_{2e} = 0$$

or $\qquad I_2^2 R_{2e} = P_c$ $\qquad\qquad\qquad\qquad\qquad\qquad\qquad$ [35.19]

To check that this condition gives the minimum and not the maximum value of the denominator in expression [35.18], $(-P_c/I_2^2 + R_{2e})$ should be differentiated with respect to I_2, thus:

$$\frac{d}{dI_2}\left(-\frac{P_c}{I_2^2} + R_{2e}\right) = \frac{2P_c}{I_2^3}$$

Since this quantity is positive, expression [35.19] is the condition for the minimum value of the denominator of equation [35.18] and therefore the maximum value of the efficiency. Hence the efficiency is a maximum when the variable I^2R loss is equal to the constant core loss.

Example 35.7 Assuming the power factor of the load to be 0.8, find the output at which the efficiency of the transformer of Example 35.6 is a maximum and calculate its value.

With the full-load output of 500 kVA, the total I^2R loss is 3.86 kW.
Let n = fraction of full-load apparent power (in kVA) at which the efficiency is a maximum.

Corresponding total I^2R loss = $n^2 \times 3.86$ kW

Hence, from equation [35.19],

$$\therefore \qquad n^2 \times 3.86 = 2.9$$

$$n = 0.867$$

and $\qquad$ Output at maximum efficiency = 0.867×500

$$= 433 \text{ kVA}$$

It will be noted that the value of the apparent power at which the efficiency is a maximum is independent of the power factor of the load.
Since the I^2R and core losses are equal when the efficiency is a maximum,

$$\therefore \qquad \text{Total loss} = 2 \times 2.9 = 5.8 \text{ kW}$$

$$\text{Output power} = 433 \times 0.8 = 346.4 \text{ kW}$$

$$\therefore \qquad \text{Maximum efficiency} = \left(1 - \frac{5.8}{346.4 + 5.8}\right) = \textbf{0.984 per unit}$$

$$= \textbf{98.4 per cent}$$

35.17 Open-circuit and short-circuit tests on a transformer

These two tests enable the efficiency and the voltage regulation to be calculated without actually loading the transformer and with an accuracy far higher than is possible by direct measurement of input and output powers and voltages. Also, the power required to carry out these tests is very small compared with the full-load output of the transformer.

(a) Open-circuit test

The transformer is connected as in Fig. 35.22 to a supply at the rated voltage and frequency, namely the voltage and frequency given on the nameplate. The ratio of the voltmeter readings, V_1/V_2, gives the ratio of the number of turns. Ammeter A gives the no-load current, and its reading is a check on the magnetic quality of the ferromagnetic core and joints. The primary current on no load is usually less than 5 per cent of the full-load current, so that the I^2R loss on no load is less than $1/400$ of the primary I^2R loss on full load and is therefore negligible compared with the core loss. Hence the wattmeter reading can be taken as the core loss of the transformer.

Fig. 35.22 Open-circuit test on a transformer

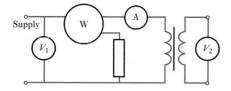

(b) Short-circuit test

The secondary is short-circuited through a suitable ammeter A_2, as shown in Fig. 35.23 and a *low* voltage is applied to the primary circuit. This voltage should, if possible, be adjusted to circulate full-load currents in the primary and secondary circuits. Assuming this to be the case, the I^2R loss in the windings is the same as that on full load. On the other hand, the core loss is negligibly small, since the applied voltage and therefore the flux are only about one-twentieth to one-thirtieth of the rated voltage and flux, and the core loss is approximately proportional to the square of the flux. Hence the power registered on wattmeter W can be taken as the I^2R loss in the windings.

Fig. 35.23 Short-circuit test on a transformer

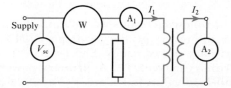

35.18 Calculation of efficiency from the open-circuit and short-circuit tests

If

P_{oc} = input power in watts on the open-circuit test

= core loss

and

P_{sc} = input power in watts on the short-circuit test with full-load currents

= total I^2R loss on full load

then

Total loss on full load = $P_{oc} + P_{sc}$

and

$$\text{Efficiency on full load} = \frac{\text{full-load } S \times \text{p.f.}}{(\text{full-load } S \times \text{p.f.}) + P_{oc} + P_{sc}} \qquad [35.20]$$

where S is the apparent power (in volt amperes). Also, for any load equal to $n \times$ full load,

Corresponding total loss = $P_{oc} + n^2 P_{sc}$

and corresponding efficiency is

$$\eta = \frac{n \times \text{full-load } S \times \text{p.f.}}{n \times \text{full-load } S \times \text{p.f.} + P_{oc} + n^2 P_{sc}} \qquad [35.21]$$

35.19 Calculation of the voltage regulation from the short-circuit test

Since the secondary voltage is zero, the whole of the applied voltage on the short-circuit test is absorbed in sending currents through the impedances of the primary and secondary windings; and since ϕ_e in Fig. 35.21 is the phase angle between the primary current and the voltage drop due to the equivalent impedance referred to the primary circuit,

$\cos \phi_e$ = power factor on short-circuit test

$$= \frac{P_{sc}}{I_1 V_{sc}}$$

where V_{sc} is the value of the primary applied voltage on the short-circuit test when *full-load* currents are flowing in the primary and secondary windings. Then, from expression [35.14],

$$\text{Per-unit voltage regulation} = \frac{V_{sc} \cos(\phi_e - \phi_2)}{V_1} \qquad [35.22]$$

Example 35.8

The following results were obtained on a 50 kVA transformer: open-circuit test – primary voltage, 3300 V; secondary voltage, 400 V; primary power, 430 W. Short-circuit test – primary voltage, 124 V; primary current, 15.3 A; primary power, 525 W; secondary current, full-load value. Calculate:

(a) the efficiencies at full load and at half load for 0.7 power factor;

(b) the voltage regulations for power factor 0.7, (i) lagging, (ii) leading;

(c) the secondary terminal voltages corresponding to (i) and (ii).

(a) $$\text{Core loss} = 430 \text{ W}$$

$$I^2R \text{ loss on full load} = 525 \text{ W}$$

$\therefore$ $$\text{Total loss on full load} = 955 \text{ W} = 0.955 \text{ kW}$$

and $$\text{Efficiency on full load} = \frac{50 \times 0.7}{(50 \times 0.7) + 0.955}$$

$$= \left(1 - \frac{0.955}{35.95}\right) = 0.973 \text{ per unit}$$

$$= 97.3 \text{ per cent}$$

$$I^2R \text{ loss on half load} = 525 \times (0.5)^2 = 131 \text{ W}$$

$\therefore$ $$\text{Total loss on half load} = 430 + 131 = 561 \text{ W} = 0.561 \text{ kW}$$

and $$\text{Efficiency on half load} = \frac{25 \times 0.7}{(25 \times 0.7) + 0.561}$$

$$= \left(1 - \frac{0.561}{18.06}\right) = 0.969 \text{ per unit}$$

$$= 96.9 \text{ per cent}$$

(b) $$\cos \phi_e = \frac{525}{124 \times 15.3} = 0.2765$$

$\therefore$ $$\phi_e = 73°57'$$

For $\cos \phi_2 = 0.7$,

$$\phi_2 = 45°34'$$

From expression [35.22], for power factor 0.7 lagging,

$$\text{Voltage regulation} = \frac{124 \cos(73°57' - 45°34')}{3300}$$

$$= 0.033 \text{ per unit} = 3.3 \text{ per cent}$$

For power factor 0.7 leading,

$$\text{Voltage regulation} = \frac{124 \cos(73°57' + 45°34')}{3300}$$

$$= -0.0185 \text{ per unit}$$

$$= -1.85 \text{ per cent}$$

(c) Secondary voltage on open circuit = 400 V. Therefore secondary voltage on full load, p.f. 0.7 lagging, is

$$400(1 - 0.033) = \textbf{387 V}$$

and secondary voltage on full load, p.f. 0.7 leading, is

$$400(1 + 0.0185) = \textbf{407 V}$$

35.20　Three-phase core-type transformers

Modern large transformers are usually of the three-phase core type shown in Fig. 35.24. Three similar limbs are connected by top and bottom yokes, each limb having primary and secondary windings, arranged concentrically. In Fig. 35.24 the primary is shown star-connected and the secondary delta-connected. Actually, the windings may be connected star/delta, delta/star, star/star or delta/delta, depending upon the conditions under which the transformer is to be used.

Fig. 35.24 Three-phase core-type transformer

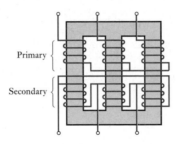

Primary

Secondary

Example 35.9

A three-phase transformer has 420 turns on the primary and 36 turns on the secondary winding. The supply voltage is 3300 V. Find the secondary line voltage on no load when the windings are connected

(a) star/delta,
(b) delta/star.

(a)　　　Primary phase voltage = 3300/1.73 = 1908 V

∴　　　Secondary phase voltage = 1908 × 36/420 = **164 V**

　　　　　　　　　　= secondary line voltage

(b)　　　Primary phase voltage = 3300 V

∴　　　Secondary phase voltage = 3300 × 36/420 = 283 V

　　　Secondary line voltage = 283 × 1.73 = **490 V**

35.21　Auto-transformers

An auto-transformer is a transformer having a part of its winding common to the primary and secondary circuits; thus, in Fig. 35.25 winding AB has a tapping at C, the load being connected across CB and the supply voltage applied across AB.

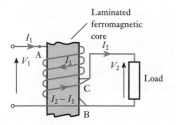

Fig. 35.25 An auto-transformer

I_1 and I_2 = primary and secondary currents respectively

N_1 = no. of turns between A and B

N_2 = no. of turns between C and B

and n = ratio of the *smaller* voltage to the *larger* voltage

Neglecting the losses, the leakage reactance and the magnetizing current, we have for Fig. 35.25,

$$n = \frac{V_2}{V_1} = \frac{I_1}{I_2} = \frac{N_2}{N_1}$$

The nearer the ratio of transformation is to unity, the greater is the economy of conductor material. Also, for the same current density in the windings and the same peak values of the flux and of the flux density, the I^2R loss in the auto-transformer is lower and the efficiency higher than in the two-winding transformer.

Auto-transformers are mainly used for (a) interconnecting systems that are operating at roughly the same voltage and (b) starting cage-type induction motors (section 39.9). Should an auto-transformer be used to supply a low-voltage system from a high-voltage system, it is essential to earth the common connection, for example, B in Fig. 35.25, otherwise there is a risk of serious shock. In general, however, an auto-transformer should not be used for interconnecting high-voltage and low-voltage systems.

35.22 Current transformers

It is difficult to construct ammeters and the current coils of wattmeters, energy (kW h) meters and relays to carry alternating currents greater than about 100 A. Furthermore, if the voltage of the system exceeds 500 V, it is dangerous to connect such instruments directly to the high voltage. These difficulties are overcome by using current transformers. Figure 35.26(a) shows an ammeter A supplied through a current transformer. The ammeter is usually arranged to give full-scale deflection with 5 A, and the ratio of the primary to secondary turns must be such that full-scale ammeter reading is obtained with full-load current in the primary. Thus, if the primary has four turns and the full-load primary current is 50 A, the full-load primary ampere-turns are 200; consequently, to circulate 5 A in the secondary, the number of secondary turns must be 200/5, namely 40.

Fig. 35.26 A current transformer

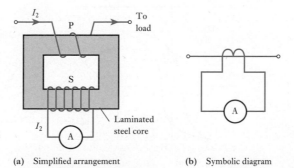

(a) Simplified arrangement

(b) Symbolic diagram

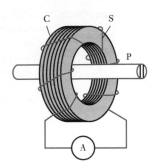

Fig. 35.27 A bar-primary current transformer

If the number of primary turns were reduced to *one* and the secondary winding had 40 turns, the primary current to give full-scale reading of 5 A on the ammeter would be 200 A. Current transformers having a single-turn primary are usually constructed as shown in Fig. 35.27, where P represents the primary conductor passing through the centre of a laminated steel ring C. The secondary winding S is wound uniformly around the ring.

The secondary circuit of a current transformer must on no account be opened while the primary winding is carrying a current, since all the primary ampere-turns would then be available to produce flux. The core loss due to the high flux density would cause excessive heating of the core and windings, and a dangerously high e.m.f. might be induced in the secondary winding. Hence if it is desired to remove the ammeter from the secondary circuit, the secondary winding must first be short-circuited. This will not be accompanied by an excessive secondary current, since the latter is proportional to the primary current; and since the primary winding is in *series* with the load, the primary current is determined by the value of the load and not by that of the secondary current.

35.23   Waveform of the magnetizing current of a transformer

In Fig. 35.4 the phasor for the no-load current of a transformer is shown leading the magnetic flux. A student may ask: is the flux not a maximum when this current is a maximum? The answer is that they are at their maximum values at the same instant (assuming the eddy-current loss to be negligible); but if the applied voltage is sinusoidal, then the magnetizing current of a ferromagnetic-core transformer is not sinusoidal, and a non-sinusoidal quantity cannot be represented by a phasor.

Suppose the relationship between the flux and the magnetizing current for the ferromagnetic core to be represented by the hysteresis loop in Fig. 35.28(a). Also, let us assume that the waveform of the flux is sinusoidal as shown by the dotted curve in Fig. 35.28(b). It was shown in section 35.4 that when the flux is sinusoidal, the e.m.f. induced in the primary is also sinusoidal and lags the flux by a quarter of a cycle. Hence the voltage applied to the primary must be sinusoidal and leads the flux by a quarter of a cycle as shown in Fig. 35.28(b).

At instant O in Fig. 35.28(b), the flux is zero and the magnetizing current is OA. At instant B, the flux is OB in Fig. 35.28(a), and the current is BC. When the flux is at its maximum value OF, the current is also at its maximum

Fig. 35.28 Waveform of magnetizing current

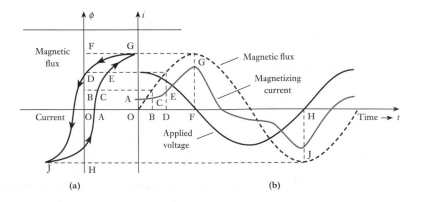

value FG. Thus, by projecting from Fig. 35.28(a) to the flux curve in Fig. 35.28(b) and erecting ordinates representing the corresponding values of the current, the waveform of the magnetizing current can be derived. It can be shown that this waveform contains a sinusoidal component having the same frequency as the supply voltage and referred to as the *fundamental* component, together with other sinusoidal components having frequencies that are odd multiples of the supply frequency. The fundamental current component lags the applied voltage by an angle ϕ that is a little less than 90°, and

$$\text{Hysteresis loss} = \text{r.m.s. fundamental current component}$$

$$\times \text{ r.m.s. voltage} \times \cos \phi$$

Most power transformers are operated near to full load, in which case the magnetizing current has negligible effect. Thus, although the magnetizing current is not sinusoidal, the total input current including the magnetizing current is almost sinusoidal. Therefore the effect of the non-sinusoidal component of the current can generally be ignored in most power applications.

However, if a transformer is designed for small loads, the effect of the magnetizing current on the total input current becomes more pronounced. This is particularly important when the transformer is incorporated into electronic circuits or even the power supply to electronic equipment. In these applications, we are normally involved with analogue signals in which it is essential not to introduce distortion – distortion would of course result from the magnetizing current component.

There are two methods responding to the needs of electronic circuits. The first is to use a better material for the core so that the magnetizing current is almost sinusoidal – this involves much more expensive material which is acceptable in small transformers. This method is not perfect, but bearing in mind that the transistors also do not have absolutely linear characteristics, we can look on the effect of the transformer as being just another small source of distortion. The transformer, however, has the advantage that it can be relatively compact and has little leakage of its magnetic field.

The alternative method is to do away with the ferromagnetic core; this arrangement is the air-cored transformer.

35.24   Air-cored transformer

The air-cored transformer may literally consist of two concentric coils which have nothing but air within the coils. However, the intent of the name infers that there is no ferromagnetic core. This immediately has the advantage that the magnetizing current has exactly the same waveform as the voltage to which it is related. However, it has the very significant disadvantage that it is difficult to produce the necessary magnetic flux to generate the appropriate e.m.f.

There is a second disadvantage. With a ferromagnetic core, the flux linking the windings is largely contained within the core. Even so there is some flux leakage. We have looked at the effect this has on equivalent impedance of the winding. However, it can also have a side effect in that it produces stray induced e.m.f.s in nearby circuits. These are insignificant in power systems, but are most unwelcome in electronic circuits where they cause distortion.

This problem becomes much worse when there is no ferromagnetic core to help contain the flux. Air-cored transformers, therefore, have to be contained in ferromagnetic shields which limit the effect of the stray fields – but then reintroduce a small non-linearity in the magnetizing current. The effect is almost negligible as the shield is remote from the transformer coils, but this no longer permits the size of the transformer to remain small.

Occasionally we come across power air-cored transformers which literally consist of two concentric coils held in space within a large steel mesh container. Even for relatively small ratings, the devices are large, but this can be acceptable in order to avoid the production of any form of distortion being introduced.

Summary of important formulae

For an idealized transformer,

$$\frac{V_2}{V_1} = \frac{N_2}{N_1} \qquad\qquad [35.1]$$

$$\frac{I_1}{I_2} = \frac{N_2}{N_1} \qquad\qquad [35.2]$$

$$E_1 = 4.44 N_1 f \Phi_m \text{ (volts)} \qquad\qquad [35.4]$$

The no-load current relates to its power and magnetizing components using

$$I_0 = \sqrt{(I_{0l}^2 + I_{0m}^2)} \qquad\qquad [35.7]$$

Voltage regulation,

$$= \frac{V_1 - V_{2'}}{V_1} \qquad\qquad [35.13]$$

Terms and concepts

Transformers effect changes of voltage with virtually no loss of power. The input is called the **primary** and the output is termed the **secondary**.

The primary and secondary systems are connected by magnetic flux linkage.

The winding terminals are so connected to their respective windings that the primary and secondary voltages are normally in phase with one another.

A **no-load current** is required to magnetize the core of a transformer. The no-load current has two components – one to supply the power losses incurred by the core and the other to create the magnetic flux. Normally the no-load current is almost insignificant in relation to the full-load current caused by a secondary load.

Not all the flux links the two windings although the leakage can be minimized by placing the low-voltage winding inside the high-voltage winding.

Losses occur in a transformer due to the I^2R losses in the windings, plus the hysteresis and eddy-current losses in the core. The losses usually are sufficiently small under full-load conditions that the efficiency is in excess of 98 per cent.

Auto-transformers have a common primary and secondary winding. **Current transformers** are intended to strictly relate the primary and secondary currents for the purposes of measurement.

Exercises 35

1. The design requirements of an 11 000 V/400 V, 50 Hz, single-phase, core-type transformer are: approximate e.m.f./turn, 15 V; maximum flux density, 1.5 T. Find a suitable number of primary and secondary turns, and the net cross-sectional area of the core.

2. The primary winding of a single-phase transformer is connected to a 230 V, 50 Hz supply. The secondary winding has 1500 turns. If the maximum value of the core flux is 0.002 07 Wb, determine: (a) the number of turns on the primary winding; (b) the secondary induced voltage; (c) the net cross-sectional core area if the flux density has a maximum value of 0.465 T.

3. An 11 000 V/230 V, 50 Hz, single-phase, core-type transformer has a core section 25 cm × 25 cm. Allowing for a space factor of 0.9, find a suitable number of primary and secondary turns, if the flux density is not to exceed 1.2 T.

4. A single-phase 50 Hz transformer has 80 turns on the primary winding and 400 turns on the secondary winding. The net cross-sectional area of the core is 200 cm². If the primary winding is connected to a 230 V, 50 Hz supply, determine: (a) the e.m.f. induced in the secondary winding; (b) the maximum value of the flux density in the core.

5. A 50 kVA single-phase transformer has a turns ratio of 300/20. The primary winding is connected to a 2200 V, 50 Hz supply. Calculate: (a) the secondary voltage on no load; (b) the approximate values of the primary and secondary currents on full load; (c) the maximum value of the flux.

6. A 200 kVA, 3300 V/230 V, 50 Hz, single-phase transformer has 80 turns on the secondary winding. Assuming an ideal transformer, calculate: (a) the primary and secondary currents on full load; (b) the maximum value of the flux; (c) the number of primary turns.

7. The following data apply to a single-phase transformer:

peak flux density in the core	= 1.41 T,
net core area	= 0.01 m²,
current density in conductors	= 2.55 MA/m²,
conductor diameter	= 2.0 mm,
primary supply (assume sinusoidal)	= 230 V, 50 Hz.

Calculate the rating (in kVA) of the transformer and the number of turns on the primary winding.

8. A transformer for a radio receiver has a 230 V, 50 Hz primary winding, and three secondary windings as follows: a 1000 V winding with a centre tapping; a 4 V winding with a centre tapping; a 6.3 V winding. The net cross-sectional area of the core is 14 cm². Calculate the number of turns on each winding if the maximum flux density is not to exceed 1 T.

Note. Since the number of turns on a winding must be an integer, it is best to calculate the number of turns on the low-voltage winding first. In this question, the 4 V winding must have an even number of turns since it has a centre tapping.

9. The primary of a certain transformer takes 1 A at a power factor of 0.4 when connected across a 230 V, 50 Hz supply and the secondary is on open circuit. The number of turns on the primary is twice that on the secondary. A load taking 50 A at a lagging power factor of 0.8 is now connected across the secondary. Sketch, and explain briefly, the phasor diagram for this condition, neglecting voltage drops in the transformer. What is now the value of the primary current?

10. A 4:1 ratio step-down transformer takes 1 A at 0.15 power factor on no load. Determine the primary current and power factor when the transformer is supplying a load of 25 A at 0.8 power factor lag. Ignore internal voltage drops.

11. A 3300 V/230 V, single-phase transformer, on no load, takes 2 A at power factor 0.25. Determine graphically, or otherwise, the primary current and power factor when the transformer is supplying a load of 60 A at power factor 0.9 leading.

12. A three-phase transformer has its primary winding delta-connected and its secondary winding star-connected. The number of turns per phase on the primary is four times that on the secondary, and the secondary line voltage is 400 V. A balanced load of 20 kW, at power factor 0.8, is connected across the secondary terminals. Assuming an ideal transformer, calculate the primary voltage and the phase and line currents on the secondary and primary sides. Sketch a circuit diagram and indicate the values of the voltages and currents on the diagram.

13. A 50 Hz, three-phase, core-type transformer is connected star–delta and has a line voltage ratio of 11 000/400 V. The cross-section of the core is square with a circumscribing circle of 0.6 m diameter. If the maximum flux density is about 1.2 T, calculate the number of turns per phase on the low-voltage and on the high-voltage windings. Assume the insulation to occupy 10 per cent of the gross core area.

14. If three transformers, each with a turns ratio of 12:1, are connected star–delta and the primary line voltage

is 11 000 V, what is the value of the secondary no–load voltage? If the transformers are reconnected delta–star with the same primary voltage, what is the value of the secondary line voltage?

15. A 400 V, three-phase supply is connected through a three-phase loss-free transformer of 1:1 ratio, which has its primary connected in mesh and secondary in star, to a load comprising three 20 Ω resistors connected in delta. Calculate the currents in the transformer windings, in the resistors and in the lines to the supply and the load. Find also the total power supplied and the power dissipated by each resistor.

16. The no-load current of a transformer is 5.0 A at 0.3 power factor when supplied at 230 V, 50 Hz. The number of turns on the primary winding is 200. Calculate: (a) the maximum value of the flux in the core; (b) the core loss; (c) the magnetizing current.

17. Calculate: (a) the number of turns required for an inductor to absorb 230 V on a 50 Hz circuit; (b) the length of airgap required if the coil is to take a magnetizing current of 3 A (r.m.s.); (c) the phase difference between the current and the terminal voltage. Mean length of steel path, 500 mm; maximum flux density in core, 1 T; sectional area of core, 3000 mm^2; maximum magnetic field strength for the steel, 250 A/m; core loss, 1.7 W/kg; density of core, 7800 kg/m^3. Neglect the resistance of the winding and any magnetic leakage and fringing. Assume the current waveform to be sinusoidal.

18. Calculate the no-load current and power factor for the following 60 Hz transformer: mean length of core path, 700 mm; maximum flux density, 1.1 T; maximum magnetic field strength for the core, 300 A/m; core loss, 1.0 W. All the joints may be assumed equivalent to a single airgap of 0.2 mm. Number of primary turns, 120; primary voltage, 2.0 V. Neglect the resistance of the primary winding and assume the current waveform to be sinusoidal.

19. The ratio of turns of a single-phase transformer is 8, the resistances of the primary and secondary windings are 0.85 Ω and 0.012 Ω respectively, and the leakage reactances of these windings are 4.8 Ω and 0.07 Ω respectively. Determine the voltage to be applied to the primary to obtain a current of 150 A in the secondary when the secondary terminals are short-circuited. Ignore the magnetizing current.

20. A single-phase transformer operates from a 230 V supply. It has an equivalent resistance of 0.1 Ω and an equivalent leakage reactance of 0.5 Ω referred to the

primary. The secondary is connected to a coil having a resistance of 200 Ω and a reactance of 100 Ω. Calculate the secondary terminal voltage. The secondary winding has four times as many turns as the primary.

21. A 10 kVA single-phase transformer, for 2000 V/400 V at no load, has resistances and leakage reactances as follows. Primary winding: resistance, 5.5 Ω; reactance, 12 Ω. Secondary winding: resistance, 0.2 Ω; reactance, 0.45 Ω. Determine the approximate value of the secondary voltage at full load, 0.8 power factor (lagging), when the primary supply voltage is 2000 V.

22. Calculate the voltage regulation at 0.8 lagging power factor for a transformer which has an equivalent resistance of 2 per cent and an equivalent leakage reactance of 4 per cent.

23. A 75 kVA transformer, rated at 11 000 V/230 V on no load, requires 310 V across the primary to circulate full-load currents on short-circuit, the power absorbed being 1.6 kW. Determine: (a) the percentage voltage regulation; (b) the full-load secondary terminal voltage for power factors of (i) unity, (ii) 0.8 lagging and (iii) 0.8 leading. If the input power to the transformer on no load is 0.9 kW, calculate the per-unit efficiency at full load and at half load for power factor 0.8 and find the load (in kVA) at which the efficiency is a maximum.

24. The primary and secondary windings of a 30 kVA, 11 000/230 V transformer have resistances of 10 Ω and 0.016 Ω respectively. The total reactance of the transformer referred to the primary is 23 Ω. Calculate the percentage regulation of the transformer when supplying full-load current at a power factor of 0.8 lagging.

25. A 50 kVA, 6360 V/230 V transformer is tested on open and short-circuit to obtain its efficiency, the results of the test being as follows. Open circuit: primary voltage, 6360 V; primary current, 1 A; power input, 2 kW. Short-circuit: voltage across primary winding, 180 V; current in secondary winding, 175 A; power input, 2 kW. Find the efficiency of the transformer when supplying full load at a power factor of 0.8 lagging and draw a phasor diagram (neglecting impedance drops) for this condition.

26. A 230 V/400 V single-phase transformer absorbs 35 W when its primary winding is connected to a 230 V, 50 Hz supply, the secondary being on open circuit. When the primary is short-circuited and a 10 V, 50 Hz supply is connected to the secondary winding, the power absorbed is 48 W when the current has the

full-load value of 15 A. Estimate the efficiency of the transformer at half load, 0.8 power factor lagging.

27. A 1 kVA transformer has a core loss of 15 W and a full-load I^2R loss of 20 W. Calculate the full-load efficiency, assuming the power factor to be 0.9.

An ammeter is scaled to read 5 A, but it is to be used with a current transformer to read 15 A. Draw a diagram of connections for this, giving terminal markings and currents.

28. Discuss fully the energy losses in single-phase transformers. Such a transformer working at unity power factor has an efficiency of 90 per cent at both one-half load and at the full load of 500 W. Determine the efficiency at 75 per cent of full load.

29. A single-phase transformer is rated at 10 kVA, 230 V/100 V. When the secondary terminals are open-circuited and the primary winding is supplied at normal voltage (230 V), the current input is 2.6 A at a power factor of 0.3. When the secondary terminals are short-circuited, a voltage of 18 V applied to the primary causes the full-load current (100 A) to flow in the secondary, the power input to the primary being 240 W. Calculate: (a) the efficiency of the transformer at full load, unity power factor; (b) the load at which maximum efficiency occurs; (c) the value of the maximum efficiency.

30. A 400 kVA transformer has a core loss of 2 kW and the maximum efficiency at 0.8 power factor occurs when the load is 240 kW. Calculate: (a) the maximum efficiency at unity power factor; (b) the efficiency on full load at 0.71 power factor.

31. A 40 kVA transformer has a core loss of 450 W and a full-load I^2R loss of 850 W. If the power factor of the load is 0.8, calculate: (a) the full-load efficiency; (b) the maximum efficiency; (c) the load at which maximum efficiency occurs.

32. Each of two transformers, A and B, has an output of 40 kVA. The core losses in A and B are 500 and 250 W respectively, and the full-load I^2R losses are 500 and 750 W respectively. Tabulate the losses and efficiencies at quarter, half and full load for a power factor of 0.8. For each transformer, find the load at which the efficiency is a maximum.

33. The following table gives the relationship between the flux and the magnetizing current in the primary winding of a transformer:

Current (A)	0	0.25	0.5	0.75	1.0	1.25	1.5
Flux (mWb)	−1.76	−1.48	−0.8	1.0	1.56	1.88	2.08
Current (A)	1.75	2.0	1.5	1.0	0.5	0	
Flux (mWb)	2.17	2.24	2.2	2.14	2.04	1.76	

Plot the above values and derive the waveform of the magnetizing current of the transformer, assuming the waveform of the flux to be sinusoidal and to have a peak value of 2.24 mWb.

Chapter thirty-six

Introduction to Machine Theory

Contents

This chapter introduces the energy-conversion process whereby electrical energy is converted to mechanical energy and vice versa by means of an electromagnetic system. It explains how mechanical forces can be developed by magnetic attraction and alignment and how rotary motion can therefore be realized. This is the basis of the electrical motor by which mechanical force is produced.

The simplest motor, the reluctance motor, is initially considered to develop an understanding of how rotational torque is produced as a rotor tries to align with stator poles. This facilitates an introduction to the analysis of more complex doubly excited machines such as synchronous and commutator machines.

By the end of the chapter, you will be familiar with the principles of electrical machine operation and be able to undertake exercises in linear machine operation.

36.1 The role of the electrical machine

In the previous chapters, we have been introduced to the transmission of electrical power by means of three-phase circuits and the use of transformers by which system voltages can be raised and lowered. Having transmitted the power, we have to consider to what purpose we intend to apply the power. Much electrical energy is used to provide simple heat, either for heating a room or for cooking. Some heat conversion is used to produce light as in the common light bulb. But much of electrical energy is used to drive machines.

All electrical machines operate on a common set of principles which will be introduced in this chapter. Later chapters will consider specific families of electrical machine. The principles apply whether the device operates with alternating current or with direct current.

The most simple (so far as construction is concerned) involve linear movement; these include the relay and the contactor. The former was by far the most common electrical machine of all, but now is being replaced by the electronic switch which has no moving parts. The contactor remains in common use for switching power circuits on and off.

Simply made rotating machines have few applications, although we will briefly consider the clock and stepper motors. Most rotating machines are complicated to make as we could see by looking inside a typical DIY electric drill.

Let us then consider the basic principles.

36.2 Conversion process in a machine

An electromagnetic machine is one that links an electrical energy system to another energy system by providing a reversible means of energy flow in its magnetic field. The magnetic field is therefore the coupling between the two systems and is the mutual link. The energy transferred from the one system to the other is temporarily stored in the field and then released to the other system.

Usually the energy system coupled to the electrical energy system is a mechanical one; the function of a motor is to transfer electrical energy into mechanical energy while a generator converts mechanical energy into electrical energy. Converters transfer electrical energy from one system to another as in the transformer. Not all machines deal with large amounts of energy – those operating at very low power levels are often termed *transducers*, particularly when providing 'signals' with which to activate electronic control devices.

Discussion of a mechanical energy system implies that a mechanical force is associated with displacement of its point of action. An electromagnetic system can develop a mechanical force in two ways:

1. By alignment.
2. By interaction.

The force of alignment can be illustrated by the arrangements shown in Fig. 36.1. In the first, two poles are situated opposite one another; each is made of a ferromagnetic material and a flux passes from the one to the other. The surfaces through which the flux passes are said to be magnetized surfaces and they are attracted towards one another as indicated in Fig. 36.1.

Fig. 36.1 Force of alignment.
(a) Force of attraction; (b) lateral
force of alignment

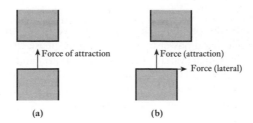

It will be shown that the force of alignment acts in any direction that will increase the magnetic energy stored in the arrangement. In the first case, it will try to bring the poles together since this decreases the reluctance of the air-gap in the magnetic circuit and hence will increase the flux and consequently the stored energy. This principle of increasing the stored magnetic energy is a most important one and it is the key to the unified machine theory that follows.

In the second case, shown in Fig. 36.1(b), the poles are not situated opposite one another. The resultant force tries to achieve greater stored magnetic energy by two component actions:

1. By attraction of the poles towards one another as before.
2. By aligning the poles laterally.

If the poles move laterally, the cross-sectional area of the air-gap is increased and the reluctance is reduced with consequent increase in the stored magnetic energy as before.

Both actions attempt to align the poles to point of maximum stored energy, i.e. when the poles are in contact with maximum area of contact. It should be noted that the force of alignment does not necessarily act in the direction of the lines of flux.

Fig. 36.2 Electromagnetic relay

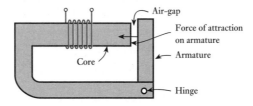

Many devices demonstrate the principle of the force of alignment. Probably the one with which the reader is most familiar is that of the permanent magnet and its attraction of ferromagnetic materials. Electromagnetic devices such as the relay shown in Fig. 36.2 demonstrate the force of alignment giving rise to linear motion.

When the coil is energized, a flux is set up in the relay core and the air-gap. The surfaces adjacent to the air-gap become magnetized and are attracted, hence pulling the armature plate in the direction indicated. The relay's function is to operate switches and it is used extensively in telephone exchanges.

The force of alignment can also be used to produce rotary motion, as in the reluctance motor shown in Fig. 36.3. In this case, the rotating piece, termed the *rotor*, experiences radial forces in opposite directions, thereby cancelling each other out. The rotor also experiences a torque due to the magnetized rotor and pole surfaces attempting to align themselves. This alignment

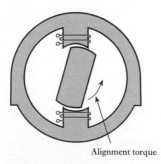

Fig. 36.3 Reluctance motor

torque occurs in any rotating machine which does not have a cylindrical rotor, i.e. in a rotor which is salient as in the case shown in Fig. 36.3.

The force of interaction has the advantage of simplicity in its application. To calculate, or even to estimate, the energy stored in the magnetic fields of many arrangements is difficult if not impossible. Many of these cases, however, can be dealt with by the relationship $F = Bli$. This includes the case of a beam of electrons being deflected by a magnetic field in, say, a cathode-ray tube.

There are many applications involving the force of interaction to give rise to rotary motion. These include the synchronous and induction machines as well as the commutator machines, all of which are given further consideration in this and subsequent chapters. All are variations on the same theme, each giving a different characteristic suitable for the various industrial drives required.

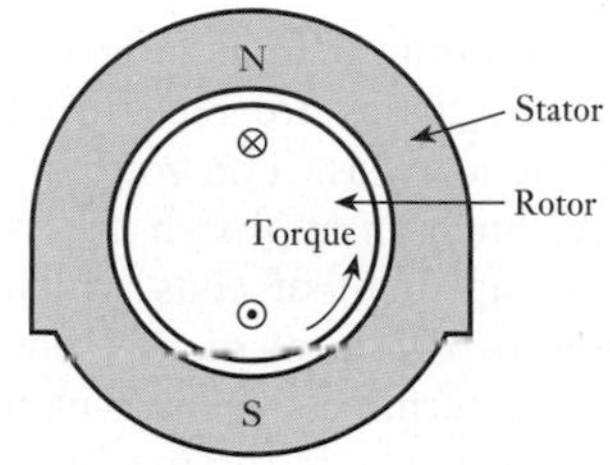

A simple machine illustrating the principle involved is shown in Fig. 36.4. By passing a current through the coil, it experiences a force on each of the coil sides resulting in a torque about the axis of rotation. A practical machine requires many conductors in order to develop a sufficient torque; depending on how the conductors are arranged, the various machines mentioned above and many others are created.

Fig. 36.4 Rotary machine illustrating force of interaction

These examples only serve to illustrate the field of study involved in machine theory. Each is a system-linking device. At the one end there is the electrical system, at the other end is the mechanical (or other) system. In between there is a magnetic field forming a two-way link between them. If there is to be a flow of energy, all three will be involved simultaneously. Note that the reaction in the electrical system, apart from the flow of current, is the introduction of an e.m.f. into that system; the product of e.m.f. and current gives the rate of electrical energy conversion.

Before progressing to analyse the energy-conversion process, consideration must be given as to how to approach the analysis. There are three methods of approach, each of which has to allow for the imperfections of a machine. No machine gives out as much energy as it takes in. The difference is termed the *losses* – the losses in the electrical system, the mechanical system and the magnetic system.

36.3	Methods of analysis of machine performance

The basic energy-conversion process is one involving the magnetic coupling field and its action and reaction on the electrical and mechanical systems. We will consider two possible approaches to analysing the energy conversion.

(a) The so-called classical approach

This dates back to the end of the nineteenth century when it was found that the operation of a machine could be predicted from a study of the machine losses. Such an approach is generally simple, but it has two major disadvantages. First it deals almost exclusively with the machine operating under steady-state conditions, thus transient response conditions are virtually ignored, i.e. when it is accelerating and decelerating. Second, the losses of each machine are different. It follows that each type of machine requires to be separately analysed. As a consequence, much attention has been given in the past to the theory of d.c. machines, three-phase induction machines and three-phase synchronous machines to the exclusion of most others. Such a process is both tedious and repetitive, but it is simple.

We will use the classical approach in Chapters 38, 39, 42 and 44. This limits our consideration of the various motors to steady-state conditions, but that will be sufficient at this introductory stage.

(b)   The generalized-machine approach

This approach depends on a full analysis of the coupling field as observed from the terminals of the machine windings. The losses are recognized as necessary digressions from the main line of the analysis. The coupling field is described in terms of mutual inductance. Such an approach there-fore considers any machine to be merely an arrangement of coils which are magnetically linked. No attention is paid as to the form of machine construction initially and it exists as a box with terminals from which measurements may be made to determine the performance of the machine. The measured quantities are voltage, current, power, frequency, torque and rotational speed, from which may be derived the resistance and inductance values for the coils. In the light of the derived values, it is possible to analyse to performance both under steady-state and transient conditions.

It should not be taken from this description of the generalized theory that the design arrangement of a machine is considered to be unimportant. However, if too much emphasis is given to the constructional detail, the principle of the energy transfer may well become ignored.

The generalized approach has been universally appreciated only since the late 1950s. Since the approach has so much to offer, one may well enquire as to why it has taken so long to attain recognition. The difficulty that pre-cluded its earlier use is the complexity of the mathematics involved. The initial complexity of the approach remains, but once the appropriate math-ematical manipulations have been carried out, they need not be repeated. The results of this preliminary step can then be separately modified to analyse not only most types of machines but also the different modes of operation. Thus an analysis is being made of the electromechanical machine in general. The individual machine is considered by making simplifying assumptions at the end of the analysis, but the effect of the theory is to concentrate attention upon the properties common to all machines.

This book is intended to provide a basic introduction to electrical prin-ciples and therefore cannot give extensive coverage to the transient operation of electrical machines.

36.4   **Magnetic field energy**

It was shown in section 8.10 that the energy in a magnetic field is given by

$$W_f = \tfrac{1}{2} L i^2$$

However, there are a number of ways in which the inductance can be expressed.

$$L = \frac{N\phi}{i} = \frac{\psi}{i} = \frac{N^2}{S} = \frac{N^2 \mu_0 \mu_r A}{l}$$

These expressions can be substituted in the energy relation to give

$$W_f = \frac{1}{2} i\psi = \frac{1}{2} F\phi = \frac{1}{2} S\phi^2$$

It should be noted that all of these expressions for the energy depend on the flux and the m.m.f. being directly proportional, i.e. the inductance is constant. This is generally limited to the case of air, which is the most important one in electrical machines.

Sometimes the energy density can be of greater importance. In the proof of the relationship of hysteresis loss in a magnetic core, it was shown that the energy stored was proportional to the shaded area due to the B/H curve. This is shown in Fig. 36.5.

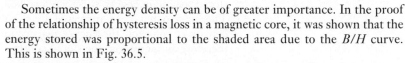

Fig. 36.5 Stored-energy diagram

In the case of an air-gap, the B/H characteristic is straight and the energy stored is given by

$$W_f - \tfrac{1}{2} BH \times \text{volume of air-gap}$$

If the air-gap has a cross-sectional area A and length l,

$$W_f = \tfrac{1}{2} BH \times Al$$

$$= \tfrac{1}{2} F\phi$$

The stored energy density is thus given by

$$w_f = \tfrac{1}{2} BH$$

$$= \frac{\mu_0 H^2}{2}$$

$$w_f = \frac{B^2}{2\mu_0} \qquad [36.1]$$

It will be noted that all the expressions have been shown for instantaneous values and not for steady-state conditions. However, this does not always draw attention to some quantities which can be variable. In particular, the flux density B, the magnetizing force H and the inductance L are all variable quantities and so instantaneous quantities are implied in these cases.

36.5 Simple analysis of force of alignment

Consider the force of alignment between two poles of a magnetic circuit as shown in Fig. 36.6.

Let there be a flux Φ in the air-gap and let there be no fringing of the flux. The uniform flux density in the air-gap is then given by

$$B = \frac{\Phi}{A}$$

Fig. 36.6 Force of alignment between two poles

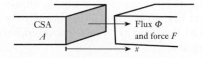

Suppose that the poles can be separated by a small distance $\mathrm{d}x$ without there being a change in the flux and the flux density. This can be effectively realized in practice. Because there is a mechanical force experienced by the poles, the mechanical work done is

$$\mathrm{d}W_\mathrm{M} = F \cdot \mathrm{d}x$$

Assume now that the magnetic core is ideal, i.e. it is of infinite permeability and therefore requires no m.m.f. to create a magnetic field in it. The stored magnetic energy will therefore be contained entirely in the air-gap. The air-gap has been increased by a volume $A \cdot \mathrm{d}x$, yet, since the flux density is constant, the energy density must remain unchanged. There is therefore an increase in the stored energy

$$\mathrm{d}W_\mathrm{f} = \frac{B^2}{2\mu_0} \times A \cdot \mathrm{d}x$$

Since the system is ideal and the motion has taken place slowly from one point of rest to another, this energy must be due to the input of mechanical energy, i.e.

$$\mathrm{d}W_\mathrm{M} = \mathrm{d}W_\mathrm{f}$$

$$F \cdot \mathrm{d}x = \frac{B^2 A}{2\mu_0} \cdot \mathrm{d}x$$

$$F = \frac{B^2 A}{2\mu_0} \qquad\qquad [36.2]$$

A more important relationship, however, may be derived from the above argument, i.e.

$$\mathrm{d}W_\mathrm{M} = F \cdot \mathrm{d}x = \mathrm{d}W_\mathrm{f}$$

$$F = \frac{\mathrm{d}W_\mathrm{f}}{\mathrm{d}x} \qquad\qquad [36.3]$$

It will be shown later that this relation only holds true provided the flux (or flux-linkage) remains constant. However, the principle demonstrated is more important, i.e. the force is given by the rate of change of stored field energy with distortion of the arrangement of the ferromagnetic poles.

In order to develop this principle, more attention must now be given to the energy-transfer process. In the ideal situation described above, all of the mechanical energy was converted into stored field energy. In order to obtain this arrangement, several conditions had to be met, e.g. the ferromagnetic core was infinitely permeable. In practice, such conditions cannot be laid down and it follows that the energy balance must be modified.

36.6 Energy balance

Consider the operation of a simple attracted–armature relay such as that shown in Fig. 36.7. Assume that initially the switch is open and that there is no stored field energy. These conditions are quite normal and in no way unreasonable.

Fig. 36.7 Attracted-armature
relay

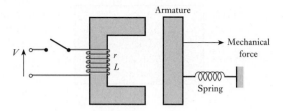

After the switch is closed, the sequence of events falls into four distinct groups as follows:

1.   After the switch is closed, the current rises exponentially in the manner described in Chapter 8. If L_1 is the inductance of the coil for the initial position of the armature, the initial rate of rise of current is given by V/L_1. The electrical energy from the source is partly dissipated in i^2R loss in the magnetizing coil while the remainder is converted into stored energy in the magnetic field. During this period the armature experiences an attractive force, but the various mechanical restraints prevent it from moving. It is not unusual to find that the steady-state value of coil current has almost been reached before the armature starts to move.

2.   At some appropriate value of current, the armature begins to move. This occurs when the force of attraction f_E balances the mechanical force f_M. During the motion of the armature, there are many changes of energy in the system. On the mechanical side, energy is required to stretch the spring, drive the external load and to supply the kinetic energy required by the moving parts. At the same time, the air-gaps are being reduced with consequent increase in the inductance of the arrangement. This causes a reaction in the electrical system in the form of an induced e.m.f.; this e.m.f. tends to reduce the coil current and also permits the conversion of electrical energy, i.e. it is the reaction to the action.

3.   The armature cannot continue to move indefinitely, but, instead, it hits an end stop. This causes the kinetic energy of the system to be dissipated as noise, deformation of the poles and vibration.

4.   Now that there is no further motion of the system, the inductance becomes constant at a new higher value L_2. The current increases exponentially to a value V/r. It should be noted that the rate of rise is less than the initial rate of rise since the inductance is now much greater.

The energy-flow processes are therefore many, and yet they are typical of many machines – in rotating machines there is no sudden stop, but otherwise the processes are similar. To be able to handle so many at one time it is necessary to set up an energy balance involving the convention which will remain during all the subsequent machine analysis.

Consider first the external systems, i.e. the electrical and mechanical systems. Because the conversion process can take place in either direction, let both electrical and mechanical energies be input energies to the system. It follows that an output energy is mathematically a negative input energy. The electrical and mechanical energies are W_E and W_M.

Second there is the internal system comprising the stored magnetic field energy W_f, the stored mechanical energy W_s and the non-useful thermal energy (due to i^2R loss, friction, etc.) which is a loss W_l. The arrangement may be demonstrated diagrammatically as in Fig. 36.8.

Fig. 36.8   Energy balance diagram

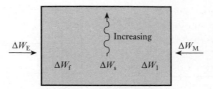

Between any two states of the system, the energy balance may be expressed as

$$\Delta W_\mathrm{E} + \Delta W_\mathrm{M} = \Delta W_\mathrm{f} + \Delta W_\mathrm{s} + \Delta W_\mathrm{l} \qquad [36.4]$$

Alternatively the energy rates of flow may be expressed as

$$p_\mathrm{E} + p_\mathrm{M} = \frac{\mathrm{d}W_\mathrm{f}}{\mathrm{d}t} + \frac{\mathrm{d}W_\mathrm{s}}{\mathrm{d}t} + \frac{\mathrm{d}W_\mathrm{l}}{\mathrm{d}t} \qquad [36.5]$$

In this chapter, the basic machine will be idealized to a limited extent by separating some of the losses as indicated in Fig. 36.9.

Fig. 36.9   Idealized energy balance diagram

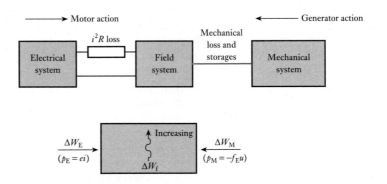

It follows that the ideal and essential energy balance may be expressed as

$$\mathrm{d}W_\mathrm{E} + \mathrm{d}W_\mathrm{M} = \mathrm{d}W_\mathrm{f} \qquad [36.6]$$

and hence the power balance may be expressed as

$$p_\mathrm{E} + p_\mathrm{M} = \frac{\mathrm{d}W_\mathrm{f}}{\mathrm{d}t} \qquad [36.7]$$

Finally there are the actions and reactions to consider. These are indicated in Fig. 36.10.

Fig. 36.10   Actions and reactions in practical conversion system

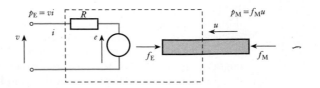

On the electrical side the applied voltage is v and this is opposed by the reaction in the form of the back e.m.f. e. The electrical power is

$$p_E = vi$$

while the rate of electrical energy conversion is ei. These two terms are only equal when the i^2R loss is either neglected or considered external to the conversion process as in the idealized system of Fig. 36.9.

On the mechanical side, the mechanical input force f_M acts towards the conversion system and moves in a similar direction, say with a velocity u. The reaction to this is the magnetically developed force f_E. These two forces are equal and opposite only when the mechanical system is at rest or moves with uniform velocity. The difference would otherwise give rise to acceleration and hence there could not be steady-state conditions. There is also a slight difference between the forces when the mechanical system is moving due to friction. However, it has been noted that the effect of friction is to be neglected for the time being.

36.7	**Division of converted energy and power**

There are several methods of demonstrating the division of energy and power of which the following case is sufficient. Consider the attracted-armature relay shown in Fig. 36.11. As before, it is assumed that the relay is to be energized with the result that the relay closes.

The movement of the armature is in the direction $-x$. Because the lengths of the air-gaps are decreasing, the inductance of the system increases; it can also be noted that the velocity is

$$u = -\frac{\mathrm{d}x}{\mathrm{d}t}$$

Having noted these points, consider the general solution of the power balance. At any instant

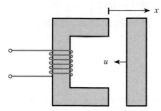

Fig. 36.11 Attracted-armature relay operation

$$e = \frac{\mathrm{d}\psi}{\mathrm{d}t}$$

$$= \frac{\mathrm{d}}{\mathrm{d}t}(Li)$$

$$= L \cdot \frac{\mathrm{d}i}{\mathrm{d}t} + i \cdot \frac{\mathrm{d}L}{\mathrm{d}t}$$

$$p_E = ei$$

$$= Li \cdot \frac{\mathrm{d}i}{\mathrm{d}t} + i^2 \cdot \frac{\mathrm{d}L}{\mathrm{d}t}$$

and $\qquad W_f = \tfrac{1}{2}Li^2$

$$\frac{\mathrm{d}W_f}{\mathrm{d}t} = Li \cdot \frac{\mathrm{d}i}{\mathrm{d}t} + \tfrac{1}{2}i^2 \cdot \frac{\mathrm{d}L}{\mathrm{d}t}$$

and $\qquad p_E + p_M = \dfrac{\mathrm{d}W_f}{\mathrm{d}t}$

hence $\quad p_M = -\frac{1}{2}i^2 \cdot \dfrac{dL}{dt} = -\frac{1}{2}i^2 \cdot \dfrac{dL}{dx} \cdot \dfrac{dx}{dt}$

$\qquad\qquad = -\frac{1}{2}i^2 \cdot \dfrac{dL}{dx} \cdot (-u) = \frac{1}{2}i^2 \cdot \dfrac{dL}{dx} \cdot u$

$\qquad\qquad = -f_E u$

$$f_E = -\frac{1}{2}i^2 \cdot \frac{dL}{dx} \qquad\qquad\qquad [36.8]$$

Relation [36.8] is an expression for the force of reaction developed by the magnetic field and not for the mechanical force. The possible difference between these forces has been noted above, but is again emphasized in view of the importance in this and other subsequent relations.

It will be noted that the expression for the rate of mechanical energy conversion p_M is negative, i.e. the machine acts as a motor and p_M is an output. This requires that the mechanical force acts away from the machine and is negative. Relation [36.8] does not readily indicate that the force has a negative value, but it should be remembered that the inductance decreases with increase of the air-gaps.

The output mechanical power accounts for both any mechanical storages as well as the useful power.

36.8 Force of alignment between parallel magnetized surfaces

To find the force of alignment between the core members of the arrangement shown in Fig. 36.12, consider again relation [36.3]:

$$f_E = \frac{dW_f}{dx}$$

From relation [36.1], where V is the volume of the air-gap,

$$W_f = w_f V$$

$$\qquad = w_f A x$$

$$\qquad = \frac{B^2}{2\mu_0} A x$$

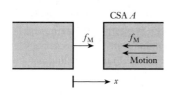

Fig. 36.12 Force of alignment between parallel magnetized surfaces

This assumes $\mu_r = 1$ for air and x is the variable distance measured along the air-gap

$$f_E = \frac{dW_f}{dx}$$

$\therefore \qquad$
$$f_E = \frac{B^2 A}{2\mu_0} \qquad\qquad\qquad [36.9]$$

This is the same relation as [36.2], but the force has been interpreted in the effect it will have, i.e. attraction.

The development of the theory has now completed a full cycle. Now consider a different arrangement and analyse it using the same principles. If the poles had been laterally displaced, a force of alignment would be experienced trying to align the poles. This arrangement is shown in Fig. 36.13.

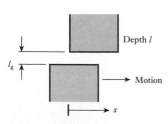

Fig. 36.13 Lateral force between magnetized surfaces

Ignoring the effect of leakage flux, let the area covered by the gap be xl, and the gap length be l_g. The air-gap volume is given by

$$V = ll_g x$$

and $\qquad W_f = w_f ll_g x$

$$= \frac{B^2}{2\mu_0} ll_g x$$

$\therefore \qquad f_E = \dfrac{\mathrm{d}W_f}{\mathrm{d}x}$

and $\qquad f_E = \dfrac{B^2 ll_g}{2\mu_0}$ [36.10]

The polarity of this expression indicates that the force tries to align the poles by increasing the cross-sectional area of the air-gap, thereby decreasing the reluctance. This expression should be used with great care since it is not a continuous function; for instance it is not immediately obvious that, if the two poles are aligned, then the force drops to zero and reverses in its direction of action thereafter.

Finally it should be noted that relations [36.9] and [36.10] can be applied to the action at any air-gap. This is because these expressions are related to the field energy at the gap and this was defined in terms of reluctance and flux. The energy stored in the core of the magnetic circuit, when the core is not ideal, does not affect the validity of the relations.

Example 36.1

An electromagnet is made using a horseshoe core as shown in Fig. 36.14. The core has an effective length of 600 mm and a cross-sectional area of 500 mm². A rectangular block of steel is held by the electromagnet's force of alignment and a force of 20 N is required to free it. The magnetic circuit through the block is 200 mm long and the effective cross-sectional area is again 500 mm². The relative permeability of both core and block is 700. If the magnet is energized by a coil of 100 turns, estimate the coil current.

Fig. 36.14

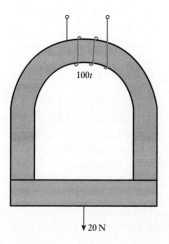

There are two air-gaps in the magnetic circuit, hence the force to part the circuit members is double that at any one gap.

$$f_M = 2 \cdot \frac{B^2 A}{2\mu_0} = \frac{B^2 A}{\mu_0} = 20 \text{ N}$$

$$B = \left(\frac{20 \times 4\pi \times 10^{-7}}{500 \times 10^{-6}} \right)^{\frac{1}{2}} = 0.222 \text{ T}$$

$$H = \frac{B}{\mu_0 \mu_r} = \frac{0.222}{4\pi \times 10^{-7} \times 700} = 250 \text{ At/m}$$

$$F = Hl = 250 \times (600 + 200) \times 10^{-3} = 200 \text{ At} = NI = 100I$$

$$I = \frac{200}{100} = \textbf{2.0 A}$$

Example 36.2　The poles of an electromagnet are shown in Fig. 36.15 relative to a steel bar. If the effects of leakage flux may be neglected, estimate the forces of alignment that act laterally on the steel bar due to each pole.

Fig. 36.15

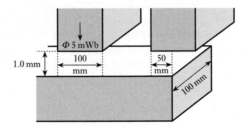

There is no force of alignment at the left-hand pole since motion in either direction would not change the reluctance of the air-gap and hence there can be no change in the field energy stored in that gap.

There is a force of alignment at the right-hand gap given as follows:

$$B = \frac{\Phi}{A} = \frac{5 \times 10^{-3}}{5 \times 10^{-3} \times 100 \times 10^{-3}} = 1.0 \text{ T}$$

$$f_E = \frac{B^2 l_g l}{2\mu_0} = \frac{1.0^2 \times 1 \times 10^{-3} \times 100 \times 10^{-3}}{2 \times 4\pi \times 10^{-7}}$$

$$= \textbf{39.7 N}$$

Example 36.3　A solenoid relay shown in Fig. 36.16 is operated from a 110 V d.c. supply and the 5000-turn coil resistance is 5.5 kΩ. The core diameter of the relay is 20 mm and the gap length is 1.5 mm, the armature being stationary. The gap faces may be taken as parallel and the permeability of the ferromagnetic parts as very high. Estimate:

(a) the gap flux density;
(b) the coil inductance;
(c) the pull on the armature.

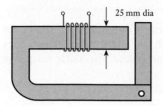

25 mm dia

Fig. 36.16

(a) $I = \dfrac{V}{R} = \dfrac{110}{5.5 \times 10^3} = 20 \times 10^{-3} \text{ A}$

$F = IN = 20 \times 10^{-3} \times 5000 = 100 \text{ At}$

$H = \dfrac{F}{l_g} = \dfrac{100}{1.5 \times 10^{-3}} = 0.67 \times 10^5 \text{ At/m}$

$B = \mu_0 H = 4\pi \times 10^{-7} \times 0.67 \times 10^5 = 84 \times 10^{-3} \text{ T} = \textbf{84 mT}$

(b) $\Phi = BA = 84 \times 10^{-3} \times 10^2 \times \pi \times 10^{-6} = 26.3 \times 10^{-6} \text{ Wb}$

$L = \dfrac{\Phi N}{I} = \dfrac{26.3 \times 10^{-6} \times 5000}{20 \times 10^{-3}} = \textbf{6.56 H}$

(c) The inductance L is inversely proportional to the gap length l_g, hence in general:

$$L = 6.56 \times \dfrac{1.5}{x}$$

$$\dfrac{\mathrm{d}L}{\mathrm{d}x} = -\dfrac{9.82}{x^2} = -\dfrac{9.82}{1.5^2} = -4.37 \text{ H/mm}$$

(negative sign as indicated in section 36.7)

$$= -4370 \text{ H/m}$$

$$f_E = -\tfrac{1}{2} i^2 \cdot \dfrac{\mathrm{d}L}{\mathrm{d}x} = \tfrac{1}{2} \times 20^2 \times 10^{-6} \times 44\,370 = \textbf{0.88 N}$$

The positive polarity indicates that it is a force of attraction.

36.9   Rotary motion

The analysis of the electromechanical machine has concentrated on the linear machine – however, the principles also apply to the rotating machine by replacing x by λ (the angle of rotational distortion) and u by ω (the angular velocity of the rotor).

Angular velocity   Symbol: ω_r Unit: **radian per second (rad/s)**

In electrical machines, the speed is often measured in revolutions per second or per minute, hence

Angular speed   Symbol: n (or n_r)   Unit: **revolution per second (r/s)**

The alternative symbol is used in this book wherever it gives continuity to ω_r, i.e.

$$\omega_r = 2\pi n_r$$

The torque of a rotating machine is given by

$$M_E = \dfrac{\mathrm{d}W_f}{\mathrm{d}\lambda} \qquad\qquad [36.11]$$

Remember that M, rather than T, is used to symbolize the torque of a rotating machine.

36.10   Reluctance motor

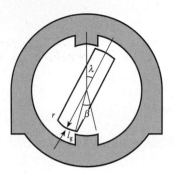

Fig. 36.17 Simple reluctance torque machine

A simple machine that demonstrates a torque of alignment is shown in Fig. 36.17. If the rotor is displaced through an angle λ, it experiences a torque which tries to align it with the stator poles. The static torque is calculated from relation [36.11].

The field energy density in the air-gaps is given by

$$w_f = \frac{B^2}{2\mu_0}$$

It will be assumed that only the gap energies need by considered. The total energy is therefore given by

$$W_f = 2\left(\frac{B^2 A l_g}{2\mu_0}\right) = \frac{B^2 l r l_g(\beta - \lambda)}{\mu_0}$$

where A is the gap area through which the pole flux passes and hence is equal to $lr(\beta - \lambda)$, l is the pole length (into the page), r is the rotor radius, l_g is the gap length between the stator poles and rotor.

$$M_E = \frac{dW_f}{d\lambda}$$

$$M_E = \frac{B^2 l r l_g}{\mu_0} \qquad\qquad [36.12]$$

Such a machine would not produce continuous rotation, but it can be adapted to do so. The resulting machine is termed a *reluctance motor*. It is not particularly important – one common application is clock motors – but its significance should be borne in mind. Any machine with saliency of the type illustrated by the rotor in Fig. 36.17 will produce a significant torque in this way.

For continuous motion, the assumption is required that the reluctance varies sinusoidally with the rotation of the rotor. This compares quite favourably with normal practice. A sinusoidal voltage is applied to the stator coil giving rise to a flux that may be defined by

$$\phi = \Phi_m \cos \omega t$$

As the rotor rotates, the reluctance will vary. Minimum reluctance occurs when the rotor centre line is coincident with the direct axis of Fig. 36.18. The minimum reluctance is S_d. This occurs when

$$\lambda = 0, \pi, 2\pi, \text{ etc.}$$

The maximum reluctance occurs with the rotor centre line coincident with the quadrature axis. The maximum reluctance is S_q. This occurs when

$$\lambda = \frac{\pi}{2}, \frac{3\pi}{2}, \frac{5\pi}{2}, \text{ etc.}$$

From the reluctance curves showing the sinusoidal variation between these limits:

Fig. 36.18 Reluctance motor with operating curves

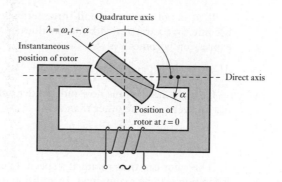

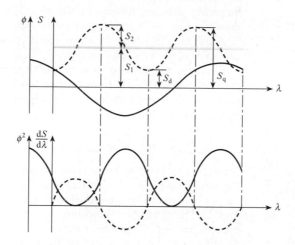

$$S = S_1 - S_2 \cos 2\lambda$$
$$= S_d + \tfrac{1}{2}(S_q - S_d) - \tfrac{1}{2}(S_q - S_d)\cos 2\lambda$$
$$= \tfrac{1}{2}(S_d + S_q) - \tfrac{1}{2}(S_q - S_d)\cos 2\lambda$$

but
$$\phi = \Phi_m \cos \omega t$$
$$\phi^2 = \Phi_m^2 \cos^2 \omega t = \tfrac{1}{2}\Phi_m^2(1 + \cos 2\omega t)$$

also
$$\frac{\mathrm{d}S}{\mathrm{d}\lambda} = (S_q - S_d)\sin 2\lambda$$

$$M_E = \tfrac{1}{2}\phi^2 \cdot \frac{\mathrm{d}S}{\mathrm{d}\lambda}$$
$$= \tfrac{1}{4} \cdot \Phi_m^2(1 + \cos 2\omega t)(S_q - S_d)\sin 2\lambda$$
$$= \tfrac{1}{4} \cdot \Phi_m^2(S_q - S_d)(1 + \cos 2\omega t)\sin(2\omega_r t - 2\alpha)$$
$$= \tfrac{1}{4} \cdot \Phi_m^2(S_q - S_d)(\sin(2\omega_r t - 2\alpha) + \sin(2\omega_r t - 2\alpha)\cos 2\omega t)$$
$$= \tfrac{1}{4} \cdot \Phi_m^2(S_q - S_d)(\sin(2\omega_r t - 2\alpha) + \tfrac{1}{2}\sin(2\omega_r t - 2\omega t - 2\alpha)$$
$$+ \tfrac{1}{2}\sin(2\omega_r t - 2\omega t - 2\alpha))$$

If ω_r is not equal to ω, all three terms within the last group of brackets become time variables with mean values of zero. If ω_r is equal to ω then the expression becomes

$$M_E = \tfrac{1}{4} \cdot \Phi_m^2 (S_q - S_d)(\sin(2\omega t - 2\alpha) + \tfrac{1}{2}\sin(4\omega t - 2\alpha) + \tfrac{1}{2}\sin(-2\alpha))$$

Only the last term is now independent of time. Consequently the torque has an average value other than zero:

$$M_{Eav} = \tfrac{1}{8}\Phi_m^2(S_q - S_d)\sin 2\alpha \qquad [36.13]$$

The rotor can thus rotate at a speed determined by the supply frequency if rotation is to be maintained. In addition, there are double- and quadruple-frequency pulsating torques produced which have no net effect. The curves shown in Fig. 36.18 are drawn to satisfy the condition of equal frequencies.

A machine in which the rotor speed is exactly fixed by the supply frequency is of the synchronous type.

Finally it will be noted that torque depends on the rotor position which does not follow directly from relation [36.13]. This relation concerns static torque and assumes a uniform air-gap which is not the case with the reluctance motor.

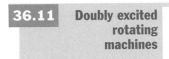

36.11   Doubly excited rotating machines

A machine of the reluctance type discussed above has many disadvantages, the most important of which are the fixed speed and the weak pulsating torque. These can be attributed to the manner in which the torque is produced. The reluctance machine, and also the relays, are singly excited systems, i.e. only the stator or the rotor is excited by a current-carrying coil. In either case, motion is caused by a movable part changing its position to reduce the reluctance of the magnetic circuit. Because of the physical construction in the rotary case, the axis of the rotor tries to align itself with the axis of the field.

In order to strengthen the attraction towards alignment, both the rotor and the stator can be excited. A simple arrangement is shown in Fig. 36.19. This is termed a doubly excited system.

With this arrangement, the stator and the rotor each have two magnetic poles. Such a machine is described as a two-pole machine. Machines can be made with greater numbers of poles but, for the purposes of this introductory chapter, only the two-pole machine will be considered.

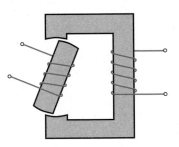

Fig. 36.19 Simple doubly excited rotary system

The windings and the magnetic circuit components give rise to clearly defined stator and rotor fields. By the symmetry of the fields, each has an axis. The axes indicate the directions of the mean magnetizing forces across the air-gaps for each of the fields. Since the gap lengths are constant, the field axes can have complexor properties ascribed to them, their magnitudes being related to the respective m.m.f.s.

It is now possible to describe the torque as being created by the m.m.f. axes trying to align themselves. These axes are functions of the coil constructions and hence become less dependent on the construction of the core members, which only serve to distribute the field although they are of course required to ensure that a sufficient field strength is available, and it follows that specially shaped rotors are no longer required as in the reluctance machine. The coil on the rotor ensures that the rotor has a well-defined axis

Fig. 36.20 Machines with cylindrical rotors and stators

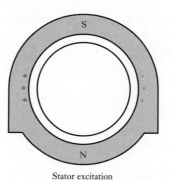
Stator excitation

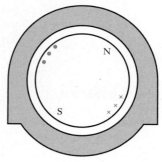
Rotor excitation

and thus cylindrical rotors can be used. This has a considerable advantage with regard to the mechanical design of the rotor, since the cylinder is inherently a very robust structure.

By removing the salient parts of the rotor shape, the reluctance torque is removed, the rotor now offering the same reluctance regardless of its position.

The stator magnetic circuit as shown in Fig. 36.19 is unnecessarily long. Instead the stator can be made in a similar form to the rotor in that it consists of a winding set into a cylinder; however, in the stator, the winding is about the inside surface of a cylinder as shown in Fig. 36.20. It will be noted that, in each case, the windings are set into the surface and are not generally wound on to the surface. This gives greater mechanical support to them and it minimizes the air-gap between the core members.

The current in the rotor winding produces its own m.m.f. and hence a flux that is termed the *armature reaction*. The total flux in the air-gap, therefore, is the result of the combination of the field due to the rotor winding and the field due to the stator winding. The torque is created by the desire for alignment of the fields.

Generally machines which are constructed with cylindrical rotors and stators are energized by alternating currents. We shall consider these arrangements in greater detail in Chapters 37, 38 and 39. Such machines provide useful general-purpose motors as well as the majority of generators. However, there are a number of specialist situations which are better satisfied by using machines which retain an element of saliency.

Most machines with salient poles do not take the form illustrated in Fig. 36.17. Rather, the salient poles appear on the stators as shown in Fig. 36.21. This arrangement creates a significant field under the salient poles and the rotor introduces currents carrying conductors which create the torque. Most machines constructed in this manner are generally excited by direct current, although small alternating currents can be used. We will deal with salient-pole machines in Chapters 41 and 42.

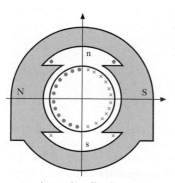
Arrows show directions of rotor and stator fields

Fig. 36.21 Salient-pole machine

There are therefore three important families of doubly excited rotating machines as follows:

1. *Synchronous machines.*   Stator flux – alternating current
   Rotor flux – direct current.

2. *Asynchronous machines.*   Stator flux – alternating current
   Rotor flux – alternating current.

3. *Commutator machines.*   Stator flux – direct current
   Rotor flux – direct current.

Synchronous and asynchronous machines usually have cylindrical stators, while the commutator machines have salient-pole stators. Most machines have cylindrical rotors.

Summary of important formulae

Magnetic stored energy density in free space

$$w_f = \frac{B^2}{2\mu_0} \tag{36.1}$$

Force of attraction between magnetic surfaces

$$F = \frac{B^2 A}{2\mu_0} \tag{36.2}$$

and

$$F = \frac{dW_f}{dx} \tag{36.3}$$

The basic energy balance

$$dW_E + dW_M = dW_f \tag{36.6}$$

$$p_E + p_M = \frac{dW_f}{dt} \tag{36.7}$$

Force of alignment between magnetized surfaces

$$F_E = -\frac{1}{2}i^2 \cdot \frac{dL}{dx} \tag{36.8}$$

The general torque is

$$M_E = \frac{dW_f}{d\lambda} \tag{36.11}$$

Terms and concepts

Magnetic systems try to optimize the stored energy by distorting the magnetic core either by closing air-gaps or by aligning poles. The former is associated with forces of attraction and the latter with forces of alignment.

Few machines are based on the force of attraction principle. The most common is the **relay** or **contactor**.

Rotating machines are based on the force of alignment principle.

The most simple are the **reluctance motors** which are singly excited systems.

Doubly excited systems can either incorporate cylindrical rotors and stators or be salient.

Doubly excited machines fall into three principal categories – **synchronous, asynchronous** and **commutator**.

Exercises 36

1. A coil of fixed inductance 4.0 H and effective resistance 30 Ω is suddenly connected to a 100 V, d.c. supply. What is the rate of energy storage in the field of the coil at each of the following instants: (a) when the current is 1.0 A; (b) when the current is 2.0 A; (c) when the current is at its final steady value?

2. In question 1, the value of the inductance of the coil was not required in the solution. Why was this the case?

3. A simple relay has an air-gap of length 1.0 mm and effective cross-sectional area 1000 mm². The magnetizing coil consists of 1000 turns of wire carrying a current of 200 mA. Calculate the energy stored in the air-gap. The reluctance of the ferromagnetic part of the magnetic circuit may be neglected.

4. The semi-circular electromagnet shown in Fig. A is to be used to lift a rectangular bar. The electromagnet and the bar have the same square cross-section of 0.01 m². The electromagnet is made of core steel; the bar, of mild steel material, has a bulk density of 7500 kg/m³. The electromagnet is energized by a 500-turn coil connected across a d.c. supply. With the aid of the magnetization characteristics given in Fig. 7.4 find the current required just to lift the bar. Neglect leakage and fringing in the air-gap.

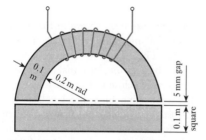

Fig. A

5. Two halves of a steel ring are placed together and a conductor is passed axially through the centre as shown in Fig. B. The ring has a mean diameter of 250 mm and a cross-sectional area of 500 mm². When a current is passed through the conductor, a force F of 25 N is required to separate the halves of the ring. For this condition the relative permeability of the iron is 500. Calculate: (a) the flux density in the ring; (b) the mean magnetizing force in the ring; (c) the current in the conductor.

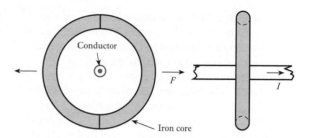

Fig. B

6. The electromagnet illustrated in Fig. C supports a mass of 11 kg. The cross-section of the magnetic parts of the magnetic circuit is 600 mm², and the excitation current is 2 A. Find the number of turns in the magnetizing coil to achieve the required supporting force. The permeability of the core material varies with flux density as in the following table:

B (T)	0.36	0.44	0.48	0.60	0.72
u_r	3300	3000	2900	2600	2300

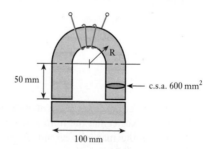

Fig. C

7. An electromagnet is shown in Fig. D to be in close proximity to a bar of steel. The flux in the magnetic

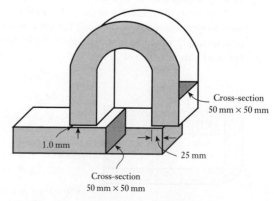

Fig. D

circuit is 2.5 mWb. Assuming that the reluctance of the magnetic circuit is concentrated in the air-gap and that fringing is negligible, calculate the horizontal and vertical forces experienced by the steel bar.

8. The cylindrical pot magnet shown in Fig. E has its axis vertical. The maximum length of the gap l_g is 15 mm. The minimum gap is limited by a stop to 5 mm. The reluctance of the steel parts of the magnetic circuit is negligible. The exciting coil is energized by a constant current of 3.0 A from a 60 V d.c. source. For maximum, mid and minimum gap lengths, find: (a) the gap flux density; (b) the coil inductance; (c) the magnetic force on the plunger; (d) check that the static magnetic force is given by $\frac{1}{2}I^2 \, \mathrm{d}l/\mathrm{d}x$.

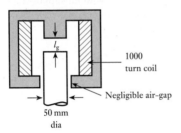

Fig. E

9. The essentials of an electromagnetic relay mechanism are shown in Fig. F. Neglecting the m.m.f. required by the steel parts of the magnetic circuit, estimate the inductance of the coil for: (a) the given gap length 3.0 mm; (b) the gap length increased to 3.6 mm. Hence estimate the average force developed on the armature between these two positions when the coil carries a steady current of 20 mA.

   Given that the resistance of the coil is 1000 Ω, find the voltage required across the coil for the current to remain steady when the armature moves between the two positions in a time of 11 ms.

10. A cylindrical pot magnet is shown in Fig. G. The coil, which has 1000 turns and a resistance of 20 Ω, is connected to a constant-voltage source of 60 V d.c. The plunger passes the point at which the air-gap is 10 mm travelling upwards with a uniform velocity of 2.0 m/s. Make an estimate of the value of the current at this instant, assuming that the initial air-gap length was 15 mm. Apportion the instantaneous input power into heat, magnetic-energy storage-rate and mechanical components. Hence estimate the magnetic force on the plunger.

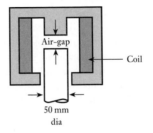

Fig. G

11. The attracted-armature relay shown in Fig. H has an initial air-gap of 3.0 mm. The reluctance of the steel part is negligible. The coil has 1000 turns. The armature is held open by a spring, such that an initial force of 0.8 N is required to close it. Explain what happens following the application of a direct voltage to the coil terminals, stating the energy balances concerned. Estimate the initial inductance of the coil, the initial rate of change of inductance with gap length and the current required to close the relay.

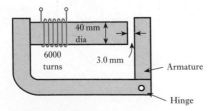

Fig. F

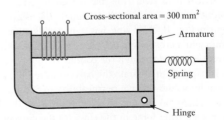

Fig. H

Exercises 36 continued

12. A simple electromagnetic relay is shown in Fig. I. The relay coil has a constant resistance R, an inductance L that varies with armature position x, and a constant

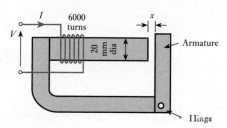

Fig. I

applied voltage V. Derive an expression relating the converted mechanical power to the rate of change of coil inductance.

Considering the armature to be stationary, and given that $V = 50$ V, $R = 2.0$ kΩ, $x = 2.0$ mm and the diameter of the core is 20 mm, estimate: (a) the gap flux density; (b) the gap flux; (c) the coil inductance; (d) the rate of change of coil inductance with gap length; (e) the magnetic pull on the armature. Assume that the gap faces are parallel and that the permeability of the iron parts is very high.

If, at this gap length of 2.0 mm, the armature were moving to close the gap at the rate 1.0 m/s, how would the current be affected? Give a brief explanation.

Chapter thirty-seven

AC Synchronous Machine Windings

Objectives

When you have studied this chapter, you should
- be familiar with simple synchronous machine construction
- be familiar with salient pole and cylindrical rotors
- have an understanding of stator windings
- be capable of calculating stator winding e.m.f.s
- understand the production of a rotating magnetic field
- understand the term synchronous speed
- be capable of analysing the air-gap flux due to three-phase currents
- understand the reversal of direction of magnetic flux

Contents

It has been observed that there are three important families of rotating machine. One of these is the synchronous machine which is commonly found in the form of the a.c. synchronous generator. Such machines are widely used in power stations throughout the world for electrical power generation. Individual generator ratings of 500 to 600 megawatts are commonplace.

In this chapter, the two common forms of rotor construction, namely the salient pole rotor and the cylindrical rotor, are discussed. Thereafter, stator windings and the e.m.f.s generated in them are explored, and the speed of rotation related to the frequency of these e.m.f.s. This leads to an understanding of the production of a rotating magnetic field in the air-gap between rotor and stator and the flux due to three-phase currents can then be analysed.

By the end of the chapter, the principles of a.c. synchronous machine windings will have been developed in preparation for the study of synchronous and asynchronous (or induction) machines.

37.1 General arrangement of synchronous machines

Let us first consider why synchronous machines, whether motors or generators, are usually constructed with stationary armature windings and rotating poles. Suppose we have a 20 MVA, 11 kV, three-phase synchronous machine; then from expression [34.6]

$$20 \times 10^6 = 1.73 \times I_L \times 11 \times 10^3$$

$$\therefore \quad \text{Line current} = I_L = 1050 \text{ A}$$

Hence, if the machine was constructed with stationary poles and a rotating three-phase winding, three slip-rings would be required, each capable of dealing with 1050 A, and the insulation of each ring together with that of the brushgear would be subjected to a working voltage of 11/1.73, namely 6.35 kV. Further, it is usual to connect synchronous generator windings in star and to join the star-point through a suitable resistor to a metal plate embedded in the ground so as to make good electrical contact with earth; consequently a fourth slip-ring would be required.

By using a stationary a.c. winding and a rotating field system, only two slip-rings are necessary and these have to deal with the exciting current. Assuming the power required for exciting the poles of the above machine to be 150 kW and the voltage to be 400 V,

$$\text{Exciting current} = 150 \times \frac{1000}{400} = 375 \text{ A}$$

In other words, the two slip-rings and brushgear would have to deal with only 375 A and be insulated for merely 400 V. Hence, by using a stationary a.c. winding and rotating poles, the construction is considerably simplified and the slip-ring losses are reduced.

Further advantages of this arrangement are:

1. The extra space available for the a.c. winding makes it possible to use more insulation and to enable operating voltages of up to 33 kV.
2. With the simpler and more robust mechanical construction of the rotor, a high speed is possible, so that a greater output is obtainable from a machine of given dimensions.

The slots in the laminated stator core of a synchronous machine are usually semi-enclosed, as shown in Fig. 37.1, so as to distribute the magnetic flux as uniformly as possible in the air-gap, thereby minimizing the ripple that would appear in the e.m.f. waveform if open slots were used.

In section 9.4 it was explained that if a synchronous machine has p pairs of poles and the speed is n revolutions per second,

$$\text{Frequency} = f = np \qquad\qquad [37.1]$$

Hence for a 50 Hz supply, a two-pole synchronous machine must operate at 3000 r/min, a four-pole synchronous machine at 1500 r/min, etc.

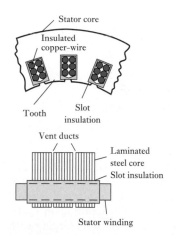

Fig. 37.1 Portion of stator of a.c. synchronous machine

37.2 Types of rotor construction

Synchronous machines can be divided into two categories: (1) those with salient or projecting poles; and (2) those with cylindrical rotors.

The salient-pole construction is used in comparatively small machines and machines driven at a relatively low speed. For instance, if a 50 Hz synchronous machine is to operate at, say, 375 r/min, then, from expression [37.1], the machine must have 16 poles; and to accommodate all these poles, the synchronous machine must have a comparatively large diameter. Since the output of a machine is roughly proportional to its volume, such a synchronous machine would have a small axial length. Figure 37.2 shows one arrangement of salient-pole construction. The poles are made of fairly thick steel laminations, L, riveted together and bolted to a steel yoke wheel Y, a bar of mild steel B being inserted to improve the mechanical strength. The exciting winding W is usually an insulated copper strip wound on edge, the coil being held firmly between the pole tips and the yoke wheel. The pole tips are well rounded so as to make the flux distribution around the periphery nearer a sine wave and thus improve the waveform of the generated e.m.f. Copper rods D, short-circuited at each end by copper bars E, are usually inserted in the pole shoes.

Fig. 37.2   Portion of a salient-pole rotor

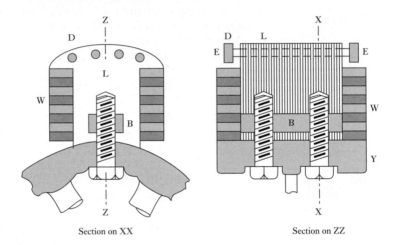

Section on XX   Section on ZZ

Most synchronous machines are essentially high-speed machines. The centrifugal force on a high-speed rotor is enormous: for instance, a mass of 1 kg on the outside of a rotor of 1 m diameter, rotating at a speed of 3000 r/min, has a centrifugal force ($= mv^2/r$) of about 50 kN acting upon it. To withstand such a force the rotor is usually made of a solid steel forging with longitudinal slots cut as indicated in Fig. 37.3, which shows a two-pole rotor with eight slots and two conductors per slot. In an actual rotor there are more slots and far more conductors per slot; and the winding is in the form of insulated copper strip held securely in position by phosphor-bronze wedges. The regions forming the centres of the poles are usually left unslotted. The horizontal dotted lines joining the conductors in Fig. 37.3 represent the end connections. If the rotor current has the direction represented by the dots and crosses in Fig. 37.3, the flux distribution is indicated by the light dotted lines. In addition to its mechanical robustness, this cylindrical construction has the advantage that the flux distribution around the periphery is nearer a sine wave than is the case with the salient-pole machine. Consequently, a better e.m.f. waveform is obtained.

Fig. 37.3   A cylindrical rotor

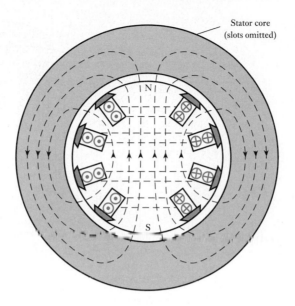

Stator core
(slots omitted)

Stator windings

In this book we shall consider only two types of three-phase winding, namely
(1) single-layer winding, and (2) double-layer winding, and of these types we
shall consider only the simplest forms. In a synchronous machine, the stator
winding is the armature winding, being that winding in which the operating
e.m.f. is induced.

(a)   Single-layer winding

The main difficulty with single-layer windings is to arrange the end connections so that they do not obstruct one another. Figure 37.4 shows one of the

Fig. 37.4   End connections of a
three-phase single-layer winding

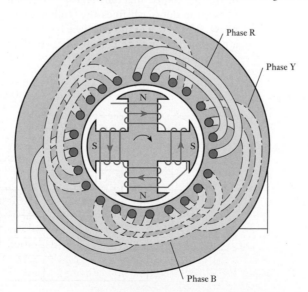

Phase R

Phase Y

Phase B

Fig. 37.5 End connections of a three-phase single-layer winding

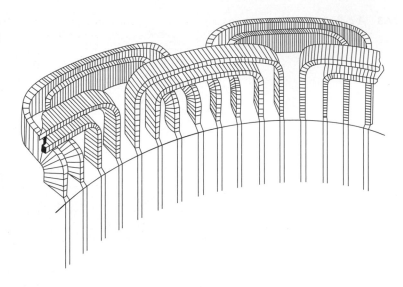

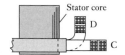

Fig. 37.6 Sectional view of end connections

most common methods of arranging these end connections for a four-pole, three-phase synchronous machine having two slots per pole per phase, i.e. six slots per pole or a total of 24 slots. In Fig. 37.4, all the end connections are shown bent outwards for clearness, but in actual practice the end connections are usually shaped as shown in Fig. 37.5 and in section in Fig. 37.6. This method has the advantage that it requires only two shapes of end connections, namely those marked C in Fig. 37.6, which are brought straight out of the slots and bent so as to lie on a cylindrical plane, and those marked D. The latter, after being brought out of the slots, are bent outwards roughly at right angles, before being again bent to form an arch alongside the core.

The connections of the various coils are more easily indicated by means of the developed diagram of Fig. 37.7. The solid lines represent the Red phase, the dot/dash lines the Yellow phase and the dashed lines the Blue phase. The width of the pole face has been made two-thirds of the pole pitch, a pole pitch being the distance between the centres of adjacent poles. The

Fig. 37.7 Three-phase single-layer winding

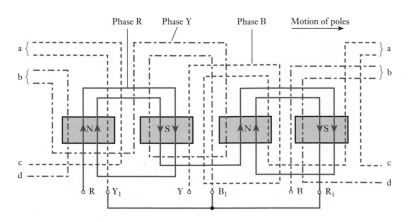

poles in Fig. 37.7 are assumed to be *behind* the winding and moving towards the right. From the right-hand rule (section 6.9) – bearing in mind that the thumb represents the direction of *motion of the conductor relative to the flux*, namely towards the left in Fig. 37.7 – the e.m.f.s in the conductors opposite the poles are as indicated by the arrowheads. The connections between the groups of coils forming any one phase must be such that all the e.m.f.s are assisting one another.

Since the stator has six slots per pole and since the rotation of the poles through one pole pitch corresponds to half a cycle of the e.m.f. wave or 180 electrical degrees, it follows that the spacing between two adjacent slots corresponds to 180°/6, namely 30 electrical degrees. Hence, if the wire forming the beginning of the coil occupying the first slot is taken to the 'red' terminal R, the connection to the 'yellow' terminal Y must be a conductor from a slot four slot-pitches ahead, namely from the fifth slot, since this allows the e.m.f. in phase Y to lag the e.m.f. in phase R by 120°. Similarly, the connection to the 'blue' terminal B must be taken from the ninth slot in order that the e.m.f. in phase B may lag the e.m.f. in phase Y by 120°. Ends R_1, Y_1 and B_1 of the three phases can be joined to form the neutral point of a star-connected system. If the windings are to be delta connected, end R_1 of phase R is joined to the beginning of Y, end Y_1 to the beginning of B and end B_1 to the beginning of R, as shown in Fig. 37.8.

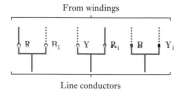

Fig. 37.8 Delta connection of windings

(b)   Double-layer winding

Let us consider a four-pole, three-phase synchronous machine with two slots per pole per phase and two conductors per slot. Figure 37.9 shows the simplest arrangement of the end connections of one phase, the thick lines representing the conductors (and their end connections) forming, say, the outer layer and the thin lines representing conductors forming the inner layer of the winding. The coils are assumed full-pitch, i.e. the spacing between the two sides of each turn is exactly a pole pitch. The main feature of the end connections of a double-layer winding is the strap X, which enables the coils of any one phase to be connected so that all the e.m.f.s of that phase are assisting one another.

Since there are six slots per pole, the phase difference between the e.m.f.s of adjacent slots is 180°/6, namely 30 electrical degrees; since there is a phase difference of 120° between the e.m.f.s of phases R and Y, there must be four

Fig. 37.9 One phase of a three-phase double-layer winding

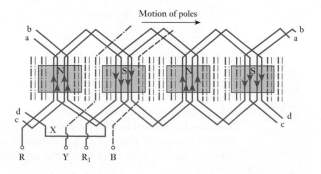

slot pitches between the first conductor of phase R and that of phase Y. Similarly, there must be four slot pitches between the first conductors of phases Y and B. Hence, if the outer conductor of the third slot is connected to terminal R, the corresponding conductor in the seventh slot, in the direction of rotation of the poles, is connected to terminal Y and that in the eleventh slot to terminal B.

In general, the single-layer winding is employed where the machine has a large number of conductors per slot, whereas the double-layer winding is more convenient when the number of conductors per slot does not exceed eight.

37.4	Expression for the e.m.f. of a stator winding

Let Z = no. of conductors in series per *phase*

Φ = useful flux per pole, in webers

p = no. of pairs of poles

and n_1 = speed in revolutions per second

Magnetic flux cutting a conductor in 1 revolution is $\Phi \times 2p$.

Magnetic flux cutting a conductor in 1 s = $2\Phi p \times n_1$

$\therefore$ Average e.m.f. generated in one conductor = $2\Phi p n_1$ volt

If the stator of a three-phase machine had only three slots per pole, i.e. one slot per phase, and if the coils were full-pitch, the e.m.f.s generated in all the conductors of one phase would be in phase with one another and could therefore be added arithmetically. Hence, for a winding concentrated in one slot per phase,

Average e.m.f. per phase = $Z \times 2\Phi p n_1$ volts

If the e.m.f. wave is assumed to be sinusoidal, r.m.s. value of e.m.f. per phase for one slot per phase is

$1.11 \times 2Z \times n_1 p \times \Phi$

e.m.f. per phase = $2.22 Z f \Phi$ [37.2]

A winding concentrated in one slot per pole per phase would have two disadvantages:

1.   The size of such slots and the number of conductors per slot would be so great that it would be difficult to prevent the insulation on the conductors in the centre of the slot becoming overheated, since most of the heat generated in the slots has to flow radially outwards to the steel core.

2.   The waveform of the e.m.f. would be similar to that of the flux distribution around the inner periphery of the stator; and, in general, this would not be sinusoidal.

37.5	Production of rotating magnetic flux by three-phase currents

Let us again consider a two-pole machine and suppose the three-phase winding to have only one slot per pole per phase, as shown in Fig. 37.10. The end connections of the coils are not shown, but are assumed to be similar to those already shown in Fig. 37.4; thus R and R_1 represent the 'start' and the 'finish' of the 'red' phase, etc. It will be noted that R, Y and B are displaced 120 electrical degrees relative to one another. Let us also assume that the

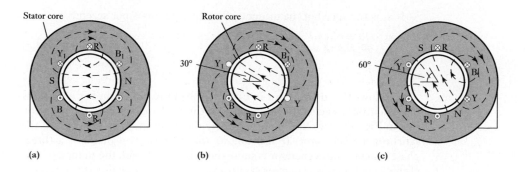

Fig. 37.10 Distribution of magnetic flux due to three-phase currents

current is positive when it is flowing inwards in conductors R, Y and B, and therefore outwards in R_1, Y_1 and B_1. As far as the present discussion is concerned, the rotor core need only consist of circular steel laminations to provide a path of low reluctance for the magnetic flux.

Suppose the currents in the three phases to be represented by the curves in Fig. 37.11; then at instant *a* the current in phase R is positive and at its maximum value, whereas in phases Y and B the currents are negative and each is half the maximum value. These currents, represented in direction by dots and crosses in Fig. 37.10(a), produce the magnetic flux represented by the dotted lines. At instant *b* in Fig. 37.11, the currents in phases R and B are each 0.866 of the maximum; and the distribution of the magnetic flux due to these currents is shown in Fig. 37.10(b). It will be seen that the axis of this field is in line with coil YY_1 and therefore has turned clockwise through 30° from that of Fig. 37.10(a). At instant *c* in Fig. 37.11, the current in phase B has attained its maximum negative value, and the currents in R and Y are both positive, each being half the maximum value. These currents produce the magnetic flux shown in Fig. 37.10(c), the axis of this flux being displaced clockwise by another 30° compared with that in Fig. 37.10(b).

Fig. 37.11 Three-phase currents

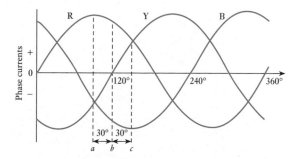

These three cases are sufficient to prove that for every interval of *time* corresponding to 30° along the horizontal axis of Fig. 37.11, the axis of the magnetic flux in a two–pole stator moves forward 30° in *space*. Consequently, in one cycle, the flux rotates through one revolution or two pole pitches. If the stator is wound for *p* pairs of poles, the magnetic flux rotates through $1/p$ revolutions in one cycle and therefore through f/p revolutions in 1 s.

If n_1 is the speed of the magnetic flux in revolutions per second

$$n_1 = f/p \qquad\qquad [37.3]$$

or $\qquad f = n_1 p$

It follows that if the stator in Fig. 37.10 had the same number of poles as the synchronous generator supplying the three-phase currents, the magnetic flux in Fig. 37.10 would rotate at *exactly* the same speed as the poles of the synchronous generator. It also follows that when a two-phase or a three-phase synchronous generator is supplying a balanced load, the stator currents of that synchronous generator set up a resultant magnetic flux that rotates at exactly the same speed as the poles. Hence the magnetic flux is said to rotate at *synchronous speed*, n_1.

37.6 Analysis of the resultant flux due to three-phase currents

We shall proceed by assuming the distribution of the flux density due to each phase to be sinusoidal over a pole pitch. Hence, the full-line curves in Fig. 37.12(a) represent the distribution of the flux densities due to the three phases at instant *a* in Fig. 37.11 and the dotted curve represents the resultant flux density on the assumption that the magnetic circuit is unsaturated. If B_m represents the maximum flux density due to the maximum current in one phase alone, it will be seen from Fig. 37.12(a) that the contributions made to the maximum resultant flux density by phases R, Y and B are B_m, $0.25B_m$ and $0.25B_m$ respectively, and that the peak value of the resultant flux density is therefore $1.5B_m$.

Figure 37.12(b) represents the distribution of the flux densities due to phases R and B at instant *b* in Fig. 37.11 and again it will be seen that the

Fig. 37.12 Distribution of flux densities due to three-phase currents

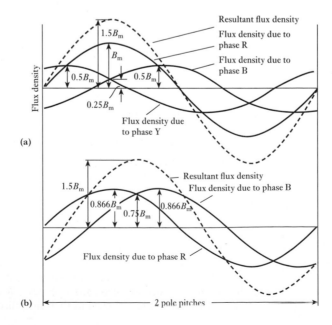

Fig. 37.13 Distribution of flux densities due to three-phase currents

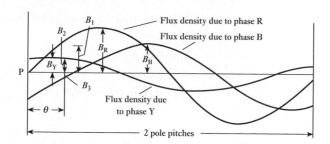

peak value of the resultant flux density is $1.5B_m$ and that its position has shifted a sixth of a pole pitch in a twelfth of a cycle.

Similarly, if the distribution of the resultant flux density were derived for instant c of Fig. 37.11, it would be found that the peak value of the resultant flux density would again be $1.5B_m$ and that it would have moved through a further sixth of a pole pitch in a twelfth of a cycle and therefore through two pole pitches in one cycle.

It will now be shown that the peak value of the resultant flux density remains constant at *every* instant and that its position rotates at a *uniform* speed.

Suppose the curves in Fig. 37.13 to represent the distributions of the flux densities due to the three phases at an instant t seconds after the flux due to phase R has passed through zero from negative to positive values, and suppose B_m to be the peak density due to one phase when the current in that phase is at its maximum value. Then, at the instant under consideration, maximum flux density due to phase R is

$$B_R = B_m \sin \omega t$$

maximum flux density due to phase Y is

$$B_Y = B_m \sin\left(\omega t - \frac{2\pi}{3}\right)$$

and maximum flux density due to phase B is

$$B_B = B_m \sin\left(\omega t - \frac{4\pi}{3}\right)$$

If P is a point of zero flux density for phase R, then for a point θ electrical radians from P, flux density due to phase R is

$$B_1 = B_R \sin \theta = B_m \sin \omega t \cdot \sin \theta$$

flux density due to phase Y is

$$B_2 = B_m \sin\left(\omega t - \frac{2\pi}{3}\right) \cdot \sin\left(\theta - \frac{2\pi}{3}\right)$$

and flux density due to phase B is

$$B_3 = B_m \sin\left(\omega t - \frac{4\pi}{3}\right) \cdot \sin\left(\theta - \frac{4\pi}{3}\right)$$

If the magnetic circuit is unsaturated, the resultant flux density at a point θ electrical radians from P is

$$B_1 + B_2 + B_3$$

$$= B_m \left\{ \sin \omega t \cdot \sin \theta + \sin\left(\omega t - \frac{2\pi}{3} \right) \cdot \sin\left(\theta - \frac{2\pi}{3} \right) \right.$$

$$\left. + \sin\left(\omega t - \frac{4\pi}{3} \right) \cdot \sin\left(\theta - \frac{4\pi}{3} \right) \right\}$$

$$= 1.5 B_m (\sin \omega t \cdot \sin \theta + \cos \omega t \cdot \cos \theta)$$

$$= 1.5 B_m \cos(\omega t - \theta)$$

and is constant at $1.5 B_m$ when $\theta = \omega t = 2\pi f t$.

For a value of t equal to $1/f$, namely the duration of one cycle,

$$\theta = 2\pi \text{ electrical radians}$$

i.e. the position of the peak value of the resultant flux density rotates through two pole pitches in one cycle.

Since the distribution of the resultant flux density remains sinusoidal and the peak value remains constant, it follows that the total flux over a pole pitch is constant and rotates at a uniform speed through two pole pitches in one cycle. If the machine has p pairs of pole, the resultant magnetic flux rotates through $1/p$ revolution in one cycle, therefore speed of rotating magnetic flux is

$$n_1 = f/p \text{ revolutions per second}$$

or $$f = n_1 p \qquad\qquad\qquad\qquad\qquad\qquad [37.4]$$

which is the same as expression [37.3] derived in section 37.5 for the frequency of the e.m.f. generated in the stator winding.

37.7   **Reversal of direction of rotation of the magnetic flux**

Suppose the stator winding to be connected as in Fig. 37.14(a) and that this arrangement corresponds to that already shown in Fig. 37.4 and discussed in section 37.5. The resultant magnetic flux was found to rotate clockwise. Let us interchange the connections between two of the supply lines, say Y and B, and the stator windings, as shown in Fig. 37.14(b). The distribution of currents at instant a in Fig. 37.11 will be exactly as shown in Fig. 37.10(a);

Fig. 37.14 Reversal of direction of rotation

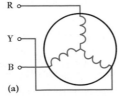

(a)

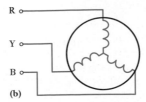

(b)

but at instant *b*, the current in the winding that was originally the 'yellow' phase is now 0.866 of the maximum and the distribution of the resultant magnetic flux is as shown in Fig. 37.15. From a comparison of Figs 37.10(a) and 37.15 it is seen that the axis of the magnetic flux is now rotating anti-clockwise. The same result may be represented thus:

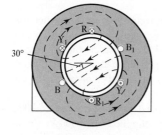

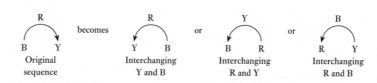

Fig. 37.15 Distribution of magnetic flux at instant b of Fig. 37.11

From this it will be seen that the direction of rotation of the resultant magnetic flux can be reversed by reversing the connections to any two of the three terminals of the motor. The ease with which it is possible to reverse the direction of rotation constitutes one of the advantages of three-phase motors.

Summary of important formulae

The rotational speed per second,

$$f = np \tag{37.1}$$

Terms and concepts

The speed of a synchronous machine depends on the frequency and the number of pole pairs.

The rotor can be salient or cylindrical and is excited by direct current.

The stator has three phase windings which if excited by a three-phase supply can produce a rotating magnetic field. The direction of rotation can be reversed by interchanging two of the phase supplies.

Exercises 37

1. Explain why the e.m.f. generated in a conductor of an alternating-current generator is seldom sinusoidal.

   A rectangular coil of 55 turns, carried by a spindle placed at right angles to a magnetic field of uniform density, is rotated at a constant speed. The mean area per turn is 300 cm². Calculate: (a) the speed in order that the frequency of the generated e.m.f. may be 60 Hz; (b) the density of the magnetic field if the r.m.s. value of the generated e.m.f. is 10 V.

2. The flux density in the air-gap of a synchronous machine at equal intervals is as follows:

Angle (degrees)	0	15	30	45	60	75	90
Flux density (T)	0	0.1	0.4	0.9	1.0	1.0	1.0

   Derive the waveform of the resultant e.m.f. of a single-phase synchronous motor having six slots per pole when the winding is: (a) concentrated in one slot per pole; (b) distributed in two adjacent slots per pole; (c) distributed in four adjacent slots per pole.

3. The distribution of flux density in a synchronous machine is trapezoidal, being uniform under the pole faces and decreasing uniformly to zero at points midway between the poles. The ratio of pole arc to pole pitch is 0.6. The machine has four slots per pole. Derive curves representing the waveform of the generated e.m.f. when the winding is concentrated in one slot per pole.

Exercises 37 continued

4. In a synchronous motor, the flux density may be assumed uniform under the poles and zero between the poles. The ratio of pole arc to pole is 0.7. Calculate the form factor of the e.m.f. generated in a full-pitch coil.

5. The field-form of a synchronous machine taken from the pole centre line, in electrical degrees, is given below, the points being joined by straight lines. Determine the form factor of the e.m.f. generated in a full-pitch coil.

Distance from pole centre (degrees)

0	20	45	60	75	90	105	120	135

Flux density (T)   etc.

0.7	0.7	0.6	0.15	0	0	0	−0.15	−0.6

6. A stator has two poles of arc equal to two-thirds of the pole pitch, producing a uniform radial flux of density 1 T. The length and diameter of the armature are both 0.2 m and the speed of rotation is 1500 r/min. Neglecting fringing, draw to scale the waveform of e.m.f. induced in a single fully pitched armature coil of 10 turns. If the coil is connected through slip-rings to a resistor of a value which makes the total resistance of the circuit 4 Ω, calculate the mean torque on the coil. Explain with the aid of sketch how a torque is produced on a coil carrying a current in a magnetic field.

7. A three-phase, star-connected synchronous generator, driven at 900 r/min, is required to generate a line voltage of 460 V at 60 Hz on open circuit. The stator has two slots per pole per phase and four conductors per slot. Calculate: (a) the number of poles; (b) the useful flux per pole.

8. A star-connected balanced three-phase load of 30 Ω resistance per phase is supplied by a 400 V, three-phase generator of efficiency 90 per cent. Calculate the power input to the generator.

9. The stator of an a.c. machine is wound for six poles, three-phase. If the supply frequency is 25 Hz, what is the value of the synchronous speed?

10. A stator winding supplied from a three-phase, 60 Hz system is required to produce a magnetic flux rotating at 1800 r/min. Calculate the number of poles.

11. A three-phase, two-pole motor is to have a synchronous speed of 9000 r/min. Calculate the frequency of the supply voltage.

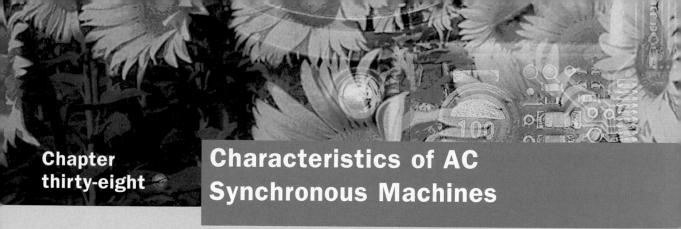

Chapter thirty-eight

Characteristics of AC Synchronous Machines

Objectives

When you have studied this chapter, you should

- understand armature reaction in a synchronous generator
- be familiar with the term voltage regulation
- be familiar with the term synchronous impedance
- understand the parallel operation of generators
- be aware of synchronous motor principles

Contents

In Chapter 37, an understanding of a.c. synchronous machine windings was established. In this chapter, the operational characteristics of the machines themselves are considered so that we know what happens as the load on a motor is increased, or what happens when the rotor excitation is varied.

Most synchronous machines operate within large power systems so that the terminal voltages are relatively constant. Armature reaction determines the relationship between the induced e.m.f. and the terminal voltage, with the result that the current in the machine can lag or lead the voltage. We shall develop an equivalent circuit involving the synchronous impedance. This equivalent circuit is a powerful tool in the analysis of machine performance.

Many synchronous machines operate as generators. In power stations, several generators are connected in parallel to supply the connected load so parallel operation will be considered. We will also introduce the synchronous motor.

By the end of the chapter, exercises analysing the performance of both synchronous generators and motors can be undertaken.

38.1 Armature reaction in a three-phase synchronous generator

'Armature reaction' is the influence of the stator m.m.f. upon the value and distribution of the magnetic flux in the air-gaps between the poles and the stator core. It has already been explained in Chapter 37 that balanced three-phase currents in a three-phase winding produce a resultant magnetic flux of constant magnitude rotating at synchronous speed. We shall now consider the application of this principle to a synchronous generator.

(a) When the current and the generated e.m.f. are in phase

Consider a two-pole, three-phase synchronous generator with two slots per pole per phase. If the machine is on open circuit there is no stator current, and the magnetic flux due to the rotor current is distributed symmetrically as shown in Fig. 38.1. If the direction of rotation of the poles is clockwise, the e.m.f. generated in phase RR_1 is at its maximum and is towards the paper in conductors R and outwards in R_1.

Let us next consider the distribution of flux (Fig. 38.2) due to the stator currents alone at the instant when the current in phase R is at its maximum positive value (instant *a* in Fig. 37.11) and when the rotor (unexcited) is in the position shown in Fig. 38.1. This magnetic flux rotates clockwise at synchronous speed and is therefore stationary relative to the rotor.

We can now derive the resultant magnetic flux due to the rotor and stator currents by superimposing the fluxes of Figs 38.1 and 38.2 on each other. Comparison of these figures shows that over the leading half of each pole face the two fluxes are in opposition, whereas over the trailing half of each pole face they are in the same direction. Hence the effect is to distort the magnetic flux as shown in Fig. 38.3. It will be noticed that the direction of most of the lines of flux in the air-gaps has been skewed and thereby lengthened. But lines of flux behave like stretched elastic cords, and consequently in Fig. 38.3 they exert a backward pull on the rotor; and to overcome the tangential component of this pull, the engine driving the generator has to exert a larger torque than that required on no load. Since the magnetic flux due to the stator currents rotates synchronously with the rotor, the flux distortion shown in Fig. 38.3 remains the same for all positions of the rotor.

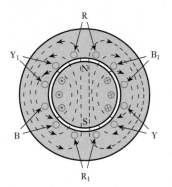

Fig. 38.1 Magnetic flux due to rotor current alone

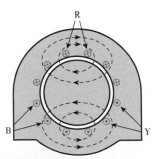

Fig. 38.2 Magnetic flux due to stator currents alone

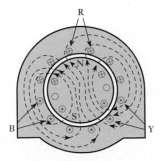

Fig. 38.3 Resultant magnetic flux for case (a)

(b) When the current lags the generated e.m.f. by a quarter of a cycle

When the e.m.f. in phase R is at its maximum value, the poles are in the position shown in Fig. 38.1. By the time the current in phase R reaches its maximum value, the poles will have moved forward through half a pole pitch to the position shown in Fig. 38.4. Reference to Fig. 38.2 shows that the stator m.m.f., acting alone, would send a flux from right to left through the rotor, namely in direct opposition to the flux produced by the rotor m.m.f. Hence it follows that the effect of armature reaction due to a current lagging the e.m.f. by 90° is to reduce the flux. The resultant distribution of the flux, however, is symmetrical over the two halves of the pole face, so that no torque is required to drive the rotor, apart from that to overcome losses.

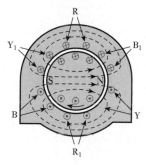

Fig. 38.4 Resultant magnetic flux for case (b)

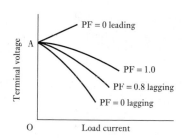

Fig. 38.5 Variation of terminal voltage with load

(c) When the current leads the generated e.m.f. by a quarter of a cycle

In this case the current in phase R is a positive maximum when the N and S poles of the rotor are in the positions occupied by the S and N poles respectively in Fig. 38.4. Consequently the flux due to the stator m.m.f. is now in the same direction as that due to the rotor m.m.f., so that the effect of armature reaction due to a leading current is to increase the flux.

The influence of armature reaction upon the variation of terminal voltage with load is shown in Fig. 38.5, where it is assumed that the field current is maintained constant at a value giving an e.m.f. OA on open circuit. When the power factor of the load is unity, the fall in voltage with increase of load is comparatively small. With an inductive load, the demagnetizing effect of armature reaction causes the terminal voltage to fall much more rapidly. The graph for 0.8 power factor is roughly midway between those for unity and zero power factors. With a capacitive load, the magnetizing effect of armature reaction causes the terminal voltage to increase with increase of load.

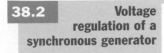

38.2 Voltage regulation of a synchronous generator

An a.c. generator is always designed to give a certain terminal voltage when supplying its rated current at a specified power factor – usually unity or 0.8 lagging. For instance, suppose OB in Fig. 38.6 to represent the full-load current and OA the rated terminal voltage of a synchronous generator. If the field current is adjusted to give the terminal voltage OA when the generator is supplying current OB at unity power factor, then when the load is removed but with the field current and speed kept unaltered, the terminal voltage rises to OC. This variation of the terminal voltage between full load and no load, expressed as a per-unit value or a percentage of the full-load voltage, is termed the per-unit or the percentage *voltage regulation* of the generator; thus:

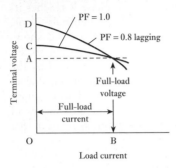

Fig. 38.6 Variation of terminal voltage with load

$$
\text{Per-unit voltage regulation} \\
= \frac{\text{change of terminal voltage when full load is removed}}{\text{full-load terminal voltage}} \quad [38.1]
$$

$$
= \frac{\text{AC}}{\text{OA}} \text{ for unity power factor}
$$

$$
= \frac{\text{AD}}{\text{OA}} \text{ for p.f. of 0.8 lagging}
$$

The voltage regulation for a power factor of 0.8 lagging is usually far greater than that at unity power factor, and it is therefore important to include the power factor when stating the voltage regulation. (See Example 38.1.)

38.3 Synchronous impedance

In Figs 38.3 and 38.4, the resultant flux was shown as the combination of the flux due to the stator m.m.f. alone and that due to the rotor m.m.f. alone. For the purpose of deriving the effect of load upon the terminal voltage, however, it is convenient to regard these two component fluxes as if they existed independently of each other and to consider the cylindrical-rotor type rather than the salient-pole type of generator (see section 37.2). Thus the flux due to the rotor m.m.f. may be regarded as generating an e.m.f., E, due to the rotation of the poles, this e.m.f. being a maximum in any one phase when the conductors of that phase are opposite the centres of the poles. On the other hand, the rotating magnetic field due to the stator currents can be regarded as generating an e.m.f. lagging the current by a quarter of a cycle. For instance, in Fig. 37.10(a), the current in R is at its maximum value flowing towards the paper, but the e.m.f. induced in R by the rotating flux due to the stator currents is zero at that instant.

A quarter of a cycle later, this rotating flux will have turned clockwise through 90°; and since we are considering a generator having a cylindrical rotor and therefore a uniform air-gap, R is then being cut at the maximum rate by flux passing from the rotor to the stator. Hence the e.m.f. induced in R at that instant is at its maximum value acting towards the paper. Since the e.m.f. induced by the rotating flux in any one phase lags the current in that phase by a quarter of a cycle, the effect is exactly similar to that of inductive reactance, i.e. the rotating magnetic flux produced by the stator currents can be regarded as being responsible for the reactance of the stator winding. Furthermore, since the rotating flux revolves synchronously with the poles, this reactance is referred to as the *synchronous reactance* of the winding.

By combining the resistance with the synchronous reactance of the winding, we obtain its *synchronous impedance*. Thus, if

$$X_s = \text{synchronous reactance per phase}$$

$$R = \text{resistance per phase}$$

and $\quad Z_s = \text{synchronous impedance per phase}$

then $\quad Z_s = \sqrt{(R^2 + X_s^2)}$

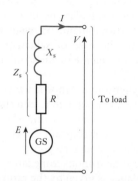

Fig. 38.7 Equivalent circuit of a synchronous generator

In generator windings, R is usually very small compared with X_s, so that for many practical purposes, Z_s can be assumed to be the same as X_s.

The relationship between the terminal voltage V of the generator and the e.m.f. E generated by the flux due to the rotor m.m.f. alone can now be derived. Thus in Fig. 38.7, GS represents *one* phase of the stator winding, and R and X_s represent the resistance and synchronous reactance of that phase. If the load takes a current I at a lagging power factor $\cos \phi$, the various quantities can be represented by phasors as in Fig. 38.8, where

$$OI = \text{current per phase}$$

$$OV = \text{terminal voltage per phase}$$

$OA = IR = $ component of the generated e.m.f. E absorbed in sending current through R

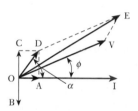

Fig. 38.8 Phasor diagram for a synchronous generator

$OB = $ e.m.f. per phase induced by the rotating flux due to stator currents and lags OI by $90°$

$OC = $ component of the generated e.m.f. E required to neutralize OB

$\quad = $ voltage drop due to synchronous reactance X_s

$\quad = IX_s$

$OD = $ component of the generated e.m.f. absorbed in sending current through the synchronous impedance Z_s

$\alpha = $ phase angle between OI and OD $\tan^{-1}(X_s/R)$

and $\quad OE = $ resultant of OV and OD

$\quad = $ e.m.f. per phase generated by the flux due to the rotor

From Fig. 38.8

$$OE^2 = OV^2 + OD^2 + 2 \cdot OV \cdot OD \cdot \cos(\alpha - \phi)$$

i.e. $\quad E^2 = V^2 + (IZ_s)^2 + 2V \cdot IZ_s \cos(\alpha - \phi)$

from which the e.m.f. E generated by the flux due to the rotor m.m.f., namely the open-circuit voltage, can be calculated. It follows that if V is the rated terminal voltage per phase of the synchronous generator and I is the full-load current per phase, the terminal voltage per phase obtained when the load is removed, the exciting current and speed remaining unaltered, is given by E; and from expression [38.1]

$$\text{Per-unit voltage regulation} = \frac{E - V}{V}$$

The synchronous impedance is important when we come to deal with the parallel operation of generators, but numerical calculations involving synchronous impedance are usually unsatisfactory, owing mainly to magnetic saturation in the poles and stator teeth and, in the case of salient-pole machines, to the value of Z_s varying with the power factor of the load and with the excitation of the poles.

One method of estimating the value of the synchronous impedance is to run the generator on open circuit and measure the generated e.m.f., and then short-circuit the terminals through an ammeter and measure the short-circuit current, the exciting current and speed being kept constant. Since the e.m.f. generated on open circuit can be regarded as being responsible for circulating the short-circuit current through the synchronous impedance of the winding, the value of the synchronous impedance is given by the ratio of the open-circuit voltage per phase to the short-circuit current per phase.

Example 38.1

A three-phase, 600 MVA generator has a rated terminal voltage of 22 kV (line). The stator winding is star-connected and has a resistance of 0.014 Ω/phase and a synchronous impedance of 0.16 Ω/phase. Calculate the voltage regulation for a load having a power factor of

(a) unity;
(b) 0.8 lagging.

(a) Since

$$600 \times 10^6 = 1.73 I_L \times 22 \times 10^3$$

∴   Line current = I_L = 15.7 kA = phase current

Terminal voltage per phase on full load is

$$\frac{22\ 000}{1.73} = 12\ 700 \text{ V}$$

Voltage drop per phase on full load due to synchronous impedance is

$$15\ 700 \times 0.16 = 2540 \text{ V}$$

From Fig. 38.8 it follows that at unity power factor,

$$OV = 12\ 700 \text{ V} \qquad OD = 2540 \text{ V} \quad \text{and} \quad \phi = 0$$

Also $\cos \alpha = \dfrac{OA}{OD} = \dfrac{R}{Z_s} = \dfrac{0.014}{0.16} = 0.088$

∴ $OE^2 = (12\ 700)^2 + (2540)^2 + 2 \times 12\ 700 \times 2540 \times 0.088$

$$= 1.734 \times 10^8$$

∴ $OE = 13\ 170 \text{ V}$

and voltage regulation at unity power factor is

$$\frac{13\ 170 - 12\ 700}{12\ 700} = 0.037 \text{ per unit} = 3.7 \text{ per cent}$$

(b) Since the rating of the generator is 600 MVA, the full-load current is the same whatever the power factor.

For power factor of 0.8,

$$\phi = 36.87°$$

Also $\alpha = \cos^{-1} 0.088 = 84.95°$

so that

$$(\alpha - \phi) = 48.08°   \text{ and }   \cos 48.08° = 0.668$$

Hence

$$OE^2 = (12\ 700)^2 + (2540)^2 + 2 \times 12\ 700 \times 2540 \times 0.668$$

$$= 2.111 \times 10^8$$

$$OE = 14\ 530\ V$$

and voltage regulation for power factor 0.8 lagging is

$$\frac{14\ 530 - 12\ 700}{12\ 700} = 0.144\ \text{per unit} = 14.4\ \text{per cent}$$

38.4	Parallel operation of synchronous generators

It is not obvious why two generators continue running in synchronism after they have been paralleled; so let us consider the case of two similar generators, A and B, connected in parallel to the bus-bars, as shown in Fig. 38.9, and let us, for simplicity, assume that there is no external load connected across the bus-bars. We will consider two separate cases:

1. The effect of varying the torque of the driving engine, e.g. by varying the steam supply.
2. The effect of varying the exciting current.

Fig. 38.9 Two generators in parallel

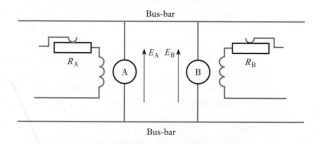

(a)   Effect of varying the driving torque

If we were to draw the complete circuit diagram for two three-phase generators operating in parallel, the diagram would become quite large. In order to simplify the diagram, let us consider a representative phase for each machine as shown in Fig. 38.9. Since all phases are identical, it does not matter which phase is shown.

Let us first assume that each engine is exerting exactly the torque required by its own generator and that the field resistors R_A and R_B are adjusted so that the generated e.m.f.s E_A and E_B are equal. Suppose the arrows in Fig. 38.9 to represent the *positive* directions of these e.m.f.s. It will be seen that, *relative to the bus-bars*, these e.m.f.s are acting in the same direction, i.e. when the e.m.f. generated in each machine is positive, each e.m.f. is making the top bus-bar positive in relation to the bottom bus-bar. But in *relation to each other*, these e.m.f.s are in opposition, i.e. if we trace the closed circuit formed by the two generators we find that the e.m.f.s oppose each other.

In the present discussion we want to find out if any current is being circulated in this closed circuit. It is therefore more convenient to consider these e.m.f.s in relation to each other rather than to the bus-bars. This condition is shown in Fig. 38.10(a) when E_A and E_B are in exact phase opposition relative to each other; and since they are equal in magnitude, their resultant is zero and consequently no current is circulated.

Let us next assume the driving torque of B's engine to be reduced, e.g. by a reduction of its steam supply. The rotor of B falls back in relation to that of A, and Fig. 38.10(b) shows the conditions when B's rotor has fallen back by an angle θ. The resultant e.m.f. in the closed circuit formed by the generator windings is represented by E_Z, and this e.m.f. circulates a current I lagging E_Z by an angle α, where

$$I = \frac{E_Z}{2Z_s} \quad \text{and} \quad \alpha = \tan^{-1} X_s / R$$

where R is the resistance of each generator, X_s the synchronous reactance of each generator and Z_s the synchronous impedance of each generator.

Since the resistance is very small compared with the synchronous reactance, α is nearly 90°, so that the current I is almost in phase with E_A and in phase opposition to E_B. This means that A is generating and B is motoring, and the power supplied from A to B compensates for the reduction of the power supplied by B's engine. If the frequency is to be maintained constant, the driving torque of A's engine has to be increased by an amount equal to the decrease in the driving torque of B's engine.

The larger the value of θ (so long as it does not exceed about 80°), the larger is the circulating current and the greater is the power supplied from A to B. Hence machine B falls back in relation to A until the power taken from the latter exactly compensates for the reduction in the driving power of B's engine. Once this balance has been attained, B and A will run at exactly the same speed.

(b) Effect of varying the excitation

Let us again revert to Fig. 38.10(a) and assume that each engine is exerting the torque required by its generator and that the e.m.f.s, E_A and E_B, are equal. The resultant e.m.f. is zero and there is therefore no circulating current.

Suppose the exciting current of generator B to be increased so that the corresponding open-circuit e.m.f. is represented by E_B in Fig. 38.11. The resultant e.m.f., E_Z (= $E_B - E_A$), circulates a current I through the synchronous impedances of the two generators, and since the machines are assumed similar, the impedance drop per machine is $\frac{1}{2}E_Z$, so that

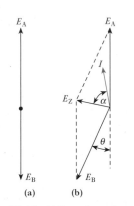

Fig. 38.10 Effect of varying the driving torque

(a)   (b)

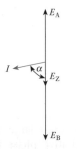

Fig. 38.11 Effect of varying the excitation

$$\text{Terminal voltage} = E_{\text{B}} - \tfrac{1}{2}E_{\text{Z}}$$
$$= E_{\text{A}} + \tfrac{1}{2}E_{\text{Z}}$$

Hence one effect has been to increase the terminal voltage. Further, since angle α is nearly 90°, the circulating current I is almost in quadrature with the generated e.m.f.s, so that very little power is circulated from one machine to the other.

In general, we can therefore conclude:

1. The distribution of load between generators operating in parallel can be varied by varying the driving torques of the engines and only slightly by varying the exciting currents.
2. The terminal voltage is controlled by varying the exciting currents.

38.5 **Three-phase synchronous motor: principle of action**

In section 38.4 it was explained that when two generators, A and B, are in parallel, with *no load* on the bus-bars, a reduction in the driving torque applied to B causes the latter to fall back by some angle θ (Fig. 38.10) in relation to A, so that power is supplied from A to B. Machine B is then operating as a *synchronous motor*.

It will also be seen from Fig. 38.10 that the current in a synchronous motor is approximately in phase opposition to the e.m.f. generated in that machine. This effect is represented in Fig. 38.12 where the rotor poles are shown in the same position relative to the three-phase winding as in Fig. 38.3. The latter represented a synchronous generator with the current in phase with the generated e.m.f., whereas Fig. 38.12 represents a synchronous motor with the current in phase opposition to this generated e.m.f. A diagram similar to Fig. 38.2 could be drawn showing the flux distribution due to the stator currents alone, but a comparison with Figs 38.1, 38.2 and 38.3 can be sufficient to indicate that in Fig. 38.12 the effect of armature reaction is to increase the flux in the leading half of each pole and to reduce it in the trailing half. Consequently the flux is distorted in the direction of rotation and the lines of flux in the gap are skewed in such a direction as to exert a clockwise torque on the rotor. Since the resultant magnetic flux due to the stator currents rotates at synchronous speed, the rotor must also rotate at exactly the same speed for the flux distribution shown in Fig. 38.12 to remain unaltered.

Fig. 38.12 Principle of action of a three-phase synchronous motor

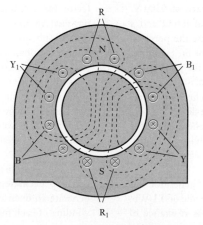

38.6   **Advantages and disadvantages of the synchronous motor**

The principal advantages of the synchronous motor are:

1. The ease with which the power factor can be controlled. An over-excited synchronous motor having a leading power factor can be operated in parallel with induction motors having a lagging power factor, thereby improving the power factor of the supply system.

   Synchronous motors are sometimes run on *no load* for power-factor correction or for improving the voltage regulation of a transmission line. In such applications, the machine is referred to as a *synchronous capacitor*.
2. The speed is constant and independent of the load. This characteristic is mainly of use when the motor is required to drive another generator to generate a supply at a different frequency, as in frequency-changers.

The principal disadvantages are:

1. The cost per kilowatt is generally higher than that of an induction motor.
2. A d.c. supply is necessary for the rotor excitation. This is usually provided by a small d.c. shunt generator carried on an extension of the shaft.
3. Some arrangement must be provided for starting and synchronizing the motor.

Terms and concepts

The magnetic flux in the air-gap between the rotor and the stator is due to the sum of the effects created by the rotor excitation and by the stator excitation. The stator component is called the **armature reaction**.

The rotor excitation varies the stator current and power factor.

The stator reactance gives rise to the synchronous impedance.

Synchronous machines can operate either as motors or as generators. The change of operation is effected by variation of the excitation current in the rotor.

Exercises 38

1. A single-phase synchronous generator has a rated output of 500 kVA at a terminal voltage of 3300 V. The stator winding has a resistance of 0.6 Ω and a synchronous reactance of 4 Ω. Calculate the percentage voltage regulation at a power factor of: (a) unity; (b) 0.8 lagging; (c) 0.8 leading. Sketch the phasor diagram for each case.
2. A single-phase generator having a synchronous reactance of 5.5 Ω and a resistance of 0.6 Ω delivers a current of 100 A. Calculate the e.m.f. generated in the stator winding when the terminal voltage is 2000 V and the power factor of the load is 0.8 lagging. Sketch the phasor diagram.
3. A three-phase, star-connected, 50 Hz generator has 96 conductors per phase and a flux per pole of 0.1 Wb. The stator winding has a synchronous reactance of

5 Ω/ph and negligible resistance. The distribution factor for the stator winding is 0.96. Calculate the terminal voltage when three non-inductive resistors, of 10 Ω/ph, are connected in star across the terminals. Sketch the phasor diagram for one phase.
4. A 1500 kVA, 6.6 kV, three-phase, star-connected synchronous generator has a resistance of 0.5 Ω/ph and a synchronous reactance of 5 Ω/ph. Calculate the percentage change of voltage when the rated output of 1500 kVA at power factor 0.8 lagging is switched off. Assume the speed and the exciting current remain unaltered.
5. Two single-phase generators are connected in parallel, and the excitation of each machine is such as to generate an open-circuit e.m.f. of 3500 V. The stator winding of each machine has a synchronous reactance

of 30 Ω and negligible resistance. If there is a phase displacement of 40 electrical degrees between the e.m.f.s, calculate: (a) the current circulating between the two machines; (b) the terminal voltage; (c) the power supplied from one machine to the other. Assume that there is no external load. Sketch the phasor diagram.

6. Two similar three-phase star-connected generators are connected in parallel. Each machine has a synchronous reactance of 4.5 Ω/ph and negligible resistance, and is excited to generate an e.m.f. of 1910 V/ph. The machines have a phase displacement of 30 electrical degrees relative to each other. Calculate: (a) the circulating current; (b) the terminal voltage/phase; (c) the active power supplied from one machine to the other. Sketch the phasor diagram for one phase.

7. If the two generators of Q. 6 are adjusted to be in exact phase opposition relative to each other and if the excitation of one machine is adjusted to give an open-circuit voltage of 2240 V/ph and that of the other machine adjusted to give an open-circuit voltage of 1600 V/ph, calculate: (a) the circulating current; (b) the terminal voltage.

8. A generator supplying 2800 kW at power factor 0.7 lagging is loaded to its full capacity, i.e. its maximum rating in kVA. If the power factor is raised to unity by means of an over-excited synchronous motor, how much more active power can the generator supply and what must be the power factor of the synchronous motor, assuming that the latter absorbs all extra power obtainable from the generator? Sketch the phasor diagram.

9. A three-phase, 50 Hz, star-connected generator, driven at 1000 r/min, has an open-circuit line voltage of 460 V when the field current is 16 A. The stator winding has a synchronous reactance of 2 Ω/ph and negligible resistance. Calculate the value of M and from it determine the driving torque when the machine is supplying 50 A/ph at a p.f. of 0.8 lagging, assuming the field current and the speed to remain 16 A and 1000 r/min respectively. Neglect the core and friction losses.

   Note: $E^2 = (V \cos \phi)^2 + (V \sin \phi + IX)^2$ and $\cos \alpha = (V \cos \phi)/W$, where $\cos \phi = $ p.f. of load.

Chapter
thirty-nine

Induction Motors

Objectives

When you have studied this chapter, you should
- be familiar with induction motor action
- understand the principle of slip
- understand the variation of torque with slip
- have an understanding of starting torque
- be familiar with methods of starting three-phase motors
- be familiar with cage and wound rotors
- be familiar with different single–phase motor constructions

Contents

The synchronous motor has limited application. However, the asynchronous machine, the induction motor, has widespread industrial and domestic application such that about 50 per cent of electric power consumption is due to induction motor loads.

In this chapter, the construction and principles of induction motor operation are first considered. Thereafter, the principle of slip and the variation of torque with slip are studied. We shall then discuss the starting of induction machines and study different starting techniques for three-phase motors. The two principal forms of rotor construction, the cage and wound rotor types, display different starting characteristics which can be matched to the demands of particular loads.

In conclusion, we will consider the single-phase motor which has widespread small power application, for example in the home. Special winding configurations are necessary in order to start such motors from single-phase supplies. Three different machine winding arrangements will be considered: the capacitor-run motor, the split-phase motor and the shaded-pole motor.

<table>
<tr><td>**39.1**</td><td>**Principle of action**</td></tr>
</table>

The stator of an induction motor is similar to that of a synchronous machine, and in the case of a machine supplied with three-phase currents, a rotating magnetic flux is produced, as already explained in section 37.3. The rotor core is laminated and the conductors often consist of uninsulated copper or aluminium bars in semi-enclosed slots, the bars being short-circuited at each end by rings or plates to which the bars are brazed or welded. In motors below about 50 kW, the aluminium rotor bars and end rings are often cast in one operation. This type is known as the *cage* or *short-circuited* rotor. The air-gap between the rotor and the stator is uniform and made as small as is mechanically possible. For simplicity, the stator slots and winding have been omitted in Fig. 39.1.

Fig. 39.1 Induction motor with a cage rotor

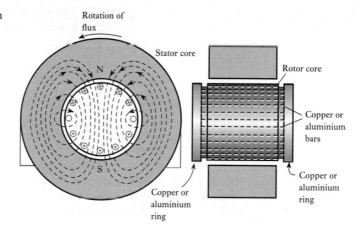

If the stator is wound for two poles, the distribution of the magnetic flux due to the stator currents at a particular instant is shown in Fig. 39.1. The e.m.f. generated in a rotor conductor is a maximum in the region of maximum flux density; if the flux is assumed to rotate anticlockwise, the directions of the e.m.f.s generated in the stationary rotor conductors can be determined by the right-hand rule and are indicated by the crosses and dots in Fig. 39.1. The e.m.f. generated in the rotor conductor shown in Fig. 39.2 circulates a current, the effect of which is to strengthen the flux density on the right-hand side and weaken that on the left-hand side, i.e. the flux in the gap is distorted as indicated by the dotted lines in Fig. 39.2. Consequently, a force is exerted on the rotor tending to drag it in the direction of the rotating flux.

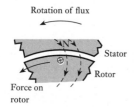

Fig. 39.2 Force on rotor

The higher the speed of the rotor, the lower is the speed of the rotating field relative to the rotor winding and the smaller is the e.m.f. generated in the latter. Should the speed of the rotor attain the synchronous value, the rotor conductors would be stationary in relation to the rotating flux. There would therefore be no e.m.f. and no current in the rotor conductors and consequently no torque on the rotor. Hence the latter could not continue rotating at synchronous speed. As the rotor speed falls more and more below the synchronous speed, the values of the rotor e.m.f. and current and therefore of the torque continue to increase until the latter is equal to that required by the rotor losses and by any load there may be on the motor.

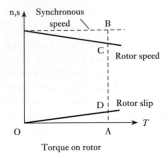

Fig. 39.3 Slip and rotor speed of an induction motor

The speed of the rotor relative to that of the rotating flux is termed the *slip*; thus for a torque OA in Fig. 39.3, the rotor speed is AC and the slip is AD, where

$$AD = AB - AC = CB$$

For torques varying between zero and the full-load value, the slip is practically proportional to the torque. It is usual to express the slip either as a per-unit or fractional value or as a percentage of the synchronous speed; thus in Fig. 39.3

Slip   Symbol: s Unit: none

$$\text{Per-unit slip} = \frac{\text{slip in r/min}}{\text{synchronous speed in r/min}} = \frac{AD}{AB}$$

$$s = \frac{\text{synchronous speed} - \text{rotor speed}}{\text{synchronous speed}}$$

$$s = \frac{n_1 - n_r}{n_1} \qquad\qquad [39.1]$$

and   Percentage slip = per-unit or fractional slip $\times$ 100

$$= \frac{AD}{AB} \times 100$$

The value of the slip at full load varies from about 6 per cent for small motors to about 2 per cent for large machines. The induction motor may therefore be regarded as practically a constant-speed machine; the difficulty of varying its speed economically constitutes one of its main disadvantages.

39.2   Frequency of rotor e.m.f. and current

It was shown in section 37.3 that for a three-phase winding with p pairs of poles supplied at a frequency of f hertz the speed of the rotating flux is given by n_1 revolutions per second, where

$$f = n_1 p$$

If n_r is the rotor speed in revolutions per second, the speed at which the rotor conductors are being cut by the rotating flux is $(n_1 - n_r)$ revolutions per second,

$$\text{Frequency of rotor e.m.f.} = f_r = (n_1 - n_r)p$$

If s = per-unit or fractional slip = $(n_1 - n_r)/n_1$

then $n_1 - n_r = sn_1$

and $f_r = sn_1 p = sf \qquad\qquad [39.2]$

Polyphase currents in the stator winding produce a resultant magnetic field, the axis of which rotates at synchronous speed, n_1 revolutions per second, relative to the stator. Similarly the polyphase currents in the rotor winding produce a resultant magnetic field, the axis of which rotates (by expression [39.2]) at a speed sn_1 revolutions per second relative to the rotor surface, *in the direction of rotation of the rotor*. But the rotor is revolving at a speed n_r revolutions per second relative to the stator core; hence the speed of the resultant rotor magnetic field relative to the stator core is

$$sn_1 + n_r = (n_1 - n_r) + n_r = n_1 \text{ revolutions per second}$$

i.e. the axis of the resultant rotor field m.m.f. is travelling at the same speed as that of the resultant stator field m.m.f., so that they are stationary relative to each other. Consequently the polyphase induction motor can be regarded as being equivalent to a transformer having an air-gap separating the steel portions of the magnetic circuit carrying the primary and secondary windings.

Because of this gap, the magnetizing current and the magnetic leakage for an induction motor are large compared with the corresponding values for a transformer of the same apparent power rating. Also, the friction and windage losses contribute towards making the efficiency of the induction motor less than that of the corresponding transformer. On the other hand, the stator field m.m.f. has to balance the rotor field m.m.f. and also provide the magnetizing and no-load loss components of the stator current, as in a transformer. Hence, an increase of slip due to increase of load is accompanied by an increase of the rotor currents and therefore by a corresponding increase of the stator currents.

39.3 Rotor e.m.f. and current

Let V_p be the voltage per phase applied to stator winding and Z_s be the number of stator conductors in series per phase. If E_0 = rotor e.m.f. generated per phase at standstill, and if E_r is the rotor e.m.f. generated per phase when the per-unit slip is s and the rotor frequency is $f_r = sf$, then

$$E_r = sE_0 \qquad\qquad [39.3]$$

If R = resistance per phase of the rotor winding

and X_0 = leakage reactance per phase of rotor winding at standstill

= $2\pi f \times$ leakage inductance per phase of rotor winding

then for per-unit or fractional slip s, corresponding reactance per phase is

$$X_r = sX_0$$

and corresponding impedance per phase is

$$Z_r = \sqrt{\{R^2 + (sX_0)^2\}} \qquad\qquad [39.4]$$

If I_0 = rotor current per phase at standstill

and I_r = rotor current per phase at slip s

$$I_0 = \frac{E_0}{\sqrt{(R^2 + X_0^2)}}$$

and $$I_r = \frac{E_r}{\sqrt{\{R^2 + (sX_0)^2\}}} = \frac{sE_0}{\sqrt{\{R^2 + (sX_0)^2\}}} \qquad\qquad [39.5]$$

If ϕ_r is the phase difference between E_r and I_r

$$\tan \phi_r = \frac{X_r}{R} = \frac{sX_0}{R} \qquad\qquad [39.6]$$

and

$$\cos \phi_r = \frac{R}{\sqrt{(R^2 + X_r^2)}} \qquad\qquad [39.7]$$

Example 39.1 A three-phase induction motor is wound for four poles and is supplied from a 50 Hz system. Calculate:

(a) the synchronous speed;
(b) the speed of the rotor when the slip is 4 per cent;
(c) the rotor frequency when the speed of the rotor is 600 r/min.

(a) From equation [37.1]

$$\text{Synchronous speed} = \frac{60f}{p} = \frac{60 \times 50}{2}$$

$$= \mathbf{1500\ r/min}$$

(b) From equation [39.1]

$$0.04 = \frac{1500 - \text{rotor speed}}{1500}$$

$$\therefore \qquad \text{Rotor speed} = \mathbf{1440\ r/min}$$

(c) Also from equation [39.1]

$$\text{Per-unit slip} = \frac{1500 - 600}{1500} = 0.6$$

Hence, from equation [39.2],

$$\text{Rotor frequency} = 0.6 \times 50 = \mathbf{30\ Hz}$$

Example 39.2 The stator winding of the motor of Example 39.1 is delta-connected with 240 conductors per phase and the rotor winding is star-connected with 48 conductors per phase. The rotor winding has a resistance of 0.013 Ω/ph and a leakage reactance of 0.048 Ω/ph at standstill. The supply voltage is 400 V. Calculate:

(a) the rotor e.m.f. per phase at standstill with the rotor on open circuit;
(b) the rotor e.m.f. and current per phase at 4 per cent slip;
(c) the phase difference between the rotor e.m.f. and current for a slip of (i) 4 per cent and (ii) 100 per cent.

Assume the impedance of the stator winding to be negligible.

(a) $\qquad E_0 = 400 \times \dfrac{48}{240} = \mathbf{80\ V}$

(b) From equation [39.3]

Rotor e.m.f. for 4 per cent slip $= 80 \times 0.04 = 3.2$ V

From [39.4] impedance per phase for 4 per cent slip is

$$\sqrt{\{(0.013)^2 + (0.04 \times 0.048)^2\}} = 0.013\ 14\ \Omega$$

$\therefore$ Rotor current $= 3.2/0.013\ 14 = \textbf{244 A}$

(c) From equation [39.6] it follows that for 4 per cent slip

$$\tan \phi_r = \frac{0.04 \times 0.048}{0.013} = 0.1477$$

$$\phi_r = 8°24'$$

For 100 per cent slip

$$\tan \phi_r = 0.048/0.013 = 3.692$$

$$\phi_r = \textbf{74°51'}$$

Example 39.2 shows that the slip has a considerable effect upon the phase difference between the rotor e.m.f. and current – a fact that is very important when we come to discuss the variation of torque with slip.

39.4 Relationship between the rotor I^2R loss and the rotor slip

The following diagram indicates concisely what becomes of the power supplied to the stator of the induction motor:

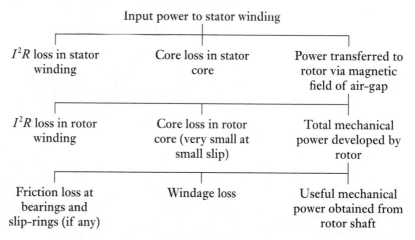

If M = torque, in newton metres, exerted on the rotor by the rotating flux and n_1 = synchronous speed in revolutions per second, power transferred from stator to rotor is

$2\pi M n_1$ watts

If n_r = rotor speed in revolutions per second, total mechanical power developed by rotor is

$2\pi M n_r$ watts

But from the diagram above it is seen that total I^2R loss in rotor

$\simeq$ power transferred from stator to rotor

$-$ total mechanical power developed by rotor

$= 2\pi M(n_1 - n_r)$ watts

$\therefore$
$$\frac{\text{Total rotor } I^2R \text{ loss}}{\text{Input power to rotor}} = \frac{2\pi M(n_1 - n_r)}{2\pi Mn_1} = s \qquad [39.8]$$

or total rotor I^2R loss (in watts) is

$s \times$ input power to rotor (in watts)

39.5 Factors determining the torque

If m = number of rotor phases, then, using the symbols given in section 39.3, we have

Electrical power generated in rotor $= mI_r E_r \cos \phi_r$ watts

$$= \frac{ms^2 E_0^2 R}{R^2 + (sX_0)^2}$$

All this power is dissipated as I^2R loss in the rotor circuits.

Since input power to rotor $= 2\pi Mn_1$ watts, hence, from equation [39.8], we have

$$s \times 2\pi Mn_1 = \frac{ms^2 E_0^2 R}{R^2 + (sX_0)^2}$$

Consequently, for given synchronous speed and number of rotor phases,

$$M \propto \frac{sE_0^2 R}{R^2 + (sX_0)^2} \propto \frac{s\Phi^2 R}{R^2 + (sX_0)^2} \qquad [39.9]$$

since $E_0 \propto \Phi$.

39.6 Variation of torque with slip, other factors remaining constant

If the impedance of the stator winding is assumed to be negligible, then for a given supply voltage, Φ and E_0 remain constant,

$\therefore$
$$\text{Torque } M \propto \frac{sR}{R^2 + (sX_0)^2} \qquad [39.10]$$

The value of X_0 is usually far greater than the resistance of the rotor winding, so let us for simplicity assume $R = 1\,\Omega$ and $X_0 = 8\,\Omega$, and calculate the value of $sR/(R^2 + s^2X_0^2)$ for various values of the slip between 1 and 0. The results are represented by curve A in Fig. 39.4. It will be seen that, for small values of the slip, the torque is almost directly proportional to the slip, whereas for slips between about 0.2 and 1, the torque is almost inversely proportional to the slip. These relationships can be easily deduced from

Fig. 39.4   Torque/slip curves
for an induction motor

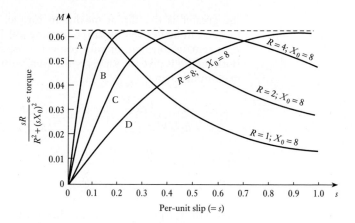

expression [39.10]. Thus, in the case of the cage rotor, R is small compared with X_0, but for values of the slip less than about 0.1 per unit, $(sX_0)^2$ is very small compared with R^2, so that

$$\text{Torque } M \propto \frac{sR}{R^2} \propto \frac{s}{R} \qquad\qquad [39.11]$$

i.e. the torque is directly proportional to the slip when the latter is very small.

For large values of the slip, R^2 is very small compared with $(sX_0)^2$ for the cage rotor and for the slip-ring rotor with no external resistance (section 39.8)

$$\therefore \qquad \text{Torque } M \propto \frac{sR}{(sX_0)^2} \propto \frac{R}{s} \qquad\qquad [39.12]$$

since X_0 is constant for a given motor, i.e. the torque is inversely proportional to the slip when the latter is large. The term R has been left in the above expressions as it is referred to in the next section.

<table>
<tr><td>**39.7**</td><td>**Effect of rotor resistance upon the torque/slip relationship**</td></tr>
</table>

From expression [39.12] it is seen that when R is small compared with sX_0, the torque for a given slip is directly proportional to the value of R, whereas from expression [39.11] it follows that when R is large compared with sX_0, the torque for a given slip is inversely proportional to the value of R. The simplest method of demonstrating this effect is to repeat the calculation of $sR/(R^2 + s^2X_0^2)$ with $R = 2\,\Omega$, $R = 4\,\Omega$ and $R = 8\,\Omega$. The results are represented by curves B, C and D respectively in Fig. 39.4. It will be seen that for a slip of, say, 0.05 per unit, the effect of doubling the rotor resistance is to reduce the torque by about 0.45 per unit, whereas for a slip of 1, the torque is nearly doubled when the resistance is increased from $1\,\Omega$ to $2\,\Omega$. Hence, if a large starting torque is required, the rotor must have a relatively high resistance.

It will also be noticed from Fig. 39.4 that the maximum value of the torque is the same for the four values of R and that the larger the resistance the greater is the slip at maximum torque. The condition for maximum torque can be derived by differentiating [39.10] with respect to s, assuming R to remain constant, or with respect to R, assuming s to remain constant. Both methods give the same result; thus, with the first method, the torque is maximum when

$$\frac{\mathrm{d}}{\mathrm{d}s}\left(\frac{sR}{R^2 + s^2 X_0^2}\right) = \frac{(R^2 + s^2 X_0^2)R - sR \times 2sX_0^2}{(R^2 + s^2 X_0^2)^2} = 0$$

i.e.
$$R^2 - s^2 X_0^2 = 0$$

so that
$$sX_0 = R \qquad\qquad [39.13]$$

Hence the torque is a maximum when the reactance is equal to the resistance. For instance, with $R = 1\ \Omega$ and $X_0 = 8\ \Omega$, maximum torque occurs when $s = 0.125$ per unit, whereas with $R = 8\ \Omega$ and $X_0 = 8\ \Omega$, maximum torque occurs when $s = 1$, namely when the rotor is at standstill.

Substituting R for sX_0 in expression [39.10], we have

$$\text{Maximum torque } M_m \propto \frac{sR}{2R^2}$$

$$\propto \frac{1}{2X_0}$$

But X_0 is the leakage reactance at standstill and is a constant for a given rotor; hence the maximum torque is the same whatever the value of the rotor resistance.

39.8 Starting torque

At the instant of starting, $s = 1$, and it will be seen from Fig. 39.4 that with a motor having a low-resistance rotor, such as the usual type of cage rotor, the starting torque is small compared with the maximum torque available. On the other hand, if the bars of the cage rotor were made with sufficiently high resistance to give the maximum torque at standstill, the slip for full-load torque – usually about one-third to one-half of the maximum torque – would be relatively large and the I^2R loss in the rotor winding would be high, with the result that the efficiency would be low; if this load was maintained for an hour or two, the temperature rise would be excessive. Also, the variation of speed with load would be large. Hence, when a motor is required to exert its maximum torque at starting, the usual practice is to insert extra resistance into the rotor circuit and to reduce the resistance as the motor accelerates. Such an arrangement involves a three-phase winding on the rotor, the three ends of the winding being connected via slip-rings on the shaft to external star-connected resistors R, as shown in Fig. 39.5. The three arms, A, are mechanically and electrically connected together.

The starting procedure is to close the triple-pole switch S and then move arms A clockwise as the motor accelerates, until, at full speed, the arms are in the ON position shown dotted in Fig. 39.5, and the starting resistors have been cut out of the rotor circuit. Large motors are often fitted with a short-circuiting and brush-lifting device which first short-circuits the three

Fig. 39.5 Induction motor with slip-ring rotor

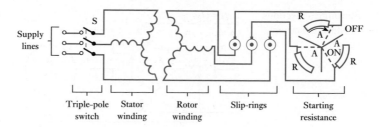

Triple-pole switch   Stator winding   Rotor winding   Slip-rings   Starting resistance

slip-rings and then lifts the brushes off the rings, thereby eliminating losses due to the brush-contact resistance and the brush friction and reducing the wear of the brushes and of the slip-rings.

39.9   Starting of a three-phase induction motor fitted with a cage rotor

If this type of motor is started up by being switched directly across the supply, the starting current is about four to seven times the full-load current, the actual value depending upon the size and design of the machine. Such a large current can cause a relatively large voltage drop in the cables and thereby produce an objectionable momentary dimming of the lamps in the vicinity. Consequently it is usual to start cage motors – except small machines – with a reduced voltage, using one of the methods given under the headings below.

(a)   Star–delta starter

The two ends of each phase of the stator winding are brought out to the starter which, when moved to the 'starting' position, connects the winding in star. After the motor has accelerated, the starter is quickly moved to the 'running' position, thereby changing the connections to delta. Hence the voltage per phase at starting is $1/\sqrt{3}$ of the supply voltage, and the starting torque is one-third of that obtained if the motor were switched directly across the supply with its stator winding delta-connected. Also, the starting current per phase is $1/\sqrt{3}$, and that taken from the supply is one-third of the corresponding value with direct switching.

(b)   Auto-transformer starter

In Fig. 39.6, T represents a three-phase star-connected auto-transformer with a mid-point tapping on each phase so that the voltage applied to motor M is half the supply voltage. With such tappings, the supply current and the starting torque are only a quarter of the values when the full voltage is applied to the motor.

After the motor has accelerated, the starter is moved to the 'running' position, thereby connecting the motor directly across the supply and opening the star-connection of the auto-transformer.

Taking the general case where the output voltage per phase of the auto-transformer is n times the input voltage per phase, we have

Fig. 39.6 Starting connections of an auto-transformer starter

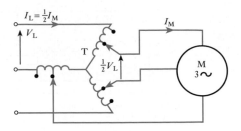

$$\frac{\text{Starting torque with auto-transformer}}{\text{Starting torque with direct switching}} = \left(\frac{\text{output voltage}}{\text{input voltage}}\right)^2 = n^2$$

Also, if Z_0 is the equivalent standstill impedance per phase of a star-connected motor, referred to the stator circuit, and if V_P is the star voltage of the supply, then, with direct switching,

$$\text{Starting current} = \frac{V_P}{Z_0}$$

With auto-transformer starting,

$$\text{Starting current in motor} = \frac{nV_P}{Z_0}$$

and   Starting current from supply $= n \times$ motor current

$$= n^2 V_P / Z_0$$

$\therefore$ $\dfrac{\text{Starting current with auto-transformer}}{\text{Starting current with direct switching}}$

$$= \left(n^2 \frac{V_P}{Z_0}\right) \div \left(\frac{V_P}{Z_0}\right) = n^2$$

The auto-transformer is usually arranged with two or three tappings per phase so that the most suitable ratio can be selected for a given motor, but an auto-transformer starter is more expensive than a star–delta starter.

(c)   Solid-state soft starter

In order to understand this form of starting, we require to understand solid-state switching which will be introduced in Chapter 45. In order to appreciate the uses of solid-state switching in respect of induction motors, sections 45.11 and 45.12 will explore this further.

39.10	Comparison of cage and slip-ring rotors

The cage rotor possesses the following advantages:

1. Cheaper and more robust.
2. Slightly higher efficiency and power factor.
3. Explosion-proof since the absence of slip-rings and brushes eliminates the risk of sparking.

The advantages of the slip–ring rotor are:

1. The starting torque is much higher and the starting current much lower.
2. The speed can be varied especially by means of solid-state switching.

39.11 Braking

We have considered the torque/speed and torque/slip characteristics for induction motors within the slip range 0–1. However, what if the slip were outside these limits? The results of varying the slip are shown in Fig. 39.7.

Fig. 39.7 Torque/slip for an induction machine

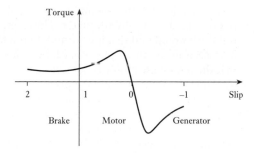

The slip can be quite simply increased above 1 by reversing any two of the phase supplies while operating the machine as a motor. The effect of reversing two supply phases is to make the stator field rotate in the opposite direction; thus at the instant of switch-over, the rotor is rotating almost at synchronous speed in one direction and now the stator field is rotating at synchronous speed in the opposite direction. The difference is almost twice the synchronous speed and hence the slip is almost 2.

The effect is that the rotor now attempts to reverse its direction which, given time, it eventually does. However, initially the effect is one of braking the rotor in order to bring it to a standstill prior to commencing rotation in the opposite direction. This braking effect is known as *plugging* and we can either have plug braking or plug reversal.

For plug braking, we reverse two phase supplies causing the braking. As soon as the machine stops, the power supply is switched off and the machine remains at standstill. However, if we were to leave the power supply switched on, the motor would accelerate the load up to full speed with a direction reversed from that previously experienced.

The action of plugging is very demanding on the supply since the current is very high and care has to be taken when designing the supply system for a plugged motor that the peak currents can be sustained. Also the mechanical effects on both motor and load are severe and they have to be sufficiently robust to withstand the effects. Fortunately the induction motor (normally we use cage rotor motors for plugging), as has already been observed, is very robust and can readily accept the stresses imposed by plugging.

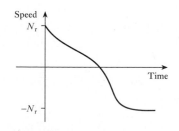

Fig. 39.8 Speed/time characteristic of a plugging motor

The change of speed is illustrated in Fig. 39.8. The time taken for small motors to complete reversal is about 1 or 2 s while larger motors take a little longer.

Finally, returning to the characteristic shown in Fig. 39.7, if the slip were negative, the machine would operate as a generator. There are a limited number of applications in which we can find induction generators, but most significant generators are synchronous machines.

| 39.12 | Single-phase induction motors |

We have considered the three-phase induction motor at some length, but it is not always convenient to have three-phase supplies available. The most common situation is in the house where almost universally we have only one-phase supplies available. It follows that we need motors which can operate from such supplies.

If there is only the possibility of one phase winding, we cannot have a rotating field set up by the stator. However, if we could somehow start the rotor, we would find that it continues to rotate in much the same way that we have seen in the three-phase machine. Therefore the problems we find are that the one-phase motor is not self-starting and we cannot be sure in which direction the rotation will take place. The latter is important – just think if the motor in a vacuum cleaner started in the wrong direction!

Let us consider an explanation of the action of a one-phase motor. Because there is only one phase winding, the field can only act in one direction, varying in magnitude with that of the excitation current. However, we can also represent it by the combined effects of two rotating fields of equal magnitude, but rotating in opposite directions. These are illustrated in Fig. 39.9.

In position (a), the two fields are aligned and their sum gives the peak magnetic field created by the one-phase system. In position (b), the vertical components of the rotating fields are aligned and their sum gives the magnetic field created by the one-phase system; however, the horizontal components are equal and opposite and therefore cancel one another. We

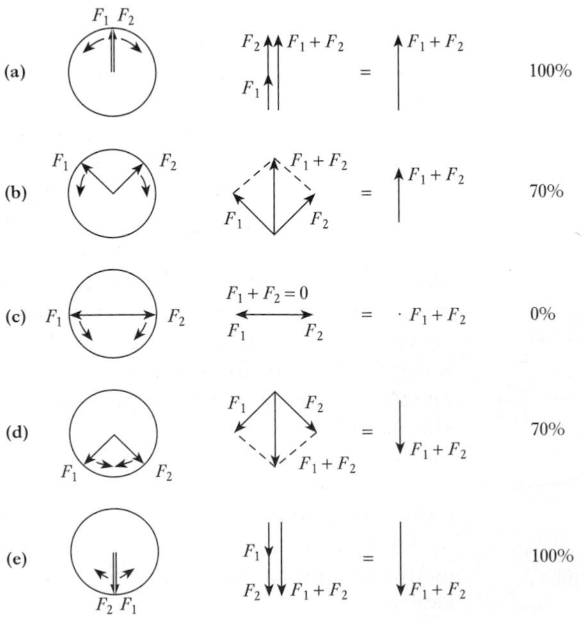

Fig. 39.9 Representation of a one-phase field

can apply the comments for positions (c) and (d); in position (c) there are no vertical components hence the total field is zero. Finally, at position (e), we have almost returned to the situation associated with (a) except that the field acts in the opposite direction.

Therefore we can consider the one-phase winding effectively to set up two rotating magnetic fields, equal in magnitude but opposite in direction. If the rotor is moving, it must be moving in the direction of one of these fields and rotating in the opposite direction to that of the other field. When related to the field moving in the same direction, we have a motor action with a value of slip between 0 and 1. However, when related to the field moving in the opposite direction, we have a plugging or braking action with a value of slip between 2 and 1. The total action is therefore the sum of these two actions – fortunately the motor action is stronger than the plugging action and therefore the device acts as a motor. The summing of the actions is illustrated in Fig. 39.10.

Fig. 39.10 Summation of motor and plugging actions

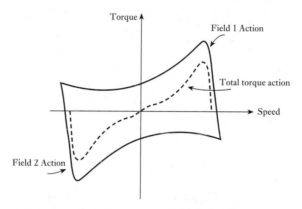

However, if the slip is 1 then the two actions are equal and opposite, hence the machine has no ability to start itself. From Fig. 39.10, we can observe that the torque characteristic of a one-phase induction motor is similar to that of its three-phase cousin, but relatively the torque cannot be quite so strong. It is for this reason that most one-phase induction motors produce outputs up to 1 kW and most have lower ratings.

There remains only the problem of how to start such a motor. It should be appreciated that as soon as movement commences the characteristic shown in Fig. 39.10 indicates that there is a torque which will continue to accelerate the load up to operating speed. All we require is a small torque sufficient to overcome the load torque to start the motor turning.

We know that a three-phase supply creates a rotating field, but this could also be achieved by a two-phase system. Normally we would think of a two-phase system as being a conventional two-phase system with the phases displaced by 90°. However, even if the displacement was less, there still would be a rotating field of some form. This means that we require two stator windings carrying currents which are phase displaced, say by 60°.

The means by which we achieve this gives rise to a variety of one-phase motors.

39.13 Capacitor-run induction motors

In the capacitor-run one-phase induction motor, the stator has two windings physically displaced by about 90°. Both windings can only be energized from the same supply, and to ensure that the windings have currents which are phase displaced, we introduce a capacitor in series with one of the windings. The circuit diagram is shown in Fig. 39.11(a).

Fig. 39.11 Single-phase, capacitor-run induction motor

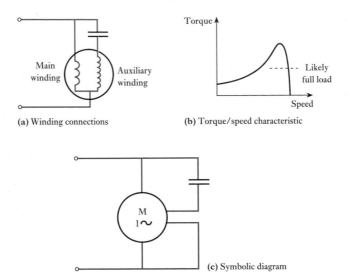

(a) Winding connections   (b) Torque/speed characteristic

(c) Symbolic diagram

The winding connected across the supply is termed the *main winding* and the winding in series with the capacitor is called the *auxiliary winding*. Usually the auxiliary winding is smaller and set deeper into the surface of the stator. Basically this is to ensure that the windings have different impedances which causes the winding currents to have phase separation.

In practice, some capacitor-run machines cause the auxiliary winding to be disconnected as the machine speed increases. This can be arranged using a centrifugal switch. Such motors are called capacitor-start motors.

The torque characteristic is shown in Fig. 39.11(b) from which it can be seen that the starting torque is small. Typical applications with small starting torques include fans where the load torque is proportional to the $(speed)^2$, and therefore there is hardly any load torque (other than friction) at starting.

39.14 Split-phase motors

In the capacitor-run induction motor, the capacitor ensured that the current in the auxiliary winding would lead the current in the main winding. If the capacitor were omitted and the auxiliary winding were made in such a manner that its resistance were emphasized, then again phase difference would occur. Thus in a split-phase motor, the main winding has relatively thick wire and high inductance while the auxiliary winding is relatively thin (giving high resistance) and low inductance.

The circuit diagram and torque characteristic are shown in Fig. 39.12. The starting torque is much improved and is usually about 50 per cent greater than the designed load torque.

Both the capacitor-run and split-phase motors can be made to reverse by changing the polarity of the supply to the auxiliary windings.

Fig. 39.12 Split-phase induction motor

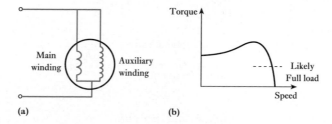

(a)   (b)

39.15 Shaded-pole motors

The shaded-pole motor is one of the most common motors to be found. The reason it is common is that it is probably the cheapest to make. The basic construction is shown in Fig. 39.13. The motor has only one winding, and that is a simple coil wound round the laminated stator. The coil is not inserted into slots in the stator, hence the cheapness of manufacture. The rotor is normally made in the form of a cast cage, again ensuring cheapness of manufacture.

Fig. 39.13 Shaded-pole motor

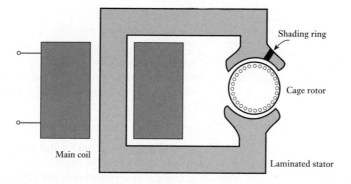

The production of the second field is again most simple. A heavy ring of copper or aluminium is mounted around the shoe of a pole. This short-circuiting ring is the shading ring. The alternating flux passing through the ring induces an e.m.f. which causes a current in the ring. Because it has almost no resistance, the current is very large and therefore it causes a second field to be created. The e.m.f. is 90° displaced from the flux which induced it, but the current in the ring is in phase with the e.m.f. as is the field which it creates. Therefore the second field is approximately 90° out of phase with the main field.

Such motors are among the least efficient and therefore are limited to very small ratings. The low efficiency arises from the losses in the shading ring and in the magnetic circuit.

Summary of important formulae

The synchronous speed,

$$n_1 = \frac{f}{p}$$

hence slip $s = \dfrac{n_1 - n_r}{n_1}$ [39.1]

Rotor frequency,

$$f_r = sf \tag{39.2}$$

Rotor e.m.f. per phase

$$E_r = sE_0 \tag{39.3}$$

Rotor impedance per phase,

$$Z_r = \sqrt{(R^2 + (sX_0)^2)} \tag{39.4}$$

For small slip,

$$\text{Torque } M \propto \frac{s}{R} \tag{39.11}$$

For large slip,

$$\text{Torque } M \propto \frac{R}{s} \tag{39.12}$$

For maximum torque,

$$R = sX_0 \tag{39.13}$$

Terms and concepts

The rotating field of the stator induces e.m.f.s and hence currents in the rotor conductors. The rotor conductors can take the form of either windings as in the wound-rotor machine or short-circuited bars as in the cage-rotor machine.

The speed of the rotor relative to the rotating field is termed the **slip**.

The torque developed varies during the acceleration period and also depends on the ratio of the rotor reactance to the rotor resistance. The reactance also varies during the acceleration period.

The torque must exceed the load torque for the machine to accelerate. Eventually the motor torque falls to balance that of the load at which point the speed stabilizes.

Three-phase induction motors can be started by the star–delta method, the auto-transformer starter and by a soft starter.

Induction motors can be used to brake the load, a procedure called **plugging**.

One-phase motors can be used for small power applications. The common forms are the **capacitor-run motor**, the **split-phase motor** and the **shaded-pole motor**.

Exercises 39

1. Explain how slip–frequency currents are set up in the rotor windings of a three–phase induction motor.

A two–pole, three–phase, 50 Hz induction motor is running on load with a slip of 4 per cent. Calculate the actual speed and the synchronous speed of the machine. Sketch the speed/load characteristic for this type of machine and state with which kind of d.c. motor it compares.

2. Explain why an induction motor cannot develop torque when running at synchronous speed. Define the slip speed of an induction motor and deduce how the frequency of rotor currents and the magnitude of the rotor e.m.f. are related to slip.

An induction motor has four poles and is energized from a 50 Hz supply. If the machine runs on full load at 2 per cent slip, determine the running speed and the frequency of the rotor currents.

3. Give a clear explanation of the following effects in a three–phase induction motor: (a) the production of the rotating field; (b) the presence of an induced rotor current; (c) the development of the torque.

4. If a six–pole induction motor supplied from a three–phase 50 Hz supply has a rotor frequency of 2.3 Hz, calculate: (a) the percentage slip; (b) the speed of the rotor in revolutions per minute.

5. Show how a rotating magnetic field can be produced by three–phase currents.

A 14–pole, 50 Hz induction motor runs at 415 r/min. Deduce the frequency of the currents in the rotor winding and the slip.

6. Explain the principle of action of a three–phase induction motor and the meaning of the term *slip*. How does slip vary with the load?

A centre–zero d.c. galvanometer, suitably shunted, is connected in one lead of the rotor of a three–phase, six–pole, 50 Hz slip–ring induction motor and the pointer makes 85 complete oscillations per minute. What is the rotor speed?

7. Describe, in general terms, the principle of operation of a three–phase induction motor.

The stator winding of a three–phase, eight–pole, 50 Hz induction motor has 720 conductors, accommodated in 72 slots. Calculate the flux per pole of the rotating field in the air–gap of the motor, needed to generate 230 V in each phase of the stator winding.

8. A three–phase induction motor, at standstill, has a rotor voltage of 100 V between the slip–rings when they are open–circuited. The rotor winding is star–connected and has a leakage reactance of 1 Ω/ph at standstill and a resistance of 0.2 Ω/ph. Calculate: (a) the rotor current when the slip is 4 per cent and the rings are short–circuited; (b) the slip and the rotor current when the rotor is developing maximum torque. Assume the flux to remain constant.

9. A three–phase, 50 Hz induction motor with its rotor star–connected gives 500 V (r.m.s.) at standstill between the slip–rings on open circuit. Calculate the current and power factor at standstill when the rotor winding is joined to a star–connected external circuit, each phase of which has a resistance of 10 Ω and an inductance of 0.04 H. The resistance per phase of the rotor winding is 0.2 Ω and its inductance is 0.04 H. Also calculate the current and power factor when the slip–rings are short–circuited and the motor is running with a slip of 5 per cent. Assume the flux to remain constant.

10. If the star–connected rotor winding of a three–phase induction motor has a resistance of 0.01 Ω/ph and a standstill leakage reactance of 0.08 Ω/ph, what must be the value of the resistance per phase of a starter to give the maximum starting torque? What is the percentage slip when the starting resistance has been reduced to 0.02 Ω/ph, if the motor is still exerting its maximum torque?

11. A three–phase, 50 Hz, six–pole induction motor has a slip of 0.04 per unit when the output is 20 kW. The frictional loss is 250 W. Calculate: (a) the rotor speed; (b) the rotor I^2R loss.

12. Sketch the usual form of the torque/speed curve for a polyphase induction motor and explain the factors which determine the shape of this curve.

In a certain eight–pole, 50 Hz machine, the rotor resistance per phase is 0.04 Ω and the maximum torque occurs at a speed of 645 r/min. Assuming that the air–gap flux is constant at all loads, determine the percentage of maximum torque: (a) at starting; (b) when the slip is 3 per cent.

13. Describe briefly the construction of the stator and slip–ring rotor of a three–phase induction motor, explain the action of the motor and why the rotor is provided with slip–rings.

A three–phase, 50 Hz induction motor has four poles and runs at a speed of 1440 r/min when the total torque developed by the rotor is 70 N m. Calculate: (a) the total input (in kilowatts) to the rotor; (b) the rotor I^2R loss in watts.

14. Explain how a rotating magnetic field may be produced by stationary coils carrying three–phase currents.

Determine the efficiency and the output kilowatts of a three-phase, 400 V induction motor running on load with a fractional slip of 0.04 and taking a current of 50 A at a power factor of 0.86. When running light at 400 V, the motor has an input current of 15 A and the power taken is 2000 W, of which 650 W represent the friction, windage and rotor core loss. The resistance per phase of the stator winding (delta-connected) is 0.5 Ω.

15. Calculate the relative values of (a) the starting torque and (b) the starting current of a three-phase cage-rotor induction motor when started by: (i) direct switching; (ii) a star–delta starter; and (iii) an auto-transformer having 40 per cent tappings.

16. A cage-rotor induction motor is loaded to operate at 1470 r/min when the connections to two of its three phases are interchanged. Explain the subsequent action of the motor.

17. When a basic one-phase motor is switched on, it is possible that it might not start, or, if it does start, we cannot be certain of the direction of rotor rotation. Explain the reason for this uncertainty and hence describe a development that will ensure that the rotor starts in a predetermined direction.

18. An engineer has two one-phase induction motors, one a capacitor-run and the other a split-phase, and two loads, a fan and one which is driven through a gear train. Which motor should be used for each application? Explain your choice.

19. You have been asked to design a cheap fan to fit in a refrigerator. Which motor would you select? Justify your choice.

20. If a basic one-phase motor were encouraged to operate in the opposite direction, it would not act as a generator. Explain why we cannot obtain a one-phase generator.

Power Systems

All industrialized countries rely heavily on electricity supply systems. Such power systems are complex interconnected networks of overhead lines, underground cables and transformers for the transmission and distribution of electrical energy over long distances from power station to consumers. Power system analytical techniques have been developed in order to design and operate efficient power systems. This chapter introduces some of these methods and explains their purpose.

One of the main design, and operation, issues involves voltage drop along long current-carrying transmission lines. In order to carry large load currents economically but without excessive loss, circuit impedance has to be carefully 'managed' and therefore conductor cross-sections need to be appropriately sized. This chapter introduces the calculation of voltage drops in electricity supply circuits.

The other major issue discussed is the calculation of fault currents in power systems. It is the circuit impedance, between generators and a fault location, which limits the short-circuit current that flows. This, in turn, determines the rating of circuit breaking equipment, installed to minimize damage to system equipment. This chapter will enable you to make simple calculations of fault current.

Power system engineers use a per-unit method for these analyses. It considerably simplifies such calculations in complex, interconnected networks operating at many different generation, transmission and distribution voltages. This chapter compares the 'ohmic value' and the 'per-unit' methods of calculation. By the end of the chapter, you will be able to carry out volt-drop and fault current calculations using the per-unit method.

40.1   System representation

In preceding chapters, individual components of power systems have been considered. Generators supply the electrical energy. Transmission and distribution networks of overhead lines and underground cables deliver the energy to consumers where all manner of electrical loads, from motors to TVs, use it. Transformers initially raise the generated voltage for efficient transmission over long distances and thereafter decrease the system voltage for local distribution and utilization.

As has also been discussed, electrical energy is most efficiently supplied by three-phase systems. The electrical loads are arranged to ensure that the currents in each of the three individual phases are roughly equal. In this condition, a power system is said to be balanced. In fact, power systems are normally well balanced and can, for many purposes, be both more simply and accurately represented by a single-phase diagram. It is often the practice also to omit the neutral line – no current flows in it in a balanced three-phase system – leaving a single-line diagram. Both the single-phase representation with neutral and the single-line diagram are commonly used.

A simple three-phase supply system is depicted in Fig. 40.1. The voltages shown are typical UK network line to line voltages.

Fig. 40.1   Simple supply system

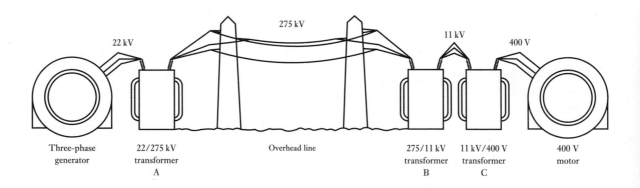

22 kV   275 kV   11 kV   400 V

| Three-phase generator | 22/275 kV transformer A | Overhead line | 275/11 kV transformer B | 11 kV/400 V transformer C | 400 V motor |

It is necessary to represent this as a single-phase diagram. In order to do this, we have to consider the representation of the transformer and of the transmission line.

(a)   Representation of the transformer

This has been discussed in sections 35.10–35.13. We will ignore the parallel magnetizing branch of Fig. 35.17 since load and fault currents through series impedance elements are of most concern to us in the analysis to be carried out in this chapter. We can therefore use the simplified equivalent circuit of Fig. 35.20. The equivalent impedance of the primary and secondary windings can be referred to either the primary or the secondary of the transformer, whichever is most convenient for the power system under consideration.

(b) Representation of the transmission line

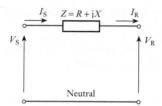

Fig. 40.2 Equivalent circuit of short transmission line. V_S is the sending-end voltage, V_R is the receiving-end voltage

A transmission line has series resistance, series inductive reactance, shunt capacitance and leakage resistance which are distributed evenly along its length. Except for long lines, the total resistance, inductance, capacitance and leakage resistance of the line can be concentrated to give a 'lumped-constant' circuit which simplifies calculation. The particular 'lumped-constant' circuit used depends on the length of the line and the required accuracy of the calculations. For the purposes of this introduction to power system calculations, we will consider a representation which is accurate for short transmission lines up to about 80 km in length. For this length of line, the shunt capacitance and leakage resistance can be ignored, as shown in Fig. 40.2. It should be noted that this assumption is not valid for unloaded lines when the shunt capacitance dominates.

Fig. 40.3 Equivalent circuit of simple power system. Z_S is the generator or source impedance, Z_A is the transformer impedance referred to 22 kV, $Z_{o/h}$ is the 275 kV transmission line impedance, Z_B is the transformer impedance referred to 275 kV, Z_C is the transformer impedance referred to 11 kV, and Z_M is the motor impedance

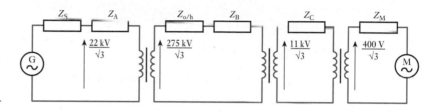

We can now represent the power system of Fig. 40.1, assuming a balanced condition, by the single-phase circuit diagram of Fig. 40.3. Voltages are now phase to neutral values.

As it represents a balanced three-phase system, the neutral can be omitted altogether and the network can be drawn as a single line diagram as in Fig. 40.4. Note however that the impedances have been left out for simplicity.

Fig. 40.4 Single line diagram of simple power system

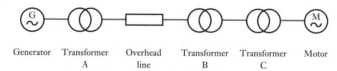

Generator   Transformer   Overhead   Transformer   Transformer   Motor
                  A   line   B   C

40.2 Power system analysis

Power system analysis is required for a large number of different purposes. These include:

- System design and control to maintain consumer voltage at statutory levels. (Effected by conductor sizing and transformer tap changer position.)
- Fault calculations to ensure that the maximum fault current can be interrupted by circuit breakers or fuses and that large fault currents cause the minimum of damage to the power system.
- Design of protection systems to ensure faulty circuits are switched off rapidly (<20 ms) to prevent damage and to ensure only the faulty circuit is switched off to minimize supply disruption.
- System design and control to maintain frequency within ±0.5 per cent.

- To ensure sufficient generation is available to meet the expected demand – load forecasting.
- To ensure that loads are supplied by the most efficient arrangement of generators – load scheduling.

We shall consider the first two of these topics – the calculation of voltage drops and the calculation of fault currents in supply networks.

| 40.3 | Voltage-drop calculations |

Consider the equivalent circuit of Fig. 40.2 used to represent a single phase of a balanced three-phase system. V_S and V_R are phase to neutral voltages. Since there are no parallel branches, the current at the sending end (I_S) and that at the receiving end (I_R) are the same
Hence the two circuit equations are

$$V_S = V_R + ZI_R$$ [40.1]

$$I_S = I_R$$ [40.2]

The current is calculated from the three-phase power delivered:

$$P = 3V_R I_R \cos \phi$$

The phasor representation is shown in Fig. 40.5.

Fig. 40.5   Phasor diagram for short line with inductive load

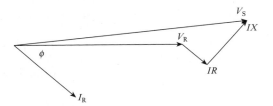

It is clear from these ideas that the impedance of networks has a considerable bearing on power system design and operation. Supply authorities have a statutory obligation to consumers to maintain voltage levels to within a few per cent of the declared supply voltage. (In Europe this is now 230 V (phase to neutral) with a permissible variation of +10 per cent and −6 per cent.) This can be problematic. Firstly, voltage drops down long transmission lines have to be minimized by choice of conductor cross-section and impedance. At the same time, it has to be realized that lines and cables having conductors of larger cross-section are much more expensive, so economic design is important. Secondly, the variation of load current throughout the 24 hour day means that the voltage drop in a network also varies. In order to support the voltage to consumers during periods of heavy demand and to reduce it during periods of light loading, network transformers have tap-changing gear which can automatically vary the transformer turns ratio, and hence maintain the secondary voltage output according to system load requirements. It should be understood though that tap changers are only designed to accommodate a small range of voltage variation and do not compensate for poor system design.

Voltage-drop theory can be applied to the following two examples:

Example 40.1

An 11 kV, three-phase, 50 Hz line of resistance 3 Ω/ph and reactance j7 Ω/ph supplies an 11 kV/400 V transformer having negligible resistance and reactance of j2 Ω/ph referred to 11 kV. The transformer supplies a 400 V feeder of resistance 0.01 Ω/ph and reactance j0.005 Ω/ph. If V_R, the receiving-end voltage, is 400 V, calculate V_S, the sending-end voltage, when the three-phase load delivered is 250 kW at unity power factor.

The supply network is represented by the single-phase diagram of Fig. 40.6.

Fig. 40.6 Network diagram for Example 40.1

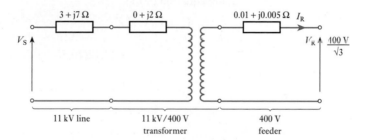

11 kV line   11 kV/400 V transformer   400 V feeder

Since we are required to calculate the sending-end voltage, nominally 11 kV, we refer all impedances to the 11 kV side of the transformer:

$$\therefore \qquad 0.01\ \Omega \qquad \text{becomes} \qquad 0.01 \times \left(\frac{11\,000}{400}\right)^2 = 7.56\ \Omega$$

$$j0.005\ \Omega \quad \text{becomes} \quad j0.005 \times \left(\frac{11\,000}{400}\right)^2 = j3.78\ \Omega$$

The network diagram simplifies to that of Fig. 40.7.

Fig. 40.7 Simplified diagram for Example 40.1, impedances referred to 11 kV

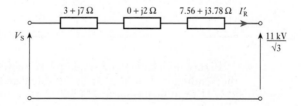

Hence:

$$Z_{\text{total}} = 10.56 + j12.78\ \Omega$$

The next step is to calculate I_R (and I_S) from the three-phase power delivered, 250 kW:

$$\text{Power delivered} = 3V_R I_R \cos\phi$$

$$250 \times 10^3 = 3 \cdot \left(\frac{400}{\sqrt{3}}\right) \cdot I_R$$

$$\therefore \qquad\qquad\qquad I_R = 360.8 \text{ A}$$

However, in order to calculate the voltage drop down the equivalent network referred to 11 kV, Fig. 40.7, we have to refer this current to 11 kV as well:

$$\text{i.e.} \qquad I'_R = 360.8 \times \frac{400}{11\,000} = 13.12 \text{ A}$$

Using equation [40.1],

$$V_S = V_R + I'_R Z_t$$

$$= \frac{11\,000}{\sqrt{3}} + 13.12(10.56 + j12.78)$$

$$= 6351 + 138.5 + j167.7$$

$$= 6489.5 + j167.7$$

$$= \mathbf{6491.7 \text{ V/ph}}$$

$$\text{or } \mathbf{11.24 \text{ kV line}}$$

| **Example 40.2** | A 400 V, three-phase induction motor takes a current of 300 A per phase at a power factor of 0.8 lag. It is supplied by a length of 275 kV overhead line of impedance $(1 + j10)$ Ω/ph, a 275/11 kV transformer of impedance $(2 + j15)$ Ω/ph referred to its primary and an 11 kV/400 V transformer of impedance $(2 + j10)$ Ω/ph referred to its primary. Determine the sending-end voltage if the voltage at the motor is 400 V. |

Fig. 40.8 Circuit diagram for Example 40.2

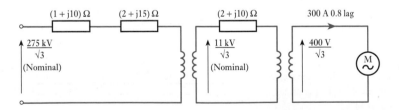

Figure 40.8 shows the single-phase representation of the system. Since the sending-end voltage is to be calculated, refer the impedance of the 275/11 kV transformer to 275 kV:

$$Z' = (2 + j10)\left(\frac{275}{11}\right)^2 \text{ Ω}$$

$$Z' = 1250 + j6250 \text{ Ω}$$

Fig. 40.9 Simplified network for Example 40.2

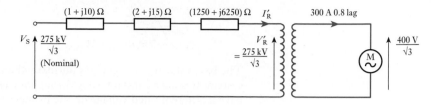

The network becomes that shown in Fig. 40.9.

The total impedance referred to 275 kV is now

$$Z_t = 1 + j10 + 2 + j15 + 1250 + j6250 \ \Omega$$

$$Z_t = 1253 + j6275 \ \Omega$$

The system volt drop per phase is calculated using equation [40.1]. However, we first need to refer the current drawn by the motor to 275 kV:

Actual motor current = 300 A at 400 V at 0.8 pf

$$\text{At} \quad 275 \text{ kV}, \ I'_R = 300(0.8 - j0.6) \frac{400}{275\,000} \text{ A}$$

$$I'_R = 0.35 - j0.26 \text{ A}$$

$$V_S = V'_R + I'_R Z_t \quad \text{where} \quad V'_R = \frac{275 \text{ kV}}{\sqrt{3}}$$

$$\therefore \qquad V_S = 158.8 \times 10^3 + (0.35 - j0.26)(1253 + j6275) \text{ V}$$

$$= 158.8 \times 10^3 + 2079 + j1872 \text{ V}$$

$$= 160.88 + j1.87 \text{ kV}$$

$$= \textbf{160.9 kV/ph} \quad \text{or} \quad \textbf{278.6 kV line}$$

In order to maintain a supply voltage of 400 V to the motor, the sending-end voltage has to be 278.6 kV.

40.4 The per-unit method

The two examples, 40.1 and 40.2, illustrate the use of the 'ohmic value method' for solving power system calculations. However, it should be recognized that a typical power system is a complex, interconnected network of parallel and series circuits containing many transformation steps. To calculate the voltage drops in such a circuit using amps, volts and ohms would be very laborious. For example, if we wanted to carry out a calculation at 400 V, a power station generator impedance might have to be referred through four or five transformations.

The common technique used by power system engineers is to use a per-unit system of currents, voltages, impedances and power. In this system, each value is expressed as a fraction of its **own** nominal or rated value allowing us to ignore the many different transformation steps.

Voltages and currents are expressed as a fraction of the nominal or base values:

$$V_{pu} = \frac{V_{actual}}{V_{base}} \tag{40.3}$$

$$I_{pu} = \frac{I_{actual}}{I_{base}} \qquad\qquad [40.4]$$

The base value of the voltage is normally the rated line to line voltage of the system. It is not a constant value throughout a system as a transformer would have a different rated voltage on the primary to that of the secondary. Thus a 33/11 kV transformer would have a base voltage on the 11 kV side of 11 kV and on the 33 kV side, of 33 kV. In a similar way, the base current on each side of the transformer would change in such a manner as to have the power input = the power output. In general, therefore:

$$\text{per-unit value (pu)} = \frac{\text{actual value}}{\text{base value}} \qquad\qquad [40.5]$$

and this applies to voltage, current, impedance and apparent power (VA).

40.5	Per-unit impedance

The base value of voltage in volts (rated voltage) is defined as 1 pu voltage and the base value of current in amps (full load current) is defined as 1 pu current.

$$\therefore \qquad \text{1 pu impedance } (\Omega) = \frac{V_{base}}{I_{base}}$$

and any other impedance in the system is a fraction of 1 pu $Z\,(\Omega)$. Thus if $V_{base} = 3464$ V (a line to line voltage) and $I_{base} = 500$ A, then

$$\text{1 pu } Z = \frac{3464}{\sqrt{3}} \Big/ 500 = 4\ \Omega$$

Note that $\sqrt{3}$ is required because V_B is a line voltage.

If a transmission line in this system had an impedance of 0.5 Ω per phase then its per-unit impedance would be 0.5/4 = 0.125 pu or

$$Z_{pu} = Z(\Omega)\frac{\sqrt{3}I_{base}}{V_{base}} \qquad\qquad [40.6]$$

Z_{pu} has no units; it is simply a number.

A transmission line having a per-unit impedance of 0.125 pu means that the magnitude of the voltage drop along the line when full load current is flowing is 0.125 pu or 12.5 per cent. This method of specifying an 'impedance' in terms of a per-unit (or percentage) voltage drop of rated voltage may seem puzzling at first glance but in fact, as is evident, it provides a ready basis for calculating voltage drops – and later short-circuit currents. It is the series impedance which limits these currents during faults.

We can now compare the ohmic and per-unit impedance of two different transformers in an example.

| Example 40.3 | Calculate the ohmic impedance of two 0.1 pu transformers: one is a 11 kV/400 V transformer having a rating of 2 MVA, the other is a 33 kV/400 V transformer having a rating of 10 MVA. |

2 MVA transformer

$$Z_{pu} = Z_\Omega \frac{\sqrt{3}\, I_{base}}{V_{base}} \quad \text{or} \quad Z_\Omega = Z_{pu} \frac{V_{base}}{\sqrt{3}\, I_{base}}$$

$$V_{base} = 11 \text{ kV}$$

$$I_{base} = \frac{S_B}{\sqrt{3} \cdot V_B} \tag{40.7}$$

where S_B is the rated MVA = 2 MVA

$$I_{base} = \frac{2 \cdot 10^6}{\sqrt{3} \cdot 11 \cdot 10^3} = 105 \text{ A}$$

$$Z_\Omega = j0.1 \cdot \frac{11 \cdot 10^3}{\sqrt{3} \cdot 105} = j6.05\Omega \quad \begin{array}{l}\text{referred to 11 kV since this voltage}\\ \text{was used in the calculation of } I_B\end{array}$$

It can also be demonstrated that the value of Z_Ω referred to the low-voltage side of the transformer is also 0.1 pu impedance:

$$Z_\Omega \text{ (referred to 400 V)} = j6.05 \times \left(\frac{400}{11\,000}\right)^2 = j0.08 \ \Omega$$

We also need to recalculate I_B at 400 V:

$$I_B = 105 \cdot \frac{11\,000}{400} = 2887.5 \text{ A}$$

$$\therefore \qquad Z_{pu} = j0.008 \cdot \sqrt{3} \cdot \frac{2887.5}{400} \text{ pu}$$

$$= j0.1 \text{ pu}$$

10 MVA transformer

An alternative method of expressing Z_{pu} can be obtained by multiplying both top and bottom of equation [40.6] by V_B:

i.e. $\qquad Z_\Omega = Z_{pu} \dfrac{V_B^2}{\sqrt{3} I_B V_B}$

$$= Z_{pu} \frac{(V_B)^2}{S_B}$$

and

$$Z_{pu} = \frac{Z_\Omega \cdot S_B}{(V_B)^2} \tag{40.8}$$

This equation is extremely useful in per-unit calculations. It makes it unnecessary to calculate I_B. In this case V_{base} is 33 kV.

$$\therefore \qquad Z_\Omega = j0.1 \cdot \frac{33^2}{10}$$

Note that when the voltage is in kV and S_B is in MVA, the powers of 10 cancel, as above:

$$Z_\Omega = j10.89 \ \Omega \quad \text{referred to 33 kV}$$

40.6 Base power – S_B or MVA$_B$

Every item of power system plant has a nominal rating – often in MVA – and a per-unit impedance value referred to that rated VA value. This has been illustrated in the previous example. For a complex circuit, when the per-unit method is used, as a first step in any solution, all per-unit values must be referred to a common VA base, S_{base} (S_B). Any convenient VA base may be chosen. Thereafter, equations [40.6] and [40.8] are commonly used.

We will illustrate the per-unit technique of solution by the following examples, the first of which compares the per-unit solution with the ohmic method.

Example 40.4

For the circuit of Fig. 40.10 calculate the sending-end voltage V_S, when a load of 1 MVA, 0.8 pf is supplied at 400 volts. Use both (i) the ohmic value method and (ii) the per-unit method.

Fig. 40.10 Single line diagram for Example 40.4

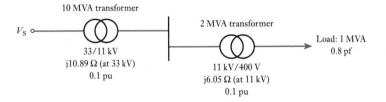

(i) Ohmic value method

Load current is obtained from $VA = \sqrt{3} V_L I_L$

$$\therefore \qquad I_L = \frac{1 \cdot 10^6}{\sqrt{3} \cdot 400} = 1443 \ A$$

If 400 V is taken as the reference phasor then $I = 1443(0.8 - j0.6)$ A. The answer expected for V_S will be around 33 kV, therefore refer all values to 33 kV by multiplying by the turns ratio:

$$\therefore \qquad I_{33} = 1443(0.8 - j0.6) \cdot \frac{0.400}{33}$$

$$= 17.5(0.8 - j0.6) \ A$$

2 MVA transformer

$$Z_\Omega = j6.05 \ \Omega \text{ referred to } 11 \text{ kV}$$

$$Z_\Omega \text{ (referred to } 33 \text{ kV)} = j6.05 \times \left(\frac{33}{11}\right)^2 = j54.45 \ \Omega$$

10 MVA transformer

$$Z_\Omega \text{ (referred to } 33 \text{ kV)} = j10.89 \ \Omega$$

To calculate V_S, use equation [40.1]:

$$V_S = V_R\angle 0 + I\angle\phi \cdot Z_{\text{total}}$$

$$Z_t = j54.45 + j10.89$$

$$V_S = \frac{33\,000}{\sqrt{3}} + 17.5(0.8 - j0.6)j65.3$$

$$= 19\,052 + j1143(0.8 - j0.6)$$

$$= 19\,052 + 686 + j915$$

$$= 19\,759 \text{ volts (phase voltage)}$$

$$= \mathbf{34.22 \ kV} \quad \text{(line to line voltage)}$$

(ii) Per-unit method

We begin by choosing the base value of MVA, S_B or MVA_B. This can be any value, but it would be sensible to choose a base MVA of either 1 MVA, 2 MVA or 10 MVA to minimize work!

Choose 2 MVA as the base power.

The per-unit impedance (reactance) of the 2 MVA transformer remains at $j0.1$ pu as the voltage drop within the transformer, when the base power flows through it, will be 0.1 pu.

What is the voltage drop within the 10 MVA transformer when the base power flows through it?

Since its voltage drop is 0.1 pu when the load is 10 MVA then when the load is 2 MVA the voltage drop will be 0.02 pu

or
$$Z_{\text{pu}} = Z_{\text{pu}} \text{ actual} \times \frac{MVA_{\text{base}}}{MVA_{\text{actual}}} \qquad [40.9]$$

$$= j0.1 \cdot \frac{2}{10}$$

$$= j0.02 \text{ pu}$$

Similarly a load of 2 MVA absorbs a current of 1 pu. Therefore the actual load of 1 MVA must absorb a current of 0.5 pu. Consequently, the load current is also 0.5 pu but the phase angle should not be forgotten. In this case

$$I_{\text{pu}} = 0.5(0.8 - j0.6)$$

Fig. 40.11

The per-unit network for a base of 2 MVA is thus as shown in Fig. 40.11. We are now in a position to complete the example using per-unit techniques:

Circuit for base power of 2 MVA:

$$\therefore \qquad V_S = V_R + IZ$$

$$= 1\angle 0 + 0.5(0.8 - j0.6)j0.12$$

$$= 1 + j0.06(0.8 - j0.6)$$

$$= 1.036 + j0.048 = 1.037 \text{ pu}$$

$$V_{\text{actual}} = V_{\text{pu}} \times V_B$$

$$= 1.037 \times 33 \text{ kV}$$

$$= \mathbf{34.22 \text{ kV}} \quad \text{(as before)}$$

Example 40.5 A supply system is shown in Fig. 40.12. The receiving-end voltage at the load is 33 kV. Determine the sending-end, input voltage at the nominal 275 kV grid.

Fig. 40.12

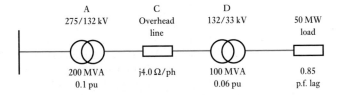

Choose 100 MVA as S_B.

For transformer A, $Z_{\text{Apu}} = Z_A \dfrac{S_B}{S_A}$ (from equation [40.9])

$$Z_{\text{Apu}} = 0.1 \times \frac{100}{200} = j0.05 \text{ pu}$$

For transformer D, $Z_{\text{Dpu}} = j0.06 \text{ pu}$

For overhead line C, the operating voltage is 132 kV. This is V_B.

$$Z_{\text{Cpu}} = \frac{Z_C S_B}{(V_B)^2} \quad \text{(from equation [40.8])}$$

$$= \frac{j4 \times 100}{(132)^2} = j0.023 \text{ pu}$$

The network can now be reduced to that shown in Fig. 40.13.

Fig. 40.13

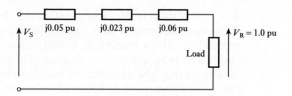

The total supply system impedance (all reactance) is

$$Z_t = j0.05 + j0.06 + j0.023 = j0.133 \text{ pu}$$

For the load,

$$S_L = \frac{P_L}{\cos \phi} = \frac{50}{0.85} = 58.8 \text{ MVA}$$

$$I_{Lpu} = \frac{S_L}{S_B} = \frac{58.8}{100} = 0.588 \text{ pu}$$

The network reduces further to that of Fig. 40.14.

Fig. 40.14

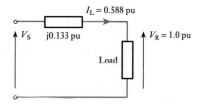

$$I_{Lpu} = I_1 \cos \phi - jI_L \sin \phi$$
$$= (0.588 \times 0.85) - j(0.588 \times 0.527)$$
$$= (0.50 - j0.31) \text{ pu}$$

Hence, using equation [40.1],

$$V_S = V_R + IZ$$
$$= (1 + j0) + (0.50 - j0.31)j0.133$$
$$= 1.0412 + j0.0665 \text{ pu}$$
$$|V_S| = (1.0412^2 + 0.0665^2)^{\frac{1}{2}} = 1.0433 \text{ pu}$$
$$V_S = 1.0433 \times 275 = \textbf{287 kV}$$

40.7   **Faults in a power system**

There are five types of fault which can arise in a three-phase system:

1. a single line connected directly to earth;
2. two lines directly connected to earth;
3. all three lines directly connected to earth;
4. two lines directly connected to one another;
5. all three lines directly connected to one another.

The first of these is the most common but the last is the most serious since it involves the greatest fault current (and fault VA) which has to be interrupted by a circuit breaker. It should be noted that in large power systems, fault currents are often many thousands of amperes. When a fault occurs, it is essential that it be detected and swiftly interrupted by the circuit breakers. Often the interruption is achieved in less than a tenth of a second. If such faults are not interrupted, the currents can severely damage power system equipment.

Interruption of such large currents requires sophisticated circuit breaking equipment which must be sized appropriately so switchgear is designed and rated according to the apparent power or current which it is to break.

As has already been indicated, a symmetrical three-phase fault, one in which the currents are equal in magnitude in each of the three phases, imposes the severest condition on circuits and circuit breakers. It is therefore most important to examine this condition. However, it is only necessary to examine one phase, other phases having identical fault current magnitudes.

It is the impedance of the transmission and distribution network between the source of the supply and the fault which determines the fault current available, and hence the rating of circuit breaking equipment. Figure 40.15 shows a single-phase representation of a power system with a fault.

Fig. 40.15

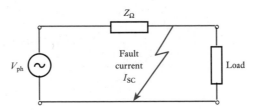

Note that we can ignore load current, the fault current being always much larger. The supply source is, of course, a generator, or in a large power system, many generators operating in parallel in many different power stations. The impedance, Z_Ω, is made up of the complex network of series and parallel connections of lines and cables and the many circuit transformers.

Hence,

$$I_{SC} = \frac{V_{ph}}{Z_\Omega}$$

For the reasons discussed earlier, power system engineers use the per-unit method for fault calculations, so we have to convert this expression, in volts, amps and ohms, to per-unit quantities.

$$\therefore \qquad I_{SC} = \frac{V_B}{\sqrt{3} \cdot Z_\Omega}$$

We can substitute for Z_Ω obtained by using equation [40.6]:

$$Z_\Omega = Z_{pu} \cdot \frac{V_B}{\sqrt{3} \cdot I_B}$$

$$\therefore \qquad I_{SC} = \frac{I_B}{Z_{pu}} \qquad\qquad [40.10]$$

The short-circuit or fault apparent power, by definition, is the three-phase apparent power at the fault:

$$VA_{SC} = \sqrt{3} \cdot V_B \cdot I_{SC}$$ [40.11]

This quantity is termed the *fault level*. Equation [40.11] can be rewritten by substituting for I_{SC}, from equation [40.10],

$$VA_{SC} = \frac{S_B}{Z_{pu}}$$ [40.12]

This is a very simple, but powerful expression. Power system engineers know the per-unit impedance of every item of plant in the network. Consequently, the fault level can readily be calculated at any point in the network and circuit breakers sized appropriately.

The rating of switchgear must always more than match that fault level. Circuit breakers and fuses for all electrical installations have a short-circuit VA rating or a short-circuit current and voltage rating. Fuses can be used where the fault level is low, typically on low-voltage networks where the short-circuit current is limited by the high value of impedance between power station generators and fault locations. When the fault level is high, it is highest on high-voltage circuits close to power stations; compressed gas circuit breakers with ratings of many MVA are typically required.

Let us now consider a simple fault level calculation. The term 'busbar' is used for a heavy conductor which interconnects, for example, generators operating in parallel in power stations or circuits arriving at and leaving network substations. Clearly, they have to be capable of conducting large currents.

Example 40.6

A 30 MVA transformer is connected to busbars which are supplied by two generators via cables as shown in Fig. 40.16. Determine the fault level at the secondary terminals of the transformer.

Fig. 40.16

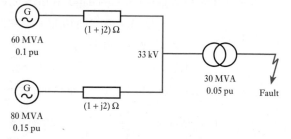

Choose a base of 120 MVA. This is S_B. We need a per-unit diagram to this V_{base}.

For the generators:

$$Z_1 = 0.1 \times \frac{120}{60} = j0.2 \text{ pu}$$

$$Z_2 = 0.15 \times \frac{120}{80} = j0.225 \text{ pu}$$

For the cables:

$$Z_C = Z_\Omega \frac{S_B}{V_B^2} = (1 + j2) \times \frac{120}{33^2}$$

$$Z_C = (0.11 + j0.22) \text{ pu}$$

For the transformer:

$$Z_3 = 0.05 \times \frac{120}{30} = j0.20 \text{ pu}$$

The network can now be reduced to that shown in Fig. 40.17.

Fig. 40.17

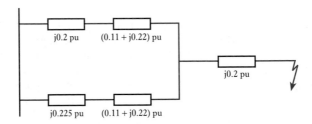

By adding the branch impedances, the network further reduces to that shown in Fig. 40.18.

Fig. 40.18

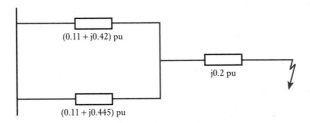

The total per-unit impedance is therefore

$$Z_t = (0.11 + j0.42) \parallel (0.11 + j0.445) + j0.2 \text{ pu}$$

$$Z_t = 0.055 + j0.416 \text{ pu}$$

$$|Z_t| = (0.55^2 + 0.416^2)^{\frac{1}{2}} = 0.42 \text{ pu}$$

Using equation [40.12], the short-circuit apparent power is

$$\text{MVA}_{SC} = \frac{S_B}{Z_t} = \frac{120}{0.42} = \textbf{286 MVA}$$

40.8	Representation of a grid connection

In the last example, the two sources are shown as generators but it is more likely that they are equivalent to the inputs from two different parts of the electricity supply network or 'grid'. In the UK the transmission network to which the generators are connected is termed the grid. In the example, effectively, we have one input, with a reactance of 0.1 pu, supplying 60 MVA and the other input, with a reactance of 0.15 pu, supplying 80 MVA.

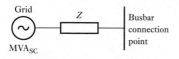

Fig. 40.19   Representation of grid supply network connection

As mentioned earlier, the fault level, anywhere in the supply network, is known so that any connection made at a particular location can be represented by a single generator in series with a reactance as shown in Fig. 40.19. The generator will have a rating equivalent to the fault VA (VA_{SC}), the fault level, at that connection location.

This connection can then be incorporated in a per-unit calculation by using equation [40.12] to calculate the per-unit value of the series reactance appropriate to the chosen value of S_B. Example 40.7 illustrates this technique.

Example 40.7

In Fig. 40.20, a grid connection having a fault level of 2000 MVA is made to 275 kV busbars which are connected to two 100 MVA, 275/33 kV transformers, each having 0.05 pu reactance. Calculate the rating of circuit breakers situated on the outgoing 33 kV circuits connected to the 33 kV busbars. Calculate the fault current available at the 33 kV busbars.

Fig. 40.20   Circuit diagram for Example 40.7

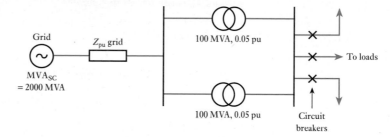

Choose S_B or MVA_B = 100 MVA.
Using equation [40.12],

$$Z_{pu}\,(grid) = \frac{MVA_B}{MVA_{SC}\,(grid)}$$

$$= \frac{100}{2000} = 0.05 \text{ pu}$$

The circuit thus reduces to that of Fig. 40.21.

Fig. 40.21

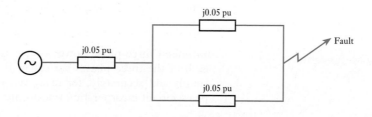

Total $Z_{pu} = 0.05 + 0.05 \parallel 0.05 = 0.075$ pu

$$MVA_{SC} \text{ at 33 kV busbars} = \frac{100}{0.075} = \textbf{1333 MVA}$$

A circuit breaker rating in excess of 1333 MVA is required.

The fault current is, from equation [40.11],

$$I_{SC} = \frac{1333 \cdot 10^6}{\sqrt{3} \cdot 33 \times 10^3}$$

$$= 23.3 \text{ kA}$$

Summary of important formulae

$$V_{pu} = \frac{V_{actual}}{V_B} \qquad [40.3]$$

$$I_{pu} = \frac{I_{actual}}{I_B} \qquad [40.4]$$

$$Z_{pu} = \frac{Z_\Omega}{Z_B} = \frac{Z_\Omega \sqrt{3} I_B}{V_B} \qquad [40.6]$$

Note V_B is the *line to line voltage*.

$$I_B = \frac{S_B}{\sqrt{3} V_B} \qquad [40.7]$$

$$Z_{pu} = \frac{Z_\Omega \cdot S_B}{(V_B)^2} \qquad [40.8]$$

$$I_{SC} = \frac{I_B}{Z_{pu}} \qquad [40.10]$$

$$VA_{SC} = \frac{S_B}{Z_{pu}} \qquad [40.12]$$

Rules for per-unit calculations

Rule 1: Choose a base VA value, S_B, and use it throughout the calculation.

Rule 2: Any per-unit impedance given for an actual equipment rating must be converted to the VA base chosen for the particular calculation.

$$\text{i.e.} \qquad Z_{pu} = Z_{pu} \,(\text{actual}) \frac{S_B \,(\text{chosen})}{S_B \,(\text{actual})} \qquad [40.9]$$

Terms and concepts

A **balanced** three-phase power system is one in which the currents in each of the three phases are equal. Balanced power systems can be simply and accurately, for many purposes, represented by a single-phase circuit incorporating transformer and line impedances.

The network impedance has a considerable bearing on power system design and operation. It determines the **voltage drop** in current-carrying conductors. It also determines the short–circuit or **fault current** that will flow between generators and any fault location.

Power system engineers use a **per-unit method** of calculation of voltage drops and fault currents. It simplifies such calculations in complex,

Terms and concepts continued

interconnected networks operating at many different generation, transmission and distribution voltages.

In the per-unit method, each value of current, voltage, impedance and apparent power (VA) is expressed as a fraction of its own **nominal or rated value**. **A common base VA value**, S_B, is chosen and used throughout a per-unit calculation.

A **symmetrical** three-phase fault, one in which the currents in each of the three phases are equal in magnitude, imposes the severest condition on circuits and circuit breakers. This situation can be represented by a single-phase circuit incorporating the network impedance between the source and the fault location.

The value of the short-circuit or fault apparent power available at any location on the power system network, in the event of a fault at that location, is termed the **fault level**. The rating of switchgear must always exceed the fault level for the successful protection of power system equipment.

Exercises 40

1. An 11 kV, 50 Hz, three-phase overhead line has a cross-sectional area of 30 mm² and an inductance per km of 1 mH. If the line is 8 km long, calculate the voltage required at the sending end when delivering to the load 500 kW at 11 kV and 0.9 lagging power factor. If 500 kW at 11 kV and 0.9 lagging power factor were supplied at the sending end what would be the voltage, power and power factor at the receiving end?

$$\rho = 1.7 \ \mu\Omega \ \text{cm}$$

2. A substation is supplied by two feeders, one having an impedance of $(1 + j2) \ \Omega/\text{ph}$, and the other $(1.5 + j1.5) \ \Omega/\text{ph}$. The total current delivered to the substation is 700 A at 0.8 lagging p.f. Find the current and its p.f. at the receiving end in each feeder.

3. The three-phase busbars of a power station are divided into two sections A and B by a section switch. Each section is connected by a feeder to a substation C at which the voltage is 6.6 kV. Feeder AC has an impedance of $(0.5 + j0.6) \ \Omega/\text{ph}$ and delivers 9 MW at 1.0 pf to C. Feeder BC has an impedance of $(0.3 + j0.4) \ \Omega/\text{ph}$ and delivers 6 MW at 0.8 pf to C. Determine the voltage across the section switch.

4. A substation is fed at 11 kV by two parallel feeders. One feeder is 50 per cent longer than the other but has a conductor of twice the cross-sectional area. The reactance per km is the same for both and equal to 1/3 of the resistance per km of the shorter feeder. If the load on the substation is 4 MW at 0.9 pf lagging find the current and power in each feeder.

5. A 33 kV feeder with impedance $(2 + j10) \ \Omega/\text{ph}$ supplies a 33/6.6 kV substation containing two parallel connected transformers of 10 and 5 MVA each with a 0.1 pu reactance. The 6.6 kV side of the substation supplies a 6.6 kV feeder with an impedance of $(0.05 + j0.2) \ \Omega/\text{ph}$. 9.6 MW at a p.f. of 0.8 lag is supplied to the 33 kV feeder at 33 kV. Determine the voltage at the end of the 6.6 kV feeder. Use both the ohmic value and the per-unit method.

6. A transmission system consists of a generator, a step-up transformer, transmission line, step-down transformer and a load. The following parameters apply to each system component.

(i) Generator:	130 MVA, 11 kV, 0.16 pu
(ii) Step-up transformer:	135 MVA, 11/132 kV, j0.15 pu
(iii) 132 kV transmission line:	$4 + j12 \ \Omega$
(iv) Step-down transformer:	125 MVA, 132/11 kV, j0.112 pu

   If we require to deliver 125 MW at 1 p.f. and 11 kV at the load, what voltage and p.f. exist at the supply?

7. A factory is supplied from a substation containing two 500 kVA, 11 kV/400 V transformers connected in parallel, each having a reactance of 0.08 pu. A 750 kVA generator with 0.12 pu reactance also supplies the 400 V busbars which in turn supply a 400 V feeder having an impedance of $(0.01 + j0.015) \ \Omega/\text{ph}$.

Calculate the short-circuit current in the feeder and the voltage of the 400 V busbars when a symmetrical three-phase fault occurs at the far end of the feeder. That fault level of the 11 kV system is 100 MVA.

8. A three-phase generating station contains a 50 MVA generator of 0.1 pu reactance and a 30 MVA generator of 0.08 pu reactance feeding the same 11 kV busbars. An outgoing 33 kV feeder having an impedance of $(0.6 + j1.5)\,\Omega/\text{ph}$ is supplied from these busbars through an 11/33 kV, 20 MVA transformer of reactance 0.1 pu. Calculate, by (a) the ohmic value method and (b) the pu method, the short-circuit current and MVA when a symmetrical three-phase short circuit occurs (i) on the feeder side of the transformer and (ii) at the far end of the feeder. Calculate also the busbar voltage during the short circuit for each case.

9. A large industrial complex has four 5 MVA generators each of reactance 0.08 pu supplying the busbars of its power station. Determine the rating of any circuit breakers connected to the busbars.

Due to increased electricity demand in the factory it is proposed to make a connection to the grid through two 2 MVA transformers in parallel each with a reactance of 0.1 pu. The fault level at the grid connection point is 1000 MVA. What will now be the required rating of circuit breakers connected to the busbars?

10. A power station contains three 15 MVA generators each having a reactance of 0.1 pu and connected to a set of 33 kV busbars. The outgoing circuit breakers connected to these busbars have a breaking capacity of 750 MVA. It is proposed to make a connection to the grid through a 30 MVA transformer having a reactance of 0.06 pu and the short-circuit MVA of the grid at the point of connection is 1500 MVA. Determine the reactance of a reactor (in ohms) to be inserted between the busbars and the transformer if the existing switchgear in the power station is to be retained.

11. A 33 kV substation has three sets of busbars A, B and C. Each of the busbars A and B is connected to C by 15 MVA reactors of 0.1 pu reactance. Busbar A is supplied by a power system having a fault level of 1000 MVA and busbar B is supplied by a 50 MVA generator of 0.3 pu reactance. Busbar B supplies an 11 kV transmission line of reactance $5\,\Omega/\text{ph}$ via a 10 MVA 33/11 kV transformer of reactance 0.1 pu.

If a symmetrical three-phase fault occurs at the far end of the transmission line determine the fault current and voltages on busbars A, B and C.

12. The busbars of an 11 kV generating station are divided into three sections A, B and C, to each of which is connected a 20 MVA generator of 0.3 pu reactance. Each of the three busbars A, B and C is connected to a common busbar D via three 15 MVA reactors of 0.1 pu reactance. Busbar A feeds a 33 kV transmission line of reactance $15\,\Omega/\text{ph}$ via an 11/33 kV, 10 MVA transformer of reactance 0.1 pu.

If a symmetrical three-phase fault occurs at the far end of the transmission line find the fault current and the voltage on busbar A during the fault. (You could also attempt this question using the ohmic value method.)

Chapter forty-one

Direct-current Machines

Direct-current machines were the first electrical machines invented; a simple d.c. motor drove an electric locomotive in Edinburgh in 1839, although it took another 40 years before the d.c. motor became widespread. It is still used today to power trains and cranes.

In this chapter, we introduce the construction of the d.c. machine, and in particular different rotor (armature) winding arrangements and the rotor switching arrangement – the commutator. Armature reaction, as in the a.c. machine, is carefully discussed because it is the interaction between the magnetic field due to the armature reaction and the field due to the stator that produces the starting torque. Armature reaction also causes problems with commutation. Sparking can occur at the brushes – as anyone overloading a DIY drill will have witnessed.

By the end of the chapter, you will be familiar with the principles of d.c. machine construction and operation. You will also be able to undertake exercises to calculate the e.m.f. induced in the armature windings.

41.1 General arrangement of a d.c. machine

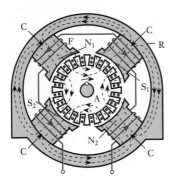

Fig. 41.1 General arrangement of a four-pole d.c. machine

Figure 41.1 shows the general arrangement of a four-pole d.c. motor or generator. The fixed part consists of four steel cores C, referred to as *pole cores*, attached to a steel ring R, called the *yoke*. The pole cores are usually made of steel plates riveted together and bolted to the yoke, which may be of cast steel or fabricated rolled steel. Each pole core has pole tips, partly to support the field winding and partly to increase the cross-sectional area and thus reduce the reluctance of the air-gap. Each pole core carries a winding F so connected as to excite the poles alternately N and S.

The armature core A consists of steel laminations, about 0.4–0.6 mm thick, insulated from one another and assembled on the shaft in the case of small machines and on a cast-steel spider in the case of large machines. The purpose of laminating the core is to reduce the eddy-current loss. Slots are stamped on the periphery of the laminations, partly to accommodate and provide mechanical security to the armature winding and partly to give a shorter air-gap for the magnetic flux to cross between the pole face and the armature 'teeth'. In Fig. 41.1, each slot has two circular conductors, insulated from each other.

The term *conductor*, when applied to armature windings, refers to the active portion of the winding, namely that part which cuts the flux, thereby generating an e.m.f.; for example, if an armature has 40 slots and if each slot contains 8 wires, the armature is said to have 8 conductors per slot and a total of 320 conductors.

The dotted lines in Fig. 41.1 represent the distribution of the *useful* magnetic flux, namely that flux which passes into the armature core and is therefore cut by the armature conductors when the armature revolves. It will be seen from Fig. 41.1 that the magnetic flux which emerges from N_1 divides, half going towards S_1 and half towards S_2. Similarly, the flux emerging from N_2 divides equally between S_1 and S_2.

Suppose the armature to revolve clockwise, as shown by the curved arrow in Fig. 41.1. Applying Fleming's right-hand rule (section 6.9), we find that the e.m.f. generated in the conductors is towards the paper in those moving under the N poles and outwards from the paper in those moving under the S poles. If the air-gap is of uniform length, the e.m.f. generated in a conductor remains constant while it is moving under a pole face, and then decreases rapidly to zero when the conductor is midway between the pole tips of adjacent poles.

Figure 41.2 shows the variation of the e.m.f. generated in a conductor while the latter is moving through two pole pitches, a *pole pitch* being the distance between the centres of adjacent poles. Thus, at instant O, the conductor is midway between the pole tips of, say, S_2 and N_1, and CD represents the e.m.f. generated while the conductor is moving under the pole face of N_1, the e.m.f. being assumed positive when its direction is towards the paper in Fig. 41.1. At instant E, the conductor is midway between the pole tips of N_1 and S_1, and portion EFGH represents the variation of the e.m.f. while the conductor is moving through the next pole pitch. The variation of e.m.f. during interval OH in Fig. 41.2 is repeated indefinitely, so long as the speed is maintained constant.

A d.c. machine, however, has to give a voltage that remains constant in direction and in magnitude, and it is therefore necessary to use a *commutator* to enable a steady or direct voltage to be obtained from the alternating e.m.f. generated in the rotating conductors.

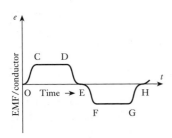

Fig. 41.2 Waveform of e.m.f. generated in a conductor

Fig. 41.3 Commutator of a d.c. machine

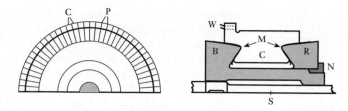

Figure 41.3 shows a longitudinal or axial section and an end elevation of half of a relatively small commutator. It consists of a large number of wedge-shaped copper segments or bars C, assembled side by side to form a ring, the segments being insulated from one another by thin mica sheets P. The segments are shaped as shown so that they can be clamped securely between a V–ring B, which is part of a cast-iron bush or sleeve, and another V–ring R which is tightened and kept in place by a nut N. The bush is keyed to shaft S.

The copper segments are insulated from the V–rings by collars of mica-based insulation M, a composite of mica flakes bonded with epoxy resin, fabricated to the exact shape of the rings. These collars project well beyond the segments so as to reduce surface leakage of current from the commutator to the shaft. At the end adjacent to the winding, each segment has a milled slot to accommodate two armature wires W which are soldered to the segment.

41.2   Double-layer drum windings

The term *armature* is generally associated with the rotating part of the d.c. machine. It essentially refers only to the rotating winding into which an e.m.f. is induced, thus we have the armature winding mounted on the armature core. By usage, the term armature, however, is frequently used to describe the entire rotating arrangement, i.e. the rotor. This can be misleading because, in a.c. machines, the e.m.f.s are induced into the fixed windings on the stator, i.e. the yoke, in which case the armature windings are the static windings. 'Armature' therefore tends to have a rather specialized interpretation when used in respect of a d.c. machine.

Let us consider a four-pole armature with, say, 11 slots, as in Fig. 41.4. In order that all the coils may be similar in shape and therefore may be wound to the correct shape before being assembled on the core, they have to be made such that if side 1 of a coil occupies the outer half of one slot, the other side 1′ occupies the inner half of another slot. This necessitates a kink in the end connections in order that the coils may overlap one another as they are being assembled. Figure 41.5 shows the shape of the end connections of a single coil consisting of a number of turns, and Fig. 41.5(b) shows how three coils, 1–1′, 2–2′ and 3–3′, are arranged in the slots so that their end connections overlap one another, the end elevation of the end connections of coils 1–1′ and 3–3′ being as shown in Fig. 41.6(a). The end connection of coil 2–2′ has been omitted from Fig. 41.6(a) to enable the shape of the other end connections to be shown more clearly. In Fig. 41.5, the two ends of the coil are brought out to P and Q, and as far as the connections to the commutator segments are concerned, the number of turns on each coil is of no consequence.

From Fig. 41.4 it is evident that if the e.m.f.s generated in conductors 1 and 1′ are to assist each other, 1′ must be moving under a S pole when 1 is

Fig. 41.4 Arrangement of a double-layer winding

Fig. 41.5 An armature coil

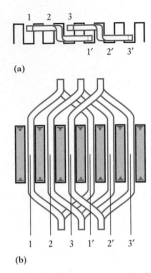

Fig. 41.6 Arrangement of overlap of end connections

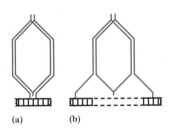

Fig. 41.7 (a) Coil of a lap winding; (b) coil of a wave winding

moving under an N pole; thus, by applying the right-hand rule (section 6.9) to Fig. 41.4 and assuming the armature to be rotated clockwise, we find that the direction of the e.m.f. generated in conductor 1 is towards the paper, whereas that generated in conductor 1′ is outwards from the paper. Hence, the distance between coil sides 1 and 1′ must be approximately a pole pitch. With 11 slots it is impossible to make the distance between 1 and 1′ exactly a pole pitch, and in Fig. 41.4 one side of coil 1–1′ is shown in slot 1 and the other side is in slot 4. The coil is then said to have a *coil span* of 4–1, namely 3. In practice, the coil span must be a whole number and is approximately equal to

$$\frac{\text{Total number of slots}}{\text{Total number of poles}}$$

In the example shown in Fig. 41.4 a very small number of slots has for simplicity been chosen. In actual machines the number of slots per pole usually lies between 10 and 15 and the coil span is slightly less than the value given by the above expression.

Let us now return to the consideration of the 11-slot armature. The 11 coils are assembled in the slots with a coil span of 3, and we are now faced with the problem of connecting to the commutator segments the 22 ends that are projecting from the winding.

Apart from a few special windings, armature windings can be divided into two groups, depending upon the manner in which the wires are joined to the commutator, namely:

1. Lap windings.
2. Wave windings.

In lap windings the two ends of any one coil are taken to adjacent segments as in Fig. 41.7(a), where a coil of two turns is shown, whereas in wave windings the two ends of each coil are bent in opposite directions and taken to segments some distance apart, as in Fig. 41.7(b).

A lap winding has as many paths in parallel between the negative and positive brushes as there are of poles; for instance, with an eight-pole lap winding, the armature conductors form eight parallel paths between the negative and positive brushes. A wave winding, on the other hand, has only two paths in parallel, irrespective of the number of poles. Hence, if a machine has p pairs of poles

$$\text{No. of parallel paths with a lap winding} = 2p$$

and $\text{No. of parallel paths with a wave winding} = 2$

For a given cross-sectional area of armature conductor and a given current density in the conductor, it follows that the total current from a lap winding is p times that from a wave winding. On the other hand, for a given number of armature conductors, the number of conductors in series per path in a wave winding is p times that in a lap winding. Consequently, for a given generated e.m.f. per conductor, the voltage between the negative and positive brushes with a wave winding is p times that with a lap winding. Hence it may be said that, in general, lap windings are used for low-voltage, heavy-current machines.

Example 41.1 An eight-pole armature is wound with 480 conductors. The magnetic flux and the speed are such that the average e.m.f. generated in each conductor is 2.2 V, and each conductor is capable of carrying a full-load current of 100 A. Calculate the terminal voltage on no load, the output current on full load and the total power generated on full load when the armature is

(a) lap-connected;
(b) wave-connected.

(a) With the armature lap-connected, number of parallel paths in the armature winding = number of poles = 8.

$$\therefore \qquad \text{No. of conductors per path} = \frac{480}{8} = 60$$

Terminal voltage on no load = e.m.f. per conductor × number of conductors per path which is

$$2.2 \times 60 = \textbf{132 V}$$

Output current on full load is

Full-load current per conductor

$$\times \text{ no. of parallel paths}$$

$$= 100 \times 8 = \textbf{800 A}$$

Total power generated on full load is

Output current

$$\times \text{ generated e.m.f.}$$

$$= 800 \times 132 = 105\ 600 \text{ W}$$

$$= \textbf{105.6 kW}$$

(b) With the armature wave-connected,

No. of parallel paths = 2

$$\therefore \qquad \text{No. of conductors per path} = \frac{480}{2} = 240$$

Terminal voltage on no load = 2.2 × 240 = **528 V**

Output current on full load = 100 × 2 = **200 A**

Total power generated on full load is

$$200 \times 528 = 105\ 600 \text{ W}$$

$$= \textbf{105.6 kW}$$

It will be seen from Example 41.1 that the total power generated by a given machine is the same whether the armature winding is lap- or wave-connected.

41.3 Calculation of e.m.f. generated in an armature winding

When an armature is rotated through one revolution, each conductor cuts the magnetic flux emanating from all the N poles and also that entering all the S poles. Consequently, if Φ is the useful flux per pole, in webers, entering or leaving the armature, p the number of *pairs* of poles and N_r the speed in revolutions per minute,

$$\text{Time of one revolution} = \frac{60}{N_r} \text{ seconds}$$

and time taken by a conductor to move one pole pitch is

$$\frac{60}{N_r} \cdot \frac{1}{2p} \text{ seconds}$$

Therefore average rate at which conductor cuts the flux is

$$\Phi \div \left(\frac{60}{N_r} \cdot \frac{1}{2p} \right) = \frac{2\Phi N_r p}{60} \text{ webers per second}$$

and average e.m.f. generated in each conductor is

$$\frac{2\Phi N_r p}{60} \text{ volts}$$

If Z is the total number of armature conductors, and c the number of parallel paths through winding between positive and negative brushes (2 for a wave winding, and $2p$ for a lap winding)

$$\therefore \qquad \frac{Z}{c} = \text{number of conductors in series in each path}$$

The brushes are assumed to be in contact with segments connected to conductors in which no e.m.f. is being generated, and the e.m.f. generated in each conductor, while it is moving between positions of zero e.m.f., varies as shown by curve OCDE in Fig. 41.2. The number of conductors in series in each of the parallel paths between the brushes remains practically constant; hence total e.m.f. between brushes is

Average e.m.f. per conductor

$\times$ no. of conductors in series per path

$$= \frac{2\Phi N_r p}{60} \times \frac{Z}{c}$$

i.e.

$$E = 2\frac{Z}{c} \times \frac{N_r p}{60} \times \Phi \text{ volts} \qquad\qquad [41.1]$$

Example 41.2

A four-pole wave-connected armature has 51 slots with 12 conductors per slot and is driven at 900 r/min. If the useful flux per pole is 25 mWb, calculate the value of the generated e.m.f.

Total number of conductors $= Z = 51 \times 12 = 612$; $c = 2$; $p = 2$; $N = 900$ r/min; $\Phi = 0.025$ Wb.

Using expression [41.1], we have

$$E = 2 \times \frac{612}{2} \times \frac{900 \times 2}{60} \times 0.025$$

$$= 459 \text{ V}$$

Example 41.3

An eight-pole lap-connected armature, driven at 350 r/min, is required to generate 260 V. The useful flux per pole is about 0.05 Wb. If the armature has 120 slots, calculate a suitable number of conductors per slot.

For an eight-pole lap winding, $c = 8$. Hence

$$260 = 2 \times \frac{Z}{8} \times \frac{350 \times 4}{60} \times 0.05$$

$$\therefore \qquad Z = 890 \text{ (approximately)}$$

and number of conductors per slot = 890/120 = 7.4 (approx.).

This value must be an even number; hence **eight conductors per slot** would be suitable.

Since this arrangement involves a total of $8 \times 120 = 960$ conductors, and since a flux of 0.05 Wb per pole with 890 conductors gave 260 V, then with 960 conductors, the same e.m.f. is generated with a flux of $0.05 \times (890/960)$ = 0.0464 Wb per pole.

41.4 Armature reaction

Armature reaction is the effect of armature ampere-turns upon the value and the distribution of the magnetic flux entering and leaving the armature core.

Let us, for simplicity, consider a two-pole machine having an armature with eight slots and two conductors per slot, as shown in Fig. 41.8. The curved lines between the conductors and the commutator segments represent the front end connections of the armature winding and those on the outside of the armature represent the back end connections. The armature winding – like all modern d.c. windings – is of the double-layer type, the end connections of the outer layer being represented by full lines and those of the inner layer by dotted lines.

Fig. 41.8 A two-pole armature winding

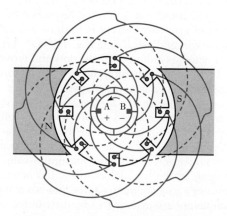

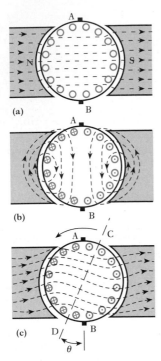

Fig. 41.9 Flux distribution due to (a) field current alone, (b) armature current alone, (c) field and armature currents of a d.c. motor

Brushes A and B are placed so that they are making contact with conductors which are moving midway between the poles and have therefore no e.m.f. induced in them. If the armature moves anticlockwise, the direction of the e.m.f.s generated in the various conductors is opposite to that of the currents, which are indicated in Fig. 41.9(b) by the dots and crosses.

In diagrams where the end connections are omitted, it is usual to show the brushes midway between the poles, as in Fig. 41.9.

In general, an armature has 10 to 15 slots per pole, so that the conductors are more uniformly distributed around the armature core than is suggested by Fig. 41.8, and for simplicity we may omit the slots and consider the conductors uniformly distributed as in Fig. 41.9(a). The latter shows the distribution of flux when there is no armature current, the flux in the gap being practically radial and uniformly distributed.

Figure 41.9(b) shows the distribution of the flux set up by current flowing through the armature winding in the direction that it will actually flow when the machine is loaded as a motor. It will be seen that at the centre of the armature core and in the pole shoes the direction of this flux is at right angles to that due to the field winding; hence the reason why the flux due to the armature current is termed *cross flux*.

The pole tip which is first met during revolution by a point on the armature or stator surface is known as the *leading tip* and the other as the *trailing pole tip*.

Figure 41.9(c) shows the resultant distribution of the flux due to the combination of the fluxes in Fig. 41.9(a) and (b); thus over the trailing halves of the pole faces the cross flux is in opposition to the main flux, thereby reducing the flux density, whereas over the leading halves the two fluxes are in the same direction, so that the flux density is strengthened. Apart from the effect of magnetic saturation, the increase of flux over one half of the pole face is the same as the decrease over the other half, and the total flux per pole remains practically unaltered. Hence, in a motor, the effect of armature reaction is to twist or distort the flux against the direction of rotation.

One important consequence of this distortion of the flux is that the magnetic neutral axis is shifted through an angle θ from AB to CD; in other words, with the machine on no load and the flux distribution of Fig. 41.9(a), conductors are moving parallel to the magnetic flux and therefore generating no e.m.f. when they are passing axis AB. When the machine is loaded as a motor and the flux distorted as in Fig. 41.9(c), conductors are moving parallel to the flux and generating no e.m.f. when they are passing axis CD.

An alternative and in some respects a better method of representing the effect of armature current is to draw a developed diagram of the armature conductors and poles, as in Fig. 41.10(a). The direction of the current in the conductors is indicated by the dots and crosses.

In an actual armature, the two conductors forming one turn are situated approximately a pole pitch apart, as in Fig. 41.8, but *as far as the magnetic effect of the currents in the armature conductors is concerned*, the end connections could be arranged as shown by the dotted lines in Fig. 41.10(a). From the latter, it will be seen that the conductors situated between the vertical axes CC_1 and DD_1 act as if they formed concentric coils producing a magnetomotive force having its maximum value along axis AA_1. Similarly, the currents in the conductors to the left of axis CC_1 and to the right of DD_1 produce a magnetomotive force that is a maximum along axis BB_1. Since the conductors are assumed to be distributed uniformly around the armature

Fig. 41.10 Distribution of armature m.m.f.

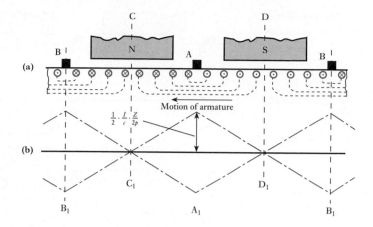

periphery, the distribution of the m.m.f. is represented by the chain-dotted line in Fig. 41.10(b). These lines pass through zero at points C_1 and D_1 midway between the brushes.

If I is the *total* armature current, in amperes, Z the number of armature conductors, c the number of parallel paths and p the number of pairs of poles

$$\text{Current per conductor} = \frac{I}{c}$$

and

$$\text{Conductors per pole} = \frac{Z}{2p}$$

∴

$$\text{Ampere-conductors per pole} = \frac{I}{c} \cdot \frac{Z}{2p}$$

Since two armature conductors constitute one turn,

$$\text{Ampere-turns per pole} = \frac{1}{2} \cdot \frac{I}{c} \cdot \frac{Z}{2p} \qquad \text{[41.2]}$$

This expression represents the armature m.m.f. at each brush axis.

The effect of the armature ampere-turns upon the distribution of the magnetic flux is represented in Fig. 41.11. The dotted lines in Fig. 41.11(a) represent the distribution of the magnetic flux in the air-gap on *no load*. The corresponding variation of the flux density over the periphery of the armature is represented by the ordinates of Fig. 41.11(b). Figure 41.11(c) and (d) represent the cross flux due to the armature ampere–turns alone, the armature current being assumed in the direction in which it flows when the machine is loaded as a generator. It will be seen that the flux density in the gap increases from zero at the centre of the pole face to a maximum at the pole tips and then decreases rapidly owing to the increasing length of the path of the fringing flux, until it is a minimum midway between the poles.

Figure 41.11(e) represents the machine operating as a motor, and the distribution of the flux density around the armature core is approximately the resultant of the graphs of Fig. 41.11(b) and (d), and is represented in Fig. 41.11(f). The effect of magnetic saturation would be to reduce the flux density at the leading pole tips, as indicated by the shaded areas P, and thereby to reduce the total flux per pole.

Fig. 41.11 Distribution of main flux, cross flux and resultant flux

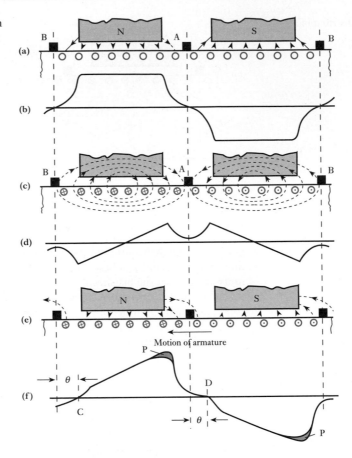

It will also be seen from Fig. 41.11(f) that the points of zero flux density, and therefore of zero generated e.m.f. in the armature conductors, have been shifted through an angle θ against the direction of rotation to points C and D.

If the machine had been operated as a generator instead of as a motor, the field patterns illustrated in Fig. 41.11 would still apply provided that the direction of the armature were reversed. The result of this observation is that the effect of magnetic saturation would be to reduce the flux density at the trailing pole tips, as indicated by the shaded areas P, and again thereby to reduce the total flux per pole.

It would also be seen from Fig. 41.11(f) that, with the armature rotating in the opposite (clockwise) direction, the points of zero flux density, and therefore of zero generated e.m.f. in the armature conductors, have been shifted through an angle θ in the direction of rotation to points C and D.

41.5	Armature reaction in a d.c. motor

The direction of the armature current in a d.c. motor is *opposite* to that of the generated e.m.f., whereas in a generator the current is in the *same* direction as the generated e.m.f. It follows that in a d.c. motor the flux is distorted backwards; and the brushes have to be shifted backwards if they are to be on the magnetic neutral axis when the machine is loaded. A backward shift in a motor gives rise to demagnetizing ampere-turns, and the reduction of flux tends to cause an increase of speed; in fact, this method – commutation permitting – may be used to compensate for the effect of the IR drop in the armature, thereby maintaining the speed of a shunt motor practically constant at all loads.

41.6   Commutation

The e.m.f. generated in a conductor of a d.c. armature is an alternating e.m.f. and the current in a conductor is in one direction when the conductor is moving under a N pole and in the reverse direction when it is moving under a S pole. This reversal of current in a coil has to take place while the two commutator segments to which the coil is connected are being short-circuited by a brush, and the process is termed *commutation*. The duration of this short-circuit is usually about 0.002 s. The reversal of, say, 100 A in an inductive circuit in such a short time is likely to present difficulty and might cause considerable sparking at the brushes.

For simplicity, in considering the variation of current in the short-circuited coil, we can represent the coils and the commutator segments as in Fig. 41.12 where the two ends of any one coil are connected to adjacent segments, as in a lap winding.

Fig. 41.12   Portion of armature winding

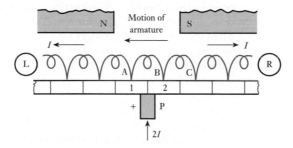

If the current per conductor is I and if the armature is moving from right to left, then – assuming the brush to be positive – coil C is carrying current from right to left (R to L), whereas coil A is carrying current from L or R. We shall therefore examine the variation of current in coil B which is connected to segments 1 and 2.

The current in coil B remains at its full value from R to L until segment 2 begins to make contact with brush P, as in Fig. 41.13(a). As the area of contact with segment 2 increases, current i_1 flowing to the brush via segment 2 increases and current $(I - i_1)$ through coil B decreases. If the current distribution between segments 1 and 2 were determined by the areas of contact only, the current through coil B would decrease linearly, as shown by line M in Fig. 41.14. It follows that when the brush is making equal areas of contact with segments 1 and 2, current through B would be zero, and further movement of the armature would cause the current through B to grow in the reverse direction.

Fig. 41.13 Coil B near the beginning and the end of commutation

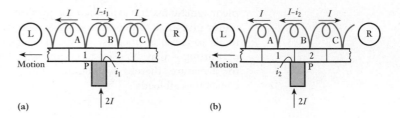

(a)   (b)

Fig. 41.14 Variation of current in the short-circuited coil

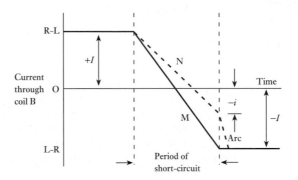

Figure 41.13(b) represents the position of the coils near the end of the period of short-circuit. The current from segment 1 to P is then i_2 and that flowing from left to right through B is $(I - i_2)$. The short-circuit is ended when segment 1 breaks contact with P, and the current through coil B should by that instant have attained its full value from L to R. Under these conditions there should be no sparking at the brush, and this linear variation of the current in the short-circuited coil is referred to as *straight line* or *linear commutation*.

It was explained in section 41.4 that the armature current gives rise to a magnetic field; thus, Fig. 41.9(b) shows the flux set up by the armature current alone. From the direction of this cross flux and assuming anticlockwise rotation, we can deduce that an e.m.f. is generated towards the paper in a conductor moving in the vicinity of brush A, namely in the direction in which current was flowing in the conductor before the latter was short-circuited by the brush. The same conclusion may be derived from a consideration of the resultant distribution of flux given in Fig. 41.11(e), where a conductor moving in the region of brush A is generating an e.m.f. in the same direction as that generated when the conductor was moving under the preceding main pole. This generated e.m.f. – often referred to as the *reactance voltage* – is responsible for delaying the reversal of the current in the short-circuited coils as shown by curve N in Fig. 41.14. The result is that when segment 1 is due to break contact with the brush, as in Fig. 41.15, the current through coil B has grown to some value i (Fig. 41.14); and the remainder, namely $(I - i)$, has to pass between segment 1 and the brush in the form of an arc. This arc is rapidly drawn out and the current through B grows quickly from i to I, as shown in Fig. 41.14.

It is this reactance voltage that is mainly responsible for sparking at the brushes of d.c. machines, and most methods of reducing sparking are directed towards the reduction or neutralization of the reactance voltage.

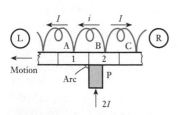

Fig. 41.15 Arcing when segment leaves brush

Summary of important formulae	EMF induced in an armature winding

$$E = 2\frac{Z}{c} \times \frac{N_r p}{60} \times \Phi \text{ volts} \qquad [41.1]$$

Terms and concepts

A d.c. machine normally has a wound rotor and a salient-pole stator with two, four, six or more poles.

The winding on the rotor is termed the **armature winding** and as a consequence it is common to refer to the rotor as the armature.

The connections to the rotor are made through carbon brushes which are held under tension against the commutator.

The armature windings are either **lap windings** or **wave windings**.

The e.m.f. induced in the armature winding is proportional to the speed of rotation and to the pole flux.

The current in the armature winding creates a second field which results in the armature reaction.

The process of switching the connections on the rotor by means of the commutator is known as **commutation**.

Exercises 41

1. A six-pole armature is wound with 498 conductors. The flux and the speed are such that the average e.m.f. generated in each conductor is 2 V. The current in each conductor is 120 A. Find the total current and the generated e.m.f. of the armature if the winding is connected: (a) wave; (b) lap. Also find the total power generated in each case.

2. A four-pole armature is wound with 564 conductors and driven at 800 r/min, the flux per pole being 20 mWb. The current in each conductor is 60 A. Calculate the total current, the e.m.f. and the electrical power generated in the armature if the conductors are connected: (a) wave; (b) lap.

3. An eight-pole lap-connected armature has 96 slots with 6 conductors per slot and is driven at 500 r/min. The useful flux per pole is 0.09 Wb. Calculate the generated e.m.f.

4. A four-pole armature has 624 lap-connected conductors and is driven at 1200 r/min. Calculate the useful flux per pole required to generate an e.m.f. of 250 V.

5. A six-pole armature has 410 wave-connected conductors. The useful flux per pole is 0.025 Wb. Find the speed at which the armature must be driven if the generated e.m.f. is to be 485 V.

6. The wave-connected armature of a four-pole d.c. generator is required to generate an e.m.f. of 520 V when driven at 600 r/min. Calculate the flux per pole required if the armature has 144 slots with two coil slides per slot, each coil consisting of three turns.

7. The armature of a four-pole d.c. generator has 47 slots, each containing six conductors. The armature winding is wave-connected, and the flux per pole is 25 mWb. At what speed must the machine be driven to generate an e.m.f. of 250 V?

8. Develop from first principles an expression for the e.m.f. of a d.c. generator.

   Calculate the e.m.f. developed in the armature of a two-pole d.c. generator, whose armature has 280 conductors and is revolving at 100 r/min. The flux per pole is 0.03 Wb.

9. Draw a labelled diagram of the cross-section of a four-pole d.c. shunt-connected generator. What are the essential functions of the field coils, armature, commutator and brushes?

   The e.m.f. generated by a four-pole d.c. generator is 400 V when the armature is driven at 1000 r/min. Calculate the flux per pole if the wave-wound armature has 39 slots with 16 conductors per slot.

Exercises 41 continuted

10. Explain the function of the commutator in a d.c. machine.

    A six-pole d.c. generator has a lap-connected armature with 480 conductors. The resistance of the armature circuit is 0.02 Ω. With an output current of 500 A from the armature, the terminal voltage is 230 V when the machine is driven at 900 r/min. Calculate the useful flux per pole and derive the expression employed.

11. A 300 kW, 500 V, eight-pole d.c. generator has 768 armature conductors, lap-connected. Calculate the number of demagnetizing and cross ampere-turns per pole when the brushes are given a lead of 5 electrical degrees from the geometric neutral. Neglect the effect of the shunt current.

12. Write a short essay describing the effects of (a) armature reaction and (b) poor commutation on the performance of a d.c. machine. Indicate in your answer how the effects of armature reaction may be reduced, and how commutation may be improved.

13. (a) Explain how armature reaction occurs in a d.c. generator and the effect it has on the flux distribution of the machine.

    (b) Explain the difficulties of commutation in a d.c. generator. What methods are used to overcome these difficulties?

14. A four-pole motor has a wave-connected armature with 888 conductors. The brushes are displaced backwards through 5 angular degrees from the geometrical neutral. If the total armature current is 90 A, calculate: (a) the cross and the back ampere-turns per pole; (b) the additional field current to neutralize this demagnetization, if the field winding has 1200 turns per pole.

15. Calculate the number of turns per pole required for the commutating poles of the d.c. generator referred to in Q. 11, assuming the compole ampere-turns per pole to be about 1.3 times the armature ampere-turns per pole and the brushes to be in the geometric neutral.

16. An eight-pole generator has a lap-connected armature with 640 conductors. The ratio of pole arc per pole pitch is 0.7. Calculate the ampere-turns per pole of a compensating winding to give uniform air-gap density when the total armature current is 900 A.

17. Define the temperature coefficient of resistance.

    A four-pole machine has a wave-wound armature with 576 conductors. Each conductor has a cross-sectional area of 5 mm^2 and a mean length of 800 mm. Assuming the resistivity for copper to be 0.0173 $\mu\Omega$ m at 20 °C and the temperature coefficient of resistance to be 0.004/°C, calculate the resistance of the armature winding at its working temperature of 50 °C.

Chapter forty-two

Direct-current Motors

Objectives

When you have studied this chapter, you should
- be familiar with armature/field connections of d.c. machines
- be familiar with the d.c. machine as a generator or motor
- understand speed characteristics of d.c. motors
- understand torque characteristics of d.c. motors
- be familiar with speed control of d.c. motors
- have an understanding of power electronic controllers
- be able to analyse d.c. machine performance

Contents

The ability to control the speed with great accuracy is an attractive feature of the d.c. motor. This chapter will explain how this can be achieved.

There are several methods of connecting the stator (field) winding of a d.c. machine. Especially important are the series and shunt connections. These terms indicate, respectively, that the field winding is connected in series and parallel with the armature winding. The speed of rotation can be related to the magnitude of the field current for each connection arrangement. Further, the torque can be related to the field and armature currents. Hence, quite different (and useful) characteristics result for each different connection. For example, where a large starting torque is required, such as in cranes and trains, the series motor is used. Shunt motors are suitable where the speed has to remain approximately constant over a wide range of load variation.

In this chapter, we will consider the derivation of the speed and torque characteristics of d.c. machines and the methods of speed control. We will also consider modern power electronic circuit arrangements which provide a means of controlling the speed of d.c. machines with greater efficiency. By the end of the chapter, you will be able to undertake exercises in the analysis of the performance of d.c. machines and be familiar with their operational characteristics.

42.1 Armature and field connections

The general arrangement of the brush and field connections of a four-pole machine is shown in Fig. 42.1. The four brushes B make contact with the commutator. The positive brushes are connected to the positive terminal A and the negative brushes to the negative terminal A_1. From Fig. 41.8 it will be seen that the brushes are situated approximately in line with the centres of the poles. This position enables them to make contact with conductors in which little or no e.m.f. is being generated since these conductors are then moving between the poles.

The four exciting or field coils C are usually joined in series and the ends are brought out to terminals F and F_1. These coils must be so connected as to produce N and S poles alternately. The arrowheads in Fig. 42.1 indicate the direction of the field current when F is positive.

In general, we may divide the methods used for connecting the field and armature windings into the following groups:

1. *Separately excited machines* – the field winding being connected to a source of supply other than the armature of its own machine.
2. *Self-excited machines*, which may be subdivided into:
   (a) *shunt-wound machines* – the field winding being connected across the armature terminals;
   (b) *series-wound machines* – the field winding being connected in series with the armature winding;
   (c) *compound-wound machines* – a combination of shunt and series windings.

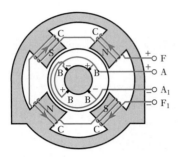

Fig. 42.1 Armature and field connections

Before we discuss the above systems in greater detail, let us consider the relationship between the magnetic flux and the exciting ampere-turns of a machine on no load. From Fig. 42.1 it will be seen that the ampere-turns of one field coil have to maintain the flux through one air-gap, a pole core, part of the yoke, one set of armature teeth and part of the armature core. The number of ampere-turns required for the air-gap is directly proportional to the flux and is represented by the straight line OA in Fig. 42.2. For low values of the flux, the number of ampere-turns required to send the flux through the ferromagnetic portion of the magnetic circuit is very small, but when the flux exceeds a certain value, some parts – especially the teeth – begin to get saturated and the number of ampere-turns increases far more rapidly than the flux, as shown by curve B. Hence, if DE represents the number of ampere-turns per pole required to maintain flux OD across the air-gap and if DF represents the number of ampere-turns per pole to send this flux through the ferromagnetic portion of the magnetic circuit, then total ampere-turns per pole to produce flux OD is

$$DE + DF = DG$$

By repeating this procedure for various values of the flux, we can derive the *magnetization curve* C representing the relationship between the useful magnetic flux per pole and the total ampere-turns per pole.

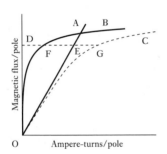

Fig. 42.2 Magnetization curve of a machine

42.2 A d.c. machine as generator or motor

There is no difference of construction between a d.c. motor and a d.c. generator. In fact, the only difference is that in a motor the generated e.m.f. is less than the terminal voltage, whereas in a generator the generated e.m.f. is greater than the terminal voltage.

Fig. 42.3 Shunt-wound
machine as generator or motor

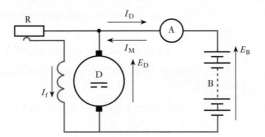

For instance, suppose a shunt generator D shown in Fig. 42.3 to be driven
by an engine and connected through a centre-zero ammeter A to a battery B.
If the field regulator R is adjusted until the reading on A is zero, the e.m.f.,
E_D, generated in D is then exactly equal to the e.m.f., E_B, of the battery. If R
is now reduced, the e.m.f. generated in D exceeds that of B, and the excess
e.m.f. is available to circulate a current I_D through the resistance of the armature circuit, the battery and the connecting conductors. Since I_D is in the
same direction as E_D, machine D is a generator of electrical energy.

Next, suppose the supply of steam or oil to the engine driving D to be cut
off. The speed of the set falls, and as E_D decreases, I_D becomes less, until,
when $E_D = E_B$, there is no circulating current. But E_D continues to decrease
and becomes less than E_B, so that a current I_M flows in the reverse direction.
Hence B is now supplying electrical energy to drive D as an electric motor.

The speed of D continues to fall until the difference between E_D and E_B
is sufficient to circulate the current necessary to maintain the rotation of D.
It will be noticed that the direction of the field current I_f is the same whether
D is running as a generator or a motor.

The relationship between the current, the e.m.f., etc. for machine D may
be expressed thus. If E is the e.m.f. generated in armature, V the terminal
voltage, R_a the resistance of armature circuit and I_a the armature current,
then, when D is operating as a generator,

$$E = V + I_a R_a \qquad\qquad [42.1]$$

When the machine is operating as a motor, the e.m.f., E, is less than the
applied voltage V, and the direction of the current I_a is the reverse of that
when the machine is acting as a generator; hence

$$E = V - I_a R_a$$

or $$V = E + I_a R_a \qquad\qquad [42.2]$$

Since the e.m.f. generated in the armature of a motor is in opposition to
the applied voltage, it is sometimes referred to as a *back e.m.f.*

Example 42.1   The armature of a d.c. machine has a resistance of 0.1 Ω and is connected to a 250 V supply. Calculate the generated e.m.f. when it is
running

(a) as a generator giving 80 A;
(b) as a motor taking 60 A.

(a) Voltage drop due to armature resistance $= 80 \times 0.1 = 8$ V. From equation [42.1]

$$\text{Generated e.m.f.} = 250 + 8 = \textbf{258 V}$$

(b) Voltage drop due to armature resistance $= 20 \times 0.1 = 6$ V. From equation [42.2]

$$\text{Generated e.m.f.} = 250 - 6 = \textbf{244 V}$$

42.3　Speed of a motor

Equation [41.1] showed that the relationship between the generated e.m.f., speed, flux, etc. is represented by

$$E = 2\frac{Z}{c} \cdot \frac{N_r p}{60} \cdot \Phi$$

For a given machine, Z, c and p are fixed; in such a case we can write

$$E = k N_r \Phi$$

where $\quad k = 2\dfrac{Z}{c} \cdot \dfrac{p}{60}$

Substituting for E in expression [42.2] we have

$$V = k N_r \Phi + I_a R_a$$

$$\therefore \qquad N_r = \frac{V - I_a R_a}{k \Phi} \qquad\qquad [42.3]$$

The value of $I_a R_a$ is usually less than 5 per cent of the terminal voltage V, so that

$$N_r \propto \frac{V}{\Phi} \qquad\qquad [42.4]$$

In words, this expression means that the speed of an electric motor is approximately proportional to the voltage applied to the armature and inversely proportional to the flux; all methods of controlling the speed involve the use of either or both of these relationships.

Example 42.2

A four-pole motor is fed at 440 V and takes an armature current of 50 A. The resistance of the armature circuit is 0.28 Ω. The armature winding is wave-connected with 888 conductors and the useful flux per pole is 0.023 Wb. Calculate the speed.

From expression [42.2] we have

$$440 = \text{generated e.m.f.} + 50 \times 0.28$$

$$\therefore \qquad \text{Generated e.m.f.} = 440 - 14 = 426 \text{ V}$$

Substituting in the e.m.f. equation [41.1], we have

$$426 = 2 \times \frac{888}{2} \times \frac{N_r \times 2}{60} \times 0.023$$

$$N_r = \textbf{626 r/min}$$

Example 42.3

A motor runs at 900 r/min off a 460 V supply. Calculate the approximate speed when the machine is connected across a 200 V supply. Assume the new flux to be 0.7 of the original flux.

If Φ is the original flux, then from expression [42.4]:

$$900 = \frac{460}{k\Phi}$$

$\therefore \qquad k\Phi = 0.511$

and $\qquad$ new speed $= \dfrac{\text{new voltage}}{k \times \text{original flux} \times 0.7}$ (approximately)

$$N_r = \frac{200}{0.511 \times 0.7} = 559\,\text{r/min}$$

42.4 Torque of an electric motor

If we start with equation [42.2] and multiply each term by I_a, namely the total armature current, we have

$$VI_a = EI_a + I_a^2 R_a$$

But VI_a represents the total electrical power supplied to the armature, and $I_a^2 R_a$ represents the loss due to the resistance of the armature circuit. The difference between these two quantities, namely EI_a, therefore represents the mechanical power developed by the armature. All of this mechanical power is not available externally, since some of it is absorbed as friction loss at the bearings and at the brushes and some is wasted as hysteresis loss (section 44.9) and in circulating eddy currents in the ferromagnetic core (section 44.10).

If M is the torque, in newton metres, exerted on the armature to develop the mechanical power just referred to, and if N_r is the speed in revolutions per minute, then from expression [1.8],

$$\text{Mechanical power developed} = \frac{2\pi M N_r}{60}\ \text{watts}$$

Hence $\qquad \dfrac{2\pi M N_r}{60} = EI_a$ $\hfill$ [42.5]

$$= 2\frac{Z}{c} \cdot \frac{N_r p}{60} \cdot \Phi \cdot I_a$$

$$M = 0.318\frac{I_a}{c} \cdot Zp\Phi \quad \text{newton metres} \hfill [42.6]$$

For a given machine, Z, c and p are fixed, in which case

$$M \propto I_a \times \Phi \hfill [42.7]$$

Or, in words, the torque of a given d.c. motor is proportional to the product of the armature current and the flux per pole.

Example 42.4

A d.c. motor takes an armature current of 110 A at 480 V. The resistance of the armature circuit is 0.2 Ω. The machine has six poles and the armature is lap-connected with 864 conductors. The flux per pole is 0.05 Wb. Calculate

(a) the speed;
(b) the gross torque developed by the armature.

(a)   Generated e.m.f. $= 480 - (110 \times 0.2)$

$$= 458 \text{ V}$$

Since the armature winding is lap-connected, $c = 6$.
Substituting in expression [41.1], we have

$$458 = 2 \times \frac{864}{6} \times \frac{N_r \times 3}{60} \times 0.05$$

$$N_r = \textbf{636 r/min}$$

(b) Mechanical power developed by armature is

$$110 \times 458 = 50\ 380 \text{ W}$$

Substituting in expression [42.5] we have

$$2\pi M \times \frac{636}{60} = 50\ 380$$

$$= \textbf{756 N m}$$

Alternatively, using expression [42.6] we have

$$M = 0.318 \times \frac{110}{6} \times 864 \times 3 \times 0.05$$

$$= \textbf{756 N m}$$

Example 42.5

The torque required to drive a d.c. generator at 15 r/s is 2 kN m. The core, friction and windage losses in the machine are 8.0 kW. Calculate the power generated in the armature winding.

Driving torque $= 2$ kN m $= 2000$ N m

From expression [1.8], power required to drive the generator is

$$2\pi \times 2000 \text{ [N m]} \times 15 \text{ [r/s]}$$

$$= 188\ 400 \text{ W} = 188.4 \text{ kW}$$

Since core, friction and windage losses are 8.0 kW,

$\therefore$ Power generated in armature winding $= 188.4 - 8.0$

$$= \textbf{180.4 kW}$$

42.5	**Speed characteristics of electric motors**

With very few exceptions, d.c. motors are shunt-, series- or compound-wound. The connections of a shunt motor are given in Fig. 42.4, and Figs 42.5 and 42.6 show the connections for series and compound motors respectively, the starter in each case being shown in the ON position. It is good practice to include starters in diagrams of motor connections. In compound motors, the series and shunt windings almost invariably assist each other, as indicated in Fig. 42.6.

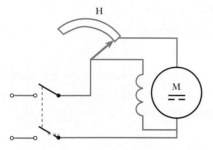

Fig. 42.4 Shunt-wound motor

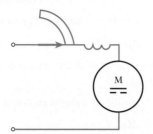

Fig. 42.5 Series-wound motor

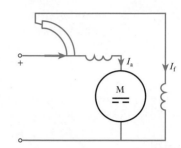

Fig. 42.6 Compound-wound motor

The speed characteristic of a motor usually represents the variation of speed with input current or input power, and its shape can be easily derived from expression [42.3], namely

$$N_r = \frac{V - I_a R_a}{k\Phi}$$

In shunt motors, the flux Φ is only slightly affected by the armature current and the value of $I_a R_a$ at full load rarely exceeds 5 per cent of V, so that the variation of speed with input current may be represented by curve A in Fig. 42.7. Hence shunt motors are suitable where the speed has to remain approximately constant over a wide range of load.

In series motors, the flux increases at first in proportion to the current and then less rapidly owing to magnetic saturation (Fig. 42.2). Also R_a in the above expression now includes the resistance of the field winding. Hence the speed is roughly inversely proportional to the current, as indicated by curve B in Fig. 42.7. It will be seen that if the load falls to a very small value, the speed may become dangerously high. A series motor should therefore not be

Fig. 42.7 Speed characteristics of shunt, series and compound motors

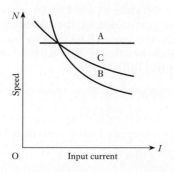

employed when there is any such risk; for instance, it should never be belt-coupled to its load except in very small machines such as vacuum cleaners.

Since the compound motor has a combination of shunt and series excitations, its characteristic (curve C in Fig. 42.7) is intermediate between those of the shunt and series motors, the exact shape depending upon the values of the shunt and series ampere-turns.

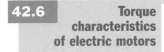

42.6 Torque characteristics of electric motors

In section 42.4 it was shown that for a given motor:

torque ∝ armature current × flux per pole

$$M \propto I_a \Phi$$

Since the flux in a shunt motor is practically independent of the armature current

∴   torque of a shunt motor ∝ armature current

$$M \propto I_a$$

Fig. 42.8 Torque characteristics of shunt, series and compound motors

and is represented by the straight line A in Fig. 42.8.

In a series motor the flux is approximately proportional to the current up to full load, so that

torque of a series motor ∝ (armature current)², approx.

$$M \propto I_a^2$$

Above full load, magnetic saturation becomes more marked and the torque does not increase so rapidly.

Curves A, B and C in Fig. 42.8 show the relative shapes of torque curves for shunt, series and compound motors having the same full-load torque OQ with the same full-load armature current OP, the exact shape of curve C depending upon the relative value of the shunt and series ampere-turns at full load.

From Fig. 42.8 it is evident that for a given current below the full-load value the shunt motor exerts the largest torque, but for a given current above that value the series motor exerts the largest torque.

The maximum permissible current at starting is usually about 1.5 times the full-load current. Consequently where a large starting torque is required, such as for hoists, cranes, electric trains, etc., the series motor is the most suitable machine.

Example 42.6   A series motor runs at 600 r/min when taking 110 A from a 250 V supply. The resistance of the armature circuit is 0.12 Ω and that of the series winding is 0.03 Ω. The useful flux per pole for 120 A is 0.024 Wb and that for 50 A is 0.0155 Wb. Calculate the speed when the current has fallen to 50 A.

Total resistance of armature and series windings is

0.12 + 0.03 = 0.15 Ω

therefore e.m.f. generated when current is 110 A is

250 − 110 × 0.15 = 232 V

In section 42.3 it was shown that for a given machine:

Generated e.m.f. = a constant (say k) × speed × flux

Hence with 120 A,

$$232 = k \times 600 \times 0.024$$

$$\therefore \qquad k = 16.11$$

With 50 A,

Generated e.m.f. $= 250 - 50 \times 0.15 = 242.5$ V

But the new e.m.f. generated is

$$k \times \text{new speed} \times \text{new flux}$$

$$\therefore \qquad 242.5 = 14.82 \times \text{new speed} \times 0.0155$$

$$\therefore \qquad \textbf{Speed for 50 A = 971 r/min}$$

42.7	Speed control of d.c. motors

It has already been explained in section 42.3 that the speed of a d.c. motor can be altered by varying either the flux or the armature voltage or both; the methods most commonly employed are as follows:

1.   A variable resistor, termed a *field regulator*, in series with the shunt winding – *only applicable to shunt and compound motors*. Such a field regulator is indicated by H in Fig. 42.4. When the resistance is increased, the field current, the flux and the generated e.m.f. are reduced. Consequently more current flows through the armature and the increased torque enables the armature to accelerate until the generated e.m.f. is again nearly equal to the applied voltage (see Example 42.7).

With this method it is possible to increase the speed to three or four times that at full excitation, but it is not possible to reduce the speed below that value. Also, with any given setting of the regulator, the speed remains approximately constant between no load and full load.

2.   A resistor, termed a *controller*, in series with the armature. The electrical connections for a controller are exactly the same as for a starter, the only difference being that in a controller the resistor elements are designed to carry the armature current indefinitely, whereas in a starter they can only do so for a comparatively short time without getting excessively hot.

For a given armature current, the larger the controller resistance in circuit, the smaller is the p.d. across the armature and the lower, in consequence, is the speed.

This system has several disadvantages: (a) the relatively high cost of the controller; (b) much of the input energy may be dissipated in the controller and the overall efficiency of the motor considerably reduced thereby; (c) the speed may vary greatly with variation of load due to the change in the p.d. across the controller causing a corresponding change in the p.d. across the motor; thus, if the supply voltage is 250 V, and if the current decreases so that the p.d. across the controller falls from, say, 100 to 40 V, then the p.d. across the motor increases from 150 V to 210 V.

The principal advantage of the system is that speeds from zero upwards are easily obtainable, and the method is chiefly used for controlling the speed of cranes, hoists, trains, etc. where the motors are frequently started and stopped and where efficiency is of secondary importance.

3.   When an a.c. supply is available the voltage applied to the armature can be controlled by thyristors, the operation of which is explained in section 45.2. Briefly, the thyristor is a solid-state rectifier which is normally non-conducting in the forward and reverse directions. It is provided with an extra electrode, termed the *gate*, so arranged that when a pulse of current is introduced into the gate circuit, the thyristor is 'fired', i.e. it conducts in the forward direction. Once it is fired, the thyristor continues to conduct until the current falls below the holding value.

Figure 42.9 shows a simple arrangement for controlling a d.c. motor from a single-phase supply. Field winding F is separately excited via bridge-connected rectifiers J (section 21.4) and armature A is supplied via thyristor T. Gate G of the thyristor is connected to a firing circuit which supplies a current pulse once every cycle. The arrangement of the firing circuit is not shown as it is too involved for inclusion in this diagram. In Fig. 42.9, R and L represent the resistance and inductance respectively of the armature winding and an external inductor that may be inserted to increase the inductance of the circuit. A diode D is connected across the armature and the inductor.

The sine wave in Fig. 42.10(a) represents the supply voltage and the wavy line MN represents the e.m.f., e_r, generated by the rotation of the armature. The value of e_r is proportional to the speed, which in turn varies with the armature current. Thus, when the current exceeds the average value, the armature accelerates, and then decelerates when the current falls below the average value. The speed fluctuation indicated in Fig. 42.10(a) is exaggerated to illustrate the effect more clearly.

Fig. 42.9   Thyristor system of speed control

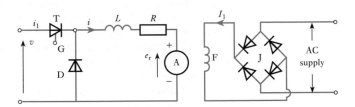

Fig. 42.10   Waveforms for Fig. 42.9

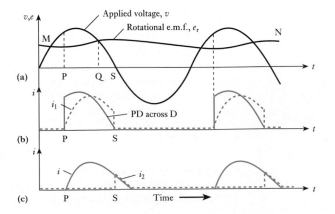

Suppose a current pulse to be applied to gate G at instant P in Fig. 42.10(a). The resulting current through the thyristor and the armature grows at a rate depending upon the difference between the applied voltage, v, and the rotational e.m.f., e_r, and upon the ratio L/R for the circuit. Thus at any instant:

$$v = e_r + iR + L \cdot \mathrm{d}i/\mathrm{d}t$$

$\therefore$ $$i = (v - e_r - L \cdot \mathrm{d}i/\mathrm{d}t)/R \qquad\qquad [42.8]$$

While the thyristor is conducting, the p.d. across the armature and inductor, and therefore across diode D, varies as shown by the full-line waveform in Fig. 42.10(b).

At instant Q, $e_r = v$, so that $i = -\dfrac{L}{R} \cdot \dfrac{\mathrm{d}i}{\mathrm{d}t}$. The current is now decreasing so that $\mathrm{d}i/\mathrm{d}t$ is negative, hence the current is still positive and is therefore continuing to exert a *driving* torque on the armature.

At instant S, the supply voltage is reversing its direction. A reverse current from the supply flows through D, thereby making the cathode of the thyristor *positive* relative to its anode. Consequently, the current i_1 through the thyristor falls to zero, as indicated by the dotted line in Fig. 42.10(b), so that the thyristor reverts to its non-conducting state.

Current is now confined to the closed circuit formed by the armature and diode D. Since the p.d. across D is practically zero, equation [42.8] can now be written

$$i = -(e_r - L \cdot \mathrm{d}i/\mathrm{d}t)/R$$

Current i is now decreasing at a sufficiently high rate for the value of $L \cdot \mathrm{d}i/\mathrm{d}t$ to exceed e_r. Since the value of $\mathrm{d}i/\mathrm{d}t$ is negative, the direction of the current is unaltered. Hence the current still continues to exert a driving torque on the armature, the energy supplied to the load being recovered partly from that stored in the inductance of the circuit while the current was growing and partly from that stored as kinetic energy in the motor and load during acceleration.

The dotted waveform in Fig. 42.10(c) represents current i_2 through diode D. The latter is often referred to as a free-wheeling diode since it carries current when the thyristor *ceases* to conduct.

The later the instant of firing the thyristor, the smaller is the average voltage applied to the armature, and the lower the speed in order that the motor may take the same average current from the supply to enable it to maintain the *same* load torque. Thus the speed of the motor can be controlled over a wide range.

An increase of motor load causes the speed to fall, thereby allowing a larger current pulse to flow during the conducting period. The fluctuation of current can be reduced by the following:

1. Using two thyristors to give full-wave rectification when the supply is single-phase.
2. Using three or six thyristors when the supply is three-phase.

An important application of the thyristor is the speed control of series motors in battery-driven vehicles. The principle of operation is that pulses of the battery voltage are applied to the motor, and the *average* value of the voltage across the motor is controlled by varying the ratio of the ON and OFF durations of the pulses. Thus, if the ON period is t_1 and the OFF period is t_2,

$$\text{Average motor voltage} = \text{battery voltage} \times \frac{t_1}{t_1 + t_2}$$

Figure 42.11 shows the essential features of this method of speed control. The series motor M is connected in series with thyristor T across the battery. A free-wheeling diode D is connected in parallel with the motor, and a switch S is used to short-circuit the thyristor at the end of each ON period.

With S open, the thyristor is fired by a current pulse applied to gate G, causing a current i_1 to flow through T and M, as shown in Fig. 42.12. After an interval t_1, switch S is closed for sufficient time to allow the thyristor current to fall to zero, thereby enabling the thyristor to revert to its non-conducting state.

After S is opened, the current through M decreases at a rate such that the e.m.f. induced in the inductance of the field and armature windings exceeds the rotational e.m.f. by an amount sufficient to circulate a current i_2 around the closed circuit formed by the motor and diode D. After an interval t_2, the operation is repeated, as indicated in Fig. 42.12.

For a low speed, the ratio t_1/t_2 is small, as shown in Fig. 42.12(a). A higher speed is obtained by increasing t_1 and reducing t_2, as in Fig. 42.12(b). The dotted horizontal lines in Fig. 42.12 represent the average values of the rotational e.m.f. E_r generated in the armature. The average value of the current is determined by the torque requirement, exactly as described in earlier sections.

The battery is supplying power only during the ON intervals t_1. During the OFF intervals t_2, current i_2 is exerting a *driving* torque on the armature, the energy supplied to the mechanical load during these intervals being derived from the magnetic fields of the series and armature windings and from the kinetic energy of the motor and load.

The frequency of the pulses may be as high as 3000 per second, and the arrangement used to control the firing of the thyristor is referred to as a *chopper circuit*, but the details of this circuit are too complex for inclusion here.

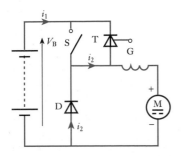

Fig. 42.11 Chopper speed control of a series motor

Fig. 42.12 Waveforms for Fig. 42.11

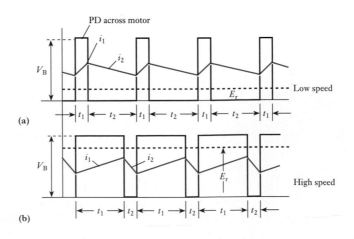

Finally let us recall the calculation of form factor as demonstrated in Example 9.6. The repeated switching of the supply in a chopper circuit gives rise to quite high form factors. The significance of this effect is that although we can control the direct currents and voltages to have low values, this is achieved at the expense of relatively high r.m.s. values. And high r.m.s. values mean high I^2R losses. Therefore we have to note that chopper circuits incur higher losses and care has to be taken when choosing the rating of the wiring.

Example 42.7 A shunt motor is running at 626 r/min (Example 42.2) when taking an armature current of 50 A from a 440 V supply. The armature circuit has a resistance of 0.28 Ω. If the flux is suddenly reduced by 5 per cent, find:

(a) the maximum value to which the current increases momentarily and the ratio of the corresponding torque to the initial torque;

(b) the ultimate steady value of the armature current, assuming the torque due to the load to remain unaltered.

(a) From Example 42.2:

Initial e.m.f. generated $= 440 - 50 \times 0.28 = 426$ V

Immediately after the flux is reduced 5 per cent, i.e. before the speed has begun to increase:

New e.m.f. generated $= 426 \times 0.95 = 404.7$ V

Therefore corresponding voltage drop due to armature resistance is

$$440 - 404.7 = 35.3 \text{ V}$$

and corresponding armature current is

$$\frac{35.3}{0.28} = \textbf{126 A}$$

From expression [42.7]:

Torque of a given machine $\propto$ armature current $\times$ flux

$$\therefore \quad \frac{\text{New torque}}{\text{Initial torque}} = \frac{\text{new current}}{\text{initial current}} \times \frac{\text{new flux}}{\text{initial flux}}$$

$$= \frac{126}{50} \times 0.95 = \textbf{2.394}$$

Hence the sudden reduction of 5 per cent in the flux is accompanied by more than a twofold increase of torque; this is the reason why the motor accelerated.

(b) After the speed and current have attained steady values, the torque will have decreased to the original value, so that

New current $\times$ new flux $=$ original current $\times$ original flux

$$\text{New armature current} = 50 \times \frac{1}{0.95} = \textbf{52.6 A}$$

Example 42.8

A shunt motor runs on no load at 700 r/min off a 440 V supply. The resistance of the shunt circuit is 240 Ω. The following table gives the relationship between the flux and the shunt current:

Shunt current (A)	0.5	0.75	1.0	1.25	1.5	1.75	2.0
Flux per pole (mWb)	6.0	8.0	9.4	10.2	10.8	11.2	11.5

Calculate the additional resistance required in the shunt circuit to raise the no-load speed to 1000 r/min.

When the motor is on no load, the voltage drop due to the armature resistance is negligible.

Initial shunt current = 440/240 = 1.83 A.

From a graph (not shown) representing the data given in the above table, corresponding flux per pole = 11.3 mWb.

Since the speed is inversely proportional to the flux,

$$\text{Flux per pole at 1000 r/min} = 11.3 \times \frac{700}{1000}$$

$$= 7.91 \text{ mWb}$$

From the graph it can be found that the shunt current to produce a flux of 7.91 mWb/pole is 0.74 A,

$$\text{Corresponding resistance of shunt circuit} = \frac{440}{0.74}$$

$$= 595 \ \Omega$$

and additional resistance required in shunt circuit is

$$595 - 240 = \textbf{355} \ \Omega$$

Example 42.9

A shunt motor, supplied at 250 V, runs at 900 r/min when the armature current is 30 A. The resistance of the armature circuit is 0.4 Ω. Calculate the resistance required in series with the armature to reduce the speed to 600 r/min, assuming that the armature current is then 20 A.

Initial e.m.f. generated is

$$250 - 30 \times 0.4 = 238 \text{ V}$$

Since the excitation remains constant, the generated e.m.f. is proportional to the speed.

$$\therefore \quad \text{EMF generated at 600 r/min} = 238 \times \frac{600}{900} = 158.7 \text{ V}$$

Hence voltage drop due to total resistance of armature circuit is

$$250 - 158.7 = 91.3 \text{ V}$$

and total resistance of armature circuit is

$$\frac{91.3}{20} = 4.567 \ \Omega$$

therefore additional resistance required in armature circuit is

$$4.567 - 0.4 = \textbf{4.17} \ \Omega$$

Summary of important formulae

For a d.c. generator,

$$E = V + I_a R_a \qquad [42.1]$$

For a d.c. motor,

$$V = E + I_a R_a \qquad [42.2]$$

$$N_r \propto \frac{V}{\Phi} \qquad [42.4]$$

and $M \propto I_a \times \Phi$ $\qquad [42.7]$

Terms and concepts

DC machines can be separately excited or self-excited. Separately excited machines are often used in control systems.

Self-excited machines can be **shunt-wound**, **series-wound** or **compound-wound**.

DC machines can readily act both as motors and as generators.

The torque developed is proportional to the pole flux and to the armature current.

The speed characteristic of a shunt motor is almost constant whilst that of a series motor is inversely proportional to the current. The consequence is that the shunt motor is useful where speed control over a limited range is required whereas the series motor gives exceptional starting torque.

The control of d.c. motors is increasingly achieved using power electronic circuits. A typical system would be based on a **thyristor chopper arrangement**.

Exercises 42

1. A shunt machine has armature and field resistances of 0.04 Ω and 100 Ω respectively. When connected to a 460 V d.c. supply and driven as a generator at 600 r/min, it delivers 50 kW. Calculate its speed when running as a motor and taking 50 kW from the same supply. Show that the direction of rotation of the machine as a generator and as a motor under these conditions is unchanged.

2. Obtain from first principles an expression for the e.m.f. of a two-pole d.c. machine, defining the symbols used.

A 100 kW, 500 V, 750 r/min, d.c. shunt generator, connected to constant-voltage bus-bars, has field and armature resistances of 100 Ω and 0.1 Ω respectively. If the prime-mover fails, and the machine continues to run, taking 50 A from the bus-bars, calculate its speed. Neglect brush-drop and armature reaction effects.

3. Derive the e.m.f. equation for a d.c. machine.

A d.c. shunt motor has an armature resistance of 0.5 Ω and is connected to a 200 V supply. If the armature current taken by the motor is 20 A, what is the e.m.f. generated by the armature? What is the effect of: (a) inserting a resistor in the field circuit; (b) inserting a resistor in the armature circuit if the armature current is maintained at 20 A?

4. Explain clearly the effect of the back e.m.f. of a shunt motor. What precautions must be taken when starting a shunt motor?

A four-pole d.c. motor is connected to a 500 V d.c. supply and takes an armature current of 80 A. The resistance of the armature circuit is 0.4 Ω. The armature is wave-wound with 522 conductors and the useful flux per pole is 0.025 Wb. Calculate: (a) the back e.m.f. of the motor; (b) the speed of the motor; (c) the torque in N m developed by the armature.

5. A shunt machine is running as a motor off a 500 V system, taking an armature current of 50 A. If the field current is suddenly increased so as to increase the flux by 20 per cent, calculate the value of the current that would momentarily be fed back into the mains. Neglect the shunt current and assume the resistance of the armature circuit to be 0.5 Ω.

6. A shunt motor is running off a 220 V supply taking an armature current of 15 A, the resistance of the armature circuit being 0.8 Ω. Calculate the value of the generated e.m.f. If the flux were suddenly reduced by 10 per cent, to what value would the armature current increase momentarily?

7. A six-pole d.c. motor has a wave-connected armature with 87 slots, each containing six conductors. The flux per pole is 20 mWb and the armature has a resistance of 0.13 Ω. Calculate the speed when the motor is running off a 240 V supply and taking an armature current of 80 A. Calculate also the torque, in N m, developed by the armature.

8. A four-pole, 460 V shunt motor has its armature wave-wound with 888 conductors. The useful flux per pole is 0.02 Wb and the resistance of the armature circuit is 0.7 Ω. If the armature current is 40 A, calculate: (a) the speed; (b) the torque in N m.

9. A four-pole motor has its armature lap-wound with 1040 conductors and runs at 1000 r/min when taking an armature current of 50 A from a 250 V d.c. supply. The resistance of the armature circuit is 0.2 Ω. Calculate: (a) the useful flux per pole; (b) the torque developed by the armature in N m.

10. A d.c. shunt generator delivers 5 kW at 250 V when driven at 1500 r/min. The shunt circuit resistance is 250 Φ and the armature circuit resistance is 0.4 Ω. The core, friction and windage losses are 250 W. Determine the torque (in N m) required to drive the machine at the above load.

11. Explain why a d.c. shunt-wound motor needs a starter on constant-voltage mains. A shunt-wound motor has a field resistance of 350 Ω and an armature resistance of 0.2 Ω and runs off a 250 V supply. The armature current is 55 A and the motor speed is 1000 r/min. Assuming a straight-line magnetization curve, calculate: (a) the additional resistance required in the field circuit to increase the speed to 1100 r/min for the same armature current; (b) the speed with the original field current and an armature current of 100 A.

12. Explain the necessity for using a starter with a d.c. motor.

A 240 V d.c. shunt motor has an armature of resistance of 0.2 Ω. Calculate: (a) the value of resistance which must be introduced into the armature circuit to limit the starting current to 40 A; (b) the e.m.f. generated when the motor is running at a constant speed with this additional resistance in circuit and with an armature current of 30 A.

13. Calculate the torque, in N m, developed by a d.c. motor having an armature resistance of 0.25 Ω and running at 750 r/min when taking an armature current of 60 A from a 480 V supply.

14. A six-pole, lap-wound, 220 V, shunt-excited d.c. machine takes an armature current of 2.5 A when unloaded at 950 r/min. When loaded, it takes an armature current of 54 A from the supply and runs at 950 r/min. The resistance of the armature circuit is 0.18 Ω and there are 1044 armature conductors. For the loaded condition, calculate: (a) the generated e.m.f.; (b) the useful flux per pole; (c) the useful torque developed by the machine in N m.

15. A d.c. shunt motor runs at 900 r/min from a 480 V supply when taking an armature current of 25 A. Calculate the speed at which it will run from a 240 V supply when taking an armature current of 15 A. The resistance of the armature circuit is 0.8 Ω. Assume the flux per pole at 240 V to have decreased to 75 per cent of its value at 480 V.

16. A shunt motor, connected across a 440 V supply, takes an armature current of 20 A and runs at 500 r/min. The armature circuit has a resistance of 0.6 Ω. If the magnetic flux is reduced by 30 per cent and the torque developed by the armature increases by 40 per cent, what are the values of the armature current and of the speed?

17. A d.c. series motor connected across a 460 V supply runs at 500 r/min when the current is 40 A. The total resistance of the armature and field circuits is 0.6 Ω. Calculate the torque on the armature in N m.

18. A d.c. series motor, having armature and field resistances of 0.06 Ω and 0.04 Ω respectively, was tested by driving it at 200 r/min and measuring the open-circuit voltage across the armature terminals, the field being supplied from a separate source. One of the readings taken was: field current, 350 A; armature p.d., 1560 V. From the above information, obtain a point on the speed/current characteristic, and one on the torque/current characteristic for normal operation at 750 V

and at 350 A. Take the torque due to rotational loss as 50 N m. Assume that brush drop and field weakening due to armature reaction can be neglected.

19. A d.c. series motor, connected to a 440 V supply, runs at 600 r/min when taking a current of 50 A. Calculate the value of a resistor which, when inserted in series with the motor, will reduce the speed to 400 r/min, the gross torque being then half its previous value. Resistance of motor = 0.2 Ω. Assume the flux to be proportional to the field current.

20. A series motor runs at 900 r/min when taking 30 A at 230 V. The total resistance of the armature and field circuits is 0.8 Ω. Calculate the values of the additional resistance required in series with the machine to reduce the speed to 500 r/min if the gross torque is: (a) constant; (b) proportional to the speed; (c) proportional to the square of the speed. Assume the magnetic circuit to be unsaturated.

Control System Motors

Two forms of control system were introduced in Chapter 29, the regulator and the servosystem, or remote position control (r.p.c.) system. The former is concerned with varying operational speed whilst servomechanisms are associated with varying position.

It has been observed that the speed of a d.c. motor can be adjusted accurately. The field winding can be even better controlled if it is excited from a separate source such as an electronic amplifier. In this chapter, we not only will consider the separately excited d.c. motor, but will also be introduced to the electronically supplied induction motor. The variation of supply frequency possible with electronic controllers facilitates speed control.

The r.p.c. system most commonly uses a stepper motor although it is possible to use a mechanism such as a Geneva cam. Both will be discussed. Finally, we will consider drive circuits for stepper motors.

By the end of the chapter, you will be familiar with the motors used and with the control systems of which they are an integral part.

43.1 Review

In Chapter 29, we were introduced to control systems in which we found that control engineering requires both an understanding of electronics and also an understanding of electrical machines. We also found that control systems were divided into two main categories:

1. Regulators.
2. Remote position control (r.p.c.) systems.

In this chapter, we will consider the motors and associated mechanisms which fulfil the requirements of such control systems.

43.2 Motors for regulators

In regulators, we are seeking to control the speed of the motor generally with a high degree of accuracy. The motor which has a good torque/speed characteristic for such applications is the d.c. shunt motor which has been described in Chapter 42. However, the shunt motor is so called on the assumption that the field winding is excited from the same source as the armature winding. This is not necessarily the case and we can have separately excited motors, i.e. motors in which the field winding is excited from a completely separate source.

In control systems, we can arrange for the output of the control amplifier to be the source of excitation for the field winding of the motor as shown in Fig. 43.1.

Fig. 43.1 Separately excited d.c. motor in a regulator

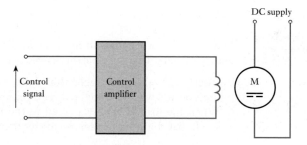

The d.c. motor is a relatively expensive motor to manufacture and it is not so robust as the a.c. induction motor. When we are considering control systems, the degree of sophistication means that high investment cost is acceptable and the relative cost of the d.c. motor is unlikely to be a major obstacle to its application (although every engineer is always cost conscious). The fact that the motor could be more robust is of some concern since down time for maintenance or repair is an important consideration. For this reason, engineers have sought alternatives and the cage-rotor induction motor is increasingly more acceptable. It will be remembered that the speed of the induction motor is effectively set by the supply frequency. If we could vary the supply frequency, it follows that we could control the speed.

In Chapter 45, we will find that there are power electronic devices which allow us to vary the supply frequency. If these electronic devices are controlled by the control system amplifier, we find that we have introduced a robust speed-controlled motor. Initially, though the motor was robust, the electronic power devices were no better than the d.c. motor which was being replaced. Now, however, the reliability of the electronic power devices such as thyristors has improved, hence the increasing number of variable-frequency cage-rotor induction motors being used in regulators. A typical variable-frequency control system is shown in Fig. 43.2.

There are other regulator motors such as the variable-frequency synchronous motor, the brushless d.c. motor and the switched reluctance motor to be found in control systems, but none of these carry the general acceptance of the two considered above.

Fig. 43.2　Regulator incorporating a variable-frequency cage-rotor induction motor

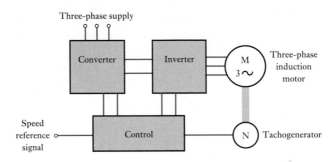

43.3　RPC system requirements

The r.p.c. motor requires to move its load to a position determined by the control system. The system experiences two limitations:

1. There is a limit to the mechanical ability of a motor to move its load – this limit occurs when the angle of required rotation becomes too small.
2. When the load is almost aligned to the input objective, the error is so small that the amplifier can no longer drive the motor.

It will be recalled that in electronic systems it was possible to generate analogue signals from digital signals provided that the digital bits were significantly small compared with the analogue wave. In much the same manner, we can arrange for a motor to turn through large values of rotation, yet the load when driven through a reduction gear will appear to move through a small and apparently precise angle of rotation.

Yet even here we find limitations. First of all, the reduction gear introduces an additional load and the greater the accuracy of positioning the load, the greater the additional load on the motor. Also the greater the gearing the more slack in the gear train, introducing further misalignment. Second, the more we reduce the gearing, the more difficult to stop the motor at exactly the required position. This can be overcome by applying a mechanical brake, but such brakes experience wear.

These limitations have been avoided by two arrangements:

1. Mechanisms such as the Geneva cam.
2. The stepping motor.

43.4	Geneva cam

The Geneva cam permits the motor to move without moving the load. However, during a quarter of every rotation, the motor engages with the cam and moves the load through a defined angle – this is the mechanical equivalent of a digital bit. During the remainder of the motor rotation there is no engagement, so it does not matter that we cannot stop the motor in exactly any given position. Rather we can stop the motor anywhere that it is not engaged.

The basic Geneva cam is shown in Fig. 43.3. Assuming that the striker arm rotates clockwise, Fig. 43.3(a) shows that the arm has just engaged in one of the slots on the cam. In order to show the action, one of the edges of the slot is marked for identification.

Fig. 43.3 Geneva cam

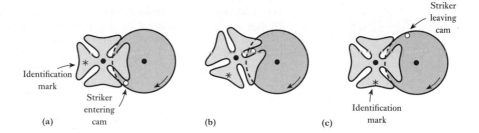

If the striker is rotated through 60°, the system arrives at the position shown in Fig. 43.3(b), i.e. the cam also has rotated through 60°. A further rotation of the striker by 60° brings the system to the position shown in Fig. 43.3(c). Here the striker has disengaged, leaving the cam rotated through exactly 90°. Further movement by the motor will provide no further cam rotation until the striker continues to position (a). The system therefore ensures exact progressions of the load and, so long as we can accept such 'digital' progressions, it is possible to obtain accurate positioning of the load.

The Geneva cam is one of a number of such mechanisms. They have been largely superseded by the stepper motor except in systems which require very large torques (and preferably can accept quite large 'digital' movements).

43.5	The stepping (or stepper) motor

Let us consider the simple reluctance motor shown in Fig. 36.3 in which we observed that the salient-pole rotor would align itself to the stator poles when they were energized by the field windings. Subsequently we became aware that the dependence of the reluctance motor on the saliency of the rotor did not provide a large torque, but if we then also excited the rotor, strong rotational torques could be established, this being the basis of the a.c. and d.c. machines which we have subsequently investigated.

However, let us consider what would happen if the rotor of the reluctance motor was permanently magnetized as indicated in Fig. 43.4. In this arrangement, when the stator poles are excited by direct current, the rotor will align itself – and stop in the exact position of alignment.

If the current in the windings is reversed, it is possible that the rotor will rotate through 180° to be aligned in the opposite direction – but equally

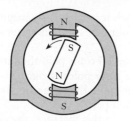

Fig. 43.4 Reluctance motor with permanent magnet rotor

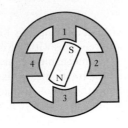

Fig. 43.5   Simple stepping motor

nothing might happen because the rotor is confused as to the direction in which to rotate, both directions being equally attractive. To avoid this difficulty, we can develop the motor to have four salient stator poles as shown in Fig. 43.5.

In this arrangement, let us excite poles 1 and 3 so that the rotor aligns accordingly. Once stability has been achieved, let us now seek to align the rotor to poles 3 and 1. To do this, we first excite poles 2 and 4; this causes the rotor to rotate through 90°, a movement which is quite decisive and causes no confusion. Subsequently let us excite poles 3 and 1 which causes a rotation of a further 90°, thus resulting in a total 180° rotation which is the objective we were seeking.

This basic arrangement might appear rather cumbersome with one movement followed by another. However, when we consider the speed of digital transmission, we can appreciate that the switching of the excitation can be very fast and that the desired rotation can be achieved in a very short time. In fact by suitable control of the windings, we could obtain continuous rotation with speed control finally stopping in any one of four desired positions. Every pulse of excitation sent to a winding steps the rotor round by 90° (hence the name of the motor – the stepping motor). By counting the pulses, we can determine the rotation achieved. This basic control is illustrated in Fig. 43.6.

Fig. 43.6   Rotational control of a stepping motor

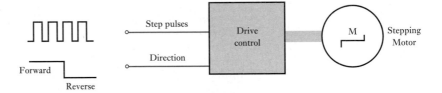

It will be noted that the control has two elements – the number of pulses which determine the angle of rotation, and the direction data which determine the order in which the poles are excited.

Now that we understand the manner in which a stepping motor may operate, let us look at practical stepping motors which come in two forms:

1. The variable–reluctance (VR) motor.
2. The hybrid motor.

43.6   **The variable-reluctance motor**

The VR motor avoids the need to use a rotor which is permanently magnetized by the simple expedient of the rotor having four poles and the stator six, as shown in Fig. 43.7.

If the No. 1 stator poles are excited, let the marked rotor pole align itself to the No. 1 pole marked N. In this position, the rotor poles at right angles to the marked pair have no function whatsoever since they are effectively neutral. Let us refer to these poles as the unmarked pair.

Now let us excite the next pair of stator poles. In this instance, the next pair is selected by moving clockwise around the stator; these are the No. 2 poles. The salient poles of the unmarked pair are nearest to the position of

Fig. 43.7 Simple VR motor

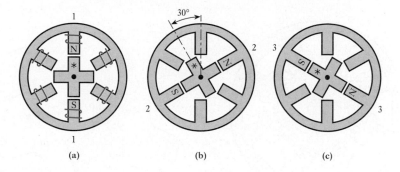

(a) (b) (c)

alignment and therefore the rotor moves to align the unmarked rotor poles with the No. 2 stator poles. This causes a rotation of 30° *anticlockwise*.

Again, let us move on to exciting the No. 3 stator poles moving round the stator in the clockwise direction. Once more, the nearest rotor poles to alignment come from the marked pair which align themselves by rotating a further 30° anticlockwise.

A complete rotation of the stator excitation only causes the rotor to rotate through 180°, hence we are required to complete two cycles of stator excitation to rotate the rotor through one revolution. This arrangement is not dependent on the direction of the excitation current. Instead only the order in which the pole pairs are excited is important.

By varying the numbers of poles, the angle of motion (the step angle) can be varied from the 30° demonstrated to values such as 15°, 22.5°, 45°, etc. but all of the values are quite large. In order to obtain smaller angles of alignment, we have to turn to the hybrid stepping motor which typically has step angles of 1.8° and 2.5°.

43.7 The hybrid stepping motor

The basic construction of this motor is illustrated in Fig. 43.8. Even with the help of diagrams, this is a difficult machine to envisage. Let us start by highlighting the significant features of its construction.

Instead of having just four, six or eight rotor salient poles, the hybrid has increased the number significantly. In Fig. 43.8, the rotor has 36 saliencies which have become so small that they appear as rotor teeth. The stator poles also have teeth which are set at the same pitch. This is emphasized by the top pole in Fig. 43.8(b) where the rotor and stator teeth are seen to match at the point of alignment. However, the teeth are not aligned at the adjacent pole, which is the same situation as we observed in the VR motor but with a much smaller displacement.

Because the displacement is small, the saliency effect requires to be strengthened and this is achieved by reintroducing the permanent magnet which we used in section 43.5. However, it is not possible to have a permanent magnet for each of the teeth, so we develop the design in a most dramatic manner.

Instead of having just one rotor with teeth, we introduce a second rotor mounted further along the same axis. The two arrangements are identical in construction, but between them is mounted a single permanent magnet as

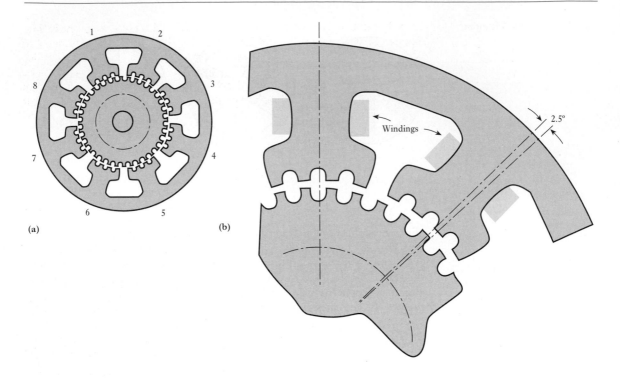

(a)   (b)

Fig. 43.8 Construction of a hybrid stepping motor (144 steps per revolution)

shown in Fig. 43.9. The result is that all the teeth on the left-hand rotor are effectively N poles while all the teeth on the right-hand rotor are S poles.

To balance the double rotor, we require a double stator. It follows that all the stator poles on the left-hand stator will act as S poles while all those on the right-hand stator will act as N poles. This means that the magnetic circuit has to be completed between the two stators, hence there is an external yoke. (In practice the two sets of stator poles are mounted directly on to this yoke.)

Owing to the symmetry of the construction, in the left-hand arrangement, poles 1 and 5 of Fig. 43.8 can both be excited as S poles to cause alignment with the N teeth of the rotor while the opposite applies in the right-hand arrangement.

Fig. 43.9 Hybrid motor rotor and stator construction

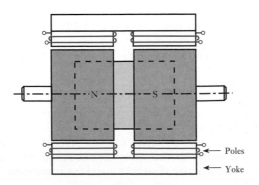

When the next pole pairs are excited, the respective teeth move into alignment which causes a shift of 2.5°. Also, because of the symmetry of the arrangement, it can be seen that when poles 1 and 5 are excited, it is also possible to excite poles 3 and 7, since they are in the neutral position. The significance of this is that we can alternately excite the odd poles and the even poles to produce motion. This avoids the need for complex switching, but it will be noted that the direction of the current is significant in terms of the direction of rotation.

A useful relationship when considering the step angle is

$$\text{Step angle} = \frac{360°}{(\text{rotor teeth}) \times (\text{stator phases})}$$

In the case which we have considered there were 36 rotor teeth and 4 stator phases, hence the step angle was 2.5°.

| 43.8 | Drive circuits |

We have previously introduced many complexities into the control of stepping motors. We can make them run in either direction, we can start them, stop them, vary their speed and determine where they stop. It follows that the drive circuits are a complete study in themselves.

However, we can anticipate the basic requirements by considering the VR motor once again. The general control system is shown in Fig. 43.10. The VR motor to which we were introduced had three stator phases and

Fig. 43.10 Drive system for a VR motor

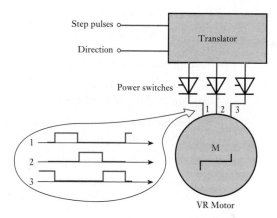

therefore the control system has to arrange for each phase to be supplied in turn. The device which gives this control is called a translator. The translator is only required to direct the incoming pulses to the phase windings. The pulses are generally low-power signals from a control unit. The control unit determines the pulse rate and the number of pulses if we are seeking a definite angle of rotation. The translator controls the supply of power from another source to the motor phases. This can be achieved by power transistors.

A typical drive circuit for one of the motor phases is shown in Fig. 43.11. The diodes are used to ensure that the output waveforms are reasonably rectangular. Without the diodes, there would be a significant transient effect

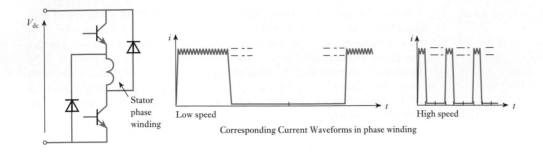

Corresponding Current Waveforms in phase winding

Fig. 43.11   Drive circuit for a VR motor

during both switch-on and switch-off. The higher the switching frequency, the more significant the transient effects, so it is essential to ensure rectangular waveforms.

Also, rectangular waveforms increase the pull-out torque performance. The pull-out torque can usually be maintained up to motional rates of 1000 steps per second.

Terms and concepts

Both d.c. and a.c. motors can be used in **regulators**. Both depend heavily on power electronic arrangements. These arrangements may control the field current of a d.c. motor or control the three-phase supply currents to an induction motor.

RPC systems can be effected either by mechanisms such as a **Geneva cam** or by a **stepping motor**.

Stepping (or stepper) motors are developed from the reluctance motor except that the input usually is pulsed.

There are two forms of stepping motor – the **variable-reluctance motor** and the **hybrid motor**.

The variable-reluctance motor has a rotor which is obviously based on a reluctance motor.

The hybrid motor has a cylindrical rotor which has teeth.

Stepping motors can rotate through given angles of rotation or they can rotate continuously. In the latter case, the speed can be controlled.

Exercises 43

1. A high-quality tape recorder requires that the speed of the tape be controlled by a regulated motor. It has been decided that the motor should be a separately excited d.c. motor. Explain why this choice is an appropriate one.

2. A paper-production plant requires that the tension on the paper be maintained by the accurate control of the speed of one of the system motors. It has been decided that the motor should be a variable-frequency cage-rotor induction motor. Explain why this choice is an appropriate one.

3. In Q.s 1 and 2, the motors could have been interchanged. Explain why such choices would be inappropriate.

4. A variable-reluctance stepper motor is required to produce 22.5° steps. Determine the number of rotor and stator poles and hence draw a diagram of its construction.

5. For the motor produced in Q. 4, describe the sequence of events that result in the rotor rotating: (a) clockwise; (b) anticlockwise.

6. A hybrid stepping motor has a step angle of 1.8°. Determine the number of rotor teeth if the stator has eight poles.

7. Describe how a variable-reluctance motor could be made to rotate at 10 r/min.

8. A cam-driver requires a vary large torque to move it from one position to another. Describe the reasons that a Geneva cam might be used in such a situation.

Chapter forty-four

Motor Selection and Efficiency

Objectives

When you have studied this chapter, you should
- understand factors influencing motor selection and application:
  - speed
  - power rating
  - duty cycle
  - load torque
  - operating environment
- have an understanding of losses due to:
  - hysteresis
  - eddy currents
- be capable of calculating motor efficiency
- have an understanding of the measurement of efficiency

Contents

In this chapter, the factors which influence the choice of motor for a particular application will be considered. These factors include speed, power rating, duty cycle, load torque and the environment in which the motor has to operate.

Efficiency will also be considered. This requires further study of electromagnetism. In particular, the concepts of hysteresis and eddy–current losses will be developed and their influence on the losses examined. By careful analysis of all the losses, the operating efficiency of a motor can be more accurately determined.

By the end of the chapter, the selection of a motor for a particular application will be better understood. It will also be possible to estimate motor efficiency more accurately.

44.1 Selecting a motor

The selection of a motor is a surprisingly complex process. Perhaps we should explore the reason for surprise. And to do that, we have to ask ourselves – what do we expect from a motor? In most cases, the answer is a fairly bland one – we are looking for a motor to drive some device.

In this answer, we have not indicated any particular requirement about speed, nor have we said anything about the starting condition, nor have we said anything about speed control. In fact we are simply content that the device is made to rotate.

If we consider many of the applications which we come across in everyday life, we find that we do not require anything special from a motor. For instance, within reason, does the speed of a food mixer really matter, or are there any limitations to starting an electric drill, or do we need to control the speed of a vacuum cleaner? For the food mixer, 300 r/min is probably just as effective as 270 r/min. For the electric drill, it is normally starting off load so there are no problems getting it started. For the vacuum cleaner, we do not require to adjust the speed from 900 r/min to 899 r/min.

Coming from a domestic introduction to the applications of electric motors, it therefore comes as something of a surprise that engineers can be concerned as to whether a factory lathe should operate at 300 r/min or 400 r/min. Or whether a motor will be able to start a passenger lift when carrying 10 people. Or whether a roller in a strip mill will ensure that the strip expansion is taken up (wrong speed and you have a large amount of expensive scrap steel).

Given the demands imposed by the applications, what are our difficulties in selecting a motor? There are three problems to be addressed as follows:

1. In most instances there are two or more types of motor which are suitable for our needs, so we have to make a choice rather than have it made for us.

2. In some instances, it is not simply a matter of matching the motor to the load when running; there is also the matter of starting the load. Thus, is the starting torque able to accelerate the load and is the acceleration time sufficiently short?

3. There are limitations imposed on motor selection which apply irrespective of the type of motor. For instance, can the motor operate under water, or in a hot surrounding, or in a dusty atmosphere?

Therefore, in selecting a motor, there is a large variety of factors to be considered. The following sections consider some of the more important factors.

44.2 Speed

Most electric motors operate at quite high speeds, say between 500 r/min and 3000 r/min (with 3600 r/min in countries such as the USA). Special motors, usually in association with power electronic systems which we will consider in Chapter 45, can operate at lower speeds or at higher speeds, but these are not especially common. Often it is easier to change speed by means of mechanical gears.

As already observed, most motors operate within a small speed range, e.g. an induction motor might operate between 1430 and 1470 r/min according to load. At the extreme condition, a comparable synchronous motor would run at 1500 r/min regardless of the load. On the other hand, commutator motors with series windings are able to accept wide speed variation, as instanced by railway trains which seldom operate at constant speed, such is the nature of railway operation.

In a number of specialist applications, precise speed control is essential. In the past, this requirement gave rise to a wide range of specially designed motors which were extremely complicated in their design: they were a tribute to the ingenuity of engineers in times gone by. However, improvements in power electronic systems have consigned such motors to the scrap yard and we will meet methods of controlling the speeds of d.c. shunt motors and cage-rotor induction motors.

44.3	Power rating and duty cycles

Basically, a motor is expected to operate continuously with a rated power output. The rating is dependent on the ability of the motor to dissipate waste heat, i.e. the heat which comes from the I^2R losses in the windings, the eddy-current losses in the rotor and stator cores and windage and friction. The losses cause the windings to become warmer, and if the insulation gets too warm it will break down. Thus the rating depends on limiting the load such that it will not overheat the winding insulation. This is a problem which we met back in Chapter 3.

Typical power ranges are illustrated in Fig. 44.1. The motors instanced are those which provide simple drives.

Fig. 44.1 Power ranges for basic motors

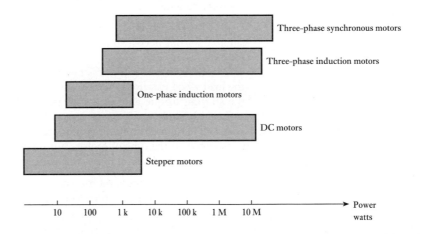

Let us consider what would happen if a motor were cold and it were loaded to its rated value. It would not immediately heat to the maximum permitted temperature – rather there would be a considerable period of time until that temperature was reached. The time would be in excess of an hour.

Alternatively we could, at the time of being cold, have increased the load. This would have caused the motor to heat more quickly, but generally this is acceptable so long as we do not continue with the overload beyond the time when the maximum operating temperature is reached. This approach leads to short-term ratings which assume higher loading followed by periods of switch-off to allow cooling. Lift motors operate in similar load cycles.

It is possible to use a 10 kW motor to deliver 20 kW for a very short period of time (say 1 or 2 min). Or for that matter to deliver, say, 11 kW for 1 h. Motors for such cyclical duties are often available in standard forms for duty periods such as 10, 30 or 60 min. Where there are shorter duty periods, motors are generally specifically designed for the intended applications.

There is quite a range of controlled-speed motors, but the two most common are the conventional d.c. motors and the variable-frequency cage-rotor induction motor. The d.c. motor has ratings between 100 W and 10 MW, while the induction motor covers almost the same range.

44.4   Load torques

Load torques can be considered under two categories:

1. Constant torque.
2. Fan or pump-type torque (so-called square law).

The constant-torque load can be instanced by the passenger lift. If we consider the arrangement shown in Fig. 44.2, let us assume a load causing a force of 5000 N in the lifting cable. If the lift were to rise at a speed of 1 m/s then the power (ignoring friction) required is 5000 W. This requires that the motor produces $2\pi n_1 T$, where n is the motor rotational speed and T the torque required. If the speed were to be doubled to 2 m/s then the power would be $10\ 000\ \text{W} = 2\pi n_2 T$. However, the rotational speed will be double to provide twice the rising speed for the lift – hence we find that the torque remains the same, i.e. it is constant.

Fig. 44.2 A motor-driven passenger lift

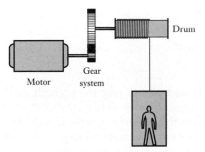

This description has not addressed the torque requirements during the acceleration and deceleration of the lift. To start the lift rising, extra torque will be required – the greater the desired acceleration, the greater the torque needed. Also, we have the problem that the torque has to be produced, even at standstill. Thus when we consider a so-called constant-torque load, we are not taking into account the acceleration and deceleration periods.

The torque requirements for the lift are shown in Fig. 44.3.

We could have considered similar arrangements which involve horizontal motion, but most of these involve higher speeds which introduce windage effects. For example, a train quickly reaches speeds at which the torque required is a mixture of the friction and the drive against the air through which it is moving.

Fig. 44.3   Torque/displacement characteristic for a lift

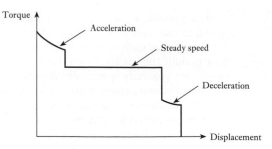

Fig. 44.4   Typical fan-type and pump-type torque/speed characteristics

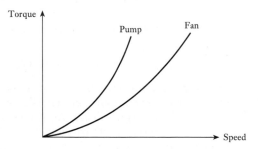

Fig. 44.5   Motor and load torque characteristics for a fan

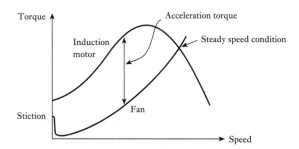

Fans and pumps take the example of the train to the extreme, and the total load is only due to the drive against air (in the case of a fan) or against a fluid (in the case of a pump). Usually the drive against air is proportional to the $(speed)^2$, as shown in Fig. 44.4. Pumps have a similar characteristic, but the proportionality of the drive requires a higher power for the speed; usually something approaching $(speed)^3$.

Both characteristics have the advantage that the motor hardly requires any torque at all at starting. In practice there is a static friction (called stiction) to overcome, so a practical load characteristic for a fan and that for an induction motor driving it take the forms shown in Fig. 44.5. The difference between the motor and load torques shows that there is a good advantage causing rapid acceleration to the operating condition.

44.5	The motor and its environment

When choosing a motor, there is one factor which is independent of the circuitry of the motor – that factor is the environment in which it is designed to operate. Towards this end, there are four ranges of environmental activity which are considered:

1. Ingress of materials.
2. Ingress of water.
3. Cooling arrangement.
4. Cooling circuit power.

Not all motors operate in a clean atmosphere, although most operate in an atmosphere which contains nothing more than a little dust. There are International Electrotechnical Commission (IEC) standards which provide for a range of levels of protection, starting with the need to keep out solid bodies about the size of a tennis ball. In such a motor, clearly we are not concerned about dust entering the motor, but we are concerned with keeping stray hands out. Progressively the range allows for smaller and smaller bodies until we wish to keep out even dust. Apart from the build-up of dust on moving parts, there is the hazard of igniting the dust.

Water could affect the insulation, but if we expect the motor normally to operate in a dry place, e.g. driving a cassette player, there is no need to protect it from water. However, some motors might experience dripping water, or the occasional jet of water, and in extreme situations be immersed in water. These situations, therefore, give rise to different casing designs for the motors.

Motors are generally cooled by a fan mounted on the end of the rotor causing air to pass between the rotor and stator. However, if we have sealed out dust then we probably could not pass air (unless it were filtered) through the motor and the heat would simply have to be released through the surface of the casing. Motors immersed in water may very well be in colder situations and surface cooling is thereby made easier.

Finally the fan need not be mounted on the rotor shaft, but could be a separate unit with its own motor. Therefore, we need to consider the manner in which the cooling circuit is powered.

There are, therefore, a number of ranges to be considered and any of the motors which we have described are capable of experiencing any of the environmental factors listed above.

44.6 Machine efficiency

In section 44.3, we observed the losses which can arise in an electrical machine. Ideally, we would hope that the power into a machine would equal the power out. When considering a machine in a dynamic state, e.g. accelerating, the power need not be equal because the magnetic and mechanical systems will be changing their stored energies and therefore absorbing some of the power. It is necessary to be able to predict these changes if we wish to anticipate the performance of a dynamic machine. In particular, we wish to predict the response to a control system demand such as a step change.

On the other hand, we can learn a lot about the practical operation of a machine by considering its imperfections. Thus the input power less the predictable losses will give its output power. Let us therefore consider the losses – and to do this, we need to reflect first on the core losses of electrical machines.

It will be recalled that in section 35.2 we noted that the core losses consist of hysteresis and eddy-current losses. At this point we need to explore the nature of these losses.

44.7　Hysteresis

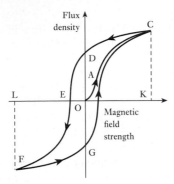

Fig. 44.6 Hysteresis loop

If we take a closed steel ring which has been completely demagnetized and measure the flux density with increasing values of the magnetic field strength, the relationship between the two quantities is represented by curve OAC in Fig. 44.6. If the value of H is then reduced, it is found that the flux density follows curve CD, and that when H has been reduced to zero, the flux density remaining in the steel is OD and is referred to as the *remanent flux density*.

If H is increased in the reverse direction, the flux density decreases, until at some value OE, the flux has been reduced to zero. The magnetic field strength OE required to wipe out the residual magnetism is termed the *coercive force*. Further increase of H causes the flux density to grow in the reverse direction as represented by curve EF. If the reversed magnetic field strength OL is adjusted to the same value as the maximum value OK in the initial direction, the final flux density LF is the same as KC.

If the magnetic field strength is varied backwards from OL to OK, the flux density follows a curve FGC similar to curve CDEF, and the closed figure CDEFGC is termed the *hysteresis loop*.

If hysteresis loops for a given steel ring are determined for different maximum values of the magnetic field strength, they are found to lie within one another, as shown in Fig. 44.7. The apexes A, C, D and E of the respective loops lie on the B/H curve determined with increasing values of H. It will be seen that the value of the remanent flux density depends upon the value of the peak magnetization; thus, for loop A, the remanent flux density is OX, whereas for loop E, corresponding to a maximum magnetization that is approaching saturation, the remanent flux density is OY. The value of the remanent flux density obtained when the maximum magnetization reaches the saturation value of the material is termed the *remanence* of that material. Thus for the material having the hysteresis loops of Fig. 44.7, the remanence is approximately OY.

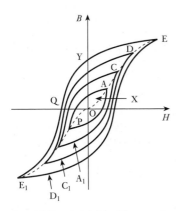

Fig. 44.7 A family of hysteresis loops

The value of the coercive force in Fig. 44.7 varies from OP for loop AA_1 to OQ for loop EE_1; and the value of the coercive force when the maximum magnetization reaches the saturation value of the material is termed the *coercivity* of that material. Thus, for Fig. 44.7 the coercivity is approximately OQ. The value of the coercivity varies enormously for different materials, being about 40 000 A/m for Alnico (an alloy of iron, aluminium, nickel, cobalt and copper, used for permanent magnets) and about 3 A/m for Mumetal (an alloy of nickel, iron, copper and molybdenum).

44.8　Current-ring theory of magnetism

It may be relevant at this point to consider why the presence of a ferro-magnetic material in a current-carrying coil increases the value of the magnetic flux and why magnetic hysteresis occurs in ferromagnets. As long ago as 1823, André-Marie Ampère – after whom the unit of current was named – suggested that the increase in the magnetic flux might be due to electric currents circulating within the molecules of the ferromagnet. Subsequent discoveries have confirmed this suggestion, and the following brief explanation may assist in giving some idea of the current-ring theory of magnetism.

An atom consists of a nucleus of positive electricity surrounded, at distances large compared with their diameter, by electrons which are charges of negative electricity. The electrons revolve in orbits around the nucleus, and each electron also spins around its own axis – somewhat like a gyroscope – and the magnetic characteristics of ferromagnetic materials appear to be due mainly to this electron spin. The movement of an electron around a circular path is equivalent to a minute current flowing in a circular ring. In a ferromagnetic atom, e.g. iron, four more electrons spin round in one direction than in the reverse direction, and the axes of spin of these electrons are parallel with one another; consequently, the effect is equivalent to four current rings producing magnetic flux in a certain direction.

The ferromagnetic atoms are grouped together in *domains*, each about 0.1 mm in width, and in any one domain the magnetic axes of all the atoms are parallel with one another. In an unmagnetized bar of ferromagnetic material, the magnetic axes of different domains are in various directions so that their magnetizing effects cancel one another out. Between adjacent domains there is a region or 'wall', about 10^{-4} mm thick, within which the direction of the magnetic axes of the atoms changes gradually from that of the axes in one domain to that of the axes in the adjacent domain.

When an unmagnetized bar of ferromagnetic material, e.g. steel, is moved into a current-carrying solenoid, there are sudden tiny increments of the magnetic flux as the magnetic axes of the various domains are orientated so that they coincide with the direction of the m.m.f. due to the solenoid, thereby increasing the magnitude of the flux. This phenomenon is known as the *Barkhausen effect* and can be demonstrated by winding a search coil on the steel bar and connecting it through an amplifier to a loudspeaker. The sudden increments of flux due to successive orientation of the various domains, while the steel bar is being moved into the solenoid, induce e.m.f. impulses in the search coil and the effect can be heard as a rustling noise.

It follows that when a current-carrying solenoid has a ferromagnetic core the magnetic flux can be regarded as consisting of two components:

1. The flux produced by the solenoid without a ferromagnetic core.
2. The flux due to ampere-turns equivalent to the current rings formed by the spinning electrons in the orientated domains. This component reaches its maximum value when all the domains have been orientated so that their magnetic axes are in the direction of the magnetic flux. The core is then said to be saturated.

These effects are illustrated in Fig. 44.8 where graph P represents the variation of the actual flux density with magnetic field strength for a ferromagnetic material. The straight line Q represents the variation of flux density with magnetic field strength if there were no ferromagnetic material present; thus for magnetic field strength OA

$$AD = \mu_0 \times OA$$

Fig. 44.8 Actual and intrinsic flux densities

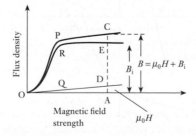

For the same magnetic field strength OA, the actual flux density B is represented by AC. Hence the flux density due to the presence of the ferromagnetic core = AC − AD = AE. This flux density is referred to as the *intrinsic magnetic flux density* or *magnetic polarization* and is represented by symbol B_i, i.e.

$$B_i = B - \mu_0 H$$

The variation of the intrinsic flux density with magnetic field strength is represented by graph R in Fig. 44.8, from which it will be seen that when the magnetic field strength exceeds a certain value, the intrinsic flux density remains constant, i.e. the ferromagnetic material is magnetically saturated, or, in other words, all the domains have been orientated so that their axes are in the direction of the magnetic field.

This alignment of the domains has a certain amount of stability, depending upon the quality of the ferromagnetic material and the treatment it has received during manufacture. Consequently, the core may retain much of its magnetism after the external magnetomotive force has been removed, as already discussed in section 44.7, the remanent flux being maintained by the m.m.f. due to the electronic current rings in the iron. A permanent magnet can therefore be regarded as an electromagnet, the relatively large flux being due to the high value of the *inherent m.m.f.* (i.e. coercive force × length of magnet) retained by the steel after it has been magnetized.

A disturbance in the alignment of the domains necessitates the expenditure of energy in taking a specimen of a ferromagnetic material through a cycle of magnetization. This energy appears as heat in the specimen and is referred to as *hysteresis loss*.

44.9   Hysteresis loss

Suppose Fig. 44.9 to represent the hysteresis loop obtained on a steel ring of mean circumference l metres and cross-sectional area A square metres. Let N be the number of turns on the magnetizing coil.

Let dB = increase of flux density when the magnetic field strength is increased by a very small amount MN in dt seconds, and i = current in amperes corresponding to OM, i.e.

Fig. 44.9   Hysteresis loss

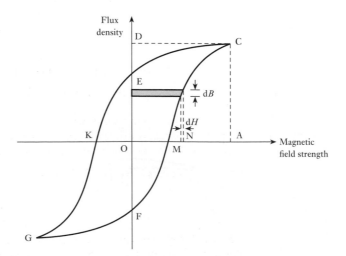

$$\text{OM} = \frac{Ni}{l} \qquad\qquad\qquad [44.1]$$

Instantaneous e.m.f. induced in winding is

$$A \times \mathrm{d}B \times \frac{N}{\mathrm{d}t} \text{ volts}$$

and component of applied voltage to neutralize this e.m.f. equals

$$AN \times \frac{\mathrm{d}B}{\mathrm{d}t} \text{ volts}$$

Therefore instantaneous power supplied to magnetic field is

$$i \times AN \times \frac{\mathrm{d}B}{\mathrm{d}t} \text{ watts}$$

and energy supplied to magnetic field in time dt seconds is

$$iAN \times \mathrm{d}B \text{ joules}$$

From equation [44.1]

$$i = l \times \frac{\text{OM}}{N}$$

hence energy supplied to magnetic field in time dt is

$$l \times \frac{\text{OM}}{N} \times AN \times \mathrm{d}B \text{ joules}$$

$$= \text{OM} \times \mathrm{d}B \times lA \text{ joules}$$

$$= \text{area of shaded strip, joules per metre}^3$$

Thus energy supplied to magnetic field when H is increased from zero to OA = area FJCDF joules per metre^3. Similarly, energy returned from magnetic field when H is reduced from OA to zero = area CDEC joules per metre^3. Then net energy absorbed by magnetic field is

$$\text{Area FJCEF joules per metre}^3$$

Hence: Hysteresis loss for a complete cycle =

Area of loop GFCEG joules per metre^3 [44.2]

If the hysteresis loop is plotted to scales of 1 cm to x amperes per metre along the horizontal axis and 1 cm to y teslas along the vertical axis, and if a represents the area of the loop in square centimetres, then:

$$\text{Hysteresis loss per cycle} = axy \text{ joules per metre}^3$$

$$= a \times \frac{NI}{l} \times \frac{\Phi}{A} \text{ joules per metre}^3$$

where NI/l = magnetic field strength per centimetre of scale and Φ/A = flux density per centimetre of scale.

Hence, total hysteresis loss per cycle in joules equals

$$lA \times a \times \frac{NI}{l} \times \frac{\Phi}{A} = a \times NI \times \Phi \qquad [44.3]$$

$$= (\text{area of loop in centimetres}^3)$$
$$\times (\text{amperes per centimetre of scale})$$
$$\times (\text{webers per centimetre of scale})$$

The total hysteresis loss cycle in a specimen can therefore be determined from the area of a loop drawn by plotting the total flux, in webers, against the total m.m.f. in amperes.

From the areas of a number of loops similar to those shown in Fig. 44.7, the hysteresis loss per cycle was found by Charles Steinmetz (1865–1923) to be proportional to $(B_{max})^{1.6}$ for a certain quality of steel, where B_{max} represents the maximum value of the flux density. It is important to realize that this index of 1.6 is purely empirical and has no theoretical basis. In fact, indices as high as 3.5 have been obtained, and in general we may say that

$$\text{Hysteresis loss per cubic metre per cycle} \propto (B_{max})^x$$

where $x = 1.5$–2.5, depending upon the quality of the steel and the range of flux density over which the measurement has been made.

From the above proof, it is evident that the hysteresis loss is proportional to the volume and to the number of cycles through which the magnetization is taken. Hence, if $f =$ frequency of alternating magnetization in hertz and $v =$ volume of steel in cubic metres,

$$\boxed{\text{Hysteresis loss} = kvf(B_{max})^x \text{ watts}} \qquad [44.4]$$

where k is a constant for a given specimen and a given range of flux density.

The magnitude of the hysteresis loss depends upon:

1. The composition of the specimen, e.g. nickel–iron alloys such as Mumetal and Permalloy have very narrow hysteresis loops due to their low coercive force, and consequently have very low hysteresis loss.
2. The heat treatment and the mechanical handling to which the specimen has been subjected. For instance, if a material has been bent or cold-worked, the hysteresis loss is increased, but the specimen can usually be restored to its original magnetic condition by a strain-relieving heat treatment.

Example 44.1

The area of the hysteresis loop obtained with a certain specimen of steel was 9.3 cm². The coordinates were such that 1 cm = 1000 A/m and 1 cm = 0.2 T. Calculate:

(a) the hysteresis loss per cubic metre per cycle;
(b) the hysteresis loss per cubic metre at a frequency of 50 Hz.
(c) If the maximum flux density was 1.5 T, calculate the hysteresis loss per cubic metre for a maximum density of 1.2 T and a frequency of 30 Hz, assuming the loss to be proportional to $(B_{max})^{1.8}$.

(a) Area of hysteresis loop in BH units is

$$9.3 \times 1000 \times 0.2 = 1860$$

Then hysteresis loss per cubic metre per cycle is

1860 J

(b) Hysteresis loss per cubic metre at 50 Hz is

$$1860 \times 50 = \textbf{93 000 W}$$

(c) From expression [44.4]:

$$93\,000 = k \times 1 \times 50 \times (1.5)^{1.8}$$

$$\log(1.5)^{1.8} = 1.8 \times 0.176 = 0.3168 = \log 2.074$$

$$k = \frac{93\,000}{50 \times 2.074} = 896$$

For $B_{max} = 1.2$ T and $f = 30$ Hz, hysteresis loss per cubic metre is

$$896 \times 30 \times (1.2)^{1.8} = \textbf{37 350 W}$$

Or, hysteresis loss at 1.2 T and 30 Hz is

$$93\,000 \times \frac{30}{50} \times \left(\frac{1.2}{1.5}\right)^{1.8} = \textbf{37 350 W}$$

44.10 Losses in motors and generators

Now that we are fully aware of the implications of hysteresis, we can proceed to consider the losses which occur in motors and generators. AC machines are more complicated due to the effects of reactance in the circuits, so we will simplify matters by considering only d.c. machines. Even here the components experience flux reversal and therefore the a.c. effects such as hysteresis and eddy-current losses are experienced, but on a limited scale. The losses in d.c. machines can be classified under the headings below.

(a) Armature losses

1. I^2R loss in armature winding. The resistance of an armature winding can be measured by the voltmeter–ammeter method. If the resistance measurement is made at room temperature, the resistance at normal working temperature should be calculated. Thus, if the resistance is R_1 at room temperature of, say, 15 °C and if 50 °C is the temperature rise of the winding after the machine has been operating on full load for 3 or 4 h, then from expression [3.17], we have

$$\text{Resistance at 65 °C} = R_1 \times \frac{1 + (0.004\,26 \times 65)}{1 + (0.004\,26 \times 15)} = 1.2R_1$$

2. Core loss in the armature core due to hysteresis and eddy currents. Hysteresis loss has been discussed in section 44.9 and is dependent upon the quality of the steel. It is proportional to the frequency and is approximately proportional to the square of the flux. The eddy-current loss is due to

Fig. 44.10   Eddy currents

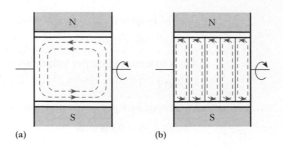

(a)   (b)

circulating currents set up in the steel laminations. Had the core been of solid steel, as shown in Fig. 44.10(a) for a two-pole machine, then, if the armature were rotated, e.m.f.s would be generated in the core in exactly the same way as they are generated in conductors placed on the armature, and these e.m.f.s would circulate currents – known as *eddy currents* – in the core as shown dotted in Fig. 44.10(a), the rotation being assumed clockwise when the armature is viewed from the right-hand side of the machine. Owing to the very low resistance of the core, these eddy currents would be considerable and would cause a large loss of power in, and excessive heating of, the armature.

If the core is made of laminations insulated from one another, the eddy currents are confined to their respective sheets, as shown dotted in Fig. 44.10(b), and the eddy-current loss is thereby reduced. Thus, if the core is split up into five laminations, the e.m.f. per lamination is only a fifth of that generated in the solid core. Also, the cross-sectional area per path is reduced to about a fifth, so that the resistance per path is roughly five times that of the solid core. Consequently the current per path is about one-twenty-fifth of that in the solid core. Hence:

$$\frac{I^2R \text{ loss per lamination}}{I^2R \text{ loss in solid core}} = \left(\frac{1}{25}\right)^2 \times 5 = \frac{1}{125} \text{ (approx.)}$$

Since there are five laminations,

$$\frac{\text{Total eddy-current loss in laminated core}}{\text{Total eddy-current loss in solid core}} = \frac{5}{125} = \left(\frac{1}{5}\right)^2$$

It follows that the eddy-current loss is approximately proportional to the square of the thickness of the laminations. Hence the eddy-current loss can be reduced to any desired value, but if the thickness of the laminations is made less than about 0.4 mm, the reduction in the loss does not justify the extra cost of construction. Eddy-current loss can also be reduced considerably by the use of silicon–iron alloy – usually about 4 per cent of silicon – due to the resistivity of this alloy being much higher than that of ordinary steel.

Since the e.m.f.s induced in the core are proportional to the frequency and the flux, therefore the eddy-current loss is proportional to (frequency × flux)^2.

(b)   Commutator losses

1.   Loss due to the contact resistance between the brushes and the segments. This loss is dependent upon the quality of the brushes. For carbon brushes,

the p.d. between a brush and the commutator, over a wide range of current, is usually about 1 V per positive set of brushes and 1 V per negative set, so that the total contact-resistance loss, in watts, is approximately 2 × total armature current.

2. Loss due to friction between the brushes and the commutator. This loss depends upon the total brush pressure, the coefficient of friction and the peripheral speed of the commutator.

(c) Excitation losses

1. Loss in the shunt circuit (if any) equal to the product of the shunt current and the terminal voltage. In shunt generators this loss increases a little between no load and full load, since the shunt current has to be increased to maintain the terminal voltage constant; but in shunt and compound motors, it remains approximately constant.

2. Losses in series, compole and compensating windings (if any). These losses are proportional to the square of the armature current.

(d) Bearing friction and windage losses

The bearing friction loss is roughly proportional to the speed, but the windage loss, namely the power absorbed in setting up circulating currents of air, is proportional to the cube of the speed. The windage loss is very small unless the machine is fitted with a cooling fan.

(e) Stray load loss

It was shown in section 41.4 that the effect of armature reaction is to distort the flux, the flux densities at certain points of the armature being increased; consequently the core loss is also increased. This stray loss is usually neglected as it is difficult to estimate its value.

44.11 Efficiency of a d.c. motor

If R_a = total resistance of armature circuit (including the brush–contact resistance and the resistance of series and compole windings, if any), I = input current, I_s = shunt current and I_a = armature current = $I - I_s$, then total loss in armature circuit = $I_a^2 R_a$.

If V = terminal voltage, loss in shunt circuit = $I_s V$. This includes the loss in the shunt regulating resistor.

If C = sum of core, friction and windage losses,

$$\text{Total losses} = I_a^2 R_a + I_s V + C$$

$$\text{Input power} = IV$$

$$\therefore \quad \text{Output power} = IV - I_a^2 R_a - I_s V - C$$

and Efficiency $\eta = \dfrac{IV - I_a^2 R_a - I_s V - C}{IV}$ [44.5]

44.12 Approximate condition for maximum efficiency

Let us assume:

1. That the shunt current is negligible compared with the armature current at load corresponding to maximum efficiency.
2. That the shunt, core and friction losses are independent of the load.

Then from expression [44.5]:

$$\text{Efficiency of a motor, } \eta = \frac{IV - I^2 R_a - I_s V - C}{IV}$$

$$\eta = \frac{V - IR_a - \dfrac{1}{I}(I_s V + C)}{V} \qquad [44.6]$$

This efficiency is a maximum when the numerator of equation [44.6] is a maximum, namely when

$$\frac{d}{dI}\left\{ V - IR_a - \frac{1}{I}(I_s V + C) \right\} = 0$$

i.e.

$$R_a - \frac{1}{I^2}(I_s V - C) = 0 \qquad [44.7]$$

∴

$$I^2 R_a = I_s V + C \qquad [44.8]$$

The condition for the numerator in expression [44.6] to be a maximum, and therefore the efficiency to be a maximum, is that the left-hand side of expression [44.7], when differentiated with respect to I, should be positive; thus

$$\frac{d}{dI}\{R_a - I^{-2}(I_s V + C)\} = 2I^{-3}(I_s V + C)$$

Since this quantity is positive, it follows that expression [44.8] represents the condition for maximum efficiency, i.e. the efficiency of the motor is a maximum when the load is such that the variable loss is equal to the constant loss.

Precisely the same conclusion can be derived for a generator.

44.13 Determination of efficiency

(a) By direct measurement of input and output powers

In the case of small machines the output power can be measured by some form of mechanical brake as that shown in Fig. 44.11 where the belt (or ropes) on an air- or water-cooled pulley has one end attached via a spring balance S to the floor and has a mass of m kilograms suspended at the other end.

$$\text{Weight of the suspended mass} = W$$

$$\simeq 9.81m \text{ newtons}$$

If reading on spring balance $= S$ newtons, therefore net pull due to friction $= (W - S)$ newtons.

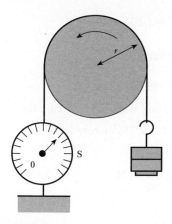

Fig. 44.11 Brake test on a motor

If r = effective radius of brake, in metres, and N = speed of pulley, in revolutions per minute,

Torque due to brake friction = $(W - S)r$ newton metres

and

$$\text{Output power} = \frac{2\pi(W - S)rN}{60} \text{ watts} \qquad [44.9]$$

If V = supply voltage, in volts, and I = current taken by motor, in amperes,

Input power = IV watts

and

$$\text{Efficiency} = \frac{2\pi(W - S)rN}{60 \times IV} = \frac{2\pi(9.81m - S)rN}{60 \times IV} \qquad [44.10]$$

The size of machine that can be tested by this method is limited by the difficulty of dissipating the heat generated at the brake.

The principal methods of testing larger machines are the Swinburne and Hopkinson methods. The main advantages of the Swinburne method are that the power required is small compared with the machine's rating and the test data enable the efficiency to be determined at any load, but the method makes no allowance for stray losses and there is no check on the performance. These disadvantages are overcome by the Hopkinson method which has only the drawback that it requires to test two identical machines simultaneously. These methods are too specialized for further consideration in this text.

Example 44.2

In a load test on a 240 V, d.c. series motor, the following data were recorded at one value of load:

Terminal voltage	238 V
Line current	6.90 A
Belt brake	See Fig. 44.12
Speed	755 r/min

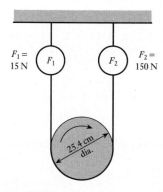

$F_1 = 15\,\text{N}$ $F_2 = 150\,\text{N}$

25.4 cm dia.

Fig. 44.12

Evaluate:

(i) the torque;
(ii) the brake power;
(iii) the efficiency.

The resistance between terminals is 2.0 Ω, and it has been found for this machine that the windage and friction losses are equal to $(1.8 \times 10^{-4})N^2$ watts, where N is the speed in revolutions per minute. Estimate the core loss for the load.

(i)   Torque $T = (F_2 - F_1)r = (150 - 15) \times \dfrac{25.4 \times 10^{-2}}{2}$

$$= \textbf{17.2 N m}$$

(ii)   Brake power $P_o = T\omega_r = 17.15 \times \dfrac{2\pi \times 755}{60} = \textbf{1355 W}$

(iii)   Input power $P_i = VI = 238 \times 6.9 = 1642.2$ W

$$\eta = \frac{1355.5}{1642.2} \equiv \textbf{82.5 per cent}$$

$$\text{Losses} = P_i - P_o = 1642.2 - 1355.5 = 286.7 \text{ W}$$

$$= I^2R \text{ loss} + \text{windage and friction loss} + \text{core loss}$$

$$= 6.9^2 \times 2 + (1.8 \times 10^{-4}) \times 755^2 + \text{core loss}$$

$$= 95.2 + 102.6 + \text{core loss}$$

$\therefore$ Core loss $= 286.7 - 95.2 - 102.6 = \textbf{89 W}$

Example 44.3   A 100 kW, 460 V shunt generator was run as a motor on no load at its rated voltage and speed. The total current taken was 9.8 A, including a shunt current of 2.7 A. The resistance of the armature circuit (including compoles) at normal working temperature was 0.11 Ω. Calculate the effciencies at

(a) full load;
(b) half load.

(a)   Output current at full load $= \dfrac{100 \times 1000}{460} = 217.5$ A

$\therefore$ Armature current at full load $= 217.5 + 2.7 = 220.2$ A

The I^2R loss in armature circuit at full load is

$$(220)^2 \times 0.11 = 5325 \text{ W}$$

$$\text{Loss in shunt circuit} = 2.7 \times 460 = 1242 \text{ W}$$

$$\text{Armature current on no load} = 9.8 - 2.7 = 7.1 \text{ A}$$

Thus input power to armature on no load is

$$7.1 \times 460 = 3265 \text{ W}$$

But loss in armature circuit on no load is

$$(7.1)^2 \times 0.11 = 5.5 \text{ W}$$

This loss is less than 0.2 per cent of the input power and could therefore have been neglected.

Hence, core, friction and windage losses $= 3260$ W, and

$$\text{Total losses at full load} = 5325 + 1242 + 3260$$

$$= 9827 \text{ W} = 9.83 \text{ kW}$$

$$\text{Input power at full load} = 100 + 9.83 = 109.8 \text{ kW}$$

and   Efficiency at full load $= \dfrac{100}{109.8} = 0.911$ per unit

$$= \textbf{91.1 per cent}$$

(b) Output current at half load = 108.7 A, therefore armature current at half load is

$$108.7 + 2.7 = 111.4 \text{ A}$$

and I^2R loss in armature circuit at half load is

$$(111.4)^2 \times 0.11 = 1365 \text{ W}$$

∴   Total losses at half load = $1365 + 1242 + 3260$

$$= 5867 \text{ W} = 5.87 \text{ kW}$$

Input power at half load = $50 + 5.87 = 55.87 \text{ kW}$

and   Efficiency at half load = $\dfrac{50}{55.87} = 0.895$ per unit

$$= \textbf{89.5 per cent}$$

| Terms and concepts | Motor selection requires knowledge of the required speed, power rating, duty cycles and load torques. |

Motor selection requires knowledge of the required speed, power rating, duty cycles and load torques.

To match a motor to a load, it is necessary to know the **torque/speed characteristic** for the load.

Machines experience core losses which include hysteresis and eddy-current losses.

Hysteresis is the variation of flux density with magnetic field strength. Owing to the nature of the domain structure of the ferromagnetic material, the variation produces a hysteresis loop which is indicative of the energy required every cycle. The repetition of these cycles gives rise to a power loss.

Eddy-current losses are due to the circulating currents in the core and the effect is reduced by core lamination.

Machines incur losses due to the currents in the windings and, in commutator machines, the brushes release waste energy due to **contact resistance**.

The efficiency of a machine can be estimated by deducting the losses from the input power.

Exercises 44

1. A hysteresis loop is plotted against a horizontal axis which scales 1000 A/m = 1 cm and a vertical axis which scales 1 T = 5 cm. If the area of the loop is 9 cm² and the overall height is 14 cm, calculate: (a) the hysteresis loss in joules per cubic metre per cycle; (b) the maximum flux density; (c) the hysteresis loss in watts per kilogram, assuming the density of the material to be 7800 kg/m³.

2. (a) Explain briefly the occurrence of hysteresis loss and eddy-current loss in a transformer core and indicate how these losses may be reduced.

(b) The area of a hysteresis loop for a certain magnetic specimen is 160 cm² when drawn to the following scales: 1 cm represents 25 A/m and 1 cm represents 0.1 T. Calculate the hysteresis loss in watts per kilogram when the specimen is carrying an alternating flux at 50 Hz, the maximum value of which corresponds to the maximum value attained on the hysteresis loop. The density of the material is 7600 kg/m^3.

3. Explain how the energy losses in a sample of ferromagnetic material subjected to an alternating magnetic field depend on the frequency and the flux density. What particular property of the material can be used as a measure of the magnitude of each type of loss?

   The area of a hysteresis loop plotted for a sample of steel is 67.1 cm^2, the maximum flux density being 1.06 T. The scales of B and H are such that 1 cm = 0.12 T and 1 cm = 7.07 A/m. Find the loss due to hysteresis if 750 g of this steel were subjected to an alternating magnetic field of maximum flux density 1.06 T at a frequency of 60 Hz. The density of the steel is 7700 kg/m^3.

4. The hysteresis loop for a certain ferromagnetic ring is drawn to the following scales: 1 cm to 300 A and 1 cm to 100 μWb. The area of the loop is 37 cm^2. Calculate the hysteresis loss per cycle.

5. In a certain transformer, the hysteresis loss is 300 W when the maximum flux density is 0.9 T and the frequency 50 Hz. What would be the hysteresis loss if the maximum flux density were increased to 1.1 T and the frequency reduced to 40 Hz? Assume the hysteresis loss over this range to be proportional to $(B_{max})^{1.7}$.

6. The armature of a four-pole d.c. motor has an armature eddy-current loss of 500 W when driven at a given speed and field excitation. If the speed is increased by 15 per cent and the flux is increased by 10 per cent, calculate the new value of the eddy-current loss.

7. A d.c. shunt motor has an output of 8 kW when running at 750 r/min off a 480 V supply. The resistance of the armature circuit is 1.2 Ω and that of the shunt circuit is 800 Ω. The efficiency at that load is 83 per cent. Determine: (a) the no-load armature current; (b) the speed when the motor takes 12 A; (c) the armature current when the gross torque is 60 N m. Assume the flux to remain constant.

8. In a brake test on a d.c. motor, the effective load on the brake pulley was 265 N, the effective diameter of the pulley 650 mm and the speed 720 r/min. The motor took 35 A at 220 V. Calculate the output power, in kW, and the efficiency at this load.

9. In a test on a d.c. motor with the brake shown in Fig. 44.11, the mass suspended at one end of the belt was 30 kg and the reading on the spring balance was 65 N. The effective diameter of the brake wheel was 400 mm and the speed was 960 r/min. The input to the motor was 23 A at 240 V. Calculate: (a) the output power; (b) the efficiency.

10. A d.c. shunt machine has an armature resistance of 0.5 Ω and a field-circuit resistance of 750 Ω. When run under test as a motor, with no mechanical load, and with 500 V applied to the terminals, the line current was 3 A. Allowing for a drop of 2 V at the brushes, estimate the efficiency of the machine when it operates as a generator with an output of 20 kW at 500 V, the field-circuit resistance remaining unchanged. State the assumptions made.

11. Describe the various losses which take place in a shunt motor when it is running on no load.

   A 230 V shunt motor, running on no load and at normal speed, takes an armature current of 2.5 A from 230 V mains. The field-circuit resistance is 230 Ω and the armature-circuit resistance is 0.3 Ω. Calculate the motor output, in kW, and the efficiency when the total current taken from the mains is 35 A. If the motor is used as a 230 V shunt generator, find the efficiency and input power for an output current of 35 A.

Chapter forty-five

Power Electronics

In this chapter, we will discuss power electronic systems. We will be concerned principally with the thyristor, which has wide application because of its ability to switch and control high alternating currents. Some thyristor circuits will be discussed, including thyristor bridge circuits for a.c./d.c. conversion.

The limitations of thyristor operation will also be mentioned. In particular, a thyristor cannot be switched off, although it switches off at the current zero of an alternating current waveform. This leads us to the consideration of a thyristor which can be turned off, the gate turn–off thyristor (GTO) and another important power semiconductor device, the insulated–gate bipolar transistor (IGBT). The use of GTOs and IGBTs in inverter circuits will also be presented.

We will finally consider some other important power electronic circuits: a three–phase fully controlled rectifier network can be used to supply and control a large d.c. load like a motor; a three–phase inverter can be used as part of a variable-frequency, speed-control system for a large induction motor.

Although transistors can be used as a switch, generally their current-carrying capacity is small. There are many applications in which it would be advantageous to have a high-speed switch which could handle up to 1000 A: such a device is the thyristor, which also has the advantages of having no moving parts and no arcing.

The basic parts of the thyristor are its four layers of alternate p-type and n-type silicon semiconductors forming three p–n junctions, A, B and C, as shown in Fig. 45.1(a). The terminals connected to the n_1 and p_2 layers are the cathode and anode respectively. A contact welded to the p_1 layer is termed the *gate*. The CENELEC Standard graphical symbol for the thyristor is given in Fig. 45.1(b). The direction of the arrowhead on the gate lead indicates that the gate contact is welded to a p-region and shows the direction of the gate current required to operate the device. If the gate contact is welded to an n-region, the arrowhead should point outwards from the rectifier.

Fig. 45.1 Thyristor arrangement and symbols

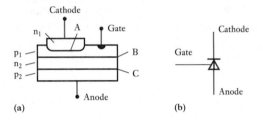

When the anode is positive with respect to the cathode, junctions A and C are forward-biased and therefore have a very low resistance, whereas junction B is reverse-biased and consequently presents a very high resistance, of the order of megohms, to the passage of a current. On the other hand, if the anode terminal is made *negative* with respect to the cathode terminal, junction B is forward-biased while A and C act as two reverse-biased junctions in series.

Let us now consider the effect of increasing the voltage applied across the thyristor, *with the anode positive relative to the cathode*. At first, the forward leakage current reaches saturation value due to the action of junction B. Ultimately, a breakover value is reached and the resistance of the thyristor instantly falls to a very low value, as shown in Fig. 45.2. The forward voltage drop is of the order of 1–2 V and remains nearly constant over a wide

Fig. 45.2 Thyristor characteristic

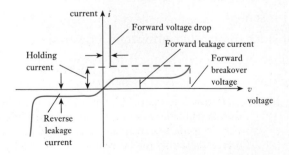

Fig. 45.3 Gate control of thyristor breakover voltage

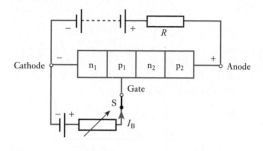

Fig. 45.4 Variation of breakover voltage with bias current

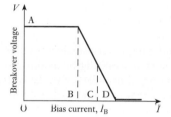

Fig. 45.5 Variation of breakover voltage with bias current, I_8

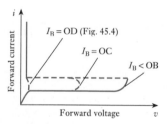

variation of current. A resistor is necessary in series with the thyristor to limit the current to a safe value.

We shall now consider the effect upon the breakover voltage of applying a positive potential to the gate as in Fig. 45.3. When switch S is closed, a bias current, I_B, flows via the gate contact and layers p_1 and n_1, and the value of the breakover voltage of the thyristor depends upon the magnitude of the bias current in the way shown in Fig. 45.4. Thus, with $I_B = 0$, the breakover voltage is represented by OA, and remains practically constant at this value until the bias current is increased to OB. For values of bias current between OB and OD, the breakover voltage falls rapidly to nearly zero. An alternative method of representing this effect is shown in Fig. 45.5.

In the triggered condition, the thyristor approximates a single p–n diode, the anode current being limited only by the resistance R of the external circuit. Once the gate has triggered the device, *it loses control over the anode current*, and the only method of restoring the device to its high-resistance condition is to reduce the anode current below the *holding* value indicated in Fig. 45.2. The value of the holding current is usually very small compared with the rated forward current; for instance, the holding current may be about 10 mA for a thyristor having a forward–current rating of 40 A.

If the thyristor is connected in series with a non-reactive load, of resistance R, across a supply voltage having a sinusoidal waveform and if it is triggered at an instant corresponding to an angle ϕ after the voltage has passed through zero from a negative to a positive value, as in Fig. 45.6(a), the value of the applied voltage at that instant is given by

$$v = V_m \sin \phi$$

Up to that instant, the voltage across the thyristor has been growing from zero to v. When triggering occurs, the voltage across the thyristor instantly falls to about 1–2 V and remains approximately constant while current flows, as shown by the slightly curved line in Fig. 45.6(a). Also, at the instant of triggering, the current increases immediately from zero to i, where

$$i = \frac{v - \text{p.d. across thyristor}}{R}$$

$$= \frac{v}{R} \text{ when the p.d. across thyristor} \ll v$$

If ϕ is less than $\pi/2$, the current increases to a maximum I_m and then decreases to the holding value, when it falls instantly to zero, as shown in Fig. 45.6(b). The average value of the current over one cycle is the shaded area enclosed by the current wave divided by 2π.

Fig. 45.6 Phase-controlled
half-wave rectification

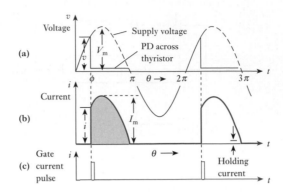

If the p.d. across the thyristor, when conducting, is very small compared with the supply voltage, and if the holding current is negligible compared with I_m, the waveform of the current is practically sinusoidal for values of θ between ϕ and π. Therefore average value of current over one cycle is

$$I_{av} = \frac{1}{2\pi} \int_{\phi}^{\pi} I_m \sin \theta \cdot d\theta$$

$$= \frac{I_m}{2\pi} \left[-\cos \theta \right]_{\phi}^{\pi} = \frac{I_m}{2\pi} (1 + \cos \phi)$$

$$I_{av} = \frac{I_m}{2\pi} (1 + \cos \phi) \qquad\qquad [45.1]$$

When $\phi = 0$

$$I_{av} = \frac{I_m}{\pi} = 0.3185 I_m$$

When $\phi = \pi/2$

$$I_{av} = 0.159 I_m$$

When $\phi = \pi$

$$I_{av} = 0$$

Hence, by varying the instant at which the thyristor is triggered, it is possible to control the output current over a wide range. Triggering signals for this purpose are usually in the form of a positive pulse of short duration compared with the time of one cycle of the alternating voltage. A gate current of about 50 mA applied for about 5 μs is all that is required to trigger a device having current ratings between a few milliamperes and hundreds of amperes.

45.3   Some thyristor circuits

A simple switch-on arrangement incorporating a thyristor is shown in Fig. 45.7. In this case, when the switch S1 is closed, the thyristor immediately blocks the passage of current through the lamp. When switch S2 is temporarily closed, some milliamperes of current flow into the gate and switch on the thyristor. Once the thyristor is switched on, it is no longer necessary to keep switch S2 closed.

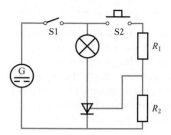

Fig. 45.7 Simple d.c. thyristor switch-on circuit

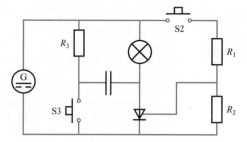

Fig. 45.8 Simple d.c. thyristor on–off circuit

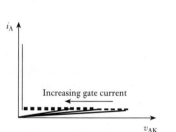

Fig. 45.9 Forward conduction characteristic of thyristor

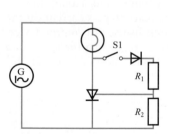

Fig. 45.10 Simple half-wave a.c. switch circuit

In order to switch off the lamp in Fig. 45.7, we could open switch S1, but this raises the question – is switch S1 not doing the same work as the thyristor and have we not duplicated our switching? Really we should be looking to S1 being replaced by the thyristor, and this can be achieved if we modify the circuit to that shown in Fig. 45.8. The closure of switch S3 connects the capacitor so that it tries to discharge in opposition to the passage of current through the thyristor, with the result that for an instant there is no current passing through the thyristor. This short interruption is sufficient to stop the thyristor conducting, and it therefore returns to the OFF condition.

When operating from an a.c. supply, the thyristor does not always receive sufficient gate current to switch it on. Instead we find that the forward conduction characteristic changes as the gate current is increased; thus if we have insufficient gate current to switch on the thyristor, it is nevertheless noticeable from the characteristic shown in Fig. 45.9 that less anode–cathode voltage is required to raise the leakage current to a sufficient level that the avalanche breakdown takes place and the thyristor switches on of its own accord.

For the simple circuit shown in Fig. 45.10, this form of switch-on action can be employed. When switch S1 is open, there is no gate current and the thyristor is able to block the applied alternating voltage even at its peak value. When S1 is closed, the applied voltage causes a current to pass through R_1 and R_2, causing volt drops across each. The voltage across R_1 causes some current to pass through the gate, thus reducing the breakdown voltage of the thyristor. Eventually a point is reached at which either there is sufficient gate current to directly switch on the thyristor, or the gate current sufficiently reduces the breakdown voltage of the thyristor that it again switches on – usually we experience this latter form of action.

Once the thyristor commences to conduct, the volt drop across it falls almost to zero; thus the gate current almost disappears. When the half-cycle of the supply is completed, the anode current falls to zero and the thyristor switches off.

During the subsequent half-cycle, the thyristor cannot conduct; thus the lamp receives approximately a half-wave supply, depending on how soon the thyristor switches on after the commencement of the positive half-wave. By varying R_1, the instant of switch-on can be delayed; thus the thyristor can be used to control the voltage developed across the lamp and consequently the power dissipated by the lamp. The diode in the circuit shown prevents reverse bias being applied to the thyristor gate during the negative half-cycles of the supply.

45.4 Limitations to thyristor operation

Because of the nature of the construction of a thyristor, there is some capacitance between the anode and the gate. If a sharply rising voltage is applied to the thyristor, then there is an inrush of charge corresponding to the relation $i = C(\mathrm{d}v/\mathrm{d}t)$. This inrush current can switch on the thyristor, and it can arise in practice due to surges in the supply system, for example due to switching or to lightning. Thus thyristors may be inadvertently switched on, and such occurrences can be avoided by providing C–R circuits in order to divert the surges from the thyristors.

The leakage current in any p–n junction depends on the temperature of the junction. It follows that if we were to raise the temperature of a thyristor, the leakage current would rise, and it approximately doubles for every 8 °C rise in temperature. If the temperature is permitted to rise too much, again the leakage current could inadvertently switch on the thyristor; thus precautions must be taken in order to maintain the operating temperature of a thyristor at a reasonably low level. Alternatively, it is possible to make use of this observation and to use the thyristor as a switch which will complete a circuit should a predetermined temperature be exceeded.

45.5 The thyristor in practice

The thyristor has therefore been seen to be a switch, and you may wonder why we should use this form instead of the normal mechanical device, especially since we see from the simple circuits that we still must retain mechanical switches. The advantage of the thyristor is that it can operate without involving arcing; thus there are no parts being worn out as a result either of motion or of the burning of the arc. The switches used to turn the thyristor on and also off do not involve the interruption of large currents, so they are less prone to wear and tear.

Apart from the aspects of reliability and safety due to the lack of moving parts, the thyristor as a switch is much more definite in its action; thus we can determine the instant at which it will commence to pass current during each cycle of an alternating current. This permits its use as a control device, thus making the thyristor available as a means of regulating the speed of a machine, regulating a voltage supply and regulating a host of other variable quantities necessary to the control engineer.

45.6 The fully controlled a.c./d.c. converter

Consider the circuit shown in Fig. 45.10 which was a basic a.c./d.c. converter. Like the half-wave rectifier, it is not making the best use of the supply available, hence a thyristor bridge network is preferable. The power circuit is obtained by replacing the diodes of a bridge rectifier with thyristors, as shown in Fig. 45.11. The circuits required to fire the thyristors have been omitted for simplicity.

Let us assume that the load is resistive, hence the current in the load is directly proportional to the applied voltage. The thyristors have to be fired in pairs and usually each pair is fired at the same point in the appropriate half-wave. This is the firing angle. The greater the firing angle, the smaller the average voltage which appears across the load. Typical waveforms are shown in Fig. 45.12.

Fig. 45.11 Full-wave
thyristor converter

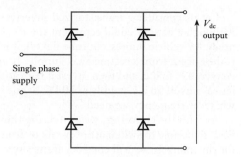

Fig. 45.12 Voltage waveforms
for varying firing angles

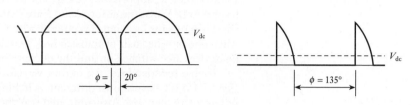

The output voltage is given by

$$V_{dc} = \frac{V_m}{2\pi}(1 + \cos \phi)$$

[45.2]

which follows from equation [45.1].

In practice it is more common that the load has an inductive element, which is what we expect of motors. In an inductive circuit, the current cannot change instantaneously. The consequence is that although the supply voltage has passed through zero switching off the respective thyristors, the load current continues driving the output voltage negative. An output waveform is shown in Fig. 45.13.

Fig. 45.13 Voltage waveform
for an inductive load

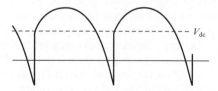

45.7　　**AC/DC
inversion**

The process of changing a.c. to d.c. is known as *conversion*, whereas the process of changing d.c. to a.c. is known as *inversion*. The significant point of inversion is that we are able to choose the frequency of the alternating signal – and for that matter we can vary the frequency at will. This is most important when controlling the speed of an a.c. motor, but can also be significant when coupling to a system in which the frequency is varying.

The problem with inversion is that the output is not sinusoidal. In fact it could hardly be more different since the output waveform is rectangular. Fortunately, most loads such as a.c. motors and fluorescent tubes can accept such waveforms.

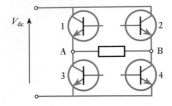

Fig. 45.14 Inverter network for one-phase supply

Let us consider a transistorized inverter bridge as shown in Fig. 45.14. Although a battery could represent the d.c. supply (which might be appropriate for vehicle fluorescent tube lights), it is more usual that the d.c. supply is the output from a rectified a.c. supply. It may seem strange that we should convert a.c. to d.c. and then invert it back to a.c., but this arrangement could let us convert an a.c. supply at 50 Hz to one operating at 100 Hz or 83 Hz or whatever frequency we need.

As with the converter, we switch the transistors on and off in pairs. The need to be able to switch off prevents us from readily using thyristors, hence for the moment we will consider transistors. When we switch on transistors 1 and 4, the d.c. supply voltage appears across the load with A positive with respect to B. When transistors 2 and 3 are switched on, the d.c. supply voltage appears across the load, but B is positive with respect to A. Between the periods when one or other pair of transistors is switched on, there are periods when no devices are switched on and the output voltage across the load is zero.

Given a resistive load, the output waveform takes the form illustrated in Fig. 45.15. We have shown different switching periods whereby we generate differing frequencies. Also indicated are the equivalent sine waves which indicate the considerable differences between sinusoidal and rectangular waveforms. Nevertheless both are alternating in nature, and also the output waveform is symmetrical.

Fig. 45.15 Inverter output voltage waveforms

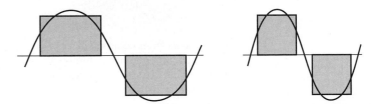

In a bridge rectifier, we do not anticipate that both diodes connected to one end of the load will conduct at the same time as this would short-circuit the supply. In the same way, we do not anticipate that the equivalent transistors in an inverter will be switched on at the same time. This could happen if we control them without due caution. If both are switched on at the same time, the system experiences shoot-through and the d.c. supply is short-circuited. In all probability, this would burn out the transistors.

Apart from varying the output frequency, it is also possible to vary the output voltage by the following methods:

1. Control of the input d.c. voltage.
2. Pulse width modulation.

In the control of the input voltage, it is necessary to assume that the d.c. supply comes from a converter which is fully controlled. This is the system considered in section 45.6 and is usually reserved for high-power applications mainly because of the expense of the converter.

For lower-power applications, the d.c. input can be derived from a simple rectifier system which is consequently cheaper. The control of the output voltage, therefore, has to be achieved within the inverter by pulse width modulation (PWM). This is a term we might readily associate with data transmission systems and it is interesting to find the same principles being applied in power situations.

Fig. 45.16 Simplistic output voltage control

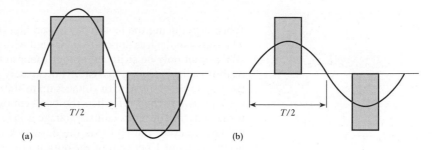

(a)   (b)

Fig. 45.17 PWM voltage control

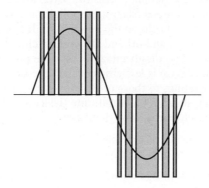

For PWM systems, we have to remember that the supply voltage is fixed, thus the magnitude of the output voltage will depend primarily on the mark–space ratio. Let us consider the waveforms shown in Fig. 45.16(a) and (b). The first shows a rectangular waveform with its equivalent sine wave and is similar to that shown in Fig. 45.15. Now let us consider a sine wave of lesser magnitude. This could be achieved by lowering the mark–space ratio as shown in Fig. 45.16(b).

There is a practical difficulty in that the difference between the rectangular wave and the equivalent sine wave is quite appreciable. The basic requirement is that the area under the rectangular waveform should remain the same, but we can distribute that area so that it more closely reflects the activity of the sine wave. This is the point at which PWM becomes effective as shown in Fig. 45.17.

In practice, we do not obtain the pristine rectangular pulses, but rather the edges are ragged due to a form of resonance. This takes the form of a high-frequency pulse which quickly decays. The resonance is due to the transfer of energy between stray capacitance and stray inductance. These effects can be minimized by snubber circuits which take the form of a string of resistors and capacitors. The fact that a string is used indicates the implication of the pulse being at a high frequency compared with the switching frequencies used for PWM.

The ringing effect is demonstrated in Fig. 45.18.

Snubber networks are rated for up to 500 A r.m.s. when used with standard 400 V loads. They are also identified by the rate of voltage change which they can suppress, e.g. 100 V/μs. As switching devices improve, the ringing effect is minimized and snubber circuits can be omitted. Such high-quality switching devices are said to be snubberless.

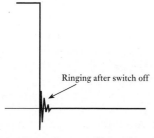

Ringing after switch off

Fig. 45.18 The ringing effect before snubbing

45.8   Switching devices in inverters

When introducing the inverter, we used bipolar junction transistors (BJTs) which are not devices normally associated with power systems. However, the BJT cannot only be switched on but can also be switched off, a facility not readily available from a thyristor. Fortunately BJTs can be used in applications up to about 5 kW with voltages up to 400/230 V. It will be recalled that the BJT is switched off when the base–emitter current is zero and, when turned on, the collector–emitter voltage is low. Apart from the relatively low ratings achieved, the BJT has the drawback that it dissipates a significant heat loss (about 1 per cent of the output power).

Just as junction transistors have given way to field-effect transistors (FETs) in general electronic applications, so the MOSFET has in a number of instances displaced the power BJT. It has the advantages that it can easily be turned on and off and the control circuitry is less complex and hence is cheaper. The disadvantage is that the power dissipation is even higher so that the device is less efficient than the BJT. The MOSFET inverters have similar ratings to those of BJT inverters.

This leaves two significant devices which have enabled engineers to design better inverters; these are:

1. The insulated-gate bipolar transistor (IGBT).
2. The gate turn-off thyristor (GTO).

The IGBT brings together the advantages of the MOSFET and the BJT. This means we have a device in which the gate is easily switched on and off, yet there is low power dissipation in the collector–emitter circuit. These advantages have attracted inverter manufacturers especially since the power ratings are approaching 1 MW while the voltage ratings have also increased to about 1 kV.

The equivalent circuit for an IGBT is shown in Fig. 45.19 as well as its symbol when appearing in circuit diagrams. From the equivalent circuit, it will be observed that the IGBT is basically equivalent to a p–n–p transistor unlike the power BJT which is usually n–p–n. The switch-on time is of the order of 0.15 μs, but the switch-off time is about 1 μs.

Fig. 45.19   IGBT equivalent circuit and symbol

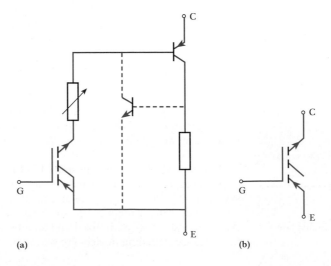

(a)   (b)

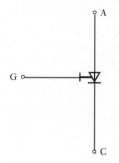

Fig. 45.20 GTO symbol

The GTO has overcome the basic drawback associated with the conventional thyristor – it can be switched off. As with the conventional thyristor, the device is switched on by a pulse of current in the gate–cathode circuit. However, a negative gate–cathode current causes the device to switch off. This gate reversal is implied by the symbol shown in Fig. 45.20.

The GTO is especially useful when we are dealing with high power ratings. It can be used in inverters up to a few megawatts. However, compared with the conventional thyristor, the equivalent GTO requires bigger gate pulses and the forward volt drops are double.

45.9 Three-phase rectifier networks

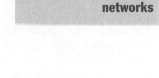

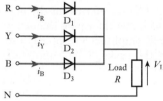

Fig. 45.21 Simple three-phase rectifier network

So far we have considered the use of thyristors and transistors to control power in one-phase circuits. However, when we were first introduced to one-phase circuits, we became aware that there was a limitation in the power ratings which could readily be achieved. These limitations were circumvented by introducing three-phase systems.

When we require to supply large d.c. loads, usually motors, it is almost certain that the supply system will be a three-phase a.c. one. We therefore require a rectifier system which will convert three-phase power to d.c. power.

Diodes are used with polyphase supplies to give a rectified output which has basically less ripple than rectifiers operating from single-phase supplies. Figure 45.21 shows a simple three-phase rectifier network. Each diode will conduct for one-third of a cycle, the conduction path changing instantaneously from one diode to another as one phase voltage becomes more positive than another. For example when v_{RN} becomes less positive than v_{YN} diode D_1 ceases to conduct while D_2 starts to conduct.

The mean load current can be determined as follows:

$$I_{dc} = \frac{1}{2\pi/3} \int_{\pi/6}^{5\pi/6} I_m \sin \omega t \, d(\omega t)$$

$$= \frac{3I_m}{2\pi} \left[-\cos \omega t \right]_{\pi/6}^{5\pi/6} = \frac{3I_m}{2\pi} \left[-\cos \frac{5\pi}{6} + \cos \frac{\pi}{6} \right]$$

$$= \frac{3I_m}{2\pi} \left[\frac{\sqrt{3}}{2} + \frac{\sqrt{3}}{2} \right]$$

$$I_{dc} = \frac{3\sqrt{3}I_m}{2\pi} \qquad [45.3]$$

The mean load voltage will be given by

$$V_{dc} = I_{dc} R$$

$\therefore$
$$V_{dc} = \frac{3\sqrt{3}I_m R}{2\pi} \qquad [45.4]$$

Fig. 45.22 Waveforms for
network shown in Fig. 45.21

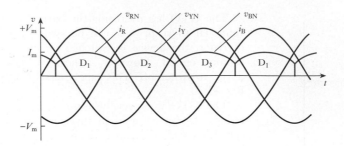

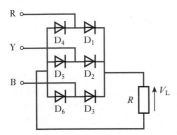

Fig. 45.23 Three-phase bridge
rectifier network

where $I_m R = V_m$, the peak value of the phase voltage if the diode forward
resistance is neglected. Figure 45.22 shows waveforms for the network.

The rectifier circuit shown in Fig. 45.23 has the advantage that there is no
necessity to have a neutral point available. Diodes D_1, D_2, D_4 and D_5 form a
bridge rectifier network, similar to that described in section 21.4, between
the R and Y lines, similarly D_2, D_3, D_5 and D_6 between the B and R lines
and D_3, D_1, D_6 and D_4 between the B and R lines. When v_{RY} is at its
maximum, positive value diodes D_1 and D_5 will be conducting. These two
diodes will continue conducting until v_{RB} is more positive than v_{RY} and
diodes D_1 and D_6 will then take over the conduction.

In this way a pair of diodes will conduct for one-sixth of a cycle at a time,
each diode conducting for one-third of a cycle. The waveforms for the network are shown in Fig. 45.24 where it can be seen that the ripple frequency
is twice that obtained in the previous circuit. The amplitude of ripple is also
decreased since the conduction periods about the peaks of the supply are
decreased.

Fig. 45.24 Waveforms for
network shown in Fig. 45.23

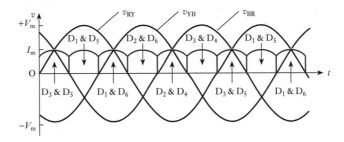

The mean load current can be determined as follows:

$$I_{dc} = \frac{1}{\pi/3} \int_{\pi/3}^{2\pi/3} I_m \sin \omega t \, d(\omega t)$$

$$= \frac{3I_m}{\pi} \left[-\cos \omega t \right]_{\pi/3}^{2\pi/3}$$

$$= \frac{3I_m}{\pi} \left[-\cos \frac{2\pi}{3} + \cos \frac{\pi}{3} \right] = \frac{3I_m}{\pi} \left[\frac{1}{2} + \frac{1}{2} \right]$$

$$\therefore \qquad I_{dc} = \frac{3I_m}{\pi} \qquad\qquad\qquad [45.5]$$

45.10 The three-phase fully controlled converter

We have already seen in section 45.6 that we could produce a fully controlled converter supplied from a one-phase source by replacing the rectifiers in a bridge by thyristors. It is a simple step forward to anticipate that replacing the rectifiers in a three-phase rectifier by thyristors produces a three-phase fully controlled converter. Examination of the waveforms in Fig. 45.25 shows how the delay in the firing angles causes the d.c. output to be varied.

This arrangement can be used to vary the supply voltage to a d.c. motor. For instance, if we consider a d.c. series motor, the starting involved energy loss in a series resistor used to regulate the voltage. Using a converter, we can obtain the same voltage control with virtually no energy loss which is significant – 50 per cent of the energy used during the period when the series resistor is connected in the circuit is lost by the resistor!

Fig. 45.25 Output voltage waveforms from a three-phase fully controlled converter

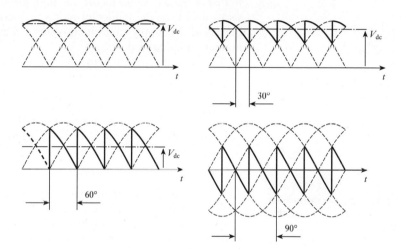

45.11 Inverter-fed induction motors

Having produced the three-phase fully controlled converter, we can also anticipate that the one-phase inverter can be developed into a three-phase inverter. As with the one-phase inverter, this has the significant advantage that the output frequency is variable.

When we consider the three-phase induction motor, we find that we have a robust machine which is mainly limited by the difficulties of controlling its speed. However, if the supply frequency is variable, we find that the torque/speed characteristic varies as shown in Fig. 45.26.

Fig. 45.26 Torque/speed characteristic with varying frequency

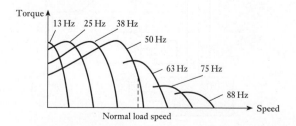

Fig. 45.27 Inverter-fed variable-frequency induction motor

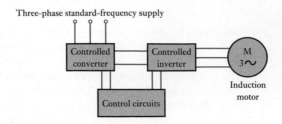

Three-phase standard-frequency supply

Controlled converter

Controlled inverter

M 3~

Induction motor

Control circuits

This variation shows that the form of the torque/speed characteristic varies and, assuming that the voltage is compensated at low frequencies, the torque at low speeds is improved. This effect can make the operation similar to that of the d.c. series motor, i.e. high torque at standstill. Unlike the d.c. series motor, as the induction motor accelerates, we can increase the frequency of the supply from the inverter and maintain the high torque up to normal running speed.

To obtain the voltage boost, we require a controlled converter as well as a controlled inverter. The arrangement is shown in Fig. 45.27. This basic arrangement is useful for starting high-torque loads, but with a little modification it could be applied to precision speed-control systems. A feedback control loop from a tachogenerator could be applied to the input control signal so that the speed could be both maintained and varied at will.

45.12	Soft-starting induction motors

Although the converter/inverter arrangement can be used to start a three-phase induction motor, it is an expensive method for starting a motor and can only be justified for high-torque loads. However, an arrangement of thyristors somewhat similar to the converter can control the starting currents. When we recall the starting problems considered in section 39.9, it will be remembered that arrangements such as the star/delta starter were intended to limit the initial current. The thyristor arrangement shown in Fig. 45.28 also controls the initial current, but there are two advantages:

Fig. 45.28 Soft starting of induction motor

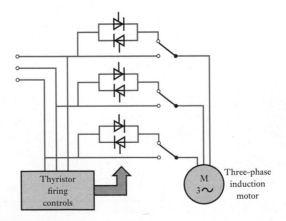

Thyristor firing controls

M 3~

Three-phase induction motor

Fig. 45.29 Soft-starting current waveforms

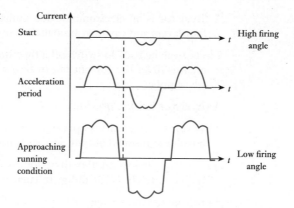

1. The star/delta starter causes a jerk when the switches convert the connections from star to delta. The continuous variation of the thyristor soft starting avoids this snatching effect.

2. The control of the current ensures that maximum current is maintained during the acceleration period ensuring that good acceleration is obtained.

The effect of varying the thyristor firing control on the motor current waveforms is shown in Fig. 45.29. Once the motor reaches full speed, the thyristors are cut out and the motor is connected directly to the supply. However, during the acceleration period, the current waveforms are not sinusoidal. The motor can accept this, but it can cause interference to other equipment connected to the same supply system.

Soft starting is taking over from other forms of induction motor starting and could eventually become predominant in new systems. Electrical power equipment tends to be long-lasting, so the more conventional starters will be operational for many years in the future.

Summary of important formulae

For a single-thyristor control circuit,

$$I_{av} = \frac{I_m}{2\pi}(1 + \cos \phi) \qquad [45.1]$$

For a thyristor-bridge control network,

$$V_{dc} = \frac{V_m}{\pi}(1 + \cos \phi) \qquad [45.2]$$

For a three-phase simple rectifier network,

$$I_{dc} = \frac{3\sqrt{3}I_m}{2\pi} \qquad [45.3]$$

Terms and concepts

A **thyristor** is an electronic device similar to a transistor switch. It has four layers and can only be switched on; it cannot be switched off.

Circuits can be used to switch off a thyristor but the most simple arrangement is to let the current fall to zero which arises when used with an a.c. supply.

A **thyristor bridge** provides a controllable a.c./d.c. converter.

Thyristor systems can also provide d.c./a.c. inverters. To obtain the best possible sinusoidal output, inverters use **pulse width modulation**.

Alternative devices to the thyristor include **insulated-gate bipolar transistors** (IGBTs) and **gate turn-off thyristors** (GTOs).

Three-phase rectifier networks provide d.c. outputs with relatively little ripple.

If the thyristors replace the diodes in a three-phase rectifier network, the result is a **fully controlled converter**.

By coupling converters and inverters, a speed-control system can be provided for three-phase induction motors.

Thyristors coupled back-to-back provide **soft-starting** arrangements for three-phase induction motors.

Exercises 45

1. A thyristor is connected in series with a 100 Ω resistor to a 230 V sinusoidal supply. If the thyristor is controlled to switch on at a firing angle of 30°, determine the average current in the resistor.
2. For the circuit arrangement in Q. 1, determine the firing angle if the average current is 0.25 A.
3. For the circuit arrangement in Q. 1, the thyristor is upgraded to a fully controlled converter. Determine: (a) the average resistor current when the firing angle is 25°; (b) the firing angle when the average resistor current is 0.30 A.

4. A simple three-phase rectifier network supplies a 50 Ω resistive load from a 400 V, 50 Hz, three-phase source. Determine: (a) the mean load current; (b) the mean load voltage.
5. The rectifier network in Q. 4 is replaced by a three-phase bridge rectifier. Again determine: (a) the mean load current; (b) the mean load voltage.
6. Describe the effect achieved by replacing the bridge rectifier by a fully controlled converter; in particular describe the effect of varying the firing angle.

Section four Measurements

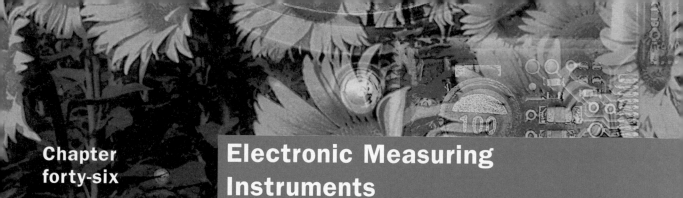

Chapter forty-six

Electronic Measuring Instruments

Objectives

When you have studied this chapter, you should
- have an understanding of the operation of digital electronic voltmeters, ammeters and wattmeters
- be able to describe the main functional elements of a cathode-ray oscilloscope
- be aware of different ways that a cathode-ray oscilloscope may be used

Contents

Throughout this book it has been assumed that it is possible to measure the electrical quantities to which reference has been made. There are two families of measuring devices: digital instruments and analogue instruments.

In this chapter electronic instruments which generally display measurements by digital means will be considered. These instruments have little effect on the circuits in which they are making measurements. Many of them depend on A/D converters and such a system will be considered in detail.

The oscilloscope will also be considered. To understand its action, thermionic emission and control will be considered. The procedures that are used to interpret the oscilloscope displays and the methods of connecting the oscilloscope in a circuit are described.

By the end of the chapter, you will be familiar with electronic measuring instruments including the oscilloscope. You will also be able to undertake exercises in the interpretation of oscilloscope displays.

46.1   Introduction to analogue and electronic instruments

Throughout this book, we have introduced all sorts of electrical quantities such as volts, amperes, ohms and farads. We have used more generally established quantities such as metres, seconds and kilograms. We have defined these units, but we have not addressed the ways in which they might be measured.

It is possible to measure all quantities with considerable accuracy. However, before introducing the means of such measurement, we should be clear about our expectations.

An example of a measurement with which we are all familiar is that of time. When we ask about time, it is usually in order to determine the time remaining until some event, say attending a meeting or catching a train (10 min so we can walk, 5 min so we start running). However, mention of running makes us think of a sprinter running 100 m; to the sprinter, time is measured to 0.01 s and here we are seeking very accurate measurement.

In electrical engineering, we find equivalent situations arising. For instance, if we are repairing some of the electric circuits on a car, we might use a meter to determine if the circuit was delivering the battery voltage of 12 V. In practice, the actual voltage is about 11–13 V and even more if the engine is running, but that does not matter since any voltage of that order is satisfactory. More or less, we are seeking all or nothing.

However, there are many situations in which our measurements must be accurate. For instance, setting the frequency in electronic data transmission equipment requires almost absolute accuracy. Equally, determining the position of a fault in an underground cable has to be exact, otherwise how big a hole do we dig? And, of course, there is the time measurement for the sprinter – or a skier.

Therefore, in practice we find that measuring devices fall into two categories – those which give graded indications, e.g. a petrol gauge, and those which give the best possible measurement, e.g. the stop-clock operating to 0.01 s accuracy. Until recently, the measuring devices were all analogue devices. An analogue instrument is one in which an action takes place directly representing the quantity which we wish to measure.

Typically we can pass a current through a wire causing a force and hence a torque to stretch a spring. The greater the current then the greater the torque and the greater the displacement of the spring. If we attach a pointer to the mechanism then its deflection across a scale represents the current, i.e. the displacement is an analogue of the current.

Such instruments were all that was available, so they were used both for graded indications and the best possible measurements, although the latter was limited. Significantly, the visual display of a needle is very popular and such analogue devices remain supreme for graded-indication applications. However, their uses for accurate measurement left much to be desired – often an accuracy of 1 or 2 per cent was all that could be achieved. Where that is satisfactory, then the analogue instrument is still in common use. We shall leave further consideration of analogue devices to Chapter 47.

In recent years, electronic measuring devices have been developed which can achieve two significant outcomes. They can be extremely accurate and they can provide graphical displays. The accurate measuring devices generally have digital displays of the measurement data, hence such devices are often referred to as digital instruments (the electronic element remains inferred). The graphical displays commonly show the waveforms to which we have been introduced and the displays appear on the screens of oscilloscopes.

It follows that most instruments now manufactured for the purpose of measurement are electronic and will be considered in the remainder of this chapter.

46.2	Digital electronic voltmeters

We were introduced to A/D conversion in Chapter 25. One of the most common applications of this conversion process is to be found in the digital voltmeter (sometimes given the abbreviation DVM). In this instrument, the analogue input is converted to a BCD-code representation which is then decoded and displayed on a digital display.

Let us take care with the term 'analogue input'. In our introduction, we used 'analogue' in the measurements context to mean a mechanical represent-entation of an electrical or other quantity. However, in the electronic instru-ment, the analogue signal is one which represents an electrical or other signal. Thus if a voltage were sufficiently small, it could be applied directly to the converter – but if it were too large, we could apply it to a preset poten-tiometer, the output of which could be applied to the converter. The poten-tiometer output being a direct fraction of the input voltage is in fact an analogue of the input.

In the same way we can measure currents by passing them through a resistor and measuring the volt drops across the resistor, these voltages being analogues of the currents.

A typical digital voltmeter has a four-digit display. It is possible to increase the number of digits, but the costs rise rapidly with every extra digit so we would need to be careful that such extra implied accuracy was justified. Figure 46.1 shows a simple continuous-reading digital voltmeter circuit arrangement in block diagram form.

The four cascaded BCD counters provide the digital inputs to a BCD-type D/A converter. Let us assume that the converter produces a step voltage of 10 mV at its output V'_A, i.e. if there are six steps then $V'_A = 60$ mV and so on. Each BCD counter feeds a decoder/driver and associated display. It follows that each digit of the count is continuously being displayed as the counters run up from 0000 to 9999. When the counting has ended, the display is held for a time so that we can observe it. (The counting period is so rapid that we can scarcely observe the changing display and normally the first effect we can observe is the steady display.)

The clock pulses are gated into the counters along with the comparator output. So long as $V_A > V'_A$, the output from the comparator (COMP) is 1 and the counter continues to receive pulses from the clock. With each clock pulse, the counter advances a step and V'_A goes up another 10 mV (this is inferred by the step waveform on the diagram). Eventually $V'_A > V_A$ by a value not more than 10 mV, but that is sufficient to make COMP go to 0 (or go LOW) and the AND gate is disabled. This prevents further clock pulses entering the counter which therefore cannot advance further.

At this point, we have the counter holding data equivalent to the unknown analogue input V_A. However, the data are slightly in excess of V_A by a value not more than one step relative to the number of steps which were counted. Provided we can arrange it so that the input analogue requires a

Fig. 46.1 Simple continuous-reading digital voltmeter

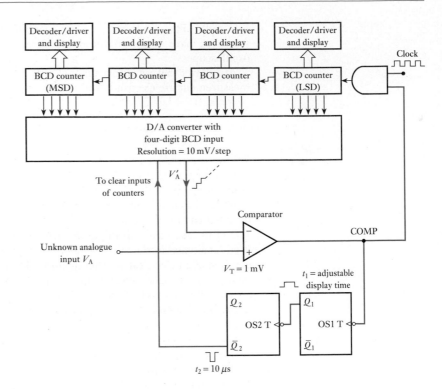

count in excess of 1000 steps then the error will only be 1 step in 1000 which is excellent for most purposes.

As soon as the comparator output COMP goes LOW to 0, it also triggers a form of set–reset relay. This form is called a one-shot device and is so called because it only requires one input pulse to set it, but subsequently it resets itself ready to repeat the cycle of events.

Our voltmeter uses two such devices, the first of which is OS1. Its function is to hold the display so that we can observe it. To do this, OS1 effectively locks out the clock for a period of time t_1. This time is preset, but is usually about 1 or 2 s. While OS1 is operated, the Q_1 output is HIGH.

The D/A converter now requires to be completely cleared so that the count can be repeated. This is achieved by the second one-shot sending a pulse to clear the D/A converter and counters. When the Q_1 input to OS2 goes LOW, OS2 is triggered and this causes $\bar{Q}_2$ to clear the BCD counters to the 0 state. The pulse is very short and a time of approximately 10 μs is sufficient.

As soon as the D/A converter is cleared, $V_A > V'_A$ and the whole counting process is repeated. If we use such a digital voltmeter, we become aware of this cyclical operation as the display varies from time to time (rather like a digital clock which includes seconds in its display).

Finally let us consider the above process by attaching values to the number of steps and the voltages measured. If the voltage to be measured is 238.46 V, it is clear that our four-digit display would be overwhelmed; at 10 mV per step, we would require 23 846 steps which is beyond the upper limit (9999) of our arrangement. We are therefore required to produce an analogue voltage $V_A = 23.846$ V with which our converter can cope.

With $V_A = 23.846$ V, then when the comparator output goes HIGH, counting commences until $V'_A > V_A$. For this to happen, the count will be $23.846/0.01 = 2385$ steps. At this point, $V'_A = 23.850$ V and $V_A = 23.846$ so that the display is a mere 4 mV high relative to 23.85 V. This is equivalent to 40 mV high relative to 238.5 V, which is a very small error indeed.

It will be noted that we have to state the discrepancy relative to the display value rather than the actual value. We have to do this because the displayed value is the only data to which we can relate.

Although the digital display shows the numbers 2385, the decimal also has to be shown, thus the display would be 238.5. The switching of the device which provided the 23.85 V from the 238.5 V input usually ensures that the decimal point is located correctly.

Returning to the last digit, it is in error because of the 10 mV resolution of the D/A converter. This quantization error can be reduced by adding further BCD counters (or it can be increased by reducing the number of counters, thus cutting the cost of the meter).

46.3 Digital electronic ammeters and wattmeters

It has already been noted that the passage of a current through a resistor produces a volt drop which can be measured in the manner described above. Usually 10 mV/step counting requires too high a volt drop and therefore we require to introduce either smaller resolutions or amplifiers to increase the analogue input voltage signals.

The development of the basic instrument circuitry would take us into more advanced electronic systems than lie within the scope of this book, and it must remain sufficient to note that ammeters and wattmeters (which measure real power) are merely sophisticated variations on the basic digital voltmeter already described.

Apart from the accuracy of display compared with the analogue instruments of Chapter 47, digital meters have considerable advantages in that introducing them into electrical circuits does not significantly change the circuits. This effect on circuits is known as *circuit disturbance* and will be described in section 47.15.

46.4 Graphical display devices

At this stage we could consider such devices to be specialized applications of the display which we see in television receivers or computer screens. Generally, for the purposes of measurement, we are seeking waveform displays to determine either their shape or their position in time relative to another waveform. The first of these usually relates to analogue transmission signals and the latter to digital signals.

In order to understand the operation of such display instruments, which we call oscilloscopes or CROs (which is the abbreviation for cathode-ray oscilloscopes), we need to consider the actions which take place within the display tubes. This also enables us to understand what happens inside the display tubes of our television sets.

First we need to consider what happens to an electron in a vacuum. We can find out by considering a vacuum diode, a device which is no longer commonly used.

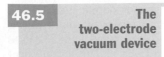

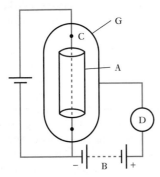

Fig. 46.2 A vacuum diode

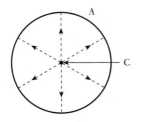

Fig. 46.3 Electron paths from cathode to anode

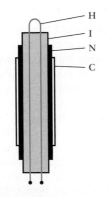

Fig. 46.4 An indirectly heated cathode

If a metal cylinder A surrounds an incandescent filament C in an evacuated glass bulb G, as shown in Fig. 46.2, and a battery B is connected in series with a milliammeter D between the cylinder and the negative end of the filament, it is found that an electric current flows through the milliammeter when the cylinder is made positive relative to the filament, but when the connections to battery B are reversed, so as to make A negative, there is no current through D. Let us now consider the reason for this behaviour.

An electrical conductor contains a large number of mobile or free electrons that are not attached to any particular atom of the material, but move at random from one atom to another within the boundary of the conductor; the higher the temperature of the conductor, the greater is the velocity attained by these electrons. In the case of an incandescent tungsten filament, for instance, some of these free electrons may acquire sufficient momentum to overcome the forces tending to hold them within the boundary of the filament. Consequently they escape outwards; but if there is no p.d. between the filament and the surrounding cylinder, the electrons emitted from the filament form a negatively charged cloud or *space charge* around the wire, the latter being left positively charged. Hence the electrons near the surface experience a force urging them to re-enter the filament, and a condition of equilibrium is established in which electrons re-enter the surface at nearly the same rate as they are being emitted, the difference being the electrons which succeed in passing through to cylinder A and represent a current of the order of microamperes. This current is reduced to zero if the potential of the cylinder is made about 1 V negative relative to the filament.

If cylinder A is made positive in relation to the filament, electrons are attracted outwards from the space charge, as indicated by the dotted radial lines in Fig. 46.3, and fewer of the electrons emitted from the filament are repelled back into the latter. The number of electrons reaching the cylinder increases with increase in the positive potential on the cylinder.

If the cylinder is made negative in relation to the filament, the electrons of the space charge are repelled towards the filament, so that none reaches the cylinder and the reading on milliammeter D (Fig. 46.2) is zero.

Since the cylinder is normally positive in relation to the filament, the former is termed the *anode* and the latter the *cathode*, and since current can flow in one direction only, the arrangement is termed a *thermionic valve* or merely a *valve*. The liberation of electrons from an electrode by virtue of its temperature is referred to as *thermionic emission*. When a valve contains only an anode and a cathode, it is referred to as a *diode*. The p.d. between the anode and the cathode is termed the *anode voltage* and the rate of flow of electrons from cathode to anode constitutes the *anode current*. The *conventional* direction of the anode current is from the positive of battery B via the anode to the cathode, as indicated by the arrowhead in Fig. 46.2; however, it is important to realize that the anode current is actually a movement of electrons in the reverse direction.

The anode is stamped out of nickel sheet and the cathode can be either of the directly heated or of the indirectly heated type.

Most diodes incorporate cathodes of the indirectly heated type, in which the cathode C consists of a mixture of barium and strontium oxides sprayed on a hollow nickel cylinder N, as in Fig. 46.4. The cathode is heated by a tungsten filament H, known as the *heater*, embedded in an insulator I to prevent the heater making electrical contact with the cathode. Oxide emitters must be activated by special heat treatment to produce a layer of metallic molecules of barium and strontium on the surface of the oxide.

46.6 Control of the anode current

Although the current between cathode and anode can be controlled by varying the anode–cathode voltage, it is easier to introduce a control grid. The resultant device is called a *triode* which is short for three electrodes.

Figure 46.5 shows the general arrangement of a triode having a directly heated cathode C. The anode cylinder A is shown cut to depict more clearly the internal construction. Grid G is usually a wire helix attached to one or two supporting rods. The pitch of this helix and the distance between the helix and the cathode are the main factors that determine the characteristics of the triode.

In a triode, the potential of the anode A is always positive with respect to the filament, so that electrons tend to be attracted towards A from the space charge surrounding cathode C. The effect of making the grid G positive with respect to C is to attract more electrons from the space charge. Most of these electrons pass through the gaps between the grid wires, but some of them are caught by the grid as shown in Fig. 46.6(a) and return to the cathode via the grid circuit. On the other hand, the effect of making the grid negative is to neutralize, partially or wholly, the effect of the positive potential of the anode. Consequently, fewer electrons reach the anode, the paths of these electrons being as shown in Fig. 46.6(b). No electrons are now reaching the grid, i.e. there is no grid current when the grid is negative by more than about 1 V with respect to the cathode. The paths of the electrons which are repelled back from the space charge into the cathode are not indicated in Fig. 46.6.

It will be seen that the magnitude of the anode current can be controlled by varying the p.d. between the grid and the cathode; since the grid is in close proximity to the space charge surrounding the cathode, a variation of, say, 1 V in the grid potential produces a far greater change of anode current than that due to 1 V variation of anode potential.

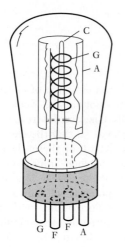

Fig. 46.5 A vacuum triode

Fig. 46.6 Influence of grid potential upon electron paths. (a) Grid positive; (b) grid negative

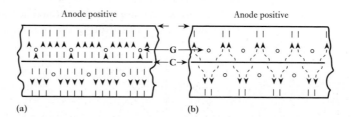

46.7 Cathode-ray tube

The CRO is almost universally employed to display the waveforms of alternating voltages and currents and has very many applications in electrical testing – especially at high frequencies. The cathode-ray tube is an important component of both the CRO and the television receiver.

Figure 46.7 shows the principal features of the modern cathode-ray tube. C represents an indirectly heated cathode and G is a control grid with a

Fig. 46.7   A cathode-ray tube

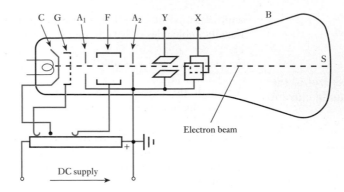

Electron beam

DC supply

variable negative bias by means of which the electron emission of C can be controlled, thereby varying the brilliancy of the spot on the fluorescent screen S. The anode discs A_1 and A_2 are usually connected together and maintained at a high potential relative to the cathode, so that the electrons passing through G are accelerated very rapidly. Many of these electrons shoot through the small apertures in the discs and their impact on the fluorescent screen S produces a luminous patch on the latter. This patch can be focused into a bright spot by varying the potential of the focusing electrode F, thereby varying the distribution of the electrostatic field in the space between discs A_1 and A_2. Electrode F may consist of a metal cylinder or of two discs with relatively large apertures. The combination of A_1, A_2 and F may be regarded as an *electron lens* and the system of electrodes producing the electron beam is termed an *electron gun*. The glass bulb B is thoroughly evacuated to prevent any ionization.

46.8   Deflecting systems of a cathode-ray tube

The electrons after emerging through the aperture in disc A_2 pass between two pairs of parallel plates, termed the X- and Y-plates and arranged as in Fig. 46.8. One plate of each pair is usually connected to anode A_2 and to earth, as in Fig. 46.7.

Suppose a d.c. supply to be applied across the Y-plates, as in Fig. 46.9, then the electrons constituting the beam are attracted towards the positive

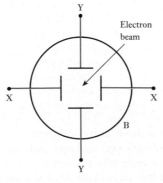

Fig. 46.8   Deflecting plates of a cathode-ray tube

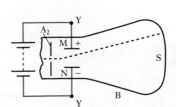

Fig. 46.9   Electrostatic deflection of an electron beam

plate M and the beam is deflected upwards. If an alternating voltage were applied to the Y-plates, the beam would be deflected alternately upwards and downwards and would therefore trace a *vertical* line on the screen. Similarly, an alternating voltage applied to the X-plates would cause the beam to trace a horizontal line.

46.9 Cathode-ray oscilloscope

Most oscilloscopes are general-purpose instruments and the basic form of their operation is illustrated in Fig. 46.10. For simplicity, we shall restrict our interest to displaying one signal, although most oscilloscopes are capable of displaying two.

Fig. 46.10 Basic schematic diagram of a CRO

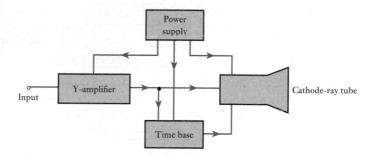

The input signal is amplified by the Y-amplifier, so called because it causes the beam to be driven up and down the screen of the cathode-ray tube in the direction described as the Y-direction by mathematicians.

The time base serves to move the beam across the screen of the tube. When the beam moves across the screen, it is said to move in the X-direction. It would not be appropriate if the movements in the X- and Y-directions were not coordinated; hence the time base may be controlled by the output of the Y-amplifier. This interrelationship is quite complex and therefore requires further explanation.

However, before proceeding, we require another major component which is the power supply. This serves to energize the grid and anode systems of the cathode-ray tube, as well as to energize the brilliance, focus and astigmatism controls of the beam. The power supply also energizes the amplifiers for the control of the beam.

Assuming that you are already familiar with the operation of the cathode-ray tube, it remains to consider those parts of the overall instrument which give rise to controls that we must operate in order to use the oscilloscope. A more complete schematic diagram therefore is shown in Fig. 46.11.

The circuitry of an oscilloscope has to be capable of handling a very wide range of input signals varying from a few millivolts to possibly a few hundred volts, while the input signal frequency may vary from zero (d.c.) up to possibly 1 GHz, although an upper limit of 10–50 MHz is more common in general-purpose instruments.

We have already noted that the input signal drives the display beam in the Y-direction. The height of the screen dictates the extent of the possible deflection. The output of the Y-amplifier therefore has to be of a sufficient

Fig. 46.11   Schematic diagram
of a cathode-ray oscilloscope

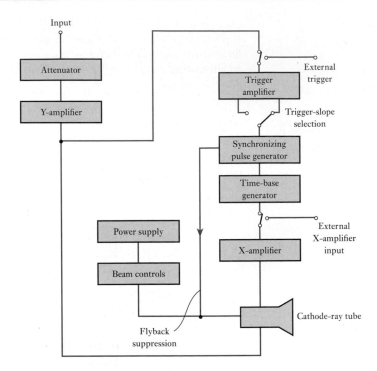

magnitude to drive the beam up and down the screen in order to give as large
a display as possible without the display disappearing off the edge of the
screen. Nevertheless, the Y-amplifier operates with a fixed gain and it is
necessary to adjust the magnitude of the input signal to the amplifier, this
being done with an attenuator. An attenuator is a network of resistors and
capacitors, and its function is to reduce the input signal.

This may seem to be a peculiar function, but the hardest task required of
the amplifier is to increase the voltage of a small signal in order to obtain a
display across the greatest extent of the screen. This determines the amplifier
gain, but, unless something is done about it, greater input signals would
cause the display to extend beyond the screen. These greater signals,
however, can readily be cut down to size by an attenuator, thus leaving the
Y-amplifier to continue to operate with its gain fixed to suit the smallest
signal.

The attenuator has a number of switched steps, the lowest normally being
0.1 V/cm and the highest being 50 V/cm. For instance, if we set the control
to 10 V/cm and apply an input signal of 50 V peak-to-peak, it follows that
the height of the display on the screen is $50/10 = 5$ cm. As many screens give
a display 8 cm high, this is the best possible scale, and such a display would
disappear at the top and bottom of the screen.

Associated with the vertical scale control, we also have a Y-shift control
which permits us to centralize the display vertically on the screen.

In oscilloscopes, the Y-amplifier may consist of a single stage in a very
basic model, but generally a number of stages are incorporated, especially in
those oscilloscopes used for measurements as opposed to simple waveform
displays.

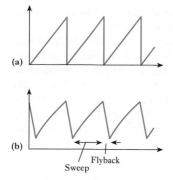

Fig. 46.12 X-amplifier sawtooth waveforms: (a) idealized form; (b) practical form

The X-amplifier is normally identical to the Y-amplifier and has an associated X-shift control comparable to the Y-shift control. In each case, the shift can be achieved by adjusting the bias voltage to the amplifier, thus causing a shift of the mean output voltage.

The most usual mode of oscilloscope operation has the X-amplifier fed from a time-base generator, the function of which is to drive the beam at a steady speed across the screen and, when it reaches the right-hand side of the screen, the beam is then made to fly back to the left-hand side and start out again across the screen. To produce such an output, the input signal to the X-amplifier must take the form of a sawtooth waveform, which is illustrated in Fig. 46.12. As the signal steadily increases, the beam is moved across the screen. When the signal reaches its peak, ideally it should drop to zero, thus instantaneously returning the beam to the beginning of its travel. In practice, there are two (possibly three) differences between the ideal waveform and that actually experienced.

The ramp of the waveform is not linear but is derived from an R–C circuit transient. Provided that the time constant of the R–C circuit is very much greater than that of the time required for the beam to scan across the screen, the ramp is almost linear. The difference has been exaggerated in Fig. 46.12(b) for clarity.

When the waveform reaches its peak, it is not possible, for reasons which we have observed in our studies of transients, for the signal to suddenly return to zero; thus there is a short period during which the voltage decays. This is called the flyback time, because it is the period during which the beam flies back to the start of its travel.

The flyback time is made as short as possible, partly to save the display time lost and partly to reduce the trace of the beam returning across the screen. To help eliminate this unwanted display, the beam current is reduced during the flyback time by means of a flyback suppression pulse.

There may be a short interval between the end of the flyback period and the following scanning period. This is necessary when the time base is controlled from an external trigger source, which is described later. The short delay ensures that the scan starts at the same point in the display waveform, thus causing the display to appear stationary on the screen. In most applications, the time base is controlled by a pulse generator synchronized to the signal from the Y-amplifier, and there is no need to have a delay between the end of the flyback period and the beginning of the scanning period.

The signal to the X-amplifier comes from the time-base generator which may operate in any of the following modes:

1. self-oscillating
2. self-oscillating and synchronized
3. externally triggered.

In the purely self-oscillating arrangement, the time-base voltage rises to a preset value, at which instant the beam has reached the right-hand extremity of its travel. When the preset value is reached, the flyback is automatically initiated and, as soon as the original value at the beginning of the scan is obtained, the generator starts generating the next sweep across the screen. The problem with the self-oscillating arrangement is that it works independently of the input signal; thus in the first sweep it may start when the input signal is at a positive maximum value, yet the next sweep starts at a negative maximum, thus giving a completely different trace. This sort of variation at best gives rise to an apparently moving display and at worst to two or three displays which are superimposed on one another.

To overcome this problem, it is necessary to synchronize the time-base generator to the frequency of the supply. In a self-oscillating and synchronized system, the initiation of the flyback is controlled by a synchronization signal from the Y-amplifier. Because the flyback is controlled by the synchronizing signal, it follows that the synchronizing signal also controls the start of the sweep of the beam. In some oscilloscopes, this arrangement is fully automatic, but in many of the cheaper general-purpose oscilloscopes, there is a stabilizing control which sets the level of signal display at which the flyback is initiated.

It is necessary to appreciate the reason why this arrangement operates from the finish of the display and not from the beginning. The time-base generator causes the beam to sweep across the display at regular intervals. Let us assume that this is taking place with a frequency 50 Hz and also let us assume that the frequency of the signal to be displayed is 150 Hz. During the sweep time, the input signal undergoes three cycles; thus we would hope to see these three cycles being displayed. In practice, a bit of one cycle would be lost because not all of the time is available for display, the remaining time being taken up by the flyback period. However, the main problems are to commence the trace at the same point in the input signal each time. Let us assume that we wish to start when the input signal is positive and rising. Ideally this would coincide with the instant at which the sweep was due to commence; thus three cycles (almost) would be displayed, followed by the flyback, and everything would be ready for the next sweep to commence displaying the following three input signal waves.

However, what happens if we just miss the start of an input signal wave? If we wait for the next instant of the signal being positive and increasing, then we have to wait for almost a complete cycle, which would be lost to the display. And, even more awkward, what happens if the frequency of the signal to be displayed is 152 Hz? After all, as the signal frequency increases we expect to see more than three cycles, so that if the frequency is 200 Hz, for example, we expect the display of four cycles.

The answer is not to wait for the chosen instant but rather to get on with the display up to the time of the chosen instant. In this way, we do not miss anything by waiting (although we shall miss that short period of display during the flyback) but, having reached the chosen instant, the beam is caused to fly back and to recommence the sweep with the minimum delay. It now starts no matter what is happening, and continues again up to the chosen point at which the flyback is again initiated. In this way we can display any number of cycles or fractions of a cycle in excess of one cycle.

The stabilizing control has to be adjusted appropriately to synchronize the flyback of the time-base generator to the output of the Y-amplifier. In most cases, this can be readily achieved, but sometimes the quality of the signal to be displayed is not sufficiently reliable, in which case the time-base generator must be controlled from an external source, which provides a suitable trigger. In this case, the trigger initiates each individual time-base sweep, and the flyback then follows automatically when the time-base signal has reached a preset value. In this case, the time-base generator remains inactive until the trigger releases another sweep. This means that possibly a significant part of the display can be omitted. For this reason, some oscilloscopes are provided with gain controls to the X-amplifier whereby the display can be expanded and we can examine the display in greater detail.

If we wish to make time or frequency measurements, the time-base control must have a calibration setting at which the display time coincides with the control markings. For instance, if the time-base control is set to 10 μs/cm, then the X-amplifier control is set to the calibration mark and we know that each centimetre of the display in the X-direction represents 10 μs of time. A typical range of time-base control settings is 0.5 s/cm to 1 μs/cm.

There are several applications of the oscilloscope in which we do not require the time-base generation at all, but instead we drive the X-amplifier from another signal source in a similar manner to the operation of the Y-amplifier. For this reason, many oscilloscopes afford direct access to the X-amplifier and we shall look at such an application in section 46.10.

The input impedance of most general-purpose oscilloscopes is 1 MΩ shunted by a capacitance of 20–50 pF according to the model used. The effect of the capacitance becomes significantly effective only at high frequencies, i.e. in excess of 1 GHz. Such a high input impedance makes the oscilloscope suitable for many measurement techniques, since the oscilloscope scarcely modifies the network into which it has been introduced.

Mention has already been made of the calibration of oscilloscopes, and many have built-in calibration circuits. Generally these give a square or trapezoidal waveform of known peak-to-peak magnitude and cycle duration. This calibration signal is fed into the oscilloscope and the gain of the Y-amplifier is adjusted to give the appropriate vertical display. Similarly, the gain of the X-amplifier is adjusted to give a signal display of appropriate length. These adjustments are usually made by potentiometers with a screw adjustment operated by a screwdriver. In this way, calibration adjustment cannot be confused with the other controls of the oscilloscope.

This brief description of the operation of the principal components has indicated the main controls that we require to use when displaying and measuring waveforms and phase differences by means of the oscilloscope.

| 46.10 | Use of the cathode-ray oscilloscope in waveform measurement |

A discussion of the use of the oscilloscope falls naturally into two parts: the use of the instrument itself, and the methods of connecting the instrument to the circuits in which the measurements are to be made. For ease of introduction, let us assume that the signals applied to the oscilloscope are suitable.

Once the connections between the source of the signal and the oscilloscope have been made, the oscilloscope should be switched on and given time to warm up. Generally a trace will appear on the screen, but should this not occur, some useful points to check are that the vertical and horizontal shift controls are centralized, that the brilliance control is centralized, that the trigger is set to the automatic position (where appropriate) and that the stabilizing control is varied to ensure that the display time base is operated. Normally these checks ensure that the display appears, but if these do not work, then you have to check out the full procedure in accordance with the manufacturer's operating manual.

Once the display has been established, adjust the brilliance to obtain an acceptable trace which is not brighter than necessary. Too bright a trace, especially if permitted to remain in the one position for a considerable period of time, can damage the fluorescent material on the screen, hence the reason for minimizing the brilliance. It also does no harm to check that the beam is focused and that there is minimum astigmatism. These do not vary much with operation but sometimes the controls are adjusted incorrectly.

The display is next centralized vertically and the scale control adjusted to give the highest possible display that can be contained within the screen.

Having the display clearly in view, it may be that the trace is stationary, but it could also be slipping slowly in a horizontal direction. In the latter case, the stabilizing control requires to be adjusted until the trace is locked in position and remains stationary.

Unless you have some unusual observations to make on the waveform, the X-gain amplifier should be set to the calibration position and the X-shift control readjusted to centralize the display horizontally. This display may contain only part of the waveform or a great many waveforms; this is changed by adjusting the time-base control until the desired number of waveforms are displayed.

This again is a brief description of the setting-up procedure of the oscilloscope and it serves only to highlight the common form of operation. Different models of oscilloscope vary in detail, but the procedure is essentially that indicated. However, words cannot substitute for the practical experience of operating oscilloscopes, and you will readily obtain a better appreciation of the oscilloscope from a few minutes' experimentation with one in a laboratory.

To aid observation of the display on an oscilloscope, a set of squares is marked on the transparent screen cover. This marking is termed a graticule and is illustrated in Fig. 46.13.

Graticules are marked out with a 1 cm grid and are presently 10 cm across by 8 cm high. Older models had graticules 8 cm by 8 cm or sometimes 10 cm by 10 cm. To avoid parallax error, you should always observe the trace directly through the graticule and not from the side.

Let us now consider the interpretation of the basic forms of display, which are sine waves, square waves and pulses. A typical sine waveform display as seen through a graticule is shown in Fig. 46.14. To obtain this display, let us assume that the vertical control is set to 2 V/cm and the time-base control to 500 μs/cm.

The peak-to-peak height of the display is 4.8 cm, hence the peak-to-peak voltage is $4.8 \times 2 = 9.6$ V. This may be a direct measurement of a voltage or the indirect measurement of, say, a current. In the latter case, if a current is passed through a resistor of known resistance, then the current value is obtained by dividing the voltage by the resistance.

You will note that the voltage measured is the peak-to-peak value. If the signal is sinusoidal, then the r.m.s. value is obtained by dividing the peak-to-peak value by $2\sqrt{2}$, which in this instance gives an r.m.s. value of 3.4 V.

The oscilloscope, therefore, can be seen to suffer the disadvantage when compared with electronic voltmeters that it is more complex to operate and to interpret. However, we are immediately able to determine whether we are dealing with sinusoidal quantities, which is not possible with other meters and which is essential to interpreting the accuracy of the measurement of alternating quantities. The oscilloscope is therefore an instrument whereby we observe the waveform in detail and measurements of magnitude of the signal are essentially those of peak-to-peak values.

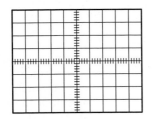

Fig. 46.13 Cathode-ray oscilloscope graticule

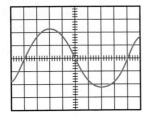

Fig. 46.14 Sine waveform display on an oscilloscope

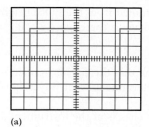

(a)

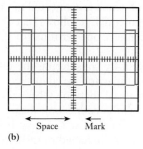

← Space → ← Mark

(b)

Fig. 46.15 Cathode-ray oscilloscope displays for square waves and pulses: (a) square wave; (b) pulse

Returning to the display shown in Fig. 46.14, the length of one cycle of the display is 8.0 cm, hence the period of the waveform is

$$8.0 \times 500 \times 10^{-6} = 4.0 \times 10^{-3} \text{ s} = 4.0 \text{ ms}$$

It follows that the frequency of the signal is

$$\frac{1}{4.0 \times 10^{-3}} = 250 \text{ Hz}$$

In each case, the accuracy of measurement is not particularly good. At best, we cannot claim an accuracy of measurement on the graticule that is better than to the nearest millimetre; thus the accuracy at best is about 2 per cent.

If we wish to determine the values of a waveform, such as the average and r.m.s. values of non-sinusoidal waveforms or the mark-to-space ratio of a pulse waveform, it is better to take a photograph of the trace. This is easily done as most oscilloscopes have camera attachments which take photographs of the type that are developed within a minute. Such photographs can be examined at leisure, whereas maintaining a trace for such a length of time on the cathode-ray tube could damage the screen, a point that has already been mentioned.

Typical traces for square waveforms and pulses are shown in Fig. 46.15.

Example 46.1

The trace displayed by a CRO is shown in Fig. 46.15(a). The signal amplitude control is set to 0.5 V/cm and the time-base control to 100 μs/cm. Determine the peak-to-peak voltage of the signal and its frequency.

Height of display is 4.6 cm. This is equivalent to $4.6 \times 0.5 = 2.3$ V. The peak-to-peak voltage is therefore **2.3 V**.

The width of the display of one cycle is 7.0 cm. This is equivalent to a period of $7.0 \times 100 \times 10^{-6} = 700 \times 10^{-6}$ s. It follows that the frequency is given by $1/(700 \times 10^{-6}) = $ **1430 Hz**.

Example 46.2

An oscilloscope has a display shown in Fig. 46.15(b). The signal amplitude control is set to 0.2 V/cm and the time-base control to 10 μs/cm. Determine the mark-to-space ratio of the pulse waveform and the pulse frequency. Also determine the magnitude of the pulse voltage.

The width of the pulse display is 0.8 cm and the width of the space between pulses is 3.2 cm. The mark-to-space ratio is therefore $0.8/3.2 = $ **0.25**.

The width of the display from the commencement of one pulse to the next is 4.0 cm. This is equivalent to

$$4.0 \times 10 \times 10^{-6} = 40 \times 10^{-6} \text{ s}$$

being the period of a pulse waveform. The pulse frequency is therefore given by

$$\frac{1}{40 \times 10^{-6}} = 25\ 000 \text{ Hz} = \mathbf{25\ kHz}$$

The magnitude of the pulse voltage is determined from the pulse height on the display, this being 4.2 cm. The pulse voltage is therefore $4.2 \times 0.2 = $ **0.84 V**.

At the start of this section, we had to assume that the signals applied to the oscilloscope were suitable. Now we must determine in what way a signal may be thought of as suitable for an oscilloscope.

Most oscilloscopes operate with the body or chassis of the instrument at earth potential. Also, most oscilloscopes are connected to the signal source by means of a coaxial cable, the outer conductor of which is connected to the body of the oscilloscope and is therefore at earth potential. It follows that one of the connections from the oscilloscope will connect one terminal of the signal source to earth.

The effect of this observation can be illustrated by considering the test arrangement shown in Fig. 46.16. A resistor and a capacitor are connected in series and supplied from a signal generator. It is usual that one terminal of the signal generator is also at earth potential; thus the connection diagram shown in Fig. 46.16 is suitable for the circuit and for the oscilloscope. It is suitable because the earth point of the oscilloscope is connected to the earth point of the generator and they are therefore at the same potential.

Fig. 46.16 An experiment involving an oscilloscope

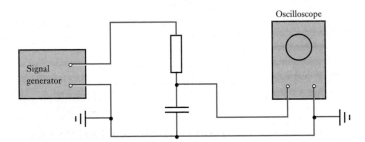

The oscilloscope displays the waveform of the voltage across the capacitor. However, what if we wished to display the waveform of the voltage across the resistor? We appear to have two choices: either we reconnect the oscilloscope across the resistor, or we reconnect the test circuit to the signal generator. Let us reconnect the oscilloscope as shown in Fig. 46.17.

Although the signal into the oscilloscope is now that of the voltage across the resistor, the test circuit has been seriously changed. The earth connection from the oscilloscope to the junction between the resistor and the capacitor causes the capacitor to be short-circuited, i.e. the current from the signal generator passes through the resistor, into the earth connection of

Fig. 46.17 Unsuitable reconnection of the oscilloscope

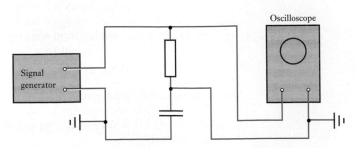

Fig. 46.18 Suitable
reconnection of the oscilloscope

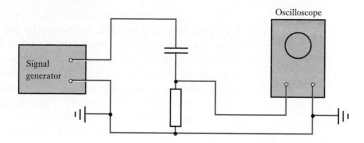

the oscilloscope and back to the generator through the earth connection of
the generator.

In this particular test, this problem can easily be overcome by interchanging the components of the test circuit as shown in Fig. 46.18. You
should also notice that the connection from the earth terminal of the
oscilloscope to the test circuit is not actually required, since it duplicates the
connection available through earth. As the connection also provides a screen
for the coaxial cable, thereby minimizing interference from other sources, it
is good practice to retain the second connection.

The reconnection of the test circuit was possible in this instance because
of the simplicity of the circuit. If the circuit had been more complicated,
such as that shown in Fig. 46.19, reconnection would not have been possible.
For instance, the amplifier transistor could not be reconnected in order to
observe the voltage across the base–collector junction.

In such cases, there are four possible methods whereby this form of
difficulty may be overome.

Fig. 46.19 Transistor amplifier
investigation by an oscilloscope

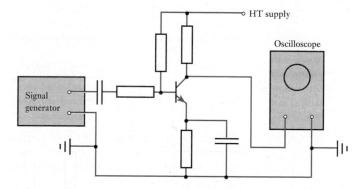

(a) Isolation of the source from earth

This is generally the most simple solution to achieve. If the source is energized from a battery, no connection is made to earth; hence the network
that is supplied can take up any potential it wishes, and the only connection
to earth is that of the oscilloscope. In such an instance, the battery source is
isolated from earth and there is no return path for current trying to leak away
from the network.

Many signal sources are energized from the mains supply, in which case they are connected to earth for safety by the third wire of the supply flex. They are also indirectly connected to earth by the neutral wire, since the neutral wire of any mains supply is connected to earth back at the supply substation. However, when such instruments are used in a laboratory, special supply arrangements may be provided whereby the frame of the source is not earthed and the 240 V supply is isolated from earth by means of an isolating transformer. In effect this has the same result as that of the source, which was battery operated, there being no return path to the source for any current trying to leak away from the network.

Such isolation procedures make for good safety practice in laboratories and workshops. If there is a return path through earth and two pieces of equipment in close proximity during testing are at different potentials, there is a risk of shock to a person touching both. The discontinuity of the earth connection minimizes such a risk in these operating conditions.

For the arrangement shown in Fig. 46.19, if the signal generator were not connected to earth, it would be quite possible to connect the oscilloscope across the base–collector junction of the transistor without affecting the operation of the circuit.

(b)   Isolation of the load from the source

This can be achieved by an isolating transformer, but the use of a transformer limits the applications to those of alternating current. An example of an experiment involving an isolating transformer is shown in Fig. 46.20. The transformer windings each have the same number of turns, thus the input and output of the transformer are essentially the same. However, the secondary winding and the load which it supplies are free of connection to earth; therefore the introduction of the earth-connected oscilloscope does not interfere with the operation of the network.

Fig. 46.20 Experiment involving the isolation of the load from the source

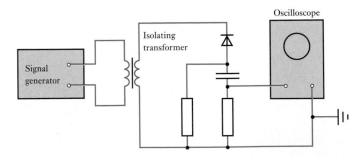

(c)   Double-beam oscilloscope with difference-of-signals facility

This facility is not available in most general-purpose oscilloscopes, although most oscilloscopes are double-beam instruments. A double-beam oscilloscope has two beams and therefore can display two traces at the one time. Thus, going back to the experiment shown in Fig. 46.16, one beam could have displayed the voltage across the capacitor while the other displayed the supply voltage. It is the difference of these two displays which gives the voltage across the resistor, and better oscilloscopes have this facility built into them.

It follows that for the test arrangement in Fig. 46.19, the base–collector voltage could be obtained by connecting one input to receive the collector-to-earth voltage and the other input to receive the base-to-earth voltage. The controls could then be set to show the difference of these signals, which would be the base–collector voltage.

(d) Isolation of the oscilloscope from earth

This arrangement is not generally favoured, partly because of the high voltages involved within an oscilloscope, but can be employed when none of the other methods is suitable. For instance, if we wish to observe the voltage across a component of a circuit which cannot be isolated readily from earth, e.g. part of a power circuit, and which is at a considerable potential to earth at both terminals, it is better to isolate the oscilloscope and to set it up prior to the application of power to the principal circuit. Under such conditions, it is not possible to adjust the oscilloscope during the test as it then takes up the potential of the power circuit. It is possible to take records of its display by camera, this being remotely controlled. This form of operation is an advanced part of engineering technology and should only be carried out under the supervision of an experienced electrical engineer.

Terms and concepts	The most common measuring instrument is based on the **electronic indicating instrument**.
	Electronic instruments make little demand on the circuit being measured and therefore are of relatively high accuracy.
	Most vacuum devices such as diodes are obsolete but one significant device remains – the **cathode-ray tube**. This is incorporated into oscilloscopes.
	Oscilloscopes provide visual displays of voltage and current waveforms.

Exercises 46

1. A four-digit voltmeter incorporates an A/D converter. Describe its operation when measuring a voltage of 148.5 V.

2. Describe how such a digital electronic voltmeter could be modified to indicate currents in the range 0 to 100 mA.

3. Describe the control of electrons in a vacuum triode by means of a grid.

4. Describe how the Y-plates of a cathode-ray oscilloscope would be connected to give a trace of: (a) the alternating voltage applied to a circuit component (such as a coil); (b) the current through the

component. What voltage waveform would normally be applied to the X-plates of the oscilloscope for this purpose?

5. Draw a block diagram showing the principal parts of a cathode-ray oscilloscope amplifier arrangement. What is the purpose of synchronization in an oscilloscope and why is it essential to the process of waveform display in a cathode-ray oscilloscope?

6. Explain with the aid of a circuit diagram showing the connections made to the cathode-ray oscilloscope, how it may be used to determine the r.m.s. value of an a.c. signal which is: (a) sinusoidal; (b) non-sinusoidal.

Exercises 46 continued

An oscilloscope is used to display a sinusoidal alternating voltage. The display shows a sine wave of amplitude 3.3 cm and the Y-amplifier sensitivity is set to 5 V/cm. Determine the r.m.s. value of the voltage.

7. Draw a block schematic diagram of a cathode-ray oscilloscope. Include in your diagram a switch for the selection of internal/external time base, and describe an application of the oscilloscope in which the external time base would be used.

   Explain the limitation of the Y-amplifier with regard to the accurate observation of rectangular pulses.

8. The display given by a cathode-ray oscilloscope is shown in Fig. A. Given that the sensitivity of the Y-amplifier is set to 10 V/cm and the time-base control to 5 ms/cm, determine: (a) the peak-to-peak value of the voltage; (b) the period of the signal; (c) the frequency.

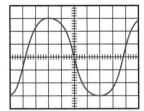

Fig. A

9. The display of a cathode-ray oscilloscope is given in Fig. B. The figure is produced by the application of two signals of equal amplitude to the X- and Y-plates. Determine the phase angle between the signals.

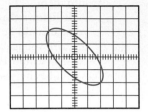

Fig. B

10. The display of a cathode-ray oscilloscope is shown in Fig. C. The signal applied to the X-amplifier is a sinusoidal signal of frequency 600 Hz and voltage 20 V peak-to-peak. The sensitivities of the X- and Y-amplifiers are set to the same values. Determine: (a) the frequency of the voltage applied to the Y-amplifier; (b) its peak-to-peak value; (c) the phase angle between the signals.

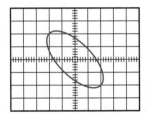

Fig. C

11. The display of a cathode-ray oscilloscope is shown in Fig. D. Describe the harmonic content of the signal.

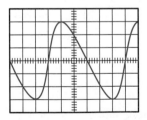

Fig. D

Chapter forty-seven

Analogue Measuring Instruments

Prior to the development of electronic instruments a wide range of analogue instruments was developed. They are generally robust, cheap and ergonomically easy to use, so most of us will use them at some time.

The most common analogue instrument is the permanent-magnet moving-coil meter. This will be considered and then extended for application either as an ammeter or as a voltmeter over a variety of ranges.

Some cumbersome measuring devices including the bridge and potentiometer are described. These involve some important principles, which may be of use in a range of areas.

Analogue instruments tend to have a greater effect than digital devices on the circuits in which they are used. Measurement errors and their associated effects are discussed.

By the end of the chapter, you will be familiar with the analogue instruments still found in regular use and be able to undertake exercises in respect of errors in measurements.

47.1   Introduction

Although most measuring instruments are now manufactured with digital electronic displays, most measuring devices still are based on analogue principles. This seems a most ambiguous statement and can be explained by considering the most common measuring device of all – the watch.

We use a watch to tell the time and most watches are extremely accurate; for example, a variation of a minute or two in a month is quite normal and is much more accurate than many of the instruments described in the last chapter or later in this chapter. Yet for all the accuracy of watches, it is strange that we can use watches with dials indicating only the quarter hours or even only the hour. Also, we do not require the time displayed in numbers but accept the position of indicators (the hands) as a suitable analogue.

It is relatively seldom that we seek to measure time with great accuracy. This can be achieved by using a stop-watch or one of a variety of electronic instruments which can measure minute fractions of a second.

Many indicating devices use the movement of an indicator across a scale to represent a quantity which we require. Car speedometers, rev counters, etc. are included in a list of such devices. In each application, we are not seeking a precise measurement but rather a reasonable indication of its value. For example, is the car engine running at 2500 or 2600 r/min? More importantly, are we speeding in a built-up area?

Among the uses of such analogue devices, we still find a variety of measuring and test instruments, i.e. devices intended to indicate precise values. They are to be found for a variety of reasons:

1. They are cheaper.
2. People still often prefer an analogue display rather than a digital one.
3. The additional cost of digital instruments can only be justified when accuracy is essential.

However, we will see in this chapter that electrical analogue instruments have to be used with care since they can distort the circuit in which they are being applied. Let us start by considering the basic features of the construction of analogue indicating instruments.

47.2   Electrical analogue indicating instruments

The moving system of an analogue instrument is usually carried by a spindle of hardened steel, having its ends tapered and highly polished to form pivots which rest in hollow-ground bearings, usually of sapphire, set in steel screws. In some instruments, the moving system is attached to two thin ribbons of spring material such as beryllium–copper alloy, held taut by tension springs mounted on the frame of the movement. This arrangement eliminates pivot friction and the instrument is less susceptible to damage by shock or vibration.

Indicating analogue instruments possess three essential features:

1. a *deflecting device* whereby a mechanical force is produced by the electric current, voltage or power
2. a *controlling device* whereby the value of the deflection is dependent upon the magnitude of the quantity being measured
3. a *damping device* to prevent oscillation of the moving system and enable the latter to reach its final position quickly.

The action of the deflecting device depends upon the type of instrument, and the principle of operation of each of the instruments most commonly used in practice will be described in later sections of this chapter.

47.3 Controlling devices

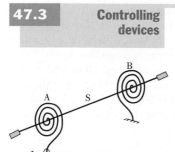

Fig. 47.1   Spring control

The most common arrangement of spring control utilizes two spiral hairsprings, A and B (Fig. 47.1), the inner ends of which are attached to the spindle S. The outer end of B is fixed, whereas that of A is attached to one end of a lever L, pivoted at P, thereby enabling zero adjustment to be easily effected. The hairsprings are of non-magnetic alloy such as phosphor–bronze or beryllium–copper.

The two springs, A and B, are wound in opposite directions so that when the moving system is deflected one spring winds up while the other unwinds, and the controlling torque is due to the combined torsions of the springs. Since the torsional torque of a spiral spring is proportional to the angle of twist, the controlling torque is directly proportional to the angular deflection of the pointer.

47.4 Damping devices

The combination of the inertia of the moving system and the controlling torque of the spiral springs or of gravity gives the moving system a natural frequency of oscillation. Consequently, if the current through an under-damped ammeter were increased suddenly from zero to OA (Fig. 47.2), the

Fig. 47.2   Damping curves

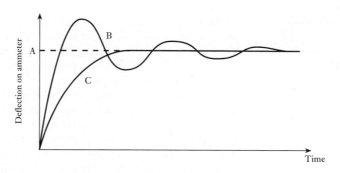

pointer would oscillate about its mean position, as shown by curve B, before coming to rest. Similarly, every fluctuation of current would cause the pointer to oscillate and it might be difficult to read the instrument accurately. It is therefore desirable to provide sufficient damping to enable the pointer to reach its steady position without oscillation, as indicated by curve C. Such an instrument is said to be *dead-beat*.

The method of damping commonly employed is eddy-current damping.

One form of eddy-current damping is shown in Fig. 47.3, where a copper or aluminium disc D, carried by a spindle, can move between the poles of a permanent magnet M. If the disc moves clockwise, the e.m.f.s induced in the disc circulate eddy currents as shown dotted. It follows from Lenz's law that

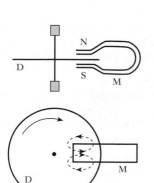

Fig. 47.3   Eddy-current damping

these currents exert a force opposing the motion producing them, namely the clockwise movement of the disc.

Another arrangement, used in moving-coil instruments (section 47.5), is to wind the coil on an aluminium frame. When the latter moves across the magnetic field, eddy current is induced in the frame; and, by Lenz's law, this current exerts a torque opposing the movement of the coil.

Apart from the electrostatic and electronic types of voltmeter, all voltmeters are in effect milliammeters connected in series with a non-reactive resistor having a high resistance. For instance, if a milliammeter has a full-scale deflection with 10 mA and has a resistance of 10 Ω and if this milliammeter is connected in series with a resistor of 9990 Ω, then the p.d. required for full-scale deflection is $0.01 \times 10\,000$, namely 100 V, and the scale of the milliammeter can be calibrated to give the p.d. directly in volts.

47.5   Permanent-magnet moving-coil ammeters and voltmeters

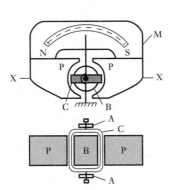

Fig. 47.4   Permanent-magnet moving-coil instrument

The high coercive force (section 44.7) of modern steel alloys, such as Alcomax (iron, aluminium, cobalt, nickel and copper), allows the use of relatively short magnets and has led to a variety of arrangements of the magnetic circuit for moving-coil instruments. The front elevation and sectional plan of one arrangement are shown in Fig. 47.4 where M represents a permanent magnet and PP are soft-iron pole pieces. The hardness of permanent-magnet materials makes machining difficult, whereas soft iron can be easily machined to give exact air-gap dimensions. In one form of construction, the anisotropic magnet and the pole pieces are of the sintered type, i.e. powdered magnet alloy and powdered soft iron are compressed in a die to the required shape and heat-treated so that the magnet and the pole pieces become alloyed, thereby eliminating air-gaps at the junctions of the materials.

An alternative arrangement is to cast the magnet and attach the soft-iron plate, in one piece, to the two surfaces of M which have been rendered flat by grinding, the joints being made by a resin-bonding technique. This construction enables the drilling of the cylindrical hole and the machining of the gaps to be done with precision.

The rectangular moving-coil C in Fig. 47.4 consists of insulated copper wire wound on a light aluminium frame fitted with polished steel pivots resting on jewel bearings. Current is led into and out of the coil by spiral hairsprings AA, which also provide the controlling torque. The coil is free to move in air-gaps between the soft-iron pole pieces PP and a soft-iron cylinder B supported by a brass plate (not shown). The functions of core B are:

1. to intensify the magnetic field by reducing the length of air-gap across which the magnetic flux has to pass;
2. to give a radial magnetic flux of uniform density, thereby enabling the scale to be uniformly divided.

The manner in which a torque is produced when the coil is carrying a current may be understood more easily by considering a single turn PQ, as in Fig. 47.5. Suppose P to carry current outwards from the paper; then Q is carrying current towards the paper. Current in P tends to set up a magnetic field in a counterclockwise direction around P and thus strengthens the magnetic field on the lower side and weakens it on the upper side. The

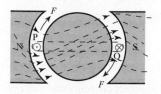

Fig. 47.5 Distribution of resultant magnetic flux

current in Q, on the other hand, strengthens the field on the upper side while weakening it on the lower side. Hence, the effect is to distort the magnetic flux as shown in Fig. 47.5. Since the flux behaves as if it was in tension and therefore tries to take the shortest path between poles NS, it exerts forces *FF* on coil PQ, tending to move it out of the magnetic field.

The deflecting torque varies as

Current through coil × flux density in gap

$$= kI \text{ for uniform flux density}$$

where k = a constant for a given instrument and I = current through coil. The controlling torque of the spiral springs

∝ angular deflection

$$= c\theta$$

where c = a constant for given springs and θ = angular deflection. For a steady deflection

Controlling torque = deflecting torque

hence $c\theta = kI$

∴ $$\theta = \frac{k}{c}I$$

i.e. the deflection is proportional to the current and the scale is therefore uniformly divided.

A numerical example on the calculation of the torque on a moving coil is given in Example 47.2.

As already mentioned in section 47.4 damping is effected by eddy currents induced in the metal frame on which the coil is wound.

Owing to the delicate nature of the moving system, this type of instrument is only suitable for measuring currents up to about 50 mA directly. When a larger current has to be measured, a *shunt* S (Fig. 47.6), having a low resistance, is connected in parallel with the moving coil MC, and the instrument scale may be calibrated to read directly the total current *I*. Shunts are made of a material such as manganin (copper, manganese and nickel) having negligible temperature coefficient of resistance. A 'swamping' resistor *r*, of material having negligible temperature coefficient of resistance, is connected in series with the moving coil. The latter is wound with copper wire and the function of *r* is to reduce the error due to the variation of resistance of the moving coil with variation of temperature. The resistance of *r* is usually about three times that of the coil, thereby reducing a possible error of, say, 4 per cent to about 1 per cent.

The shunt shown in Fig. 47.6 is provided with four terminals, the milli-ammeter being connected across the potential terminals. If the instrument were connected across the current terminals, there might be considerable error due to the contact resistance at these terminals being appreciable compared with the resistance of the shunt.

The moving-coil instrument can be made into a voltmeter by connecting a resistor *R* of manganin or other similar material in series, as in Fig. 47.7. The scale may be calibrated to read directly the voltage applied to the terminals TT.

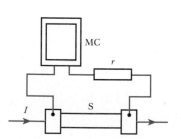

Fig. 47.6 Moving-coil instrument as an ammeter

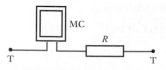

Fig. 47.7 Moving-coil instrument as a voltmeter

The main advantages of the moving-coil instrument are:

1. high sensitivity
2. uniform scale
3. well shielded from any stray magnetic field.

Its main disadvantage is that it is only suitable for direct currents and voltages.

Example 47.1   A moving-coil instrument gives full-scale deflection with 15 mA and has a resistance of 5 Ω. Calculate the resistance required:

(a) in parallel to enable the instrument to read up to 1 A;
(b) in series to enable it to read up to 10 V.

(a) Current through coil (Fig. 47.6) is

$$\frac{\text{p.d. across coil}}{\text{resistance of coil}}$$

$$\therefore \qquad \frac{15}{1000} = \frac{\text{p.d. (in volts) across coil}}{5}$$

so that   p.d. across coil = 0.075 V

For Fig. 47.6

$$\text{Current through S} = \text{total current} - \text{current through coil}$$

$$= 1 - 0.015 = 0.985 \text{ A}$$

$$\text{Current through S} = \frac{\text{p.d. across S}}{\text{resistance of S}}$$

$$\therefore \qquad 0.985 = \frac{0.075}{\text{resistance of S (in ohms)}}$$

and $$\text{Resistance of S} = \frac{0.075}{0.985} = \textbf{0.076 16 Ω}$$

(b) For Fig. 47.7

$$\text{Current through coil} = \frac{\text{p.d. across TT}}{\text{resistance between TT}}$$

$$\frac{15}{1000} = \frac{10}{\text{resistance between TT}}$$

so that

$$\text{Resistance between TT} = 666.7 \text{ Ω}$$

Hence, resistance of resistor R required in series with coil is

$$\text{Total resistance between TT} - \text{resistance of coil}$$

$$= 666.7 - 5 = \textbf{661.7 Ω}$$

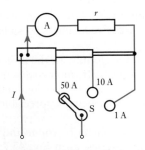

Fig. 47.8 Multi-range moving-coil ammeter

The moving-coil instrument can be arranged as a multi-range ammeter by making the shunt of different sections as shown in Fig. 47.8, where A represents a milliammeter in series with a 'swamping' resistor r of material having negligible temperature coefficient of resistance.

With the selector switch S on, say, the 50 A stud, a shunt having a very low resistance is connected across the instrument, the value of its resistance being such that full-scale deflection is produced when $I = 50$ A. With S on the 10 A stud, the resistance of the two sections of the shunt is approximately five times that of the 50 A section, and full-scale deflection is obtained when $I = 10$ A. Similarly, with S on the 1 A stud, the total resistance of the three sections is such that full-scale deflection is obtained with $I = 1$ A. Such a multi-range instrument is provided with three scales so that the value of the current can be read directly.

A multi-range voltmeter is easily arranged by using a tapped resistor in series with a milliammeter A, as shown in Fig. 47.9. For instance, with the data given in Example 47.1, the resistance of section BC would be 661.7 Ω for the 10 V range. If D is the tapping for, say, 100 V, the total resistance between O and D = 100/0.015 = 6666.7 Ω, so that the resistance of section CD = 6666.7 − 666.7 = 6000 Ω.

Similarly, if E is to be 500 V tapping, section DE must absorb 400 V at full-scale deflection; hence the resistance of DE = 400/0.015 = 26 667 Ω. With the aid of selector switch S, the instrument can be used on three voltage ranges, and the scales can be calibrated to enable the value of the voltage to be read directly.

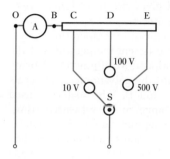

Fig. 47.9 Multi-range moving-coil voltmeter

Example 47.2

The coil of a moving-coil instrument is wound with $42\frac{1}{2}$ turns. The mean width of the coil is 25 mm and the axial length of the magnetic field is 20 mm. If the flux density in the air-gap is 0.2 T, calculate the torque, in newton metres, when the current is 15 mA.

Since the coil has $42\frac{1}{2}$ turns, one side has 42 wires and the other side has 43 wires.

From expression [6.1], force on the side having 42 wires is

$$0.2\,[\text{T}] \times 0.02\,[\text{m}] \times 0.015\,[\text{A}] \times 42$$

$$= 2520 \times 10^{-6}\,\text{N}$$

$\therefore$ Torque on that side of coil

$$= (2520 \times 10^{-6})\,[\text{N}] \times 0.0125\,[\text{m}]$$

$$= 31.5 \times 10^{-6}\,\text{N m}$$

Similarly, torque on side of coil having 43 wires is

$$31.5 \times 10^{-6} \times \frac{43}{42}$$

$$= 32.2 \times 10^{-6}\,\text{N m}$$

$\therefore$ Total torque on coil = $(31.5 + 32.2) \times 10^{-6}\,\text{N m}$

$$= \mathbf{63.7 \times 10^{-6}\,N\,m}$$

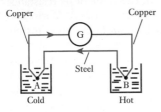

Fig. 47.10 A thermocouple

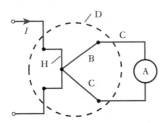

Fig. 47.11 A thermocouple ammeter

This type of instrument utilizes the thermoelectric effect observed by Thomas Seebeck (1770–1831) in 1821, namely that in a closed circuit consisting of two different metals, an electric current flows when the two junctions are at different temperatures. Thus, if A and B in Fig. 47.10 are junctions of copper and steel wires, each immersed in water, then if the vessel containing B is heated, it is found that an electric current flows from the steel to the copper at the cold junction and from the copper to the steel at the hot junction, as indicated by the arrowheads. A pair of metals arranged in this manner is termed a *thermocouple* and gives rise to a *thermo-e.m.f.*, when the two junctions are at different temperatures.

This *thermoelectric* effect may be utilized to measure temperature. Thus, if the reading on galvanometer G is noted for different temperatures of the water in which junction B is immersed, the temperature of junction A being maintained constant, it is possible to calibrate the galvanometer in terms of the difference of temperature between A and B. The materials used in practice depend upon the temperature range to be measured; thus, copper–constantan couples are suitable for temperatures up to about 400 °C and steel–constantan couples up to about 900 °C, constantan being an alloy of copper and nickel. For temperatures up to about 1400 °C, a couple made of platinum and platinum–iridium alloy is suitable.

A thermocouple can be used to measure the r.m.s. value of an alternating current by arranging for one of the junctions of wires of dissimilar material, B and C (Fig. 47.11), to be placed near or welded to a resistor H carrying the current *I* to be measured. The current due to the thermo-e.m.f. is measured by a permanent-magnet moving-coil microammeter A. The heater and the thermocouple can be enclosed in an evacuated glass bulb D, shown dotted in Fig. 47.11, to shield them from draughts. Ammeter A may be calibrated by noting its reading for various values of direct current through H, and it can then be used to measure the r.m.s. value of alternating currents of frequencies up to several megahertz.

Thermocouple instruments are good for recording effective values, e.g. the current in a cable. They ignore instantaneous or short-term values and concentrate on the underlying continuous r.m.s. value.

The action of this type of instrument depends upon the electromagnetic force exerted between fixed and moving coils carrying current. The upper diagram in Fig. 47.12 shows a sectional elevation through fixed coils FF and the lower diagram represents a sectional plan on XX. The moving coil M is carried by a spindle S and the controlling torque is exerted by spiral hairsprings H, which also serve to lead the current into and out of M.

The deflecting torque is due to the interaction of the magnetic fields produced by currents in the fixed and moving coils; thus Fig. 47.13(a) shows the magnetic field due to current flowing through F in the direction indicated by the dots and crosses, and Fig. 47.13(b) shows that due to current in M. By combining these magnetic fields, it will be seen that when currents flow simultaneously through F and M, the resultant magnetic field is distorted as shown in Fig. 47.13(c) and the effect is to exert a clockwise torque on M.

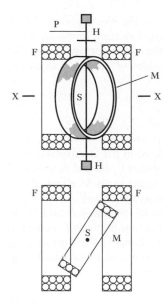

Fig. 47.12 Electrodynamic or dynamometer instrument

Fig. 47.13 Magnetic fields due to fixed and moving coils

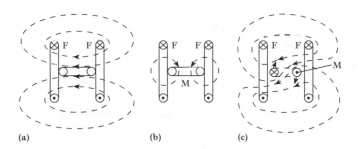

(a)   (b)   (c)

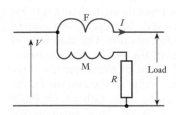

Fig. 47.14 Wattmeter connections

Since M is carrying current at right angles to the magnetic field produced by F

Deflecting force on each side of M

$\propto$ (current in M)

$\times$ (density of magnetic field due to current in F)

$\propto$ current in M $\times$ current in F

In dynamometer ammeters, the fixed and moving coils are connected in parallel, whereas in voltmeters they are in series with each other and with the usual resistor. In each case, the deflecting force is proportional to the square of the current or the voltage. Hence, when the dynamometer instrument is used to measure an alternating current or voltage, the moving coil – due to its inertia – takes up a position where the average deflecting torque over one cycle is balanced by the restoring torque of the spiral springs. For that position, the deflecting torque is proportional to the mean value of the square of the current or voltage, and the instrument scale can therefore be calibrated to read the r.m.s. value.

Owing to the higher cost and low sensitivity of dynamometer ammeters and voltmeters, they are seldom used commercially, but *electrodynamic* or *dynamometer wattmeters* are very important because they are commonly employed for measuring the power in a.c. circuits. The fixed coils F are con-

nected in series with the load, as shown in Fig. 47.14. The moving coil M is connected in series with a non-reactive resistor R across the supply, so that the current through M is proportional to and practically in phase with the supply voltage V; hence:

Instantaneous force on each side of M

$\propto$ (instantaneous current through F)

$\times$ (instantaneous current through M)

$\propto$ (instantaneous current through load)

$\times$ (instantaneous p.d. across load)

$\propto$ instantaneous power taken by load

$\therefore$ Average deflecting force on M

$\propto$ average value of the power over a complete number of cycles

When the instrument is used in an a.c. circuit, the moving coil – due to its inertia – takes up a position where the average deflecting torque over one cycle is balanced by the restoring torque of the spiral springs; hence the instrument can be calibrated to read the mean value of the power in an a.c. circuit.

47.8 Electrostatic voltmeters

In Chapter 5 we were introduced to the mutual attraction between positive and negative charges. This phenomenon is utilized in the electrostatic voltmeter. This instrument consists of fixed metal plates F, shaped as indicated in the lower part of Fig. 47.15, and very light metal vanes M attached to a spindle controlled by spiral springs S and carrying a pointer P.

The voltage to be measured is applied across terminals A and B. In section 5.20 it was shown that the force of attraction between F and M is proportional to the square of the applied voltage; hence this instrument can be used to measure either direct or alternating voltage, and when used in an a.c. circuit it reads the r.m.s. value.

The main advantages of the electrostatic voltmeter are:

1. It takes no current from a d.c. circuit (apart from the small initial charging current) and the current taken from an a.c. circuit is usually negligible. Hence, it can be used to measure the p.d. between points in a circuit where the current taken by other types of voltmeter might considerably modify the value of that p.d.
2. It is particularly suitable for measuring high voltages, since the electrostatic forces are then so large that its construction can be greatly simplified.

The most significant application is the measurement of static voltage induced in metal objects near to electrical equipment, e.g. fence wires near overhead lines. As an example, a static voltage of 18 kV was experienced in an umbrella on a station platform of an electrified railway: there was, however, no danger due to the charge being very small.

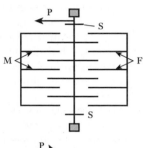

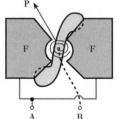

Fig. 47.15 Electrostatic voltmeter

47.9 Rectifier ammeters and voltmeters

In this type of instrument a rectifier such as a copper-oxide rectifier is used to convert the alternating current into a unidirectional current, the mean value of which is measured on a permanent-magnet moving-coil instrument.

Rectifier ammeters usually consist of four rectifier elements arranged in the form of a bridge, as shown in Fig. 47.16 where the apex of the black triangle indicates the direction in which the resistance is low, and A represents a moving-coil ammeter. During the half-cycles that the current is flowing from left to right in Fig. 47.16 current flows through elements B and D, as shown by the full arrows. During the other half-cycles, the current flows through C and E, as shown by the dashed arrows. The waveform of the current through A is therefore as shown in Fig. 47.17. Consequently, the deflection of A depends upon the average value of the current, and the scale of A can be calibrated to read the r.m.s. value of the current on the assumption that the waveform of the latter is sinusoidal with a form factor of 1.11.

In a rectifier voltmeter, A is a milliammeter and the bridge circuit of Fig. 47.16 is connected in series with a suitable non-reactive resistor.

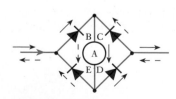

Fig. 47.16 Bridge circuit for full-wave rectification

Fig. 47.17   Waveform of current through moving-coil ammeter

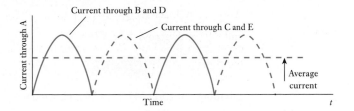

The main advantage of the rectifier voltmeter is that it is far more sensitive than other types of voltmeter suitable for measuring alternating voltages. Also, rectifiers can be incorporated in universal instruments, such as the Avometer, thereby enabling a moving-coil milliammeter to be used in combination with shunt and series resistors to measure various ranges of direct current and voltage, and in combination with a bridge rectifier and suitable resistors to measure various ranges of alternating current and voltage.

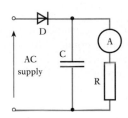

Fig. 47.18   A simple form of electronic voltmeter

If a diode or a solid-state rectifier D is connected in series with a capacitor C across an a.c. supply, as shown in Fig. 47.18, the capacitor is charged until its p.d. is practically equal to the peak value of the alternating voltage. The value of this voltage can be measured by connecting across the capacitor a microammeter A in series with a resistor R having a high resistance.

The capacitor C is charged during the very small fraction of each cycle that the applied voltage exceeds the p.d. across C. During the remainder of each cycle, the capacitor supplies the current flowing through A and R. The microammeter can be calibrated to read either the peak or the r.m.s. value of the voltage. In the latter case, it has to be assumed that the voltage is sinusoidal.

47.10   Measurement of resistance by the Wheatstone bridge

The four branches of the network CDFEC, in Fig. 47.19, have two known resistances P and Q, a known variable resistance R and the unknown resistance X. A battery B is connected through a switch S_1 to junctions C and F; and a galvanometer G, a variable resistor A and a switch S_2 are in series across D and E. The function of A is merely to protect G against an excessive current should the system be seriously out of balance when S_2 is closed.

With S_1 and S_2 closed, R is adjusted until there is no deflection on G even with the resistance of A reduced to zero. Junctions D and E are then at the same potential, so that the p.d. between C and D is the same as that between C and E, and the p.d. between D and F is the same as that between E and F.

Suppose I_1 and I_2 to be the currents through P and R respectively when the bridge is balanced. From Kirchhoff's first law it follows that since there is no current through G, the currents through Q and X are also I_1 and I_2 respectively

But   p.d. across $P = PI_1$

and   p.d. across $R = RI_2$

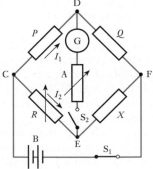

Fig. 47.19   Wheatstone bridge

$\therefore$ $$PI_1 = RI_2$$ [47.1]

Also   p.d. across $Q = QI_1$

and   p.d. across $X = XI_2$

$$\therefore \qquad QI_1 = XI_2 \qquad\qquad\qquad\qquad [47.2]$$

Dividing equation [47.2] by [47.1], we have

$$\frac{Q}{P} = \frac{X}{R}$$

and $$X = R \times \frac{Q}{P} \qquad\qquad\qquad\qquad [47.3]$$

The resistances P and Q may take the form of the resistance of a slide-wire, in which case R may be a fixed value and balance obtained by moving a sliding contact along the wire. If the wire is homogeneous and of uniform section, the ratio of P to Q is the same as the ratio of the lengths of wire in the respective arms. A more convenient method, however, is to arrange P and Q so that each may be 10, 100 or 1000 Ω. For instance, if $P = 1000\ \Omega$ and $Q = 10\ \Omega$, and if R has to be 476 Ω to give a balance, then from equation [47.3]:

$$X = 476 \times \frac{10}{1000} = 4.76\ \Omega$$

On the other hand, if P and Q had been 10 and 1000 Ω respectively, then for the same value of R,

$$X = 476 \times \frac{1000}{10} = 47\ 600\ \Omega$$

Hence it is seen that with this arrangement it is possible to measure a wide range of resistance with considerable accuracy and to derive very easily and accurately the value of the resistance from that of R.

47.11   The potentiometer

One of the most useful instruments for the accurate measurement of p.d., current and resistance is the potentiometer, the principle of action being that an unknown e.m.f. or p.d. is measured by balancing it, wholly or in part, against a known difference of potential.

In its simplest form, the potentiometer consists of a wire MN (Fig. 47.20) of uniform cross-section, stretched alongside a scale and connected across a secondary cell B of ample capacity. A standard cell SC of known e.m.f. E_1, for example a cadmium cell having an e.m.f. of 1.018 59 V at 20 °C, is connected between M and terminal a of a two-way switch S, care being taken that the corresponding terminals of B and SC are connected to M.

Slider L is then pressed momentarily against wire MN and its position adjusted until the galvanometer deflection is zero when L is making contact with MN. Let l_1 be the corresponding distance between M and L. The fall of potential over length l_1 of the wire is then the same as the e.m.f. E_1 of the standard cell.

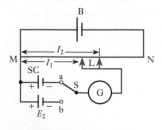

Fig. 47.20   A simple potentiometer

Switch S is then moved over to terminal b, thereby replacing the standard cell by another cell, such as a Leclanché cell, the e.m.f. E_2 of which is to be measured. Slider L is again adjusted to give zero deflection on G. If l_2 is the new distance between M and L, then

$$\frac{E_1}{E_2} = \frac{l_1}{l_2}$$

$$\therefore \qquad E_2 = E_1 \times l_2/l_1 \qquad\qquad\qquad\qquad [47.4]$$

47.12 A commercial form of potentiometer

The simple arrangement described in the preceding section has two disadvantages:

1. The arithmetical calculation involved in expression [47.4] can introduce an error, and in any case takes an appreciable time.
2. The accuracy is limited by the length of slide-wire that is practicable and by the difficulty of ensuring exact uniformity over a considerable length.

In the commercial type of potentiometer shown in Fig. 47.21, these disadvantages are practically eliminated. Here R consists of 14 resistors in series, the resistance of each resistor being equal to that of the slide-wire S. The value of the current supplied by a lead–acid cell B is controlled by a slide-wire resistor W. A double-pole change-over switch T, closed on the upper side as in Fig. 47.21, connects the standard cell SC between arm P and galvanometer G. A special key K, when slightly depressed, inserts a resistor F having a high resistance in series with G; but when K is further depressed, F is short-circuited. The galvanometer is thereby protected against an excessive current should the potentiometer be appreciably out of balance when K is first depressed.

Fig. 47.21 A commercial potentiometer

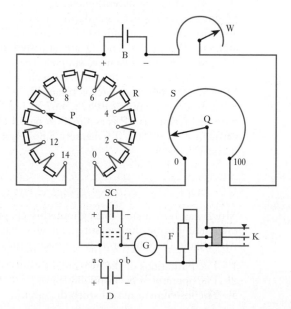

47.13	Standardization of the potentiometer

Suppose the standard cell SC to be of the cadmium type having an e.m.f. of 1.018 59 V at 20 °C. Arm P is placed on stud 10 and Q on 18.59 – assuming the scale alongside S to have 100 divisions. The value of W is then adjusted for zero deflection on G when K is fully depressed. The p.d. between P and Q is then exactly 1.018 59 V, so that the p.d. between two adjacent studs of R is 0.1 V and that corresponding to each division of S's scale is 0.001 V. Consequently, if P is moved to, say, stud 4 and Q to 78.4 on the slide-wire scale, the p.d. between P and Q is

$$(4 \times 0.1) + (78.4 \times 0.001) = 0.4784 \text{ V}$$

It is therefore a simple matter to read the p.d. directly off the potentiometer.

Since most potentiometers have 14 steps on R, it is usually not possible to measure directly a p.d. exceeding 1.5 V.

47.14	Calibration of an ammeter by means of a potentiometer

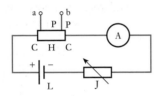

Fig. 47.22 Calibration of an ammeter

The ammeter A to be calibrated is connected in series with a standard resistor H and a variable resistor J across a cell L of ample current capacity, as in Fig. 47.22. The standard resistor H is usually provided with four terminals, namely two heavy current terminals CC and two potential terminals PP. The resistance between the potential terminals is known with a high degree of accuracy and its value must be such that with the maximum current through the ammeter, the p.d. between terminals PP does not exceed 1.5 V. For instance, suppose A to be a 10 A ammeter; then the resistance of H must not exceed 1.5/10, namely 0.15 Ω. Further, the resistance of H should preferably be a round figure, such as 0.1 Ω in this case, in order that the current may be quickly and accurately deduced from the potentiometer readings.

Terminals PP of the standard resistor are connected to terminals ab (Fig. 47.21) of the potentiometer (cell D having been removed). After the potentiometer has been standardized, switch T is changed over to terminals ab; with the current adjusted to give a desired reading on the scale of ammeter A, arms P and Q are adjusted to give zero deflection on the galvanometer. For instance, suppose the current to be adjusted to give a reading of, say, 6 A on the ammeter scale, and suppose the readings on P and Q, when the potentiometer is balanced, to be 5 and 86.7 respectively, then the p.d. across terminals PP is 0.5867 V, and since the resistance between the potential terminals PP is assumed to be 0.1 Ω, the true value of the current through H is 0.5867/0.1, namely 5.867 A. Hence, the ammeter is reading high by 0.133 A.

47.15	Calibration accuracy and errors

Much has been made throughout this chapter concerning the errors that may arise during the measurement of voltage, current and resistance. But why should there be error? Error occurs for three reasons:

1. The limitations of the instrument used.
2. The operator is never infallible.
3. The instrument may disturb the circuit.

1. First there are the limitations of the instrument. These may arise from incorrect calibration of the instrument, which does not necessarily mean that the scale indications have been put in the wrong place. It could be that the meter has changed its deflection with age, for instance the springs may not be as stiff as when they were made. The friction of the bearings may have changed with time. So in a number of ways it is quite possible that the meter may not indicate exactly the quantity that it is supposed to measure. Because of this, every meter is permitted a margin of error which is stated as a percentage of the indication.

This is unusual because it is the general case in engineering that you state the desired measurement and the error permitted about that quantity. Thus you may wish to have a shaft made to the diameter of 100 mm with a tolerance of 1 mm, i.e. the error of manufacture is ±1 per cent about the desired diameter.

In meter measurement, however, it is usual to calculate errors on the incorrect basis of the indicated quantity and not the actual quantity. This procedure has no undue effect provided the error is less than about 2 per cent, and makes the method of calculation very much easier. However, this method of error calculation must not be used when the error exceeds 3 per cent.

The reason that this change of basis is employed is that it would be difficult to set a supply to, say, 120 V and then read accurately the indicated voltage from the meter whereby the error could be ascertained. Instead we set the supply so that the meter indicates 120 V and then determine by potentiometer or similar means the correct supply voltage. Let this correct voltage be, say, 120.84 V; thus the difference between the indicated voltage and the correct voltage is 0.84 V, which is $0.84 \times 100/120 = 0.7$ per cent. However, the meter gave an indication that was lower than the correct voltage so the error is stated as −0.7 per cent.

Now let the supply voltage be varied so that the meter indicates 60 V. This time, we may find that the correct voltage as measured by the potentiometer is 59.94 V which means that the meter has overestimated the voltage by 0.06 V, which relative to 60 V means that the error is +0.1 per cent.

Thus we have the following rules:

(a) If the instrument error is positive then the indicated quantity is higher than the true quantity.
(b) If the instrument error is negative then the indicated quantity is lower than the true quantity.

It would be much more difficult if we were to calculate the errors relative to the correct quantities and the difference would scarcely be noticeable when the errors are so small.

Apart from the causes described, there are other sources of instrument error. Knowing that a range of error is acceptable (and inevitable) it follows that the manufacturer need not seek perfection during construction and can therefore use components which have a range of value, e.g. a multiplier for a voltmeter may acceptably have a resistance of, say, ±1 per cent about its stated value. Also the meter scales can be produced in number and therefore are not exactly matched to the individual instrument. However, even with such sources of possible error, it is relatively simple to make instruments of ±2.5 per cent error and even these tend to be much better than the range

Fig. 47.23   Variation of error
with deflection

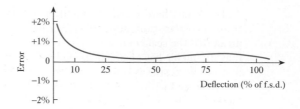

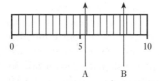

Fig. 47.24   Scale indications

suggests. A typical range of error is shown in Fig. 47.23 and it will be seen that the error changes over the range of indication.

2.   However, no matter how well a meter performs, it remains only as good as the operator who can be careless about reading the scale and thereby cause an error, or there may be a limit as to how accurately the meter may be read to the nearest division with certainty although usually we try to divide up a division into 10 parts so that we may give a reading including a fraction of a division. This leads to an unexpected error; let us look at the scale shown in Fig. 47.24.

In position A, the pointer is indicating 5.4 which is what many people would take it to be. However, without being careless, some will read it as 5.5 while others will see it as 5.3. This is a defect in their judgement and for this reason, no one should place too much reliance on such estimated figures.

Position B illustrates another common reading error. The pointer indicates approximately half-way between 8.2 and 8.3. This leads to some reading the indication as being 8.2 while others take it as 8.3. It is the human desire to be helpful that causes the error, and not the meter, and we must be aware that this introduces such a source of error.

3.   Finally, there is the error due to circuit disturbance. Meters require a certain amount of power to cause operation. Provided this power is small relative to the power in the measured circuit, then little error will result. However, if the meter power is comparable to the power in the circuit a serious error will result. Before giving an example to illustrate this, another point to bear in mind is that any meter will, within the limitations of calibration, indicate the conditions at its terminals correctly. Thus a voltmeter will indicate, within the limitations of its normal accuracy, the terminal voltage and similarly an ammeter will indicate the current passing through it.

To illustrate these remarks consider the following example.

Example 47.3   A voltage of 100 V is applied to a circuit comprising two 50 kΩ resistors in series. A voltmeter, with an FSD of 50 V and a figure of merit 1 kΩ/V, is used to measure the voltage across one of the 50 kΩ resistors. Calculate:

(a) the voltage across the 50 kΩ resistor;
(b) the voltage measured by the voltmeter.

(a) Let V_1 be the voltage across a 50 kΩ resistor when the voltmeter is not in circuit; thus

$$V_1 = \frac{R}{2R} = \frac{50 \times 10^3}{100 \times 10^3} \times 100 = \mathbf{50\ V}$$

(b) Let R_V be the resistance of the voltmeter:

$$R_V = 50 \times 1000 = 50\,000\ \Omega = 50\ \text{k}\Omega$$

When the voltmeter is connected in circuit, it shunts the 50 kΩ resistor. If R_e is the resistance of the parallel networks, then for two parallel 50 kΩ resistances, $R_e = 25$ kΩ. The network is thus effectively changed into an equivalent resistance of 25 kΩ in series with a resistance of 50 kΩ. The voltage thus measured by the voltmeter is given by

$$\frac{R_e \cdot V}{R + R_e} = \frac{25 \times 100}{50 + 25} = 33.3\ \text{V}$$

In Example 47.3 clearly the voltage as indicated by the voltmeter is quite erroneous. The error has been caused by the effect of the voltmeter on the circuit. Because of the values chosen, the voltmeter takes the same current as the load whose voltage is being measured. The power taken by the voltmeter is equal to the power in the measured load.

Even if the resistance of the voltmeter had been 10 times as great, the error would still have been almost 2 per cent. It can therefore be seen that the meter can affect the circuit to which it is being applied, i.e. the circuit has been disturbed.

While the sources of error discussed are general, there are further sources of error that have been introduced, but which are specific to the practice of alternating current, such as change of waveform causing error in certain a.c. meters. Such problems cause further errors due to limitations of the instrument in coping with extreme conditions.

47.16 Determination of error due to instrument errors

When using instruments to measure electrical quantities, you cannot avoid the introduction of error. It follows, therefore, that you ought to develop an appreciation of the quality of the measurements you make on the measurements you determine. For instance, if you have found the resistance of a resistor to be 100 Ω, you then have to decide if this is the correct value or if it is a value with a possible error of, say, ±5 per cent. In the second case, you know that the resistance has a value between 95 Ω and 105 Ω, although probably it has a value near to 100 Ω.

Using other measurement techniques, we may determine the same resistance with a better (or a poorer) degree of accuracy. The accuracy you require depends on the application to which you intend to put the resulting information. In the case of a 1000 m underground cable which has developed a break in its insulation, you could determine the resistance or the capacitance from one end of the cable up to the point of breakdown. If this were done to an accuracy of 5 per cent, you would require to dig up 100 m of the cable to be sure of finding the fault! If your accuracy of determination were 0.5 per cent then the hole would shrink to 10 m long and if the accuracy were 0.1 per cent, the hole would only be 2 m long, which seems much more reasonable. In this case, therefore, we wish to ensure an accuracy of 0.1 per cent.

Ideally we would measure any electrical quantity directly by one instrument. For instance, we would wish to measure voltage directly by means of a voltmeter, an oscilloscope or a potentiometer. The voltmeter is the simplest to operate, so let us suppose that on a voltmeter operating with a full-scale deflection of 100 V we obtain an indication of 82.6 V. For the particular model of voltmeter, we note that the limit of error over the effective range expressed as a percentage of the scale range is 2.0 per cent; thus the greatest error on any indication is 2.0 per cent of 100 V which is 2.0 V. We may therefore conclude that for the voltmeter operating under the given conditions, the voltage across its terminals is 82.6 ± 2.0 V. It follows that the error may be expressed as ± 2.4 per cent of the indicated value.

As the indication relative to the full-scale deflection becomes lower, the specified limit of error becomes relatively more important; thus for the given voltmeter, the limit of error is still 2.0 V when the voltmeter indicates 50.0 V, in which case the percentage error limit is ± 4.0 per cent of the indicated voltage. By the time the voltage indicated falls to 20.0 V, the percentage error limit rises to 10.0 per cent.

From these observations, we appreciate that a meter of apparently reasonable accuracy can in fact give indications with relatively large errors when operating under conditions which do not demand deflection of the meter approaching full-scale deflection. It is for this reason that you should always choose a meter scale that gives the greatest possible deflection within the scale of the meter.

By comparison, circuit components can be manufactured to have tolerances which remain the same at all settings. For instance, in a bridge network, we use decade resistors. If the decade resistor is manufactured to have a percentage error limit of, say, 1.0 per cent, then this is the percentage error limit at all settings.

Many experimental techniques involve the combination of two or more observations. The resistance of a circuit component is the ratio of the voltage across it to the current passing through it, and these quantities could be measured by a voltmeter and an ammeter. Each meter has an error, and this raises the question of the effect of combining the two errors. Similarly, a bridge determines the resistance, say, by comparing it with three other resistance values, each of which has an error. Again there is the question of the effect of combining the possible errors in each of the three values. Suppose that

$$X = \frac{AB}{C}$$

then $\ln X = \ln A + \ln B - \ln C$

and $\dfrac{\mathrm{d}X}{X} = \dfrac{\mathrm{d}A}{A} + \dfrac{\mathrm{d}B}{B} - \dfrac{\mathrm{d}C}{C}$

Each of the small changes can be positive or negative; thus the right-hand side of the relation is greatest when all the small terms produce errors of the same kind. Thus the maximum possible error occurs when

$$\frac{\mathrm{d}X}{X} = \pm\left(\frac{\mathrm{d}A}{A} + \frac{\mathrm{d}B}{B} + \frac{\mathrm{d}C}{C} \right) \qquad [47.5]$$

This curious looking expression becomes more readily understood when we apply it to an example, as in Example 47.4.

Example 47.4

The current in a circuit is measured as 235 μA and the accuracy of measurement is ±0.5 per cent. The current passes through a resistor of resistance 35 kΩ ± 0.2 per cent. Estimate the voltage across the resistor.

Estimate of voltages:

$$V = IR = 235 \times 10^{-6} \times 35 \times 10^{3} = 8.225 \text{ V}$$

Estimate of errors from relation [47.5]: the maximum relative error is given by the sum of the relative errors; thus the maximum relative error = 0.5 + 0.2 = 0.7 per cent.

The voltage may therefore be expressed as 8.225 V ± 0.7 per cent, but the basic value is given to a greater degree of accuracy than the error. The voltage should therefore be

8.23 V ± 0.7 per cent

It may be that we would wish to express the answer to this problem in volts, in which case we require to determine the maximum possible error. The error is ϵ and

$$\epsilon = \frac{0.7}{100} \times 8.225 = 0.058 \text{ V}$$

The voltage across the resistor may therefore be expressed as

8.225 ± 0.058 V

Again it would be consistent to round off the figures and express the voltage as

8.23 ± 0.06 V

In Example 47.4, the relative error values were already determined. If we return to the instance of the voltmeter in which the limit of error was referred to the scale range, the determination of the final result becomes more complicated.

Example 47.5

The voltage across a resistor is measured by a voltmeter which gives an indication of 75.5 V when operating on a scale of range 0–100 V. The current in the resistor is measured by an ammeter which gives an indication of 3.45 A when operating on a scale of range 0.5 A. Both instruments have a limit of error of 1.0 per cent of the scale range. Determine the resistance of the resistor and express the value both in error form and in percentage error form.

Estimate of resistance:

$$R = \frac{V}{I} = \frac{75.5}{3.45} = 21.88 \ \Omega$$

Estimate of error: error of voltage is 1.0 per cent of 100 V, which is 1.0 V. Relative to the indication, the voltage error is

$$\epsilon_V = \frac{1.0}{75.5} \times 100 = 1.32 \text{ per cent}$$

Error of current is 1.0 per cent of 5 A, which is 0.05 A. Relative to the indication, the current error is

$$\epsilon_I = \frac{0.05}{3.45} \times 100 = 1.45 \text{ per cent}$$

The maximum relative error is therefore given by

$$\epsilon = \epsilon_V + \epsilon_I = 1.32 + 1.45 = 2.77 \text{ per cent}$$

The error in the determination of the resistance can therefore be as great as

$$\frac{2.77}{100} \times 21.88 = 0.61 \ \Omega$$

The resistance of the resistor is therefore

$$21.9 \pm 0.6 \ \Omega$$

or   **21.9 $\Omega \pm 2.8$ per cent**

When attempting to obtain the best possible accuracy from a number of related measurements, it is usual to determine the error characteristics of the instruments. A typical error characteristic relates percentage error to indication as shown in Fig. 47.25. From such characteristics, we can relate the error appropriate to each indication and hence determine a corrected value. An instance of this operation is given in the following example.

Fig. 47.25   Error characteristic of an indicating instrument

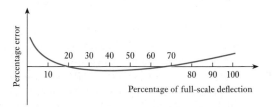

Example 47.6

A voltmeter indicates that the voltage applied to a resistor is 21.8 V and an ammeter indicates that the current in the resistor is 0.68 A. The error characteristics for the instruments are shown in Fig. 47.26. Determine the resistance of the resistor.

Fig. 47.26

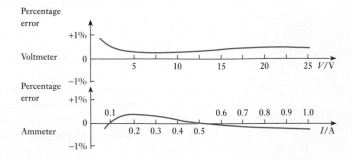

From the characteristics, the percentage error of the voltmeter is $+0.6$ when indicating 21.8 V and that of the ammeter -0.2 when indicating 0.68 A.

The estimated resistance is

$$R = \frac{V}{I} = \frac{21.8}{0.68} = 32.06 \ \Omega$$

The total error is given by

$$\epsilon = \epsilon_V - \epsilon_I = +0.6 - (-0.2) = +0.8$$

The resistance value is therefore 0.8 per cent high and the corrected value is therefore

$$32.06 \times \frac{100 - 0.8}{100} = \mathbf{31.80 \ \Omega}$$

To be consistent with the accuracy of the data supplied, the resistance should be given as 31.8 Ω.

One of the most common mistakes when dealing with instrument indications, and particularly when dealing with the results of calculations, is to attribute to them an accuracy that is not justified. For instance an instrument with an accuracy of 1.0 per cent should indicate to that accuracy and no more. Thus if it were a voltmeter of full-scale deflection, a reasonable indication might be given as 75.6 V, but certainly not 75.64 V. Often the extra decimal figure is given as an indication on the part of the observer that the reading was between 75.6 and 75.7 and he or she does not intend to make out that the indication was quite so accurate – but that is not what is stated by the figure quoted.

This false accuracy arises from a failure to appreciate that the number 2.5 is not the same as 2.50. Whereas 2.5 indicates a value between 2.45 and 2.55, 2.50 indicates a number between 2.495 and 2.505. While this may sometimes be appreciated in part, it is surprising how often an experimental table of results shows such figures as 1.65, 1.75, 1.9, 2, 2.15, 2.25, etc. At one extreme the figures are given to a maximum error of 0.005, yet at the other extreme the maximum error can rise to 0.5. This has possibly arisen in part from the observer being too lazy to write 2.0 when satisfied with a mere 2 – but he or she has conveyed completely different information as a result.

Another difficulty arises when a figure such as 234 000 is given, suggesting that the number is correct to a maximum error of 0.5, i.e. the actual value lies between 233 999.5 and 234 000.5. More likely the intention is that the third significant figure is the last reliable figure, in which case the value should have been given as 234×10^3.

Finally, the result of a calculation cannot produce a value of greater accuracy than the accuracy of the information used to formulate the calculation. For instance, if a current of 11 A passes through a 23 Ω resistor, then each figure has been given to an accuracy of two significant places. The voltage across the resistor should therefore be determined as 250 V and not 253 V, which suggests three-figure accuracy. With the advent of calculators, such false accuracy can run riot, so that we find such extreme mistakes as a voltage of 65 V associated with a current of 3.2 A giving rise to a determined value of resistance of 20.3125 Ω. Alas, accuracy cannot be achieved so painlessly, but instead it must be won with great care in the choice of instrumentation and its application.

Terms and concepts

Analogue instruments depend generally on a pointer making an indication by moving across a scale.

The moving mechanism requires a controlling device, usually two counterwound springs, and a clamping device, usually eddy-current damping.

Permanent-magnet moving-coil instruments are driven by direct current but by the use of rectifier bridges they provide the basis of most cheap indicating instruments.

The ranges of application are extended by introducing series resistors (**multipliers**) into voltmeters or parallel resistors (**shunts**) into ammeters.

Other measuring instruments include **electrodynamic meters** and **electrostatic voltmeters**.

Reference measurements are ones of high accuracy and can be derived from **Wheatstone (and other) bridges** and from **potentiometers**.

Errors in measurement occur due to the limitations of the instrument used, the fallibility of the operator and circuit disturbance.

Exercises 47

1. Sketch and describe the construction of a moving-coil ammeter and give the principle of operation.

   A moving-coil instrument gives full-scale deflection with 15 mA and has a resistance of 5 Ω. Calculate the resistance of the necessary components in order that the instrument may be used as: (a) a 2 A ammeter; (b) a 100 V voltmeter.

2. Why is spring control to be preferred to gravity control in an electrical measuring instrument?

   The coil of a moving-coil meter has a resistance of 5 Ω and gives full-scale deflection when a current of 15 mA passes through it. What modification must be made to the instrument to convert it into: (a) an ammeter reading to 15 A; (b) a voltmeter reading to 15 V?

3. If the shunt for Q. 2(a) is to be made of manganin strip having a resistivity of 0.5 $\mu\Omega$ m, a thickness of 0.6 mm and a length of 50 mm, calculate the width of the strip.

4. Draw a diagram to show the essential parts of a modern moving-coil instrument. Label each part and state its function.

   A moving-coil milliammeter has a coil of resistance 15 Ω and full-scale deflection is given by a current of 5 mA. This instrument is to be adapted to operate: (a) as a voltmeter with a full-scale deflection of 100 V; (b) as an ammeter with a full-scale deflection of 2 A. Sketch the circuit in each case, calculate the value of any components introduced and state any precautions regarding these components. (c) Explain how the moving-coil instrument can be adapted to read alternating voltage or current.

5. (a) A moving-coil galvanometer, of resistance 5 Ω, gives a full-scale reading when a current of 15 mA passes through the instrument. Explain, with the aid of circuit diagrams, how its range could be altered so as to read up to: (i) 5 A; (ii) 150 V. Calculate the values of the resistors required.

   (b) A uniform potentiometer wire, AB, is 4 m long and has resistance 8 Ω. End A is connected to the negative terminal of a 2 V cell of negligible internal resistance, and end B is connected to the positive terminal. An ammeter of resistance 5 Ω has its negative terminal connected to A and its positive terminal to a point on the wire 3 m from A. What current will the ammeter indicate?

6. A moving-coil instrument, which gives full-scale deflection with 15 mA, has a copper coil having a resistance of 1.5 Ω at 15 °C, and a temperature coefficient of 1/234.5 at 0 °C, in series with a swamp resistor of 3.5 Ω having a negligible temperature coefficient. Determine: (a) the resistance of shunt required for a full-scale deflection of 20 A; (b) the resistance required for a full-scale deflection of 250 V. If the instrument reads correctly at 15 °C, determine the percentage error in each case when the temperature is 25 °C.

7. Describe, with the aid of sketches, the effect on a current-carrying conductor lying in and at right angles to a magnetic field.

The coil of a moving-coil instrument is wound with $40\frac{1}{2}$ turns. The mean width of the coil is 4 cm and the axial length of the magnetic field is 5 cm. If the flux density in the gap is 0.1 T, calculate the torque in N m when the coil is carrying a current of 10 mA.

8. Explain, with the aid of a circuit diagram, how a d.c. voltmeter may be calibrated by means of a potentiometer method.

   A moving-coil instrument, used as a voltmeter, has a coil of 150 turns with a width of 3 cm and an active length of 3 cm. The gap flux density is 0.15 T. If the full-scale reading is 150 V and the total resistance of the instrument is 100 000 Ω, find the torque exerted by the control springs at full scale.

9. A rectangular moving coil of a milliammeter is wound with $30\frac{1}{2}$ turns. The effective axial length of the magnetic field is 20 mm and the effective radius of the coil is 8 mm. The flux density in the gap is 0.12 T and the controlling torque of the hairsprings is 0.5×10^{-6} N m/degree of deflection. Calculate the current to give a deflection of 60°.

10. If a rectifier-type voltmeter has been calibrated to read the r.m.s. value of a sinusoidal voltage, by what factor must the scale readings be multiplied when it is used to measure the r.m.s. value of: (a) a square-wave voltage; (b) a voltage having a form factor of 1.15?

11. A permanent-magnet moving-coil milliammeter, having a resistance of 15 Ω and giving full-scale deflection with 5 mA, is to be used with bridge-connected rectifiers (Fig. 45.16), and a series resistor to measure sinusoidal alternating voltages. Assuming the 'forward' resistance of the rectifier units to be negligible and the 'reverse' resistance to be infinite, calculate the resistance of the series resistor if the instrument is to give full-scale deflection with 10 V (r.m.s.).

12. Give a summary of four different types of voltmeters commonly used in practice. State whether they can be used on a.c. or d.c. circuits. In *one* case, give a sketch showing the construction, with the method of control and damping employed.

   A d.c. voltmeter has a resistance of 28 600 Ω. When connected in series with an external resistor across a 480 V d.c. supply, the instrument reads 220 V. What is the value of the external resistance?

13. The resistance of a coil is measured by the ammeter–voltmeter method. With the voltmeter connected across the coil, the readings on the ammeter and voltmeter are 0.4 A and 3.2 V respectively. The resistance of the voltmeter is 500 Ω. Calculate: (a) the true value

of the resistance; (b) the percentage error in the value of the resistance if the voltmeter current were neglected.

14. A voltmeter is connected across a circuit consisting of a milliammeter in series with an unknown resistor R. If the readings on the instruments are 0.8 V and 12 mA respectively and if the resistance of the milliammeter is 6 Ω, calculate: (a) the true resistance of R; (b) the percentage error had the resistance of the milliammeter been neglected.

15. Describe the principle of the Wheatstone bridge and derive the formula for balance conditions.

   The ratio arms of a Wheatstone bridge are 1000 and 100 Ω respectively. An unknown resistor, believed to have a resistance near 800 Ω, is to be measured, using a resistor adjustable between 60 and 100 Ω. Sketch an appropriate circuit. If the bridge is balanced when the adjustable resistor is set to 77.6 Ω, calculate the value of the unknown resistor.

   State, with reasons, the direction in which the current in the detector branch would flow when the bridge is slightly off balance due to the adjustable resistor being set at too *low* a value.

16. Describe with the aid of a circuit diagram the principle of the Wheatstone bridge, and hence deduce the balance condition giving the unknown in terms of known values of resistance.

   In a Wheatstone bridge ABCD, a galvanometer is connected between B and D, and a battery of e.m.f. 10 V and internal resistance 2 Ω is connected between A and C. A resistor of unknown value is connected between A and B. When the bridge is balanced, the resistance between B and C is 100 Ω, that between C and D is 10 Ω and that between D and A is 500 Ω. Calculate the value of the unknown resistance and the total current supplied by the battery.

17. The arms of a Wheatstone bridge have the following resistances: AB, 10 Ω; BC, 20 Ω; CD, 30 Ω; and DA, 10 Ω. A 40 Ω galvanometer is connected between B and D and a 2 V cell, of negligible internal resistance, is connected across A and C, its positive end being connected to A. Calculate the current through the galvanometer and state its direction.

18. Describe, with the aid of a circuit diagram, how a simple potentiometer can be used to check the calibration of a d.c. ammeter.

   The current through an ammeter connected in series with a standard resistor of 0.1 Ω was adjusted to 8 A. A standard cell, of e.m.f. 1.018 V, gives a balance

Exercises 47 continued

at 78 cm, while the potential difference across the standard resistor gives a balance at 60 cm when measured with a simple potentiometer. Calculate the percentage error of the ammeter.

19. The resistance of a resistor is to be determined by measuring the voltage drop across it when passing a given current, as shown in Fig. A. The instrument indications are 2 mA and 12.0 V, the ammeter having a resistance of 10 Ω and the voltmeter a resistance of 10 kΩ. Calculate: (a) the resistance of the resistor as indicated by the instrument readings; (b) the correct resistance of the resistor allowing for instrument error; (c) the percentage error of the uncorrected resistance with respect to the corrected resistance.

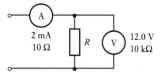

Fig. A

20. With reference to Q. 19, repeat the calculation assuming that the voltmeter and the ammeter had been connected as shown in Fig. B but had indicated the same readings as before.

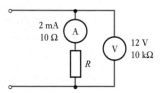

Fig. B

21. A voltage of 80.0 V is applied to a circuit comprising two resistors of resistance 105 Ω and 55 Ω respectively. The voltage across the 55 Ω resistor is to be measured by a voltmeter of internal resistance 100 Ω/V. Given that the meter is set to a scale of 0–50 V, determine the voltage indicated. (Give your answer to the third significant figure.)

22. An industrial grade ammeter (0–50 mA) was compared with a precision grade ammeter and the following test data, after the indications of the precision grade ammeter had been corrected, were obtained:

Industrial (mA)	0	10.0	20.0	30.0	40.0	50.0
Precision (mA)	0	9.6	18.4	28.9	39.2	49.5

Determine whether or not the industrial grade ammeter is within the required ±1.0 per cent error of full-scale range.

23. For the network shown in Fig. C, determine the readings indicated by the ammeter and the wattmeter. The supply voltage may be assumed sinusoidal. A rectifier diode is connected into the circuit between the ammeter and the resistor; again determine the readings indicated by the ammeter and the wattmeter. The diode may be assumed ideal.

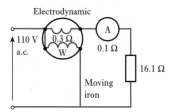

Fig. C

24. The voltage applied to the simple transistor network shown in Fig. D is 6.0 V and the ammeter indicates exactly 1.7 mA. Calculate: (a) the apparent collector–emitter voltage; (b) the maximum and minimum values of the collector–emitter voltage, given that the accuracy of the ammeter is ±2 per cent and the tolerance of the resistor R_c is ±10 per cent; (c) the percentage error of the collector–emitter voltage in each case.

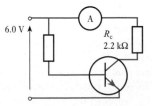

Fig. D

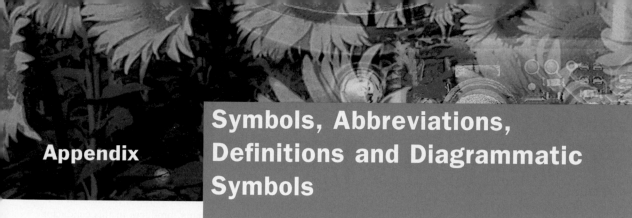

Based upon PD 5685: 1972 (BSI) and The International System of Units (1970), prepared jointly by the National Physical Laboratory, UK, and the National Bureau of Standards, USA, and approved by the International Bureau of Weights and Measures.

Notes on the use of symbols and abbreviations

1. A unit symbol is the same for the singular and the plural: for example, 10 kg, 5 V; and should be used only after a numerical value: for example, m kilograms, 5 kg.
2. Full point should be omitted after a unit symbol: for example, 5 mA, 10 μF, etc.
3. Full point should be used in a multi-word abbreviation: for example, e.m.f., p.d.
4. In a compound unit symbol, the product of two units is preferably indicated by a dot, especially in manuscript. The dot may be dispensed with when there is no risk of confusion with another unit symbol: for example, N · m or N m, but not mN. A solidus (/) denotes division: for example, J/kg, r/min.
5. Only one multiplying prefix should be applied to a given unit: for example, 5 megagrams, not 5 kilokilograms.
6. A prefix should be applied to the numerator rather than the denominator: for example, MN/m^2 rather than N/mm^2.
7. The abbreviated forms a.c. and d.c. should be used only as adjectives: for example, d.c. motor, a.c. circuit.

Abbreviations for Multiples and Sub-multiples

Y	yotta	10^{24}
Z	zetta	10^{21}
E	exa	10^{18}
P	peta	10^{15}
T	tera	10^{12}
G	giga	10^{9}
M	mega or meg	10^{6}
k	kilo	10^{3}
h	hecto	10^{2}
da	deca	10
d	deci	10^{-1}
c	centi	10^{-2}
m	milli	10^{-3}
μ	micro	10^{-6}
n	nano	10^{-9}
p	pico	10^{-12}
f	femto	10^{-15}
a	atto	10^{-18}
z	zepto	10^{-21}
y	yocto	10^{-24}

Miscellaneous

Term	Symbol
Approximately equal to	$\simeq$
Proportional to	$\propto$
Infinity	∞
Sum of	Σ
Increment or finite difference operator	Δ, δ
Greater than	$>$
Less than	$<$
Much greater than	$\gg$
Much less than	$\ll$
Base of natural logarithms	e
Common logarithm of x	$\log x$
Natural logarithm of x	$\ln x$
Complex operator $\sqrt{(-1)}$	j
Temperature	θ
Time constant	T
Efficiency	η
Per unit	p.u.

Greek letters used as symbols

Letter	Capital	Small
Alpha	—	α (angle, temperature coefficient of resistance, current amplification factor for common-base transistor)
Beta	—	β (current amplification factor for common-emitter transistor)
Delta	Δ (increment, mesh connection)	δ (small increment)
Epsilon	—	ε (permittivity)
Eta	—	η (efficiency)
Theta	—	θ (angle, temperature)
Lambda	—	λ (wavelength)
Mu	—	μ (micro, permeability, amplification factor)
Pi	—	π (circumference/diameter)
Rho	—	ρ (resistivity)
Sigma	Σ (sum of)	σ (conductivity)
Phi	Φ (magnetic flux)	ϕ (angle, phase difference)
Psi	Ψ (magnetic flux linkage, electric flux)	ψ
Omega	Ω (ohm)	ω (solid angle, angular velocity, angular frequency)

Definitions of electric and magnetic SI units

The *ampere* (A) is that constant *current* which, if maintained in two straight parallel conductors of infinite length, of negligible circular cross-section, and placed 1 m apart in vacuum, would produce between these conductors a force equal to 2×10^{-7} N per metre of length.

The *coulomb* (C) is the *quantity of electricity* transported in 1 s by 1 A.

The *volt* (V) is the *difference of electrical potential* between two points of a conductor carrying a constant current of 1 A, when the power dissipated between these points is equal to 1 W.

The *ohm* (Ω) is the *resistance* between two points of a conductor when a constant difference of potential of 1 V, applied between these points, produces in this conductor a current of 1 A, the conductor not being a source of any electromotive force.

The *henry* (H) is the *inductance* of a closed circuit in which an e.m.f. of 1 V is produced when the electric current in the circuit varies uniformly at a rate of 1 A/s.

(*Note*: this also applies to the e.m.f. in one circuit produced by a varying current in a second circuit, i.e. mutual inductance.)

The *farad* (F) is the *capacitance* of a capacitor between the plates of which there appears a difference of potential of 1 V when it is charged by 1 C of electricity.

The *weber* (Wb) is the *magnetic flux* which, linking a circuit of one turn, produces in it an e.m.f. of 1 V when it is reduced to zero at a uniform rate in 1 s.

The *tesla* (T) is the *magnetic flux density* equal to 1 Wb/m^2.

Definitions of other derived SI units

The *newton* (N) is the *force* which, when applied to a mass of 1 kg, gives it an acceleration of 1 m/s^2.

The *pascal* (Pa) is the *stress* or *pressure* equal to 1 N/m^2.

The *joule* (J) is the *work done* when a force of 1 N is exerted through a distance of 1 m in the direction of the force.

The *watt* (W) is the *power* equal to 1 J/s.

The *hertz* (Hz) is the unit of *frequency*, namely the number of cycles per second.

The *lumen* (lm) is the *luminous flux* emitted within unit solid angle by a point source having a uniform intensity of 1 candela.

The *lux* (lx) is an *illuminance* of 1 lm/m^2.

Units of length, mass, volume and time

Quantity	Unit	Symbol	Quantity	Unit	Symbol
Length	metre	m	Volume	cubic metre	m^3
	kilometre	km		litre (0.001 m^3)	l
Mass	kilogram	kg	Time	second	s
	megagram	Mg		minute	min
	or tonne (1 Mg)	t		hour	h

Electricity and Magnetism

Quantity	Quantity symbol	Unit	Unit symbol
Admittance	Y	siemens	S
Angular velocity	ω	radian per second	rad/s
Capacitance	C	farad	F
		microfarad	μF
		picofarad	pF
Charge on Quantity of electricity	Q	coulomb	C
Conductance	G	siemens	S
Conductivity	σ	siemens per metre	S/m
Current			
Steady or r.m.s. value	I	ampere	A
		milliampere	mA
		microampere	μA
Instantaneous value	i		
Maximum value	I_m		
Current density	J	ampere per square metre	A/m^2
Difference of potential			
Steady or r.m.s. value	V	volt	V
		millivolt	mV
		kilovolt	kV
Instantaneous value	v		
Maximum value	V_m		
Electric field strength	E	volt per metre	V/m
Electric flux	Ψ	coulomb	C
Electric flux density	D	coulomb per square metre	C/m^2
Electromotive force			
Steady or r.m.s. value	E	volt	V
Instantaneous value	ε		
Maximum value	E_m		
Energy	W	joule	J
		kilojoule	kJ
		megajoule	MJ
		watt hour	W h
		kilowatt hour	kW h
		electronvolt	eV
Force	F	newton	N
Frequency	f	hertz	Hz
		kilohertz	kHz
		megahertz	MHz
Impedance	Z	ohm	Ω
Inductance, self	L	henry (plural, henrys)	H
Inductance, mutual	M	henry (plural, henrys)	H
Magnetic field strength	H	ampere per metre	A/m
		ampere-turns per metre	At/m
Magnetic flux	Φ	weber	Wb
Magnetic flux density	B	tesla	T

Quantity	Quantity symbol	Unit	Unit symbol
Magnetic flux linkage	Ψ	weber	Wb
Magnetomotive force	F	ampere	A
		ampere-turns	At
Permeability of free space or Magnetic constant	μ_0	henry per metre	H/m
Permeability, relative	μ_r		
Permeability, absolute	μ	henry per metre	H/m
Permittivity of free space or Electric constant	ε_0	farad per metre	F/m
Permittivity, relative	ε_r		
Permittivity, absolute	ε	farad per metre	F/m
Power	P	watt	W
		kilowatt	kW
		megawatt	MW
Power, apparent	S	voltampere	VA
Power, reactive	Q	var	var
Reactance	X	ohm	Ω
Reactive voltampere	Q	var	var
Reluctance	S	ampere per weber	A/Wb
Resistance	R	ohm	Ω
		microhm	$\mu\Omega$
		megohm	$M\Omega$
Resistivity	ρ	ohm metre	$\Omega\,m$
Speed, linear	u	metres per second	m/s
Speed, rotational	ω	radians per second	rad/s
	n	revolutions per second	r/s
	N	revolutions per minute	r/min
		microhm metre	$\mu\Omega\,m$
Susceptance	B	siemens	S
Torque	T, M	newton metre	N m
Voltampere	—	voltampere	VA
		kilovoltampere	kVA
Wavelength	λ	metre	m
		micrometre	μm

Light

Quantity	Quantity symbol	Unit	Unit symbol
Illuminance	E	lux	lx
Luminance (objective brightness)	L	candela per square metre	cd/m^2
Luminous flux	Φ	lumen	lm
Luminous intensity	I	candela	cd
Luminous efficacy	—	lumen per watt	lm/W

Selection of graphical symbols from BS 3939

Description	Symbol
Direct current or steady voltage	—
Alternating	~
Positive polarity	+
Negative polarity	–
Primary or secondary cell	
Battery of primary or secondary cells	
Fixed resistor	
Variable resistor	
Resistor with moving contact	
Filament lamp	
Crossing of conductor symbols on a diagram (no electrical connection)	
Junction of conductors	
Double junction of conductors	
Earth	
Capacitor: general symbol	
Polarized capacitor	
Winding	
Inductor and core	
Transformer	
Ammeter	Ⓐ
Voltmeter	Ⓥ
Wattmeter	Ⓦ

Description	Symbol
Galvanometer	
Motor	Ⓜ
Generator	Ⓖ
Make contact (normally open)	
Break contact (normally closed)	
Rectifier	
Zener diode	
p–n–p transistor	
n–p–n transistor	
n-channel JUGFET	
p-channel JUGFET	
n-channel IGFET	
p-channel IGFET	
Amplifier	
Thyristor	
MOSFET	
IGBT	
GTO	
Binary logic units — AND	
Binary logic units — OR	
Binary logic units — NOT	
Binary logic units — NAND	
Binary logic units — NOR	

Answers to Exercises

Exercises 1

1. 0.4 m/s^2
2. 2 kN
3. 98.1 N
4. 9.81 m/s, 19.62 m/s, 29.43 m/s
5. 50 s
6. 625 kN
7. 98.1 N, 128.1 N
8. 0.98 m/s^2 downwards
9. 14.7 kN; 221 MJ, 61.3 kW h; 368 kW; 26 kW h
10. $16\,370$ N m

Exercises 2

1. $5\ \mu$A
2. 3 MV
3. $10\ \Omega$
4. 40 V
5. 0.25 A
6. $2.9\ k\Omega$
7. 28.2 kA
8. 5.1 mA
9. 110 V
10. 1 mA, 0.67 mA
11. $1\ k\Omega$
12. $1\ k\Omega$
13. See Fig. A.13
14. See Fig. A.14
15. 20 C
16. 8 C
17. 80 V
18. 212 N m
19. 94.7 A, 1091 p
20. 83.8 per cent, 146 kJ; $67.2\ \Omega$
21. 12.82 kW, 26.7 A, 102.6 kWh
22. $59.5\ \Omega$, 0.65 p
23. 13.93 kW, 2229 p

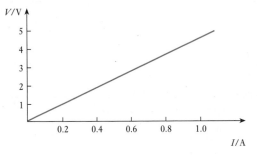

Fig. A.13

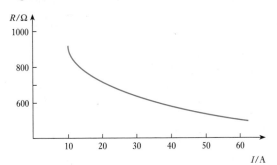

Fig. A.14

Exercises 3

1. 20 V
2. $500\ \Omega$, 0.48 A
3. $20\ \Omega$
4. $60\ \Omega$
5. 30 V, 70 V, 100 V, 70 V
6. -35 V, -5 V, 35 V, 50 V, 10 V
7. 1.5 A, 1.0 A, 0.6 A, 3.1 A, $2.9\ \Omega$
8. $24\ \Omega$, $36\ \Omega$
9. 6 A, 1.5 A
10. (a), (d)
11. (b), (c), (e)
12. 4.04 V, 780 J

13. 26.57 V, 13.8 W
14. 5 Ω
15. A: 2.4 A, 14.4 V, 9.6 V; B: 1.6 A, 16 V, 8 V; 96 W
16. 10.96 V; 121.7 A, 78.3 A; 1.33 kW, 0.86 kW
17. 11.33 Ω, 208 W
18. 40 Ω, 302.5 W
19. 0.569 mm
20. 70°C
21. 30 Ω
22. 0.0173 $\mu\Omega$ m, 0.398 Ω
23. 78.1°C

Exercises 4

1. 1.069 A, discharge; 0.115 A, discharge; 94.66 V
2. 5; if A is positive with respect to B, load current is from D to C.
3. 72.8 mA; 34.3 mA, 1.93 V
4. 11.55 A, 3.86 A
5. I_A, −0.183 A, charge; I_H, 5.726 A, discharge; I_C, 5.543 A; 108.5 V
6. 0.32 A
7. 1 V
8. 2.84 Ω, 1.45 Ω
9. 20.6 mA from B to E
10. 0.047 A
11. 0.192 A
12. 40 Ω; 6 V, 0.9 W; 0.8 W
13. 1 A, 12 A, 13 A; 104 V
14. 2.295 A, 0.53 A, 1.765 A, both batteries discharging; 4.875 A
15. AB, 183.3 Ω; BC, 550 Ω; CA, 275 Ω
16. A, 4.615 Ω; B, 12.31 Ω; C, 18.46 Ω
17. $R_{AB} = 247.8$ Ω, $R_{BC} = 318.6$ Ω, $R_{CA} = 223$ Ω; 80 Ω
18. 56.2 mA
19. 17.5 V; 12.5 Ω

Exercises 5

1. 150 V, 3 mC
2. 45 μF, 4.615 μF
3. 3.6 μF, 72 μC, 288 μJ
4. 1200 μC; 120 V, 80 V; 6 μF
5. 500 μC; 250 V, 167 V, 83 V; 0.0208 J
6. 15 μF in series
7. 2.77 μF, 18.46 μC
8. 4.57 μF, 3.56 μF
9. 3.6 μF, 5.4 μF
10. 200 V; 1.2 mC, 2 mC, 3.2 mC on A, B, C respectively
11. 360 V, 240 V; 40 μF, 0.8 J
12. 267 V; 0.32 J; 0.213 J
13. 150 V, 100 V, 600 μC; 120 V, 480 μC; 720 μC
14. 62.5 V
15. 120 V, 0.06 J, 0.036 J
16. 590 pF, 0.354 μC, 200 kV/m, 8.85 μC/m^2
17. 1594 pF, 0.797 μC, 167 kV/m, 8.86 μC/m^2
18. 664 pF, 0.2656 μC, 100 kV/m, 4.43 μC/m^2
19. 177 pF, 1062 pF, 33.3 V, 0.0354 μC
20. 619.5 pF, 31 ms, 0.062 μC
21. 2212 pF
22. 1.416 mm
23. 1288 pF; 1.82 kV/mm (paper), 0.454 kV/mm; 0.0161 J
24. 0.245 m^2, 2.3 MV/m in air-gap
25. 857 kV/m in air-gap, 143 kV/m
26. 20 μC, 15 μC; 1000 μJ, 2250 μJ; 35 μC; 140 V; 2450 μJ
27. 8.8×10^{-12} F/m; 100 kV/m, 0.88 μC/m^2
28. 2.81; 30 kV/m, 0.7425 μC/m^2; 0.4455 μJ
29. See Fig. A.29
30. 0.8 S, 125 V/s, 12.5 mA, 0.01 C, 0.5 J
32. 18.4 μA, 4000 μJ
33. 100 V, 105 μS
34. 50 mA, 500 mA, 1.25 J
35. 1344 s
36. 100 MΩ

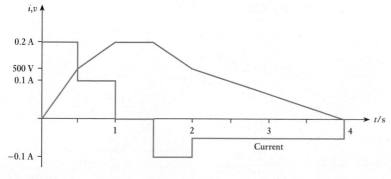

Fig. A.29

37. 39.7 m (approx), if wound spirally
38. 4.75 kV, 1 ms, 0.4 N
39. 3.75 μF, 1500 μC
40. 300 V; 4 μJ, 16 μJ
41. 2510 μJ/m^3
42. 20.8 N/m^2, 20.8 J/m^3
43. 15 940 pF, 3000 kV/m, 159.4 μC/m^3, 71.7 N

Exercises 6

1. 60 N/m
2. 50 A
3. 0.333 T
4. 0.452 N
5. 0.667 T
6. 10 m/s, 0.8 N, 0.48 J
7. 0.375 T, 0.141 N
8. 12 N, 1.2 V, 120 W
9. 2.33 mV
10. 0.333 V
11. 4.42 μV
12. 37.5 V
13. 0.32 V
14. −33.3 V, 0, 100 V
15. 24 V, 224 V

Exercises 7

1. 880 A, 2.2×10^6/H, 605
2. 720 A/m, 663; 2400 A/m, 398
3. 3000 A, 346 μWb
4. 1.5×10^6/H, 144
5. 4.5 A; 0.438×10^6/H, 306 A
6. 121 A, 582
7. 82.5 mA
8. 1100 A
9. 1015 A
10. 5.85 A
11. 1520 A
12. 0.64 T, 358 A/m
13. 308 A/m, 0.593 T;
    616 A/m, 1.04 T;
    924 A/m, 1.267 T;
    1232 A/m, 1.41 T;
    1540 A/m, 1.49 T
14. 1.22 T, 1620
15. 1.23×10^{-9} C/div
16. 995 A/m, 1.25 mT

17. 6000 A, 500 000/H; 2650
18. 125 N/m
19. A, 4780 A/m, 6 mT; B, 6370 A/m, 8 mT; C, 2390 A/m, 3 mT; A, 0, 0; B, 3185 A/m, 4 mT; C, 2390 A/m, 3 mT

Exercises 8

1. 0.375 H
2. 0.15 H
3. 160 A/s
4. 3.5 V
5. 1.25 μH, 0.1 V
6. 530 A/m, 354 A/m; 2.66 A; 18.8 μH
7. 0.24 H, 0.06 H
8. 1492, 47.1 mH, 1.884 V
9. 157 μH, 94.2 mH
10. 15 H; 300 V
11. 0.15 H, 750 V, 0.122 H
12. 2.68 A; 13.32 mH, 7.14 V
13. 0.09 H, 180 V
14. 26.67 μH
15. 100 μH, 6 mV, 0.527
16. 2.23 mm, 0.125 H
17. See Fig. A.17
18. See Fig. A.18

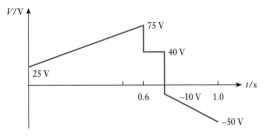

Fig. A.17

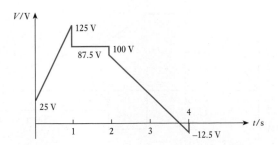

Fig. A.18

19. 83.3 mH, 41.7 ms; 1.667 s
20. 20 A/s, 1.55, 0.23 s
21. 20 A/s, 0.1 s, 1.5 A
22. 0.316 A, 632 Ω 1.896 H, 94.6 mJ
23. 1 A, 2.5 A, 0, 200 V; 2.5 A, 2.5 A, 700 V, 500 V
24. 0.316 A, 155.5 V
25. −40 A/s, 10 J
26. 12 V
27. 62.5 mH
28. −10 V
29. 15.06 μH
30. 1.35 mH
31. 150 μC
32. 740 μH
33. 12.8×10^{-10} C, 10.65 μH
34. 1.612 mH, 0.417
35. 27.5 mJ, 17.5 mJ; 0.354
36. 32 mH, 48 mH, 160 mJ, 96 mJ
37. 290 μH, 440 μH

Exercises 9

1. 125.6 V, 33.3 Hz
2. 20 Hz, 35.5 V, 38.5 V
3. 28.27 sin 157t volts; 8.74 Ω
4. 14.2 A, 16.4 A, 1.155
5. 81.5 V, 1.087
6. 5.0 V, 5.77 V, 1.154
7. 30 A r.m.s.; 20 A average
8. 10 V, 11.55 V, 1.155
9. 0.816; none
10. 0.637 A, 1.0 A
11. 2.61 kV, 3.0 kV
12. 5.06 A
13. 500 sin(314t + 0.93) V, 10 sin(314t + 0.412) A; −300 V, −9.15 A; 29.5°, current lagging voltage
14. 150 Hz; 6.67 ms; 0.68 ms, 7.35 ms; 0.5 kW h
15. 35 A
16. 68.0 V, 48.1 V
17. 55.9 A, 1.2° lead
18. 23.1 sin(314t + 0.376) A; 16.3 A, 50 Hz
19. 262 sin(314t + 0.41) V; 185.5 V, 50 Hz
20. 121 V, 9.5° leading: 101.8 V, 101° leading
21. 72 sin(ωt + 0.443)
22. 208 sin(ωt − 0.202), 11°34′ lagging, 18°26′ leading; 76 sin(ωt + 0.528)
23. 10.75 A, 8.5° lagging

Exercises 10

1. 125.6 V (max), 1.256 A (max), 6.03 N m (max)
2. 500 W
3. Zero
4. 2.65 A, 0.159 A
5. 6.23 A, 196 V, 156 V, 51.5°
6. 21.65 Ω, 39.8 mH, 2165 W
7. 25.1 Ω, 29.3 Ω, 8.19 A, 0.512, 1008 W
8. 2.64 A, 69.7 W, 0.115
9. 43.5 mH, 16 Ω, 58.6°
10. 170 V, 46°; 3.07 Ω, 60 mH
11. 254 Ω, 1.168 mH
12. 5.34 A, 0.857
13. 77.5 W, 5.5 Ω
14. 41.4 A, 4.5°; 41.6 sin(314t − 0.078) A; 4.83 Ω, 4.81 Ω, 0.38 Ω
15. 61.1 Ω, 0.229 H, 34°18′
16. 78.6 W, 37.6 mH, 0.716
17. 174 V, 75.6 V; 109.5 W, 182.5 W; 0.47
18. 0.129 H, 81.8 mH, 0.71 A or 1.72 A
19. 49.6 mA, 0.0794; 189 mA, 0.303
20. 9.95 A
21. 133 sin(ωt − 0.202) V; 6.65 sin(ωt + 0.845) A

Exercises 11

1. 25.8 μF, 0.478 leading, 145.5 V
2. 1.54 A, 1.884 A, 2.435 A, 50°42′, 308 W, 0.633 leading
3. 11.67 Ω, 144 μF; 14.55 A, 0.884 leading
4. 10.3 A, 103 V, 515 V, 309 V
5. 7.27 A, 5.56 A, 468 V
6. 7.07 A, 5 A, 5 A
7. 10 A, 13.4 A, 5.6 A; 20.8 A, 0.95 lagging
8. 10.3 A, 10.2 A, 13.8 A, 3160 W
9. 2 A, 4 A, 2 A, 4.56 A
10. 127.4 μF, 56 A
11. 8.5 A, 13.53 Ω, 10.74 Ω, 8.25 Ω
12. 6.05 A, 15.2° lagging

Exercises 12

1. 28.7 μF
2. 5.02 kVA
3. 10.3 kVA, 0.97 leading; 20.6 kVA, 0.97 lagging
4. 324 μF
5. 17.35 A, 3967 W, 375 var, 0.996 lagging

6. 59 kW, 75.5 kVA, 0.78 lagging; 47 kvar in parallel
7. 60.8 A, 6.7 kvar, 0.643 lag

Exercises 13

1. $10 + j0$, $10\angle 0°$; $2.5 - j4.33$, $5\angle -60°$; $34.64 + j20$, $40\angle 30°$
2. $20 + j0$, $0 + j40$, $26 - j15$, $5 - j8.66$, $51 + j16.34$; 53.5, $17°46'$ lead
3. $11.2\angle 26°34'$, $8.54\angle -69°27'$
4. $10 + j17.32$, $28.28 - j28.28$
5. $13 - j3$, $13.35\angle -13°$
6. $7 + j13$, $14.77\angle 61°42'$
7. $38.28 - j10.96$, $39.8\angle -16°$
8. $-18.28 + j45.6$, $49.1\angle 111°48'$
9. 50 Ω, 95.6 mH; 30 Ω, 63.7 μF; 76.6 Ω, 0.2045 H; 20 Ω, 92 μF
10. $(0.0308 - j0.0462)$ s, $0.0555\angle -56°18'$ s; $(0.04 + j0.02)$ s, $0.0447\angle 26°34'$ s; $(0.0188 - j0.006\ 84)$ s, $0.02\angle -20°$ s; $(0.0342 + j0.094)$ s, $0.1\angle 70°$ s
11. $(0.69 - j1.725)$ Ω, $1.86\angle -68°12'$ Ω; $(10.82 + j6.25)$ Ω, $12.5\angle 30°$ Ω
12. 4 Ω, 1910 μF; 20 Ω, 31.85 mH; 1.44 Ω, 1274 μF; 3.11 Ω, 8.3 mH
13. $(91.8 + j53)$ V, $(2.53 - j1.32)$ A; $106\angle 30°$ V, $2.85\angle 27°30'$ A
14. $(1.46 - j0.83)$ A; $2.82\angle 87°53'$ A
15. 25 Ω, 50 Ω
16. 24 Ω, 57.3 mH; 12 Ω, 145 μF; $0.0482\angle 18°10'$ s, $18°10'$, current leading
17. 7.64 A, $8°36'$ lagging
18. $(13.7 - j3.2)$ Ω, capacitive; 2250 W; $13°9'$
19. $(1.192 - j0.538)$ A
20. 20 Ω, 47.8 mH; 10 Ω, 53 μF; $(0.0347 - j0.0078)$ s, $12°40'$ lagging
21. 2.83 A, $30°$ lagging
22. 2.08 A, lagging $34°18'$ from supply; 4.64 A, lagging $59°$ through coil; 2.88 A, leading $103°30'$ through capacitor
23. 18.75 A, 0.839 lagging
24. 940 W; 550 var lagging
25. 1060 W; 250 var leading
26. 4 kVA, 3.46 kW; 8 kVA, 8 kW; 11.64 kVA, 11.46 kW; $(44.64 + j37.32)$ A

Exercises 14

1. 159 pF, 5 mA; 145 pF, 177 pF
2. 920 Hz, 173 V

3. 15.8
4. 400 V, 401 V, 10 V, 0.5 A, 15 W
5. 50 Hz, 12.95 Ω, 0.154 H, 208 W; $21.87 \sin(314t + \pi/4)$ A
6. 31.5 V, 60 Ω, 0.167 H, 4220 pF
7. 84.5 μH, 30 Ω, 17.7
8. 57.2 Hz, 167 Ω
9. 125 kΩ
10. 76.5 pF, 0.87 MHz

Exercises 15

1. $5.7\angle -58.6°$
2. $5.85\angle -38°$ A
3. $(4 + j2)$ A
4. $(22.2 + j15.8)$ V or 27.2 V
5. $3.58\angle -3.4°$ mA, $17.9\angle 86.4°$ V
6. $22.4\angle 26.6°$ V, $(3.2 - j4.4)$ Ω
7. $(-3.2 + j8.0)$ V, $(12.8 + j1.6)$ Ω
8. $(22.2 + j15.8)$ V or 27.2 V
9. $(4 + j2)$ A
10. 0.83 mA
11. $(-0.2 + j0.1)$ A, $(19.23 - j3.84)$ Ω
12. $(-0.17 + j0.65)$ A, $(12.8 + j1.6)$ Ω
13. $3.58\angle -3.4°$ mA
14. $(1.46 - j0.83)$ A
15. $(0.30 - j0.13)$ A
16. $(3.2 + j4.4)$ Ω
17. 12.8 Ω resistance, 1.6 Ω capacitive reactance
18. 36.24 W
19. 9.16 W
20. 569 W
21. $(4 - j14)$ Ω, $(5.2 + j13.6)$ Ω, $(17 - j12)$ Ω
22. $(6.1 - j5.1)$ Ω, $(6.9 + j2.0)$ Ω, $(5.1 + j6.1)$ Ω
23. $(18.16 - j2.07)$ Ω
24. 0.393 A

Exercises 18

1. 5.653, 2.314, −4
2. 1000, 1.097, 125.9
3. 1.68
4. 54 dB
5. 10 dB
6. 37.5 dB
7. 250 rad/s, 8.5 V, $10.7\angle -26.6°$, $5.36\angle -63.4°$, $2.9\angle -76°$
8. 10 kHz, $0.995\angle -5.7°$, $0.894\angle -26.6°$, $0.707\angle -45°$, $0.196\angle -78.6°$, $0.1\angle -84°$

9. 50 krad/s, 0.894∠−26.6°, 0.45∠−63.4°, 0.45∠63.4°, 0.894∠26.6°

10. 25 krad/s, 0.1∠84°, 0.45∠63.4°, 0.707∠45°, 0.894∠26.6°, 0.95∠5.7°

11. 0 to 237 Hz

12. −3 dB at 1592 Hz, −20 dB, −7 dB, −0.97 dB

13. −3 dB∠45° at 6772 Hz (f_1), −20 dB∠84° (at $f_1/10$), −7 dB∠63° (at $f_1/2$), 1 dB∠26.6° (at $2f_1$), 0 dB∠5.7° (at $10f_1$)

4. 8.1 A, 1.7 Ω

5. 325 mA, 103 mA, 162 mA, 26.2 W

6. 26 Ω, 833 V, 0.78

8. 250 Ω, 1.54 W

10. 0.35 A, 0.59 A

11. 1.6 Ω, 0.89

12. 2.6 Ω, 13.8 A

13. 40 Ω, 0.07

14. 22.2 Ω, 100 Ω, about 20–70 mA, 11–17 V, 0.17

Exercises 19

2. 25, 25, 625

3. 488, 130 000, 63 400 000, 156 000

4. 470

5. 750, 2000, 8000, 250 Ω, 40 mA

6. 20.6, 38

7. 632 mV

8. 20 dB

9. 0.775 V, 2.45 V, 0.245 V

10. 54 dB, 115 dB, 64.5 dB, 9 W

11. 2.4 V, 75.8 dB

12. 1.39 kΩ, 1610

13. 4.5 μV, 72 μV, 24.1 dB

14. −0.013, 26.1

15. 41.5

16. 36

17. 97, −0.0916

18. 108.8, 28.26, 35.86, 40.7 dB, 29 dB, 31.1 dB

19. −0.0071, −4.10 dB

20. 4.87, 13.75 dB

21. −0.008, 97.8, 101.9

22. 51.3, 20 000, 5.3 per cent

23. −0.053, 123.9 mV

Exercises 22

10. 74 μA; 32.0 dB; 33.6 dB

11. 2.06 mV; 1290; **27.5**

12. 25 Ω; 10 kΩ

13. 4.9 mA; 8.6 V; 3.36 V, 21.3

14. 4.6 V; 1.9 mA; 0.9 mA; 45

15. 113 kΩ; 1.08 V, 1.16 mW

16. 30; 30; 900

17. 28.2, 38.3, 128, 188, 4900, 5300

18. 63.6; 42.4; 2700

19. 50; 250; 12 500; 10 mW

20. 41.7; 20.9; 870

21. 1.046 V; 5970; 37.8 dB

22. 0.33 V

23. 2.07 mA; 3.74 V; 1396; 31.4 dB

24. 476; 1455; 693 000

25. 3.19 mA, 2.28 mA, 4.56 V, 88

26. 0.95 V; 0°; 60

27. −4.26 dB

28. 1.33 V; 44.2 dB

29. 80; 320; 44.1 dB

30. 101; 40.5; 4100

31. 81; 48; 2304

Exercises 20

1. (c)   2. (b)   3. (b)   4. (b)   5. (a)   6. (d)

Exercises 21

1. 0, 20 mA, 220 mA

2. 0, 5.7 mA, 63 mA

3. 800 Hz

Exercises 23

3. 220 kΩ; 2.5 mA/V

4. 50 kΩ; 2.4 mA/V; 120; for G, 8 mA; 10 V; −0.49 V

5. 95; 57.2

6. 1.19 V; 2.85 V

7. 499 MΩ, 795 Ω

8. 2 mA/V

9. 3.66 nF; 11

10. 389 mV

11. 200 Ω; 36.4 V

12. 24.2 V

Exercises 24

1. 3750; 20 000
2. 1200; 288 000
3. 33.3; 33.3; 1110; 1 230 000
4. 150; 4
5. 100 mV; 1 V; 0.16 μW
6. 0.125 mV; 7.94 mV; 26 pW
7. 60 dB; 58.75 dB
9. 9 V
11. 1 MΩ
12. 2.8 V
13. 33.3 kΩ
14. 0.5 sin ωt; 2.12 sin($\omega t + \pi/4$)
15. -6.44; 7.44
16. -34.2 V
17. 0.9 mV

Exercises 25

1. 450 kΩ
2. 2.25 V
4. 0.5 V, 1.0 V, 2.0 V, 4.0 V
5. 0 to 7.5 V in 0.75 V steps
6. 0.79 per cent; 0.006 per cent
10. 30 kΩ, 60 kΩ, 120 kΩ
11. 25 kΩ, 50 kΩ, 100 kΩ

Exercises 26

1. 1111
2. 11011
3. 1010110
4. 1, 1
5. 25
6. 43
7. 127
8. 001 111 010
9. 001 011 000 000 000
10. 001 011 000 000 001
11. 1111
14. 01010
16. 01010
17. 10111101
20. 10
23. 0101010

Exercises 27

1. 1; A + B; A + B
2. A + B; 0; A
3. A + B; A $\cdot$ B; A + B $\cdot$ $\overline{C}$
4. 0; 1; 1
6. A $\cdot$ B $\cdot$ C
7. $\overline{A}$ + C $\cdot$ B
8. A
9. $\overline{B}$; $\overline{A} \cdot B + \overline{B} \cdot C$; $\overline{B} \cdot \overline{C} + \overline{A} \cdot B \cdot C$; no simplification; $B \cdot \overline{D} + \overline{B} \cdot D$
12. B $\cdot$ (A $\cdot$ $\overline{C}$ + $\overline{A}$ $\cdot$ C)

14.

A	B	C	F
0	0	0	0
0	0	1	0
0	1	0	1
0	1	1	1
1	0	0	0
1	0	1	1
1	1	0	0
1	1	1	1

15. A $\cdot$ (B + C); $\overline{A}$ $\cdot$ B $\cdot$ C; A $\cdot$ $\overline{B}$ + B $\cdot$ $\overline{A}$
16. A $\cdot$ C + $\overline{A}$ $\cdot$ $\overline{C}$ + B
17. $\overline{A}$ $\cdot$ B $\cdot$ $\overline{C}$
18. X = $\overline{A} \cdot \overline{B} \cdot \overline{C}$ + A $\cdot$ $\overline{B}$ $\cdot$ C + $\overline{A}$ $\cdot$ B $\cdot$ C + A $\cdot$ B $\cdot$ $\overline{C}$; X = A; X = E $\cdot$ $\overline{F}$ + $\overline{D}$ $\cdot$ $\overline{F}$
19. F = A $\cdot$ C + $\overline{A}$ $\cdot$ $\overline{B}$ $\cdot$ $\overline{C}$; G = $\overline{B}$ $\cdot$ C + $\overline{A}$ $\cdot$ B + B $\cdot$ $\overline{C}$
20. X = C $\cdot$ $\overline{B}$ + $\overline{A}$ $\cdot$ $\overline{B}$ $\cdot$ D + A $\cdot$ C $\cdot$ D; X = $\overline{A}$ $\cdot$ $\overline{B}$ $\cdot$ $\overline{C}$ + A $\cdot$ $\overline{C}$ $\cdot$ D + A $\cdot$ B $\cdot$ $\overline{D}$ + B $\cdot$ C $\cdot$ $\overline{D}$; X = A $\cdot$ C + $\overline{A}$ $\cdot$ $\overline{C}$

Exercises 28

The exercises permit alternative programs hence only a few are illustrated below.

```
 4. LD   0260
    ADD  0261
    ST   0262
    BRK
14. LD   0266
    ST   0267
    SHL  0267
    SHL  0267
```

LD   0267
ADD   0266
ST   0268
BRK
16. LD   #1E
    ST   0290
    OR   0293
    ST   0292
    BRK

7. = 94.8 per cent
8. 01010110; 11010110
9. Yes
10. Even numbers of errors. No difference
11. $x_1 = 0$, $x_2 = k$, $x_3 = 10$; 100 per cent; 75 per cent
13. F = 1, O = 010, M = 0000, B = 0001, G = 0010,
    E = 0011, C = 0110, A = 0111, entropy = 2.366,
    average length = 2.37, efficiency = 99 per cent

Exercises 29

3. 9.8 V per volt
4. 1600 r/min
5. 96.94 V
7. 0.44 s
8. 31.6 per cent

Exercises 30

2. periodic, 2 s; aperiodic
4. $u(t) = r(t) - r(t-4) + 2\gamma(t-6) - 6\gamma(t-8)$;
   $u(t) = {}^3/_4 r(t) - {}^3/_4 r(t-4) - 3r(t-6) + 3r(t-7)$
5. (a) (i) non-periodic (ii) non-periodic (iii) non-periodic
   (iv) periodic (v) non-periodic (vi) non-periodic
   (b) (i) energy (ii) power (iii) energy (iv) power
   (v) neither (vi) power
6. 6 10 12 12 10 6 6 6 . . .
7. 1 2 2 1 0 0 0 0 0 . . .
9. $y(t) = (0.48 - 0.48 \mathrm{e}^{-4t} \cos 3t + 0.36 \mathrm{e}^{-4t} \sin 3t) \gamma(t)$

Exercises 31

3. 125 kHz, 6 μs

Exercises 32

1. = 15.4 kbits/s
2. 3.32 bits; 4.68 bits
3. 5 bits, 1.5 bits
4. kbits
5. 7 bits; 8 bits
6. 3 bits

Exercises 33

1. b
2. a
3. a
4. b
5. b
6. c
7. 250×10^{12} Hz
10. 133.3×10^6 m/s

Exercises 34

1. 346 V; 346 V, 200 V, 200 V
2. 11.56 kW, 34.7 kW
3. 96.2 A, 55.6 A
4. 8.8 A, 416 V, 15.25 A, 5810 W
5. 4.3 Ω, 66.4 mH, 18.8 A, 32.6 A
6. 1.525 A; 2.64 A, 2.64 A, 4.57 A; 1.32 A, 1.32 A, 0;
   210 V
7. 59.5 A, 0.858, 35.5 kW
8. 90.8 A, 240 V
9. 110.7 Ω, 0.264 H, 1.67 A, 926 W; 1.445 A, 1.445 A, 0
10. 462 V, 267 V, 16.7 kvar, 5.87 Ω, 8.92 Ω, 75 A, 33 kW
11. 43.5 A, 26.1 A, 17.4 A, 23 A
12. 17.3 A, 31.2 kW
13. 4.00 A, 6.67 A, 3.08 A, 6.85 A, 10.33 A, 5.8 A
14. 21.6 A in R, 49.6 A in Y, 43.5 A in B
15. 1.1 kW, 0.715
16. 8 kW
17. 0.815
18. 13.1 kW, 1.71 kW
19. 21.8 A, 36°35' lagging; 7 kW; 13 kW
20. 0.655, 0.359
21. 15 kW, 0.327
23. 11.7 kW, 5.33 kW
24. Line amperes $\times$ line volts $\times$ sin ϕ = 8750 var
25. 75 μF/ph
26. 3.81 kvar, 70.5 μF, 23.5 μF, 0.848 lagging

Exercises 35

1. 27 turns, 742 turns, 44 500 mm^2
2. 523, 688 V, 4450 mm^2
3. 16 turns, 765 turns
4. 1200 V, 0.675 T
5. 146.7 V; 22.73 A, 341 A; 0.033 Wb
6. 60.6 A, 833 A, 0.0135 Wb, 1100 turns
7. 74 turns, 1.84 kVA
8. 14, 22, 840, 3500 turns
9. 25.9 A
10. 7 A, 0.735 lagging
11. 4.26 A, 1.0
12. 960 V; 34.82 A, secondary; 8.71 A, 15.06 A, primary
13. 10 turns, 159 turns
14. 530 V, 1586 V
15. 34.6 A, 59.9 A, 59.9 A, 103.8 A, 72 kW, 24 kW
16. 5.4 mWb, 360 W, 4.77 A (r.m.s.)
17. 346 turns, 1.07 mm, 88.4°
18. 3.24 A, 0.154
19. 176.5 V
20. 928 V
21. 377.6 V
22. 4 per cent
23. 2.13 per cent, 225.1 V; 0.41 per cent, 223.5 V; 2.81 per cent, 228.7 V; 0.960 p.u., 0.985 p.u.; 56.25 kVA
24. 3.08 per cent
25. 0.887 p.u.
26. 0.981 p.u.
27. 0.9626 p.u.
28. 0.905 p.u.
29. 0.96 p.u., 8.65 kVA, 0.96 p.u. at unity power factor
30. 0.9868 p.u., 0.9808 p.u.
31. 0.961 p.u., 0.9628 p.u., 23.3 kW
32. A, 93.77, 96.24, 96.97 per cent; B, 96.42, 97.34, 96.97 per cent; A, 40 kVA; B, 23.1 kVA
33. 362.8 W

Exercises 36

1. 70 W; 80 W; 0
3. 25 mJ
4. 4.2 A
5. 0.25 T; 400 A/m; 314 A
6. 19 turns
7. 79.5 N; 2985 N
8.

0.25 T	0.38 T	0.75 T
164 mH	246 mH	492 mH
49 N	110 N	440 N

9. 18.9 H; 15.8 H; 1.04 N; 25.7 V
10. 80 W; −78.4 W; 118.4 W; 49.2 N
11. 126 mH, −42 H/m; 195 mA
12. 94 mT; 29.6 μWb; 7.2 H; 3600 H/m; 1.1 N

Exercises 37

1. 3600 r/min, 0.0227 T
4. 1.195
5. 1.24
6. 4.19 N m
7. 8 poles, 31.2 mWb
8. 6.4 kW
9. 500 r/min
10. 4 poles
11. 150 Hz

Exercises 38

1. 4.4 per cent, 14.0 per cent, −7.3 per cent
2. 2412 V
3. 1650 V
4. 12.5 per cent increase
5. 39.9 A, 3290 V, 131.3 kW
6. 111 A, 184.5 V, 615 kW
7. 71.1 A, 1920 V/ph
8. 1200 kW, 0.388 leading
9. 0.225 H, 223 Nm

Exercises 39

1. 2880 r/min, 3000 r/min
2. 1470 r/min, 1 Hz
4. 4.6 per cent, 954 r/min
5. 1.585 Hz, 3.17 per cent
6. 971.7 r/min
7. 9.36 mWb
8. 11.33 A; 0.2, 40.8 A
9. 10.67 A, 0.376; 21.95 A, 0.303
10. 0.07 Ω, 37.5 per cent
11. 960 r/min, 855 W
12. 27.5 per cent, 41 per cent
13. 11 kW, 440 W
14. 0.857 p.u., 25.57 kW
15. 1:0.333:0.16; 1:0.333:0.16

Exercises 40

1. 11.25 kV, 10.74 kV
2. 345 A, 0.69 MW
   363 A, 0.88 MW
3. 388 V
4. 108 A, 1.96 MW
   129 A, 2.04 MW
5. 5.56 kV
6. 12.54 kV, $\angle 25.5°$
7. 8904 A, 280 V (line)
8. (i) 163 MVA, $I_{SC} = 2848$ A, $V_{BB} = 8.95$ kV line
   (ii) 133 MVA, $I_{SC} = 2334$ A, $V_{BB} = 9.32$ kV line
9. In excess of 288 MVA
10. 0.8 Ω
11. 944 A, $V_A = 32.8$ kV (line), $V_B = 30.5$ kV (line),
    $V_C = 31.6$ kV (line)
12. 550 A, 8.20 kV

Exercises 41

1. 240 A, 498 V; 720 A, 166 V; 119.5 kW
2. 120 A, 301 V, 36.12 kW; 240 A, 150.5 V, 36.12 kW
3. 432 V
4. 0.02 Wb
5. 946 r/min
6. 0.0274 Wb
7. 1065 r/min
8. 140 V
9. 0.0192 Wb
10. 0.0333 Wb
11. 200 At, 3400 At
14. 4440 At, 555 At; 0.463 A
15. 8
16. 3150 At
17. 0.443 Ω

Exercises 42

1. 589 r/min
2. 714 r/min
3. 190 V
4. 468 V, 1075 r/min, 333 N m
5. 140 A
6. 208 V, 41 A
7. 439 r/min; 400 N m
8. 730 r/min, 226 N m
9. 13.85 mWb, 114.7 N m

10. 36.2 N m
11. 35 Ω, 962 r/min
12. 5.8 Ω, 60 V
13. 356 N m
14. 210.3 V, 0.0127 Wb, 114.3 N m
15. 595 r/min
16. 40 A, 695 r/min
17. 333 N m
18. 917 r/min, 2560 N m
19. 6.53 Ω
20. 3.05 Ω, 5.67 Ω, 9.2 Ω

Exercises 43

4. Stator −16; rotor −4 or 8
6. 50

Exercises 44

1. 1800 J, 1.4 T, 11.55 W at 50 Hz
2. 2.63 W
3. 0.333 W
4. 1.11 J
5. 337 W
6. 800 W
7. 1.87 A, 766 r/min, 10.3 A
8. 6.49 kW, 0.843 p.u.
9. 4.61 kW, 0.835 p.u.
10. 0.893 p.u.
11. 7130 W, 0.886 p.u.; 9010 W, 0.893 p.u.

Exercises 45

1. 1.008 A
2. 122.5°
3. 2.06 A, 136.2°
4. 5.61 A, 280.7 V
5. 6.48 A, 323.9 V

Exercises 46

6. 23.3 V
8. 60 V; 38 ms; 26.3 Hz
9. 137°

10. 600 Hz; 20 V; 134°
11. Fundamental plus even harmonics

Exercises 47

1. 0.0378 Ω; 6662 Ω
2. 5005 μΩ; 995 Ω
3. 8.325 mm
4. 19.985 kΩ; 0.0378 Ω
5. 0.015 04 Ω; 9.995 kΩ; 0.231 A
6. 0.003 753 Ω; 16.662 kΩ; −1.2 per cent, −0.000 36 per cent
7. 81 μN m
8. 30.4 μN m

9. 51.2 mA
10. 0.9; 1.036
11. 1786 Ω
12. 33.8 kΩ
13. 8.13 Ω; 1.6 per cent low
14. 60.67 Ω; 9.9 per cent high
15. 776 Ω
16. 5000 Ω; 21.45 mA
17. 3.08 mA from D to B
18. 2.17 per cent high
19. 6 kΩ, 15 kΩ, 60 per cent
20. 6 kΩ, 5.99 kΩ, 1.69 per cent
21. 27.3 V
22. Outwith limit
23. 6.67 A; 734 W; 4.71 A; 519 W
24. 2.26 V; 1.81 V; 2.71 V; 19.9 per cent, 19.9 per cent

Index